Safety Symbols

These safety symbols are used in laboratory and investigations in this book to indicate possible hazards. Learn the meaning of each symbol and refer to this page often. *Remember to wash your hands thoroughly after completing lab procedures.*

SAFETY SYMBOLS	HAZARD	EXAMPLES	PRECAUTION	REMEDY
DISPOSAL	Special disposal procedures need to be followed.	certain chemicals, living organisms	Do not dispose of these materials in the sink or trash can.	Dispose of wastes as directed by your teacher.
BIOLOGICAL	Organisms or other biological materials that might be harmful to humans	bacteria, fungi, blood, unpreserved tissues, plant materials	Avoid skin contact with these materials. Wear mask or gloves.	Notify your teacher if you suspect contact with material. Wash hands thoroughly.
EXTREME TEMPERATURE	Objects that can burn skin by being too cold or too hot	boiling liquids, hot plates, dry ice, liquid nitrogen	Use proper protection when handling.	Go to your teacher for first aid.
SHARP OBJECT	Use of tools or glassware that can easily puncture or slice skin	razor blades, pins, scalpels, pointed tools, dissecting probes, broken glass	Practice common-sense behavior and follow guidelines for use of the tool.	Go to your teacher for first aid.
FUME	Possible danger to respiratory tract from fumes	ammonia, acetone, nail polish remover, heated sulfur, moth balls	Make sure there is good ventilation. Never smell fumes directly. Wear a mask.	Leave foul area and notify your teacher immediately.
ELECTRICAL	Possible danger from electrical shock or burn	improper grounding, liquid spills, short circuits, exposed wires	Double-check setup with teacher. Check condition of wires and apparatus.	Do not attempt to fix electrical problems. Notify your teacher immediately.
IRRITANT	Substances that can irritate the skin or mucous membranes of the respiratory tract	pollen, moth balls, steel wool, fiberglass, potassium permanganate	Wear dust mask and gloves. Practice extra care when handling these materials.	Go to your teacher for first aid.
CHEMICAL	Chemicals that can react with and destroy tissue and other materials	bleaches such as hydrogen peroxide; acids such as sulfuric acid, hydrochloric acid; bases such as ammonia, sodium hydroxide	Wear goggles, gloves, and an apron.	Immediately flush the affected area with water and notify your teacher.
TOXIC	Substance may be poisonous if touched, inhaled, or swallowed.	mercury, many metal compounds, iodine, poinsettia plant parts	Follow your teacher's instructions.	Always wash hands thoroughly after use. Go to your teacher for first aid.
FLAMMABLE	Open flame may ignite flammable chemicals, loose clothing, or hair.	alcohol, kerosene, potassium permanganate, hair, clothing	Avoid open flames and heat when using flammable chemicals.	Notify your teacher immediately. Use fire safety equipment if applicable.
OPEN FLAME	Open flame in use, may cause fire.	hair, clothing, paper, synthetic materials	Tie back hair and loose clothing. Follow teacher's instructions on lighting and extinguishing flames.	Always wash hands thoroughly after use. Go to your teacher for first aid.

 Eye Safety Proper eye protection should be worn at all times by anyone performing or observing science activities.

 Clothing Protection This symbol appears when substances could stain or burn clothing.

 Animal Safety This symbol appears when safety of animals and students must be ensured.

 Radioactivity This symbol appears when radioactive materials are used.

 Handwashing After the lab, wash hands with soap and water before removing goggles

GLENCOE

PHYSICAL
SCIENCE
WITH
EARTH
SCIENCE

Ralph Feather
Charles William McLaughlin
Marilyn Thompson
Dinah Zike

Mc
Graw
Hill **Education**

About the Cover

Lightning—a plasma formed by charged particles in air—blazes a path between the clouds and Earth. How this happens is not entirely understood. Lightning research is just one exciting area of scientific inquiry—the process that seeks to understand the natural world.

The McGraw-Hill Companies

 Education

Send all inquiries to:
McGraw-Hill Education
8787 Orion Place
Columbus, OH 43240-4027

ISBN: 978-0-07-894582-3
MHID: 0-07-894582-8

Printed in the United States of America.

7 8 9 10 11 12 13   QVS/LEH   20 19 18 17 16 15 14

Contents in Brief

Online Guide

Get ConnectED

connectED.mcgraw-hill.com

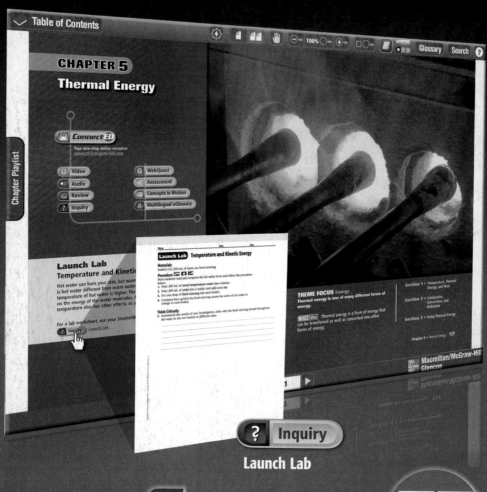

Inquiry

Launch Lab

ConnectED

▷ **Your Digital Science Portal**

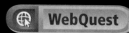

Video	Audio	Review	Inquiry	WebQuest
ee the science in al life through e exciting s.	Click the link and you can listen to the text while you follow along.	Try these interactive tools to help you review the lesson concepts.	Explore concepts through hands–on and virtual labs.	These web-based challenges relate the concepts you're learning about to the latest news and research.

The icons in your online student edition link you to interactive learning opportunities. Browse your online student book to find more.

Review

Additional Practice Problems

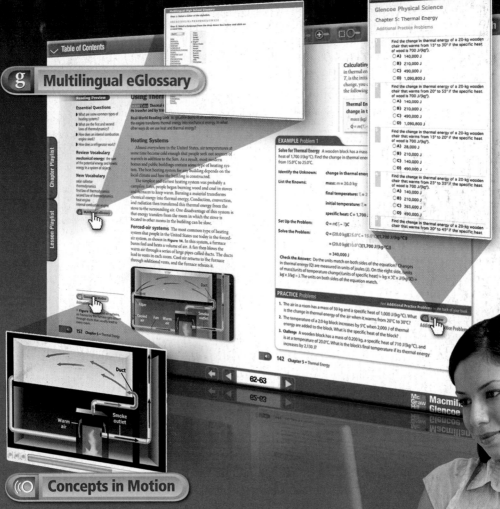

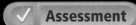
Concepts in Motion

Animation

"It's easy to do my assignments online and quick to find everything I need."

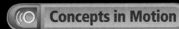
Assessment

Check how well you understand the concepts with online quizzes and practice questions.

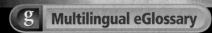
Concepts in Motion

The textbook comes alive with animated explanations of important concepts.

Multilingual eGlossary

Read key vocabulary in 13 languages.

About the Authors

Ralph Feather, Jr, PhD received a B.A. in Geology and MEd in geoscience education from Indiana University of PA. He earned a PhD in science education from the University of Pittsburgh. Dr. Feather has been awarded the Presidential Award for Excellence in Science Teaching, the Award for Excellence in Earth Science Teaching from the Geological Society of America, the Outstanding Earth Science Teacher Award from NAGT, Eastern Section, and the Kevin Burns Citation (Excellence in Science Teaching) from the Spectroscopy Society of Pittsburgh. Dr. Feather taught for 32 years at the secondary level. He was an assistant professor of astronomy and education specialist at Indiana University of PA. He is currently an assistant professor at the College of Professional Studies at Bloomsburg University in Pennsylvania.

Charles William McLaughlin, PhD holds a BS in science education from Northwest Missouri State University, an MS in chemistry from Kansas State University-Pittsburg, and a PhD in analytical chemistry from University of Nebraska. Dr. McLaughlin taught chemistry for more than 35 years including high school, science education courses, outreach programs, and as a visiting professor. He currently teaches general chemistry and serves as the Coordinator of General Chemistry at the University of Nebraska-Lincoln. He has received the Outstanding Teacher in the State award, a National Presidential Award for teaching science, and the Outstanding Campus Educator Award three times.

Marilyn Thompson, PhD received a BA in chemistry from Carleton College, an MA in science curriculum and instruction from the University of Kansas, and a PhD in science education from the University of Kansas. Dr. Thompson's career has included teaching physics, chemistry, and physical science while serving as science department chair. Dr. Thompson is currently an assistant professor at Arizona State University.

Dinah Zike is an international curriculum consultant and inventor who has designed and developed educational products and three-dimensional, interactive graphic organizers for over thirty years. Dinah has a BS and an MS in educational curriculum and instruction from Texas A&M University. She is frequently a featured speaker at national, regional, and state science teacher's conferences.

Contributing Authors

Nancy Ross-Flanigan
Science Writer
Detroit, MI

Nicholas Hainen
Science Writer
Carroll, OH

Steve Hardesty
Adjunct Professor
Valencia Community College
Orlando, Florida

Tina C Hopper
Science and Health Writer
Rockwell, TX

Sharon Nicholson, PhD
Heinz and Katherine Lettau
    Professor of Climatology
Meteorology Department
Florida State University
Tallahassee, FL

Margaret K. Zorn
Science Writer
Yorktown, VA

Reviewers and Consultants

High School Reviewers

Each teacher reviewed selected chapters of *Glencoe Physical Science with Earth Science* and provided feedback and suggestions regarding the effectiveness of the instruction.

Danielle Chirip
University High School
Orlando, FL

Dwight Dutton
East Chapel Hill High School
Chapel Hill, NC

Rob Hall
Grove City High School
Grove City, OH

Candace Hebert
Alfred M. Barbe High School
Lake Charles, LA

Dr. Carol Jones
Science Consultant
Macomb Intermediate School
   District
Clinton Township, MI

Christina McCray
Leesville Road High School
Raleigh, NC

Shelli Pace
Captain Shreve High School
Shreveport, LA

Jim Perkins
Triad High School
Troy, IL

Balmatie Sagramsingh
Cypress Creek High School
Orlando, FL

Glenn Smith
Bentworth School District
Bentleyville, PA

Tiffany Tammasini
Cypress Creek High School
Orlando, FL

Josephine K. Thirunayagam
Cypress Creek High School
Orlando, FL

Sara Tondra
Marysville High School
Marysville, OH

Elizabeth Trageser
Bentworth School District
Bentleyville, PA

Content Consultants

Content consultants each reviewed selected chapters of *Glencoe Physical Science with Earth Science* for content accuracy and clarity.

David G. Haase
Professor of Physics
North Carolina State
   University
Raleigh, NC

Michael O. Hurst
Associate Professor of Biochemistry
Georgia Southern University
Statesboro, GA

Sally Koutsoliotas
Associate Professor of Physics
Bucknell University
Lewisburg, PA

Monika Kress
Assistant Professor of Physics &
   Astronomy
San Jose State University
San Jose, CA

Nathan Niemi
Assistant Professor of Geological
   Sciences
University of Michigan
Ann Arbor, MI

Dr. Maria Pacheco
Associate Professor of Chemistry
Buffalo State College
Buffalo, NY

Contributing Writers and Consultants

Contributing Writers

Contributing writers helped develop chapter features.

Kelinda Rutan-Jorgensen	Andrew Schroeder	Karen Sottosanti	Stephen Whitt	Jenipher Willoughby
Columbus, OH	Columbus, OH	Pickerington, OH	Columbus, OH	Forest, VA

Safety Consultants

The safety consultant reviewed lab and lab materials for safety and implementation.

Kenneth R. Roy, PhD
Director of Environmental Health and Safety
Glastonbury Public Schools
Glastonbury, CT

Understanding Physical Science

Your book is organized into chapters that are based on the
BIG Ideas of Physical Science.

At the start of each chapter,
you will see two features that will preview how what you are about to investigate fits into the big picture of science.

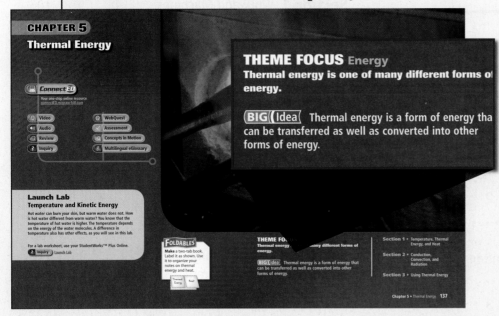

THEMES FOCUS

Themes are overarching concepts used throughout the entire book and help you connect together what you learn.

BIG Ideas

The Big Idea is the focus of the chapter. The labs, text, and other activities will build an in-depth understanding of these major concepts. Along the way, there are various activities and questions to evaluate your progress.

At the start of each section,
you will find a Reading Preview that summarizes what you will learn while exploring the section.

Essential Questions

Essential Questions reflect the important goals of the section. Together, an understanding of these questions will lead toward understanding of the chapter's Big Idea.

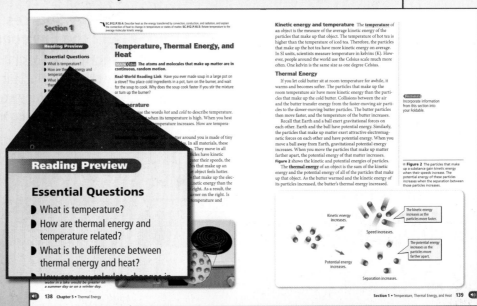

Contents

EXPLORE SCIENCE
In Action!

How SCIENCE Works

In the Field

SCIENCE & TECHNOLOGY

SCIENCE & HISTORY

For complete listings, see page **xvii**.

Contents

Contents

For complete listings, see page **xvii**.

Contents

Launch Labs

MiniLabs

Labs

Features

How SCIENCE Works

In the Field

SCIENCE & TECHNOLOGY

SCIENCE & HISTORY

Activities

EXAMPLE Problems

Apply Science

Concepts in Motion

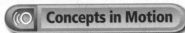
Concepts in Motion   Animation

 **FOLDABLES**®   by Dinah Zike

Folding Instructions

The following pages offer step-by-step instructions to make the Foldables study guides.

Layered-Look Book

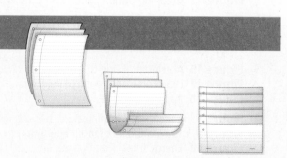

1. Collect three sheets of paper and layer them about 1 cm apart vertically. Keep the edges level.

2. Fold up the bottom edges of the paper to form ten equal tabs.

3. Fold the papers and crease well to hold the tabs in place. Staple along the fold. Label each tab.

Trifold Book

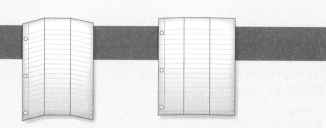

1. Fold a vertical sheet of paper into thirds.

2. Unfold and label each row.

Three-Tab Book

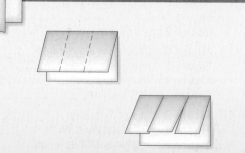

1. Fold a vertical sheet of paper from side to side. Make the front edge about 2 cm shorter than the back edge.

2. Turn length-wise and fold into thirds.

3. Unfold and cut only the top layer along both folds to make three tabs. Label each tab.

Two- and Four-Tab Books

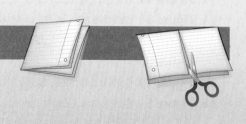

1. Fold a sheet of paper in half.

2. Fold in half. If making a four-tab book, then fold in half again to make three folds.

3. Unfold and cut only the top layer along the folds to make two or four tabs. Label each tab.

Four-Door Book

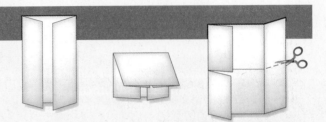

1. Find the middle of a horizontal sheet of paper. Fold both edges to the middle and crease the folds.

2. Fold the folded paper in half, from top to bottom.

3. Unfold and cut along the fold lines of the top layers to make four tabs. Label each tab.

Concept-Map Book

1. Fold a vertical sheet of paper from top to bottom. Make the top edge about 2 cm shorter than the bottom edge.

2. Turn length-wise and fold into thirds.

3. Unfold and cut only the top layer along both folds to make three tabs. Label the top and each tab.

Vocabulary Book

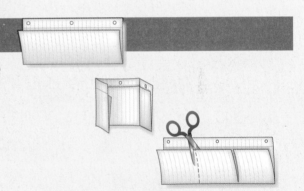

1. Fold a vertical sheet of notebook paper in half.

2. Cut along every third line of only the top layer to form tabs. Label each tab.

Folded Chart

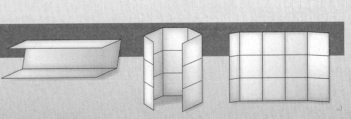

1. Fold a sheet of paper length-wise into thirds.

2. Fold the paper width-wise into fourths.

3. Unfold, lay the paper length-wise, and draw lines along the folds. Label the table.

UNIT 1

Motion and Forces

THEMES

Motion and Forces Newton's laws describe how force affects motion.

Scientific Inquiry Science methods form a creative and dynamic inquiry process that is validated by peer review and argumentation.

Technology The results of science provide technologies that improve everyday life.

WebQuest

STEM UNIT PROJECT

Model Research suspension bridges. Design and construct a model bridge that can support a toy car.

CHAPTER 1

The Nature of Science

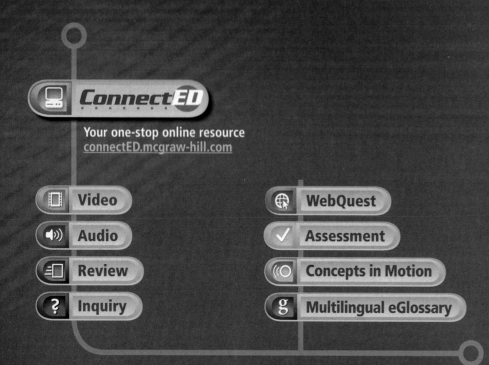

ConnectED

Your one-stop online resource
connectED.mcgraw-hill.com

- Video
- Audio
- Review
- Inquiry
- WebQuest
- Assessment
- Concepts in Motion
- Multilingual eGlossary

Launch Lab
Technology in Your Life

For nearly 10,000 years, farmers have used technology to help them optimize crop production. Today, they use things like GPS systems to guide their tractors. Technology is the application of science to help people. How much technology do you use?

For a lab worksheet, use your StudentWorks™ Plus Online.

Inquiry Launch Lab

FOLDABLES®

Make a two-tab book. Label it as shown. Use it to organize your notes on science and technology.

Science

Technology

THEME FOCUS Scientific Inquiry
Scientists ask questions and conduct investigations to learn more about the world around us.

BIG ((Idea(Science is a method of learning and communicating information about the natural world.

Reading Preview

Essential Questions

▶ What steps do scientists often use to solve problems?

▶ Why do scientists use variables?

▶ What is the difference between a scientific law and a scientific theory?

Review Vocabulary

investigation: an observation or study by close examination

New Vocabulary

scientific methods
hypothesis
experiment
variable
dependent variable
independent variable
constant
control
bias
model
theory
scientific law

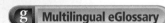 **Multilingual eGlossary**

The Methods of Science

MAIN ⟨Idea Scientific investigations don't always proceed with identical steps but do contain similar methods.

Real-World Reading Link Think of what it would be like if we still thought the world was flat or didn't have indoor plumbing or electricity. Science helps us learn about the natural world and improve our lives.

What is science?

Science is not just a subject in school. It is a method for studying the natural world. After all, science comes from the Latin word *scientia*, which means "knowledge." Science is a process based on inquiry that helps develop explanations about events in nature.

Nature follows a set of rules. Many rules, such as those concerning how the human body works, are complex. Other rules, such as the fact that Earth rotates about once every 24 hours, are much simpler. Scientists, such as the one shown in **Figure 1,** ask questions and make observations to learn about the rules that govern the natural world.

Major categories of science Science covers many different topics that can be classified according to three main categories. (1) Life science deals with living things. (2) Earth science investigates Earth and space. (3) Physical science studies matter and energy. Sometimes, though, a scientific study will overlap the categories. One scientist, for example, might study how to build better artificial limbs. Is this scientist studying energy and matter or how muscles operate? She is studying both life science and physical science.

■ **Figure 1** This scientist might use the transmitter on the turtle's back to investigate the turtle's speed, where it makes its home, gathers food, or interacts with other turtles.

Observe *What evidence in the photograph do you see of the three main branches of science?*

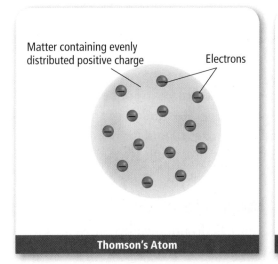

Matter containing evenly distributed positive charge

Electrons

Thomson's Atom

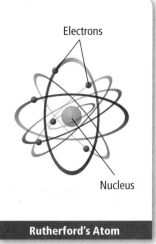

Electrons

Nucleus

Rutherford's Atom

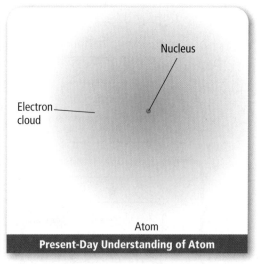

Nucleus

Electron cloud

Atom

Present-Day Understanding of Atom

■ **Figure 2** Our understanding of the atom has changed over time.

Science changes Scientific explanations help us understand the natural world. Sometimes these explanations must be modified. As more is learned, earlier explanations might be found to be incomplete or new technology might provide more accurate answers.

For example, scientists have been studying the atom for more than two centuries. Throughout this time, they have revised their thinking on what atoms might look like, how they interact, and even how they combine to form other substances.

In the early 1900s, British physicist J.J. Thomson created a model of the atom that consisted of electrons embedded in a ball of positive charge. Several years later, physicist Ernest Rutherford created a model of the atom based on new research. As shown in **Figure 2,** his model was different from Thomson's model. Instead of a solid ball, the atom consisted of a nucleus with electrons orbiting it like the planets orbit the Sun.

Later in the 20th century, scientists discovered the nucleus is not a solid ball but is made of protons and neutrons. This improved our understanding of the atom and its behavior. The present-day model of the atom is a nucleus made of protons and neutrons surrounded by an electron cloud. The electron cloud represents the space containing rapidly moving electrons.

The electron cloud model is the result of scientists pulling together evidence from many investigations. They came to an agreement that it is the best model they had at the time. Because it is the nature of science to be open to change, investigations into the model of the atom continue today.

Investigations Scientists learn new information about the natural world by performing investigations. Some investigations involve simply observing something that occurs and recording the observations. Other investigations involve setting up experiments with a control to test the effect of one thing on another.

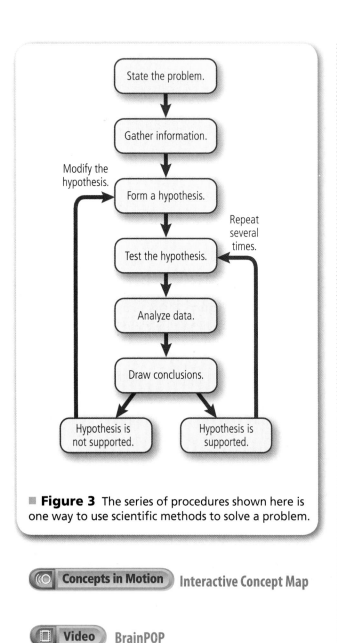

State the problem.

↓

Gather information.

↓

Form a hypothesis. ← Modify the hypothesis.

↓

Test the hypothesis. ← Repeat several times.

↓

Analyze data.

↓

Draw conclusions.

↓ ↓

Hypothesis is not supported. / Hypothesis is supported.

■ **Figure 3** The series of procedures shown here is one way to use scientific methods to solve a problem.

Concepts in Motion Interactive Concept Map

Video BrainPOP

Modeling Sometimes an investigation involves building a model that resembles something, such as a model of a new space vehicle, and then testing the model to see how it acts. Other models represent processes or objects that cannot be seen with the unaided eye, such as the models of the atom. Often, a scientist will use information from several types of investigations when attempting to learn about the natural world.

Scientific Methods

Although scientists do not always follow a rigid set of steps, investigations often follow a general pattern. The pattern of investigation procedures is called the **scientific methods.** Six common steps found in the scientific methods are shown in **Figure 3.** A scientist might add new steps, repeat some steps many times, or skip steps altogether.

State the problem To begin the process, a scientist must state what he or she is going to investigate. Many investigations begin when someone observes an event in nature and wonders why or how it occurs. The question of "why" or "how" is the problem.

Scientists once posed questions about why objects fall to Earth, what causes day and night, and how to generate electricity for daily use. Many times a statement of a problem arises when an investigation is complete and its results lead to new questions. For example, once scientists understood why we experience day and night, they wanted to know why Earth rotates.

Sometimes a new question is posed when an investigation runs into trouble. For example, some early work on guided missiles found the instruments in the nose cone did not always work properly. The original problem statement involved finding materials to protect the instruments during flight. The new statement involved how to repair the instruments.

Research and gather information Before beginning an investigation, scientists research what is already known about the problem. They gather and examine observations and interpretations from reliable sources. This background helps the scientist fine-tune his or her question and form a hypothesis.

Form a hypothesis A **hypothesis** is a possible explanation for a problem using what you know and what you observe. When trying to find a better material to protect the space shuttle, NASA scientists looked to other materials that were used in similar situations. Scientists knew that a ceramic coating had been found to solve the guided missile problem. They hypothesized that a ceramic material might work on the space shuttle.

Test a hypothesis Some hypotheses can be tested by making observations. Others can be tested by building a model and relating it to real-life situations. One common way to test a hypothesis is to perform an experiment. An **experiment** tests the effect of one thing on another using a control.

Variables An experiment usually contains at least two variables. A **variable** is a quantity that can have more than a single value. **Table 1** summarizes the types of variables. For example, as shown in **Figure 4,** numerous experiments aboard space shuttles and the *International Space Station* have studied the effects of microgravity on plants. Before these experiments could begin, scientists had to think of every factor that might affect plant growth. Each of these factors is a variable.

Independent and dependent variables In the microgravity experiment, plant growth is the **dependent variable** because its value changes according to the changes in the other variables. The variable changed to see how it will affect the dependent variable is called the **independent variable.** The microgravity is the independent variable.

Constants To be sure they were testing to see how microgravity affects growth, mission specialists kept the other possible factors the same. A factor that does not change is called a **constant.** The microgravity experiments used the same soil and type of plant. Additionally, each plant was given the same amount of light and water and was kept at the same temperature. Type of soil, type of plant, amount of light, amount of water, and temperature were constants for this experiment.

■ **Figure 4** An astronaut displays soybean plants growing in a microgravity experiment aboard the *International Space Station*. One goal of this experiment was to find out whether soybean plants would produce seeds in a microgravity environment.

Table 1	Types of Variables
Dependent Variable	changes according to the changes of the independent variable
Independent Variable	the variable that is changed to test the effect on the dependent variable
Constant	a factor that does not change when other variables change
Control	the standard by which the test results can be compared

Controls A **control** is the standard by which the test results can be compared. After the mission specialists gathered their data on the plants grown in microgravity, they could compare their results with the same types of plants grown on Earth's surface with the same constants. This comparison allowed them to analyze the data and form a conclusion about whether microgravity has an effect on plant growth.

✔ **Reading Check** **Identify** What is the purpose of a control in an experiment?

Analyze the data An important part of every investigation includes recording observations and organizing the test data into easy-to-read tables and graphs. Later in this chapter, you will study ways to display data. When you are making and recording observations, you should include all results, even unexpected ones. Many important discoveries have been made from unexpected results.

Scientific inferences are based on observations made using scientific methods. All possible scientific explanations must be considered. If the data are not organized in a logical manner, wrong conclusions can be drawn. When a scientist communicates and shares data, other scientists will examine that data, how it is analyzed, and compare it to the work of others. Scientists share their data through reports and conferences. In **Figure 5,** a student is presenting his data.

Draw conclusions Based on the analysis of the data, the next step is to decide whether the hypothesis is supported. For the hypothesis to be considered valid and widely accepted, the experiment must result in the exact same data every time it is repeated. If the experiment does not support the hypothesis, the hypothesis must be reconsidered. Perhaps the hypothesis needs to be revised, or maybe the experiment's procedure needs to be refined.

[?] **Inquiry** Virtual Lab

■ **Figure 5** An exciting and important part of investigating something is sharing your ideas with others, as this student is doing at a science fair.

Peer review Before it is made public, science-based information is reviewed by scientists' peers—scientists who are in the same field of study. Peer review is a process by which the procedures and results of an experiment are evaluated by other scientists who are in the same field as those who are conducting similar research. Reviewing other scientists' work is a responsibility that many scientists have.

Being objective Scientists also should be careful to reduce bias in their investigations. **Bias** occurs when the scientist's expectations changes how the results are analyzed or the conclusions are made. This might cause a scientist to select a result from one trial over those from other trials. Bias might also be found if the advantages of a product being tested are used in a promotion and the drawbacks are not presented.

Scientists can lessen bias by running as many trials as possible and by keeping accurate notes of each observation made. **Figure 6** shows a scientist who is researching how effective a medication is. One way to reduce bias in this case would be to conduct the experiment so that the researchers didn't know which group was the study group and which group was the control. This is called a blind experiment.

Valid experiments must also have data that are measurable. For example, a scientist performing a global warming study must base his or her data on accurate measures of global temperature. This allows others to compare the results to data they obtain from similar experiments. Most importantly, the experiment must be repeatable. Findings are supportable when other scientists around the world perform the same experiment and get the same results.

✔ **Reading Check** **Define** What is bias in science?

Visualizing with Models

Sometimes, scientists cannot see everything that they are testing. They might be observing something that is too large or small, takes too much time to see completely, or is hazardous. In these cases, scientists use models. A **model** represents an idea, event, or object to help people better understand it.

Models in history Models have been used throughout history. Lord Kelvin, a scientist who lived in England in the 1800s, was famous for making models. To model his idea of how light moves through space, he put balls into a bowl of jelly and encouraged people to move the balls around with their hands. Recall the models of the atom in **Figure 2**. Scientists used models of atoms to represent their current understanding because of the small size of an atom.

High-tech models Scientific models don't always have to be something you can touch. Another type of model is a computer simulation, like the one shown in **Figure 7.** A computer simulation uses a computer to test a process or procedure and to collect data. Computer software is designed to safely and conveniently mimic the processes under study. Today, many scientists use computers to build models.

Computer simulations enable pilots, such as the one shown in **Figure 7,** to practice all aspects of flight without ever leaving the ground. In addition, the computer simulation can simulate harsh weather conditions or other in-flight challenges that pilots might face.

■ **Figure 7** This is a computer simulation of an aircraft landing on a runway. The image on the screen in front of the pilot mimics what he would see if he were landing a real plane.

Identify *other models around your classroom.*

Scientific Theories and Laws

A scientific **theory** is an explanation of things or events based on knowledge gained from many observations and investigations. It is not a guess. If scientists repeat an investigation and the results always support the hypothesis, the hypothesis can be called a theory. Just because a scientific theory has data supporting it does not mean it will never change. As new information becomes available, theories can be modified.

A **scientific law** is a statement about what happens in nature and that seems to be true all the time. Laws tell you what will happen under certain conditions, but they don't explain why or how something happens. Gravity is an example of a scientific law. The law of gravity states that any one mass will attract another mass. To date, no experiments have been performed that disprove the law of gravity.

A theory can be used to explain a law, but theories do not become laws. For example, many theories have been proposed to explain how the law of gravity works. Even so, there are few accepted theories in science and even fewer laws.

The Limitations of Science

Science can help you explain many things about the world, but science cannot explain or solve everything. Although it's the scientist's job to make guesses, the scientist also has to make sure his or her guesses can be tested and verified.

Questions about emotions and values are not scientific, as shown in **Figure 8.** They cannot be tested. You might take a survey to gather opinions about such questions, but that would not prove the opinions are true for everyone.

■ **Figure 8** Science can't answer all questions, such as questions about emotions and values. This piece of art might look very beautiful to one person but not to another.

Analyze *Can anyone prove that you like artwork? Explain.*

Section 1 Review

Section Summary

▶ Scientists ask questions and perform investigations to learn more about the natural world.

▶ Scientists use scientific methods to test their hypotheses.

▶ Models help scientists visualize concepts.

▶ A theory is a possible explanation for observations, while a scientific law describes a pattern but does not explain why things happen.

1. **MAIN Idea Define** Summarize the steps you might use to carry out an investigation using scientific methods.

2. **Explain** why a theory cannot become a law.

3. **Think Critically** What is the dependent variable in an experiment that shows how the volume of gas changes with changes in temperature?

Apply Math

4. **Find the Average** An experiment to determine how many breaths a fish takes per minute yields this data: minute 1: 65 breaths; minute 2: 73 breaths; minute 3: 67 breaths; minute 4: 71 breaths; minute 5: 62 breaths. Calculate the average number of breaths that the fish takes per minute.

Section 2

Reading Preview

Essential Questions

▶ What is a standard of measurement?

▶ What multiple of ten does each SI prefix represent?

▶ What are the SI units and symbols for length, volume, mass, density, time, and temperature?

▶ How can related SI units be converted?

Review Vocabulary

measurement: the dimensions, capacity, or amount of something

New Vocabulary

standard
SI
volume
matter
mass
density

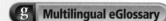 **Multilingual eGlossary**

Standards of Measurement

MAIN Idea Standard measurement units, such as centimeters and seconds, are exact quantities used to compare measurements.

Real-World Reading Link You know what time to go to school because you use seconds, minutes, and hours to tell time. The bus driver knows what speed to drive because the speedometer measures the speed of the bus. What if we all used different types of measurement?

Units and Standards

Suppose you and a friend want to find out whether a desk will fit through a doorway. You have no ruler, so you decide to use your hands as measuring tools. Using the width of his hands, your friend measures the doorway and says it is eight hands wide. Using the width of your hands, you measure the desk and find it is $7\frac{3}{4}$ hands wide. Will the desk fit through the doorway? You can't be sure. Even though you both used hands to measure, you didn't check to see whether your hands were the same width as your friend's hands. In other words, because you didn't use a measurement standard, you can't compare the measurements. A **standard** is an exact quantity that people agree to use to compare measurements.

Measurement Systems

Suppose the label on a ball of string says that the length of the string is 1. Is the length 1 meter (m), 1 foot (ft), or 1 centimeter (cm)? For a measurement to make sense, it must include a number and a unit, as shown in **Figure 9**.

Your family might buy lumber by the foot, milk by the gallon, and potatoes by the pound. These measurement units are part of the English system of measurement, which is commonly used in the United States. Most other nations use the metric system, a system of measurement based on multiples of ten.

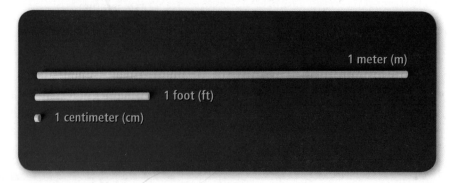

■ **Figure 9** For data collected in an investigation or experiment, you need appropriate and consistent standards of measurement. It is important that you always use a number and a unit when describing length.

International System of Units In 1960, an improved version of the metric system was devised. Known as the International System of Units, this system is often abbreviated **SI**, from the French *Le Systeme Internationale d'Unites*. All SI standards are universally accepted and understood by scientists throughout the world.

For example, the standard kilogram is kept in Sevres, France, and the standard meter equals the exact distance that light travels through a vacuum in 1/299,792,458 seconds. All kilograms and meters used throughout the world match these standards.

The kilogram is an example of a base unit. A base unit in SI is one that is based on an object or event in the physical world. The meter is the base unit of length. There are seven base units in SI. Each has its own symbol. These names and symbols for the seven base units are shown in **Table 2.** All other SI units are obtained from these seven units.

SI prefixes The SI system is easy to use because it is based on multiples of ten. Prefixes are used with the names of the units to indicate what multiple of ten should be used with the units. For example, the prefix *kilo*– means "1,000," which means that one kilometer equals 1,000 meters. Likewise, one kilogram equals 1,000 grams. Because *deci*– means "one-tenth," one decimeter equals one tenth of a meter. A decigram equals one tenth of a gram. The most frequently used prefixes are shown in **Table 3.**

✔ **Reading Check Calculate** How many meters is 1 km? How many grams is 1 dg?

Converting between SI units Sometimes, quantities are measured using different units. A conversion factor is a ratio that is equal to one. It is used to change one unit to another. For example, there are 1,000 mL in 1 L, so 1,000 mL = 1 L. If both sides in this equation are divided by 1 L, the equation becomes:

$$\frac{1{,}000 \text{ mL}}{1 \text{ L}} = 1$$

To convert units, multiply by the appropriate conversion factor. For example, to convert 1.255 L to mL, multiply 1.255 L by a conversion factor. Use the conversion factor with new units (mL) in the numerator and the old units (L) in the denominator.

$$1.255 \text{ L} \times \frac{1{,}000 \text{ mL}}{1 \text{ L}} = 1{,}255 \text{ mL}$$

Table 2	SI Base Units	
Quantity Measured	Unit	Symbol
Length	meter	m
Mass	kilogram	kg
Time	second	s
Electric current	ampere	A
Temperature	kelvin	K
Amount of substance	mole	mol
Intensity of light	candela	cd

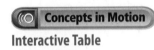 **Concepts in Motion**
Interactive Table

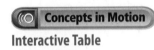 **Concepts in Motion**
Interactive Table

Table 3	Common SI Prefixes	
Prefix	Symbol	Multiplying Factor
Kilo–	k	1,000
Deci–	d	0.1
Centi–	c	0.01
Milli–	m	0.001
Micro–	μ	0.000 001
Nano–	n	0.000 000 001

Convert Units How long, in centimeters, is a 3,075-mm rope?

Identify the Unknown: rope length in cm

List the Knowns: rope length in mm = 3,075 mm

1 m = 100 cm = 1,000 mm

Set Up the Problem: length in cm = length in mm × $\dfrac{100\ cm}{1,000\ mm}$

Solve the Problem: length in cm = 3,075 mm × $\dfrac{100\ cm}{1,000\ mm}$ = 307.5 cm

Check the Answer: Millimeters are smaller than centimeters, so make sure your answer in mm is larger than the measurement in cm. Because SI is based on tens, the answer in mm should differ from the length in cm by a factor of ten.

PRACTICE Problems   Find **Additional Practice Problems** in the back of your book.

5. If your pencil is 11 cm long, how long is it in millimeters?

6. **Challenge** Some birds migrate 20,000 miles. If 1 mile equals 1.6 kilometers, calculate the distance these birds fly in kilometers.

 Review

Additional Practice Problems

Measuring Length

The word *length* is used in many ways. For example, the length of a novel is the number of pages or words it contains. In scientific measurement, however, length is the distance between two points. That distance might be the diameter of a hair or the distance from Earth to the Moon. The SI base unit of length is the meter (m). A baseball bat is about 1 m long. Metric rulers and metersticks are used to measure length. **Figure 10** compares a meter and a yard.

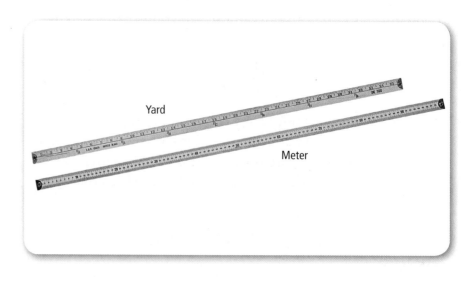

Yard

Meter

■ **Figure 10** One meter is slightly longer than 1 yard, and 100 m is slightly longer than a football field.

Predict *whether your time for a 100-m dash would be slightly more or less than your time for a 100-y dash.*

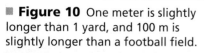

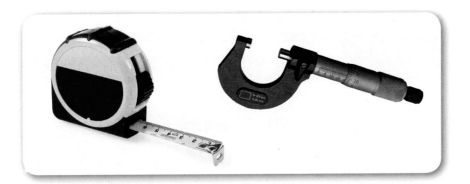

Choosing a unit of length As shown in **Figure 11,** the unit with which you measure will depend on the size of the object being measured. For example, the diameter of a shirt button is about 1 cm. You would most likely use the centimeter to measure the length of your pencil but the meter to measure the length of your classroom. What unit would you use to measure the distance from your home to school? You would probably want to use a unit larger than a meter. The kilometer (km), which is 1,000 m, is used to measure these kinds of distances.

By choosing an appropriate unit, you avoid large-digit numbers and numbers with many decimal places. Twenty-one kilometers is easier to deal with than 21,000 m. And 13 mm is easier to use than 0.013 m.

Measuring Volume

The amount of space occupied by an object is called its **volume.** If you want to know the volume of a solid rectangle, such as a brick, you measure its length, width, and height and multiply the three numbers and their units together: $V = l \times w \times h$. For a brick, your length measurements would most likely be in centimeters. The volume would then be expressed in cubic centimeters (cm^3) because when you multiply, you add the exponents. To find out how large of a load a moving van can carry, your length measurements would probably be in meters and the volume would be expressed in cubic meters (m^3).

Sometimes, liquid volumes, such as doses of medicine, are expressed in cubic centimeters. One cubic centimeter and one milliliter are the same volume.

$$1 \text{ mL} = 1 \text{ cm}^3$$

Suppose you wanted to convert a measurement in liters to cubic centimeters. You use conversion factors to convert L to mL and then mL to cm^3.

$$1.5 \text{ L} \times \frac{1{,}000 \text{ mL}}{1 \text{ L}} \times \frac{1 \text{ cm}^3}{1 \text{ mL}} = 1{,}500 \text{ cm}^3$$

VOCABULARY . . . .
SCIENCE USAGE V. COMMON USAGE
Volume
Science usage
  the amount of space occupied by an object
  To measure the volume of the cube, Raul submerged it in a beaker of water.

Common usage
  the degree of loudness
  Alisa couldn't hear the radio, so she turned up the volume. . . . .

Table 4	Densities of Some Materials at 20°C		
Material	Density (g/cm³)	Material	Density (g/cm³)
Hydrogen	0.00009	Aluminum	2.7
Oxygen	0.0014	Iron	7.9
Water	1.0	Gold	19.3

MiniLab

? Inquiry MiniLab

Determine the Density of a Pencil

Procedure

1. Read the procedure and safety information, and complete the lab form.
2. Find a **pencil** that will fit entirely in a **100-mL graduated cylinder** below the 90-mL mark.
3. Measure the mass of the pencil in grams with a **balance.**
4. Put 90 mL of **water** (initial volume) into the 100-mL graduated cylinder. Lower the pencil, eraser first, into the cylinder. Push the pencil down until it is just submerged. Hold it there, and record the final volume to the nearest tenth of a milliliter.

Analysis

1. **Determine** the water displaced by the pencil by subtracting the initial volume from the final volume.
2. **Calculate** the pencil's density by dividing its mass by the volume of water displaced.
3. **Compare** the density of the pencil to the density of water.

? Inquiry Video Lab

Measuring Mass and Density

Matter is anything that takes up space and has mass. A table-tennis ball and a golf ball have about the same volume. If you pick them up, you notice a difference. The golf ball has more mass. **Mass** is a measurement of the quantity of matter in an object. The mass of the golf ball is about 45 g. It is almost 18 times the mass of the table-tennis ball, which is about 2.5 g. A bowling ball has a mass of about 5,000 g. This makes its mass roughly 100 times greater than the mass of the golf ball and 2,000 times greater than the table-tennis ball's mass.

Density A cube of polished aluminum and a cube of silver that are the same size look similar and have the same volume, but they have different masses. The mass and volume of an object can be used to find the density of the material of which the object is made. **Density** is the mass per unit volume of a material. You find density by dividing an object's mass by the object's volume. For example, the density of an object having a mass of 10 g and a volume of 2 cm³ is 5 g/cm³. **Table 4** lists the densities of some familiar materials.

Derived units The measurement unit for density, g/cm³, is a combination of SI units. A unit obtained by combining different SI units is called a derived unit. An SI unit multiplied by itself also is a derived unit. Thus, the liter, which is based on the cubic decimeter, is a derived unit. A cubic meter, m³, is another example of a derived unit.

Measuring Time and Temperature

It is often necessary to keep track of how long it takes for something to happen. Time is the interval between two events. The SI unit for time is the second. In the laboratory, you will use a stopwatch or a clock with a second hand to measure time.

Another type of measurement common to science is temperature. You will learn the scientific meaning of the word *temperature* in a later chapter, but for now, think of temperature as a measure of how hot or how cold something is. **Table 5** summarizes measurements in SI, including time and temperature.

Table 5	SI Dimensions
Unit	**Example**
Millimeters	A dime is about 1 mm thick.
Meters	A football field is about 91 m long.
Kilometers	The distance from your house to the store can be measured in kilometers.
Milliliters	A teaspoonful of medicine is about 5 mL.
Liters	This carton holds 1.89 L of milk.
Grams/cm³	This stone sinks because it is denser—has more grams per cubic centimeter—than water.
Meters/second	The speed of a roller coaster car can be measured in meters per second.
Kelvin	Water boils at 373 K and freezes at 273 K.
Grams	The mass of a paper clip can be measured in grams.

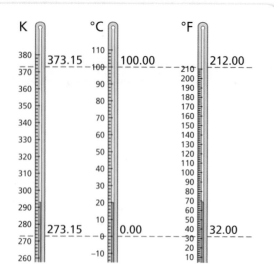

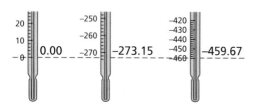

Figure 12 These three thermometers illustrate the scales of temperature from absolute zero to the boiling point of water.

Compare *the boiling points of the three scales.*

Celsius Look at **Figure 12**. For much scientific work, temperature is measured on the Celsius (C) scale. On this scale, the freezing point of water is 0°C, and the boiling point of water is 100°C. Between these points, the scale is divided into 100 equal divisions. Each one represents 1°C. On the Celsius scale, average human body temperature is 37°C, and a typical room temperature is between 20°C and 25°C.

Kelvin and Fahrenheit The SI unit of temperature is the kelvin (K). Zero on the Kelvin scale (0 K) is the coldest possible temperature, also known as absolute zero. Absolute zero is roughly equal to −273°C, which is 273°C below the freezing point of water.

Most laboratory thermometers are marked only with the Celsius scale. Because the divisions on the Celsius and Kelvin scales are the same size, the Kelvin temperature can be found by adding 273 to the Celsius reading. So, on the Kelvin scale, water freezes at 273 K and boils at 373 K. Notice that degree symbols are not used with the Kelvin scale.

The temperature measurement with which you are probably most familiar is the Fahrenheit scale. On the Fahrenheit scale, the freezing point of water is 32°F and the boiling point is 212°F. A temperature difference of 1° on the Fahrenheit scale is 5/9° on the Celsius scale.

Section 2 Review

Section Summary

▶ The International System of Units, or SI, was established to provide a standard of measurement and to reduce confusion.

▶ Conversion factors are used to change one unit to another and involve using a ratio equal to 1.

▶ The size of an object determines which unit you will use to measure it.

7. **MAIN Idea Explain** why it is important to have exact standards of measurement.

8. **Make a Table** Organize the following measurements from smallest to largest and include the multiplying factor for each: kilometer, nanometer, centimeter, meter, and micrometer.

9. **Think Critically** Explain why density is a derived unit.

Apply Math

10. **Convert Units** Make the following conversions: 27°C to Kelvin, 20 dg to milligrams, and 3 m to decimeters.

11. **Calculate Density** What is the density of an unknown metal that has a mass of 158 g and a volume of 20 mL? Use **Table 4** to identify this metal.

✓ Assessment Online Quiz

Essential Questions

▶ What are the three types of graphs, and how are they used?

▶ How are dependent and independent variables expressed in a graph?

▶ How can you analyze data using the various types of graphs?

Review Vocabulary

data: information gathered during an investigation or observation

New Vocabulary

graph

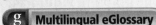

g Multilingual eGlossary

Communicating with Graphs

MAIN Idea Graphs are visual representations of numerical data that help scientists detect patterns.

Real-World Reading Link Just as a carpenter needs a variety of tools depending on the job, a scientist needs different types of graphs to communicate data.

A Visual Display

Scientists often graph the results of their experiments because they can detect patterns in the data easier in a graph than in a table. A **graph** is a visual display of information or data. **Figure 13** is a graph that shows the time and distance from home as a girl walked her dog. The horizontal axis, called the *x*-axis, measures time. Time is the independent variable because as it changes, it affects the measure of another variable. The distance from home that the girl and the dog walk is the other variable. It is the dependent variable and is measured on the vertical axis, called the *y*-axis.

Graphs are useful for displaying numerical information in business, science, sports, advertising, and many everyday situations. Graphs make it easier to understand patterns by displaying data in a visual manner.

Scientists often graph their data to detect patterns that would not have been evident in a table. Business people may graph sales dollars to determine trends. Different kinds of graphs—line, bar, and circle—are appropriate for displaying different types of information. Additional information on making and using graphs can be found in the Math Skill Handbook in the back of this book.

Concepts in Motion

Animation

■ **Figure 13** This graph tells the story of the motion that takes place when a girl takes her dog for a 10-minute walk. The dependent variable is on the *y*-axis, and the independent variable is on the *x*-axis.

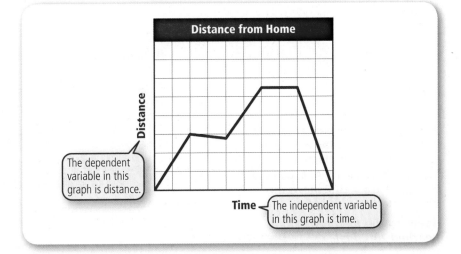

Table 6	Room Temperature		
Time* (min)	Classroom Temperature (°C)		
	A	B	C
0	16	16	16
5	17	17	16.5
10	19	19	17
15	20	21	17.5
20	20	23	18
25	20	25	18.5

*minutes after turning on heat

Line Graphs

A line graph can show any relationship where the dependent variable changes due to a change in the independent variable. Line graphs often show how a relationship between variables changes over time. You can use a line graph to track many things, such as how certain stocks perform or how the population changes over any period of time—a month, a week, or a year.

Displaying data on line graphs You can show more than one event on the same graph as long as the relationship between the variables is identical. Suppose a builder had to choose one thermostat from among three different kinds for a new school. He tested them to find out which was the best brand to install throughout the building. He installed different thermostats in classrooms A, B, and C. He set each thermostat at 20°C. He turned on the furnace and checked the temperatures in the three rooms every 5 min for 25 min. He recorded his data in **Table 6.**

The builder then plotted the data on the graph in **Figure 14.** He could see from the table that the data did not vary much for the three classrooms. So, he chose small intervals for the y-axis and left out part of the scale (the part between 0°C and 15°C). This allowed him to spread out the area on the graph where the data points lie.

You can easily see the contrast in the colors of the three lines and their relationship to the black horizontal line. The black line represents the thermostat setting and is the control. The control is what the resulting room temperature of the classrooms should be if the thermostats are working efficiently.

■ **Figure 14** The builder's graph compares the time the furnace has been running with the temperature in each of three rooms, A, B, and C. The black line at y = 20°C shows the thermostat.

Identify *the thermostat that reached 20°C first.*

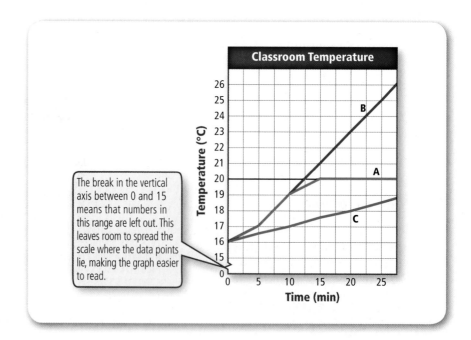

The break in the vertical axis between 0 and 15 means that numbers in this range are left out. This leaves room to spread the scale where the data points lie, making the graph easier to read.

Constructing line graphs In addition to choosing a scale that makes a graph readable, other factors are involved in constructing useful graphs. The most important factor in making a line graph is always using the *x*-axis for the independent variable. The *y*-axis is always used for the dependent variable. Recall that the dependent variable changes in response to the changes that you make to the independent variable, and the independent variable is the variable that you change to see how it will affect the dependent variable.

Another factor in constructing a graph involves units of measurement. You must use consistent units when graphing data. For example, you might use a Celsius thermometer for one part of your experiment and a Fahrenheit thermometer for another. But you must first convert your temperature readings to the same unit of measurement before you make your graph.

Once the data is plotted as points, a straight line or a curve is drawn based on those points. This should not be done like a connect-the-dots game. Instead, a best fit line or a most-probable smooth curve is placed among the data points, as shown in **Figure 15.**

In the past, graphs had to be made by hand, with each point plotted individually. Today, scientists, mathematicians, and students use a variety of tools, such as computer programs and graphing calculators, to help them draw and interpret graphs.

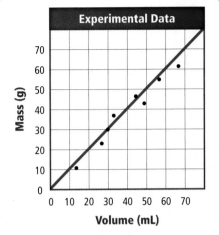

■ **Figure 15** Generally, the line or curve that you draw will not intersect all of your data points.

Apply Science

Make and Use Graphs

Line graphs are useful tools in showing the relationship between the independent and dependent variable. In an experiment, you checked the air temperature at certain hours of the day and recorded it in the data table shown here.

Time	Temp
8:00 A.M.	27°C
12:00 P.M.	32°C
4:00 P.M.	30°C

Identify the Problem

    time = independent variable
    temperature = dependent variable

Temperature is the dependent variable because it varies with time. Graph time on the *x*-axis and temperature on the *y*-axis. Mark equal increments on the graph and include all measurements. Plot each point on the graph by finding the time on the *x*-axis and moving up until you find the recorded temperature on the *y*-axis. Continue placing points on the graph. Then, connect the points from left to right.

Solve the Problem

1. Based on your graph, what was the temperature at 10:00 A.M.? What was the temperature at 2:00 P.M.?
2. What is the relationship between time and temperature?
3. Why is a line graph a useful tool in viewing this data?
4. For what other types of data might a line graph be useful?

Table 7	Classroom Size
Number of Students	Number of Classrooms
20	1
21	3
22	3
23	2
24	3
25	5
26	5
27	3

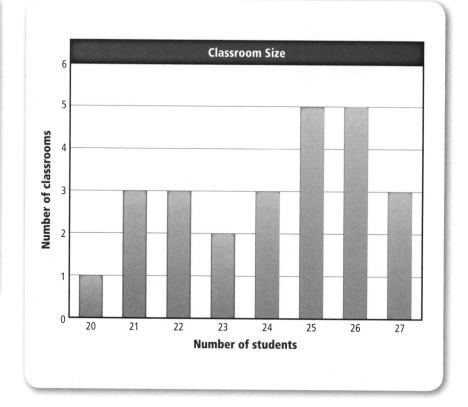

■ **Figure 16** The height of each bar corresponds to the number of classrooms having a particular number of students.

Bar Graphs

A bar graph is useful for comparing information or displaying data that do not change continuously. Suppose you counted the number of students in every classroom in your school and organized your data in **Table 7.** You could show these data in a bar graph like the one in **Figure 16.** Notice you can easily determine which classrooms have the largest and least numbers of students. You can also easily see that there are the same numbers of classrooms with 21 and 22 students and with 25 and 26 students.

Bar graphs can be used to compare oil or crop production, to compare costs of different products, or as data in promotional materials. Similar to a line graph, the independent variable is plotted on the *x*-axis and the dependent variable is plotted on the *y*-axis.

Recall that you might need to place a break in the scale of the graph to better illustrate your results. For example, if your data set included the points 1,002, 1,010, 1,030, and 1,040 and the intervals on the scale were every 100 units, you might not be able to see the difference from one bar to another. If you had a break in the scale and started your data range at 1,000 with intervals of ten units, you could make a more accurate comparison.

Reading Check **Describe** possible data where using a bar graph would be better than using a line graph.

Circle Graphs

A circle graph, sometimes called a pie chart, is used to show how some fixed quantity is broken into parts. The circular pie represents the total. The slices represent the parts and usually are represented as percentages of the total.

Figure 17 illustrates how a circle graph could be used to show the percentage of buildings in a neighborhood using each of a variety of heating fuels. You easily can see that more buildings use gas heat than any other kind of system. What other information does the graph provide?

To create a circle graph, you start with the total of what you are analyzing. Suppose the survey of heating fuels counted 72 buildings in the neighborhood. For each type of heating fuel, you divide the number of buildings using each type of fuel by the total (72). Then multiply that decimal by 360° to determine the angle that the decimal makes in the circle. For example, 18 buildings use steam. Therefore, $18 \div 72 = 0.25$ and $0.25 \times 360° = 90°$ on the circle graph. You then would measure 90° on the circle with your protractor. You can also calculate that if this graph shows 50 percent of the buildings use gas, then 36 of the buildings use gas ($0.50 \times 72 = 36$).

When you create a graph, think carefully about which type of graph you will use and how you will present your data. In addition, consider the conclusions you may draw from your graph. Make sure your conclusions are based on sound information and that you present your information clearly.

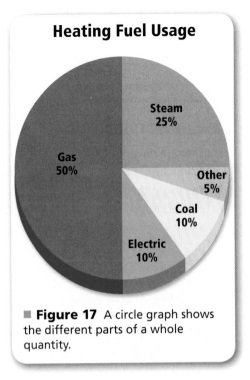

Heating Fuel Usage

Steam 25%
Gas 50%
Other 5%
Coal 10%
Electric 10%

■ **Figure 17** A circle graph shows the different parts of a whole quantity.

Section 3 Review

Section Summary

▶ Graphs are a visual representation of data.

▶ Scientists often graph their data to detect patterns.

▶ A line graph shows how a relationship between two variables changes over time.

▶ Bar graphs are best used to compare information collected by counting.

▶ A circle graph shows how a fixed quantity is broken down into parts.

12. **MAIN Idea Identify** the kind of graph that would best show the results of a survey of 144 people where 75 ride a bus, 45 drive cars, 15 carpool, and 9 walk to work.

13. **State** which type of variable is plotted on the *x*-axis and which type is plotted on the *y*-axis.

14. **Explain** why the points in a line graph can be connected.

15. **Think Critically** How are line, bar, and circle graphs similar? How are they different?

Apply Math

16. **Use Percentages** In a survey, it was reported that 56 out of 245 people would rather drink orange juice than coffee in the morning. Calculate what percentage of a circle graph orange juice drinkers would occupy.

Essential Questions

▶ What are the different types of technology?

▶ Why might the value of technology vary for different people at different times?

▶ From where do funds for science research come?

▶ How can consumers affect technological development?

Review Vocabulary

industrialized: a country that has developed industry and experiences large economic growth

New Vocabulary

technology
society

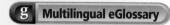 **Multilingual eGlossary**

🎬 **Video** What's PHYSICAL and EARTH SCIENCE Got To Do With It?

Science and Technology

MAIN ◀Idea Technology is the application of scientific knowledge to benefit people.

Real-World Reading Link Imagine what your life would be like without computers or cell phones. These and many other technologies have completely changed the way we live our lives.

What is technology?

The terms *science* and *technology* often are used interchangeably. However, these terms have very different definitions. Science is an exploration process. Scientific processes are used to gain knowledge to explain and predict natural occurrences. Scientists often pursue scientific knowledge for the sake of learning new information. There may or may not be a plan to use the knowledge.

When scientific knowledge is used to solve a human need or problem, as shown in **Figure 18,** the result is referred to as technology. **Technology** is the application of scientific knowledge to benefit people. Given this definition, is an aspirin tablet technology? Is a car technology? What about the national highway system? Although these examples appear to be very different, they all represent examples of technology. Technology can be

- any human-made object (such as a radio, computer, or pen),
- methods or techniques for making any object or tool (such as the process for making glass or ceramics),
- knowledge or skills needed to operate a human-made object (such as the skills needed to pilot an airplane), or
- a system of people and objects used to do a particular task (such as the Internet; a system to share information).

■ **Figure 18** The hand in this image is a technological object. The knowledge needed to interpret this image is also technology.

Technological objects The value of a technological object changes through time, as shown in **Figure 19.** What is considered new technology today might be considered an antique tomorrow. For example, special feathers called quills were used long ago to write with ink. The quill pen was the height of technology for over 1,000 years. Then, in the middle of the nineteenth century, metallic pens and writing points came into use. The modern ballpoint pen was not widely used until the 1940s.

Technological methods or techniques Just as writing instruments have changed over time, so have techniques for performing various tasks. Long ago, people would sit for hours and copy each page of a book by hand. Books were expensive and could be bought by only the very rich. Today, books can be created in different ways. They can be made on a computer, printed with a computer printer, and bound with a simple machine. Modern printing presses, like those shown in **Figure 20,** are used to produce the majority of books used today, including your textbook.

The methods used for printing books have changed over time. Each technique has its own technology. Other techniques that characterize technology include using a compact disc to store information, using a refrigerator to preserve food, and using e–mail to correspond with friends.

Technological knowledge or skills Technology also can be the knowledge or skills needed to perform a task. For example, computer skills are needed to use the software that is used to make books and other documents. Printing press operators must use their skills to print books successfully. Any time a complex machine is used to perform a task, technological skills must be used by the operator.

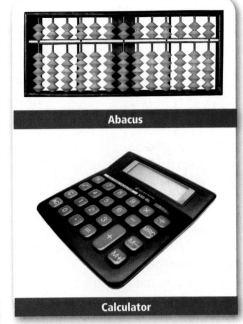
Abacus

Calculator

■ **Figure 19** Technology changes through time. As new scientific knowledge and techniques are learned, the processes used to create objects change. An abacus was once the primary way people could add large numbers. Today, we often rely on a calculator.

■ **Figure 20** Hundreds of years ago, books had to be written and published by hand. Now, printing presses can generate thousands of pages an hour.

Research the Past

Procedure

1. Read the procedure and safety information, and complete the lab safety form.
2. Prepare at least five questions to ask a senior citizen about technology when he or she was a teenager.
3. Perform the interview.
4. Record the responses to your questions with **pen** and **paper**.

Analysis

1. **Compare and contrast** at least five differences in technology between then and now.

Technological systems A network of people and objects that work together to perform a task also is technology. One example of this technology is the Internet. The Internet is a collection of computers and software that is used to exchange information. A technological system is a collection of various types of technology that are combined to perform a specific function. The airline industry is an example of a technological system. This industry is a collection of objects, methods, systems, knowledge, and procedures. The airports, pilots, fuel, and ticketing process form a technological system that is used to move people and goods.

✓ **Reading Check Identify** another example of a technological system. Explain why it fits into this category.

Global Technological Needs

The value of technology may differ for different people and at different times. The technology that is needed in the United States is not necessarily needed in other parts of the world. The needs for technology are different in developing and industrialized countries.

Developing countries The people of some countries work hard for basic needs such as food, shelter, clothing, safe drinking water, and health care. For example, the family shown in **Figure 21** from Mali, Africa, lives without electricity and running water in their home. Instead of going to the grocery store for the food their family eats, they might grow most of it themselves. Children might walk to school instead of taking the school bus.

■ **Figure 21** The technological needs of this family from Mali might be different than those of a family in another location. All people need safe water, food supplies, health care, education, and a safe place to work, but the technology involved in meeting these needs can differ from location to location.

Compare and contrast *these needs to the needs of your family.*

Meeting basic needs Technological solutions in developing countries are often limited to supplying basic needs for these families. Technology that would supply adequate and safe drinking water and food supplies would be valued before technology such as access to the Internet. Increasing the accessibility of basic health care would improve the quality of life and increase the life expectancy in developing countries. The technology valued by rural people in developing countries contrasts with the technology valued by people in those countries' industrialized cities.

Industrialized countries The United States is considered an industrialized country. Technology gives people the ability to clean polluted waters. Improving the quality of the food supply is also valued. But the level of urgency and focus differs from those in developing countries.

Most areas of the United States have adequate and safe water and food supplies. Because the needs for survival are met in industrialized nations, money often is spent on different types of technology. The technology used in industrialized countries is designed to improve the quality of life for individuals.

Look at the home in **Figure 22,** and compare it to the home on the previous page. Most homes in the United States have electricity and running water. Quality health care is available to many people. The life expectancy of Americans is the late-seventies, generally higher than in developing countries. Most homes in the United States contain many different types of technology, including computers, telephones, and televisions. Money is spent on such medical procedures as cosmetic surgery to remove wrinkles and eye surgery so that a person no longer has to wear glasses.

Contrasting needs As you can see, the human needs for developing and industrialized countries are very different. Both developing and industrialized countries value technologies that supply basic human needs. The actual technology required in each country may be very different.

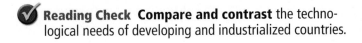 **Reading Check** **Compare and contrast** the technological needs of developing and industrialized countries.

■ **Figure 22** The basic needs for survival are available to most people in industrialized countries. Value is placed on technology that improves the quality of life, such as devices that make tasks easier and devices that provide entertainment.

Describe *three technological objects that you value that would be of less value to a family in a developing country.*

Figure 23 Consumers often decide which technologies will be developed. If consumers do not purchase a product, additional money usually will not be spent on the production or improvement of the product.

Social Forces that Shape Technology

Science and society are closely connected. **Society** is a group of people that share similar values and beliefs. Discoveries in science and technology bring about changes in society. In turn, society affects how new technologies develop. The development of technology is affected by society and its changing values, politics, and economics.

In the past 100 years, attitudes in the United States have changed toward automobiles. Many people became able to own cars due to the changes in technology and manufacturing. As car ownership increased, so did fossil fuel consumption. With rising gasoline prices, some consumers began buying more fuel-efficient cars. The automotive industry has researched and developed technologies that make cars more fuel-efficient. Today, hybrid cars use both gasoline and electricity.

Personal values People will support the development of technologies that agree with their personal values, directly and indirectly. Purchasing technology is a direct way in which people support the development of technology. For example, if consumers continue to purchase fuel-efficient cars, as shown in **Figure 23,** additional money will be spent on improving the technology. If consumers fail to buy a product, companies usually will not spend additional money on that type of technology. People also support the development of technology directly when they give their money to organizations committed to a specific project, such as cancer research.

There are also many ways to indirectly support technologies that agree with your personal values. For example, people vote for a congressional candidate based on the candidate's views on various issues. This is an indirect way in which people's personal values influence how technology projects receive funding and support.

Economic Forces that Shape Technology

Many factors influence how much money is spent on technology. Before funding is given for a project, several questions should be answered. What is the benefit for this product? What is the cost? Who will buy this product? All of these questions should be answered before money is given for a project. Various methods exist to fund new and existing technology.

Federal government One way in which funds are allocated for research and development of technology is through the federal government. Every year, Congress and the president place large amounts of money in the federal budget for scientific research and development. These funds are reserved for specific types of research, such as agriculture, defense, energy, and transportation. This money is given to companies and institutions in the form of contracts and grants to do specific types of research. Citizens can affect how the government funds technology by voting and other political activities, as shown in **Figure 24.**

Private foundations Some scientific research is funded using money from private foundations. Funds are raised for various types of disease research, such as breast cancer and muscular dystrophy through events such as races and telethons. Many private foundations focus on research for a specific cause.

Private industries Research and development is also funded by private industries. Industries budget a portion of their profits for research and development. Investing in research and development can make money for the company in the long-term. Bringing new products to the marketplace is one way companies make profits.

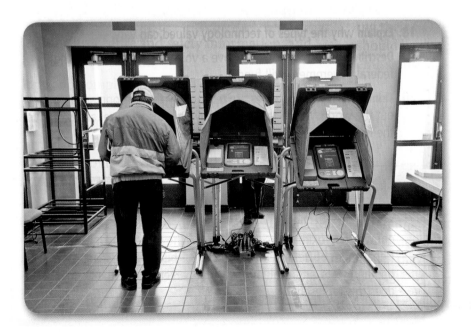

■ **Figure 24** Participating in the political process is one way in which people can influence which technologies are developed and which are not.

Explain *how voting for someone in Congress influences which technologies will be developed.*

LAB   Care Package

Objectives

- **Model** packaging and shipping a product for consumer interest.
- **Calculate** the cost to produce a consumer product.
- **Test** your packaging design to see if it keeps the snacks from being damaged.

Background: Suppose you are a young businessperson with a new idea for a product. You are going to create a care package with some popular snack foods for students to use during exams week. You must select the items for your care package, pack the items for display and shipping, and calculate the cost of the care package. You must find a way to make sure the snack foods arrive without being damaged, while keeping the cost of shipping to a minimum.

Question: *How can you package foods to survive the hazards of shipping and keep your costs to a minimum?*

Preparation

Possible Materials

packages of snack foods
shoe box
assorted snack foods
packing peanuts
shredded newspaper
packing tape
wrapping paper
markers

Safety Precautions

Make the Model

1. Read the procedure and safety information, and complete the lab form.
2. Choose the snack items that you would like to include in your care package.
3. Plan how you are going to pack your items so that they are nicely displayed and protected for shipping.
4. Plan the design of your decorative wrapper and labels. Remember to make the wrapper attractive so that customers will want to purchase your product.
5. Design a data table to keep track of all your costs to make the care package.
6. Evaluate your design plan. Determine if there are ways to cut costs to increase your profit.
7. Make sure your teacher approves your plan before you start.

Data Table

Food Item	Unit Cost	Cost of Packaging	Number of Items	Total Cost

Test the Model

1. Assemble the items that you need to carry out your plan.

2. Pack the materials according to your plan. Do not put on the decorative wrapper until after you have tested your package.

3. After you have carefully packed your box, test to make sure that everything is secure. Drop the box from the height of your lab table five times.

4. Unwrap your box to see if anything is damaged. If the contents are broken, determine changes that can be made to improve your design.

5. If needed, retest your new design. After the drop tests are successful, apply your wrapper and labels.

Analyze Your Data

1. **Calculate** the cost of your package.

2. **Determine** the limitations of your design.

Conclude and Apply

1. **Explain** how you can improve your design. Consider cost, contents, and packaging.

2. **Determine** under what conditions the contents of your package might be damaged. For example, if your package contains chocolate, the package cannot be shipped during hot summer months. What changes can you make to your design to remove these limitations?

3. **Predict** what would happen to the contents of your package if the package were dropped from heights taller than your lab table.

4. **Determine** if you could produce many of these types of packages quickly. How might that influence how widely used your design was?

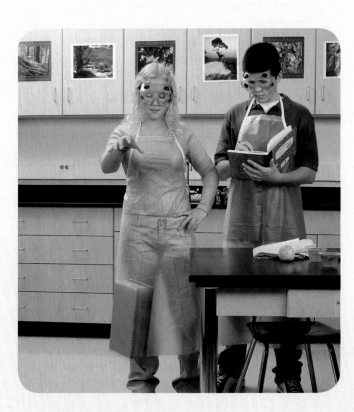

COMMUNICATE YOUR DATA

Present your design to your classmates. Compare your design to those of your classmates. Look for ways that you could improve your design.

How SCIENCE Works

Scientific Methods

Abu Ali al-Hasan ibn al-Hasan ibn al-Haytham is sometimes known as the father of the scientific method, the first true scientist, and the founder of optics. He was born in A.D. 965 in what is now modern-day Iraq, but he spent most of his life in Cairo, Egypt. As a young man, he read the works of Aristotle, the third-century B.C. Greek philosopher. He approved of Aristotle's reliance on experience, rather than just untested ideas, in pursuit of truth.

In one of his essays, Ibn al-Haytham, shown in **Figure 1,** wrote that "the seeker after truth is . . . the one who submits to argument and demonstration and not to the sayings of a human being whose nature is fraught with all kinds of imperfection and deficiency." Ibn al-Haytham studied and wrote about many scientific disciplines, including mathematics (especially geometry), astronomy, optics, physics, and medicine. He became one of the first scientists to perform experiments to test his hypotheses. In addition, he contributed to the fields of philosophy and psychology.

Experiments with optics The study of optics is the branch of physics that studies light and vision. Earlier thinkers, such as the second-century A.D. Greek scientist Ptolemy, thought that people could see because rays of light came out of their eyes and fell on the objects they were observing. Ibn al-Haytham performed experiments to test the behavior of light and the workings of the human eye. His observations led him to conclude that people could see because rays of light entered, not exited, their eyes.

Figure 1 Ibn al-Haytham is often regarded as the founder of scientific methods.

A lasting influence Ibn al-Haytham's insistence on experimentation and on making quantifiable observations helped develop scientific methods as we know them today. In A.D. 1040, centuries after his death, Ibn al-Haytham's most important work, *Kitab al-Manazir (Optics),* influenced European scientists such as Roger Bacon, a 13th-century philosopher who is sometimes given credit for laying the foundation for the use of scientific methods.

Because of Ibn al-Haytham's lifelong quest to develop a rational way of exploring his ideas, today's scientists and students of science have scientific methods, a powerful tool for testing hypotheses.

WebQuest Webquest Summarize what characterizes science and its methods. Identify a question you think can be answered through science and explain why.

Theme Focus Scientific Inquiry Scientists ask questions and conduct investigations to learn more about the world around us. Investigations should be repeatable, producing similar results when repeated around the world.

BIG Idea Science is a method of learning and communicating information about the natural world.

Section 1 The Methods of Science

bias (p. 11)
constant (p. 9)
control (p. 10)
dependent variable (p. 9)
experiment (p. 9)
hypothesis (p. 9)
independent variable (p. 9)
model (p. 12)
scientific law (p. 13)
scientific methods (p. 8)
theory (p. 13)
variable (p. 9)

MAIN Idea Scientific investigations don't always proceed with identical steps but do contain similar methods.

- Scientists ask questions and perform investigations to learn more about the natural world.
- Scientists use scientific methods to test their hypotheses.
- Models help scientists visualize concepts.
- A theory is a possible explanation for observations, while a scientific law describes a pattern but does not explain why things happen.

Section 2 Standards of Measurement

density (p. 18)
mass (p. 18)
matter (p. 18)
SI (p. 15)
standard (p. 14)
volume (p. 17)

MAIN Idea Standard measurement units, such as centimeters and seconds, are exact quantities used to compare measurements.

- The International System of Units, or SI, was established to provide a standard of measurement and to reduce confusion.
- Conversion factors are used to change one unit to another and involve using a ratio equal to 1.
- The size of an object determines which unit you will use to measure it.

Section 3 Communicating with Graphs

graph (p. 21)

MAIN Idea Graphs are visual representations of numerical data that help scientists detect patterns.

- Graphs are a visual representation of data.
- Scientists often graph their data to detect patterns.
- A line graph shows how a relationship between two variables changes over time.
- Bar graphs are best used to compare information collected by counting.
- A circle graph shows how a fixed quantity is broken down into parts.

Section 4 Science and Technology

society (p. 30)
technology (p. 26)

MAIN Idea Technology is the application of scientific knowledge to benefit people.

- Technology can be an object, a technique, a skill, or a system.
- Societal and economic forces influence which technologies will be developed and used around the world.
- The development of new technology is influenced by voting and buying habits.
- The federal government, private foundations, and private industries fund the research and development of technology.

Use Vocabulary

Match each phrase with the correct term from the Study Guide.

22. an exact quantity that people agree to use to compare measurements

23. the amount of space occupied by an object

24. application of science to help people

25. the amount of matter in an object

26. a variable that changes as another variable changes

27. a visual display of data

28. a test set up under controlled conditions

29. a variable that does NOT change as another variable changes

30. mass per unit volume

31. an educated guess using what you know and observe

Check Concepts

32. Which of the following questions CANNOT be answered by science?
A) How do birds fly?
B) Is this a good song?
C) What is an atom?
D) How does a clock work?

33. Which of the following is an example of an SI unit?
A) foot   C) pound
B) second   D) gallon

34. One one-thousandth is expressed by which prefix?
A) kilo–   C) centi–
B) nano–   D) milli–

35. On which of the following is SI based?
A) inches   C) English units
B) powers of five   D) powers of ten

36. What is the symbol for deciliter?
A) dL   C) dkL
B) dcL   D) Ld

37. Which of the following is NOT a derived unit?
A) dm^3 C) cm^3
B) m   D) g/ml

38. Which of the following is NOT considered technology?
A) car   C) leaf
B) shovel   D) book

Interpret Graphics

Use the images below to answer question 39.

39. **BIG Idea** The photos above show the items needed for an investigation. Which item is the independent variable? Which items are the constants? What might be a dependent variable?

✓ Assessment   Online Test Practice

Concepts in Motion Interactive Concept Map

40. Copy and complete this concept map on scientific methods.

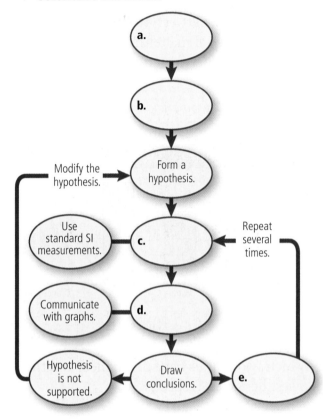

Think Critically

41. MAIN Idea **Communicate** Standards of measurement used during the Middle Ages often were based on such things as the length of the king's arm. How would you go about convincing people to use a different system of standard units?

42. **Identify** three technological items that you value that might not be important in a developing country.

43. **Identify** when bias occurs in scientific experimentation. Describe steps scientists can take to reduce bias and validate their experimental data.

44. **Demonstrate** Not all objects have a volume that is measured easily. If you were to determine the mass, volume, and density of your textbook, a container of milk, and an air-filled balloon, how would you do it?

45. **Apply** Suppose you set a glass of water in direct sunlight for 2 h and measure its temperature every 10 min. What type of graph would you use to display your data? What would the dependent variable be? What would the independent variable be?

46. THEME FOCUS **Form a Hypothesis** A metal sphere is found to have a density of 5.2 g/cm³ at 25°C and a density of 5.1 g/cm³ at 50°C. Form a hypothesis to explain this observation. How could you test your hypothesis?

47. **Apply** Suppose you are a scientist and think you have discovered the cure for skin cancer, but you need to conduct more research to confirm your discovery. Discuss the ethical issues surrounding using either animal or human subjects to support your research.

Apply Math

48. **Solve** Make the following conversions.
   a. 1,500 mL to L
   b. 2 km to cm
   c. 5.8 dg to mg
   d. 22°C to K

49. **Calculate** the density of an object having a mass of 17 g and a volume of 3 cm³.

50. **Solve** A block of wood is 0.4 m by 0.2 m by 0.7 m. Find its dimensions in centimeters. Then find its volume in cubic centimeters.

Standardized Test Practice

Multiple Choice

Record your answers on the answer sheet provided by your teacher or on a sheet of paper.

Use the graph below to answer questions 1 and 2.

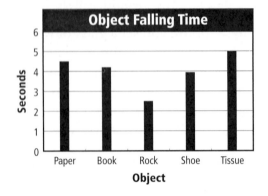

1. Students drop different objects from the same height and measure the time it takes each object to reach the ground. What is the dependent variable?
   A. falling time   C. drop height
   B. shoe   D. paper

2. What is a constant in this experiment?
   A. throwing some objects and dropping others
   B. measuring different falling times for each object
   C. dropping each object from the same height
   D. dropping a variety of objects

3. Which of the following is a statement that describes something that happens in nature that seems to be true all of the time?
   A. theory   C. hypothesis
   B. scientific law D. conclusion

4. What does the abbreviation *ns* represent?
   A. millisecond
   B. nanosecond
   C. microsecond
   D. Kelvin

5. Which of these best defines mass?
   A. the amount of space occupied by an object
   B. the distance between two points
   C. the quantity of matter in an object
   D. the interval between two events

Use the graph below to answer questions 6 and 7.

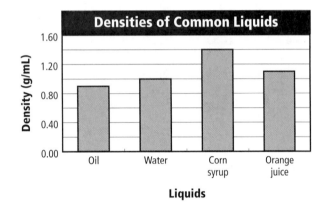

6. Which two liquids have the highest and the lowest densities, respectively?
   A. water and oil
   B. corn syrup and oil
   C. orange juice and water
   D. corn syrup and orange juice

7. What is the density of oil in units of mg/cm^3?
   A. 900 mg/cm^3
   B. 90 mg/cm^3
   C. 0.090 mg/cm^3
   D. 9,000 mg/cm^3

8. Which type of graph is most useful for showing how the relationship between the independent and dependent variables changes over time?
   A. circle graph
   B. bar graph
   C. pictograph
   D. line graph

✓ Assessment   Standardized Test Practice

Short Response

Record your answers on the answer sheet provided by your teacher or on a sheet of paper.

9. Define the term *technology*. Identify three ways that technology makes your life easier, safer, or more enjoyable.

10. Describe the three major categories into which science is classified. Which branches of science would be most important to an environmental engineer? Why?

11. Make the following conversions:
    a. 615 mg to g
    b. 75 dL to mL
    c. 0.95 km to cm

Use the illustration below to answer questions 12 and 13.

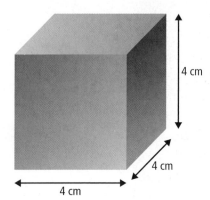

4 cm

4 cm

4 cm

12. Define the term *volume*. Calculate the volume of the cube shown above. Give your answer in cm³ and mL.

13. Define the term *density*. If the mass of the cube is 96 g, what is the density of the cube material?

Extended Response

Record your answers on a sheet of paper.

14. A friend frequently misses the morning school bus. Use the scientific method to address this problem.

Use the illustration below to answer questions 15 and 16.

15. Explain why this item is considered technology.

16. Will it always be considered technology? Why or why not?

17. You must decide what items to pack for a hiking and camping trip. Space is limited, and you must carry all items during hikes. What measurements are important in your preparation?

Use the table below to answer question 18.

Animal Life Span			
	Cow	Dog	Horse
Resting heart rate	52 beats per min	95 beats per min	48 beats per min
Average life span	18 years	16 years	27 years

18. Create a graph to display the data shown above.

NEED EXTRA HELP?																		
If You Missed Question . . .	1	2	3	4	5	6	7	8	9	10	11	12	13	14	15	16	17	18
Review Section . . .	1	1	1	2	2	2	2	3	4	1	2	2	2	1	4	4	2	3

CHAPTER 2
Motion

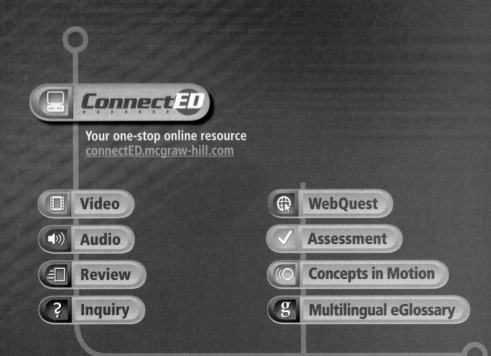

ConnectED
Your one-stop online resource
connectED.mcgraw-hill.com

- Video
- Audio
- Review
- Inquiry
- WebQuest
- Assessment
- Concepts in Motion
- Multilingual eGlossary

Launch Lab
Animal Race

A cheetah can run at a speed of almost 120 km/h. It is the fastest runner in the world. A horse can reach a speed of 64 km/h, an elephant's top speed is about 40 km/h, and a tortoise walks at a speed of about 0.3 km/h. The speed of an object is calculated by dividing the distance that the object travels by the time it takes it to move that distance. How does your speed compare to the speeds of these animals?

For a lab worksheet, use your StudentWorks™ Plus Online.

? Inquiry Launch Lab

FOLDABLES

Make a three-tab book. Label it as shown. Use it to organize your notes on motion.

Displacement	Velocity	Acceleration
	Motion	

THEME FOCUS Motion and Forces
Objects are in motion all around us.

BIG⟨⟨Idea⟩ Motion occurs when an object changes its position.

Reading Preview

Essential Questions

▶ How are distance and displacement different?

▶ How is an object's speed calculated?

▶ What information does a distance-time graph provide?

Review Vocabulary

meter: the SI unit of length, abbreviated m

New Vocabulary

motion
displacement
speed

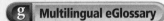 **Multilingual eGlossary**

■ **Figure 1** As the mail truck follows its route, it stops at each mailbox along the street.

Explain *How do you know the mail truck has moved?*

Describing Motion

MAIN ⟨Idea Position describes where an object is, and speed describes how fast the object is moving.

Real-World Reading Link How would you describe the trip from your home to your school? You might include whether you walked or rode a bus, how far away the school was, and how long it took to get there. Scientists describe motion in a similar, but very specific way.

Motion and Position

You do not always need to see something move to know that motion has taken place. For example, suppose you look out a window and see a mail truck stopped next to a mailbox, as shown in **Figure 1**. One minute later, you look out again and see the same truck stopped farther down the street. Although you did not see the truck move, you know it moved because its position relative to the mailbox changed.

Reference points A reference point is needed to determine the position of an object. In **Figure 1**, the reference point might be a mailbox. **Motion** is a change in an object's position relative to a reference point. How you describe an object's motion depends on the reference point that is chosen. For example, the description of the mail truck's motion in **Figure 1** would be different if the reference point were a tree instead of a mailbox.

After a reference point is chosen, a frame of reference can be created. A frame of reference is a coordinate system in which the position of the object is measured. The *x*-axis and *y*-axis of the reference frame are drawn so that they are perpendicular to each other and intersect the reference point.

9:00 A.M.

9:01 A.M.

Coordinate systems **Figure 2** shows a map of the city where the mail truck is delivering mail with a coordinate system drawn on it. The *x*-axis is in the east-west direction, the *y*-axis is in the north-south direction, and each division represents a city block. The post office is located at the origin. The mail truck is located at 3 blocks east (*x* = 3) and 2 blocks north (*y* = 2) of the post office.

Change in Position

Have you ever run a 50-m dash? Describing how far and in what direction you moved was an important part of describing your motion.

Distance In a 50-m dash, each runner travels a total distance of 50 m. The SI unit of distance is the meter (m). Longer distances are measured in kilometers (km). One kilometer is equal to 1,000 m. Shorter distances are measured in centimeters (cm) or millimeters (mm). One meter is equal to 100 cm and to 1,000 mm.

Displacement Suppose a runner jogs to the 50-m mark and then turns around and runs back to the 20-m mark, as shown in **Figure 3.** The runner travels 50 m in the original direction (east) plus 30 m in the opposite direction (west), so the total distance that she ran is 80 m. How far is she from the starting line? The answer is 20 m. Sometimes, you may want to know the change in an object's position relative to the starting point. An object's **displacement** is the distance and direction of the object's change in position. In **Figure 3,** the runner's displacement is 20 m east.

The length of the runner's displacement and the total distance traveled would be the same if the runner's motion were in a single direction. For example, if the runner ran east from the starting line to the finish line without changing direction, then the distance traveled would be 50 m and the displacement would be 50 m east.

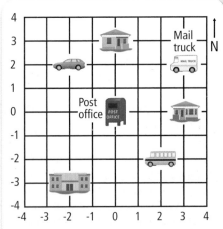

Figure 2 A coordinate system is like a map. The reference point is at the origin, and each object's position can be described with its coordinates.

Identify *the position of the orange car.*

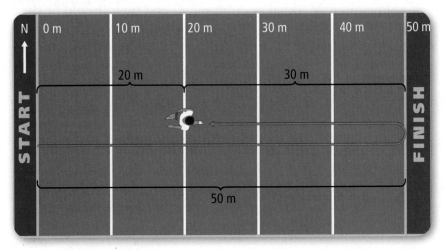

Figure 3 An object's displacement is not the same as the total distance that the object traveled. The runner's displacement is 20 m east of the starting line. However, the total distance the runner traveled is 80 m.

Describe *the difference between the total distance traveled and the displacement.*

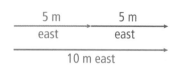
Displacements in the same direction can be added.

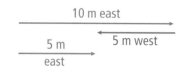
Displacements in opposite directions can be subtracted.

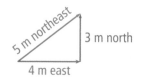
Displacements that are not in the same direction or opposite directions cannot be directly added or subtracted.

■ **Figure 4** These arrows represent the students' walks. The green arrows show the first part of the walk and the purple arrows show the second part. The red arrows show the students' displacements.

FOLDABLES
Incorporate information from this section into your Foldable.

Table 1	Rules for Adding Displacements
1.	Add displacements in the same direction.
2.	Subtract displacements in opposite directions.
3.	Displacements that are not in the same or in opposite directions cannot be directly added together.

Adding displacements

You know that you can add distances together to get the total distance. For example, 2 m + 3 m = 5 m. But how would you add the displacements 5 m east and 10 m east? Directions in math problems are much like units: you can add numbers with like directions. For example, suppose a student walks 5 m east, stops at a crosswalk, and then walks another 5 m east, as shown on the left in **Figure 4.** His displacement is

$$5 \text{ m east} + 5 \text{ m east} = 10 \text{ m east}$$

But what if the directions are not the same? Then compare the two directions. If the directions are exactly opposite, the distances can be subtracted. Suppose a student walks 10 m east, turns around, and walks 5 m west, as shown in the center of **Figure 4.** The size of the displacement would be

$$10 \text{ m} - 5 \text{ m} = 5 \text{ m}$$

The direction of the total displacement is always the direction of the larger displacement. In this case, the larger displacement is east, so the total displacement is 5 m east.

Now suppose the two displacements are not in the same direction or in opposite directions, as illustrated on the right in **Figure 4.** Here, the student walks 4 m east and then 3 m north. The student walks a total distance of 7 m, but the displacement is 5 m in a roughly northeast direction. 4 m east and 3 m north cannot be directly added or subtracted, and they should be discussed separately. The rules for adding displacements are summarized in **Table 1.**

Speed

Think back to the mail truck moving down the street. You could describe the movement by the distance traveled or by the displacement. You might also want to describe how fast the truck is moving. To do this, you need to know how far it travels in a given amount of time. To describe how fast an object moves, scientist use the object's speed. **Speed** is the distance an object travels per unit of time.

Calculating speed Any change over time is called a rate. For example, you could describe how quickly water is leaking from a tank by stating how many liters are lost each hour. This would be the rate of water leakage. If you think of distance as the change in position, then speed is the rate of change in position. Speed can be calculated from this equation.

Speed Equation

$$\text{speed (in meters/second)} = \frac{\text{distance (in meters)}}{\text{time (in seconds)}}$$

$$s = \frac{d}{t}$$

In SI units, distance is measured in meters and time is measured in seconds. Therefore, the SI unit for speed is meters per second (m/s). Sometimes, it is more convenient to express speed in other units, such as kilometers per hour (km/h). **Table 2** shows the speeds of some common objects.

Table 2	Common Speeds	
Motion	**Speed (m/s)**	
Olympic 100-m dash	10 m/s	
Car on city street (35 mph)	16 m/s	
Car on interstate highway (65 mph)	30 m/s	
Commercial airplane	250 m/s	

 **Inquiry**   Virtual Lab

---

EXAMPLE Problem 1

Calculate Speed A car traveling at a constant speed covers a distance of 750 m in 25 s. What is the car's speed?

Identify the Unknown: speed: *s*

List the Knowns: distance: $d = 750\text{ m}$
time: $t = 25\text{ s}$

Set Up the Problem: $s = \dfrac{d}{t} = \dfrac{750\text{ m}}{25\text{ s}}$

Solve the Problem: $s = \dfrac{750\text{ m}}{25\text{ s}} = 30\text{ m/s}$

Check the Answer: 30 m/s is approximately the speed limit on a U.S. interstate highway, so the answer is reasonable.

PRACTICE Problems   Find **Additional Practice Problems** in the back of your book.

1. A passenger elevator travels from the first floor to the 60th floor, a distance of 210 m, in 35 s. What is the elevator's speed?

2. A motorcycle is moving at a constant speed of 40 km/h. How long does it take the motorcycle to travel a distance of 10 km?

3. How far does a car travel in 0.75 h if it is moving at a constant speed of 88 km/h?

4. **Challenge** A long-distance runner is running at a constant speed of 5 m/s. How long does it take the runner to travel 1 km?

 **Review**

Additional Practice Problems

Figure 5 The cyclist's speed varies from 0 km/h to 30 km/h during his trip.

Explain *how you can describe the speed of an object when the speed is changing.*

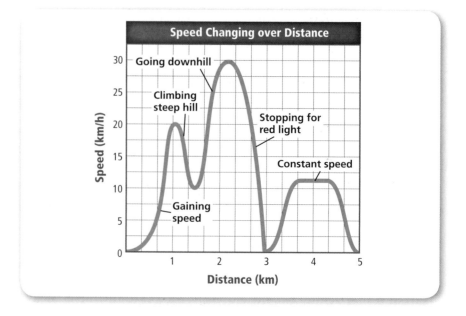

Speed Changing over Distance

Going downhill

Climbing steep hill

Stopping for red light

Constant speed

Gaining speed

Speed (km/h)

Distance (km)

Constant speed Suppose you are in a car traveling on a nearly empty freeway. You look at the speedometer and see that the car's speed hardly changes. If the car neither slows down nor speeds up, the car is traveling at a constant speed. If you are traveling at a constant speed, you can calculate your speed by dividing any distance interval over the time it took you to travel that distance. The speed you calculate will be the same regardless of the interval you choose.

Changing speed Usually, speed is not constant. Think about riding a bicycle for a distance of 5 km. The bicycle's speed will vary, as in **Figure 5.** As you start out, your speed increases from 0 km/h to 20 km/h. You slow down to 10 km/h as you pedal up a steep hill and speed up to 30 km/h going down the other side of the hill. You stop for a red light, speed up again, and move at a constant speed for a while. Finally, you slow down and come to a stop.

Checking your watch, you find that the trip took 15 min. How would you express your speed on such a trip? Would you use your fastest speed, your slowest speed, or some speed between the two? Two common ways of expressing a changing speed are average speed and instantaneous speed.

Average speed Average speed is one way to describe the speed of the bicycle trip. Average speed is the total distance traveled divided by the total time of travel. It can be calculated using the relationships between speed, distance, and time. For the bicycle trip just described, the total distance traveled was 5 km and the total time was $\frac{1}{4}$ h, or 0.25 h. Therefore, the average speed was

$$s = \frac{d}{t} = \frac{5 \text{ km}}{0.25 \text{ h}} = 20 \text{ km/h}$$

Instantaneous speed Suppose you watch a car's speedometer, like the one in **Figure 6**, go from 0 km/h to 80 km/h. A speedometer shows how fast a car is going at one point in time, or at one instant. The speed shown on a speedometer is the instantaneous speed. Instantaneous speed is the speed at a given point in time. When something is speeding up or slowing down, its instantaneous speed is changing. The speed is different at every point in time. If an object is moving with constant speed, the instantaneous speed does not change. The speed is the same at every point in time.

■ **Figure 6** A speedometer gives the car's instantaneous speed. Instantaneous speed is the speed at one instant in time.

✔ **Reading Check** **Identify** two examples of motion in which an object's instantaneous speed changes.

Graphing Motion

The motion of an object over a period of time can be shown on a distance-time graph. For example, the graph in **Figure 7** shows the distance traveled by three swimmers during a 30-minute workout. Time is plotted along the horizontal axis of the graph, and the distance traveled is plotted along the vertical axis of the graph.

Each axis must have a scale that covers the range of numbers to be plotted. In **Figure 7,** the distance scale must range from 0 to 2,400 m and the time scale must range from 0 to 30 min. Next, the *x*-axis is divided into equal time intervals, and the *y*-axis is divided into equal distance intervals.

Once the scales for each axis are in place, the data points can be plotted. In **Figure 7,** there is a data point plotted for each swimmer every two and a half minutes. After plotting the data points, a line is drawn connecting the points.

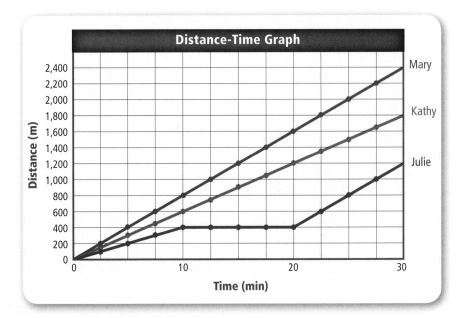

Concepts in Motion

Animation

■ **Figure 7** This graph shows how far each girl swam during a 30 minute workout. Time is divided into 2.5-minute intervals along the *x*-axis. Distance swam is divided into 200-m intervals along the *y*-axis.

Examine *the graph and determine which girl swam the farthest during the workout.*

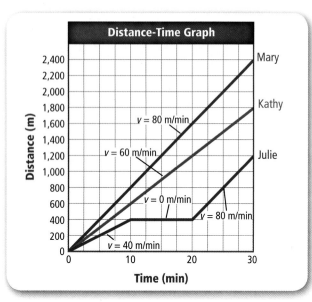

Distance-Time Graph

Mary — v = 80 m/min
Kathy — v = 60 m/min
Julie
v = 0 m/min
v = 80 m/min
v = 40 m/min

Distance (m) / Time (min)

■ **Figure 8** An object's speed is equal to the slope of the line on a distance-time graph.

Identify *the part of the graph that shows one of the swimmers resting for 10 min.*

Speed on distance-time graphs If an object moves with constant speed, the increase in distance over equal time intervals is the same. As a result, the line representing the object's motion is a straight line. For example, look at the graph of the swimmers' workouts in **Figure 8.** The straight red line represents the motion of Mary, who swam with a constant speed of 80 m/min. The green line represents the motion of Julie, who did not swim with a constant speed. She swam with a constant speed of 40 m/min for 10 minutes, rested for 10 minutes, and then swam with a constant speed of 80 m/min for 10 minutes.

The graph shows that the line representing the motion of the faster swimmer is steeper. The steepness of a line on a graph is the line's slope. The slope of a line on a distance-time graph equals the object's speed. Because Mary has a greater speed (80 m/min) than Kathy (60 m/min), the line representing her motion has a larger slope. Now look at the green line representing Julie's motion. During the time she is resting, her line is horizontal. A horizontal line on a distance-time graph has zero slope and represents an object at rest.

Section 1 Review

Section Summary

▶ Motion occurs when an object changes its position relative to a reference point.

▶ Displacement is the distance and direction of a change in position from the starting point.

▶ Speed is the rate at which an object's position changes.

▶ On a distance-time graph, time is the x-axis and distance is the y-axis.

▶ The slope of a line plotted on a distance-time graph is the speed.

5. **MAIN Idea Describe** the trip from your home to school using the words *position, distance, displacement,* and *speed.*

6. **Explain** whether the size of an object's displacement could be greater than the distance the object travels.

7. **Describe** the motion represented by a horizontal line on a distance-time graph.

8. **Describe** the difference between average speed and constant speed.

9. **Think Critically** During a trip, can a car's instantaneous speed ever be greater than its average speed? Explain.

Apply Math

10. **Calculate Speed** Michiko walked a distance of 1.60 km in 30 min. Find her average speed in m/s.

11. **Calculate Distance** A car travels at a constant speed of 30.0 m/s for 0.80 h. Find the total distance traveled in km.

Assessment Online Quiz

Reading Preview

Essential Questions

▶ What is the difference between speed and velocity?

▶ How is the motion of two objects relative to each other described?

▶ How can an object's momentum be calculated?

Review Vocabulary

speed: rate of change of position

New Vocabulary

velocity
momentum

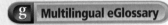

g Multilingual eGlossary

FOLDABLES
Incorporate information from this section into your Foldable.

Velocity and Momentum

MAIN ‹Idea **An object's velocity describes the object's speed and direction of motion.**

Real-World Reading Link Cars, trucks, and many other vehicles can go both backward and forward. The driver must put the car in the proper gear for each direction. What could happen if the driver put the car into reverse instead of forward?

Velocity

You turn on the radio and hear a news story about a hurricane. The storm, traveling at a speed of 20 km/h, is located 500 km east of your location. Should you worry?

Unfortunately, you do not have enough information to answer that question. Knowing only the speed of the storm is not much help. Speed describes only how fast something is moving. To decide whether you need to move to a safer area, you also need to know the direction that the storm is moving. In other words, you need to know the velocity of the storm. **Velocity** includes the speed of an object and the direction of its motion. Velocity has the same units as speed, m/s. If you had been told that the hurricane was traveling straight toward your house at 20 km/h, you would have known to evacuate.

Velocity and speed Because velocity depends on direction as well as speed, the velocity of an object can change even if the speed of the object remains constant. For example, the race cars in **Figure 9** have constant speeds through the turn. Even though the speeds remain constant, their velocities changes because they change direction thoughout the turn.

✓ **Reading Check Describe** how velocity and speed are different.

■ **Figure 9** These cars travel at constant speed, but not with constant velocity. The cars' velocities change because their direction of motion changes.

Figure 10 The two escalators move with a speed of 0.5 m/s. But the left escalator's velocity is 0.5 m/s downward and the right escalator's velocity is 0.5 m/s upward.

VOCABULARY . . . .

WORD ORIGIN
Velocity
from the Latin *veloci–* meaning fast, swift, or rapid.
The students found that the velociraptor's velocity was 10 m/s north. . . . .

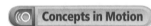

(((○ Concepts in Motion

Animation

Figure 11 Geologic evidence suggests that Earth's surface is changing. The continents have moved slowly over time and are still moving today.

Same speed, different velocities It is possible for two objects to have the same speed but different velocities. For example, the two escalators pictured in **Figure 10** are moving at the same speed but in opposite directions. The speeds of the two sets of passengers are the same, but their velocities are different because they are moving in different directions. Cars traveling in opposite directions on a road with the same speed also have different velocities.

Motion of Earth's Crust

Can you think of something that is moving so slowly that you cannot detect its motion, but you can see evidence of its motion over long periods of time? As you look around the surface of Earth from year to year, its basic structure seems the same. Mountains, plains, and oceans seem to remain unchanged. Yet, if you examined geologic evidence of what Earth's surface looked like over the past 250 million years, you would see that large changes have occurred. **Figure 11** shows how, according to the theory of plate tectonics, the positions of landmasses have changed during this time. Changes in the landscape occur constantly as continents drift slowly over Earth's surface.

These moving plates cause geologic changes, such as the formation of mountain ranges, earthquakes, and volcanic eruptions. The movement of the plates changes the size of the oceans. The Pacific Ocean is getting smaller, and the Atlantic Ocean is getting larger. The plates' movement also changes the shape of the continents as they collide and spread apart.

Plates move so slowly that their speeds are given in units of centimeters per year. Along the San Andreas Fault in California, two plates move past each other with an average speed of about 1 cm per year. The Australian Plate moves faster and pushes Australia north at an average speed of about 17 cm/y. Therefore, the velocity of the Austrailian plate is 17 cm/y north.

About 250 million years ago, the continents formed a super-continent called Pangaea.

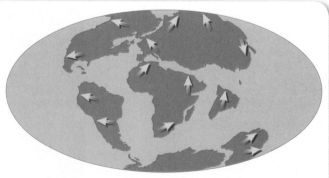

Pangaea separated into smaller pieces. About 66 million years ago, the continents looked like the figure above.

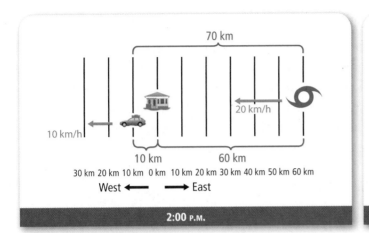

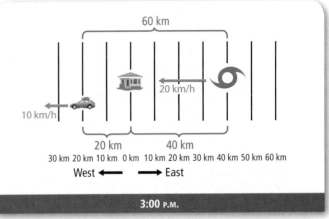

2:00 P.M.

3:00 P.M.

Relative Motion

Have you ever watched cars pass you on the highway? Cars traveling in the same direction often seem to creep by, while cars traveling in the opposite direction seem to zip by. This apparent difference in speeds is because the reference point—your vehicle—is also moving.

The choice of a moving reference point affects how you describe motion. For example, the motion of a hurricane can be described using a stationary reference point, such as a house. **Figure 12** shows the locations and velocities of a hurricane and a car relative to a house at 2:00 P.M. and 3:00 P.M. The distance between the hurricane and the house is decreasing at a rate of 20 km/h. The distance between the house and the car is increasing at a rate of 10 km/h.

How would the description of the hurricane's motion be different if the reference point were a car traveling at 10 km/h west? **Figure 13** shows the motion of the hurricane and the house relative to the car. A person in the car would say that the hurricane is approaching with a speed of 10 km/h and that the house is moving away at a speed of 10 km/h. It is important to notice that **Figure 12** and **Figure 13** show the same changes, but they use different reference points. Velocity and position always depend on the point of reference chosen.

■ **Figure 12** If the house is chosen for the reference point, the car appears to be traveling 10 km/h west and the hurricane appears to be traveling 20 km/h west.

■ **Figure 13** If the car is chosen as the reference point, the hurricane appears to be moving towards the car at 10 km/h and the house is moving away from the car at 10 km/h.

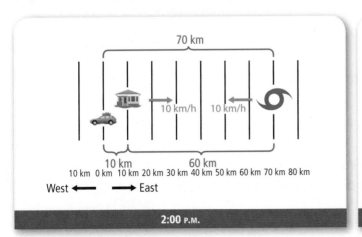

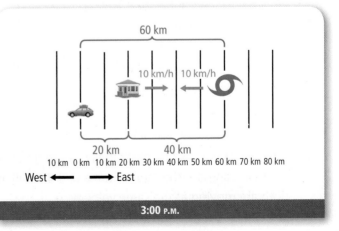

2:00 P.M.

3:00 P.M.

Table 3	Typical Momentums	
Object	Momentum (kg·m/s)	
Tossed baseball	0.15	
Person walking	100	
Car on interstate	45,000	

Inquiry Video Lab

Video What's PHYSICAL and EARTH SCIENCE Got To Do With It?

Momentum

An object is moving at 2 m/s toward a glass vase. Will the vase be damaged in the collision? If the object has a small mass, like a bug, a collision will not damage the vase. But if the object has a larger mass, like a car, a collision will damage the vase.

A useful way of describing both the velocity and mass of an object is to state its momentum. The **momentum** of an object is the product of its mass and velocity. Momentum is usually represented by the symbol p.

> **Momentum Equation**
>
> **momentum** (in kg·m/s) = **mass** (in kg) × velocity (in m/s)
>
> $$p = mv$$

The unit for momentum is kg·m/s. Like velocity, momentum has a size and a direction. An object's momentum is always in the same as the direction as its velocity. **Table 3** shows the sizes of the momentums of some common objects.

EXAMPLE Problem 2

Solve for Momentum At the end of a race, a sprinter with a mass of 80.0 kg has a velocity of 10.0 m/s east. What is the sprinter's momentum?

Identify the Unknown:	momentum: p
List the Knowns:	mass: $m = 80.0$ kg velocity: $v = 10.0$ m/s east
Set Up the Problem:	$p = mv = (80.0$ kg$) \times (10.0$ m/s$)$ east
Solve the Problem:	$p = (80.0$ kg$)(10.0$ m/s$)$ east $= 800.0$ kg·m/s east
Check the Answer:	Our answer makes sense because it is greater than the momentum of a walking person, but much smaller than the momentum of a car on the highway.

PRACTICE Problems

Find **Additional Practice Problems** in the back of your book.

12. What is the momentum of a car with a mass of 1,300 kg traveling north at a speed of 28 m/s?

13. A baseball has a momentum of 6.0 kg·m/s south and a mass of 0.15 kg. What is the baseball's velocity?

14. Find the mass of a person walking west at a speed of 0.8 m/s with a momentum of 52.0 kg·m/s west.

15. **Challenge** The mass of a basketball is three times greater than the mass of a softball. Compare the momentums of a softball and a basketball if they both are moving at the same velocity.

Review

Additional Practice Problems

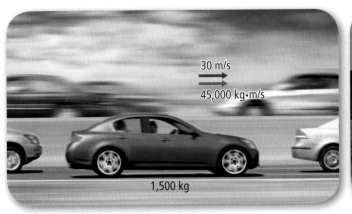

30 m/s

45,000 kg·m/s

1,500 kg

30 m/s

900,000 kg·m/s

30,000 kg

Comparing momentums Think about the car and the truck in **Figure 14**. Which has the larger momentum? The truck does because it has more mass. When two objects travel at the same velocity, the object with more mass has a greater momentum. A difference in momentums is why a car traveling at 2 m/s might damage a porcelain vase, but an insect flying at 2 m/s will not.

Now consider two 1-mg insects. One insect flies at a speed of 2 m/s, and the other flies at a speed of 4 m/s. The second insect has a greater momentum. If two objects have the same mass, the object with the larger velocity has the larger momentum.

■ **Figure 14** Both the car and the truck have a velocity of 30 m/s west, but the truck has a much larger momentum.

Section **2** Review

Section Summary

▶ The velocity of an object includes the object's speed and its direction of motion relative to a reference point.

▶ An object's motion is always described relative to a reference point.

▶ The momentum of an object is the product of its mass and velocity: $p = mv$.

16. **MAIN Idea** **Describe** a car's velocity as it goes around a track at a constant speed.

17. **Explain** why streets and highways have speed limits rather than velocity limits.

18. **Identify** For each of the following news stories, determine whether the object's speed or velocity is given: the world record for the hundred-meter dash is about 10 m/s; the wind today is 30 km/h from the north-west; a 200,000 kg train was traveling north at 70 km/h when it derailed; a car was issued a ticket for traveling at 140 km/h on the interstate.

19. **Think Critically** You are walking toward the back of a bus that is moving forward with a constant velocity. Describe your motion relative to the bus and relative to a point on the ground.

Apply Math

20. **Calculate Momentum** What is the momentum of a 100-kg football player running north at a speed of 4 m/s?

21. **Compare** the momentums of a 6,300-kg elephant walking 0.11 m/s and a 50-kg dolphin swimming 10.4 m/s.

✓ Assessment   Online Quiz

Essential Questions

▶ How are acceleration, time, and velocity related?

▶ What are three ways an object can accelerate?

▶ How can an object's acceleration be calculated?

▶ What are the similarities and differences between straight line motion, circular motion, and projectile motion?

Review Vocabulary

velocity: describes the speed and direction of a moving object

New Vocabulary

acceleration
centripetal acceleration

 **Multilingual eGlossary**

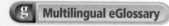 **Video** **BrainPOP**

Acceleration

MAIN Idea Acceleration describes how the velocity of an object is changing with time.

Real-World Reading Link If you have ever ridden on an airplane, you know that the plane taxis at a slow speed. Once on the runway, you are pressed back against your seat as the plane speeds up until it is going fast enough to take off. When the plane speeds up, it is accelerating.

Velocity and Acceleration

You are sitting in a car at a stoplight when the light turns green. The driver steps on the gas pedal and the car starts moving faster and faster. Just as speed is the rate of change of position, **acceleration** is the rate of change of velocity. When the velocity of an object changes, the object is accelerating.

Remember that velocity includes the speed and direction of an object. Therefore, a change in velocity can be either a change in speed or a change in direction. Acceleration occurs when an object changes its speed, its direction, or both.

When you think of acceleration, you probably think of something speeding up. However, an object that is slowing down also is accelerating, as is an object that is changing direction. **Figure 15** shows the three ways an object can accelerate.

✓ **Reading Check Identify** three ways that an object can accelerate.

Like velocity and momentum, acceleration has a direction. If you look at the car in **Figure 15,** you will see that when it is speeding up, its acceleration and velocity are in the same direction. When the car is slowing down, its acceleration is in the opposite direction of its velocity. When the car changes direction, the acceleration is not in the same direction or opposite direction as the car's velocity.

■ **Figure 15** An object, such as this car, is accelerating if it is speeding up, slowing down, or changing direction.

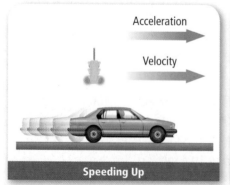

Speeding Up

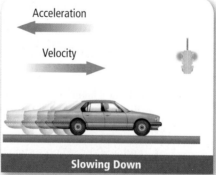

Slowing Down

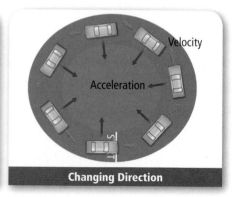

Changing Direction

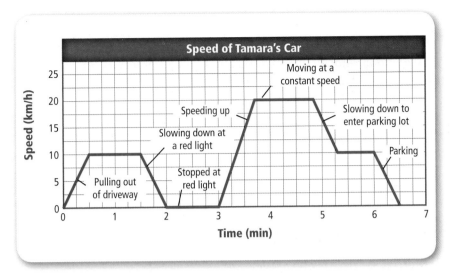

Speed of Tamara's Car

Moving at a
constant speed

Speeding up

Slowing down to
enter parking lot

Slowing down at
a red light

Parking

Pulling out
of driveway

Stopped at
red light

Speed (km/h)

Time (min)

■ **Figure 16** For objects that are speeding up and slowing down, the slope of the line on a speed-time graph is the acceleration.

Identify *the time intervals when Tamara's car is not accelerating.*

Speed-time graphs and acceleration

When an object travels in a straight line and does not change direction, a graph of speed versus time can provide information about an object's acceleration. **Figure 16** shows the speed-time graph of Tamara's car as she drives to the store. Just as the slope of a line on a distance-time graph is the object's speed, the slope of a line on a speed-time graph is the object's acceleration. For example, when Tamara pulls out of her driveway, the car's acceleration is 0.33 km/min², which is equal to the slope of the line from $t = 0$ to $t = 0.5$ min.

Calculating acceleration

Acceleration is the rate of change in velocity. To calculate the acceleration of an object, the change in velocity is divided by the length of the time interval over which the change occurred. The change in velocity is final velocity minus the initial velocity. If the direction of motion does not change and the object moves in a straight line, the size of the change in velocity can be calculated from the change in speed. Then, the acceleration of an object can be calculated from the following equation.

Acceleration Equation

$$\text{acceleration (in meters/second}^2\text{)} = \frac{\text{change in velocity (in meters/second)}}{\text{time (in seconds)}}$$

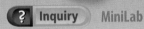

$$a = \frac{v_f - v_i}{t}$$

In SI units, velocity has units of m/s and time has units of s, so the SI unit of acceleration is m/s². In some cases, your calculations will result in a negative acceleration. The negative sign means *in the opposite direction*. For example, an acceleration of −10 m/s² north is the same as 10 m/s² south.

MiniLab

? **Inquiry** MiniLab

Determine the Direction of Acceleration

Procedure 🥽 🖐

1. Read the procedure and safety information, and complete the lab form.

2. **Tape** a **bubble level** onto the top of a **laboratory cart** and center the bubble. Tie a **string** to the front of the cart.

3. Pull the cart forward, and observe the direction of the motion of the bubble.

4. Move the cart at a constant speed, and observe the direction of the motion of the bubble.

5. Allow the cart to coast to a stop, and observe the direction of the motion of the bubble.

Analysis

1. **Relate** the motion of the bubble to the acceleration of the cart.

2. **Predict** how the bubble would move if you tie the string to the back of the cart and repeat the experiment in the opposite direction. Try it.

Calculate Acceleration A skateboarder has an initial velocity of 3 m/s west and comes to a stop in 2 s. What is the skateboarder's acceleration?

Identify the Unknown: acceleration: a

List the Knowns: initial velocity: $v_i = 3$ m/s west
final velocity: $v_f = 0$ m/s west
time: $t = 2$ s

Set Up the Problem: $a = \dfrac{(v_f - v_i)}{t} = \dfrac{(0 \text{ m/s} - 3 \text{ m/s})}{2 \text{ s}}$ west

Solve the Problem: $a = \dfrac{(0 \text{ m/s} - 3 \text{ m/s})}{2 \text{ s}} = -1.5$ m/s² west
The acceleration has a negative sign, so the direction is reversed.

$a = 1.5$ m/s² east

Check the Answer: The magnitude of the acceleration (1.5 m/s²) is reasonable for a skateboard that takes 2 s to slow from 3 m/s to 0 m/s. The acceleration is in the opposite direction of the velocity, so the skateboard is slowing down, as we expected.

PRACTICE Problems

Find **Additional Practice Problems** in the back of your book.

22. An airplane starts at rest and accelerates down the runway for 20 s. At the end of the runway, its velocity is 80 m/s north. What is its acceleration?

23. A cyclist starts at rest and accelerates at 0.5 m/s² south for 20 s. What is the cyclist's final velocity?

24. **Challenge** A ball is dropped and falls with an acceleration of 9.8 m/s² downward. It hits the ground with a velocity of 49 m/s downward. How long did it take the ball to fall to the ground?

> **Review**
> Additional Practice Problems

Motion in Two Dimensions

So far, we have only discussed motion in a straight line. But most objects are not restricted to moving in a straight line. Recall that we cannot add measurements that are not in the same or opposite directions. So, we will discuss motion in each direction separately. For example, suppose a student walked three blocks north and four blocks east. The trip could be described this way: the student walked north for three blocks at 1 m/s and then walked east for four blocks east at 2 m/s.

Recall that objects that change direction are accelerating. For an object that is changing direction, its acceleration is not in the same or opposite direction as its velocity. This means that we cannot use the acceleration equation. Just as with displacement and velocity, accelerations that are not in the same or opposite directions cannot be directly combined.

Circular motion Think about a horse's horizontal motion on a carousel, such as the one in **Figure 17.** The horse moves in a circular path. Its speed remains constant, but it is accelerating because its direction of motion changes. The change in the direction of the horse's velocity is toward the center of the carousel. The horse's velocity is perpendicular to the inward acceleration. Acceleration toward the center of a curved or circular path is called **centripetal acceleration.** In the same way, Earth experiences centripetal acceleration as it orbits the Sun in a nearly circular path.

✓ **Reading Check** **Define** the term *centripetal acceleration.*

Projectile motion If you have tossed a ball to someone, you have probably noticed that thrown objects do not travel in straight lines. They curve downward. That is why quarterbacks, dart players, and archers aim above their targets. Anything that is thrown or shot through the air is called a projectile. Earth's gravity causes projectiles to follow a curved path.

Horizontal and vertical motion When you throw or shoot an object, such as the rubber band in **Figure 18,** the force exerted by your hand gives the object a horizontal velocity. For example, after releasing the rubber band, its horizontal velocity is constant. The rubber band does not accelerate horizontally. If there were no gravity, the rubber band would move along the straight dotted line in **Figure 18.**

However, when you release a rubber band, gravity causes it to accelerate downward. The rubber band has an increasing vertical velocity. The result of these two motions is that the rubber band travels in a curve, even though its horizontal and vertical motions are completely independent of each other.

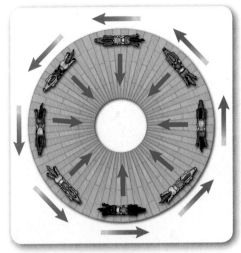

■ **Figure 17** The horizontal speed of the horses in this carousel is constant, but the horses are accelerating because their direction is changing constantly. The acceleration of each horse is toward the center of the circular carousel.

VOCABULARY
ACADEMIC VOCABULARY
Constant
not varying or changing over time; a quantity that does not vary
The constant hum of the fan made it difficult to sleep.

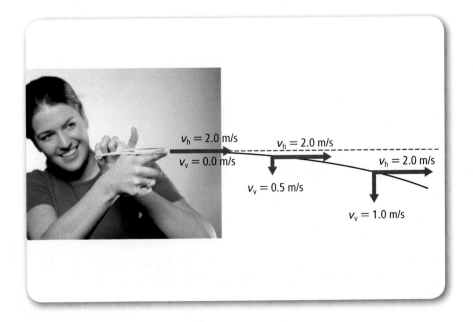

$v_h = 2.0$ m/s $v_h = 2.0$ m/s
$v_v = 0.0$ m/s $v_h = 2.0$ m/s
$v_v = 0.5$ m/s
$v_v = 1.0$ m/s

■ **Figure 18** The student gives the rubber band a horizontal velocity. The horizontal velocity of the rubber band remains constant, but gravity causes the rubber band to accelerate downward. The combination of these two motions causes the rubber band to move in a curved path.

Section 3 • Acceleration **59**

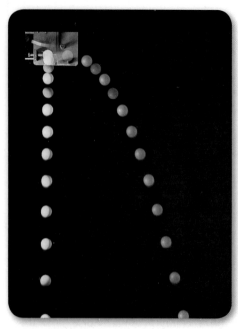

■ **Figure 19** The dropped ball and the thrown ball in this multiflash photograph have the same downward acceleration.

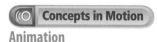
Concepts in Motion
Animation

Throwing and dropping If you were to throw a ball as hard as you could in a perfectly horizontal direction, would it take longer to reach the ground than if you dropped a ball from the same height? Surprisingly, it will not. A thrown ball and a dropped ball will hit the ground at the same time. Both balls in **Figure 19** travel the same vertical distance in the same amount of time. However, the ball thrown horizontally travels a greater horizontal distance than the ball that is dropped.

Amusement park acceleration Riding roller coasters in amusement parks can give you the feeling of danger, but these rides are designed to be safe. Engineers use the laws of physics to design amusement park rides that are thrilling but harmless. Roller coasters are constructed of steel or wood. Because wood is not as strong as steel, wooden roller coasters do not have hills that are as high and as steep as some steel roller coasters have.

The highest speeds and accelerations are usually produced on steel roller coasters. Steel roller coasters can offer multiple steep drops and inversion loops, which give the rider large accelerations. As riders move down a steep hill or an inversion loop, they will accelerate toward the ground due to gravity. When riders go around a sharp turn, they are also accelerated. This acceleration makes them feel as if a force is pushing them toward the side of the car.

Section 3 Review

Section Summary

▶ Acceleration is the rate of change of velocity.

▶ The speed of an object increases if the acceleration is in the same direction as the velocity.

▶ The speed of an object decreases if the acceleration and the velocity of the object are in opposite directions.

▶ If an object is moving in a straight line, the size of the change in velocity equals the final speed minus the initial speed.

▶ Acceleration toward the center of a curved or circular path is called centripetal acceleration.

25. **MAIN Idea** **Describe** the acceleration of your bicycle as you ride it from your home to the store.

26. **Determine** the change in velocity of a car that starts at rest and has a final velocity of 20 m/s north.

27. **Describe** the motion of an object that has an acceleration of 0 m/s².

28. **Think Critically** Suppose a car is accelerating so that its speed is increasing. First, describe the line that you would plot on a speed-time graph for the motion of the car. Then describe the line that you would plot on a distance-time graph.

Apply Math

29. **Calculate Time** A ball is dropped from a cliff and has an acceleration of 9.8 m/s². How long will it take the ball to reach a speed of 24.5 m/s?

30. **Calculate Speed** A sprinter leaves the starting blocks with an acceleration of 4.5 m/s². What is the sprinter's speed 2 s later?

✓ **Assessment** Online Quiz

LAB   Motion Graphs

Objectives

- **Measure** the position of a moving object.
- **Create** a distance-time graph.
- **Use** a distance-time graph to explain how an object is changing speeds.

Background: Cars are equipped with speedometers that allow the driver to monitor the car's speed. Toy cars, however, do not have speedometers. How would you determine the speed of a toy car? In this activity, you will graph the motion of a toy car. Your distance-time graph will allow you to determine whether the toy car is speeding up, slowing down, or moving at a constant speed.

Question: *How is the changing speed of an object represented on a distance-time graph?*

Preparation

Materials

camera with video capabilities
windup toy car
meterstick

Safety Precautions

Procedure

1. Read the procedure and safety information, and complete the lab form.
2. Make a data table to record the position and speed of the toy car every 0.1 s.
3. Mark a starting line on the lab table or the surface recommended by your teacher.
4. Place a meterstick parallel to the path that the toy car will take. Have one of the members of your group get ready to operate the camera.
5. Place the toy car at the starting line.

6. Use the camera to record a video of the toy car's motion.
7. Set the camera to play the video frame by frame. Replay the video for 0.5 s, stopping to take a measurement every 0.1 s.
8. Determine the toy car's position for each 0.1 s time interval by reading the meterstick in the video. Record it in the data table.

Conclude and Apply

1. **Draw** a distance-time graph for the toy car using the data that you collected.
2. **Calculate** the toy car's speed for each interval.
3. **Rank** the speeds of each 0.1 s interval. On your graph, label the fastest as 1 and the slowest as 5.
4. **Identify** when the toy car's speed increased, decreased, and remained constant. *(Hint: How does the slope of the line change between intervals?)*
5. **Infer** how you might use a distance-time graph to check that a car's speedometer is working.

Compare your graph to those made by your classmates. Discuss possible reasons why the graphs might be different.

The Momentum of Colliding Objects

Objectives

- **Observe** and **calculate** the momentum of different balls.
- **Compare** the results of collisions involving different amounts of momentum.

Background: In bowling, the ball's momentum is very important. The bowler must ensure that the ball will travel toward the pins and not into the gutter. The size of the momentum also matters. If the momentum is too small, the ball will knock over very few pins.

Question: *How do the mass and velocity of a moving object affect its momentum?*

Preparation

Materials

meterstick	stopwatch
softball	racquetball
tennis ball	baseball
masking tape	balance
trough	

Safety Precautions

Procedure

1. Read the procedure and safety information, and complete the lab form.
2. Copy the data table.
3. Use the balance to measure the mass of the racquetball, tennis ball, and baseball. Record these masses in your data table.
4. Measure a 2-m distance on the floor, and mark it with two pieces of masking tape. Arrange the trough so that it begins at one line of tape and extends about a meter beyond the other line of tape.
5. Place the softball in the trough over the piece of tape. Starting from the other piece of tape, slowly roll the racquetball along the trough toward the softball.

? Inquiry Lab

Data Table

Action	Time	Velocity	Mass	Momentum	Distance Softball Moved
Racquetball rolled slowly					
Racquetball rolled quickly					
Tennis ball rolled slowly					
Tennis ball rolled quickly					
Baseball rolled slowly					
Baseball rolled quickly					

6. Use a stopwatch to time how long it takes the racquetball to roll the 2-m distance and hit the softball. Record this time in your data table.

7. Measure and record the distance that the softball moved.

8. Repeat steps 5–7, rolling the racquetball quickly.

9. Repeat steps 5–7, rolling the tennis ball slowly and then quickly.

10. Repeat steps 5–7, rolling the baseball slowly and then quickly.

Analyze Your Data

1. **Calculate** the momentum of the rolled ball for each trial using the formula $p = mv$. Record your calculations in the data table.

2. **Graph** the relationship between the momentum of each ball and the distance that the softball moved. The x-axis should be momentum (kg·m/s), and the y-axis should be distance (m).

Conclude and Apply

1. **Infer** from your graph how the distance that the softball moves after each collision depends on the momentum of the ball that hits it.

2. **Describe** How do an object's velocity and mass affect the amount of momentum that it has?

3. **Explain** why bowling balls have such a large mass. What would happen if you tried to bowl with a table tennis ball? Explain.

4. **Infer** When you bowl, should you roll the ball gently? Explain.

COMMUNICATE YOUR DATA

Make a Graph As a class, make a momentum-distance graph using data from everyone in the class. Discuss how this graph is similar to and different from the graphs made by individual groups.

In the Field

No Driver? No Worries!

The diesel-powered car nicknamed Stanley faced a difficult mission—drive more than 200 km across the Mojave Desert in less than ten hours, navigating narrow tunnels, dozens of sharp turns, and a winding mountain pass with sheer drops on each side. The real challenge? Stanley, shown in **Figure 1,** was not allowed to have a driver.

Grand challenge This was the setting for the 2005 DARPA (Defense Advanced Research Projects Agency) Grand Challenge, a competition won by Stanley and the Stanford Racing Team. The competition was part of an ongoing effort to develop autonomous ground vehicles—vehicles that drive and navigate without a remote control or a human driver. Using autonomous vehicles in combat situations keeps soldiers off the battlefield and protects lives.

Physics at the wheel How can a car drive itself? To drive safely, it must interpret the environment, evaluate its position relative to the destination, navigate obstacles, and control speed and direction of motion. Light beams emitted by lasers and radio waves given off by a radar unit bounce off surrounding objects and geographical features. Reflection times determine distances between the car and these objects and help evaluate changes in position and terrain.

Additional position data is provided by the Global Positioning System (GPS) and sensors that measure wheel rotation and direction of motion. Stanley's sophisticated computer integrates all incoming data, compares it to a map of the route, and makes necessary adjustments in steering, throttling, and braking.

Figure 1 Stanley, an autonomous vehicle designed by a team from Stanford University, won the 2005 DARPA Grand Challenge.

Eliminating dangerous driving Is this technology useful in the civilian world? Human drivers face many distractions. Factors like cell phone use and unexpected traffic situations contribute to almost 40,000 deaths in U.S. traffic accidents each year. But autonomous cars would not be distracted and could make rapid course and speed adjustments and reduce accidents.

Participants in the 2007 DARPA Urban Challenge proved that autonomous cars could function in a city setting. This competition required complex maneuvers like merging into traffic, parking, negotiating stop signs, and rerouting around unexpected roadblocks. Ultimately, cars like Stanley could save lives on the battlefield and in our communities.

> **WebQuest**
>
> ## Design an Advertisement
> **Brainstorm the potential benefits of using automated cars in a city setting. Use your ideas to design a poster to advertise an automated car service. Share your poster with the class.**

The motion of an object can be described by the distance it travels, its displacement, speed, velocity, momentum, and acceleration.

BIG ⟨Idea Motion occurs when an object changes its position.

Section 1 Describing Motion

displacement (p. 45)
motion (p. 44)
speed (p. 46)

MAIN ⟨Idea Position describes where an object is, and speed describes how fast the object is moving.

- Motion occurs when an object changes its position relative to a reference point.
- Displacement is the distance and direction of a change in position from the starting point.
- Speed is the rate at which an object's position changes.
- On a distance-time graph, time is the x-axis and distance is y-axis.
- The slope of a line plotted on a distance-time graph is the speed.

Section 2 Velocity and Momentum

momentum (p. 54)
velocity (p. 51)

MAIN ⟨Idea An object's velocity describes the object's speed and direction of motion.

- The velocity of an object includes the object's speed and its direction of motion relative to a reference point.
- An object's motion is always described relative to a reference point.
- The momentum of an object is the product of its mass and velocity: $p = mv$.

Section 3 Acceleration

acceleration (p. 56)
centripetal acceleration (p. 59)

MAIN ⟨Idea Acceleration describes how the velocity of an object is changing with time.

- Acceleration is the rate of change of velocity.
- The speed of an object increases if the acceleration is in the same direction as the velocity.
- The speed of an object decreases if the acceleration and the velocity of the object are in opposite directions.
- If an object is moving in a straight line, the size of the change in velocity equals the final speed minus the initial speed.
- Acceleration toward the center of a curved or circular path is called centripetal acceleration.

Use Vocabulary

Compare and contrast the following pairs of terms.

31. speed—velocity

32. motion—displacement

33. velocity—momentum

34. acceleration—velocity

Check Concepts

35. Which of the following do you calculate when you divide the total distance traveled by the total travel time?
   A) average speed
   B) constant speed
   C) variable speed
   D) instantaneous speed

36. Which of the following is the SI unit of acceleration?
   A) s/km^2 **C)** m/s^2
   B) km/h   **D)** cm/s

37. Which of the following is not used in calculating acceleration?
   A) initial velocity
   B) average speed
   C) time interval
   D) final velocity

38. A car, a bicycle, a mouse, and a bug have the same velocity. Which has the greatest momentum?
   A) the car
   B) the bicycle
   C) the mouse
   D) the bug

39. In which of the following conditions does the car not accelerate?
   A) A car moves at 80 km/h on a flat, straight highway.
   B) The car slows from 80 km/h to 35 km/h.
   C) The car turns a corner.
   D) The car speeds up from 35 km/h to 80 km/h.

40. How is speed defined?
   A) acceleration/time
   B) change in velocity/time
   C) distance/time
   D) displacement/time

41. Which best describes why projectiles move in a curved path?
   A) They have constant horizontal velocity and vertical acceleration.
   B) They have horizontal acceleration and constant vertical velocity.
   C) They have horizontal momentum and constant vertical velocity.
   D) They have horizontal acceleration and vertical momentum.

Interpret Graphics

Use the table below to answer question 42.

Distance-Time for Runners				
Time (s)	1	2	3	4
Sally's distance (m)	2	4	6	8
Alonzo's distance (m)	1	2	2	4

42. Make a distance-time graph that shows the motion of both runners. What is the average speed of each runner? Which runner stops briefly? Over what time interval do they both have the same speed?

✓ Assessment   **Online Test Practice**

43. **THEME FOCUS** Copy and complete this concept map on motion.

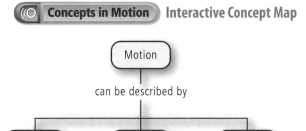

Concepts in Motion Interactive Concept Map

Think Critically

44. **Explain** Why is knowing only a hurricane's speed not enough information to be able to warn people to evacuate?

45. **Evaluate** Which of the following represents the greatest speed: 20 m/s, 200 cm/s, or 0.2 km/s?

46. **Recognize** Acceleration can occur when a car is moving at constant speed. What must cause this acceleration?

47. **BIG Idea Determine** If you walked 20 m, took a book from a table, and walked back to your seat, what are the distance you traveled and your displacement?

48. **Explain** When you are describing the rate that a race car goes around a track, should you use the term *speed* or *velocity* to describe the motion?

Apply Math

49. **Calculate Speed** A cyclist must travel 800 km. How many days will the trip take if the cyclist travels 8 h/day at an average speed of 16 km/h?

50. **Calculate Acceleration** A satellite's speed is 5,000 m/s. After 1 min, it is 10,000 m/s. What is the satellite's acceleration?

51. **Calculate Displacement** A cyclist leaves home and rides due east for a distance of 45 km. She returns home on the same bike path. If the entire trip takes 4 h, what is her average speed? What is her displacement?

52. **Calculate Velocity** The return trip of the cyclist in question 51 took 30 min longer than her trip east, although her total time was still 4 h. What was her velocity in each direction?

Use the figure below to answer question 53.

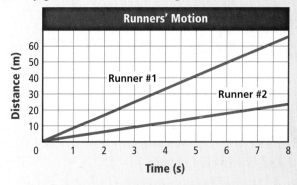

53. **Interpret a Graph** Use the graph to determine which runner had the greater speed.

54. **Calculate Mass** Find the mass of a car that has a speed of 30 m/s and a momentum of 45,000 kg·m/s.

Standardized Test Practice

Multiple Choice

Record your answers on the answer sheet provided by your teacher or on a sheet of paper.

1. If the speed of sound during a thunderstorm is 330 m/s, how long does it take for the sound of thunder to travel 1485 m?
   A. 45 s   C. 4900 s
   B. 4.5 s   D. 0.22 s

Use the figure below to answer questions 2–4.

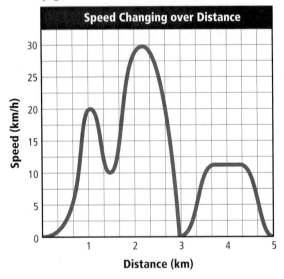

2. The graph shows how a cyclist's speed changed over a 0.25-h trip. What is the cyclist's average speed?
   A. 2 km/h   C. 20 km/h
   B. 30 km/h   D. 8 km/h

3. Once the trip was started, how many times did the cyclist stop?
   A. 0   C. 2
   B. 4   D. 5

4. What was the fastest speed that the cyclist traveled?
   A. 20 km/h
   B. 30 km/h
   C. 12 km/h
   D. 10 km/h

5. A skier is going down a hill at a speed of 9 m/s. The hill gets steeper and her speed increases to 18 m/s in 3 s. What is her acceleration?
   A. 9 m/s^2 C. 27 m/s^2
   B. 3 m/s^2 D. 6 m/s^2

6. Which of the following best describes an object with constant velocity?
   A. It is changing direction.
   B. Its acceleration is increasing.
   C. Its acceleration is zero.
   D. Its acceleration is negative.

Use the table below to answer questions 7–9.

Runner	Distance Covered (km)	Time (min)
Ling-Ling	12.5	42
LaToya	7.8	38
Bill	10.5	32
José	8.9	30

7. What is Ling-Ling's average speed?
   A. 0.3 km/min   C. 3.0 km/min
   B. 530 km/min   D. 3.4 km/min

8. Which runner has the fastest average speed?
   A. Ling-Ling   C. Bill
   B. LaToya   D. José

9. If all four runners have the same mass, who has the smallest momentum?
   A. Ling-Ling   C. Bill
   B. LaToya   D. José

10. The movement of the Australian plate pushes Australia north at an average speed of about 17 cm per year. What will Australia's displacement be in meters in 1,000 years?
    A. 170 m north   C. 1,700 m north
    B. 170 m south   D. 1,700 m south

Short Response

Record your answers on the answer sheet provided by your teacher or on a sheet of paper.

Use the figure below to answer questions 11 and 12.

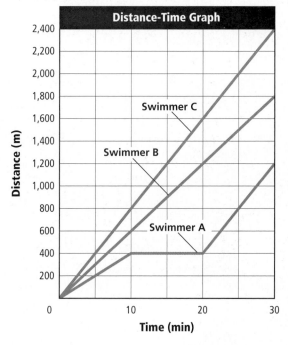

Distance-Time Graph

11. The graph shows the motion of three swimmers during a 30-min workout. Which swimmer had the highest average speed over the 30-min time interval? Explain.

12. Did all of the swimmers swim at a constant speed? Explain how you know.

13. If a car has a velocity of 40 km/h west and then comes to a stop in 5 s, what is its acceleration in m/s^2?

Extended Response

Record your answers on a sheet of paper.

14. Describe three ways that your velocity could change as you jog along a park's path.

15. Where would you place the location of a reference point in order to describe the motion of a space probe traveling from Earth to Jupiter? Explain your choice.

16. Two cars approach each other. How does the speed of one car relative to the other compare with speed of each car relative to the ground?

Use the figure below to answer question 17.

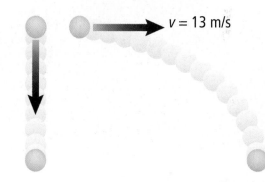

$v = 13$ m/s

17. Two balls are at the same height and released at the same time. One ball is dropped and hits the ground 5 s later. The other initially moves horizontally. When does the second ball hit the ground? How far does it travel horizontally?

NEED EXTRA HELP?																	
If You Missed Question . . .	1	2	3	4	5	6	7	8	9	10	11	12	13	14	15	16	17
Review Section . . .	1	1	1	1	3	3	1	1	2	2	1	1	3	3	2	2	3

CHAPTER 3

Forces and Newton's Laws

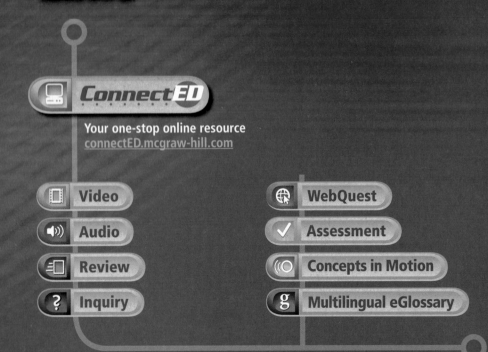

ConnectED

Your one-stop online resource
connectED.mcgraw-hill.com

- Video
- Audio
- Review
- Inquiry

- WebQuest
- Assessment
- Concepts in Motion
- Multilingual eGlossary

Launch Lab
Force, Mass, and Acceleration

You have probably noticed that it is more difficult to lift a bookbag containing three books than it is to lift a bookbag containing one book. What effect does mass have on how much force is needed to change an object's motion?

For a lab worksheet, use your StudentWorks™ Plus Online.

? Inquiry Launch Lab

FOLDABLES®

Make a three-tab book. Label it as shown. Use it to organize your notes on Newton's laws of motion.

| 1st law of motion | 2nd law of motion | 3rd law of motion |

Newton's Laws

THEME FOCUS Motion and Forces
Objects that interact exert forces on each other.

BIG Idea A force is a push or pull.

Section 1 • Forces

Section 2 • Newton's Laws of Motion

Section 3 • Using Newton's Laws

Reading Preview

Essential Questions

▶ How are force and motion related?

▶ How is the net force on an object determined?

▶ Why is there friction between objects?

▶ What is the difference between mass and weight?

Review Vocabulary

mass: amount of matter in an object

New Vocabulary

force
net force
friction
gravity
field
weight

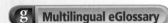 **Multilingual eGlossary**

Forces

MAIN ⟨Idea⟩ Unbalanced forces change motion.

Real-World Reading Link Have you ridden a bike down a steep hill? You might have noticed that you went faster and faster or that you had to brake very hard to stop the bike. In both cases, you and your bike were accelerating. What caused that acceleration?

What is force?

Catching a basketball and hitting a baseball with a bat are examples of applying force to an object. A **force** is a push or a pull. In both examples, the applied force changes the movement of the ball. Sometimes, it is obvious that a force has been applied. But other forces are not as noticeable. For instance, are you conscious of the force that the floor exerts on your feet? Can you feel the force of the atmosphere pushing against your body or gravity pulling on your body? Think about all of the forces that you exert in a day. Every push, pull, stretch, or bend results in a force being applied to an object.

Changing motion What happens to the motion of an object when you exert a force on it? A force can cause the motion of an object to change. Think of kicking a soccer ball, as shown in **Figure 1**. The player's foot strikes the ball with a force that causes the ball to stop and then move in the opposite direction. If you have played billiards, you know that you can force a ball at rest to roll into a pocket by striking it with another ball. The force of the moving ball causes the ball at rest to move in the direction of the force. In each case, the velocity of the ball was changed by a force.

■ **Figure 1** When the player kicks the soccer ball, she is exerting a force on the ball. This kick will cause the ball's motion to change.

Net force When two or more forces act on an object at the same time, the forces combine to form the net force. The **net force** is the sum of all of the forces acting on an object. Forces have a direction, so they follow the same addition rules as displacement, as listed in **Table 1**. Forces are measured in the SI unit of newtons (N). A force of about 3 N is needed to lift a full can of soda at a constant speed.

Unbalanced forces Look at **Figure 2A**. The students are each pushing on the box in the same direction. These forces are combined, or added together, because they are exerted on the box in the same direction. The students in **Figure 2B** are pushing in opposite directions. Here, the direction of the net force is the same as the direction of the larger force. In other words, the student who pushes harder causes the box to move in the direction of that push. The net force will be the difference between the two forces because they are in opposite directions. In **Figure 2A** and **Figure 2B,** the net force had a value that was not zero and the box moved. The forces that the students applied are considered unbalanced forces.

Balanced forces Now suppose that the students were pushing with the same size force but in opposite directions, as shown in **Figure 2C**. The net force on the box is zero because the two forces cancel each other. Forces on an object that are equal in size and opposite in direction are called balanced forces. Unbalanced forces cause changes in motion. Balanced forces do not cause a change in motion.

Table 1	Rules for Adding Forces
1.	Add forces in the same direction.
2.	Subtract forces in opposite directions.
3.	Forces not in the same direction or in opposite directions cannot be directly added together.

? Inquiry **Video Lab**

■ **Figure 2** Forces can be balanced or unbalanced.

Identify *another example of unbalanced forces and another example of balanced forces.*

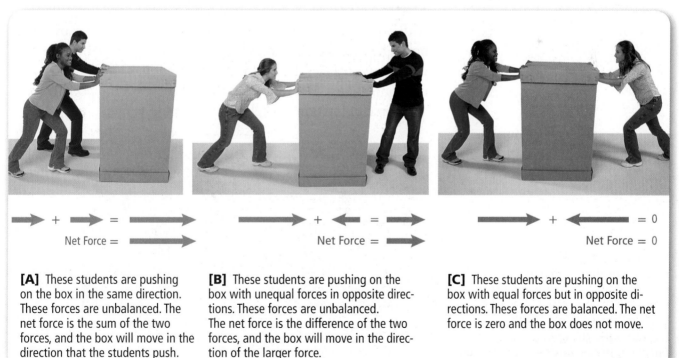

Net Force = ⟹

Net Force = ⟹

Net Force = 0

[A] These students are pushing on the box in the same direction. These forces are unbalanced. The net force is the sum of the two forces, and the box will move in the direction that the students push.

[B] These students are pushing on the box with unequal forces in opposite directions. These forces are unbalanced. The net force is the difference of the two forces, and the box will move in the direction of the larger force.

[C] These students are pushing on the box with equal forces but in opposite directions. These forces are balanced. The net force is zero and the box does not move.

Friction

Suppose you give a skateboard a push with your hand. After you let go, the skateboard slows down and eventually stops. Because the skateboard's motion is changing as it slows down, there must be a force acting on it. The force that slows the skateboard is called friction. **Friction** is the force that opposes the sliding motion of two surfaces that are touching each other.

What causes friction? Would you believe that the surface of a highly polished piece of metal is rough? Surfaces that appear smooth actually have many bumps and dips. These bumps and dips can be seen when the surface is examined under a microscope, as shown in **Figure 3.** If two surfaces are in contact, welding or sticking occurs where the bumps touch each other. These microwelds are the source of friction. To move one surface over the other, a force must be applied to break the microwelds.

Figure 3 The surface of this teapot looks and feels smooth, but it is rough at the microscopic level.

Reading Check **Describe** the source of friction.

The amount of friction between two surfaces depends on the kinds of surfaces and the force pressing the surfaces together. Rougher surfaces have more bumps and can form more microwelds, increasing the amount of friction. In addition, a larger force pushing the two surfaces together will cause more of the bumps to come into contact, as shown in **Figure 4.** The microwelds will be stronger, and a greater force must be applied to the object to break the microwelds.

Static friction Suppose you have a cardboard box filled with books, such as the one in **Figure 5,** and want to move that box. The box is resting on what seems to be a smooth floor, but when you push on the box, it does not budge. The box experiences no change in motion, so the net force on the box is zero. The force of friction cancels your push.

Figure 4 Friction is caused by microwelds that form between two surfaces. Microwelds are stronger when the two surfaces are pushed together with a larger force.

Explain *how the area of contact between the surfaces changes when they are pushed together.*

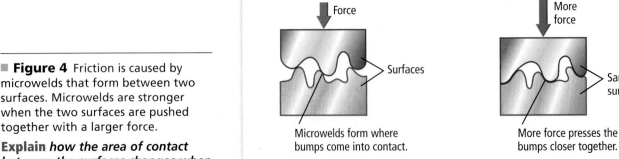

Force

Surfaces

Microwelds form where bumps come into contact.

More force

Same two surfaces

More force presses the bumps closer together.

Static friction balances the applied force. The box remains at rest and does not accelerate.

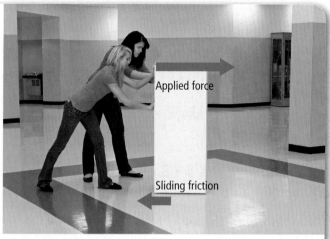

Sliding friction and the applied force are unbalanced. The box accelerates to the right.

This type of friction is called static friction. Static friction prevents two surfaces from sliding past each other and is due to the microwelds that have formed between the bottom of the box and the floor. Your push is not large enough to break the microwelds, and the box does not move, as shown in **Figure 5.**

Sliding friction If you and a friend push together, as shown on the right in **Figure 5,** the box moves. Together, you and your friend have exerted enough force to break the microwelds between the floor and the bottom of the box. But if you stop pushing, the box quickly comes to a stop. To keep the box moving, you must continually apply a force. This is because sliding friction opposes the motion of the box as the box slides across the floor. Sliding friction opposes the motion of two surfaces sliding past each other and is caused by microwelds constantly breaking and forming as the objects slide past each other. The force of sliding friction is usually smaller than the force of static friction.

Rolling friction You may think of friction as a disadvantage. But wheels, like the ones shown in **Figure 6,** would not work without friction. As a wheel rolls, static friction acts over the area where the wheel and surface are in contact. This special case of static friction is sometimes called rolling friction.

You may have seen a car that was stuck in snow, ice, or mud. The driver steps on the gas, but the wheels just spin without the car moving. The force used to rotate the tires is larger than the force of static friction between the wheels and the ground, so the tires slide instead of gripping the ground. Spreading sand or gravel on the surface increases the friction until the wheels stop slipping and begin rolling. When referring to tires on vehicles, people often use the term *traction* instead of friction.

■ **Figure 5** Friction opposes the sliding motion of two surfaces that are touching each other.

Describe *the net force on each box.*

VOCABULARY ....

WORD ORIGIN

Friction
   from the Latin *fricare* meaning "to rub"
   Oil can be used to reduce friction. ......

■ **Figure 6** Rolling friction between the in-line skate's wheels and the pavement keeps the wheels from slipping.

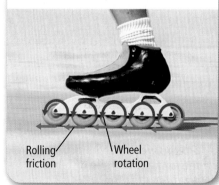

Rolling friction   Wheel rotation

Gravity

At this moment, you are exerting an attractive force on everything around you—your desk, your classmates, and even the planet Jupiter, millions of kilometers away. This attractive force acts on all objects with mass and is called gravity. **Gravity** is an attractive force between any two objects that depends on the masses of the objects and the distance between them.

Gravity is one of the four basic forces. These forces are called the fundamental forces. The other basic forces are the electromagnetic force, the strong nuclear force, and the weak nuclear force. Gravity acts on all objects with mass, and the electromagnetic force acts on all charged particles. Both gravity and the electromagnetic force have an infinite range. The nuclear forces only affect particles in the nuclei of atoms.

The law of universal gravitation

In the 1660s, British scientist Isaac Newton used data on the motions of the planets to find the relationship between the gravitational force between two objects, the objects' masses, and the distance between them. This relationship is called the law of universal gravitation and can be written as the following equation.

$$F = G \frac{m_1 m_2}{d^2}$$

In this equation, G is the universal gravitational constant, and d is the distance between the centers of the two masses, m_1 and m_2. The law of universal gravitation states that the gravitational force increases as the mass of either object increases and as the objects move closer, as shown in **Figure 7.** The force of gravity between any two objects can be calculated if their masses and the distance between them are known.

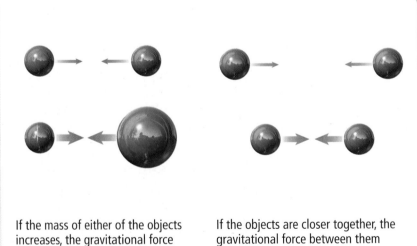

■ Figure 7 The law of universal gravitation states that the gravitational force between two objects depends on their masses and the distance between them.

If the mass of either of the objects increases, the gravitational force between them increases.

If the objects are closer together, the gravitational force between them increases.

Gravity and you The law of universal gravitation explains why you feel Earth's gravity but not the Sun's gravity or this book's gravity. While the Sun has much more mass than Earth, the Sun is too far away to exert a noticeable gravitational attraction on you. And while this book is close, it does not have enough mass to exert an attraction that you can feel. Only Earth is both close enough and has a large enough mass that you can feel its gravitational attraction.

The range of gravity According to the law of universal gravitation, the gravitational force between two masses decreases rapidly as the distance between the masses increases. For example, if the distance between two objects increases from 1 m to 2 m, the gravitational force between them becomes one-fourth as large. If the distance increases from 1 m to 10 m, the gravitational force between the objects is one-hundredth as large. However, no matter how far apart two objects are, the gravitational force between them never completely goes to zero. Because the gravitational force between two objects never disappears, gravity is called a long-range force.

✓ **Reading Check Explain** why gravity is called a long-range force.

The gravitational field Because contact between objects is not required, gravity is sometimes discussed as a field. A **field** is a region of space that has a physical quantity (such as a force) at every point. All objects are surrounded by a gravitational field. **Figure 8** shows that Earth's gravitational field is strongest near Earth and becomes weaker as the distance from Earth increases. The strength of the gravitational field, represented by the letter g, is measured in newtons per kilogram (N/kg).

▢ **Review**

Additional Practice Problems

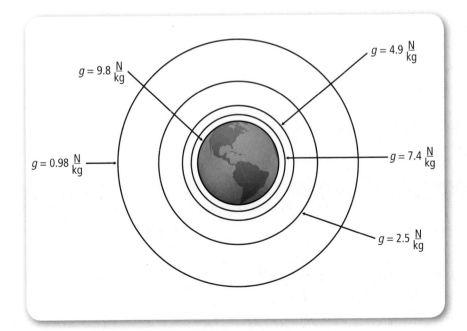

$g = 9.8 \frac{N}{kg}$

$g = 4.9 \frac{N}{kg}$

$g = 0.98 \frac{N}{kg}$

$g = 7.4 \frac{N}{kg}$

$g = 2.5 \frac{N}{kg}$

■ **Figure 8** Earth's gravitational field exists at all points in space. It is strongest near the surface and decreases in strength as one moves away from Earth.

Table 2	Weight of Common Objects on Earth	
Object	**Weight**	
Cell phone	1 N	
Backpack full of books	100 N	
Jumbo jet	3.4 million N	

Weight The gravitational force exerted on an object is the object's **weight.** The universal law of gravitation can be used to calculate weight, but scientists use a simplified version of this equation that combines m_1, d^2, and G into a single number called the gravitational strength, g.

Weight Equation

$$\text{weight (N)} = \text{mass (kg)} \times \text{gravitational strength (N/kg)}$$
$$F_g = mg$$

We use F_g for weight because weight is the force due to gravity. Weight has units of newtons (N) because it is a force. The g in the subscript stands for *gravity*. The gravitational strength, g, has the units N/kg. Recall that $1\ N = 1\ kg \cdot m/s^2$. So, g can also be written with units of m/s^2.

Weight and mass Weight and mass are not the same. Weight is a force, and mass is a measure of the amount of matter an object contains. But, according to the weight equation, weight and mass are related. Weight increases as mass increases.

Weight on Earth We often need to know an object's weight on Earth. Near Earth's surface, m_1, from the law of universal gravitation, is Earth's mass and d is Earth's radius. As a result, $g = 9.8$ N/kg. **Table 2** lists the weights of some objects on Earth.

EXAMPLE Problem 1

Solve for Weight An elephant has a mass of 5,000 kg. What is the elephant's weight?

Identify the Unknown: weight: F_g

List the Knowns: mass: m = **5,000 kg**
     gravitational strength: g = **9.8 N/kg**

Set Up the Problem: $F_g = mg$

Solve the Problem: F_g = **(5,000 kg)(9.8 N/kg) = 49,000 N**

Check the Answer: The gravitational strength is about 10 N/kg, so we would expect the elephant's weight to be about 50,000 kg. Our answer (F_g = 49,000 N) makes sense.

PRACTICE Problems

Find **Additional Practice Problems** in the back of your book.

1. A squirrel has a mass of 0.5 kg. What is its weight?

2. A boy weighs 400 N. What is his mass?

3. **Challenge** An astronaut has a mass of 100 kg and has a weight of 370 N on Mars. What is the gravitational strength on Mars?

Review

Additional Practice Problems

Weight away from Earth An object's weight usually refers to the gravitational force between the object and Earth. But the weight of an object can change, depending on the gravitational force on the object. For example, the gravitational strength on the Moon is 1.6 N/kg, about one-sixth as large as Earth's gravitational strength. As a result, a person, such as the astronaut in **Figure 9,** would weigh only about one-sixth as much on the Moon as on Earth.

Finding other planets Earth's motion around the Sun is affected by the gravitational pulls of the other planets in the solar system. In the same way, the motion of every planet in the solar system is affected by the gravitational pulls of all of the other planets.

In the 1840s, the most distant planet known was Uranus. The motion of Uranus calculated from the law of universal gravitation disagreed slightly with its observed motion. Some astronomers suggested that there must be an undiscovered planet affecting the motion of Uranus. Using the law of universal gravitation and the laws of motion, two astronomers independently calculated the orbit of this planet. As a result of these calculations, the planet Neptune was found in 1846.

■ **Figure 9** Although the astronaut has the same mass on the Moon, he weighs less than he does on Earth. He can take longer steps and jump higher than on Earth.

Section 1 Review

Section Summary

▶ A force is a push or a pull on an object.

▶ The net force on an object is the combination of all of the forces acting on the object.

▶ Unbalanced forces cause the motion of objects to change.

▶ Friction is the force that opposes the sliding motion of two surfaces that are in contact.

▶ Gravity is an attractive force between all objects that have mass.

4. **MAIN Idea** **Describe** two forces that would change the motion of a bicycle traveling along a road.

5. **Explain** Can there be forces acting on an object if the object is at rest? Must there be an unbalanced force acting on a moving object? Explain your answers.

6. **Explain** Why does coating surfaces with oil reduce friction between the surfaces?

7. **Distinguish** between the mass of an object and the object's weight.

8. **Think Critically** Suppose Earth's mass increased but Earth's diameter did not change. How would the gravitational strength near Earth's surface change?

Apply Math

9. **Calculate Weight** On Earth, what is the weight of a large-screen TV that has a mass of 75 kg?

10. **Calculate Net Force** Two students push on a box in the same direction, and one student pushes in the opposite direction. What is the net force on the box if each student pushes with a force of 50 N?

Section 2

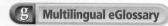

Reading Preview

Essential Questions

▶ What is inertia and how is it related to Newton's first law of motion?

▶ How can an object's acceleration be calculated using Newton's second law of motion?

▶ According to Newton's third law of motion, how are the forces between interacting objects related?

Review Vocabulary

acceleration: rate of change of velocity

New Vocabulary

Newton's first law of motion
inertia
Newton's second law of motion
Newton's third law of motion

 Multilingual eGlossary

Video BrainPOP

Figure 10 You might have seen the law of inertia without even knowing it. For example, as the car hits the block and stops, the red block continues to move forward.

Describe *another example of inertia.*

Newton's Laws of Motion

MAIN Idea Newton's laws of motion relate the change in an object's motion with the forces acting on it.

Real-World Reading Link Suppose you try to push a wagon and a car with the same amount of force. How will their motions be different? Newton's laws of motion describe how the wagon and the car will move.

Isaac Newton and the Laws of Motion

In 1665, Cambridge University in Britain closed for 18 months because of the bubonic plague. Isaac Newton, a 23-year-old student, used this time to develop the law of universal gravitation, calculus, and three laws describing how forces affect the motion of objects. Newton's laws of motion apply to the motion of everyday objects, such as cars and bicycles, as well as the motion of planets and stars.

Newton's First Law of Motion

Recall that forces change an object's motion. Newton's first law explains the relationship between force and change in motion. **Newton's first law of motion** states that an object moves at a constant velocity unless an unbalanced force acts on it. This means that a moving object will continue to move in a straight line at a constant speed unless an unbalanced force acts on it. If an object is at rest, its velocity is zero and it stays at rest unless an unbalanced force acts on it.

Inertia Newton's first law is sometimes called the law of inertia. **Inertia** (ih NUR shuh) is the tendency of an object to resist any change in its motion. **Figure 10** shows a toy car hitting a block and stopping. The block on top of the car is not attached to the car and keeps traveling. The block's forward motion demonstrates the property of inertia.

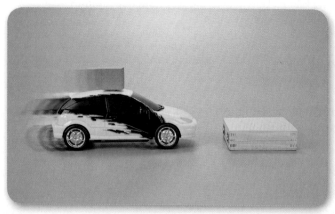
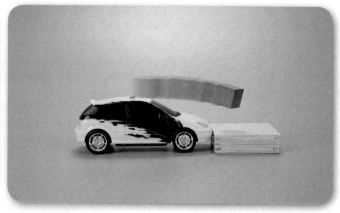

Inertia and mass Does a bowling ball have the same inertia as a table-tennis ball? Why is there a difference? You could not change the motion of a bowling ball much by swatting it with a table-tennis paddle. But you could easily change the motion of the table-tennis ball. A greater force would be needed to change the motion of the bowling ball because it has greater inertia.

Recall that mass is the amount of matter in an object. An object's inertia is related to its mass. The greater an object's mass is, the greater its inertia is. A bowling ball has more mass than a table-tennis ball has, so the bowling ball has a greater inertia.

You will sometimes hear people say that when an object begins to move, inertia is overcome. This is not true. The object still has mass when it is moving, so it still has inertia. As long as the mass is the same, the object has the same inertia.

Newton's Second Law of Motion

Newton's first law of motion states that the motion of an object changes only if an unbalanced force acts on the object. Newton's second law of motion describes how the forces exerted on an object, its mass, and its acceleration are related.

Force and acceleration How are throwing a ball as hard as you can and tossing it gently different? When you throw hard, you exert a greater force on the ball. The ball has a greater velocity when it leaves your hand. The hard-thrown ball has a greater change in velocity, and the change occurs over a shorter period of time. Recall that acceleration equals the change in velocity divided by the time it takes for the change to occur. So, a hard-thrown ball is accelerated more than a gently thrown ball, as shown in **Figure 11**.

MiniLab

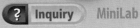

? Inquiry | MiniLab

Observe Inertia

Procedure 🥽 ✋

1. Read the procedure and safety information, and complete the lab form.
2. Create an inclined plane between 25° and 50° using a **board** and **textbooks.** Place a **stop block** (brick or other heavy object) at the end of the plane.
3. Place a **small object** in a **cart** and allow both to roll down the plane. Record the results.
4. Secure the object in the cart with **rubber bands** (safety belts). Allow both to roll down the plane again. Record the results.

Analysis

1. **Identify** the forces acting on the object in the cart in both runs.
2. **Explain** why it is important to wear safety belts in a car.

120 m/s²

17 N

360 m/s²

50 N

? Inquiry | Virtual Lab

■ **Figure 11** According to Newton's second law of motion, the harder you throw a baseball, the greater its acceleration will be.

240 m/s²
0.14 kg
34 N

170 m/s²
0.20 kg
34 N

Mass and acceleration If you throw a softball and a baseball as hard as you can, as shown in **Figure 12,** why do they not have the same speed? The difference is due to their masses. A softball has a mass of about 0.20 kg, but a baseball's mass is about 0.14 kg. The softball has less velocity after it leaves your hand than the baseball does, even though you exerted the same force. The softball has a smaller final speed because it experienced a smaller acceleration. The acceleration of an object depends on its mass as well as the force exerted on it. Force, mass, and acceleration are related.

✓ **Reading Check Identify** You apply a force of 2 N to a toy car and to a real car. Which car has the greater acceleration?

Relating force, mass, and acceleration Newton's **second law of motion** states an object's acceleration is in the same direction as the net force on the object and is equal to the net force exerted on it divided by its mass. Newton's second law can be written as the following equation.

The Second Law of Motion Equation

$$\text{acceleration (in meters/second}^2) = \frac{\text{net force (in newtons)}}{\text{mass (in kilograms)}}$$

$$a = \frac{F_{net}}{m}$$

In the above equation, acceleration has units of meters per second squared (m/s^2) and mass has units of kilograms (kg). Recall that the net force is the sum of all of the forces on an object. Remember that the SI unit for force is a newton (N) and that $1\ N = 1\ kg \cdot m/s^2$. Just like velocity and acceleration, force has a size and a direction.

FOLDABLES ▷
Incorporate information from this section into your Foldable.

Calculating net force Newton's second law can also be used to calculate the net force if mass and acceleration are known. To do this, the equation for Newton's second law must be solved for the net force, F_{net}. To solve for the net force, multiply both sides of the above equation by the mass.

$$m \times a = \cancel{m} \times \frac{F_{net}}{\cancel{m}}$$

The mass, m, on the right side cancels.

$$F_{net} = ma$$

For example, when a tennis player hits a ball, the racket and the ball might be in contact for only a few thousandths of a second. Because the ball's velocity changes over such a short period of time, the ball's acceleration could be as high as 5,000 m/s². The ball's mass is 0.06 kg, so the size of the net force exerted on the ball would be 300 N.

$$F_{net} = ma = (0.06 \text{ kg})(5,000 \text{ m/s}^2) = 300 \text{ kg} \cdot \text{m/s}^2 = 300 \text{ N}$$

VOCABULARY
ACADEMIC VOCABULARY
Period
  a length of time
  Each class period is 45 minutes.

EXAMPLE Problem 2

Solve for Acceleration You push a wagon that has a mass of 12 kg. If the net force on the wagon is 6 N south, what is the wagon's acceleration?

Identify the Unknown: acceleration: *a*

List the Knowns: mass: *m* = **12 kg**
                              net force: F_{net} = **6 N south**

Set Up the Problem: $a = \dfrac{F_{net}}{m}$

Solve the Problem: $a = \dfrac{6 \text{ N south}}{12 \text{ kg}} = $ **0.5 m/s² south**

Check the Answer: The value of the net force (6) is less than the value of the wagon's mass (12), so we would expect the acceleration's value to be less than one. Our answer (0.5 m/s²) makes sense.

PRACTICE Problems   Find **Additional Practice Problems** in the back of your book.

11. If a helicopter's mass is 4,500 kg and the net force on it is 18,000 N upward, what is its acceleration?

12. What is the net force on a dragster with a mass of 900 kg if its acceleration is 32.0 m/s² west?

13. A car pulled by a tow truck has an acceleration of 2.0 m/s² east. What is the mass of the car if the net force on the car is 3,000 N east?

14. **Challenge** What is the net force on a sky diver falling with a constant velocity of 10 m/s downward?

Review

Additional Practice Problems

Newton's Third Law of Motion

What happens when you push against a wall? If the wall is sturdy, nothing happens. But if you pushed against a wall while wearing roller skates, you would go rolling backward. This is a demonstration of Newton's third law of motion. **Newton's third law of motion** states that when one object exerts a force on a second object, the second object exerts a force on the first that is equal in strength and opposite in direction.

Sometimes, Newton's third law is written as "to every action force there is an equal and opposite reaction force." However, one force is not causing the second force. They occur at the same time. It does not matter which object is labeled Object 1 and which is labeled Object 2. Think about a boat tied to a dock with a taut rope. You could say that the action force is the boat pulling on the rope and the reaction force is the rope pulling on the boat. But it would be just as correct to say that the action force is the rope pulling on the boat and the reaction force is the boat pulling on the rope.

Forces on different objects do not cancel If these two forces are equal in size and opposite in direction, you might wonder how some things ever happen. For example, if the box in **Figure 13** pushes on the student when the student pushes on the box, why does the box move? According to the third law of motion, action and reaction forces act on different objects. Recall that the net force is the sum of the forces on a single object. The left picture in **Figure 13** shows the forces on the student. The net force is zero and the student remains at rest. The right picture shows that there is a net force of 20 N to the right on the box and the box accelerates to the right.

✓ **Reading Check** **Explain** why the action and reaction forces do not cancel.

■ **Figure 13** The student pushes on the box, and the box pushes on the student. To understand the motion of each object, the other forces on that object must be examined.

Net force = 0 N

Force from box, 30 N

Static friction, 30 N

Forces on the Student

Net force = 20 N

Applied force, 30 N

Sliding friction, 10 N

Forces on the Box

Forces are interactions Newton's third law states a very important fact: forces are interactions between objects. For example, it makes no sense to say, "The box has a force of 30 N." Is this force acting on the box? Is the box pushing on something? What is causing this force? It makes sense, however, to say, "The student applied a force of 30 N to the box."

Furthermore, both objects experience a force from the interaction. Look at the skaters in **Figure 14.** The male skater is pulling upward on the female skater, while the female skater is pulling downward on the male skater. The two forces are equal in size but opposite in direction. Both skaters feel a pull of the same size.

Even if the two objects have different masses, they will feel the same size force. For example, if a bug flies into the windshield of a truck, the bug and the truck exert forces on each other. According to Newton's third law, the bug and the truck experience the same size force even though their masses and accelerations are different.

■ **Figure 14** The skaters exert forces on each other. According to Newton's third law of motion, the two forces are equal in size but opposite in direction. Both skaters feel the same amount of force.

Section 2 Review

Section Summary

▶ Newton's first law of motion states that the motion of an object at rest or moving with constant velocity will not change unless an unbalanced force acts on the object.

▶ Inertia is the tendency of an object to resist a change in motion.

▶ Newton's second law of motion states that the acceleration of an object depends on its mass and the net force exerted on it.

▶ According to Newton's third law of motion, when an object exerts a force on a second object, the second object exerts a force on the first object.

15. **MAIN Idea Describe** Use Newton's laws of motion to describe what happens when you kick a soccer ball.

16. **Explain** why Newton's first law of motion is sometimes called the law of inertia.

17. **Determine** whether the inertia of an object changes as the object's velocity changes.

18. **Explain** why an object with a smaller mass has a larger acceleration than an object with a larger mass if the same force acts on each.

19. **Think Critically** You push a book across a table. The book moves at a constant speed, but you do not move. Identify all of the forces on you. Then, identify all of the forces on the book.

Apply Math

20. **Calculate Acceleration** A student pushes on a 5-kg box with a force of 20 N forward. The force of sliding friction is 10 N backward. What is the acceleration of the box?

21. **Calculate Mass** You push yourself on a skateboard with a force of 30 N east and accelerate at 0.5 m/s² east. Find the mass of the skateboard if your mass is 58 kg.

Essential Questions

▶ How does Newton's first law explain what happens in a car crash?

▶ How does Newton's second law explain the effects of air resistance?

▶ When is momentum conserved?

Review Vocabulary

momentum: property of a moving object that equals its mass times its velocity

New Vocabulary

air resistance
terminal velocity
free fall
centripetal force
law of conservation of momentum

g Multilingual eGlossary

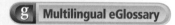

■ **Figure 15** The crash dummies have inertia and resist changes in motion.

Using Newton's Laws

MAIN ◀Idea▶ Newton's laws can be used to explain everyday events, such as falling and collisions.

Real-World Reading Link Have you ever stepped off a skateboard and seen the skateboard moved backward? You may have wondered why this happened. Newton's laws of motion explain this occurrence.

What happens in a crash?

Newton's first law of motion can explain what happens in a car crash. When a car traveling about 50 km/h collides head-on with something solid, the car crumples, slows down, and stops within approximately 0.1 s. According to Newton's first law, the passengers will continue to travel at the same velocity that the car was moving unless a force acts on them.

This means that within 0.02 s after the car stops, any unbelted passengers will slam into the windshield, dashboard, steering wheel, or the backs of the front seats, as shown on the left in **Figure 15**. These unbelted passengers are traveling at the car's original speed of 50 km/h (about 30 miles per hour).

Safety belts Passengers wearing safety belts, also shown in **Figure 15,** will be slowed down by the force of the safety belt. This prevents the person from being thrown out of the seat. Car-safety experts say that about half of the people who die in car crashes would survive if they wore safety belts. Thousands of others would suffer fewer serious injuries.

Even in a low-speed crash, inertia causes the unrestrained dummies to slam into the windshield or seat in front of them.

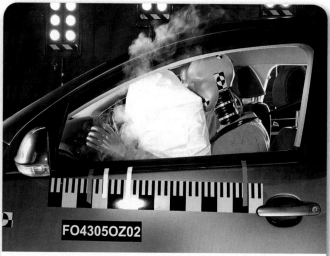

FO4305OZ02

This crash dummy was restrained safely with a safety belt and cushioned with an air bag in this low-speed crash.

The force needed to slow a person's speed from 50 km/h to 0 in 0.1 s is equal to 14 times the force of gravity on the person. Therefore, safety belts are designed to loosen a little as they restrain the passengers. The loosening increases the time it takes to slow the person down, meaning a smaller acceleration and a smaller force is exerted on the passenger.

Air bags Air bags also reduce injuries in car crashes by providing a cushion that reduces the acceleration of the passengers and prevents them from hitting the dashboard. When impact occurs, a chemical reaction occurs in the air bag that produces nitrogen gas. The air bag expands rapidly and then deflates just as quickly as the nitrogen gas escapes out of tiny holes in the bag. The entire process is completed in about 0.04 s.

Newton's Second Law and Gravitational Acceleration

Recall that the gravitational force exerted on an object is equal to the object's mass times the strength of gravity ($F_g = mg$). If gravity is the only force acting on an object ($F_{net} = F_g$), then Newton's second law states that the object's acceleration is the force of gravity divided by the object's mass. Notice that the mass cancels and the object's acceleration due to gravity is equal to the strength of gravity.

$$a = \frac{F_{net}}{m} = \frac{F_g}{m} = \frac{mg}{m} = g$$

When discussing the acceleration due to gravity, g is written with the units m/s^2. On Earth, any falling object with only gravity acting on it will accelerate at 9.8 m/s^2 toward Earth.

Air resistance The gravitational force causes objects to fall toward Earth. However, objects falling in the air experience air resistance. **Air resistance** is a friction-like force that opposes the motion of objects that move through the air. Air resistance acts in the direction opposite to the motion of an object moving through air. If the object is falling downward, air resistance exerts an upward force on the object.

The amount of air resistance on an object depends on the size, shape, and speed of the object, as well as the properties of the air. Air resistance, not the object's mass, is the reason feathers, leaves, and pieces of paper fall more slowly than pennies, acorns, and billiard ball. If there were no air resistance, then all objects, including the feather and the billiard ball, would fall with the same acceleration, as shown in **Figure 16.**

✅ **Reading Check Explain** why some objects fall faster than others.

🔘 **Concepts in Motion**
Animation
■ **Figure 16** In a vacuum, there is no air resistance. As a result, the feather and the billiard ball fall with the same acceleration in a vacuum.

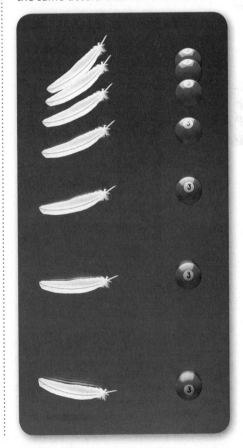

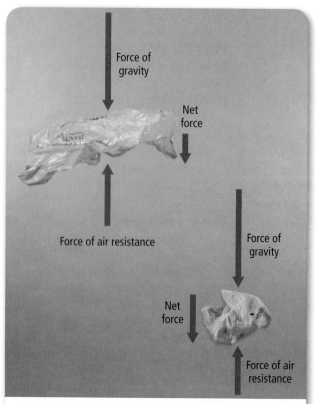

Figure 17 An object's surface size and shape determine how fast it will fall. Because of its greater surface area, the bag on the left has more air resistance acting on it as it falls.

(((O))) **Concepts in Motion** Animation

Figure 18 As the sky diver's speed increases, so does the force of air resistance acting on her. At some point, the force of the air resistance equals the force due to gravity and the sky diver falls at a constant speed called terminal velocity.

Size and shape The more spread out an object is, the more air resistance it will experience. Picture dropping two plastic bags, as shown in **Figure 17.** One is crumpled into a ball and the other is spread out. When the bags are dropped, the crumpled bag falls faster than the spread-out bag. The downward force of gravity on both bags is the same, but the upward force of air resistance on the crumpled bag is less. As a result, the net downward force on the crumpled bag is greater.

Speed and terminal velocity The amount of air resistance also increases as the object's speed increases. As an object falls, gravity causes it to accelerate downward. But, as an object falls faster, the upward force of air resistance increases. So, the net force on the object decreases as it falls, as shown with the sky diver in **Figure 18.**

Eventually, the upward air resistance force becomes large enough to balance the downward force of gravity and the net force on the object is zero. Then the acceleration of the object is zero, and the object falls with a constant speed called the terminal velocity. **Terminal velocity** is the maximum speed an object will reach when falling through a substance, such as air.

A falling object's terminal velocity depends on its size, shape, and mass. For example, the air resistance on an open parachute is much larger than the air resistance on the sky diver alone. With the parachute open, the sky diver's terminal velocity is small enough that she can land safely.

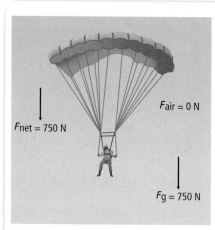

$F_{net} = 750 N$ $F_{air} = 0 N$ $F_g = 750 N$

Sky diver begins to fall.
(*v* is zero)

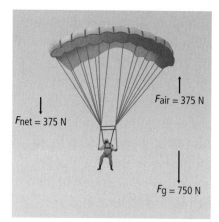

$F_{net} = 375 N$ $F_{air} = 375 N$ $F_g = 750 N$

Sky diver accelerates downward.
(*v* is increasing)

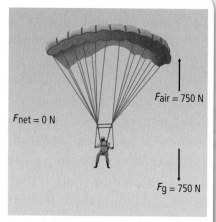

$F_{air} = 750 N$ $F_{net} = 0 N$ $F_g = 750 N$

Sky diver falls at terminal velocity.
(*v* is constant)

Free fall Suppose an object were falling and there were no air resistance. Gravity would be the only force acting on the object. If gravity is the only force acting on an object, the object is said to be in **free fall.** For example, the feather and the billiard ball falling in a vacuum, previously shown in **Figure 16,** are in free fall. Another example is an object in orbit. Earth, for example, is in free fall around the Sun. If Earth did not have a velocity perpendicular to the gravitational force, it would fall into the Sun. Similarly, satellites are in free fall around Earth.

Weightlessness You might have seen pictures of astronauts and equipment floating inside an orbiting spacecraft. They are said to be experiencing weightlessness. But, according the law of universal gravitation, the strength of Earth's gravitational field at a typical orbiting altitude is about 90 percent of its strength at Earth's surface. So, an 80-kg astronaut would weigh about 700 N in orbit and would not be weightless.

What does it mean to say that something is weightless? Think about how you measure your weight. When you stand on a scale, as shown on the left of **Figure 19,** you are at rest. The net force on you is zero, according to Newton's second law of motion. The scale exerts an upward force that balances your weight. The dial on a scale shows the size of the upward force, which is the same size as your weight.

Now suppose you stand on a scale in an elevator that is in free fall, as on the right in **Figure 19.** You would no longer push down on the scale at all. The scale dial would read zero, even though the force of gravity has not changed. An orbiting spacecraft is in free fall, and objects in it seem to float because they are all falling around Earth at the same rate.

Review
Additional Practice Problems

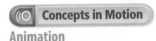

Concepts in Motion
Animation

■ **Figure 19** The reading on the scale depends on the upward force the scale exerts on the girl. If both the girl and the scale are in free fall, the girl experiences the sensation of weightlessness, even though the force of gravity has not changed.

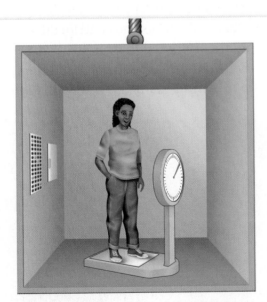

When the elevator is stationary, the scale shows the girl's weight.

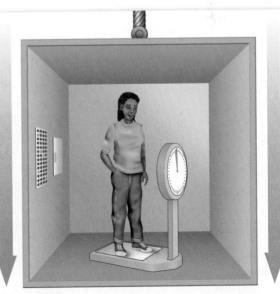

If the elevator were in free fall, the scale would show a zero weight.

■ **Figure 20** The riders on this amusement park ride are traveling in a circle because of the centripetal force acting on them.

Identify *the force that acts as the centripetal force.*

Centripetal Forces

Orbiting objects, such as space shuttles, are traveling in nearly circular paths. According to Newton's first law, this change in motion is caused by a net force acting on the object. Newton's second law states that because the object's acceleration is toward the center of the curved path, the net force is also toward the center. A **centripetal force** is a force exerted toward the center of a curved path.

Anything that moves in a circle is doing so because a centripetal force is accelerating it toward the center. Many different forces can act as a centripetal force. Gravity is the centripetal force that keeps planets orbiting the Sun. On the amusement park ride in **Figure 20,** the push of the walls of the ride on the people is the centripetal force.

When a car rounds a level curve on a highway, friction between the tires and the road acts as the centripetal force. If the road is slippery and the frictional force is small, it might not be large enough to keep the car moving around the curve. Then the car will slide in a straight line, as shown in **Figure 21.**

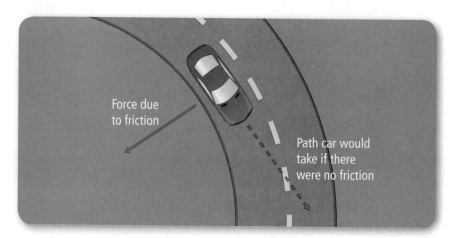

Force due to friction

Path car would take if there were no friction

■ **Figure 21** If the inward frictional force is too small, the car will continue in a straight line and not make it around the curve.

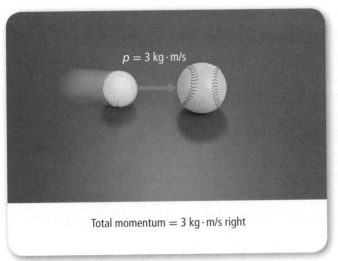

$p = 3 \ \text{kg} \cdot \text{m/s}$

Total momentum = 3 kg·m/s right

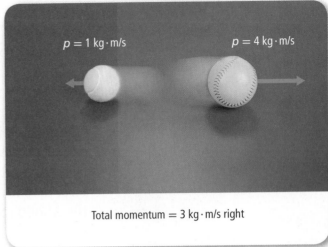

$p = 1 \ \text{kg} \cdot \text{m/s}$ $p = 4 \ \text{kg} \cdot \text{m/s}$

Total momentum = 3 kg·m/s right

Force and Momentum

Recall that acceleration is the difference between the initial and final velocities, divided by the time. Therefore, we can write Newton's second law in the following way.

$$F = ma = m \times \frac{(v_\text{f} - v_\text{i})}{t} = \frac{(mv_\text{f} - mv_\text{i})}{t}$$

Recall that an object's momentum equals its mass multiplied by its velocity. In the equation above, mv_f is the final momentum and mv_i is the initial momentum. The equation states that the net force exerted on an object equals the change in its momentum divided by the time over which the change occurs. In fact, this is how Newton originally wrote the second law of motion.

Conservation of momentum Newton's second and third laws of motion can be used to describe what happens when objects collide. For example, consider the collision of the balls in **Figure 22.** We will assume that friction is too small to cause a noticeable change in the balls' motion. In this ideal case, there are no external forces, but the balls exert forces on each other during the collision. According to Newton's third law, the forces are equal in size and opposite in direction. Therefore, the momentum lost by the first ball is gained by the second ball and the total momentum of the two balls is the same before and after the collision. This is the **law of conservation of momentum**—if no external forces act on a group of objects, their total momentum does not change.

Collisions with multiple objects When a cue ball hits the group of motionless balls, as shown in **Figure 23,** the cue ball slows down and the rest of the balls begin to move. The momentum that the group of balls gained is equal to the momentum that the cue ball lost. Momentum is conserved.

■ **Figure 22** When two objects collide and there are no external forces acting on the objects, momentum is conserved.

Inquiry   Virtual Lab

Review
Additional Practice Problems

■ **Figure 23** Momentum is conserved in collisions with more than two objects if there are no external forces acting on the objects. The momentum of the cue ball just before the collision is equal to the total of the momentums of the billiard balls (including the cue ball) just after the collision.

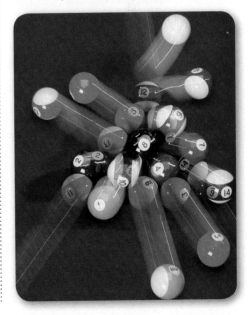

Rocket propulsion Suppose you are standing on skates holding a softball. You exert a force on the softball when you throw it. According to Newton's third law, the softball exerts a force on you. This force pushes you backward in the direction opposite the softball's motion. Rockets use the same principle to move, even in the vacuum of outer space. In the rocket engine, burning fuel produces hot gases. The rocket engine exerts a force on these gases and causes them to escape out the back of the rocket. By Newton's third law, the gases exert a force on the rocket and push it forward. **Figure 24** shows one of the Apollo rockets that traveled to the Moon. Notice that the force of the rocket on the gases is equal in size to the force of the gases on the rocket.

Momentum is conserved when a rocket ejects the hot gas. If the rocket is initially at rest, then the total momentum of the rocket and the fuel is zero. After the fuel is burned and the hot gas is expelled, the gas travels backward with a momentum of $m_{gas}v_{gas}$ and the rocket travels forward with a momentum of $m_{rocket}v_{rocket}$. These momentums are equal in size, but opposite in direction. By controlling how much gas is ejected and the gas's velocity, the rocket's motion can be controlled.

■ **Figure 24** If more gas is ejected from the rocket engine or expelled at a greater velocity, the engine will exert a larger force on the rocket.

Section **3** Review

Section Summary

▶ In a car crash, an unrestrained passenger will continue moving at the velocity of the car before the crash.

▶ Air resistance is a force that opposes an object's motion through the air.

▶ If gravity is the only force acting on an object, then the object is in free fall.

▶ Objects appear to be weightless in free fall.

▶ The force that causes an object to move in a circular path is called the centripetal force.

▶ The law of conservation of momentum states that if objects exert forces only on each other, their total momentum is conserved.

22. **MAIN Idea Describe** Use Newton's laws to describe how inertia, gravity, and air resistance affect sky divers as they fall, open their parachutes, and reach terminal velocity.

23. **Discuss** the advantages of wearing a safety belt when riding in a vehicle.

24. **Explain** why planets orbit the Sun instead of traveling off into space.

25. **Describe** what happens to the momentum of two billiard balls that collide.

26. **Explain** how a rocket can move through outer space where there is no matter for it to push on.

27. **Think Critically** Suppose you are standing on a scale in an elevator that is accelerating upward. Will the scale read your weight as larger or smaller than the weight it reads when you are stationary? Explain.

Apply Math

28. **Calculate Momentum** A fuel-filled rocket is at rest. It burns its fuel and expels hot gas. The gas has a momentum of 1,500 kg · m/s backward. What is the momentum of the rocket?

✓ **Assessment** Online Quiz

LAB

The Effects of Air Resistance

Objectives

- **Measure** the effect of air resistance on sheets of paper with different shapes.

- **Design** a shape that maximizes air resistance. Use a piece of paper to create your design.

Background: If you dropped a bowling ball and a feather from the same height on the Moon, they would both hit the surface at the same time. The objects are attracted to the Moon with the same force, but there is no atmosphere and no air resistance. But on Earth, a bowling ball and a feather will not hit the ground at the same time. Even though all objects on Earth are attracted with the same force, air resistance affects the feather more than it affects the bowling ball.

Question: *How does air resistance affect the acceleration of falling objects?*

Preparation

Materials

paper (4 sheets of equal size)
scissors
meterstick
stopwatch
masking tape

Safety Precautions

Procedure

1. Read the procedure and safety information, and complete the lab form.

2. Copy the data table, or create it on a computer.

3. Measure a height of 2.5 m on the wall, and mark the height with a piece of masking tape.

Effects of Air Resistance

Paper Type	Time
Flat paper	
Loosely crumpled paper	
Tightly crumpled paper	
Your paper design	

4. Have one group member drop the flat sheet of paper from the 2.5-m mark. Use the stopwatch to time how long it takes for the paper to reach the ground. Record the time in your data table.

5. Crumple a sheet of paper into a loose ball, and repeat step 4.

6. Crumple a sheet of paper into a tight ball, and repeat step 4.

7. Shape a piece of paper so that it will fall slowly. You may cut, tear, or fold your paper into any design that you choose. Repeat step 4.

Conclude and Apply

1. **Compare** the falling times of the different sheets of paper.

2. **Explain** why the different-shaped papers fell at different accelerations.

3. **Explain** how your design caused the force of air resistance on the paper to be greater or smaller than the air resistance on the other paper shapes.

Compare your paper design with the designs created by your classmates. As a class, compile a list of characteristics that increase air resistance.

Use Vocabulary

Complete each statement with the correct term from the Study Guide.

29. _____ opposes the sliding motion of two surfaces that are in contact.

30. The _____ of an object is different on every planet in the solar system.

31. When an object moves in a circular path, the net force is called a(n) _____.

32. The attractive force between two objects that depends on their masses and the distance between them is _____.

33. _____ relates the net force exerted on an object to its mass and acceleration.

Check Concepts

34. What is the gravitational force exerted on an object called?
   A) centripetal force   **C)** momentum
   B) friction   **D)** weight

35. Which explains why astronauts seem weightless in orbit?
   A) Earth's gravity is much less in orbit.
   B) The spacecraft is in free fall.
   C) The gravity of Earth and the Sun cancel.
   D) The centripetal force on the shuttle balances Earth's gravity.

36. Which term best describes the forces on an object with a net force of zero?
   A) balanced forces
   B) unbalanced forces
   C) inertia
   D) acceleration

37. Which body exerts the strongest gravitational force on you?
   A) the Moon   **C)** the Sun
   B) Earth   **D)** this book

38. Which term describes the tendency of an object to resist a change in motion?
   A) balanced force   **C)** net force
   B) inertia   **D)** acceleration

Use the figure below to answer question 39.

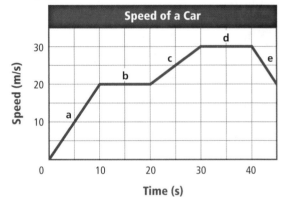

39. The graph shows the speed of a car moving in a straight line. Over which segments are the forces on the car balanced?
   A) A and C   **C)** C and E
   B) B and D   **D)** A, C, and E

40. Which of the following is true about an object in free fall?
   A) Its acceleration depends on its mass.
   B) It has no inertia.
   C) It pulls on Earth, and Earth pulls on it.
   D) Its momentum is constant.

41. An object always accelerates in the same direction as which of the following?
   A) net force   **C)** static friction
   B) air resistance   **D)** gravity

42. **BIG Idea** Which of the following is not a force?
   A) weight   **C)** momentum
   B) friction   **D)** air resistance

✓ Assessment   Online Test Practice

Interpret Graphics

43. Copy and complete the following concept map on forces.

((O **Concepts in Motion**   **Interactive Concept Map**

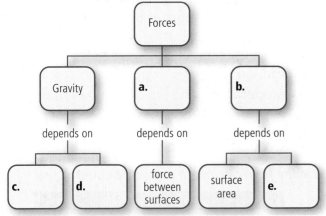

Use the table below to answer questions 44–46.

Time of Fall for Dropped Objects		
Object	Mass (g)	Time of Fall (s)
A	5.0	2.0
B	5.0	1.0
C	30.0	0.5
D	35.0	1.5

44. If the objects in the data table above all fell the same distance, which object fell with the greatest average speed?

45. Which object was aggeted the most by air resistance?

46. Explain why the four objects do not fall with the same speed.

Think Critically

47. **Predict** Suppose you are standing on a bathroom scale next to a sink. How does the reading on the scale change if you push down on the sink? Explain your prediction.

48. **Explain** whether there can be forces acting on a car that is moving in a straight line with constant speed.

49. **Explain** You pull open a door. If the force the door exerts on you is equal to the force you exert on the door, why do you not move?

50. **THEME FOCUS** **Describe** the action and reaction force pairs involved when an object falls toward Earth. Ignore the effects of air resistance.

51. **Explain** why a passenger who is not wearing a safety belt will likely hit the windshield in a head-on collision.

52. **Determine** the direction of the net force on a car if it is slowing down. Explain.

Apply Math

53. **Calculate Mass** Find your mass if a scale on Earth reads 650 N when you stand on it.

54. **Calculate Gravitational Strength** You weigh yourself at the top of a high mountain and the scale reads 728 N. If your mass is 75 kg, what is the gravitational strength at your location?

Use the figure below to answer question 55.

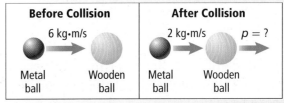

55. **Calculate Momentum** What is the momentum of the wooden ball after the collision?

56. **Calculate Sliding Friction** A box being pushed with a force of 85 N right slides along the floor with a constant speed. What is the force of sliding friction on the box?

Standardized Test Practice

Multiple Choice

Record your answers on the answer sheet provided by your teacher or on a sheet of paper.

1. The net force on an object moving with constant speed in circular motion is in which direction?
   A. downward
   B. opposite to the object's motion
   C. toward the center of the circle
   D. in the direction of the object's velocity

Use the figure below to answer questions 2–4.

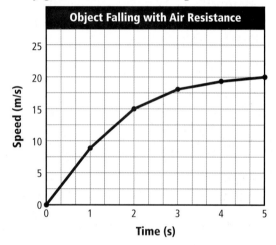

2. According to the trend in these data, which of the following values is most likely the speed of the object after falling for 6 s?
   A. 26.7 m/s   C. 20.1 m/s
   B. 15.1 m/s   D. 0 m/s

3. Over which of the following time intervals is the acceleration of the object the greatest?
   A. 0 s to 1 s
   B. 1 s to 2 s
   C. 2 s to 3 s
   D. 4 s to 5 s

4. Over which of the following time intervals is the net force on the object the smallest?
   A. 0 s to 1 s   C. 2 s to 3 s
   B. 1 s to 2 s   D. 4 s to 5 s

5. Which of the following would cause the gravitational force between object A and object B to increase?
   A. Increase the mass of object A.
   B. Increase the distance between them.
   C. Decrease the mass of object A.
   D. Decrease the mass of both objects.

6. Which of the following is a force?
   A. friction
   B. acceleration
   C. inertia
   D. velocity

Use the figure below to answer question 7.

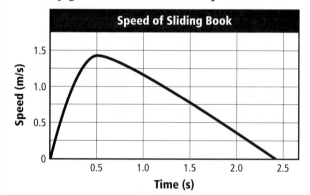

7. The graph above shows how the speed of a book changes as it slides across a table in a straight line. Over what time interval is the net force on the book in the opposite direction of the book's motion?
   A. 0 s to 0.5 s
   B. 0.5 s to 2.4 s
   C. 0 s to 2.4 s
   D. 2 s to 2.4 s

8. How does the gravitational force change as two objects move farther apart?
   A. It increases.
   B. It decreases.
   C. It remains constant.
   D. It is zero.

✓ Assessment   Standardized Test Practice

Short Response

Record your answers on the answer sheet provided by your teacher or on a sheet of paper.

9. You are pushing a 30-kg wooden crate across the floor. The force of sliding friction on the crate is 90 N. How much force must you exert on the crate to keep it moving with a constant velocity?

10. A sky diver with a mass of 60 kg jumps from an airplane. Five seconds after jumping, the force of air resistance on the sky diver is 300 N. What is the sky diver's acceleration five seconds after jumping?

11. A pickup truck is carrying a load of gravel. The driver hits a bump and gravel falls out, so that the mass of the truck is one half as large after hitting the bump. If the net force on the truck does not change, how does the truck's acceleration change?

Use the table below to answer question 12.

Car	Mass (kg)	Stopping Distance (m)
A	1,000	80
B	1,250	100
C	1,500	120
D	2,000	160

12. What is the relationship between a car's mass and its stopping distance? How can you explain this relationship?

Extended Response

Record your answers on a sheet of paper.

13. Give an example of a force applied to an object that does not change the object's velocity. Why does the object's velocity not change?

14. A tennis ball is rolling east. You apply a force to it in the north direction. How will your push affect the direction of the ball? Explain.

15. Two balls are dropped from an airplane. Both balls are the same size, but one has a mass ten times greater than the other. The force of air resistance on each ball depends on the ball's speed. Explain whether both balls will reach the same terminal velocity.

16. When the space shuttle is launched from Earth, the rocket engines burn fuel and apply a constant force on the shuttle until they burn out. Explain why the shuttle's acceleration increases while the rocket engines are burning fuel.

17. An object in motion slows down and comes to a stop. Use Newton's first law of motion to explain why this happens.

18. In an airplane flying at a constant speed, the force pushing the airplane forward is exerted by the engine and is equal to the opposite force of air resistance. Describe how these forces compare when the plane speeds up and slows down. In which direction is the net force on the airplane in each case?

NEED EXTRA HELP?																		
If You Missed Question . . .	1	2	3	4	5	6	7	8	9	10	11	12	13	14	15	16	17	18
Review Section . . .	3	3	3	3	1	1	2	1	2	3	2	2	1	2	3	2	2	2

UNIT 2
Energy

THEMES

Energy Energy can be transferred and transformed from one form into others.

Motion and Force Electric and magnetic forces can be used to do work.

Natural Resources Fossil fuels, water, and Earth's internal heat are sources of energy.

Scientific Inquiry Scientists investigate energy resources and develop new technologies.

Technology Science provides data-driven information that informs decisions about technology and alternative energy resources.

WebQuest STEM UNIT PROJECT

Model Research maglev trains. Develop a design and build your own model maglev train.

CHAPTER 4

Work and Energy

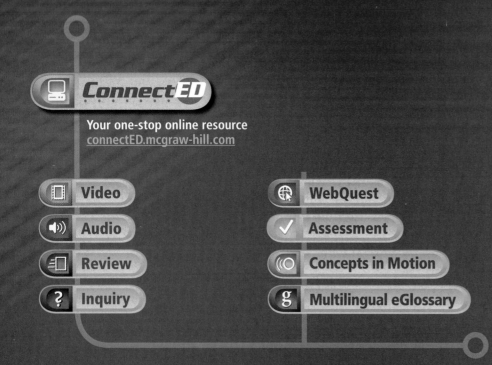

ConnectED

Your one-stop online resource
connectED.mcgraw-hill.com

- Video
- Audio
- Review
- Inquiry
- WebQuest
- Assessment
- Concepts in Motion
- Multilingual eGlossary

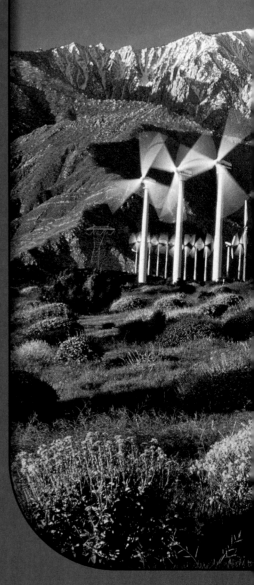

Launch Lab
Doing Work with a Simple Machine

Did you know that you can lift several times your weight with the help of a pulley? Before the hydraulic lift was invented, a car mechanic used pulleys to raise a car off the ground. In this lab, you will see how a pulley can increase a force.

For a lab worksheet, use your StudentWorks™ Plus Online.

 **Inquiry** Launch Lab

 **FOLDABLES**

Make a three-tab book. Label it as shown. Use it to organize your notes on energy.

Know? | Want to know? | Learned?

THEME FOCUS Energy
Energy can be transformed among its many forms, but energy cannot be created or destroyed.

BIG Idea Energy has many forms and can be transferred through work.

Essential Questions

▶ What is work?
▶ How can work be calculated when force and motion are parallel to each other?
▶ How do machines make doing work easier?
▶ What are mechanical advantage and efficiency?

Review Vocabulary

force: a push or pull exerted on an object

New Vocabulary

work
machine
simple machine
compound machine
efficiency
mechanical advantage

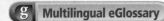

g Multilingual eGlossary

Work and Machines

MAIN ‹Idea Machines make doing work easier or faster by changing the force needed to do the work.

Real-World Reading Link When was the last time you rode a bicycle? Think of walking or running the distance that you went on the bicycle. A bicycle is a machine that allows you to travel much faster than you could on foot.

Definition of Work

To many people, the word *work* means something that people do to earn money. In that sense, work can be anything from fixing cars to designing Web sites. The word *work* might also mean exerting a force with muscles. However, in science, the word *work* is used in a different way.

Motion and work Press your hand against the surface of your desk as hard as you can. Have you done any work on the desk? The answer is no, no matter how tired you get from the effort. In science, **work** is force applied through a distance. If you push against the desk and it does not move, then you have not done any work on the desk because the desk has not moved.

Force and direction of motion Imagine that you are pushing a lawn mower, as shown in **Figure 1.** You could push on this mower in many different directions. You could push it horizontally. You could also push down on the mower or push on it at an angle. Think about how the mower's motion would be different each time. The direction of the force that you apply to the lawn mower affects how much work you do on it.

■ **Figure 1** You apply a force through a distance when you push a lawn mower over a lawn. In other words, you do work on the lawn mower when you push that mower over a lawn.

Explain *whether you could do work on the mower without moving it.*

Force parallel to motion Imagine that you push on the lawn mower in **Figure 1** with a force of 25 N and through a distance of 4 m. In what direction would you push to do the maximum amount of work on the mower? You do the maximum amount of work when you push the lawn mower in the same direction as it is moving. When force and motion are parallel, which means they are in the same direction, work is equal to force multiplied by distance.

Work Equation

work (in joules) =
    applied force (in newtons) × distance (in meters)

$$W = Fd$$

If force is measured in newtons (N) and distance is measured in meters (m), then work is measured in joules (J). You do about 1 J of work on a cell phone when you pick it up off the floor.

EXAMPLE Problem 1

Solve for Work You push a refrigerator with a horizontal force of 100 N. If you move the refrigerator a distance of 5 m while you are pushing, how much work do you do?

Identify the Unknown:	work: W
List the Knowns:	applied force: $F = 100\,N$ distance: $d = 5\,m$
Set Up the Problem:	$W = Fd$
Solve the Problem:	$W = (100\,N)(5\,m) = 500\,J$
Check the Answer:	Check to see whether the units match on both sides of the equation.
	units of W = (units of F) × (units of d) = N × m = J

PRACTICE Problems

Find **Additional Practice Problems** in the back of your book.

1. A couch is pushed with a horizontal force of 80 N and moves a distance of 5 m across the floor. How much work is done in moving the couch?

2. How much work do you do when you lift a 100-N child 0.5 m?

3. The brakes on a car do 240,000 J of work in stopping the car. If the car travels a distance of 40 m while the brakes are being applied, how large is the average force that the brakes exert on the car?

4. **Challenge** The force needed to lift an object is equal in size to the gravitational force on the object. How much work is done in lifting an object that has a mass of 5 kg a vertical distance of 2 m?

Review
Additional Practice Problems

Reading Preview

Essential Questions

▶ What is the difference between kinetic energy and potential energy?

▶ How can you calculate kinetic energy?

▶ What are some different forms of potential energy?

▶ How can you calculate gravitational potential energy?

Review Vocabulary

work: a force applied through a distance

New Vocabulary

energy
system
kinetic energy
potential energy
elastic potential energy
chemical potential energy
gravitational potential energy

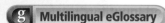 **Multilingual eGlossary**

Describing Energy

MAIN ⟨Idea Energy is the ability to cause change.

Real-World Reading Link Think about what is written on the side of a cereal box. The side of the box tells you how many Calories are in each serving. A Calorie is a unit of energy. It takes energy to run, jump, grow, and even think.

Change Requires Energy

When something is able to change its surroundings or itself, it has energy. **Energy** is the ability to cause change. Without energy, nothing would ever change. The moving tennis racket in **Figure 6** has energy. That racket causes change when it deforms the tennis ball and changes the tennis ball's motion.

Work transfers energy The tennis racket in **Figure 6** also does work on the tennis ball, applying a force to that ball through a distance. When this happens, the racket transfers energy to the ball. Therefore, energy can also be described as the ability to do work.

Because energy can be described as the ability to do work, energy can be measured with the same units as work. Energy, like work, can be measured in joules. Imagine that the tennis racket in **Figure 6** does 250 J of work on the tennis ball. Then, 250 J of energy are transferred from the racket to the ball.

Systems The tennis racket and the tennis ball in **Figure 6** are systems. A **system** is anything around which you can imagine a boundary. A system can be a single object, such as a tennis ball, or a group of objects, such as the solar system. When one system does work on a second system, energy is transferred from the first system to the second system.

■ **Figure 6** The tennis racket causes changes to occur when it hits the tennis ball.

Describe *the changes that are occurring.*

Chemical Potential Energy

Electrical Energy

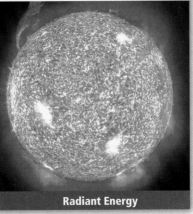
Radiant Energy

■ **Figure 7** Energy can be stored, and it can be transferred from one place to another. For example, the energy due to the chemical bonds in gasoline is easy to transport in cars. Electrical energy is transferred from a power plant to appliances in your home. Radiant energy is transferred from the Sun to Earth.

Different Forms of Energy

Turn on an electric light, and a dark room becomes bright. Turn on a portable music player, and sound comes through your headphones. In both situations, a change occurs. These changes differ from each other and the tennis racket hitting the tennis ball in **Figure 6.** This is because energy has many different forms. These forms include mechanical energy, electrical energy, chemical energy, and radiant energy.

Figure 7 shows some everyday situations in which you might notice energy. Automobiles make use of the chemical energy of gasoline. Many household appliances require electrical energy to function. Radiant energy from the Sun warms Earth. In short, energy plays a role in every activity that you do.

✔ **Reading Check** **Identify** three different forms of energy.

An energy analogy Is the chemical energy of food the same as the energy that comes from the Sun or the energy from gasoline? Money can be used in an analogy to help you understand energy. Money exists in a variety of forms, such as coins, dollar bills, and twenty-dollar bills. You can convert money from one form to another. For example, you could obtain four quarters for a dollar bill. Regardless of its form, money is money. The same is true for energy. Energy from the Sun that warms you and energy from the food that you eat are only different forms of the same thing.

VOCABULARY . . . .
SCIENCE USAGE V. COMMON USAGE
Energy
Science usage:
  the ability to cause change
  You transfer energy when you do work.

Common usage:
  the capacity of acting or being active
  That soccer player had a lot of energy on the field today. . . . .

Kinetic energy When you think of energy, you might think of objects in motion. Objects in motion can collide with other objects and cause change. Therefore, objects in motion have energy. **Kinetic energy** is energy due to motion. A car moving along a highway and a ballet dancer leaping through the air have kinetic energy. The kinetic energy from an object's motion depends on that object's mass and speed.

Kinetic Energy Equation

kinetic energy (in joules) =
$\frac{1}{2}$ mass (in kg) × [speed (in m/s)]²

$$KE = \frac{1}{2}mv^2$$

If mass is measured in kg and speed is measured in m/s, then kinetic energy is measured in joules. If you drop a softball from just above your knee, the kinetic energy from that ball's falling motion is about 1 J, just before the ball reaches the floor.

EXAMPLE Problem 4

Solve for Kinetic Energy A jogger with a mass of 60.0 kg is moving forward at a speed of 3.0 m/s. What is the jogger's kinetic energy from this forward motion?

Identify the Unknown: kinetic energy: *KE*

List the Knowns: mass: *m* = **60.0 kg**

     speed: *v* = **3.0 m/s**

Set Up the Problem: $KE = \frac{1}{2}mv^2$

Solve the Problem: $KE = \frac{1}{2}$ **(60.0 kg)**(3.0 m/s)²

     $KE = \frac{1}{2}$ **(60.0 kg)**(9.0 m²/s²)

     KE = **270 J**

Check the Answer: Check the last step by estimating. Round 9.0 m²/s² upward to 10 m²/s². Then, $\frac{1}{2}$ (60.0 kg)(10 m²/s²) = 300 J. This is close to 270 J, so the final calculation was reasonable.

PRACTICE Problems

Find **Additional Practice Problems** in the back of your book.

16. A baseball with a mass of 0.15 kg is moving at a speed of 40.0 m/s. What is the baseball's kinetic energy from this motion?

Review

Additional Practice Problems

17. **Challenge** A 1,500-kg car doubles its speed from 50 km/h to 100 km/h. By how many times does the kinetic energy from the car's forward motion increase?

Potential energy Energy does not always involve motion. Even motionless objects can have energy. **Potential energy** is energy that is stored due to the interactions between objects. One example is the energy stored between an apple hanging on a tree and Earth. Energy is stored between the apple and Earth because of the gravitational force between the apple and Earth. Another example is the energy stored between objects that are connected by a compressed spring or a stretched rubber band.

Elastic potential energy If you stretch a rubber band and let it go, it sails across the room. As it flies through the air, it has kinetic energy due to its motion. Where did this kinetic energy come from? Just as there is potential energy due to gravitational forces, there is also potential energy due to the elastic forces between the particles that make up a stretched rubber band. The energy of a stretched rubber band or a compressed spring is called elastic potential energy. **Elastic potential energy** is energy that is stored by compressing or stretching an object.

Chemical potential energy The food that you eat and the gasoline in cars also have stored energy. This stored energy is due to the chemical bonds between atoms. **Chemical potential energy** is energy that is due to chemical bonds. You might notice chemical potential energy when you burn a substance. When an object is burned, chemical potential energy becomes thermal energy and radiant energy. **Figure 8** shows the process for burning methane.

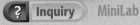

MiniLab

? Inquiry MiniLab

Interpret Data from a Slingshot

Procedure
1. Read the procedure and safety information, and complete the lab form.
2. Using two fingers, carefully stretch a **rubber band** on a **table** until it has no slack.
3. Place a **nickel** on the table, slightly touching the midpoint of the rubber band.
4. Push the nickel back 0.5 cm into the rubber band and release. Measure the distance that the nickel travels with a **metric ruler.**
5. Repeat steps 3 and 4, each time pushing the nickel back an additional 0.5 cm.

Analysis
1. **Describe** how the distance that the nickel travels depends on the distance that you stretch the rubber band.
2. **Infer** how the takeoff speed of the nickel depends on the distance that you stretch the rubber band.
3. **Infer** how the kinetic energy from the nickel's motion depends on the distance that you stretch the rubber band.

■ **Figure 8** When methane burns, it combines with oxygen to form carbon dioxide and water. In this chemical reaction, chemical potential energy is converted to other forms of energy.

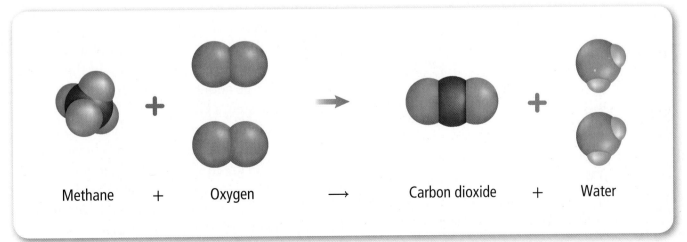

Methane + Oxygen ⟶ Carbon dioxide + Water

Gravitational potential energy Consider the blue vase in **Figure 9.** Together, the blue vase and Earth have potential energy. **Gravitational potential energy** is energy that is due to the gravitational forces between objects. Gravitational potential energy is often shortened to GPE.

Any system that has objects that are attracted to each other through gravity has gravitational potential energy. An apple and Earth have gravitational potential energy. The solar system also has gravitational potential energy. The gravitational potential energy of a system containing just Earth and another object depends on the object's mass, Earth's gravity, and the object's height. Recall that near Earth's surface, *g* is equal to 9.8 N/kg.

Gravitational Potential Energy Equation

gravitational potential energy (J) =
mass (kg) × gravity (N/kg) × **height** (m)

$$GPE = mgh$$

Height and gravitational potential energy

Look at the bookcase in **Figure 9.** Imagine that this bookcase is on the second floor of a building and that this building is at the top of a large hill.

How should you measure the heights of the objects on the shelves? You could measure the heights from the floor. You could also measure the heights from the ceiling, the ground outside, the bottom of the hill, or Earth's center.

To calculate gravitational potential energy, height is measured from a reference level. This means that gravitational potential energy varies depending on the chosen reference level.

Relative to the floor, the GPE of a system containing just the blue vase and Earth is about 90 J. Relative to the ceiling, the GPE of this same system might be about –40 J. Relative to Earth's center, this system's GPE is about 300 million J. All these statements are correct. In addition, the GPE of the blue vase-Earth system is greater than the GPE of the green vase-Earth system for every reference level. However, statements such as, "The gravitational potential energy is 100 J," are meaningless, unless a reference level is given.

■ **Figure 9** The gravitational potential energy of any system containing only an object on the bookcase and Earth depends on the object's mass, the strength of Earth's gravity, and the object's height. The object's height is measured relative to a reference level. The floor, the ground, the ceiling, and Earth's center are possible reference levels.

Concepts in Motion Animation

Inquiry Video Lab

FL Solve for Gravitational Potential Energy A 4.0-kg ceiling fan is placed 2.5 m above the floor. What is the gravitational potential energy of the Earth-ceiling fan system relative to the floor?

Identify the Unknown: gravitational potential energy: GPE

List the Knowns

mass: $m = $ **4.0 kg**

gravity: $g = $ **9.8 N/kg**

height: $h = $ **2.5 m**

Set Up the Problem: $GPE = mgh$

Solve the Problem: $GPE = $ **(4.0 kg)**(9.8 N/kg)**(2.5 m)** $ = $ **98 N · m = 98 J**

Check the Answer: Round 9.8 N/kg to 10 N/kg. Then, $GPE = $ (4.0 kg)(10 N/kg)(2.5 m) = 100 J. This is very close to the answer above. Therefore, that answer is reasonable.

PRACTICE Problems

Find **Additional Practice Problems** in the back of your book.

18. An 8.0-kg history textbook is placed on a 1.25-m high desk. How large is the gravitational potential energy of the textbook-Earth system relative to the floor?

19. **Challenge** How large is the GPE of the textbook-Earth system in problem 18, relative to the desktop?

🖥️ **Review**

Additional Practice Problems

Section 2 Review

Section Summary

▶ Forms of energy include mechanical, electrical, chemical, thermal, and radiant energy.

▶ Kinetic energy is the energy that a moving object has because of its motion.

▶ Potential energy is stored energy due to the interactions between objects.

▶ Different forms of potential energy include elastic potential energy, chemical potential energy, and gravitational potential energy.

20. **MAIN ‹Idea› Describe** a change caused by kinetic energy as well as a change that involves potential energy.

21. **Infer** whether a system can have kinetic energy and potential energy at the same time.

22. **Contrast** elastic potential energy and chemical potential energy.

23. **Think Critically** The different molecules that make up the air in a room have, on average, the same kinetic energy. How does the speed of the different molecules that make up the air depend on their masses?

Apply Math

24. **Calculate Kinetic Energy** A 0.06-kg ball is moving at 5.0 m/s. How large is the kinetic energy from this motion?

25. **Calculate GPE** A 0.50-kg apple is 2.0 m above the reference level. What is the GPE of the apple-Earth system?

Reading Preview

Essential Questions

▶ What is the law of conservation of energy?

▶ What is mechanical energy?

▶ Why is mechanical energy not always conserved?

▶ How are power and energy related?

Review Vocabulary

friction: a force that opposes the sliding motion of two surfaces that are touching each other

New Vocabulary

law of conservation of energy
mechanical energy
power

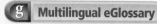 **Multilingual eGlossary**

Conservation of Energy

MAIN Idea Energy cannot be created or destroyed.

Real-World Reading Link When you toss a ball into the air, you give it kinetic energy. This kinetic energy transforms into potential energy as the ball rises and then back into kinetic energy as the ball falls. What happens to the energy when you catch the ball?

The Law of Conservation of Energy

Suppose you are riding on a roller coaster like the one in **Figure 10.** As your height above the ground changes, gravitational potential energy changes. As your speed changes, kinetic energy changes. Think about the motion of the roller-coaster cars. When the cars are high above the ground, GPE is large and kinetic energy is small. When the cars are low, GPE is small and kinetic energy is large. Energy is changing back and forth between GPE and kinetic energy. In addition, some kinetic energy is slowly converted into other forms of energy during a roller-coaster ride.

However, the total energy remains constant. The **law of conservation of energy** states that energy cannot be created or destroyed. Energy can only be converted from one form to another or transferred from one place to another.

✓ **Reading Check State** the law of conservation of energy.

Conserving resources You might have heard about energy conservation or have been asked to conserve energy. These ideas are related to using energy resources, such as coal and oil, wisely. The law of conservation of energy, on the other hand, is a universal principle that states that total energy remains constant.

■ **Figure 10** Energy can be transferred or transformed, but it cannot be created or destroyed. For the roller-coaster cars, energy is converted back and forth between kinetic energy and gravitational potential energy. In addition, some kinetic energy is converted to other forms of energy. However, the total amount of energy is constant.

Energy Transformations

You might not think that a vase on a table has any relationship with energy—until it falls. You probably associate energy more with race cars roaring by you or the Sun warming your skin on a summer day. All these situations involve energy transformations.

Mechanical energy transformations

Bicycles, roller coasters, and swings can often be described in terms of mechanical energy. **Mechanical energy** is the sum of the kinetic energy and potential energy of the objects in a system. Mechanical energy includes the kinetic energy of objects, elastic potential energy, and gravitational potential energy. It does not include nuclear energy, thermal energy, or chemical potential energy.

Mechanical energy and total energy are not the same, because there are types of energy that are not mechanical energy. As a result, mechanical energy is not necessarily conserved. However, the mechanical energy of a system often remains constant or nearly constant. When this is the case, energy is only transformed between different kinds of mechanical energy.

Falling objects Look at the apple tree in **Figure 11**. An apple-Earth system, which is a system that includes an apple and Earth, has gravitational potential energy. The apple-Earth system does not have kinetic energy while the apple is hanging from the tree because the apple is not moving.

However, when the apple falls, it gets closer to Earth, so the GPE of the apple-Earth system decreases. This potential energy is transformed into kinetic energy as the apple's speed increases.

If potential energy is being converted into kinetic energy, then the mechanical energy of the apple-Earth system does not change as the apple falls. The potential energy that the apple-Earth system loses is gained back as kinetic energy. The form of mechanical energy changes, but the total amount of mechanical energy remains the same.

Reading Check **Describe** what happens to the mechanical energy of the apple-Earth system as the apple falls from the tree.

■ **Figure 11** The gravitational potential energy of an apple-Earth system could be converted into kinetic energy. However, the mechanical energy of the apple-Earth system remains nearly constant as the apple falls.

Estimate *the gravitational potential energy of one of the apples on the tree relative to the ground.*

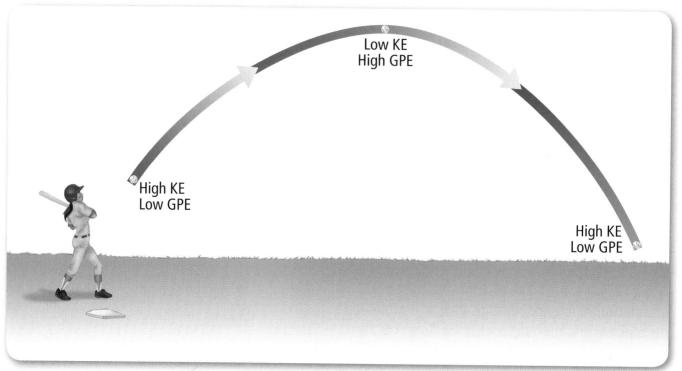

Low KE
High GPE

High KE
Low GPE

High KE
Low GPE

Figure 12 Kinetic energy is transformed into gravitational potential energy as the ball rises. As the ball falls, gravitational potential energy is transformed back into kinetic energy.

Predict *How large will the mechanical energy of the ball-Earth system be after the ball has reached the ground and rolled to a stop? Use the ground as the reference level.*

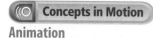
Concepts in Motion
Animation

VOCABULARY
WORD ORIGIN
Kinetic
  comes from the Greek word *kinetikos*, which means *putting in motion*
  A truck traveling on an interstate highway has a lot of kinetic energy.

Projectile motion Energy transformations also occur during projectile motion when an object moves in a curved path. Look at **Figure 12** and consider the ball-Earth system. When the ball leaves the bat, the ball is moving fast, so the system's kinetic energy is relatively large.

The ball's speed decreases as it rises, so the system's kinetic energy decreases. However, the system's gravitational potential energy increases as the ball goes higher. At the top of the ball's path, the system's GPE is larger and kinetic energy is smaller. Then, as the baseball falls, the system's GPE decreases as its kinetic energy increases. However, the mechanical energy of the ball-Earth system remains constant as the ball rises and falls.

Swings The mechanical energy transformations for a swing, like the one shown in **Figure 13,** are similar to the mechanical energy transformations for a roller coaster.

The ride starts with a push, which transfers kinetic energy to the rider. As the swing rises, the rider loses speed but gains height. In energy terms, kinetic energy changes to GPE. At the top of the rider's path, GPE is at its greatest.

Then, as the swing moves back downward, gravitational potential energy changes back to kinetic energy. At the bottom of each swing, the kinetic energy is at its maximum and the GPE is at its minimum. As the rider swings back and forth, energy is continually transformed between kinetic energy and GPE. However, the rider swings less and less on each cycle unless he or she pumps the swing or gets someone to provide a push. What is happening to the rider's mechanical energy?

FIGURE 13

Visualizing Energy Transformations

A ride on a swing illustrates how kinetic energy changes to potential energy and back to kinetic energy.

Calculate Your Power

Procedure

1. Read the procedure and safety information, and complete the lab form.
2. Find a set of **stairs** that you can safely run up.
3. Using a **stopwatch,** record how many seconds it takes you to run up the stairs.
4. Using a **meterstick,** measure the vertical height of the stairs.
5. Calculate the amount of chemical potential energy that you transformed into GPE to get up the stairs. Your mass in kilograms is your weight in pounds divided by 2.2.
6. Use the formula $P = \frac{E}{t}$ to calculate your power in running up the stairs.

Analysis

1. **Infer** how you could increase your power when you run up the stairs.

Power—how fast energy changes

Think again about the energy that your body extracts from food every day. You probably get enough energy from food in one day to jump nearly 10 km into the air. If this is true, then why can't you do this? You might have enough energy, but you don't have enough power. **Power** is the rate at which energy is converted. Power can be found using the following equation:

Power Equation

$$\text{Power (in watts)} = \frac{\text{Energy (in joules)}}{\text{time (in seconds)}}$$

$$P = \frac{E}{t}$$

Power is measured in watts; 1 watt equals 1 joule per second. A 13-W lightbulb transforms 13 J of electrical energy into radiant energy each second. A typical person can develop a power of only about 500 W for a jump. This results in a jump that is less than 1 m high for a person with average mass.

EXAMPLE Problem 6

Solve for Power You transform 950 J of chemical energy into mechanical energy to push a sofa. If it took you 5.0 s to move the sofa, what was your power?

Identify the Unknown: power: **P**

List the Knowns   Energy transformed: **E = 950 J**

time: **t = 5.0 s**

Set Up the Problem: $P = \frac{E}{t}$

Solve the Problem: $P = \frac{950 \text{ J}}{5.0 \text{ s}} = 190 \text{ W}$

Check the Answer: A typical human can develop a power of 400 W to 1,000 W for short periods of time. Developing a power of 190 W would require some exertion but would not be too difficult. The answer is reasonable.

PRACTICE Problems

Find **Additional Practice Problems** in the back of your book.

26. If a runner's power is 400 W as she runs, how much chemical energy does she convert into other forms in 10.0 minutes?

27. **Challenge** One horsepower is a unit of power equal to 746 W. How much energy can a 150-horsepower engine transform in 10.0 s?

 Review

Additional Practice Problems

Energy conversions in your body You transfer energy from your surroundings to your body when you eat. The chemical potential energy of food supplies the cells in your body with the energy that they need to function. Energy from food is often measured in Calories (C).

You have probably seen descriptions of Calories per serving on food packages, such as the sides of cereal boxes or milk cartons. One Calorie is equal to about 4,000 J.

Every gram of fat in a food supplies a person with about 10 C (40,000 J) of energy. Carbohydrates and proteins each supply about 5 C (20,000 J) of energy per gram. Everything that your body does requires energy. The number of Calories that you need for different activities depends on your weight, your body type, and your degree of physical activity. **Table 1** shows the amount of energy needed to do various activities.

Table 1	Calories Used in 1 Hour		
Type of Activity	Body Frames		
	Small	Medium	Large
Sleeping	48	56	64
Sitting	72	84	96
Eating	84	98	112
Standing	96	112	123
Walking	180	210	240
Playing tennis	380	420	460
Bicycling (fast)	500	600	700
Running	700	850	1,000

Section 3 Review

Section Summary

▶ According to the law of conservation of energy, energy cannot be created or destroyed.

▶ Energy can be transformed from one form to another.

▶ Mechanical energy is the sum of all the potential energies and kinetic energies of all the objects in a system.

▶ Power is the rate at which energy is converted from one form to another.

28. **MAIN Idea** Apply the law of conservation of energy and describe the energy transformations that occur as you coast down a long hill on a bicycle and then apply the brakes to make the bike stop at the bottom.

29. **Identify** whether each of the following is a form of mechanical energy: elastic potential energy, chemical potential energy, gravitational potential energy.

30. **Explain** how friction affects the mechanical energy of a system.

31. **Think Critically** A roller coaster is at the top of a hill and rolls to the top of a lower hill. If mechanical energy is constant, then on the top of which hill is the kinetic energy from the roller coaster's motion greater?

Apply Math

32. **Estimate Power** Approximately how much electrical energy does a 5-W lightbulb convert to radiant and thermal energy in one hour?

33. **Calculate Thermal Energy** The mechanical energy of a bicycle at the top of a hill is 6,000 J. The bicycle stops at the bottom of the hill by applying the brakes. If the gravitational potential energy of the bicycle-Earth system is 2,000 J at the bottom of the hill, how much mechanical energy was converted into thermal energy?

Objectives

- **Construct** a pendulum to compare the exchange of potential and kinetic energy when a swing is interrupted.
- **Measure** the starting and ending heights of the pendulum.

Background: Picture yourself swinging on a swing. Even without getting a push or pumping the swing, you rise to nearly the same height on each cycle. What would happen if a friend grabbed the swing's chains as you passed the lowest point? Would you come to a complete stop or continue rising to your previous maximum height, or would you rise to somewhere in between?

Question: *How does the motion and maximum height reached by a swing change if the swing is interrupted?*

Preparation

Possible Materials

ring stand
test-tube clamp
support-rod clamp, right angle
30-cm support rod
2-hole, medium rubber stopper
string (1 m)
metersticks (2)
graph paper

Safety Precautions

WARNING: *Be sure that the base is heavy enough or well anchored so the apparatus will not tip over.*

Form a Hypothesis

Examine the diagram on this page. How is it similar to the situation in the introductory paragraph? An object that is suspended so it can swing back and forth is called a pendulum. Hypothesize what will happen to the pendulum's motion and final height if its swing is interrupted.

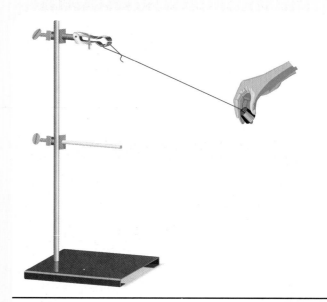

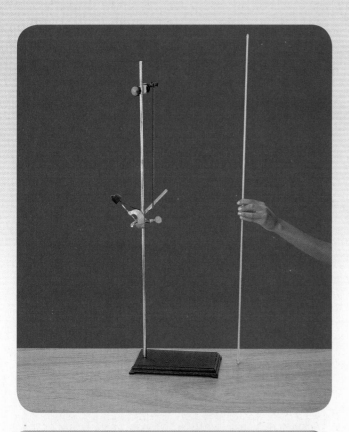

Make a Plan

1. Read the procedure and safety information, and complete the lab form.

2. As a group, write your hypothesis and identify the steps that you will take to test it. Be specific. Make sure that you test only one variable at a time. Also, state the materials that you will need.

3. Design a data table.

4. Set up an apparatus similar to the one shown in the diagram.

5. Devise a way to measure the starting and ending heights of the stopper.

6. Decide how to release the stopper from the same height each time.

7. Be sure to test your swing, starting it above the height of the crossarm, below the height of the crossarm, and even with the height of the corssarm. How many times should you repeat each starting point?

Follow Your Plan

1. Make sure your teacher approves your plan before you begin.

2. Carry out the approved experiment as planned.

3. While the experiment is going on, write any observations that you make and complete your data table.

Analyze Your Data

1. **Describe** When the stopper is released from the same height as the crossarm, is the ending height of the stopper exactly the same as its starting height?

2. **Analyze** the energy transformations. At what point along a single swing is the kinetic energy from the pendulum's motion greatest? When is the gravitational potential energy greatest?

3. **Identify** any possible sources of error. Are you confident that your measurements of the starting and ending heights are accurate?

Conclude and Apply

1. **Explain** Do the results support your hypothesis?

2. **Compare** the starting heights to the ending heights of the stopper. Is there a pattern? Can you account for the observed behavior?

3. **Discuss** Do your results support the law of conservation of energy? Why or why not?

4. **Discuss** Does the position of the crossarm affect the results of your experiment? Explain your reasoning.

5. **Infer** What would happen if the mass of the stopper were increased? How would you test it?

COMMUNICATE YOUR DATA

Compare your conclusions with those of the other lab teams in your class. Explain any differences between results.

World's Fastest Bicycle

Could a person pedaling a bicycle over flat terrain move faster than a car on a highway? With a unique type of bicycle called a recumbent, the answer is *yes*.

Mechanical disadvantage How could a bicycle convert an ordinary human into the fastest creature on Earth? The answer is mechanical *dis*advantage. The idea behind many simple machines is to increase force by decreasing distance.

On the other hand, a bicycle increases the effort of the rider, but the bicycle moves a greater distance. Look at the hand-powered cycle in **Figure 1.** The riders must exert great effort with their arms to accelerate their cycles forward. However, the payoff is a much higher top speed.

Recumbent bicycles, such as the one shown in **Figure 2,** maximize this mechanical disadvantage. At the same time, recumbent bicycles position the rider to be able to exert maximum force on the pedals.

Against the wind At the World Human Powered Speed Challenge, held every year in the Nevada wilderness, the enemy is the wind. The top competitors travel faster than 130 km/h (81 mph).

Figure 1 This hand-powered bicycle makes use of mechanical disadvantage.

Figure 2 The recumbent bike in the back is enclosed in an aerodynamic shell.

At that speed, air resistance would pummel an unprotected rider. To combat this air resistance, the fastest bikes are clad in shells, such as the one shown in **Figure 2.** These shells, constructed of Kevlar and carbon fibers for lightness and strength, deflect air around and past the rider.

The design of the shell means that the wheel can steer no more than a few degrees in either direction. The bicycles in the World Human Powered Speed Challenge are built for straight-line speed, not maneuverability.

Racing occurs in the early evening, when desert winds calm. Riders get a rolling start, sometimes taking up to 6 km to get up to speed. The timing begins at a marked spot along a closed highway and ends just 200 meters later. At top speed, riders cover this distance in less than 6 seconds.

WebQuest

Design a hand-powered bicycle or tricycle for use in a speed competition that is similar to the World Human Powered Speed Challenge.

BIG Idea Energy has many forms and can be transferred through work.

Section 1 Work and Machines

compound machine (p. 109)
efficiency (p. 110)
machine (p. 109)
mechanical advantage (p. 111)
simple machine (p. 109)
work (p. 106)

MAIN Idea Machines make doing work easier or faster by changing the force needed to do the work.

- Work is force applied through a distance.
- A machine can increase speed, change the direction of a force, or increase force.
- The output work from a machine is never as great as the input work on that machine.
- Efficiency is the ratio of output work to input work.
- Mechanical advantage is the ratio of output force to input force.

Section 2 Describing Energy

chemical potential energy (p. 117)
elastic potential energy (p. 117)
energy (p. 114)
gravitational potential energy (p. 118)
kinetic energy (p. 116)
potential energy (p. 117)
system (p. 114)

MAIN Idea Energy is the ability to cause change.

- Forms of energy include mechanical, electrical, chemical, thermal, and radiant energy.
- Kinetic energy is the energy that a moving object has because of its motion.
- Potential energy is stored energy due to the interactions between objects.
- Different forms of potential energy include elastic potential energy, chemical potential energy, and gravitational potential energy.

Section 3 Conservation of Energy

law of conservation of energy (p. 120)
mechanical energy (p. 121)
power (p. 126)

MAIN Idea Energy cannot be created or destroyed.

- According to the law of conservation of energy, energy cannot be created or destroyed.
- Energy can be transformed from one form to another.
- Mechanical energy is the sum of all the potential energies and kinetic energies of all the objects in a system.
- Power is the rate at which energy is converted from one form to another.

Use Vocabulary

Complete each sentence with the correct term from the Study Guide.

34. A combination of two or more simple machines is a(n) _____.

35. The ratio of the output force to the input force is the _____ of a machine.

36. _____ is the rate at which energy is transformed.

37. The energy due to compressing a spring is _____.

38. When a book is moved from a higher shelf to a lower shelf, _____ changes.

39. The muscles of a runner transform chemical potential energy into _____.

40. According to the _____, the universe's total amount of energy does not change.

Check Concepts

41. Using the scientific definition, which statement is always true of work?
   A) It is difficult.
   B) It involves levers.
   C) It involves a transfer of energy.
   D) It is done with a machine.

42. The gravitational potential energy of a cucumber-Earth system changes when which factor changes?
   A) the cucumber's speed
   B) the cucumber's mass
   C) the cucumber's temperature
   D) the cucumber's length

43. Which cannot be done by a machine?
   A) increase force
   B) increase work
   C) change direction of a force
   D) increase velocity

44. Which factor increases as the efficiency of a machine increases?
   A) work input   **C)** friction
   B) work output   **D)** input force

Use the following figure to answer question 45.

45. Which energy transformation is happening in the figure above?
   A) kinetic energy to GPE
   B) GPE to kinetic energy
   C) thermal energy to mechanical energy
   D) mechanical energy to chemical potential energy

46. Friction causes mechanical energy to be converted into which form?
   A) thermal energy   **C)** kinetic energy
   B) nuclear energy   **D)** potential energy

47. What term indicates the number of times a machine multiplies the input force?
   A) efficiency   **C)** mechanical advantage
   B) power   **D)** resistance

✓ **Assessment**   **Online Test Practice**

Interpret Graphics

48. Copy and complete the concept map of simple machines using the following terms: *compound machines, efficiency, mechanical advantage, output force, and work output.*

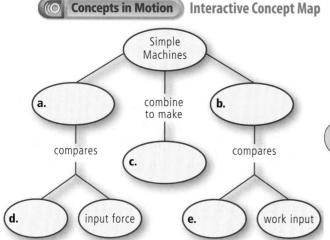

Concepts in Motion   Interactive Concept Map

Use the table below to answer question 49.

Toy Cars Rolling Down Ramps			
Ramp Height (m)	Speed at Bottom (m/s)	GPE (J)	KE (J)
0.50	3.13	**a.**	**b.**
0.75	3.83	**c.**	**d.**
1.00	4.43	**e.**	**f.**

49. THEME FOCUS **Make and Use Tables** Three toy cars, each with a mass of 0.050 kg, roll down ramps with different heights. The height of each ramp and the speed of each car at the bottom of each ramp is given in the table. Copy and complete the table by calculating the GPE for each car at the top of the ramp relative to the bottom of the ramp and the KE at the bottom from each car's motion down the ramp. Round your values two decimal places. How do the values of GPE and KE you calculate compare?

Think Critically

50. BIG Idea **Infer** What happens to the mechanical energy of a book when you push it across a table at constant speed?

51. **Diagram** On a cold day, you rub your hands together to make them warm. Diagram the energy transformations that occur, starting with the chemical potential energy that your muscles store.

Apply Math

52. **Calculate Work** Find the work needed to lift a 20-N book 2 m.

53. **Calculate Input Work** A machine has an efficiency of 20 percent. Find the input work if the output work is 140 J.

54. **Calculate Output Force** A certain lever has a mechanical advantage of 5.5. How heavy of a load could the lever lift with an input force of 20.0 N?

55. **Calculate Power** A person weighing 500 N climbs 3 m. How much power is needed to make the climb in 5 s?

56. **Calculate Kinetic Energy** What is the kinetic energy from the motion of a 0.06-kg tennis ball traveling at a speed of 50 m/s?

57. **Estimate Potential Energy** A boulder with a mass of 2,500 kg rests on a ledge that is 200 m above a canyon floor. Estimate the gravitational potential energy of the boulder-Earth system relative to the canyon floor.

58. **Estimate Speed** A boulder with a mass of 2,500 kg on a ledge 200 m above the ground falls. Estimate the speed of the boulder just before it hits the ground.

Standardized Test Practice

Multiple Choice

Record your answers on the answer sheet provided by your teacher or on a sheet of paper.

1. How much work is done lifting a 9.10-kg box straight up onto a shelf that is 1.80 m high?
   A. 5 J
   B. 15 J
   C. 50 J
   D. 160 J

Use the figure below to answer questions 2 and 3.

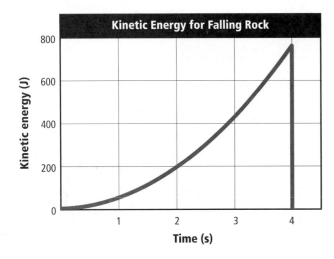

2. According to the graph, which is the best estimate for the kinetic energy from the rock's falling motion after that rock has fallen for 1 s?
   A. 0 J
   B. 50 J
   C. 100 J
   D. 200 J

3. If the rock has a mass of 1 kg, which is the speed of the rock after it has fallen for 2 s?
   A. 10 m/s
   B. 20 m/s
   C. 100 m/s
   D. 200 m/s

4. Which sequence describes the energy conversions in a car's engine?
   A. chemical to thermal to mechanical
   B. mechanical to thermal to chemical
   C. thermal to mechanical to chemical
   D. kinetic to potential to mechanical

5. A 10-kg box sits on the floor. Approximately how high would you have to lift the box to increase gravitational potential energy by 350 J?
   A. 3.5 m   C. 15 m
   B. 7.0 m   D. 40 m

6. It takes 600 J of work to lift a crate straight up but 1,200 J to push that crate up a ramp to the same height. What is the efficiency of the ramp?
   A. 30 percent   C. 75 percent
   B. 50 percent   D. 200 percent

7. What is the gravitational potential energy of an Earth-dictionary system if the dictionary has a mass of 5.0-kg and is located 2.0 m above the ground? Use the ground as the reference level.
   A. 2.5 J   C. 98 J
   B. 10 J   D. 196 J

8. The kinetic energy from an object's motion always changes when which factor changes?
   A. the object's chemical potential energy
   B. the object's volume
   C. the object's direction of motion
   D. the object's speed

9. An input force of 80 N is used to lift an object weighing 240 N with a system of pulleys. What is the mechanical advantage of the pulley system?
   A. 2   C. 4
   B. 3   D. 5

Short Response

Record your answers on the answer sheet provided by your teacher or on a sheet of paper.

10. As you throw a ball, you exert a force on that ball of 4.2 N. You exert this force on the ball while the ball moves a distance of 0.50 m. The ball then leaves your hand and travels a horizontal distance of 8.5 m to your friend. How much work have you done on the ball?

11. A student walks to school at a speed of 2.0 m/s. If the student's mass is 50 kg, what is the kinetic energy from the student's forward motion?

12. A book sliding across a horizontal table slows down and comes to a stop. The book's kinetic energy was converted into what form of energy?

Use the figure below to answer question 13.

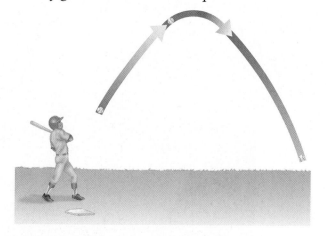

13. At what point on the ball's path is the kinetic energy from the ball's motion the least?

Extended Response

Record your answers on a sheet of paper.

14. What are three ways that simple machines can make work easier? For each one, give an example of a machine that makes work easier in that way.

15. Explain how a lubricant can increase the efficiency of a machine.

Use the figure below to answer question 16.

16. Describe how the work that the boy in the photograph does on the box is related to the energy transfer that occurs.

17. Name and describe at least two examples of how different forms of energy can be stored.

18. Explain why scientists consider conservation of energy to be a law and not a theory.

NEED EXTRA HELP?																		
If You Missed Question . . .	1	2	3	4	5	6	7	8	9	10	11	12	13	14	15	16	17	18
Review Section . . .	1	2	2	3	2	1	2	2	1	1	2	3	3	1	1	2	2	3

CHAPTER 5
Thermal Energy

Your one-stop online resource
connectED.mcgraw-hill.com

 Video

 Audio

Review

Inquiry

WebQuest

Assessment

Concepts in Motion

Multilingual eGlossary

Launch Lab
Temperature and Kinetic Energy

Hot water can burn your skin, but warm water does not. How is hot water different from warm water? You know that the temperature of hot water is higher. The temperature depends on the energy of the water molecules. A difference in temperature also has other effects, as you will see in this lab.

For a lab worksheet, use your StudentWorks™ Plus Online.

Inquiry Launch Lab

FOLDABLES®

Make a two-tab book. Label it as shown. Use it to organize your notes on thermal energy and heat.

Thermal Energy | Heat

THEME FOCUS Energy

Thermal energy is one of many different forms of energy.

BIG Idea Thermal energy is a form of energy that can be transferred as well as converted into other forms of energy.

Reading Preview

Essential Questions

▶ What is temperature?
▶ How are thermal energy and temperature related?
▶ What is the difference between thermal energy and heat?
▶ How can you calculate changes in thermal energy?

Review Vocabulary

energy: the ability to cause change; equivalent to the ability to do work

New Vocabulary

temperature
thermal energy
heat
specific heat

g Multilingual eGlossary

Temperature, Thermal Energy, and Heat

MAIN ‹Idea› The atoms and molecules that make up matter are in continuous, random motion.

Real-World Reading Link Have you ever made soup in a large pot on a stove? You place cold ingredients in a pot, turn on the burner, and wait for the soup to cook. Why does the soup cook faster if you stir the mixture or turn up the burner?

Temperature

You can use the words *hot* and *cold* to describe temperature. Something is hot when its temperature is high. When you heat water on a stove, its temperature increases. How are temperature and heat related?

Matter in motion The matter around you is made of tiny particles—atoms, ions, and molecules. In all materials, these particles are in constant, random motion. They move in all directions at different speeds. These particles have kinetic energy because they are moving. The greater their speeds, the greater their kinetic energy. If the particles that make up an object have more kinetic energy, then that object feels hotter.

For example, in **Figure 1,** the particles that make up the electric stove burner on the left have more kinetic energy than the particles that make up the burner on the right. As a result, the burner on the left feels hotter than the burner on the right. Is there a more exact relationship between temperature and kinetic energy?

■ **Figure 1** The particles that make up the left burner are moving faster than the particles that make up the right burner.

Predict *whether the kinetic energy of the particles that make up the water in a lake would be greater on a summer day or on a winter day.*

Kinetic energy and temperature The **temperature** of an object is the measure of the average kinetic energy of the particles that make up that object. The temperature of hot tea is higher than the temperature of iced tea. Therefore, the particles that make up the hot tea have more kinetic energy on average. In SI units, scientists measure temperature in kelvins (K). However, people around the world use the Celsius scale much more often. One kelvin is the same size as one degree Celsius.

Thermal Energy

If you let cold butter sit at room temperature for awhile, it warms and becomes softer. The particles that make up the room temperature air have more kinetic energy than the particles that make up the cold butter. Collisions between the air and the butter transfer energy from the faster-moving air particles to the slower-moving butter particles. The butter particles then move faster, and the temperature of the butter increases.

Recall that Earth and a ball exert gravitational forces on each other. Earth and the ball have potential energy. Similarly, the particles that make up matter exert attractive electromagnetic forces on each other and have potential energy. When you move a ball away from Earth, gravitational potential energy increases. When you move the particles that make up matter farther apart, the potential energy of that matter increases. **Figure 2** shows the kinetic and potential energies of particles.

The **thermal energy** of an object is the sum of the kinetic energy and the potential energy of all of the particles that make up that object. As the butter warmed and the kinetic energy of its particles increased, the butter's thermal energy increased.

FOLDABLES
Incorporate information from this section into your Foldable.

■ **Figure 2** The particles that make up a substance gain kinetic energy when their speeds increase. The potential energy of these particles increases when the separation between those particles increases.

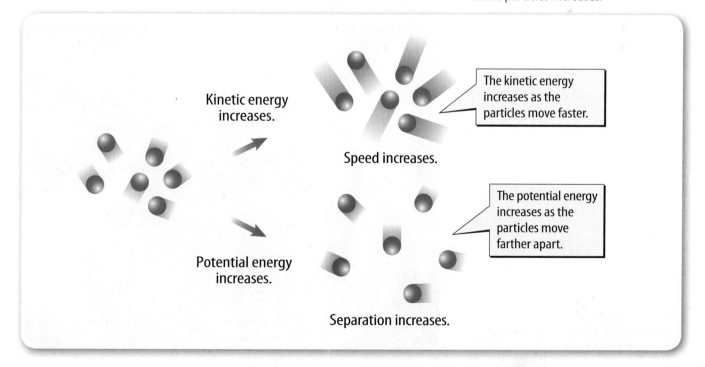

Kinetic energy increases.

Speed increases.

The kinetic energy increases as the particles move faster.

Potential energy increases.

The potential energy increases as the particles move farther apart.

Separation increases.

Temperature and thermal energy Thermal energy depends on temperature. The average kinetic energy of the particles that make up an object increases when the temperature of that object increases. Thermal energy is the total kinetic and potential energy of all of the particles that make up an object. The thermal energy of the object increases when the average kinetic energy of the particles that make up that object increases. Therefore, the thermal energy of an object increases as its temperature increases.

Mass and thermal energy Thermal energy also depends on mass. If the mass of the object increases, the thermal energy of that object also increases. Suppose you have a glass of water and a beaker of water, both at the same temperature. However, the beaker contains twice as much water as the glass. The average kinetic energy of the water molecules is the same in both containers. However, there are twice as many water molecules in the beaker. Therefore, the water in the beaker has twice as much thermal energy as the water in the glass.

Heat

Suppose you place a pot of water on a hot burner, as shown in **Figure 3**. When you do this, you transfer thermal energy from the warmer stove to the cooler pot and water. **Heat** is energy that is transferred between objects due to a temperature difference between those objects. Warmer objects always heat cooler objects, but the reverse never occurs. For example, a hot stove will heat a cold pot of water. However, a cold pot of water can never heat a hot stove.

■ **Figure 3** The warmer stove heats the cooler water.

Describe *the transfers of thermal energy in this figure.*

Specific Heat

Have you ever been to the beach during the summer? The ocean was probably cool, but the sand was probably hot. Energy from the Sun falls on the water and sand at nearly the same rate. However, the Sun's energy changes the sand's temperature more quickly than the water's temperature.

A substance's temperature changes when that substance absorbs thermal energy. This temperature change depends on the amount of thermal energy that the substance absorbs and the mass of the substance. This temperature change also depends on the nature of the substance.

The **specific heat** of a material is the amount of heat needed to raise the temperature of 1 kg of that material by 1°C. Scientists measure specific heat in joules per kilogram degree Celsius [J/(kg · °C)]. **Table 1** compares the specific heats of some familiar materials.

Compare water with iron in **Table 1**. Water has a very high specific heat. Metals, such as iron, have low specific heats. To raise equal masses of water and iron 1°C, water must absorb almost 10 times more thermal energy than iron. **Figure 4** explains why this is so.

✓ Reading Check Define specific heat.

Water as a coolant A coolant is a substance that can absorb a great amount of thermal energy with little change in temperature. Water is useful as a coolant because it can absorb thermal energy without a large change in temperature. For example, people use water as a coolant in automobile engines. Thermal energy transfers from the engine to the water as long as the water temperature is lower than the engine temperature.

Table 1	Comparison of Specific Heats*	
Substance	**Specific Heat [J/(kg·°C)]**
Water	4,200
Wood	1,700
Sand	830
Carbon (graphite)	710
Iron	450

*Values have been rounded.

■ **Figure 4** The change in temperature due to the transfer of thermal energy depends on the nature of the substance. Water molecules have strong attractions to each other. Therefore, the specific heat of water is high. The electrons in metals move through metals easily, transferring thermal energy throughout that metal quickly. As a result, the specific heats of metals are low.

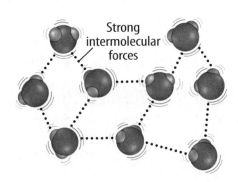

When thermal energy is added to water, some of the added thermal energy has to overcome some of the attraction between the molecules before those molecules can start moving faster.

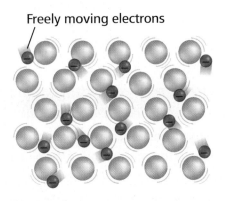

Freely moving electrons

In metals, electrons can move freely. When thermal energy is added, no strong attractions need to be overcome before the electrons can start to move faster.

Calculating thermal energy changes If Q is the change in thermal energy, C is specific heat, T_f is the final temperature, T_i is the initial temperature, and $T_f - T_i$ is the temperature change, you can calculate the change in thermal energy with the following equation.

Thermal Energy Equation

change in thermal energy (J) =

mass (kg) · temperature change(°C) · specific heat $\left(\dfrac{J}{kg \cdot °C}\right)$

$Q = m\,(T_f - T_i)\,C$

EXAMPLE Problem 1

Solve for Thermal Energy A wooden block has a mass of 20.0 kg and a specific heat of 1,700 J/(kg·°C). Find the change in thermal energy of the block as it warms from 15.0°C to 25.0°C.

Identify the Unknown:	change in thermal energy: Q
List the Knowns:	mass: m = 20.0 kg
	final temperature: T_f = 25.0°C
	initial temperature: T_i = 15.0°C
	specific heat: C = 1,700 J/(kg·°C)
Set Up the Problem:	$Q = m(T_f - T_i)C$
Solve the Problem:	Q = (20.0 kg)(25.0°C – 15.0°C)(1,700 J/(kg·°C))
	= (20.0 kg)(10.0°C)(1,700 J/(kg·°C))
	= 340,000 J
Check the Answer:	Do the units match on both sides of the equation? Changes in thermal energy (Q) are measured in units of joules (J). On the right side, (units of mass)(units of temperature change)(units of specific heat) = k̶g̶ × °C × J/(k̶g̶·°C) = J. The units on both sides of the equation match.

PRACTICE Problems

Find **Additional Practice Problems** in the back of your book

1. The air in a room has a mass of 50 kg and a specific heat of 1,000 J/(kg·°C). What is the change in thermal energy of the air when it warms from 20°C to 30°C?

   Review
   Additional Practice Problems

2. The temperature of a 2.0-kg block increases by 5°C when 2,000 J of thermal energy are added to the block. What is the specific heat of the block?

3. **Challenge** A wooden block has a mass of 0.200 kg, a specific heat of 710 J/(kg·°C), and is at a temperature of 20.0°C. What is the block's final temperature if its thermal energy increases by 2,130 J?

Measuring Specific Heat

A scientist can calculate the specific heat of a material from the measurements that he or she takes from a calorimeter, such as the one shown in **Figure 5**. To determine the specific heat of a material using a calorimeter, a scientist measures the mass of a sample of the material and the mass and initial temperature of the water in the calorimeter. The scientist then heats the sample, measures the sample's temperature, and places the sample in the water in the inner chamber of the calorimeter.

The sample cools as thermal energy is transferred to the water, and the temperature of the water increases. The transfer of thermal energy continues until the sample and the water are at the same temperature. Then the initial and final temperatures of the water are known, and the amount of heat gained by the water can be calculated from the water's mass, temperature change, and specific heat.

At this point, the scientist knows the mass of the substance, the thermal energy change of the substance, and the temperature change of the substance. From there, the specific heat of the unknown substance can be calculated using the thermal energy equation. With this information, the scientist might be able to accurately identify the substance.

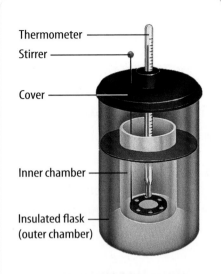

■ **Figure 5** A calorimeter can be used to determine the specific heat of materials. The sample is placed in the inner chamber.

Section 1 Review

Section Summary

▶ The temperature of an object is a measure of the average kinetic energy of the particles that make up the object.

▶ Thermal energy is the sum of the kinetic and potential energies of all of the particles in an object.

▶ Heat is a transfer of thermal energy due to a temperature difference.

▶ The specific heat of a material is the amount of heat needed to raise the temperature of 1 kg of the material by 1°C.

4. **MAIN Idea Describe** how the motions of the particles that make up an object change when the object's temperature increases.

5. **Describe** the energy transfer when you touch a block of ice with your hand.

6. **Infer** When one object heats another, does the temperature increase of one object always equal the temperature decrease of the other object? Explain.

7. **Explain** why water is often used as a coolant.

8. **Think Critically** Explain whether the following statement is true: For any two objects, the one with the higher temperature always has more thermal energy.

Apply Math

9. **Estimate** the change in the thermal energy of water in a pond with a mass of 1,000 kg and a specific heat of 4,200 J/(kg·°C) if the water cools by 1°C.

10. **Calculate** the specific heat of a metal if 0.3 kg of the metal absorb 9,000 J of heat as the metal warms by 10°C.

Reading Preview

Essential Questions

▶ What are conduction, convection, and radiation?

▶ How do thermal conductors differ from thermal insulators?

▶ How are thermal insulators used to control the transfer of thermal energy?

Review Vocabulary

density: the mass per unit volume of a substance

New Vocabulary

conduction
convection
radiation
thermal insulator

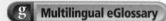 **Multilingual eGlossary**

Conduction, Convection, and Radiation

MAIN ◁ Idea Conduction, convection, and radiation are three ways to transfer energy.

Real-World Reading Link We control thermal energy transfers every day to keep from getting too hot or too cold. During the summer months, you might spend a lot of time in air-conditioned buildings. In the winter months, you might stay warm by wearing an insulated jacket when you go outside.

Conduction

Conduction, convection, and radiation transfer energy. **Conduction** is the transfer of thermal energy by collisions between the particles that make up matter. Conduction occurs because particles that make up matter are in constant motion.

Collisions transfer thermal energy If you leave a metal spoon in a pot of soup while it cooks on the stove, the spoon might get too hot to touch. As one end of the spoon heats up, the kinetic energy of the particles that make up that part of the spoon increases. These particles collide with neighboring particles. Conduction transfers thermal energy to the other end of the spoon as particles with more kinetic energy transfer kinetic energy to particles with less kinetic energy. Conduction transfers thermal energy without transferring matter. Conduction spreads thermal energy from warmer areas to cooler areas, as shown in **Figure 6.**

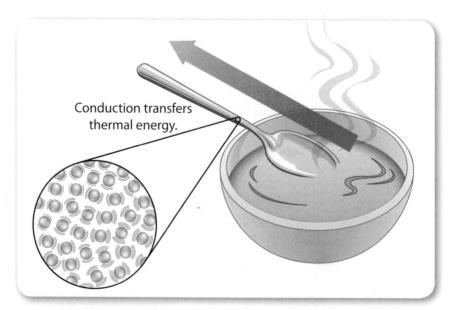

Conduction transfers thermal energy.

■ **Figure 6** Conduction occurs within a material as faster-moving particles collide with slower-moving particles.

State *If two objects are in contact, would conduction be from the cooler material to the warmer material or from the warmer material to the cooler material?*

Thermal conductors The rate at which conduction transfers thermal energy depends on the material. Conduction is faster in solids and liquids than it is in gases. In gases, particles are farther apart. Therefore, collisions among particles occur less frequently in gases.

The best conductors of thermal energy are metals. This is one reason why manufacturers often make cooking pots, like those in **Figure 7,** out of metal. In a piece of metal, some electrons are not bound to individual atoms. These electrons can move easily through the metal. Collisions between these electrons and other particles in the metal enable more rapid thermal energy transfers than in other materials. Silver, copper, and aluminum are among the best conductors of thermal energy.

Convection

Unlike solids, liquids and gases are fluids that flow. In fluids, convection can transfer thermal energy. **Convection** is the transfer of thermal energy in a fluid by the movements of warmer and cooler fluid. When conduction occurs, more energetic particles collide with less energetic particles and transfer thermal energy. When convection occurs, more energetic particles move from one place to another.

Most substances expand as their temperatures increase. That is, as the particles move faster, they tend to be farther apart. Recall that density is the mass of a material divided by its volume. When a fluid expands, its volume increases, but its mass does not change. Therefore, a fluid's density decreases when that fluid is heated.

Because fluids decrease in density as they are heated, a fluid that absorbs thermal energy also decreases in density. The density of a warmer sample of a fluid is less than the density of a cooler sample of that same fluid. The same is true for the parts of a fluid. The warmer parts of a fluid are less dense than the cooler parts of a fluid.

These differences in density within a fluid drive convection. The warmer portions of the fluid rise to the top of the fluid, and the cooler portions sink to the bottom. If a fluid is heated from below, convection currents form.

■ **Figure 7** These pots are made of metals that are excellent conductors of thermal energy.

Infer *why chefs often prefer pots that are good conductors of thermal energy when cooking.*

VOCABULARY . . . .
ACADEMIC VOCABULARY
Transfer
to convey from one place to another
She is going to transfer her pictures from her camera to the Web site.. . . . .

Convection currents How does convection occur? Look at the lamp shown in **Figure 8.** Some of these lamps contain oil and alcohol. When the oil is cool, its density is greater than the alcohol so it sits at the bottom of the lamp. When the light heats the two liquids in the lamp, the oil becomes less dense than the alcohol. The oil rises to the top of the lamp because it is less dense than the alcohol. As the oil rises, conduction transfers thermal energy from the warmer oil to the cooler alcohol. As a result, the oil cools as it rises.

By the time the oil reaches the top of the lamp, it cools and again becomes denser than the alcohol. As a result, the oil sinks. This rising-and-sinking action illustrates a convection current. Convection currents transfer thermal energy from warmer to cooler parts of the fluid. In a convection current, both conduction and convection transfer thermal energy.

✓ **Reading Check Contrast** conduction with convection.

Deserts and rainforests Earth's atmosphere is a fluid that is made of various gases. The atmosphere is warmer at the equator than it is at the North Pole and the South Pole. Also, the atmosphere is warmer at Earth's surface than it is at higher altitudes. These temperature differences produce convection currents that carry thermal energy away from the equator. Moist, warm air near the equator rises. As this air rises, it cools and loses moisture. Rain then falls over the equator.

The cooler, drier air sinks down toward the ground north and south of the equator. Desert zones form as a result. The temperature of this air increases as the air sinks. **Figure 9** shows how these convection currents result in rain forests and deserts.

■ **Figure 8** The heat from the light at the bottom of the lamp causes fluid to expand and rise. This creates convection currents in the lamp.

Explain *why the substances in the lamp rise and sink.*

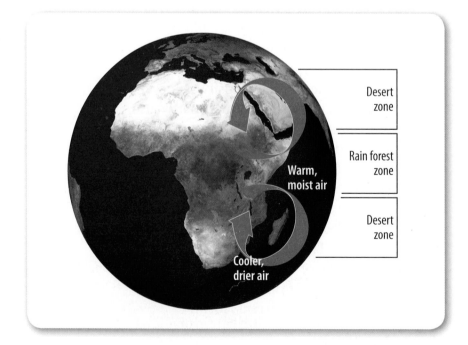

Desert zone

Rain forest zone

Desert zone

Warm, moist air

Cooler, drier air

■ **Figure 9** Sunlight on Earth is most intense at the equator. Convection currents form around Earth's equator as a result. Warm, moist air at the equator rises. As this air rises, it cools and loses its moisture as the rain that sustains rain forests near the equator. The convection currents then carry the dry air farther north and south. Deserts have formed where this air descends.

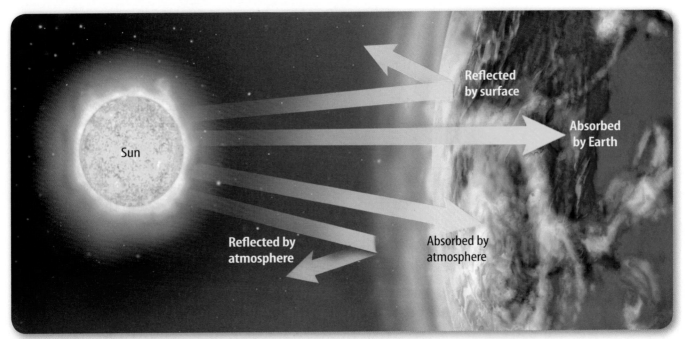

Reflected by surface

Sun

Absorbed by Earth

Reflected by atmosphere

Absorbed by atmosphere

Radiation

Energy from the Sun reaches Earth, but how does that energy travel through space? Almost no matter exists in the space between Earth and the Sun, so neither conduction nor convection could warm Earth. Instead, radiation transfers energy from the Sun to Earth.

Radiation is the transfer of energy by electromagnetic waves, such as light and microwaves. These waves travel through space even when matter is not present. Energy that is transferred by radiation is often called radiant energy. When you stand near a fire, radiation transfers energy from the fire and increases the thermal energy of your body.

Radiation and matter When radiation strikes a material, that material absorbs, reflects, and transmits some of the energy. **Figure 10** shows what happens to radiation from the Sun as it reaches Earth. The amount of energy that a material absorbs, reflects, and transmits depends on the type of material. The thermal energy of a material increases when that material absorbs radiant energy.

Radiation in solids, liquids, and gases In a solid, liquid, or gas, radiation travels through the space between particles. Particles can absorb and re-emit this radiation. This energy then travels through the space between particles, and other particles then absorb and re-emit the energy. Radiation usually passes more easily through gases than through solids or liquids because particles are much farther apart in gases than in solids or liquids. Thus, radiation transfers energy more rapidly and efficiently through gases than through liquids or solids.

Controlling Heat

You might not realize it, but you probably do a number of things every day to control thermal energy transfers between your body and your body's surroundings. For example, when it is cold outside, you put on a coat or a jacket before you leave your home. When you reach into an oven to pull out a hot dish, you might put a thick, cloth mitten over your hand.

In both cases, you have used various materials to help control transfers of thermal energy. Your jacket decreased how much thermal energy your body transferred to the surrounding air, keeping you from getting cold. The oven mitten decreased how much thermal energy the hot dish transferred to your hand, preventing that hot dish from burning your hand.

Animals and heat The animals shown in **Figure 11** have special features that help them control thermal energy transfers between their bodies and their bodies' surroundings. For example, the Antarctic fur seal's thick coat and the emperor penguin's thick layer of fat help to keep them from transferring thermal energy to their surroundings. This helps them survive in a climate where the temperature is often below freezing. In the desert, however, the scaly skin of the desert spiny lizard has just the opposite effect. It reflects the Sun's rays and keeps the animal from becoming too hot.

An animal's color also can play a role in keeping it warm or cool. The black feathers on the penguin's back, for example, allow it to absorb radiant energy.

✓ **Reading Check Identify** two animal adaptations for controlling thermal energy transfers.

■ **Figure 11** Animals have different adaptations that help them control heat. Both the Antarctic fur seal and the emperor penguin have a thick layer of fat that reduces transfers of thermal energy to those animals' surroundings. The scaly skin of the desert spiny lizard reflects radiation from the Sun to prevent overheating.

Antarctic Seal

Emperor Penguin

Spiny Lizard

■ **Figure 12** The tiny pockets of air in fleece make this jacket a good thermal insulator. They help reduce the transfer of the cyclist's thermal energy to the colder outside air.

Inquiry

Virtual Lab

Thermal insulators A **thermal insulator** is a material through which thermal energy moves slowly. Wood, fiberglass, and air are all good thermal insulators. Metals and other good conductors are poor thermal insulators. In conductors, thermal energy moves more rapidly from one place to another.

Insulated clothing Gases, such as air, are usually much better thermal insulators than solids or liquids. Some thermal insulators contain many pockets of trapped air. These air pockets conduct thermal energy poorly and also keep convection currents from forming. Fleece jackets, like the one worn by the cyclist in **Figure 12,** work in this way. When you put on a jacket like this one, the fibers in fleece trap the air and hold it next to you. The air slows the transfer of your body's thermal energy into its surroundings. Under the jacket, a blanket of warm air covers you.

✔ **Reading Check Explain** how trapped air makes a material, such as fleece, a good thermal insulator.

Insulated buildings Insulation, or materials that are thermal insulators, helps prevent thermal energy transfers out of buildings in cold weather and thermal energy transfers into buildings in warm weather. Manufacturers usually make building insulation from a fluffy material, such as fiberglass, that contains pockets of trapped air. Builders pack insulation into a structure's outer walls and attic, where it reduces thermal energy transfers between the structure and the surrounding air.

Insulation helps furnaces and air conditioners work more efficiently. In the United States, about 50 percent of the energy used in homes is for climate control. Repairing and improving insulation can dramatically reduce heating and cooling costs.

MiniLab

Inquiry   MiniLab

Compare Thermal Conductors

Procedure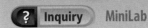

1. Read the procedure and safety information, and complete the lab form.
2. Obtain a **plastic spoon,** a **metal spoon,** and a **wooden spoon** of similar length.
3. Stick a **small, plastic bead** to the handle of each spoon with a **dab of butter or wax.** Each bead should be the same distance from the tip of the spoon.
4. Stand the spoons in a **beaker,** with the beads hanging over the edge of the beaker.
5. Carefully pour **boiling water** to a depth of about 5 cm in the beaker holding the spoons.

Analysis

1. **Describe** In what order did the beads fall from the spoons?
2. **Describe** how heat was transferred from the water to the beads.
3. **Rank** the spoons in their abilities to conduct heat.

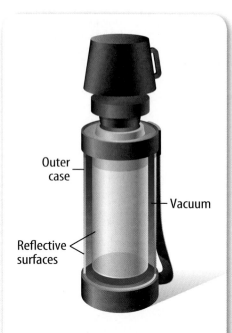

Figure 13 There is very little air between the two surfaces of a thermos bottle. This minimizes transfers of thermal energy by conduction and convection. Thermos bottles also often have reflective surfaces, which minimize energy transfers by radiation.

Labels on figure: Outer case, Vacuum, Reflective surfaces

Thermoses You might have used a thermos bottle, like the one shown in **Figure 13,** to carry hot soup or iced tea. A thermos bottle reduces energy transfers between the bottle's contents and its surroundings. As a result, the temperature of the contents changes very little, even over many hours.

In order to minimize thermal energy transfers from conduction and convection, a thermos bottle has two glass walls with very little air between those walls. Scientists call any region with very low gas density a vacuum. Glass is a good thermal insulator, and a vacuum is an extremely good thermal insulator.

To reduce energy transfers from radiation, manufacturers often coat the thermos bottle's inside and outside glass surfaces with aluminum. This makes each surface highly reflective. The inner reflective surface prevents radiation from transferring energy out of the liquid. The outer reflective surface prevents radiation from heating the liquid.

A thermos bottle keeps the liquid inside it warm or cool. Think about the things that you do to stay warm or cool. Sitting in the shade reduces energy transfers from radiation. Opening or closing windows affects thermal energy transfers from convection. Putting on a jacket reduces thermal energy transfers from conduction. In what other ways do you control heat?

Section 2   Review

Section Summary

▶ Conduction is the transfer of thermal energy by collisions between more energetic and less energetic particles.

▶ Convection is the transfer of thermal energy by the movement of warmer and cooler materials.

▶ Radiation is the transfer of energy by electromagnetic waves.

▶ Thermal insulators are used to reduce the transfer of thermal energy from one place to another.

11. **MAIN Idea Identify** which method of thermal energy transfer would be fastest through a vacuum, which would be fastest through a gas, and which would be fastest through a solid.

12. **Explain** why the air temperature near the ceiling of a room tends to be warmer than the air temperature near the floor.

13. **Predict** whether plastic foam, which contains pockets of air, would be a good thermal conductor or a good thermal insulator.

14. **Think Critically** Several days after a snowfall, the roofs of some homes on a street have almost no snow on them, while the roofs of other homes are still snow-covered. Give one reason, related to home insulation, that might cause this difference.

Apply Math

15. **Calculate Solar Radiation** Averaged over a year in the central United States, radiation from the Sun transfers about 200 W to each square meter of Earth's surface. If a house is 10 m long by 10 m wide, how much solar energy falls on the house each second?

Assessment   Online Quiz

LAB

Convection in Gases and Liquids

Objectives
- **Observe** convection currents formed in water.
- **Observe** convection currents formed in air.

Background: Birds use a large amount of energy to flap their wings. Many combine flapping flight with gliding through the air. Some birds conserve energy with soaring flight. For example, a hawk rarely flaps its wings as it circles high overhead on a warm, sunny day. Soaring birds use convection currents to stay aloft.

Question: *How do the convection currents that hawks glide on form?*

Preparation

Materials
500-mL beaker   black pepper
water   hot plate
balance   candle
small piece of paper

Safety Precautions

WARNING: *Use care when working with hot materials. Remember that hot and cold glass appear the same.*

Procedure

1. Read the procedure and safety information, and complete the lab form.
2. Pour 300 mL of water into the beaker.
3. Place a small piece of paper on the balance, and use the balance to measure 1 g of black pepper.
4. Sprinkle the pepper into the beaker of water, and let it settle to the bottom of the beaker.
5. Heat the bottom of the beaker using the hot plate. Do not boil the water.

6. Observe how the particles of pepper move as the water is heated, and make a drawing showing their motion.
7. Turn off the hot plate. Light the candle and let it burn for a few minutes.
8. Blow out the candle, and observe the motion of the smoke.
9. Make a drawing of the movement of the smoke.

Conclude and Apply

1. **Describe** how the particles of pepper moved as the water became hotter.
2. **Discuss** how the motion of the pepper particles is related to the motion of the water.
3. **Describe** how the smoke particles moved when the candle was blown out.
4. **Explain** how the convection currents that you formed in this lab are similar to the convection currents on which hawks glide.
5. **Form a hypothesis** that explains how the convection currents on which hawks glide form.

COMMUNICATE YOUR DATA

Compare your conclusions with other students in your class. State to what degree your conclusions agreed with the conclusions of your classmates. Explain any differences.

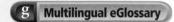

g | Multilingual eGlossary

Using Thermal Energy

MAIN ⟨Idea Thermal energy can be made useful by controlling its transfer and by transforming it into other forms of energy.

Real-World Reading Link As gasoline burns in an automobile engine, the engine transforms thermal energy into mechanical energy. In what other ways do we use heat and thermal energy?

Heating Systems

Almost everywhere in the United States, air temperatures at some time become cold enough that people seek out sources of warmth in addition to the Sun. As a result, most modern homes and public buildings contain some type of heating system. The best heating system for any building depends on the local climate and how the building is constructed.

The simplest and earliest heating system was probably a campfire. Later, people began burning wood and coal in stoves and furnaces to keep warm. Burning a material transforms chemical energy into thermal energy. Conduction, convection, and radiation then transferred this thermal energy from the stove to the surrounding air. One disadvantage of this system is that energy transfers from the room in which the stove is located to other rooms in the building can be slow.

Forced-air systems The most common type of heating system that people in the United States use today is the forced-air system, as shown in **Figure 14**. In this system, a furnace burns fuel and heats a volume of air. A fan then blows the warm air through a series of large pipes called ducts. The ducts lead to vents in each room. Cool air returns to the furnace through additional vents, and the furnace reheats it.

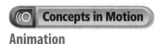

Concepts in Motion

Animation

■ **Figure 14** In forced-air systems, air heated by the furnace gets blown through ducts that usually lead to every room.

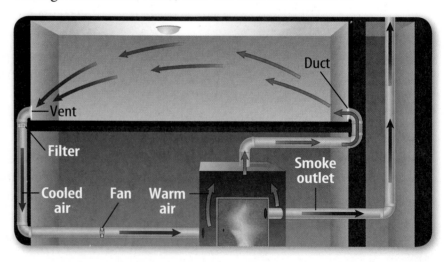

Radiator systems Before forced-air systems were widely used, radiators heated many homes and buildings. A radiator is a closed metal container that contains hot water or steam. In radiator heating systems, a central furnace heats a tank of water. A system of pipes carries the hot water or steam to radiators in other rooms.

Contrary to the name, radiators do not only transfer energy through radiation. Conduction transfers the thermal energy in the hot water or steam to the metal of the radiator and then to the surrounding air. Convection helps spread this energy throughout the room.

Electric heating systems An electric heating system has no central furnace. Instead, electric heating coils transform electrical energy into thermal energy. Portable space heaters contain such coils. Conduction transfers thermal energy from the heating coils to the surrounding air, and convection distributes thermal energy throughout the room.

 Reading Check Identify the energy transformation that occurs in an electric heating system.

Solar heating The Sun emits an enormous amount of radiant energy that strikes Earth every day. This energy can be used to help heat homes and other buildings through both passive solar heating and active solar heating.

Passive solar heating In passive solar heating systems, materials inside a building absorb radiant energy from the Sun and heat up during the day. At night when the building begins to cool, thermal energy absorbed by these materials helps keep the room warm.

Active solar heating In active solar heating, a solar collector is used. A **solar collector** is a device that transforms radiant energy from the Sun into thermal energy. Radiation from the Sun heats air or water in the solar collector. A pump circulates the hot fluid to radiators in rooms of the house. Both passive solar heating and active solar heating are shown in **Figure 15.**

Reading Check Describe the function of a solar collector.

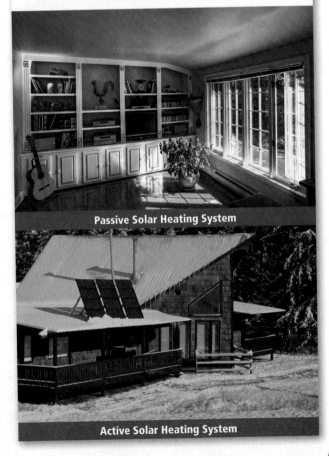

■ **Figure 15** In the passive solar heating system shown, the windows allow the Sun to transfer thermal energy into the room through radiation. Materials, such as ceramics, absorb and store some of this thermal energy. At night, these materials transfer this stored energy back to the room. In an active solar heating system, water or air is heated in a solar collector and then pumped throughout the building.

Compare and contrast *passive solar heating with active solar heating.*

Passive Solar Heating System

Active Solar Heating System

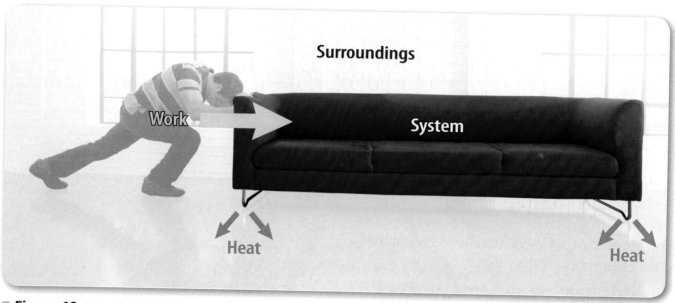

Surroundings

Work

System

Heat

Heat

Figure 16 A couch can be a system. If you define the couch in this figure as the system, then everything else is the surroundings. Work is done on this system by pushing the couch along the floor. As the couch slides along the floor, it heats the floor slightly through friction. The work done on this system is about equal to the heat from this system. The total energy of this system is nearly constant.

Describe *a system in which thermal energy is being transferred into that system.*

Thermodynamics

There is another way to increase the thermal energy of an object besides heating it. Have you ever rubbed your hands together to warm them on a cold day? Your hands get warmer and their temperature increases, even though you are not heating them near a fire or a stove. You do work to increase your hands' thermal energy when you rub them together. Thermal energy, heat, and work are related. **Thermodynamics** is the study of the relationships between thermal energy, heat, and work.

Heat and work increase thermal energy You can warm your hands by placing them near a fire because the fire heats your hands by radiation. If you rub your hands and hold them near a fire, the increase in your hands' thermal energy is even greater. Both the work you do and the heat from the fire increase your hands' thermal energy.

In the example above, your hands can be considered as a system. Recall that a system is anything around which you can draw a boundary. A system can be a group of objects, such as a galaxy, a car's engine, or something as simple as a ball. An example of a system is shown in **Figure 16**.

A system can interact with its surroundings in many ways. You increase the energy of a system whenever you do work on that system or heat that system. The work done on a system is the work done by something outside the system's boundary on something inside the system's boundary. A system can also heat its surroundings or do work on its surroundings. When a system does work on its surroundings or heats its surroundings, the total energy of the system decreases. The total energy of the surroundings increases by the same amount.

The first law of thermodynamics The **first law of thermodynamics** states that if the mechanical energy of a system is constant, the increase in thermal energy of that system equals the sum of the thermal energy transfers into that system and the work done on that system. This means that there are two ways to increase the temperature of a system. One way is to heat that system. Another way is to do work on that system. For example, you could increase the temperature of your hands both by warming them near a fire and by rubbing them together.

✓ **Reading Check** **Identify** two ways to increase the temperature of a system.

Isolated and non-isolated systems
A system is isolated if there are no energy transfers between that system and its surroundings. The total energy of an isolated system cannot change.

However, recall that energy can be converted between forms. According to the first law of thermodynamics, the thermal energy of an isolated system can still change, as long as the total energy of that system does not change.

A system is non-isolated if energy is transferred between the system and its surroundings. For example, a pan on a hot stove is a non-isolated system. This means that the total energy of a non-isolated system can change. However, remember that energy cannot be created or destroyed. Energy can only be transferred from one system to another or converted from one form to another.

The second law of thermodynamics
Energy spontaneously transfers from the warm radiator to the cat in **Figure 17.** Could the reverse ever happen? Could energy spontaneously transfer from the cat to the warmer radiator?

Energy could never spontaneously transfer from the cat to the radiator. This is due to the second law of thermodynamics. The **second law of thermodynamics** states that energy spontaneously spreads from regions of higher concentration to regions of lower concentration. It is possible to transfer energy from regions of lower concentration to regions of higher concentration, but this process does not happen spontaneously.

VOCABULARY ....

WORD ORIGIN
Thermodynamics
comes from the Greek word *thermos,* which means hot; and the Greek word *dynamis,* which means power
The women and men who design the engines for cars must have a good understanding of thermodynamics. ....

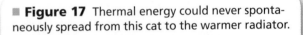

■ **Figure 17** Thermal energy could never spontaneously spread from this cat to the warmer radiator.

Converting Thermal Energy into Mechanical Energy

If you push a book sitting on a table, the book will slide and come to a stop. Friction between the book and the table converted the book's mechanical energy into thermal energy.

In the example above, mechanical energy was converted completely into thermal energy. Is it possible to do the reverse and convert thermal energy completely into mechanical energy? Remember that a system's mechanical energy is related to the motion of the objects in that system as well as the interactions between the objects of that system. A system's thermal energy is related to the motions and interactions of all the particles in that system.

Because there are far more particles than objects in a system, thermal energy is spread among more particles than objects than mechanical energy. As a result, mechanical energy tends to transform into thermal energy. Therefore, it is not possible to completely convert thermal energy into mechanical energy.

Heat engines A **heat engine** is a device that converts some thermal energy into mechanical energy. A car's engine is an example of a heat engine. A car's engine converts the chemical energy in gasoline into thermal energy. The engine then transforms some of this thermal energy into mechanical energy and rotates the car's wheels, as shown in **Figure 18.** However, only about 25 percent of the thermal energy released by the burning gasoline is converted into mechanical energy.

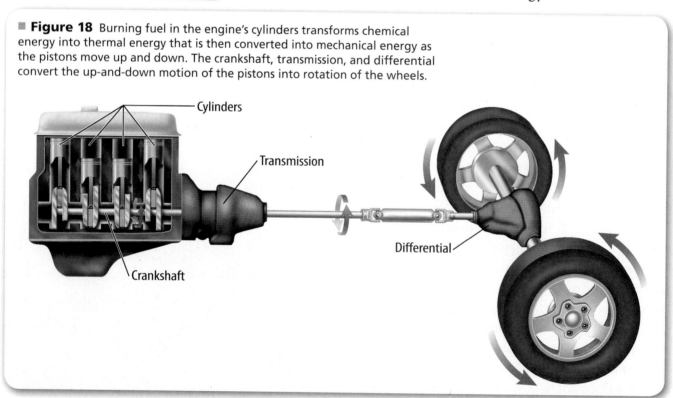

■ **Figure 18** Burning fuel in the engine's cylinders transforms chemical energy into thermal energy that is then converted into mechanical energy as the pistons move up and down. The crankshaft, transmission, and differential convert the up-and-down motion of the pistons into rotation of the wheels.

Cylinders

Transmission

Crankshaft

Differential

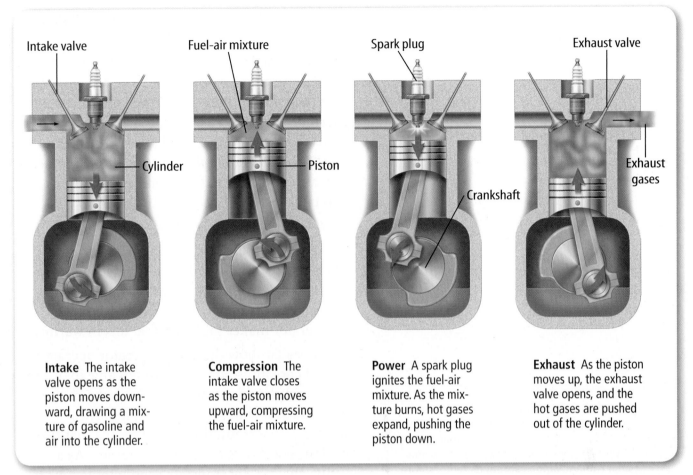

Intake The intake valve opens as the piston moves downward, drawing a mixture of gasoline and air into the cylinder.

Compression The intake valve closes as the piston moves upward, compressing the fuel-air mixture.

Power A spark plug ignites the fuel-air mixture. As the mixture burns, hot gases expand, pushing the piston down.

Exhaust As the piston moves up, the exhaust valve opens, and the hot gases are pushed out of the cylinder.

Internal combustion engines Almost all cars are powered by internal combustion engines. A heat engine that burns fuel inside a set of cylinders is an **internal combustion engine.** Each cylinder contains a piston that moves up and down. Each up or down movement of the piston is called a stroke. Automobile and diesel engines have four strokes per cycle. **Figure 19** shows the four-stroke cycle in an automobile engine.

Another type of heat engine is an external combustion engine. The cylinders in external combustion engines are heated by burning fuel outside the cylinders. Old-fashioned steam engines are external combustion engines.

Efficiency of heat engines Approximately three-fourths of the chemical energy that is transformed into thermal energy in a car engine is never converted into mechanical energy. The inefficiency of heat engines is not due only to friction. Even if friction could be completely eliminated, a typical heat engine still would not be 100 percent efficient.

Instead, the efficiency of a heat engine depends on the difference in the temperature of the burning gases in the cylinder and the temperature of the air outside the engine. Increasing the temperature of the burning gases and decreasing the temperature of the surroundings make the engine more efficient.

■ **Figure 19** The up-and-down moveme-nt of a piston in an automobile engine consists of four separate strokes. These four strokes form a cycle that is repeated many times a second by each piston.

Identify *the stroke during which the spark plug ignites.*

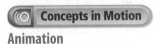

Animation

Doing Work to Transfer Thermal Energy

Is it possible to transfer thermal energy from a cooler area to a warmer area? You may think that the answer is no due to the second law of thermodynamics. However, thermal energy can be transferred from a cooler area to a warmer area if work is done in the process. Refrigerators and air conditioners function based on this principle.

Refrigerators A refrigerator does work as it transfers thermal energy from inside the cool refrigerator to the warmer room. The energy to do the work comes from the electrical energy the refrigerator obtains from an electrical outlet. A refrigerator makes the room that it is in warmer.

Figure 20 shows how a refrigerator operates. Liquid coolant is pumped through an expansion valve and changes into a gas. When the coolant expands as it changes into a gas, it pushes outward on its surroundings. This means that the coolant does work on its surroundings. As a result, the coolant transfers energy to its surroundings and the coolant cools. The cold gas is pumped through pipes inside the refrigerator, where it absorbs thermal energy from the area where you keep your food. When this happens, the inside of the refrigerator cools.

The gas then is pumped to a compressor that does work on the gas by compressing it. This makes the gas warmer than the temperature of the room. The warm gas is pumped through the condenser coils. Because the gas is warmer than the room, thermal energy spreads from the gas to the room. As the gas heats the room, it cools and changes back to a liquid and enters the expansion valve again. The cycle is then repeated.

Air conditioners An air conditioner operates like a refrigerator, except that warm air from the room is forced to pass over tubes containing the coolant. Thermal energy is transferred from the warm air to the coolant. The thermal energy that is absorbed by the coolant is then transferred to the air outdoors.

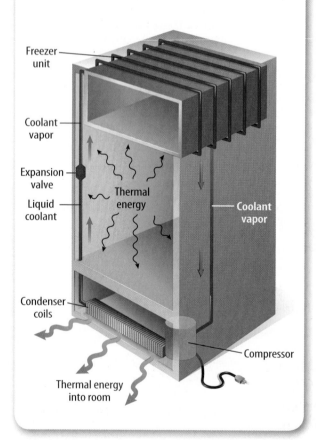

■ **Figure 20** A refrigerator must do work on the coolant in order to transfer thermal energy from inside the refrigerator to the warmer air outside. Work is done when the compressor compresses the coolant vapor, causing its temperature to increase.

Freezer unit

Coolant vapor

Expansion valve

Liquid coolant

Thermal energy

Coolant vapor

Condenser coils

Compressor

Thermal energy into room

Heat pumps A heat pump is a two-way air conditioner. In warm weather, a heat pump operates as an ordinary air conditioner. It does work to transfer thermal energy from the cooler building to the warmer outdoors. This cools the building while warming the outdoors very slightly. In cold weather, a heat pump operates like an air conditioner in reverse. It does work to transfer thermal energy from the cooler outdoors to the warmer building. This warms the building while cooling the building's surroundings very slightly.

Energy transformations and thermal energy

Many energy transformations occur around you every day that convert one form of energy into a more useful form. However, when these energy transformations occur, some of the energy is usually converted into thermal energy. For example, friction converts mechanical energy into thermal energy when the electric generators in **Figure 21** rotate. A laptop computer converts electrical energy into thermal energy. The thermal energy from these energy transformations is no longer in a useful form and is transferred to the surroundings by conduction and convection.

■ **Figure 21** Electric generators transform mechanical energy into electrical energy. However, some energy is always converted into thermal energy during any energy transformation.

Section 3 Review

Section Summary

▶ The first law of thermodynamics states that if the mechanical energy of a system is constant, then the increase in thermal energy of that system equals the work done on that system plus the thermal energy transferred into that system.

▶ The second law of thermodynamics states that energy spontaneously spreads from regions of higher concentration to regions of lower concentration.

▶ A heat engine transforms thermal energy into mechanical energy.

▶ Refrigerators, air conditioners, and heat pumps do work to transfer thermal energy from a cooler region to a warmer region.

16. **MAIN Idea Describe** a device that transforms thermal energy into another useful form.

17. **Explain** how the thermal energy of an isolated system changes with time if the mechanical energy of that system is constant.

18. **Compare and contrast** an active solar heating system with a radiator system.

19. **Predict** whether energy will ever spontaneously transfer from a cold pot of water to a hot stove.

20. **Diagram** how the thermal energy of the coolant changes as the coolant flows through the refrigerator.

21. **Think Critically** Suppose you vigorously shake a bottle of fruit juice. Predict how the temperature of the juice will change. Explain your reasoning.

Apply Math

22. **Calculate Change in Thermal Energy** Suppose you push down on the handle of a bicycle pump with a force of 20 N. The handle moves 0.3 m, and there is no heat between the pump and its surroundings. What is the change in thermal energy of the bicycle pump?

"20 DEGREES COOLER INSIDE!"

Imagine the splashy movie theater signs all across the country in the summer months of the 1920s. Many were festooned with pictures of polar bears and false icicles. The idea of cool relief from the sweltering summer heat caught on quickly, and theater owners played up the attraction that air-conditioning held for their hot and weary summertime patrons.

A simple idea The concept behind air conditioning is simple and as everyday as stepping from a swimming pool or a shower. When you leave the water, the chill that you feel is not necessarily from cold air. When liquids evaporate, they absorb thermal energy. Thermal energy transfers to the water from your body, through your skin, so you feel cooler.

Air-conditioning systems do not use water. Instead, air conditioners circulate a substance with a very low boiling point. As the substance changes from a liquid to a gas, it absorbs thermal energy and cools the surrounding air. That cool air is pumped throughout the building, whether it is in a movie theater, a shopping mall, or your own home.

The modern air conditioner was invented in 1902 by Willis Carrier. Carrier designed his air conditioner to cool machinery at a printing plant.

Theaters caught on to the air-conditioning trend in the 1920s, but it was not until the 1950s that home air-conditioning caught on with America's middle class. According to some historians, this change altered architecture as well as where and how people live.

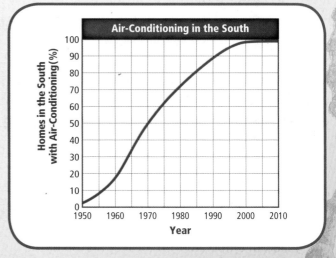

Figure 1 The graph shows how home air-conditioning grew in the South after World War II.

Architecture and air-conditioning Air-conditioning's biggest effects probably came in warmer climates. Over time, more and more houses in these climates were built with air-conditioning in mind. As a result, interior rooms got bigger, and front porches began to disappear.

Where people live The existence of air conditioning also encouraged people to move to warmer climates. Between 1910 and 1950, the South lost more than 10 million people due to migration to cooler climates. That trend reversed after 1950, as in-home air-conditioning became more common. In the 1970s alone, over 3 million more people moved to the South as moved away. **Figure 1** shows how the percentage of homes in the South using air-conditioning changed with time.

🌐 **WebQuest** **Design** Research ways architects and engineers increase the efficiency of climate control systems. Focus on building design, materials used, and advances in climate-control technology. Present a simple model or sketch of a new home designed to have an efficient climate-control system.

Theme Focus Energy

Thermal energy is one of many different forms of energy. Thermodynamics is the study of the relationships between thermal energy, heat, and work.

BIG Idea Thermal energy is a form of energy that can be transferred as well as converted into other forms of energy.

Section 1 Temperature, Thermal Energy, and Heat

heat (p. 140)
specific heat (p. 141)
temperature (p. 139)
thermal energy (p. 139)

MAIN Idea The atoms and molecules that make up matter are in continuous, random motion.

- The temperature of an object is a measure of the average kinetic energy of the particles that make up the object.
- Thermal energy is the sum of the kinetic and potential energies of all the particles in an object.
- Heat is a transfer of thermal energy due to a temperature difference.
- The specific heat of a material is the amount of heat needed to raise the temperature of 1 kg of the material by 1°C.

Section 2 Conduction, Convection, and Radiation

conduction (p. 144)
convection (p. 145)
radiation (p. 147)
thermal insulator (p. 149)

MAIN Idea Conduction, convection, and radiation are three ways to transfer energy.

- Conduction is the transfer of thermal energy by collisions between more energetic and less energetic particles.
- Convection is the transfer of thermal energy by the movement of warmer and cooler materials.
- Radiation is the transfer of energy by electromagnetic waves.
- Thermal insulators are used to reduce the transfer of thermal energy from one place to another.

Section 3 Using Thermal Energy

first law of thermodynamics (p. 155)
heat engine (p. 156)
internal combustion engine (p. 157)
second law of thermodynamics (p. 155)
solar collector (p. 153)
thermodynamics (p. 154)

MAIN Idea Thermal energy can be made useful by controlling its transfer and by transforming it into other forms of energy.

- The first law of thermodynamics states that if the mechanical energy of a system is constant, then the increase in thermal energy of that system equals the work done on that system plus the thermal energy transferred into that system.
- The second law of thermodynamics states that energy spontaneously spreads from regions of higher concentration to regions of lower concentration.
- A heat engine transforms thermal energy into mechanical energy.
- Refrigerators, air conditioners, and heat pumps do work to transfer thermal energy from a cooler region to a warmer region.

Use Vocabulary

Complete each sentence with the correct term from the Study Guide.

23. A(n) _____ is a device that converts thermal energy into mechanical energy.

24. _____ is the study of the relationships between heat, work, and thermal energy.

25. A(n) _____ is a device that transforms the Sun's radiant energy into thermal energy.

26. The energy required to raise the temperature of 1 kg of a material by 1°C is a material's _____.

27. The _____ states that thermal energy can be transferred from a cooler location to a warmer location only when work is done.

28. Thermal energy is transferred slowly through a(n) _____.

Check Concepts

Use the figure below to answer question 29.

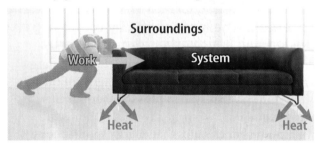

29. If work is greater than heat in the figure, then what must be true about the couch?
  A) Its total energy is decreasing.
  B) Its total energy is constant.
  C) Its total energy is increasing.
  D) Its total energy increases, then decreases.

30. In which device is fuel burned inside chambers called cylinders?
  A) internal combustion engine
  B) external combustion engine
  C) heat pump
  D) air conditioner

31. During which phase of a four-stroke engine are waste gases removed?
  A) power stroke
  B) intake stroke
  C) compression stroke
  D) exhaust stroke

32. Which material is a poor thermal insulator?
  A) iron   **C)** air
  B) feathers   **D)** plastic

33. Which device transfers thermal energy from a cooler region to a warmer region?
  A) solar panel
  B) refrigerator
  C) internal combustion engine
  D) thermal conductor

34. Which term describes the measure of the average kinetic energy of the particles that make up an object?
  A) temperature
  B) specific heat
  C) potential energy
  D) thermal energy

35. Which is energy that is transferred between objects due to a temperature difference between those objects?
  A) heat
  B) radiation
  C) conduction
  D) insulation

Assessment   Online Test Practice

Interpret Graphics

36. Complete the following events-chain concept map to show how an active solar heating system works.

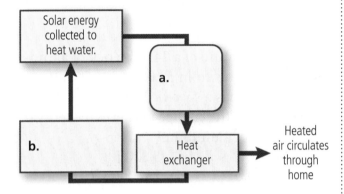

37. Copy and complete this concept map.

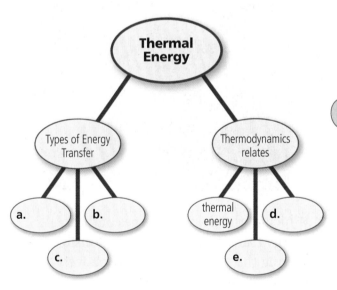

Think Critically

38. **THEME FOCUS** **Explain** On a hot day, a friend suggests that you can make your kitchen cooler by leaving the refrigerator door open. Explain whether leaving the refrigerator door open would cause the air temperature in the kitchen to decrease.

39. **Compare** Which has the greater amount of thermal energy—one liter of water at 50°C or two liters of water at 50°C?

40. **Explain** whether the following statement is true: If the thermal energy of an object increases, the temperature of the object must also increase.

41. **Predict** Suppose a beaker of water is heated from the top. Predict which is more likely to occur in the water: thermal energy transfer by conduction or convection.

42. **BIG Idea** **Order** the events that occur in the removal of thermal energy from an object by a refrigerator. Draw the complete cycle, from the placing of a warm object in the refrigerator to the changes in the coolant.

43. **Compare and contrast** the scientific definitions of *theory* and *law*. Explain why the first law of thermodynamics is not considered to be a theory.

Apply Math

44. **Calculate Thermal Energy** How much thermal energy is needed to raise the temperature of 4.0 kg of water from 25°C to 75°C?

45. **Calculate Temperature Change** Approximately how much does the temperature of 33.0 g of graphite change when it absorbs 350 J of thermal energy?

46. **Calculate Specific Heat** The temperature of a 3.0-kg material increases by 5.0°C when 6,750 J of thermal energy are added to it. What is the specific heat of the material?

47. **Calculate Final Temperature** 200 g of water in a glass is at 10.0°C when an ice cube is placed in the glass. What is the temperature of the water after it has transferred 4,200 J of thermal energy to the ice cube?

Standardized Test Practice

Multiple Choice

Record your answers on the answer sheet provided by your teacher or on a sheet of paper.

1. Which is temperature most directly related to?
   A. kinetic energy
   B. mechanical energy
   C. potential energy
   D. thermal energy

Use the table below to answer questions 2 and 3.

Material	Specific Heat [J/(kg·°C)]
Copper	385
Gold	129
Lead	130
Tin	227
Zinc	388

2. A 2-kg block of which material would require about 450 joules of thermal energy to increase its temperature by 1°C?
   A. tin   C. gold
   B. zinc   D. lead

3. Which material would require the most heat to raise a 5-kg sample of the material from 10°C to 50°C?
   A. gold
   B. lead
   C. tin
   D. zinc

4. Automobile engines are usually four-stroke engines. During which stroke does the spark from a spark plug ignite the fuel-air mixture?
   A. the intake stroke
   B. the compression stroke
   C. the power stroke
   D. the exhaust stroke

5. Which scientific principle states that energy spontaneously spreads from more concentrated regions to less concentrated regions?
   A. first law of thermodynamics
   B. second law of thermodynamics
   C. law of conservation of energy
   D. law of heat and work

6. The temperature of a 24.5-g block of aluminum decreases from 30.0°C to 21.5°C. If aluminum has a specific heat of 897 J/(kg·°C), how large is the change in thermal energy of the block of aluminum?
   A. 187 J   C. 5,820 J
   B. 2,590 J   D. 187,000 J

Use the figure below to answer question 7.

7. Which of the following types of thermal energy transfer is primarily responsible for heating the water in the figure above?
   A. conduction
   B. convection
   C. convection and radiation
   D. conduction and convection

8. Which device does work to transfer thermal energy from a cooler area to a warmer area?
   A. heat engine
   B. refrigerator
   C. solar collector
   D. stove

Short Response

*Record your answers on the answer sheet
provided by your teacher or on a sheet of paper.*

9. Define the term *heat* and state what units
   are used to measure heat.

10. Give an example of how electromagnetic
    waves transfer energy by radiation. Give an
    example of how energy is transferred by
    conduction.

11. How is the color of a material related to its
    absorption and reflection of radiant energy?

12. What property of water makes it useful as a
    coolant?

Use the figure below to answer question 13.

13. Name and describe the type of thermal
    energy transfer that occurred through the
    spoon in the above figure.

14. How is a heat pump different from an air
    conditioner?

Extended Response

*Record your answers on a sheet of paper.
Use the figure below to answer question 15.*

15. Describe the process by which the above
    object is used to measure the specific heat
    of materials.

16. Define the terms *temperature* and *thermal
    energy*. Explain how the temperature and
    thermal energy of an object are related.

17. Conduction can occur in solids, liquids,
    and gases. Explain why solids and liquids
    are better conductors than gases.

18. Explain why radiation usually passes
    through gases more easily than through
    solids or liquids.

19. Suppose you have a half-full glass of 250 mL
    of water at a temperature of 30°C. You then
    add 250 mL of water at the same tempera-
    ture to the glass. Explain any changes in the
    water's temperature and thermal energy.

NEED EXTRA HELP?																			
If You Missed Question . . .	1	2	3	4	5	6	7	8	9	10	11	12	13	14	15	16	17	18	19
Review Section . . .	1	1	1	3	3	1	2	3	1	2	2	1	2	3	1	1	2	2	1

CHAPTER 6
Electricity

ConnectED

Your one-stop online resource
connectED.mcgraw-hill.com

 Video

 Audio

 Review

 Inquiry

 WebQuest

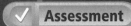 Assessment

Concepts in Motion

Multilingual eGlossary

Launch Lab
Electric Circuits

No lights! No refrigeration! No computers, video games, or TVs! Without electricity, many of the things that make your life enjoyable would not exist. For these devices to operate, electric charge must flow in the electric circuits that are part of the device. Under what conditions does electric charge flow in an electric circuit?

For a lab worksheet, use your StudentWorks™ Plus Online.

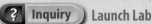 Inquiry Launch Lab

FOLDABLES®

Make a three-tab book. Label it as shown. Use it to organize your notes on electric charge, electric current, and electrical energy.

Electricity

| Electric Charge | Electric Current | Electrical Energy |

THEME FOCUS Technology
Our ability to manipulate electric charge has
allowed us to produce everything from electric
lights to digital music players to ultraportable
computers.

BIG ((Idea(Electricity consists of static and moving
electric charges.

Section 1 • Electric Charge

Section 2 • Electric Current

Section 3 • More Complex
Circuits

Reading Preview

Essential Questions

▶ How do gravitational force and electric force compare?

▶ What is the difference between conductors and insulators?

▶ How can objects become electrically charged?

Review Vocabulary

gravity: attractive force between two objects that depends on the masses of the objects and the distances between them

New Vocabulary

static electricity
law of conservation of charge
electric field
conductor
insulator
charging by contact
charging by induction
electroscope

g Multilingual eGlossary

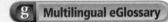

■ **Figure 1** Particles in the shoe's sole hold electrons more tightly than particles in the carpet hold electrons.

Compare *the total number of charges in the left panel with the total number of charges in the right panel.*

Electric Charge

MAIN ◁Idea **Like electric charges repel each other, and unlike charges attract each other.**

Real-World Reading Link Have you ever been shocked when you touched a metal doorknob or the outside of a car? This shocking experience was a result of electric charge.

Positive and Negative Charges

Matter is composed of atoms. These atoms are composed of protons, neutrons, and electrons. Protons have positive electric charge and electrons have negative electric charge. Neutrons have no electric charge.

The amount of positive charge on a proton equals the amount of negative charge on an electron. An atom has equal numbers of protons and electrons, so the positive and negative charges cancel out and an atom has no net electric charge. Objects with no net charge are said to be electrically neutral. The amount of electric charge is measured in coulombs (C). There are 6,250 million billion protons in 1 C of electric charge and 6,250 million billion electrons in −1 C of electric charge.

Transferring charge Electrons are bound more tightly to some atoms and molecules. For example, compared to the electrons in the atoms that make up a carpet, electrons are bound more tightly to the atoms that make up the soles of your shoes. **Figure 1** shows that when you walk on the carpet, electrons transfer from the carpet to the soles of your shoes.

As a result, the soles of your shoes gain an excess of electrons and become negatively charged. The carpet has lost electrons and has an excess of positive charge. **Static electricity** is the accumulation of excess electric charge on an object.

Before the shoe scuffs against the carpet, both the sole of the shoe and the carpet are electrically neutral.

As the shoe scuffs against the carpet, electrons transfer from the carpet to the sole of the shoe.

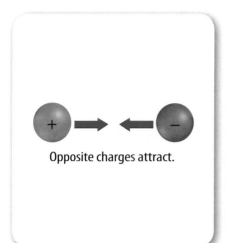

Opposite charges attract.

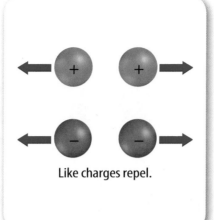

Like charges repel.

■ **Figure 2** Positive charges and negative charges exert forces on each other.

Describe *what would happen if two protons were placed near each other.*

Conservation of charge When your shoe soles become charged, they illustrate the law of conservation of charge. The **law of conservation of charge** states that charge can be transferred from object to object, but it cannot be created or destroyed. In **Figure 1,** electrons are transferred from the carpet to the shoes. However, the total charge does not change. Usually, it is the electrons that transfer from one object to another and not the protons.

✓ **Reading Check Compare** the number of electrons with the number of protons in a charged object.

Charges Exert Forces

Have you noticed how clothes sometimes cling together when removed from the dryer? These clothes cling together because of the forces that electric charges exert on each other.

Figure 2 shows that unlike charges attract each other, and like charges repel each other. Charged particles do not attract or repel particles with no charge, such as neutrons, through the electric force. The force between electric charges depends on the amount of charge as well as on the distance between charges. The force increases as the amount of charge increases and as the charges get closer together.

Just as with two electric charges, the force between any two objects that are electrically charged decreases as the objects get farther apart. This force also depends on the amount of charge on each object. As the amount of charge on either object increases, the electrical force also increases.

As clothes tumble in a dryer, the atoms that make up some clothes gain electrons and become negatively charged. The atoms that make up other clothes lose electrons and become positively charged. Electrons are transferred from some objects onto other objects. Clothes that are oppositely charged attract each other and stick together, as shown in **Figure 3.**

■ **Figure 3** The clothes from this dryer are oppositely charged, causing them to stick together.

Predict *what would happen if the clothes from the dryer were identically charged.*

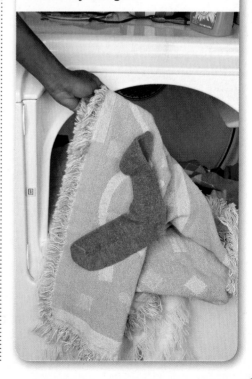

Figure 4 Surrounding every electric charge is an electric field that exerts forces on other electric charges. The arrows represent electric fields near positive and negative charges. They point in the direction a positive charge would move.

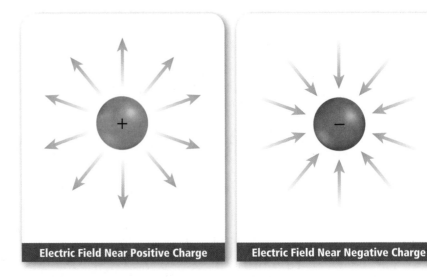

Electric Field Near Positive Charge

Electric Field Near Negative Charge

Electric fields Electric charges do not need to be touching to exert forces on each other. An **electric field** surrounds every electric charge and exerts the force that causes other electric charges to be attracted or repelled. Any charge that is placed in an electric field will be pushed or pulled by the field. Electric fields are usually represented by arrows that indicate how the electric field would make a positive charge move, as shown in **Figure 4.** Electric field arrows point away from positive charges and toward negative charges.

Comparing electric forces and gravitational forces The force of gravity between you and Earth seems to be strong. Yet, compared with electric forces, the force of gravity is very weak. For example, the electric force between a proton and an electron in a hydrogen atom is about a thousand trillion trillion trillion times larger, which is about 10^{39} times larger, than the gravitational force between these two particles.

All atoms are held together by electric forces between protons and electrons that are tremendously larger than the gravitational forces between the same particles. The chemical bonds that form between atoms in molecules are also due to the electric forces between the atoms. These electric forces are much larger than the gravitational forces between the atoms.

Reading Check **Compare** the strength of electric forces and gravitational forces between protons and electrons.

However, the electric forces between the objects around you are much less than the gravitational forces between them. Most objects that you see are nearly electrically neutral and have almost no net electric charge. There is usually no noticeable electric force between these objects. But even if a small amount of charge is transferred between objects, the electric force between those objects can be noticeable.

FOLDABLES
Incorporate information from this section into your Foldable.

Conductors and Insulators

If you reach for a metal doorknob after walking across a carpet, you might see a spark. The spark is caused by electrons moving from your hand to the doorknob, as shown in **Figure 5.** Recall that electrons were transferred from the carpet to your shoes. How did these electrons move from your shoes to your hand?

Conductors A **conductor** is a material through which electrons move easily. Electrons on your shoes repel each other, and some are pushed onto your skin. Because your skin is an effective conductor, the electrons spread over your skin, including your hand.

The best electrical conductors are metals. The atoms that make up metals have electrons that are able to move easily through the material. In **Figure 5,** the electrons move from your hand to the doorknob because the doorknob is made of metal. What would happen if the doorknob were made of glass?

Insulators Glass is an insulator. An **insulator** is a material in which electrons are not able to move easily. Electrons are held tightly to the atoms that make up insulators. Most plastics are insulators. The plastic coating around electric wires, such as the one shown in **Figure 6,** prevents dangerous electric shocks. If the doorknob in **Figure 5** were made of glass, you would not experience an electric shock from moving charges.

✓ **Reading Check Contrast** conductors with insulators.

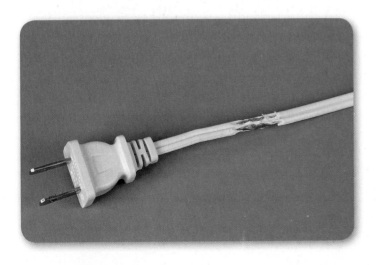

■ **Figure 5** As you walk across a carpeted floor, excess electrons can accumulate on your body. When you reach for a metal doorknob, electrons flow from your hand to the doorknob and you see a spark.

(((O **Concepts in Motion** Animation

VOCABULARY . . . .
SCIENCE USAGE V. COMMON USAGE
Conductor
Science usage
  a material in which electrons are able to move easily
  Copper is used in wiring because it is a good conductor.

Common usage
  the leader of a musical ensemble
  The conductor bowed to the audience after the orchestra's performance. . . . .

■ **Figure 6** The plastic coating around wires is an insulator. An electrical cord is hazardous when the conducting wire is exposed.

■ **Figure 7** The balloon on the left is neutral. The balloon on the right is negatively charged and produces a positively charged area on the sleeve by repelling electrons.

Determine *the direction of the electric force acting on the balloon on the right.*

? Inquiry | MiniLab

Investigate Charged Objects

Procedure

1. Read the procedure and safety information, and complete the lab form.

2. Fold over about 1 cm on the end of a **roll of transparent tape** to make a handle. Tear off a strip of tape about 10 cm long.

3. Stick the strip to a **clean, dry, smooth surface.** Make another identical strip, and stick it directly on top of the first strip.

4. Pull both pieces off the counter together, and pull them apart. Bring the nonsticky sides of both tapes together. What happens?

5. Repeat steps 1 through 3 with two new pieces of tape, but stick them side by side instead of one on top of the other.

Analysis

1. **Explain** what happened when you brought the pieces close together the first time.

2. **Explain** what happened when you brought the pieces close together the second time.

Charging by Contact

Every example of charging that has been discussed so far is an example of charging by contact. **Charging by contact** is the process of transferring charge by touching or rubbing. Rubbing two materials together can result in a transfer of electrons. Then one material is left with a positive charge and the other with an equal amount of negative charge.

Charging by Induction

Because electric forces act at a distance, charged objects brought near a neutral object will cause electrons to rearrange their positions on the neutral object. Suppose you charge a balloon by rubbing it with a cloth. If you bring the negatively charged balloon near your sleeve, the extra electrons on the balloon repel the electrons in the sleeve.

The electrons near the sleeve's surface move away from the balloon, leaving a positively charged area on the surface of the sleeve, as shown in **Figure 7.** The rearrangement of electrons on a neutral object caused by a nearby charged object is called **charging by induction.**

 Reading Check Contrast charging by contact with charging by induction.

Lightning

How is getting shocked when you touch a metal doorknob similar to lightning? Both are static discharges. A static discharge is a transfer of charge between two objects because of a buildup of static electricity.

A thundercloud is a mighty static electricity generator. As air masses move and swirl in the cloud, areas of positive and negative charge build up. Eventually, enough charge builds up to cause a static discharge between the cloud and the ground. As the electric charges move through air, they collide with atoms and molecules. These collisions cause the atoms and molecules in the air to emit light. You see this light as a flash of lightning, as shown in **Figure 8.**

 Reading Check Describe What is lightning?

FIGURE 8

Visualizing Lightning

When humid, sun-warmed air rises to meet a colder air layer, storm clouds can form. Lightning strikes occur when positive charges at the ground attract negative charges at the bottom of a storm cloud.

Through processes not yet understood, many of the particles in a storm cloud become charged. Convection currents in the storm cloud cause charge separation. The bottom of the cloud becomes negatively charged, and the top of the cloud becomes positively charged.

The negative charges at the bottom of the cloud repel the negative charges in the ground, tree, and house, inducing a positive charge on the ground below the cloud. A person standing beneath the cloud might feel a tingling sensation at this point.

The positive charges at the ground attract the negative charges in the cloud. The negative charges in the cloud move toward the ground once the bottom of the cloud has accumulated enough negative charges.

As the negative charges approach the ground, positively charged ions surge upward and meet those negative charges. The resulting connection is the spark that you see as a lightning flash.

Figure 9 Due to its shape, size, and position, a lightning rod attracts electric charges from storm clouds. A lightning rod guides the charge from a lightning bolt safely to the ground.

Thunder Not only does lightning produce a brilliant flash of light, it also generates powerful sound waves. The electrical energy in a lightning bolt rips electrons off atoms in the atmosphere and produces great amounts of heat.

The surrounding air temperature can rise to about 30,000°C, several times hotter than the Sun's surface. The heat causes air in the lightning bolt's path to expand rapidly, producing sound waves that you hear as thunder.

Grounding Lightning strikes can cause power outages, injury, fires, and loss of life. The sensitive electronics in a computer can be harmed by large static discharges. A discharge can occur any time that charge builds up in one area.

Providing a path for charge to reach Earth prevents any charge from building up. Earth is a large, neutral object that is also a conductor of charge. Any object connected to Earth by a good conductor will transfer any excess electric charge to Earth. Connecting an object to Earth with a conductor is called grounding.

✅ **Reading Check Explain** the purpose of grounding.

Buildings often have a metal lightning rod that provides a conducting path from the highest point on the building to the ground to prevent damage by lightning, as shown in **Figure 9.** Plumbing fixtures, such as metal faucets, sinks, and pipes, often provide a convenient ground connection. Look around. Do you see anything that might act as a path to the ground?

Electroscopes

An **electroscope** is a device that can detect electric charge. One kind of electroscope is shown in **Figure 10.** This electroscope is made of two thin, metal leaves attached to a metal rod with a metal knob at the top. The leaves are allowed to hang freely from the metal rod. When the device is not charged, the leaves hang straight down, as shown in **Figure 10.** When the device is charged, electric forces push the leaves apart so they are separated.

✅ **Reading Check Identify** the purpose of an electroscope.

Figure 10 The electroscope's metal leaves hang freely below. This indicates that the electroscope is not charged.

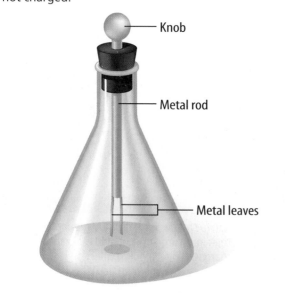

— Knob

— Metal rod

— Metal leaves

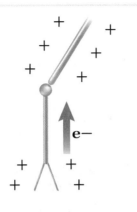

Negatively charging
an electroscope

Positively charging
an electroscope

■ **Figure 11** Notice the position of the leaves on the electroscope when they are negatively charged and positively charged.

Predict *Bring a negatively charged rod close to a positively charged electroscope. In what way will the leaves of the electroscope move?*

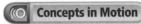 **Concepts in Motion**
Animation

Charging an electroscope Suppose a negatively charged object touches the knob. Because the metal is a good conductor, electrons travel down the rod into the leaves. Both leaves become negatively charged as they gain electrons, as shown in **Figure 11** on the left. Because the leaves have similar charges, they repel each other.

When a positively charged object touches the knob of an uncharged electroscope, electrons flow out of the metal leaves and onto the rod. Each leaf then becomes positively charged, as shown in **Figure 11** on the right. Once again, the leaves repel each other.

Section 1 Review

Section Summary

▶ There are two types of electric charge: positive charge and negative charge.

▶ Electric charge can be transferred between objects but cannot be created or destroyed.

▶ An electric charge is surrounded by an electric field that exerts forces on other charges.

▶ Charging by contact is the transfer of charge between two objects that are touching.

▶ Charging by induction is the rearrangement of charge in an object due to the influence of another charged object nearby.

1. **MAIN Idea Predict** what would happen if you touched the knob of a positively charged electroscope with a negatively charged object. Explain your prediction.

2. **Compare and contrast** the electric force with the gravitational force.

3. **Contrast** conductors and insulators.

4. **Explain** how electrically neutral objects can become electrically charged even though charge cannot be created or destroyed.

5. **Think Critically** Humid air is a better electrical conductor than dry air. Explain why you are more likely to receive a shock after walking across a carpet when the air is dry than when the air is humid.

Apply Math

6. **Calculate Electric Force** A 0.020-kg balloon is charged by rubbing and then stuck to the ceiling. If the strength of Earth's gravity on an object is 9.8 N/kg, how large is the electric force on the balloon?

Reading Preview

Essential Questions

▶ When and how does a voltage difference produce an electric current?

▶ How do batteries produce a voltage difference in a circuit?

▶ How does Ohm's law relate current, voltage difference, and resistance?

Review Vocabulary

SI: International System of Units–the improved, universally accepted version of the metric system that is based on multiples of ten and includes the meter (m), liter (L), and kilogram (kg)

New Vocabulary

electric current
voltage difference
electric circuit
resistance
Ohm's law

g Multilingual eGlossary

▢ Video BrainPOP

Electric Current

MAIN ⟨Idea An electric current is a flow of electric charge.

Real-World Reading Link You control electric circuits every day. Whether you are changing the volume on a stereo, the brightness of the lights, or how dark your toast gets, you are modifying the current, voltage, and resistance in an electric circuit.

Current and Voltage Difference

Electric devices, such as stereos, lights, and toasters, work when there is an electric current through them. **Electric current** is the net movement of electric charges in a single direction. Electric current is measured in amperes (A). One ampere is equal to one coulomb of electric charge flowing past a point every second.

In a metal wire without an electric current, electrons are in constant motion in all directions. As a result, there is no net movement of electrons in one direction. However, when there is an electric current in the wire, electrons continue their random movements but they also drift in the direction of the current. The movement of an electron in an electric current is similar to a ball bouncing down a flight of stairs. Even though the ball changes direction when it strikes a stair, the net motion of the ball is downward.

Voltage difference In some ways, the electric force that causes charges to flow is similar to the force acting on the water in a pipe. Water flows from higher pressure to lower pressure, as shown in **Figure 12.** In a similar way, electric current is from higher voltage to lower voltage. A **voltage difference** is related to the force that causes electric charges to flow. Voltage difference is measured in volts (V).

■ **Figure 12** A voltage difference in a conductor produces an electric current, just as a pressure difference in the pipe produces a water current. Water flows from higher-pressure regions to lower-pressure regions. Electric current is from higher-voltage regions to lower-voltage regions. However, as you will read on the next page, electron flow is actually from lower-voltage regions to regions of higher-voltage.

Compare *an electric current with a water current.*

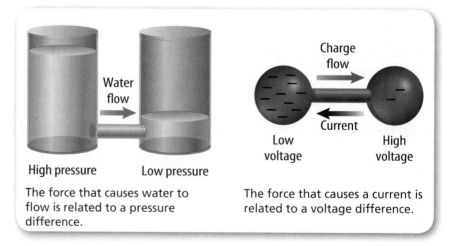

High pressure   Low pressure

The force that causes water to flow is related to a pressure difference.

Charge flow

Current

Low voltage   High voltage

The force that causes a current is related to a voltage difference.

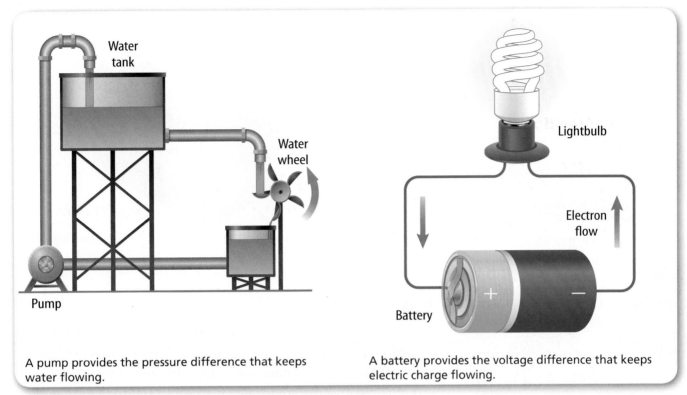

A pump provides the pressure difference that keeps water flowing.

A battery provides the voltage difference that keeps electric charge flowing.

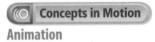

Electric circuits

A water circuit is shown in **Figure 13.** Water flows out of the tank and falls on a water wheel, causing it to rotate. A pump then lifts the water back up into the tank. The constant flow of water would stop if the pump stopped working. The flow of water would also stop if one of the pipes broke. Then water could no longer flow in a closed loop, and the water wheel would stop rotating.

Figure 13 also shows an electric circuit. Just as the water current stops if there is no longer a closed loop, the electric current stops if there is no longer a closed path to follow. An **electric circuit** is a closed path that electric current follows. If the battery, the lightbulb, or one of the wires is removed from the path in **Figure 13,** the electric circuit is broken and there will be no current.

Current and electron flow

For historical reasons, we think of current as being in the direction that positive charge flows. However, in almost all circuits, positive charge does not flow. Instead, it is the negatively charged electrons that actually move through the circuit. Because current is in the direction of positive charge flow, the flow of electrons and the current are in opposite directions. The direction of the electric current is always from higher voltage to lower voltage, but the electrons in a circuit actually flow from lower voltage to higher voltage.

✅ **Reading Check Contrast** the direction of current with the direction of electron flow.

■ **Figure 13** Water or electric charge will flow continually only through a closed loop. If any part of the loop is broken or disconnected, the flow stops.

(◎ **Concepts in Motion**
Animation

◀ **FOLDABLES**
Incorporate information from this section into your Foldable.

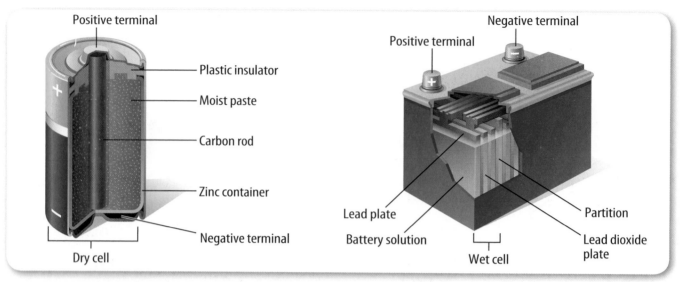

Positive terminal
Plastic insulator
Moist paste
Carbon rod
Zinc container
Negative terminal
Dry cell

Negative terminal
Positive terminal
Lead plate
Battery solution
Partition
Lead dioxide plate
Wet cell

Figure 14 Chemical reactions in batteries produce a voltage difference between the positive and negative terminals.

Infer *when these chemical reactions start and stop.*

Batteries

In order to keep water flowing continually in a water circuit, a pump is used to provide a pressure difference. In a similar way, to keep electric charges continually flowing in an electric circuit, a voltage difference needs to be maintained in the circuit. Power supplies, such as the batteries shown in **Figure 14,** can provide this voltage difference.

Dry-cell batteries You are probably most familiar with dry-cell batteries. The cylindrical batteries that are placed in flashlights are dry-cell batteries. A cell consists of two electrodes surrounded by a material called an electrolyte. The electrolyte enables charges to move from one electrode to the other.

Look at the dry cell shown in **Figure 14.** One electrode is the carbon rod, and the other is the zinc container. The electrolyte is a moist paste containing several chemicals. When the two terminals of a dry-cell battery are connected in a circuit, a chemical reaction occurs. Electrons are transferred between some of the compounds in this chemical reaction.

As a result, the carbon rod becomes positive, forming the positive terminal. Electrons accumulate on the zinc, making it the negative terminal. The voltage difference between these two terminals causes a current through a closed circuit.

Wet-cell batteries A wet-cell battery is another common type of battery. A wet cell, like the one shown in **Figure 14,** contains two connected plates made of different metals or metallic compounds in an electrolyte. Unlike for the dry cell, the electrolyte in a wet cell is a conducting liquid solution. A wet-cell battery contains several wet cells connected together. The most common type of wet-cell battery in use today is a car battery.

 **Reading Check** Contrast a dry-cell battery with a wet-cell battery.

Lead-acid batteries Most car batteries are lead-acid batteries, like the wet-cell battery shown in **Figure 14**. A lead-acid battery contains a series of six wet cells composed of lead and lead dioxide plates in a sulfuric acid solution. The chemical reaction in each cell provides a voltage difference of about 2 V, giving a total voltage difference of 12 V.

Electrical outlets A voltage difference is also provided at electrical outlets, such as a wall socket. This voltage difference is usually higher than the voltage difference provided by batteries. Most types of household devices are designed to use the voltage difference supplied by a wall socket. In the United States, the average voltage difference across the two holes in a wall socket is usually 120 V. Certain devices, such as clothes dryers, plug into special 240-V electrical outlets. In other areas of the world, the average voltage difference across the two holes of an electric outlet may be 110 V, 240 V, or another value.

Resistance

If you look inside a computer, a radio, or a telephone, you might find striped objects like those in **Figure 15**. These objects are called resistors, and they are designed to resist the flow of electrons. Electrical engineers use resistors to reduce the current through all or part of a circuit. Resistors help protect more delicate electronic components. These components can melt or break if too much current is sent through them.

Current and resistance When electrons move through a resistor, some energy is transferred to that resistor. **Resistance,** which is measured in ohms (Ω), is the tendency for a material to resist the flow of electrons and to convert electrical energy into other forms of energy, such as thermal energy. This is why cell phones can get hot during a long telephone call. Almost all materials have some resistance. For example, even copper wires have some resistance. The resistances of long copper wires can noticeably affect the currents through a circuit.

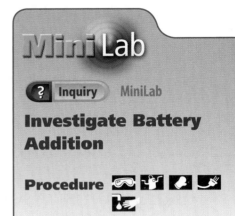

MiniLab

? Inquiry   MiniLab

Investigate Battery Addition

Procedure 🥽 🧤 ✋ 🧹 🤚

1. Read the procedure and safety information, and complete the lab form.
2. Make a circuit by using **wire** to link two **lightbulbs** and one **D-cell battery** in a loop. Observe the brightness of the bulbs.
3. Add **one more D-cell battery** to the loop. Observe the brightness of the bulbs.

Analysis

1. **Identify** the voltage difference of each D cell. Add them together to find the total voltage difference for the circuit that you tested in step 3.
2. **Infer** what you can conclude about the relationship between the voltage difference and current. Assume that a brighter bulb indicates a greater current.

■ **Figure 15** The striped objects in this photograph are resistors. The different colored stripes indicate the resistance of each resistor to engineers. Resistors are designed to reduce the current through all or part of a circuit. In sensitive electronics, it is important to prevent the current from being too high.

Explain *why desktop computers have resistors.*

Ohm's Law

The voltage difference, current, and resistance in a circuit are related. The relationship between voltage difference, current, and resistance in a circuit is known as Ohm's law. If *I* stands for electric current, Ohm's law can be written as the following equation.

Ohm's Law

$$\text{current (amperes)} = \frac{\text{voltage difference (volts)}}{\text{resistance (ohms)}}$$

$$I = \frac{V}{R}$$

According to **Ohm's law,** the current in a circuit equals the voltage difference divided by the resistance. If voltage is measured in volts (V) and resistance is measured in ohms (Ω), then current is measured in amperes (A).

EXAMPLE Problem 1

Solve for Current The voltage difference in a graphing calculator is 6 V, and the resistance is 1,200 Ω. What is the current through the batteries of the graphing calculator?

Identify the Unknown: current: *I*

List the Knowns: voltage difference: *V* = **6 V**   resistance: *R* = **1,200 Ω**

Set Up the Problem:

$$I = \frac{V}{R}$$

$$I = \frac{6\ V}{1{,}200\ \Omega}$$

Solve the Problem:

$$I = 0.005\ \frac{V}{\Omega} = 0.005\ A$$

The current through the batteries of the graphing calculator is 0.005 A.

Check the Answer: The current in a graphing calculator is probably very small, and the answer for current is very small. Therefore, the answer is reasonable.

PRACTICE Problems

Find **Additional Practice Problems** in the back of your book.

7. You measure the voltage difference of a circuit to be 5 V and the resistance to be 1,000 Ω. What is the current in the circuit?

Review
Additional Practice Problems

8. The current through a circuit is 0.0030 A. What is the resistance of this circuit if the voltage difference across the circuit is 12 V?

9. **Challenge** The current in a lamp is 0.5 A when plugged into a standard wall outlet. What is the resistance of the lamp when it is plugged into a standard wall outlet?

Alternating Current and Direct Current

Have you ever seen an AC/DC adapter like the one in **Figure 16**? This is a special piece of equipment that allows you to plug a battery-powered device into a wall socket. Most chargers for portable music players, laptops, and cell phones are AC/DC adapters.

Alternating current The AC in AC/DC stands for alternating current. This is the kind of current from household electrical outlets. In the United States, electric power grids are built so that alternating current changes direction 120 times each second. Household appliances, such as toasters and hair dryers, are built to use alternating current.

Direct current The DC in AC/DC stands for direct current. Battery-powered devices, such as flashlights, use direct current. Unlike alternating current, direct current never changes direction. An AC/DC adapter is a device that converts the alternating current of an electrical outlet into direct current. With AC/DC adapter, you can charge a cell phone battery with the current from an electrical outlet.

■ **Figure 16** The black AC/DC adapter converts the alternating current from an electrical outlet into direct current.

Section 2 Review

Section Summary

▶ Electric current is the net movement of electric charge in a single direction.

▶ A voltage difference is related to the force that causes charges to flow.

▶ A circuit is a closed conducting path.

▶ Resistance is the tendency of a material to resist the flow of electric charge.

▶ Ohm's law relates the current, resistance, and voltage difference in an electric circuit.

10. **MAIN Idea** **Compare and contrast** a current through a circuit with a static discharge.

11. **Compare and contrast** the cause of a flow of water in a pipe and the cause of a flow of electrons in a wire.

12. **Explain** how a carbon-zinc dry cell produces a voltage difference between the positive and negative terminals.

13. **Identify** two ways to increase the current in a simple circuit.

14. **Think Critically** Explain how the resistance of the heating element in an electric heater changes as it gets hotter and hotter, transferring thermal energy to the heater's surroundings at a faster and faster rate.

Apply Math

15. **Calculate** the voltage difference in a circuit that has a resistance of 24 Ω if the current is 0.50 A.

LAB

Conductors and Insulators

Objectives

- **Identify** conductors and insulators.
- **Describe** the common characteristics of conductors and insulators.

Background: The resistance of an insulator is so large that there is only a small current when the insulator is connected in a circuit. As a result, a lightbulb connected in a circuit with an insulator usually will not glow. In this lab, you will use the brightness of a lightbulb to identify conductors and insulators.

Question: *What materials are conductors, and what materials are insulators?*

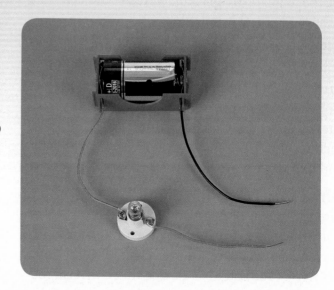

Preparation

Materials

battery
flashlight bulb
bulb holder
insulated wires (3)

Safety Precautions

Procedure

1. Read the procedure and safety information, and complete the lab form.
2. Set up an incomplete circuit as pictured in the photograph.
3. Touch the free bare ends of the wires to at least 10 different objects around the room. Record whether the lightbulb glows in each case.
4. Organize your data for this lab into a data table.

Conclude and Apply

1. **Identify** any patterns in your data.
2. **Explain** whether the materials that illuminate the lightbulb have something in common.
3. **Explain** whether the materials that do not illuminate the lightbulb have something in common.
4. **Explain** why one material might allow the lightbulb to illuminate and why another material prevents the lightbulb from illuminating.
5. **Predict** which other materials will allow the lightbulb to illuminate and which materials will prevent the lightbulb from illuminating.
6. **Classify** all of the materials that you have tested as conductors or insulators.

Expand your table by comparing your conclusions with those of other students in your class. Evaluate your conclusions with regards to these additional results.

? Inquiry Lab

Reading Preview

Essential Questions

▶ How do series circuits differ from parallel circuits?

▶ What is the function of a circuit breaker?

▶ How can you calculate electrical power?

▶ How can you calculate the cost of using an electrical appliance?

Review Vocabulary

power: the rate at which energy is converted from one form to another

New Vocabulary

series circuit
parallel circuit
electrical power

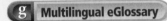 **Multilingual eGlossary**

More Complex Circuits

MAIN ◀Idea Electrical devices can be placed into series circuits as well as parallel circuits.

Real-World Reading Link Whether doing laundry or playing video games, transforming electrical energy into other forms of energy costs money. The most expensive electrical devices to run are large appliances such as refrigerators, electric stoves, and clothes dryers.

Series and Parallel Circuits

Have you ever wondered how a hair dryer works? If you were to look inside one, you would see the components of an electric circuit. You would see wires, an electric motor, a heating element, and other components.

You can see the parts of a circuit if you look inside any electrical device, whether that device is a video-disc player, a flashlight, or a computer. As a safety precaution, you should always check the instruction manual before investigating a device's circuitry. In addition, you should always unplug a device and remove its batteries before investigating that device's circuitry.

Most real circuits are not simple circuits with only one power source, one device, and one path for the current. Multiple power sources, multiple devices, and multiple paths for the current are very common. The two main types of more complex circuits are series circuits and parallel circuits.

Series circuits One kind of circuit is called a series circuit. A **series circuit** is an electric circuit with only one branch, as shown in **Figure 17.** Series circuits are used in flashlights. Because the parts of a series circuit are wired one after another, the amount of current is the same through every part. When any part of a series circuit is disconnected, there is no current through the circuit.

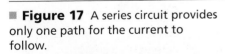 **Figure 17** A series circuit provides only one path for the current to follow.

Infer *What happens to the brightness of each bulb as more bulbs are added?*

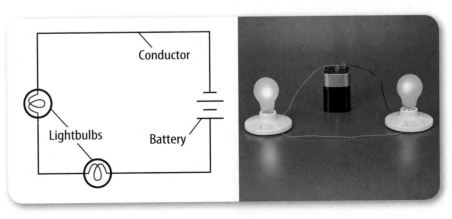

Conductor

Lightbulbs

Battery

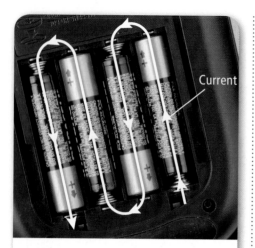

Current

■ **Figure 18** The four batteries in a graphing calculator are connected in series. The arrow shows the path of current through the batteries. If one battery is removed, then the circuit is broken and there is no current.

Open circuits A typical graphing calculator requires four AAA batteries to operate, as shown in **Figure 18**. If you remove one the batteries, the calculator will not turn on. Why does removing this battery prevent the calculator from turning on?

You might think that the other three batteries do not provide enough electrical energy to power the calculator. However, there is another reason. The batteries in a typical graphing calculator are connected in series. The arrow in **Figure 18** shows the path of electric current through the batteries.

Because the parts of a series circuit are wired one after another, the amount of current is the same through every part. When any part of a series circuit is disconnected, there is no current through the circuit. This is called an *open circuit*. Removing a battery from a typical graphing calculator results in an open circuit.

Parallel circuits What would happen if your home were wired in a series circuit and you turned off one light? All the other lights and appliances in your home would go out too. This is why houses are wired with parallel circuits. **Parallel circuits** contain two or more branches for current. Devices on each branch can be turned on or off separately.

Look at the parallel circuit in **Figure 19**. There can be a current through both or either of the branches. Because all branches connect the same two points of the circuit, the voltage difference is the same in each branch. Then, according to Ohm's law, the current is greater through the branches that have lower resistance.

When one branch of a parallel circuit is opened, such as when you turn off a light, the current continues through the other branches. Homes use parallel wiring so that individual devices can be turned off without affecting the entire circuit.

✓ **Reading Check Explain** what happens when you open one branch of a parallel circuit.

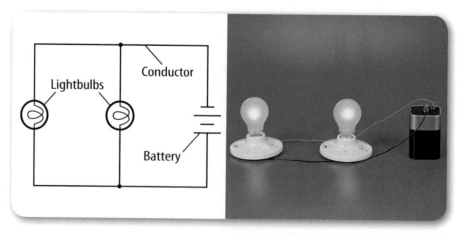

■ **Figure 19** In parallel circuits, the current follows more than one path.

Compare *the voltage differences for each of the branches.*

Lightbulbs
Conductor
Battery

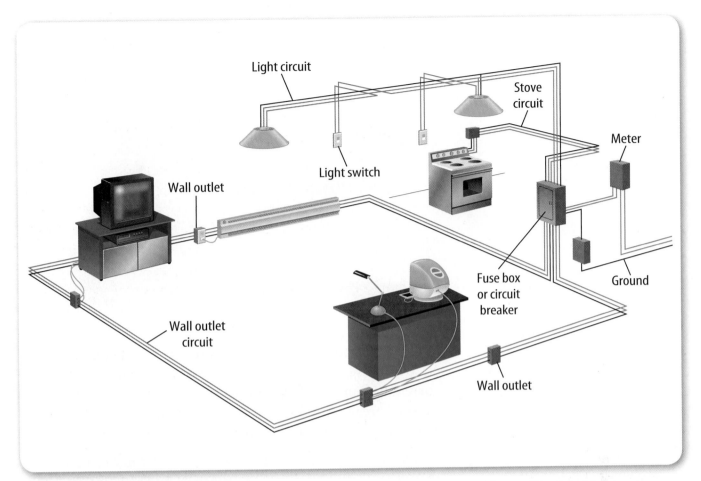

Light circuit

Stove circuit

Meter

Light switch

Wall outlet

Ground

Fuse box or circuit breaker

Wall outlet circuit

Wall outlet

Household Circuits

Count how many different things in your home require electrical energy. You do not see the wires because most of them are hidden behind the walls, ceilings, and floors. **Figure 20** shows a few rooms of a home without the walls, ceilings, or floors so that you can see these wires. This wiring is mostly a combination of parallel circuits.

In the United States, the voltage difference in most of the branches is 120 V. In some branches that are used for electric stoves or electric clothes dryers, the voltage difference is 240 V. The main switch and circuit breaker or fuse box serve as an electrical headquarters for your home. Parallel circuits branch out from the breaker or fuse box to wall sockets, major appliances, and lights.

Safety devices In a home, many appliances draw current from the same circuit. If more appliances are connected, there will be more current through the wires. As the amount of current increases, so does the amount of heat produced in the wires. If the wires get too hot, the insulation can melt and the bare wires can cause a fire. To protect against overheating of the wires, all household circuits should contain either a fuse or a circuit breaker.

■ **Figure 20** The wiring in a house must allow for the individual use of various appliances and fixtures.

Identify *the type of circuit that is most common in household wiring.*

■ **Figure 21** Fuses and circuit breakers are designed to stop the current when a device is in danger of overheating.

Evaluate *which device, a fuse or a circuit breaker, would be more convenient to have in the home.*

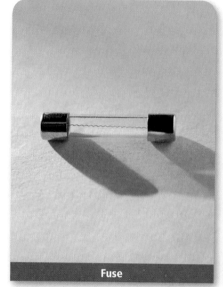

Fuse | Circuit Breaker

Fuses An electrical fuse is shown on the left in **Figure 21.** An electrical fuse contains a small piece of metal that melts if the current becomes too high. When this piece of metal melts, it causes a break in the circuit, stopping the current. To enable charge to flow again in the circuit, the fuse must be replaced.

Circuit breakers A circuit breaker is another device that prevents a circuit from overheating and causing a fire. In a circuit breaker, a switch is automatically flipped when the current becomes too great. Flipping the switch opens the circuit and stops the current. Circuit breakers usually can be reset by pushing the switch back to its "on" position. Many residences in the United States contain a box of circuit breakers similar to the one shown on the right in **Figure 21.**

✓ **Reading Check Identify** the purpose of fuses and circuit breakers in household circuits.

Electrical Power and Energy

The reason that electricity is so useful is that electrical energy is converted easily to other types of energy. For example, electrical energy is converted to mechanical energy as the blades of a fan rotate to cool you. An electric heater converts electrical energy into thermal energy.

Electrical power is the rate at which electrical energy is converted to another form of energy. The electrical power used by appliances varies. In general, heating appliances, such as electric stoves, use much more electrical power than electronic devices, such as computers. Appliances, like the microwave oven in **Figure 22,** often are labeled with a power rating that describes how much power the appliance uses.

■ **Figure 22** Many appliances are labeled with a power rating.

POWER CONSUMPTION	:	1430 W
MICROWAVE OUTPUT	:	900 W
	(IEC-705 Test Procedure	
LINE VOLTS	NORMAL TEMPERATURE R	
108 V	16°F (8.9°C) MIN.	
120V	18°F (10.0°C)-24.5°	

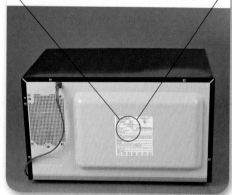

Calculate electrical power A device's electrical power depends on the voltage difference across that device and the current through that device. Electrical power can be calculated from the following equation.

Electrical Power Equation

electrical power (in watts) =

current (in amperes) × voltage difference (in volts)

$$P = IV$$

If current is measured in amperes and if voltage difference is measured in volts, then electrical power is calculated in watts (W). Because the watt is a small unit of power, electrical power is often expressed in kilowatts (kW). One kilowatt equals 1,000 watts.

EXAMPLE Problem 2

Solve for Electrical Power The current in a clothes dryer is 15 A when it is plugged into a 240-volt outlet. What is the power of the clothes dryer?

Identify the Unknown:	power: P
List the Knowns:	current: $I = 15$ A voltage difference: $V = 240$ V
Set Up the Problem:	$P = IV$
Solve the Problem:	$P = (15$ A$)(240$ V$) = 3,600$ W The clothes dryer has a power of **3,600 W**.
Check the Answer:	You can find the power rating for a clothes dryer by looking at the instruction manual, the manufacturer's web site, or the clothes dryer itself. The power rating on a typical clothes dryer is about 4,000 W. Therefore, an answer of 3,600 W seems reasonable.

PRACTICE Problems

Find **Additional Practice Problems** in the back of your book.

16. A toaster oven is plugged into an outlet where the voltage difference is 120 V. How much power does the toaster oven use if the current in the oven is 10 A?

17. A video-disc player that is not playing still uses 6.0 W of power. What is the current into the video-disc player if it is plugged into a standard 120-V outlet?

18. A flashlight bulb uses 2.4 W of power when the current in the bulb is 0.8 A. What is the voltage difference supplied by the batteries?

19. **Challenge** A hair dryer is rated at 1.2 kW. When it is plugged into a standard outlet and turned on, what is the current in the hair dryer?

[Review]

Additional Practice Problems

Calculate electrical energy Electric companies provide the electrical energy that you can then transform into other forms of energy, such as mechanical energy or thermal energy. The electrical energy that electric companies provide can be calculated from this equation.

> **Electrical Energy**
>
> electrical energy (in kWh) =
>   electrical power (in kW) × time (in h)
>
> $$E = Pt$$

Electrical energy is usually measured in units of kilowatt hours (kWh). One kilowatt hour is equal to 3.6 million joules and is enough energy to lift an average-sized car more than 200 meters.

EXAMPLE Problem 3

Solve for Electrical Energy A microwave oven with a power rating of 1,200 W is used for 0.25 h. How much electrical energy did the power company provide for the microwave?

Identify the Unknown: electrical energy provided: E

List the Knowns: electrical power used: P = 1,200 W = 1.2 kW
time: t = 0.25 h

Set Up the Problem: $E = Pt$

Solve the Problem: E = (1.2 kW) (0.25 h) = 0.30 kWh

The power company provided 0.30 kWh of electrical energy for the microwave.

Check the Answer: Use rounding to estimate the answer. The microwave oven's power is very close to 1,000 W, which is the same as 1 kW. Therefore, using the microwave for 0.25 h should require approximately 0.25 kWh of energy. Our answer, 0.30 kWh, is very close to 0.25 kWh. Therefore, our answer is reasonable.

PRACTICE Problems

Find **Additional Practice Problems** in the back of your book.

20. A refrigerator operates on average for 10.0 h a day. If the power rating of the refrigerator is 700 W, how much electrical energy does the refrigerator use in one day?

Review

Additional Practice Problems

21. A TV with a power rating of 200 W uses 0.8 kWh of electrical energy in one day. For how many hours was the TV on during this day?

22. A hair dryer has a power of 1,200 W. How much electrical energy does the hair dryer use in 3 minutes?

23. **Challenge** An electric light is plugged into a 120-V outlet. If the current in the bulb is 0.50 A, how much electrical energy does the bulb use in 15 minutes?

The cost of electrical energy The electrical energy that power companies provide costs money. The cost of using an appliance can be computed by multiplying the electrical energy used by the amount that the power company charges for each kWh. For example, if a 100-W lightbulb is left on for 5 h, the amount of electrical energy provided by the power company is

$$E = Pt = (0.1 \text{ kW})(5 \text{ h}) = 0.5 \text{ kWh}$$

If the power company charges $0.15 per kWh, the cost of using the bulb for 5 h is

cost = (kWh provided)(cost per kWh)

= (0.5 kWh)($0.15/kWh)

= $0.08

The cost of using some sample household appliances is given in **Table 1,** where the cost per kWh is assumed to be $0.15/kWh.

Table 1	Sample Costs of Using Home Appliances		
Appliance	Hair Dryer	Stereo	Color Television
Power rating (W)	1,000	100	200
Hours used daily	0.25	2.0	4.0
kWh used monthly	7.5	6.0	24.0
Cost per kWh	$0.15	$0.15	$0.15
Monthly cost	$1.13	$0.90	$3.60

Section 3 Review

Section Summary

▶ Two types of circuits are series circuits and parallel circuits.

▶ Fuses and circuit breakers are used to prevent wires from overheating.

▶ Electrical power is the rate at which electrical energy is converted into other forms of energy.

24. **MAIN Idea Compare** series circuits with parallel circuits.

25. **Explain** what determines the current in each branch of a parallel circuit.

26. **Infer** whether a circuit breaker should be connected in parallel to the circuit that it is protecting.

27. **Think Critically** A parallel circuit consisting of four branches is connected to a battery. Explain how the amount of current out of the battery is related to the amount of current in the branches of the circuit.

Apply Math

28. **Calculate** the current into a desktop computer plugged into a 120-V outlet if the power used is 180 W.

29. **Calculate Electrical Power** A circuit breaker is tripped when the current in the circuit is greater than 15 A. If the voltage difference is 120 V, what is the power being used when the circuit breaker is tripped?

30. **Estimate** the monthly cost of using a 700-W refrigerator that runs for 10 h a day if the cost per kWh is $0.20.

Use Vocabulary

Complete each sentence with the correct term from the Study Guide.

31. A(n) _____ is a circuit with only one path for current to follow.

32. An accumulation of excess electric charge is _____.

33. The electric force that makes charge flow in a circuit is related to the _____.

34. According to the _____, electric charge cannot be created or destroyed.

35. As _____ increases, current decreases.

36. Charging a balloon by rubbing it on wool is an example of _____.

Check Concepts

37. Which is a conductor?
    A) glass   C) copper
    B) wood   D) plastic

38. Resistance in wires
    A) increases current.
    B) destroys energy in a circuit.
    C) converts electrical energy to other forms of energy.
    D) converts other forms of energy to electrical energy.

39. The electric force between two charged objects depends on which of the following?
    A) their masses and their separation
    B) their speeds
    C) their charges and their separation
    D) their masses and their charges

Use the figure below to answer question 40.

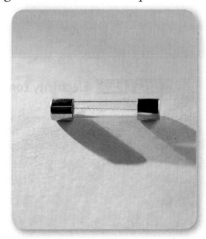

40. The object shown is a
    A) resistor.   C) circuit breaker.
    B) battery.   D) fuse.

41. Which does not provide a significant voltage difference in a circuit?
    A) wet cell   C) electrical outlet
    B) wires   D) dry cell

42. A commonly used unit for electrical energy is the
    A) kilowatt-hour.   C) ohm.
    B) ampere.   D) newton.

43. Which is the rate at which appliances convert electrical energy into other forms of energy?
    A) electrical power   C) resistance
    B) electric current   D) voltage

44. Which correctly expresses the relationship between voltage, current, and resistance?
    A) $V = I / R$ C) $V = R / I$
    B) $V = IR$ D) $V = I + R$

45. Which correctly expresses the relationship between voltage, current, and electrical power?
    A) $P = I + V$ C) $P = I / V$
    B) $P = IV$ D) $P = V / I$

✓ Assessment   Online Test Practice

Interpret Graphics

Use the table below to answer question 46.

Current in Electric Circuits	
Circuit	Current (A)
A	2.3
B	0.6
C	0.2
D	1.8

46. The table shows the current in circuits that were each connected to a 6-V dry cell. Calculate the resistance of each circuit. Graph the current versus the resistance of each circuit. Describe the shape of the line on your graph.

47. Copy and complete the following concept map on electric current.

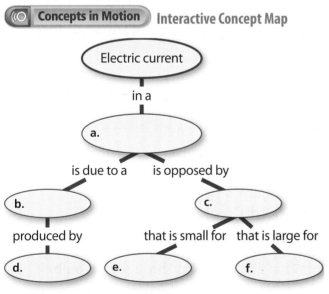

Concepts in Motion Interactive Concept Map

Think Critically

48. **BIG Idea** **Predict** You walk across a carpet on a dry day and touch a wood column. Predict whether you would receive an electric shock. Explain your reasoning.

49. **Explain** The electric force between electric charges is much larger than the gravitational force between the charges. Why, then, is the gravitational force between Earth and the Moon much larger than the electric force between Earth and the Moon?

50. **THEME FOCUS** **Diagram** Draw a circuit diagram showing how a stereo, a TV, and a computer can be connected to a single source of voltage difference, such that turning off one appliance does not turn off all the others. In your diagram, include a circuit breaker that will protect all of the appliances.

Apply Math

Use the figure below to answer question 51.

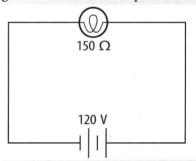

150 Ω

120 V

51. **Calculate Current** What is the current through the circuit shown above?

52. **Calculate Current** A toy car with a resistance of 20 Ω is connected to a 3-V battery. What is the current in the car?

53. **Estimate Electrical Energy** The current in an appliance connected to a 120-V source is 2A. Approximately how many kilowatt-hours of electrical energy does the power company provide for the appliance in 4 h?

54. **Calculate Electrical Power** A calculator uses a 9-V battery and draws 0.1 A of current. What is the power of the calculator?

Standardized Test Practice

Multiple Choice

Record your answers on the answer sheet provided by your teacher or on a sheet of paper.

1. Which is true about two adjacent electric charges?
   A. If both are positive, they attract.
   B. If both are negative, they attract.
   C. If one is positive and one is negative, they attract.
   D. If one is positive and one is negative, they repel.

Use the figure below to answer questions 2 and 3.

2. If a negatively charged rod is brought close to, but not touching, the knob, the two leaves will
   A. move closer together.
   B. move farther apart.
   C. not move at all.
   D. become positively charged.

3. If a positively charged rod touches the knob, the two leaves will
   A. move closer together.
   B. move farther apart.
   C. not move at all.
   D. become positively charged.

4. A kilowatt-hour is a unit of
   A. power.
   B. electrical energy.
   C. current.
   D. resistance.

5. When two objects become charged by contact, which is true?
   A. The net charge on each object does not change.
   B. Both become negatively charged.
   C. Both become positively charged.
   D. Electrons are transferred.

6. When an isolated object becomes charged by induction, which best describes the net charge on the object?
   A. The net charge increases.
   B. The net charge decreases.
   C. The net charge does not change.
   D. The net charge is negative.

Use the figure below to answer questions 7–9.

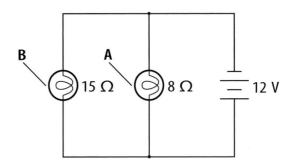

7. Which is the same for each lightbulb?
   A. current in the filament
   B. voltage difference
   C. electric resistance
   D. electrical power

8. What is the current through lightbulb B?
   A. 1.25 A   C. 0.80 A
   B. 0.67 A   D. 1.5 A

9. What is the electrical power of lightbulb A?
   A. 8 W
   B. 18 W
   C. 12 W
   D. 15 W

Short Response

Record your answers on the answer sheet provided by your teacher or on a sheet of paper.

10. The current through a lightbulb is 2.5 A. The lamp is connected to a battery supplying a voltage difference of 12 V. What is the power of the lightbulb?

11. A person spends four hours a day in a room but leaves a 100-W lightbulb on 24 hours a day. How much electrical energy would be saved if the bulb was on for only the four hours that the person was in the room?

Use the figure below to answer questions 12–14.

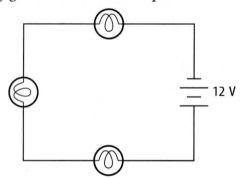

12. If the current in the circuit is 1.0 A, what is the total resistance of the circuit?

13. If the current in the circuit is 1.0 A, how much electrical energy is converted into other forms by the circuit in 2.5 h?

14. How does the current in the circuit change if one of the lightbulbs is removed and the circuit is reconnected with a wire?

Extended Response

Record your answers on a sheet of paper.

15. Explain how a refrigerator with a power rating of 650 W can convert more electrical energy into other forms of energy in a day than a hair dryer with a power rating of 1,000 W.

16. After using a laptop computer for about an hour, you notice that the laptop is getting very hot. Explain why this occurs.

17. A rubber rod rubbed on hair becomes negatively charged. A glass rod rubbed with a piece of silk becomes positively charged. Suppose the charged rubber rod is suspended so it is free to rotate. Describe how the rubber rod moves as the piece of silk is brought close to it.

18. A metal rod is charged by induction when a positively charged plastic rod is brought nearby. Describe how the distribution of charge on the metal rod has changed.

19. A lamp that is plugged in at the end of a long extension cord will be noticeably dimmer than the same lamp that is plugged in at the end of a short cord. Explain why this might be so.

20. The study of lightning is an active area of scientific research. Describe how lightning between a cloud and the ground is generated.

NEED EXTRA HELP?																			
If You Missed Question . . .	1	2	3	4	5	6	7	8	9	10	11	12	13	14	15	16	17	18	19	20
Review Section . . .	1	1	1	3	1	1	3	3	3	3	3	2	3	2	3	2	1	1	2	1

CHAPTER 7
Magnetism and Its Uses

ConnectED

Your one-stop online resource
connectED.mcgraw-hill.com

- Video
- Audio
- Review
- Inquiry
- WebQuest
- Assessment
- Concepts in Motion
- Multilingual eGlossary

Launch Lab
The Strength of Magnets

Did you know that magnets are used in computers, stereo speakers, electric motors, and many other devices? Magnets also help create images of the inside of the human body. Even Earth acts like a giant bar magnet. Where is a magnet strongest?

For a lab worksheet, use your StudentWorks™ Plus Online.

 Inquiry   Launch Lab

FOLDABLES

Make a folded table. Label it as shown. Use it to organize your notes on how magnets are used to transform energy.

Electrical to Mechanical Energy | Mechanical To Electrical Energy

THEME FOCUS Motion and Forces
The electromagnetic force changes the motion of
electric charges and magnetic materials.

BIG Idea The magnetic field surrounding a magnet
interacts with other magnets and with moving electric charges.

Section 1 • Magnetism

Section 2 • Electricity and
Magnetism

Section 3 • Producing Electric
Current

Reading Preview

Essential Questions

▶ How do magnetic poles interact?
▶ Why does a magnet exert a force on distant magnetic materials?
▶ Why are some materials magnetic but others are not?
▶ How do magnetic domains model magnetic behavior?

Review Vocabulary

electric field: surrounds an electric charge and exerts a force on other electric charges

New Vocabulary

magnetism
magnetic field
magnetic pole
magnetic domain

 **Multilingual eGlossary**

 **Video** What's PHYSICAL and EARTH SCIENCE Got to Do With It?

■ **Figure 1** Magnets can be found on your refrigerator door and in many devices that you use every day, such as TVs, headphones, video games systems, and telephones.

Magnetism

MAIN Idea A magnet is surrounded by a magnetic field that exerts a force on magnetic materials.

Real-World Reading Link Have you ever tried to stick a magnet onto an aluminum can or a steel can? If you have, you know that aluminum is not attracted by the magnet, but the steel can is. How are these two metal cans different?

Magnets

More than 2,000 years ago, the Greeks discovered deposits of a mineral that was a natural magnet. They noticed that chunks of this mineral could attract pieces of iron. This mineral was found in a region of Turkey that was known as Magnesia, so the Greeks called the mineral *magnetic stone*. This mineral is now called magnetite.

Since the discovery of magnetite, many devices have been developed that rely on magnets. In the twelfth century, Chinese sailors used magnetite to make compasses that improved navigation. **Figure 1** shows common magnets and familiar devices that use magnets today. In science, **magnetism** refers to the properties and interactions of magnets.

Magnetic force You probably have played with magnets and might have noticed that two magnets exert forces on each other. Depending on which ends of the magnets are close together, the magnets either repel or attract each other. You might have noticed that the interaction between two magnets can be felt even before the magnets touch. The strength of the force between two magnets increases as magnets move closer together and decreases as the magnets move farther apart.

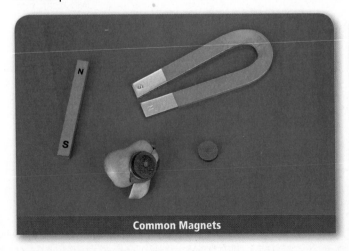
Common Magnets

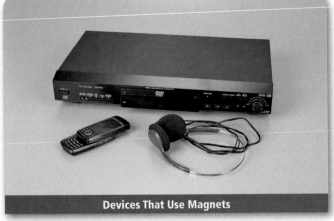
Devices That Use Magnets

Iron fillings sprinkled around a magnet line up along the magnetic field lines.

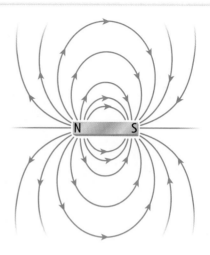

A magnetic field is represented by magnetic field lines.

Magnetic strength

You might also have noticed that some magnets are stronger than others. For example, a magnet from your refrigerator can be used to pick up paper clips, but a much stronger magnet would be needed to lift a car.

Magnetic Fields

A magnet can exert a force on a distant object because of its magnetic field. A **magnetic field** is a region of space that surrounds a magnet and exerts a force on other magnets and objects made of magnetic materials. The magnetic field and the magnetic force are related. A stronger magnetic field means that a magnetic object placed in the field will experience a stronger magnetic force. In other words, stronger magnets have stronger magnetic fields.

The strength of the magnetic field affects the use of the magnet. For example, the magnetic field used for a type of medical test called Magnetic Resonance Imaging (MRI) is about 200 times stronger than the magnetic field of a refrigerator magnet. This is why you must remove all magnetic metal objects before entering the MRI room.

Magnetic field lines

The magnetic field can be represented by lines of force called magnetic field lines. **Figure 2** shows iron filings lined up along the magnetic field lines surrounding a bar magnet. The closer together the magnetic field lines are, the stronger the magnetic field is at that point. Field lines are closer together near the magnet, as shown in **Figure 2**. This agrees with our observation that the magnetic force and the magnetic field are strongest close to the magnet. **Figure 3** shows the magnetic fields around a horseshoe magnet and a disk magnet.

■ **Figure 2** A magnet is surrounded by a magnetic field. Field lines that are closer together indicate a stronger magnetic field.

Identify *where the magnetic field is strongest.*

_____

■ **Figure 3** Horseshoe magnets and disk magnets are also surrounded by magnetic fields.

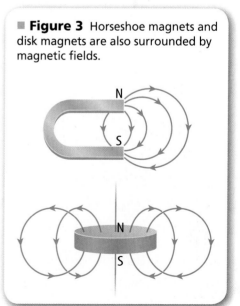

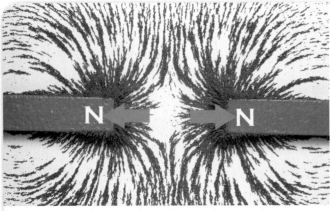

Like poles repel. Notice that in the area between the magnets, the field lines from the north pole of one magnet bend away from the field lines from the north pole of the other magnet.

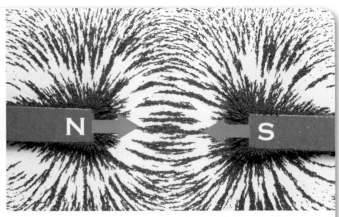

Unlike poles attract. Notice that in the area between the magnets, the field lines connect the north pole of one magnet to the south pole of the other magnet.

■ **Figure 4** When two magnetic poles are placed near each other, their magnetic fields combine. The magnets will either attract or repel each other, depending on which poles are closest together.

((◎)) **Concepts in Motion** **Animation**

■ **Figure 5** Compass needles placed around a bar magnet line up along magnetic field lines. The north poles of the compass needles are shaded red.

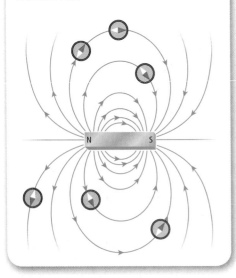

Magnetic poles You might have also noticed in **Figure 2** that the magnetic field lines are closest together at the ends of the bar magnet. This means the magnetic field is strongest at the ends. The regions of a magnet that exert the strongest force are called **magnetic poles.** All magnets have a north pole and a south pole. For a bar magnet, the north and south poles are at the opposite ends.

How magnetic poles interact Whether two magnets attract or repel each other depends on which poles are brought close together. Two north poles or two south poles repel each other. However, a north pole and a south pole always attract each other. Like magnetic poles repel each other, and unlike poles attract each other. **Figure 4** illustrates these interactions.

When two magnets are brought close to each other, their magnetic fields combine to produce a new magnetic field. If you look again at **Figure 4,** you will see iron filings illustrating the magnetic fields that result when like poles and unlike poles of bar magnets are brought close to each other.

Magnetic field direction A magnetic field also has a direction. The magnetic field always points away from north magnetic poles and toward south magnetic poles. The direction of the magnetic field around a bar magnet is shown by the arrows on the magnetic field lines in **Figure 5.** When a compass is brought near a bar magnet, the compass needle rotates. The compass needle is a small bar magnet with a north pole and a south pole. The force exerted on the compass needle by the magnetic field causes the needle to rotate until it lines up with the magnetic field lines, as shown in **Figure 5.** The north pole of a compass points in the direction of the magnetic field. Notice that in **Figure 5** the north compass needles point along the field lines toward the south magnetic pole of the magnet.

Earth's magnetic field A compass can help determine direction because the needle's north pole points to a location near Earth's geographic north pole. Earth's geographic north pole is the northernmost point on Earth. Earth acts like a giant bar magnet with a magnetic field that extends into space. A compass needle will align with Earth's magnetic field lines, as shown in **Figure 6**. Earth's magnetic field is not very strong in comparison to common magnets. The magnetic field of a refrigerator magnet is about 100 times stronger than Earth's magnetic field.

Earth's magnetic poles The north pole of a magnet is defined as the end of the magnet that points toward Earth's geographic north pole. Sometimes the north pole and south pole of magnets are called the north-seeking pole and the south-seeking pole.

Opposite magnetic poles attract, so the north pole of a compass needle is attracted to a south magnetic pole. Therefore, Earth's south magnetic pole is near its geographic north pole, as seen in **Figure 6**. However, Earth's magnetic poles move slowly with time and sometimes switch places. Measurements of magnetism in rocks show that Earth's magnetic poles have changed places over 150 times in the past 70 million years.

The source of Earth's magnetic field No one is certain what produces Earth's magnetic field, but many scientists think that it is generated by Earth's core. Earth's core is thought to be made of iron and nickel. The solid inner core is surrounded by a liquid outer core. The circulation of the molten iron and nickel in Earth's outer core is believed to produce Earth's magnetic field.

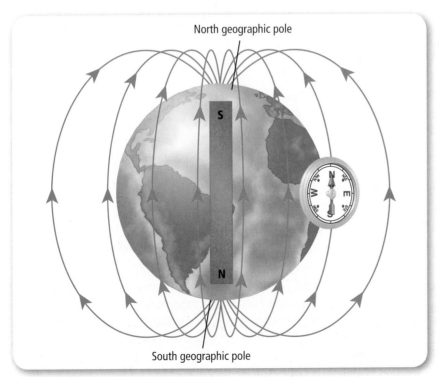

North geographic pole

South geographic pole

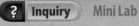

Mini Lab

? Inquiry Mini Lab

Make Your Own Compass

Procedure
1. Read the procedure and safety information, and complete the lab form.
2. Cut off the bottom of a **plastic foam cup** to make a foam disk.
3. Magnetize a **sewing needle** by continuously stroking the needle in the same direction with a **magnet** for 1 min. **WARNING:** *Use care when handling a sharp needle.*
4. **Tape** the needle to the center of the foam disk.
5. Fill **a plate** with **water** and float the disk, needle side up, in the water.
6. Bring the magnet close to the foam disk.

Analysis
1. **Discuss** Where did the needle point when you placed your compass in the water? Explain.
2. **Explain** How did the needle and disk move when the magnet was brought near them? Explain.

■ **Figure 6** Currently, Earth's south magnetic pole is located about 1,000 km from the geographic north pole.

Predict *which way a compass needle would point if Earth's magnetic poles switched places.*

Magnetic Materials

You might have noticed that a magnet will not attract all metal objects. For example, a magnet will not attract pieces of aluminum foil. Only a few metals, such as iron, cobalt, and nickel, are attracted by magnets or can be made into permanent magnets.

What makes these elements magnetic? Remember that every atom contains electrons. Electrons have magnetic properties. In the atoms of most elements, the magnetic properties of the electrons cancel out. But in the atoms of magnetic elements, these magnetic properties do not cancel out and each atom behaves like a small magnet with its own magnetic field and north and south poles.

✓ **Reading Check Explain** why the atoms of magnetic materials behave like small magnets.

Even though these atoms have their own magnetic fields, objects made from these metals are not always magnets. For example, if you hold an iron nail close to a refrigerator door and let go, the nail falls to the floor. However, you can make the nail behave like a magnet by placing it in a magnetic field. A magnetized nail would stick to the refrigerator door.

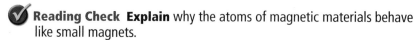

Apply Science

How can magnetic parts of a car be recovered?

Every year, over 10 million cars are scrapped. The junk car is fed into a shredder, and then large magnets are used to separate many of its metal parts from its nonmetal parts. The retrieved materials can be reused, saving both natural resources and energy. How much of the car does a magnet actually help separate? Find the answer by interpreting the circle graph.

Identify the Problem

The circle graph shows the average percent by mass of the different materials in a car. According to the graph, how much of the car can a magnet separate for recycling?

Solve the Problem

1. Cars are made of several metals, including iron, steel (a mixture of iron and carbon), aluminum, copper, and zinc. Which of these metals would the magnet attract?

Percentage Mass of Materials in a Car

- Other — 10%
- Plastic, glass, and rubber — 15%
- Magnetic metals — 65%
- Nonmagnetic metals — 10%

2. If a scrapped car has a mass of 1,500 kg, what is the mass of the materials that can be recovered using a magnet?

3. If the average mass of a scrapped car is 1,500 kg and 10 million cars are scrapped each year, what is the total mass of magnetic metals that could be recovered from scrapped cars each year?

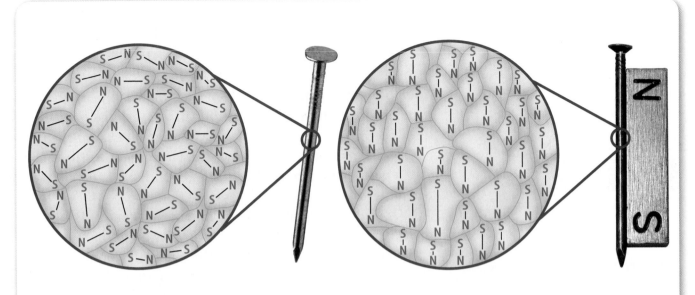

An unmagnetized iron nail is made up of microscopic domains that point in different directions.

The domains of a magnetic material will tend to align themselves along the magnetic field lines of a nearby magnet.

Magnetic domains—a model for magnetism
In magnetic materials such as iron, the magnetic field created by each atom exerts a force on other nearby atoms. These forces cause large groups of neighboring atoms to align. This means that almost all north magnetic poles in the group point in the same direction. These groups of atoms with aligned magnetic poles are called **magnetic domains.** Because the magnetic poles of the individual atoms are aligned, the domain itself behaves like a larger magnet with a north pole and a south pole.

Random arrangement of domains
An iron nail contains an enormous number of these magnetic domains. So, why does the nail not behave like a magnet? Even though each domain behaves like a magnet, the poles of the domains are arranged randomly and point in different directions, as shown on the left in **Figure 7.** As a result, the magnetic fields from the domains cancel each other and the nail is not magnetic.

Lining up domains
If you place a nail in a magnetic field, many of the domains align in the direction of the magnetic field, as shown on the right in **Figure 7.** The like poles of many of the domains point in the same direction and no longer cancel each other out. The nail itself now acts as a magnet.

Permanent magnets
When domains align, their magnetic fields add together and create a magnetic field inside the material. In a permanent magnet, this field is strong enough to prevent the constant motion of the atoms from bumping the domains out of alignment. A permanent magnet is any magnet whose domains remain aligned without an external field.

■ **Figure 7** The randomly arranged magnetic domains in an iron nail will align when placed in a magnetic field. With the domains aligned, the nail is a magnet.

   BrainPOP

VOCABULARY . . . .
SCIENCE USAGE V. COMMON USAGE
Field
Science usage
a region of space with a physical property (such as a force) that has a definite value at every point
The magnetic field is strongest close to the ends of the bar magnet.

Common usage
a yard or meadow; an area of interest
From the front porch, the farmer could see his tractor and his corn field. . . . . . .

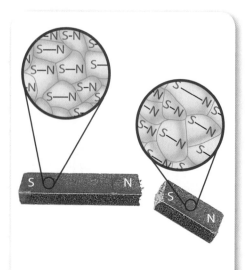

Figure 8 No matter how many pieces you break a magnet into, each piece still has a north pole and a south pole.

Making permanent magnets Permanent magnets can be made by stroking a magnetic material, as you did with the sewing needle in the MiniLab. They can also be made by heating magnetic material and letting it cool in a magnetic field.

How long does a magnet last? The domains in permanent magnets do not have to remain aligned forever. The magnetized needle that you made in the MiniLab, for example, will not remain magnetized for very long. However, it is still considered a permanent magnet because it remained magnetized when there was no external magnetic field.

Why did the sewing needle become unmagnetized? When the external magnetic field is removed, the constant motion and vibration of the atoms bump the magnetic domains out of alignment and the domains return to a random arrangement.

How quickly this happens depends on the magnet's material and its environment. For example, heating a permanent magnet causes its atoms to move faster. If the magnet is heated enough, its atoms move fast enough to bump the domains out of alignment.

Can a pole be isolated? If a magnet is broken in two, is one piece a north pole and one piece a south pole? Look at the domain model of the broken magnet in **Figure 8.** Recall that even individual atoms of magnetic materials act as tiny magnets. Because every magnet is made of many aligned smaller magnets, even the smallest pieces have both a north pole and a south pole.

Section 1 Review

Section Summary

▶ A magnet exerts a force on other magnets.

▶ A magnet has a north pole and a south pole. Like magnetic poles repel, and unlike poles attract.

▶ Atoms in magnetic materials behave like magnets.

▶ Magnetic domains are regions in a magnetic material where the magnetic poles of the atoms are aligned.

▶ In a permanent magnet, the magnetic domains remain aligned without an external magnetic field.

1. **MAIN ⟨Idea Explain** Why does a magnet exert a force on another magnet when the two magnets are not in contact?

2. **Describe** the magnetic field when two unlike magnetic poles are close together. Draw a diagram to illustrate your answer.

3. **Describe** how a compass needle moves when it is placed in a magnetic field.

4. **Explain** why only certain materials are magnetic.

5. **Explain** how heating a bar magnet would change its magnetic field.

6. **Think Critically** Use the magnetic domain model to explain why a magnet sticks to a refrigerator door.

Apply Math

7. **Calculate Number of Domains** Magnetic domains have an average volume of 0.0001 mm³. If a magnet has dimensions of 50 mm by 10 mm by 4 mm, how many domains does the magnet contain?

✓ Assessment   Online Quiz

Reading Preview

Essential Questions

◗ How do moving electric charges and magnets interact?

◗ What is the electromagnetic force?

◗ How do an electromagnet's properties affect its magnetic field strength?

◗ How does an electric motor operate?

Review Vocabulary

electric current: the flow of electric charges in a wire or any conductor

New Vocabulary

electromagnetic force
electromagnetism
solenoid
electromagnet
galvanometer
electric motor

g Multilingual eGlossary

? Inquiry   Video Lab

Electricity and Magnetism

MAIN ◖Idea An electric current in a wire is surrounded by a magnetic field.

Real-World Reading Link Electric motors are found in many devices that you use every day, including fans, cars, and moving toys. These devices work because electric currents produce magnetic fields.

Electric Current and Magnetism

In 1820, Danish physics teacher Hans Christian Oersted found that electricity and magnetism are related. While doing a demonstration involving electric current, he happened to have a compass near an electric circuit. He noticed that the electric current affected the direction of the compass needle. Oersted hypothesized that the electric current must produce a magnetic field around the wire and that the direction of the field depends on the direction of the current.

Oersted's hypothesis that an electric current creates a magnetic field was correct. It is now known that all moving charges, like those in an electric current, produce magnetic fields. Around a current-carrying wire, the magnetic field lines form circles, as shown in **Figure 9.** The direction of the magnetic field around the wire reverses when the direction of the current in the wire reverses. The strength of the magnetic field can be increased by increasing the current in the wire. Also, as you move farther away from the wire, the strength of the magnetic field decreases.

✔ **Reading Check Explain** how the strength of the magnetic field around a wire can be increased.

■ **Figure 9** When there is electric current through a wire, a magnetic field forms around the wire. The direction of the magnetic field depends on the direction of the current in the wire.

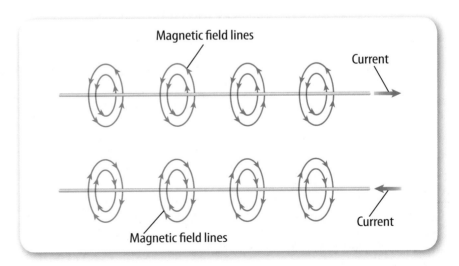

Magnetic field lines

Current

Magnetic field lines

Current

Electromagnetism

Moving electric charges are surrounded by magnetic fields. As a result, there is a force between the moving electric charge and magnets. These interactions led scientists to the realization that the electric force and the magnetic force are parts of the same force, called the electromagnetic force. The **electromagnetic force** is the attractive or repulsive force between electric charges and magnets. Like gravity, the electromagnetic force is one of the fundamental forces.

The interaction between electric charges and magnets is called **electromagnetism.** Electromagnetism is what makes magnets so useful. Many devices, including MP3 players, operate because of these interactions. Electromagnetism is also essential in producing, transmitting, and using electricity.

Electromagnets

The magnetic field around a current-carrying wire can be made stronger by changing the shape of the wire. When there is a current in a wire loop, such as the one shown on the left in **Figure 10,** the magnetic field inside the loop is stronger than the field around a straight wire. A single wire wrapped into a cylindrical wire coil is called a **solenoid.** The magnetic field inside a solenoid is stronger than the field in a single loop. The magnetic field around each loop in the solenoid adds together to form the field shown in **Figure 10.**

An **electromagnet** is a temporary magnet created when there is a current in a wire coil. Often, the current-carrying coil is wrapped around an iron core, as shown on the right in **Figure 10.** The coil's magnetic field temporarily magnetizes the iron core. As a result, the electromagnet's field can be more than 1,000 times greater than the field of the solenoid.

VOCABULARY ·····················

WORD ORIGINS

Solenoid

comes from the Greek words *solen–,* meaning "channel" or "pipe," and *–oid,* meaning "like" or "similar to" *Solenoids are used in electromagnets.* ···

■ **Figure 10** Coiling a current-carrying wire increases the strength of the magnetic field. Adding an iron core further increases the strength of the field.

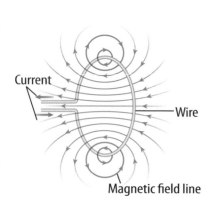

A wire loop has a magnetic field inside the loop.

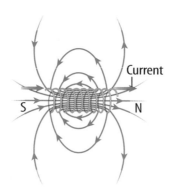

Winding the wire into a coil forms a solenoid.

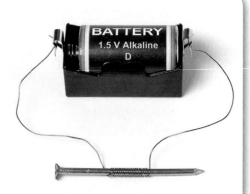

A solenoid wrapped around an iron core forms an electromagnet.

Properties of electromagnets

An electromagnet behaves like any other magnet when there is current in the solenoid. One end of the electromagnet is a north pole, and the other end is a south pole. If placed in a magnetic field, an electromagnet will align itself along the magnetic field lines. An electromagnet will also exert a magnetic force magnetic materials and other magnets.

Electromagnets are useful because their magnetic properties can be controlled by the user. The strength of the magnetic field can be increased by adding more turns of wire to the solenoid or by increasing the current in the wire. An electromagnet can even be turned off by shutting off the current to the wire.

Reading Check **Compare and contrast** permanent magnets and electromagnets.

Making an electromagnet rotate

The forces exerted on an electromagnet by another magnet can be used to make the electromagnet rotate. **Figure 11** shows an electromagnet suspended between the poles of a permanent magnet. The poles of the electromagnet are repelled by the like poles and attracted by the unlike poles of the permanent magnet.

When the electromagnet is in the position shown on the left side of **Figure 11,** there is a downward force on the left side and an upward force on the right side of the electromagnet. These forces cause the electromagnet to rotate as shown.

The electromagnet continues to rotate until its poles are next to the opposite poles of the permanent magnet, as shown on the right of **Figure 11.** Like a compass needle, the electromagnet will come to rest so that it is aligned in the magnetic field..

■ **Figure 11** An electromagnet can be made to rotate in a magnetic field.

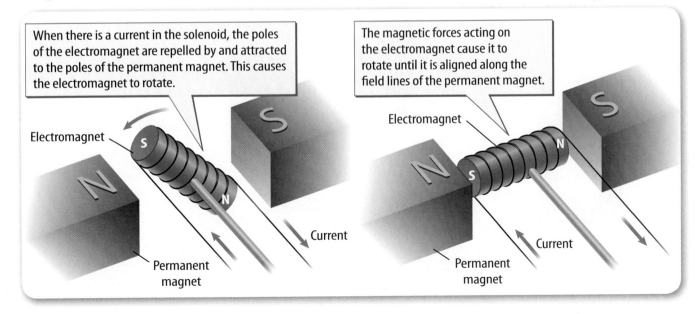

When there is a current in the solenoid, the poles of the electromagnet are repelled by and attracted to the poles of the permanent magnet. This causes the electromagnet to rotate.

Electromagnet

Current

Permanent magnet

The magnetic forces acting on the electromagnet cause it to rotate until it is aligned along the field lines of the permanent magnet.

Electromagnet

Current

Permanent magnet

Concepts in Motion Animation

Using Electromagnets

When an electromagnet rotates, electrical energy is converted into mechanical energy to do work. Electromagnets do work in various devices, such as stereo speakers and electric motors.

Speakers How does musical information stored on an MP3 player become sound that you can hear? The sound is produced by a loudspeaker, which is an electromagnet connected to a flexible speaker cone. The electromagnet changes electrical energy into the mechanical energy that vibrates the speaker cone to produce the sounds that you hear. **Figure 12** shows the parts of the loudspeaker.

When you listen to music, the MP3 player produces a voltage that changes according to the musical information in the music file. This varying voltage causes a varying electric current in the electromagnet. Both the size and the direction of the electric current change, depending on the information in the music file.

The varying electric current causes both the strength and the direction of the magnetic field in the electromagnet to change. This changing magnetic field direction causes the electromagnet to be attracted to or repelled by the surrounding permanent magnet. The permanent magnet is fixed, so the changing magnetic force makes the electromagnet move back and forth. This causes the speaker cone to vibrate and makes the surrounding air vibrate to create sound waves.

 Reading Check Summarize how a stereo speaker uses an electromagnet to produce sound.

■ **Figure 12** Changes in the electric current cause the electromagnet to move back and forth. The attached speaker cone vibrates and produces sound.

Explain *why a speaker needs a permanent magnet to produce sound.*

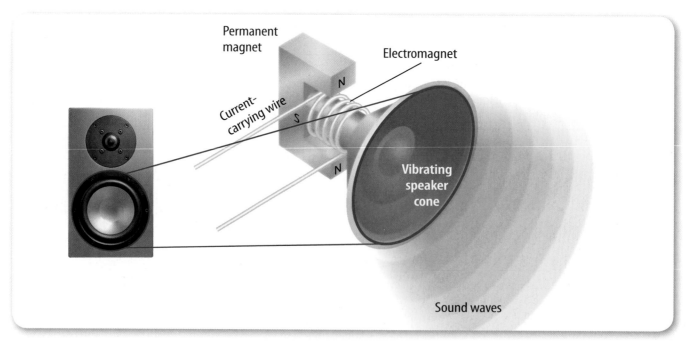

Permanent magnet

Electromagnet

N

Current-carrying wire

S

N

Vibrating speaker cone

Sound waves

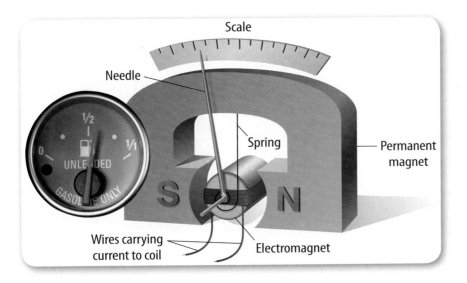

Galvanometers

You have probably noticed the gauges in the dashboard of a car. One gauge shows the amount of gasoline in the tank. How does a change in the amount of gasoline in a tank make a needle move in a gauge on the dashboard? The fuel gauge is a galvanometer. A **galvanometer** is a device that uses an electromagnet to measure electric current.

A diagram of a galvanometer is shown in **Figure 13**. In a galvanometer, the electromagnet is connected to a small spring. When there is a current in the electromagnet, it will rotate until the force exerted by the spring is balanced by the magnetic forces on the electromagnet. Changing the amount of current in the electromagnet changes the strength of the force between the electromagnet and the permanent magnet. Therefore, the amount that the needle rotates is related to the amount of current in the electromagnet. If the galvanometer is marked with a calibrated scale, it can be used to measure the current in a circuit.

In a car, a float in the fuel tank is attached to a sensor. The sensor sends a current to the fuel gauge galvanometer. As the level of the float in the tank changes, the amount of current sent by the sensor changes and rotation of the needle changes. The gauge is calibrated so that the current sent when the tank is full causes the needle to rotate to the full mark on the scale.

Electric Motors

Do you ever use an electric fan like the one in **Figure 14** to keep cool on hot summer days? A fan uses an electric motor to turn the blades. An **electric motor** is a device that changes electrical energy into mechanical energy. Electric motors are used in all types of industry, agriculture, and transportation. If you were to look carefully, you might find electric motors in every room of your house. For example, electric motors are used in DVD players, computers, and hair dryers.

FOLDABLES
Incorporate information from this section into your Foldable.

■ **Figure 14** This electric fan uses an electric motor to transform electrical energy into mechanical energy that turns the blades.

List *three additional devices that contain electric motors.*

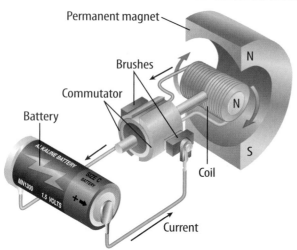

Step 1 When there is a current in the coil, the magnetic force between the permanent magnet and the coil causes the coil to rotate.

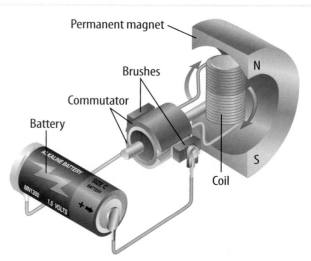

Step 2 In this position, the brushes are not in contact with the commutator and there is no current in the coil. The inertia of the coil keeps it rotating.

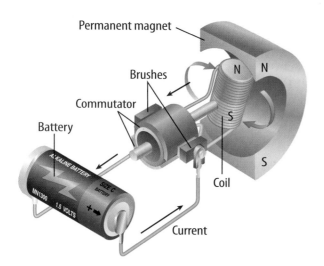

Step 3 The commutator reverses the direction of the current in the coil. This flips the north and south poles of the magnetic field around the coil.

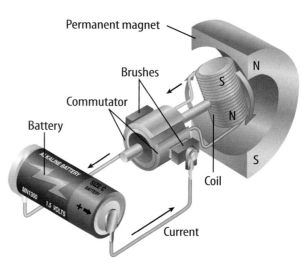

Step 4 The coil rotates until its poles are opposite the poles of the permanent magnet. The commutator reverses the current, and the coil keeps rotating.

■ **Figure 15** In a simple electric motor, a coil rotates between the poles of a permanent magnet. To keep the coil rotating, the current must change direction twice during each rotation.

Concepts in Motion

Animation

A simple electric motor A diagram of the simplest type of electric motor is shown in **Figure 15**. The main parts of a simple electric motor include a wire coil, a permanent magnet, and a source of electric current, such as a battery. The electric current makes the coil an electromagnet.

A simple electric motor also includes components called brushes and a commutator. The brushes are conducting pads connected to the battery. The brushes make contact with the commutator, which is a conducting metal ring that is split. Each half of the commutator is connected to one end of the coil so that the commutator rotates with the coil. The brushes and the commutator form a closed electric circuit between the battery and the coil.

Making the motor spin When there is a current in the coil, the forces between the coil and the permanent magnet cause the coil to rotate, as shown in step 1 of **Figure 15**. The coil rotates until it reaches the position shown in step 2. The brushes line up with the gaps in the commutator and no longer make contact with it. As a result, there is no current in the coil and no magnetic forces are exerted on the coil. However, the inertia of the coil causes it to continue rotating.

In step 3, the coil has rotated so that the brushes are again in contact with the commutator. However, the halves of the commutator that are in contact with the positive and negative battery terminals have switched. The current in the coil has reversed direction. Now the top of the electromagnet is a north magnetic pole and the bottom is a south pole. These poles are repelled by the nearby like poles of the permanent magnet, causing the magnet to continue to rotate in the same direction.

In step 4, the coil rotates until its poles are next to the opposite poles of the permanent magnet. It will continue to rotate until the brushes are not in contact with the commutator. Inertia keeps the electromagnet rotating and it returns to the position of step 1. The cycle repeats. The coil will rotate as long as the battery remains connected. **Figure 16** shows a simple motor.

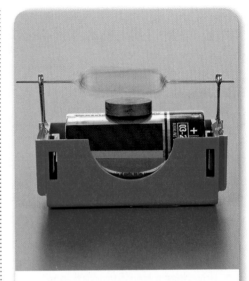

■ **Figure 16** The coil in this simple motor will continue to rotate unless the battery is removed or its energy source is depleted.

Section 2   Review

Section Summary

▶ Moving electric charges, such as electric currents, are surrounded by magnetic fields.

▶ The electromagnetic force is one of the fundamental forces.

▶ An electromagnet is a temporary magnet consisting of a current-carrying wire wrapped around an iron core.

▶ The magnetic field strength of an electromagnet can be controlled by changing the number of wire loops and by changing the current in the wire.

▶ In a simple electric motor, an electromagnet rotates between the poles of a permanent magnet.

8. **MAIN ⟨Idea Explain** what happens when a magnet is placed near a current-carrying wire.

9. **Infer** A bar magnet is repelled when an electromagnet is brought close to it. Describe how the bar magnet would have moved if the current in the electromagnet had been reversed.

10. **Define** the term *electromagnetic force.* Give an example of a machine that uses this force.

11. **Describe** two ways that you could increase the strength of the magnetic field produced by an electromagnet.

12. **Explain** how a simple electric motor transforms electrical energy into mechanical energy.

13. **Think Critically** How would an electromagnet's magnetic field change if the iron core were replaced by an aluminum core?

Apply Math

14. **Use Ratios** The magnetic field around a current-carrying wire at a distance of 1 cm is twice as strong as at 2 cm. How does the field strength at 0.5 cm compare to the field strength at 1 cm?

✓ **Assessment**   **Online Quiz**

Reading Preview

Essential Questions

▶ What is electromagnetic induction?

▶ How does a generator produce an electric current?

▶ What is the difference between alternating current and direct current?

▶ How does a transformer change the voltage of an alternating current?

Review Vocabulary

voltage difference: related to the force that causes electric charges to flow; measured in volts (V)

New Vocabulary

electromagnetic induction
generator
turbine
direct current (DC)
alternating current (AC)
transformer

g Multilingual eGlossary

Producing Electric Current

MAIN Idea A changing magnetic field produces an electric current in a wire loop.

Real-World Reading Link You might use a computer to type an essay. When it gets dark, you probably turn on a lamp. Magnets play a crucial role in producing and transmitting electricity for your lamp and computer.

Electromagnetic Induction

Working independently in 1831, Michael Faraday in Britain and Joseph Henry in the United States both found that moving a loop of wire through a magnetic field caused an electric current in the wire. They also found that moving a magnet through a loop of wire produces a current, as shown in **Figure 17.** In both cases, the mechanical energy associated with the motion of the wire loop or the magnet is converted into electrical energy associated with the current in the wire.

When the magnet and wire loop are moving relative to each other, the magnetic field inside the loop changes with time and induces, which means generates, an electric current in the wire coil. The generation of an electric current by a changing magnetic field is **electromagnetic induction.**

✓ **Reading Check** **Define** the term *electromagnetic induction.*

Electromagnetic induction can occur in other ways, too. For example, if the current in a wire changes with time, then the magnetic field around the wire is also changing. This changing magnetic field can induce a current in a nearby coil. Both types of electromagnetic induction are important to generating and transmitting the electrical energy that you use.

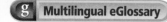
((O)) Concepts in Motion
Animation

■ **Figure 17** When the magnet is moved inside the solenoid, it induces a current in the solenoid. This solenoid is attached to a galvanometer that measures the current in microamperes.

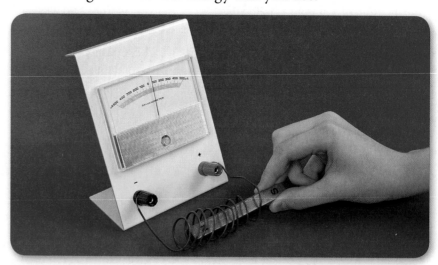

Generators

Most of the electrical energy that you use every day is provided by generators. A **generator** uses electromagnetic induction to transform mechanical energy into electrical energy. **Figure 18** shows a hand-operated generator being used to power a flashlight. The mechanical energy is provided by turning the handle on the generator. The generator transforms the mechanical energy into electrical energy, which lights the bulb.

Simple generators A diagram of a simple generator is shown in **Figure 19.** In this type of generator, a current is produced in the coil as the coil rotates between the poles of a permanent magnet. As the generator's wire coil rotates through the magnetic field of the permanent magnet, a current is induced in the coil. After the wire coil makes one-half of a revolution, the ends of the coil are moving past the opposite poles of the permanent magnet. This causes the current to change direction.

As the coil keeps rotating, the current that is produced continues to change direction. The direction of the current in the coil changes twice with each revolution. The frequency with which the current changes direction can be controlled by regulating the rotation rate of the generator. In the United States, electrical energy for residential use is produced by generators that rotate 60 times a second (60 Hz). Therefore, the current changes direction 120 times per second.

Using electric generators Generators similar to the one in **Figure 19** are used in cars, where they are called alternators. The alternator provides electrical energy to operate lights and other accessories. Spark plugs in the car's engine also use this electrical energy to ignite the fuel in the cylinders of the engine. Once the engine is running, it provides the mechanical energy that is used to turn the coil in the alternator.

■ **Figure 18** The coil in a generator is rotated by an external source of mechanical energy. Here, the student supplies the mechanical energy by turning the handle.

? **Inquiry** **Virtual Lab**

FOLDABLES
Incorporate information from this section into your Foldable.

■ **Figure 19** Moving the coil in the magnetic field of the permanent magnet induces a current in the coil. The current in the coil changes direction each time the ends of the coil move past the poles of the permanent magnet.

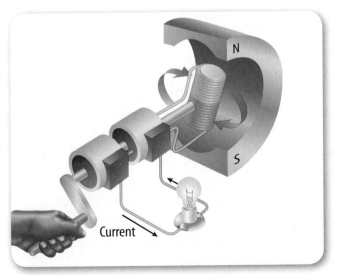

Current

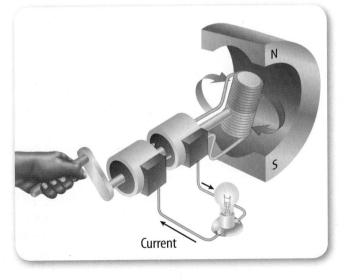

Current

These windmills are turbines. When the wind blows, the spinning turbine rotates a permanent magnet in the attached electric generator at the top of the tower.

Moving water provides the mechanical energy to rotate the large magnets in Hoover Dam's generators. Each can generate more than 100,000 kW of electrical power.

■ **Figure 20** At a power plant, an energy source, such as the wind or water, is used to rotate a turbine. The rotating turbine is attached to a permanent magnet, which induces a current in the coil.

Rotating the permanent magnet Suppose that the coil in a generator were fixed and the permanent magnet rotated. The current generated would be the same as when the coil rotates and the magnet does not move. The huge generators used in electrical power plants are made this way. Mechanical energy is used to rotate the magnet, and the current is induced in the stationary coil.

Generating electricity for your home Unless you have a generator in your home, the electrical energy needed to watch television or to wash your clothes comes from a power plant. Each power plant must have an energy source.

For example, some power plants burn fossil fuels or use nuclear reactions to produce thermal energy. This thermal energy is used to heat water and produce steam. Thermal energy is then converted into mechanical energy as the steam pushes the turbine blades. A **turbine** (TUR bine) is a large wheel that rotates when pushed by water, wind, or steam. In some areas, fields of windmills, like those shown in **Figure 20,** can be used to capture the mechanical energy in wind. Other power plants use the mechanical energy in falling water to drive the turbine.

The turbine is connected to the rotating magnet in the generator. Power plant generators, like those shown in **Figure 20,** are very large. The electromagnets in these generators have many coils of wire wrapped around huge iron cores. The generator changes the mechanical energy of the rotating turbine into the electrical energy that you use.

Generators and motors Both generators and electric motors use electromagnets to convert electrical and mechanical energy. **Figure 21** summarizes the differences and similarities between electric motors and generators.

FIGURE 21

Visualizing Motors and Generators

Motors convert electrical energy into mechanical energy. In contrast, generators convert mechanical energy into electrical energy.

The electromagnet rotates because of the attractive and repulsive forces between the coil and the permanent magnet. The rotating coil causes a shaft to rotate.

Permanent magnet

An outside source provides electrical energy. The current creates an electromagnet.

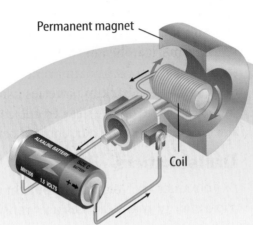

Coil

The mechanical energy from the motor is used to turn the fan's blades.

Motors
convert electrical energy into mechanical energy.

Electrical Energy

Mechanical Energy

Generators
convert mechanical energy into electrical energy.

The electrical energy from the generator is transmitted and used to light the bulb.

An outside source provides mechanical energy. This mechanical energy causes the turbine to rotate.

Permanent magnet

Coil

The rotating turbine rotates the coil or the magnet. A current is induced in the coil because it is in a changing magnetic field.

Direct and Alternating Currents

Because power outages can occur, some electric devices, such as alarm clocks, use batteries as a backup source of electrical energy. However, the current produced by a battery is different from the current produced by an electric generator. A battery produces a direct current. **Direct current** (DC) is electric current that is always in one direction through a wire.

When you plug an appliance into a wall outlet, it is receiving an alternating current. **Alternating current** (AC) is electric current that reverses the direction in a regular pattern. In North America, the alternating current in a wall socket has a frequency of 60 Hz and an average voltage of 120 V. Electronic devices that use backup batteries usually require direct current to operate. The device's electronic components reduce the voltage of the alternating current and convert it to direct current.

Transformers

To reduce the voltage without changing the amount of electrical energy, many devices use transformers. A **transformer** is a device that increases or decreases the voltage of an alternating current. A transformer is made of a primary coil and a secondary coil wrapped around the same iron core. A transformer that increases the voltage is called a step-up transformer, and a transformer that decreases the voltage is called a step-down transformer. Both types of transformers are shown in **Figure 22**.

How a transformer works As an alternating current passes through the primary coil, the coil's magnetic field magnetizes the iron core. The primary coil's magnetic field changes direction at the same frequency as the current in the primary coil. The magnetic field in the iron core also changes direction at that frequency. The changing magnetic field in the iron core induces an alternating current in the secondary coil. The input current and the output current alternate at the same frequency.

? Inquiry   Virtual Lab

VOCABULARY ....
ACADEMIC VOCABULARY
Primary
first in order of time or importance
*Elementary school is sometimes called
primary school.* ....

■ **Figure 22** Transformers increase or decrease voltage, depending on the ratio of turns in the primary coil to turns in the secondary coil.

Infer *the output voltage of the step-up transformer if the input voltage were 120 V.*

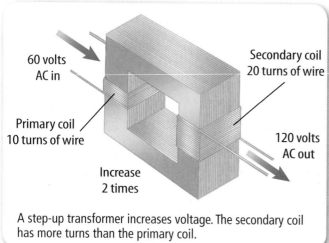

60 volts AC in

Primary coil 10 turns of wire

Secondary coil 20 turns of wire

120 volts AC out

Increase 2 times

A step-up transformer increases voltage. The secondary coil has more turns than the primary coil.

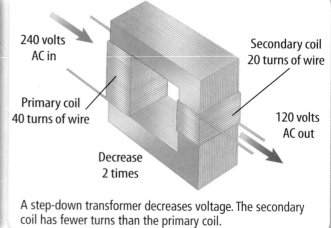

240 volts AC in

Primary coil 40 turns of wire

Secondary coil 20 turns of wire

120 volts AC out

Decrease 2 times

A step-down transformer decreases voltage. The secondary coil has fewer turns than the primary coil.

Calculate output voltage The output voltage of a transformer depends on the input voltage and the number of turns in the primary and secondary coils. This relationship can be expressed in an equation.

Transformer Voltage Equation

$$\frac{\text{output voltage (in volts)}}{\text{input voltage (in volts)}} = \frac{\text{turns in secondary coil}}{\text{turns in primary coil}}$$

$$\frac{V_{out}}{V_{in}} = \frac{N_2}{N_1}$$

In this equation, N stands for the number of turns in the coil. Because N is just a number, it has no units. Consider the step-up transformer in **Figure 22**. The secondary coil has twice as many turns as the primary coil has. So, the ratio of the output voltage to the input voltage is two to one. For this transformer, the output voltage is twice as large as the input voltage.

EXAMPLE Problem 1

Solve for Output Voltage A transformer has 150 turns in its primary coil and 50 turns in its secondary coil. If the input voltage is 12 V, what is the output voltage?

Identify the Unknown: output voltage: V_{out}

List the Knowns:
input voltage: $V_{in} = \textbf{12 V}$
turns in the primary coil: $N_1 = \textbf{150}$
turns in the secondary coil: $N_2 = \textbf{50}$

Set Up the Problem: Use the equation: $\dfrac{V_{out}}{V_{in}} = \dfrac{N_2}{N_1}$

Solve for $V_{out} = \dfrac{V_{in} N_2}{N_1}$

Solve the Problem: $V_{out} = \dfrac{12\,\text{V} \times 50}{150} = \textbf{4 V}$

Check the Answer: The number of turns in the secondary coil is one-third the number of turns in the primary coil. An output voltage (4 V) that is one-third the input voltage (12 V) makes sense.

PRACTICE Problems

Find **Additional Practice Problems** in the back of your book.

15. A transformer has 90 turns in the primary coil and 9 turns in the secondary coil. If the output voltage is 6 V, what is the input voltage?

Review
Additional Practice Problems

16. You have a transformer with 50 turns in the primary coil. If you want to increase the voltage from 5 V to 6 V, how many turns should be in the secondary coil?

17. **Challenge** High voltage power lines transmit electrical energy at 240,000 V. Before the current reaches your home, the voltage is decreased to the standard voltage for a wall socket. If a single step-down transformer is used, what is the ratio of turns in the primary coil to turns in the secondary coil?

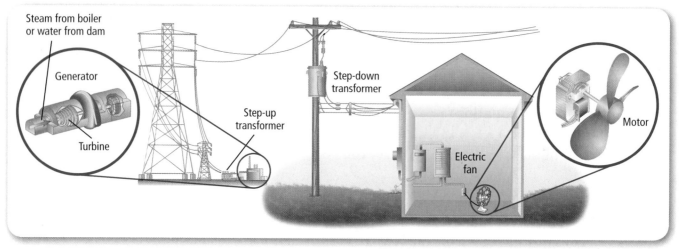

Figure 23 Many steps are involved in the production and transportation of electric current to your home.

Identify *the steps that involve electromagnetic induction.*

 **Concepts in Motion** Animation

Transmitting Electrical Energy

Electrical energy is often transmitted along wires known as power lines, as shown in **Figure 23.** During transmission, some of the electrical energy is converted into thermal energy due to the resistance of the wires and cannot be used to operate electrical devices. Electric resistance increases as the wires get longer. Power lines can be many kilometers long. As a result, large amounts of electrical energy transform into thermal energy that heats the surroundings.

To reduce this wasteful conversion, a step-up transformer increases the voltage to more than 100,000 V before transmission. After transmission, a step-down transformer decreases the voltage to 120 V for consumer use. This process is shown in **Figure 23.**

Section 3 Review

Section Summary

▶ A changing magnetic field can induce a current in a wire.

▶ A generator transforms mechanical energy into electrical energy.

▶ Direct current is current in one direction. Alternating current changes direction in a regular pattern.

▶ A transformer increases or decreases the voltage of an alternating current.

18. **MAIN Idea Explain** why a magnet sitting next to a wire does not induce a current in the wire.

19. **Define** the term *electromagnetic induction,* and explain how a generator uses electromagnetic induction.

20. **Describe** the difference between the current from a battery and the current from an electric socket.

21. **Summarize** the steps involved when a transformer changes the voltage of an alternating current.

22. **Think Critically** Why is the output voltage from a transformer zero if the current in the primary coil is a direct current?

Apply Math

23. **Use a Ratio** A transformer has 1,000 turns of wire in the primary coil and 50 turns in the secondary coil. If the input voltage is 2,400 V, what is the output voltage?

 **Assessment** Online Quiz

LAB

Electricity and Magnetism

Objectives

- **Observe** how a magnet and a wire coil produce an electric current in a wire.
- **Compare** the currents created by moving the magnet and the wire coil in different ways.

Background: Huge generators in power plants produce electricity by moving magnets past coils of wire. The moving magnet creates a changing magnetic field. This changing magnetic field induces a current in the wire coil.

Question: *How does the amount of current produced depend on how the magnet and wire coil move?*

Preparation

Materials

cardboard tube
bar magnet
insulated wire
scissors
galvanometer or ammeter

Safety Precautions

WARNING: *Do not touch bare wires when current is running through them. The ends of the wire may be sharp—handle with care.*

Procedure

1. Read the procedure and safety information, and complete the lab form.
2. Wrap the wire around the cardboard tube to make a coil of about 20 turns.
3. Remove the tube from the coil.
4. Use the scissors to cut and remove 2 cm of insulation from each end of the wire.

5. Connect the ends of the wire to a galvanometer or ammeter. These meters measure the current in the wire. Record the reading on your meter.
6. Insert one end of the magnet into the coil and then pull it out. Record the current measured by the meter.
7. Move the magnet at different speeds inside the coil, and record the current.
8. Watch the meter, and move the bar magnet in different ways around the outside of the coil. Record your observations.
9. Repeat steps 5 and 6, keeping the magnet stationary and moving the wire coil.

Conclude and Apply

1. **Explain** Is the current generated always in the same direction? How do you know?
2. **Describe** How was the largest current generated?
3. **Predict** what would happen if you used a coil with fewer turns of wire. Try it.
4. **Infer** whether a current would have been generated if the cardboard tube had been left in the coil. Why or why not? Try it.
5. **Infer** Would moving a magnet in and out of coil by hand be a practical way to produce electric current for use in your home? Why or why not?

Compare the currents generated by different members of the class. What was the value of the largest current that was generated? How was this current generated?

Locating Land Mines

The end of a war does not always make civilians safe from weapons. In nearly 70 countries, millions of buried explosive devices called land mines pose a silent, long-lasting threat. When a person steps on the land mine, the mine explodes and can cause blindness, burns, loss of limbs, and death. Humanitarian groups make neutralizing land mines a high priority.

Anatomy of a land mine Antipersonnel land mines target humans rather than heavy vehicles. They are relatively small, as shown in **Figure 1,** and are triggered by only a few kilograms of pressure. Locating them is difficult, but their components, shown in **Figure 2,** provide clues. Some of the mine's components, such as the firing pin, are often made of metal and have magnetic properties, so metal detectors are often used to find land mines.

How do metal detectors work? Basic principles of electromagnetism govern metal detector function. An alternating electric current is sent through a wire coil at the bottom of the detector. The coil is an electromagnet that creates an alternating magnetic field.

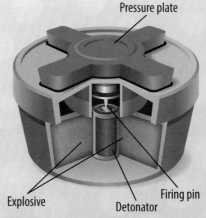

Figure 2 Antipersonnel land mines contain a firing mechanism, detonator, and an explosive material encased in plastic.

Pressure plate · Explosive · Detonator · Firing pin

Some buried metallic objects respond to this pulsating magnetic field with weak magnetic fields of their own. A receiver coil in the instrument detects this field and changes it into an audible signal, pinpointing the object's location. But, metal detection is dangerous and inefficient. Signals from buried cans, rusty nails, and land mines must all be investigated, which slows the process. In addition, manufacturers deliberately limit the metal in their mines to make locating them difficult.

A different approach The magnetic properties of the explosive material in land mines could hold a better solution to locating them. Researchers discovered that some atomic nuclei in trinitrotoluene (TNT) produce a detectable signal in the presence of a magnetic field. The next step is to construct an instrument that uses the new technique. Such a device has the potential to save lives worldwide.

Figure 1 Land mines are often difficult to find because of their small size.

> ### WebQuest
> **Research, compare, and analyze**
> Research other innovative methods for locating land mines. Create a chart to compare these methods. Choose one method, and prepare a presentation why you believe it is a safe, effective way to detect buried mines.

THEME FOCUS Motion and Forces

The electromagnetic force attracts or repels magnets, magnetic materials, and moving electric charges. These interactions are essential for producing, transmitting, and using electrical energy.

BIG Idea The magnetic field surrounding a magnet interacts with other magnets and with moving electric charges.

Section 1 Magnetism

magnetic domain (p. 207)
magnetic field (p. 203)
magnetic pole (p. 204)
magnetism (p. 202)

MAIN Idea A magnet is surrounded by a magnetic field that exerts a force on magnetic materials.

- A magnet exerts a force on other magnets.
- A magnet has a north pole and a south pole. Like magnetic poles repel, and unlike poles attract.
- Atoms in magnetic materials behave like magnets.
- Magnetic domains are regions in a magnetic material where the magnetic poles of the atoms are aligned.
- In a permanent magnet, the magnetic domains remain aligned without an external magnetic field.

Section 2 Electricity and Magnetism

electric motor (p. 213)
electromagnet (p. 210)
electromagnetic force (p. 210)
electromagnetism (p. 210)
galvanometer (p. 213)
solenoid (p. 210)

MAIN Idea An electric current in a wire is surrounded by a magnetic field.

- Moving electric charges, such as currents, are surrounded by magnetic fields.
- The electromagnetic force is one of the fundamental forces.
- An electromagnet is a temporary magnet consisting of a current-carrying wire wrapped around an iron core.
- The magnetic field strength of an electromagnet can be controlled by changing the number of wire loops and by changing the current in the wire.
- In a simple electric motor, an electromagnet rotates between the poles of a permanent magnet.

Section 3 Producing Electric Current

alternating current (p. 220)
direct current (p. 220)
electromagnetic induction (p. 216)
generator (p. 217)
transformer (p. 220)
turbine (p. 218)

MAIN Idea A changing magnetic field produces an electric current in a wire loop.

- A changing magnetic field can induce a current in a wire.
- A generator transforms mechanical energy into electrical energy.
- Direct current is current in one direction. Alternating current changes direction in a regular pattern.
- A transformer increases or decreases the voltage of an alternating current.

Review Vocabulary eGames

Use Vocabulary

Complete each statement with the correct term from the Study Guide.

24. A(n) _____ can be used to change the voltage of an alternating current.

25. A(n) _____ is the region where the magnetic field of a magnet is strongest.

26. _____ does not change direction.

27. The properties and interactions of magnets are called _____.

28. A group of atoms with aligned magnetic poles is called a(n) _____.

29. A device that uses an electromagnet to measure electric current is a(n) _____.

Check Concepts

30. **THEME FOCUS** What happens to the magnetic force as the distance between two magnetic poles decreases?
    A) It remains constant.
    B) It decreases.
    C) It increases.
    D) It decreases then increases.

31. Which describes the direction of the electric current in AC?
    A) It is constant.
    B) It is direct.
    C) It varies regularly.
    D) It varies irregularly.

32. Which is not part of a generator?
    A) wire coil   C) turbine
    B) magnet   D) battery

33. Atoms at the north pole of a magnet have
    A) only north magnetic poles.
    B) only south magnetic poles.
    C) no magnetic poles.
    D) north and south magnetic poles.

Use the figure below to answer question 34.

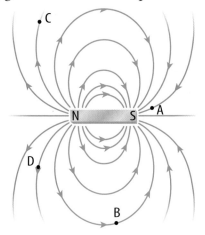

34. Where is the magnetic field strongest?
    A) point A   C) point C
    B) point B   D) point D

Interpret Graphics

35. Copy and complete the Venn diagram below. Include the functions, part names, and power sources for these devices.

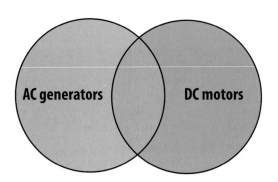

AC generators   DC motors

Assessment   Online Test Practice

Use the figure below to answer questions 36 and 37.

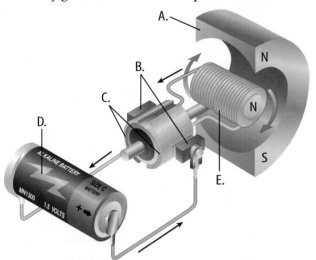

36. Fill in the missing labels on the diagram, and explain the function of each.

37. Identify the device and describe what it does.

Use the figure below to answer questions 38–41.

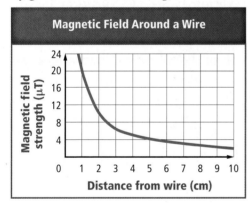

38. Use the graph to estimate the magnetic field strength 3 cm from the wire.

39. Estimate the distance from the wire where the magnetic field strength is 4 μT.

40. **BIG Idea** How much larger is the magnetic field strength 1 cm from the wire compared to 5 cm from the wire?

41. Does the magnetic field strength decrease more rapidly with distance closer to the wire or farther from the wire? Explain.

Think Critically

42. **Infer** how you could use a horseshoe magnet to find Earth's south magnetic pole.

43. **Explain** In Europe, generators produce alternating current at a frequency of 50 Hz. Would the electric appliances that you use in North America work if you plugged them into an outlet in Europe? Why or why not?

44. **Predict** Two generators are identical except for the loops of wire that rotate through their magnetic fields. One has twice as many turns of wire as the other one does. Which generator will produce the most electric current? Why?

45. **Explain** why a bar magnet will attract an iron nail to either its north pole or its south pole, but it will attract another magnet to only one of its poles.

46. **Compare and contrast** electromagnetic induction and the formation of electromagnets.

47. **Design** Describe how you might build an electromagnet. How can you control the strength of your electromagnet?

Apply Math

48. **Calculate** A step-down transformer reduces a 2,400-V current to 120 V. If the primary coil has 500 turns of wire, how many turns of wire are there on the secondary coil?

49. **Use a Ratio** A transformer increases the input voltage from 12 V to 12,000 V. What is the ratio of the number of wire turns on the primary coil to the number of turns on the secondary coil?

Standardized Test Practice

Multiple Choice

Record your answers on the answer sheet provided by your teacher or on a sheet of paper.

A student built a transformer by wrapping 50 turns of wire on one side of an iron ring to form the primary coil. The student then wrapped 10 turns of wire around the opposite side to form the secondary coil. The results are shown in the table below.

Use the table below to answer questions 1–3.

Voltage and Current in a Transformer				
Trial	Input Voltage (V)	Primary Coil Current (A)	Output Voltage (V)	Secondary Coil Current (A)
1	5	0.1	1	0.5
2	10	0.2	2	1.0
3	20	0.2	4	1.0
4	50	0.5	10	2.5

1. What is the ratio of the input voltage to the output voltage for this transformer?
   - A. 2:5
   - B. 4:1
   - C. 5:1
   - D. 1:5

2. What is the ratio of the primary coil current to the secondary coil current?
   - A. 2:5
   - B. 4:1
   - C. 5:1
   - D. 1:5

3. Describe the ratio of the current in the secondary coil to the current in the primary coil.
   - A. It equals the ratio of turns in the secondary coil to turns in the primary coil.
   - B. It equals the ratio of the output voltage to the input voltage.
   - C. It equals the ratio of turns in the primary coil to turns in the secondary coil.
   - D. It always equals one.

Use the figure below to answer questions 4 and 5.

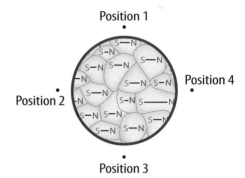

Position 1

Position 2

Position 4

Position 3

4. A steel paper clip is sitting on a desk. The figure above shows the magnetic domains in a section of the paper clip after the north pole of a magnet has been moved close to it. According to the diagram, the magnet's north pole is most likely at which of the following positions?
   - A. position 1
   - B. position 2
   - C. position 3
   - D. position 4

5. Which of the following diagrams shows the orientation of the needle of a compass that is placed at position 2?

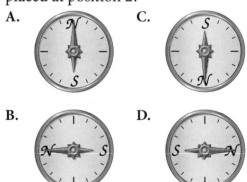

6. Which change occurs in an electric motor?
   - A. electrical energy into mechanical energy
   - B. thermal energy into wind energy
   - C. mechanical energy into electrical energy
   - D. wind energy into electrical energy

Short Response

Record your answers on the answer sheet provided by your teacher or on a sheet of paper.

7. Describe how a hydroelectric power plant converts mechanical energy into electrical energy.

8. Explain how an electromagnet can be turned on and off.

9. An electric motor rotates 60 times per second if the alternating current source is 60 Hz. How many times will an electric motor rotate in one hour if the alternating current source is changed to 50 Hz?

Use the figure below to answer questions 10 and 11.

120 V 20 turns 2 turns 8 Ω

10. A transformer is plugged into an electric outlet and a light is plugged into the transformer. What is the voltage at the output coil? What is the current in the light?

11. Is this transformer a step-up transformer or a step-down transformer?

Extended Response

Record your answers on a sheet of paper.

12. A bar magnet is placed inside a wire coil. The bar magnet and coil are then carried across a room. Explain whether an electric current will be induced in the coil as the magnet and coil are moving.

13. A student connects a battery to the primary coil of a step-up transformer in order to boost the voltage. Explain why a small electric motor does not spin when it is connected to the secondary coil of the transformer.

Use the graph below to answer question 14.

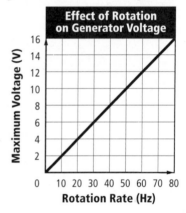

Effect of Rotation on Generator Voltage

Maximum Voltage (V) vs. Rotation Rate (Hz)

14. What voltage does the generator produce at 60 Hz? Describe how to change the voltage to 120 V for household use.

15. Describe how a permanent magnet is similar to and different from a piece of unmagnetized iron.

NEED EXTRA HELP?															
If You Missed Question . . .	1	2	3	4	5	6	7	8	9	10	11	12	13	14	15
Review Section . . .	3	3	3	1	1	2	3	2	2	3	3	3	3	3	1

CHAPTER 8

Energy Sources and the Environment

ConnectED

Your one-stop online resource
connectED.mcgraw-hill.com

- Video
- Audio
- Review
- Inquiry
- WebQuest
- Assessment
- Concepts in Motion
- Multilingual eGlossary

Launch Lab
Resources and Population Growth

Suppose you are the first one on the school bus in the morning. After a few more stops, you notice that the bus is rather noisy. By the time you get to school, every seat on the bus is taken. Like space on the school bus, space on Earth is limited. In this lab, you will model population density on Earth.

For a lab worksheet, use your StudentWorks™ Plus Online.

Inquiry Launch Lab

FOLDABLES®

Energy Sources Create a concept-map book to help you organize information about various types of energy sources.

THEME FOCUS Energy
The energy that we use daily comes from natural resources including coal, oil, wind, water, and the Sun.

BIG (Idea) Energy can be transformed from one form to another for human use.

Reading Preview

Essential Questions

▶ Which energy resources do you use daily?

▶ What is the law of conservation of energy?

▶ How do fossil fuels form?

▶ How is the chemical potential energy stored in fossil fuels converted into electrical energy?

Review Vocabulary

chemical potential energy: the energy stored in chemical bonds between atoms

New Vocabulary

fossil fuel
petroleum
nonrenewable resource

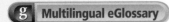
g **Multilingual eGlossary**

Fossil Fuels

MAIN ⟨Idea The burning of fossil fuels converts chemical potential energy to thermal energy, which is then converted into other useful forms.

Real-World Reading Link Lightning strikes, and your electricity goes out. You are frustrated that you cannot charge your cell phone, log on to the Internet, or use the stove to prepare dinner.

Energy Resources

How many different ways have you relied on energy resources today? You can see energy being used in many ways throughout the day, such as those shown in **Figure 1.** Furnaces and stoves use thermal energy to heat buildings and to cook food, respectively. Air conditioners use electrical energy to cool homes. Cars and other vehicles use mechanical energy to transport people and materials from one area to another.

Energy transformation According to the law of conservation of energy, energy cannot be created or destroyed. Energy can only be transformed from one form to another. To use energy means to transform energy from one form into another. For example, you use energy when the chemical potential energy from coal, oil, or natural gas is transformed into thermal energy that is used to heat your home.

Sometimes, energy is transformed into a form that is not useful. For example, when an electric current travels through power lines, about 10 percent of the electrical energy is converted to thermal energy. This reduces the amount of useful electrical energy that is delivered to homes, schools, and businesses.

■ **Figure 1** Energy is used in many ways. Automobiles use energy from the burning of gasoline.

Identify *other processes in this photo that require energy resources.*

Energy use in the United States In 2009, more energy was used annually in the United States than in any other country in the world. **Figure 2** shows energy use in the United States in 2008. About 22 percent of the energy was used in homes for heating and cooling, to run appliances, to provide lighting, and for other household needs. About 28 percent was used for transportation, powering vehicles such as cars and planes. Another 19 percent was used by businesses to heat, cool, and light shops and buildings. About 31 percent of this energy was used by industry and agriculture for manufacturing and food production.

As shown in **Figure 2,** about 85 percent of the energy used in the United States in 2008 came from burning fossil fuels. Nuclear power plants provided 8 percent, and alternative energies supplied 7 percent.

Fossil Fuel Formation

In one hour of driving, a car might use two or three gallons of gasoline. It might be hard to believe that it took millions of years to make the fuels that are used to power your car, to produce electricity, and to heat your home. Coal, natural gas, and petroleum, also called crude oil, are **fossil fuels** because they form from the remains of ancient plants and animals that were buried and altered over millions of years.

Combustion reactions When fossil fuels are burned, a combustion reaction occurs. During this reaction, carbon and hydrogen atoms combine with oxygen in the air to form carbon dioxide and water. This process converts the chemical potential energy that is stored in the bonds between atoms into thermal energy and light. Compared to wood, the energy stored in fossil fuels is much more concentrated. In fact, burning 1 kg of coal releases two to three times more energy than burning 1 kg of wood. **Figure 3** shows the energy content of different fuels.

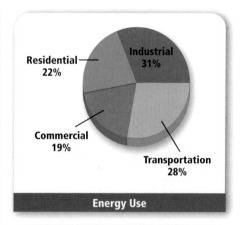

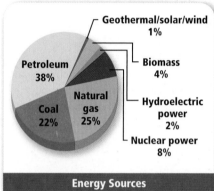

■ **Figure 2** These circle graphs show where energy was used in the United States in 2008, as well as the sources of this energy.

Interpret *Which energy source supplies the greatest amount of energy in the United States?*

 **Inquiry** Virtual Lab

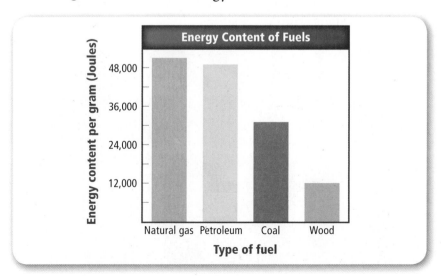

■ **Figure 3** The fuel with the greatest chemical potential energy per gram releases the greatest amount of energy.

FOLDABLES
Incorporate information from this section into your Foldable.

VOCABULARY ....

WORD ORIGIN

Petroleum
from the Middle English *petra,* meaning rock, and *oleum,* meaning oil
Petroleum is a nonrenewable resource that is limited in supply. ....

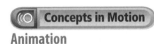

Concepts in Motion

Animation

■ **Figure 4** Petrolatum, also known as petroleum jelly, is blended with paraffin wax to make medicines, moisturizers, and other toiletries that you might find in a bathroom cabinet.

Identify *objects in your classroom that are petroleum-based products.*

Petroleum

Millions of liters of petroleum, a fossil fuel, are pumped from wells within Earth's crust every day. **Petroleum** is a flammable liquid formed from the decay of ancient organisms, such as microscopic plankton and algae. It is a mixture of thousands of chemical compounds. Most of these compounds are hydrocarbons, which means that their molecules are made of different arrangements of carbon and hydrogen atoms.

Fractional distillation Hydrocarbon compounds found in petroleum differ based on the number and arrangement of carbon and hydrogen atoms. The composition and structure of a hydrocarbon determines its chemical and physical properties.

The many different hydrocarbon compounds found in petroleum can be separated in a process called fractional distillation. This separation occurs in distillation towers at oil refineries. First, petroleum is pumped into the bottom of the tower and heated. The chemical compounds in the petroleum boil at different temperatures. Materials with the lowest boiling points rise to the top of the tower as vapor and are collected from the tower. Hydrocarbons with high boiling points, such as asphalt and waxes, remain liquid and are drained from the bottom of the tower.

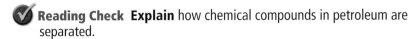

Reading Check **Explain** how chemical compounds in petroleum are separated.

Petroleum use Petroleum supplies nearly 38 percent of all energy generated in the United States each year. However, about 15 percent of petroleum-based materials in the United States are not used for fuel. Look at the materials in your home or classroom. Do you see any plastics? In addition to fuel, plastics, synthetic fabrics, cosmetics, and medicines, such as those shown in **Figure 4,** are made from petroleum. Also, lubricants such as grease and motor oil, as well as wax-based products and asphalt, are made from petroleum.

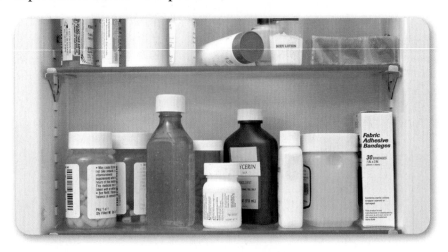

Natural Gas

The chemical processes that produced petroleum when ancient organisms decayed and were buried on the seafloor also formed natural gas. Due to differences in density, light-weight natural gas compounds are found trapped on top of petroleum deposits. Natural gas is a fossil fuel composed mostly of methane, but it also contains other gaseous hydrocarbon compounds, such as propane and butane.

Natural gas contains more chemical potential energy per kilogram than petroleum or coal does. Additionally, natural gas burns cleaner than other fossil fuels, produces fewer pollutants, and leaves no ash residue. Natural gas is burned to provide energy for cooking, heating, and manufacturing. About one-fourth of the energy used in the United States comes from the combustion of natural gas. There is a good chance that your home has a stove, a furnace, a hot-water heater, or a clothes dryer that is powered by natural gas. Some cars and buses also are powered by natural gas.

Coal

Coal is a solid fossil fuel that can be found in mines, such as the one shown in **Figure 5.** During the first half of the twentieth century, most houses in the United States were heated by the combustion of coal. In fact, during this time, coal provided more than half of the energy used in the United States. Now, almost two-thirds of the energy used comes from petroleum and natural gas, and only about one-fourth comes from coal. About 90 percent of all the coal that is used in the United States is burned by power plants to produce electricity.

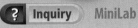

MiniLab

? Inquiry MiniLab

Design an Efficient Water Heater

Procedure 🥽 🧤 🔥

1. Read the procedure and safety information, and complete the lab form.
2. Measure and record the mass of a **candle.**
3. Measure 50 mL of **water** into a **beaker.** Record the temperature of the water.
4. Light the candle and use it to increase the temperature of the water by 10°C.
5. Put out the candle and measure its mass again.
6. Repeat steps 2–4 with an **aluminum chimney** surrounding the candle to help direct the heat upward.

Analysis

1. **Compare** the mass change in the two trials. Does a smaller or larger mass change in the candle show greater efficiency?
2. **Analyze** Natural gas burners are often used in hot-water heaters. What must be considered in the design of these heaters?

■ **Figure 5** Coal mines are common in Pennsylvania, West Virginia, and Ohio. This region of the United States was home to ancient swamps hundreds of millions of years ago. The coal formed from the remains of plants that lived in these swamps.

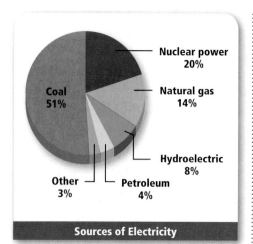

Sources of Electricity

■ **Figure 6** This circle graph shows the percentage of electrical energy that comes from different energy sources used in the United States.

Coal 51%
Nuclear power 20%
Natural gas 14%
Hydroelectric 8%
Petroleum 4%
Other 3%

Inquiry Virtual Lab

■ **Figure 7** Power plant efficiency describes how much energy is available to do work and produce electricity.

Determine *which stage in this process is the most inefficient.*

Origin of coal Coal mines were once the sites of ancient swamps. Coal formed as swampy plant material, buried beneath sediments, decayed and compacted into peat. Over millions of years, heat and pressure converted the peat into coal.

Coal is a mixture of hydrocarbons and other chemical compounds. Compared to petroleum and natural gas, coal contains more chemical impurities, such as sulfur and nitrogen-based compounds. As a result, more pollutants, including sulfur dioxide and nitrogen oxides, are produced when coal is burned.

✓ **Reading Check** **Describe** how coal forms.

Coal use Coal is the most abundant fossil fuel in the world. The amount of coal available is estimated to last between 200 and 250 years at our current rate of consumption. Because of its supply, scientists are looking for ways to make coal a cleaner energy source. For example, filters on smoke stacks and stricter government standards have reduced harmful particulates released into the atmosphere when coal is burned.

Electricity

Figure 6 shows that almost 70 percent of the electrical energy used in the United States is produced by burning fossil fuels, such as coal. How is the chemical potential energy stored in fossil fuels converted to electrical energy in a power plant? The process of energy conversion is shown in **Figure 7.**

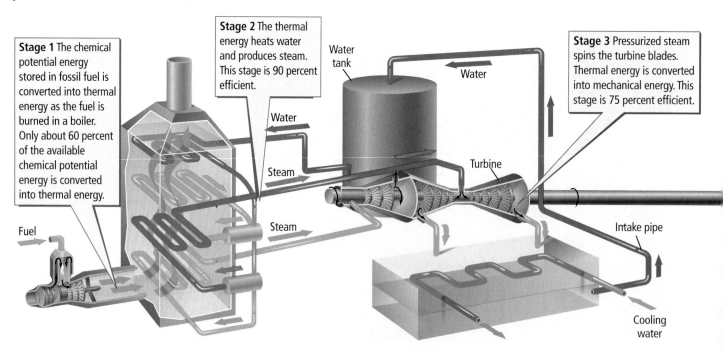

Stage 1 The chemical potential energy stored in fossil fuel is converted into thermal energy as the fuel is burned in a boiler. Only about 60 percent of the available chemical potential energy is converted into thermal energy.

Stage 2 The thermal energy heats water and produces steam. This stage is 90 percent efficient.

Stage 3 Pressurized steam spins the turbine blades. Thermal energy is converted into mechanical energy. This stage is 75 percent efficient.

Fuel

Water tank

Water

Water

Steam

Steam

Turbine

Intake pipe

Cooling water

Fuel burned in a boiler or combustion chamber converts chemical potential energy into thermal energy, which heats water and produces pressurized steam. This steam strikes the blades of a turbine, causing it to spin, converting thermal energy into mechanical energy. The shaft of the turbine connects to an electric generator, which converts mechanical energy into electrical energy. The electrical energy is then transmitted to homes, schools, and businesses through power lines.

Power plant efficiency In the power plant, not all of the chemical potential energy stored in fuel is converted into electrical energy. Some energy is converted into thermal energy. As a result, no stage of the process of electricity production is 100 percent efficient.

The overall efficiency of a fossil fuel–burning power plant is roughly 35 percent. This means that only 35 percent of the energy stored in fossil fuels is transported to homes, schools, and businesses as electrical energy. The remaining 65 percent is converted into thermal energy. Often, this heat is released into the environment.

The Cost of Fossil Fuels

Although fossil fuels are common energy resources, their uses have undesirable effects. Burning fossil fuels releases small particulates into the atmosphere that can cause breathing problems. Fossil fuels also release carbon dioxide (CO_2) when they are burned. **Figure 8** shows how the CO_2 concentration in the atmosphere has increased from 1958 to 2010. Many scientists think this increase in atmospheric CO_2 concentrations has contributed to global warming.

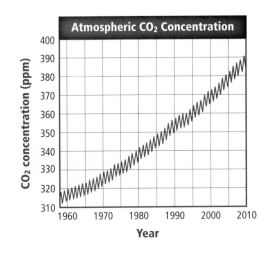

Figure 8 The carbon dioxide concentration in Earth's atmosphere has been measured at Mauna Loa in Hawaii. From 1958 to 2010, the carbon dioxide concentration has increased by 1.4 parts per million (ppm) per year.

Predict *how the concentration of carbon dioxide will change in the next several decades based on the graph trend.*

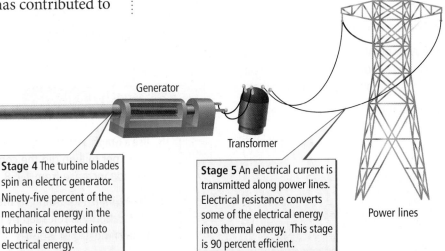

Stage 4 The turbine blades spin an electric generator. Ninety-five percent of the mechanical energy in the turbine is converted into electrical energy.

Stage 5 An electrical current is transmitted along power lines. Electrical resistance converts some of the electrical energy into thermal energy. This stage is 90 percent efficient.

Generator

Transformer

Power lines

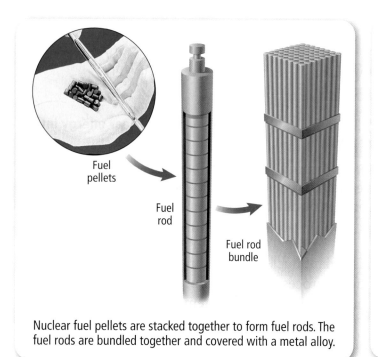

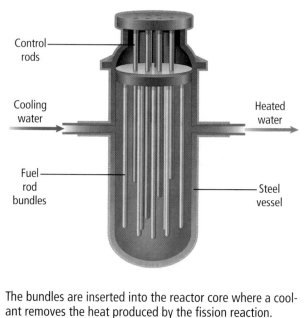

Nuclear fuel pellets are stacked together to form fuel rods. The fuel rods are bundled together and covered with a metal alloy.

The bundles are inserted into the reactor core where a coolant removes the heat produced by the fission reaction.

■ **Figure 11** The core of a nuclear reactor contains the fuel rod bundles. Control rods that absorb neutrons are inserted between the fuel rod bundles.

Animation

■ **Figure 12** The reactor core, which contains the fuel rod bundles, is submerged in a cooling chamber.

Nuclear Reactors

A **nuclear reactor** uses energy from controlled nuclear reactions to generate electricity. Although nuclear reactors vary in design, all reactors share some similarities. They contain fuel that can undergo fission and control rods that are used to control the nuclear reactions. They have a cooling system that keeps the reactor from being damaged by the enormous amount of heat produced. The actual fission of the radioactive fuel occurs in a relatively small part of the reactor known as the core, shown in **Figure 11.**

Nuclear fuel Only certain elements have nuclei that can undergo fission. Naturally occurring uranium contains the isotope U-235 with nuclei that can be split apart. Naturally occurring uranium typically contains 0.72 percent of the isotope U-235. The uranium used in a reactor is enriched so it contains 3–5 percent U-235. The fuel that is used in a nuclear reactor is usually uranium dioxide.

Fuel rods The reactor core contains uranium dioxide fuel in the form of tiny pellets like the ones shown in **Figure 11.** The pellets are about the size of a pencil eraser and are placed end-to-end in a fuel rod. The fuel rods are then bundled and covered with a metal alloy. The core of a typical reactor, shown in **Figure 12,** has about 100,000 kg of uranium contained in fuel rods. For every kilogram of uranium that undergoes fission in the core, 1 g of matter is converted into energy. You would have to burn more than 3 million kg of coal to generate an equivalent energy output.

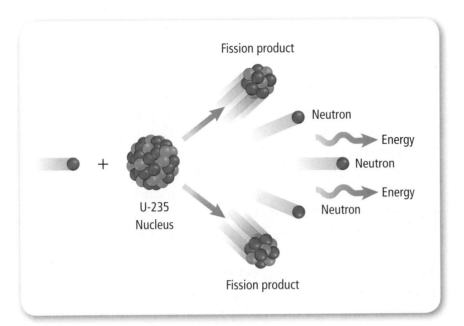

Fission product

Neutron

Energy

Neutron

U-235 Nucleus

Energy

Neutron

Fission product

■ **Figure 13** When a neutron strikes the nucleus of a U-235 atom, the nucleus splits apart into two smaller nuclei. In the process, two or three neutrons also are emitted. The smaller nuclei are called fission products.

Explain *what happens to the neutrons that are released in this reaction.*

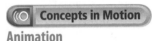
Concepts in Motion
Animation

The nuclear chain reaction How does a fission reaction proceed in the reactor core? As U-235 nuclei undergo fission, neutrons are released and are absorbed by other U-235 nuclei. When a U-235 nucleus absorbs a neutron, it splits into two smaller nuclei and two or three free neutrons, as shown in **Figure 13.** These neutrons strike other U-235 nuclei, triggering the release of more neutrons, and fission continues.

Because every uranium atom that splits apart releases free neutrons that cause other uranium atoms to split apart, this process is called a nuclear chain reaction. In the chain reaction, the number of nuclei that are split can more than double at each stage of the process. As a result, an enormous number of nuclei can be split after only a small number of stages. For example, if you start with one uranium nucleus and the number of nuclei involved doubles at each stage, after only 50 stages, more than a quadrillion nuclei might be split.

Nuclear chain reactions take place in a matter of milliseconds. If the process is not controlled, the chain reaction could release a tremendous amount of energy in the form of an explosion.

A constant rate To control the chain reaction, some of the neutrons that are released when U-235 splits apart must be prevented from colliding with other U-235 nuclei. These neutrons are absorbed by control rods containing boron or cadmium inserted into the reactor core, as shown in **Figure 11.** Moving these control rods deeper into the reactor causes them to absorb more neutrons and to slow down the chain reaction. Eventually, only one of the neutrons released in the fission of each of the U-235 nuclei strikes another U-235 nucleus, so energy is released at a constant rate.

◀ **FOLDABLES**
Incorporate information from this section into your Foldable.

Nuclear Power Plants

Nuclear power plants produce an electric current in a way that is similar to fossil fuel–burning power plants. As shown in **Figure 14,** the thermal energy released in fission is used to heat water and produce pressurized steam. To transfer thermal energy from the reactor core, the core contains a fluid coolant. The hot coolant is pumped into a heat exchanger. In the heat exchanger, thermal energy is transferred from the hot coolant to water, causing the water to boil and produce pressurized steam that spins a turbine. When the steam leaves the turbine, it enters a chamber where it is condensed back into liquid water. Cool water absorbs the thermal energy released during condensation. The thermal energy is then carried to the cooling tower where it is released into the environment. The overall efficiency of nuclear power plants is about 35 percent, which is similar to that of fossil fuel–burning power plants.

Benefits and Risks of Nuclear Power

Extracting energy from atomic nuclei has its advantages. Nuclear power plants do not produce the air pollutants that fossil fuel–burning power plants release into the atmosphere. Also, nuclear power plants do not release carbon dioxide into the atmosphere.

Nuclear power plants also have their disadvantages. For example, nuclear power plants are very expensive to build, and the building process can take 10 or more years to complete. Nuclear power plants also produce radioactive waste that can be harmful to living organisms and the environment.

Concepts in Motion

Animation

■ **Figure 14** A nuclear power plant converts water into pressurized steam that spins a turbine and generates electricity.

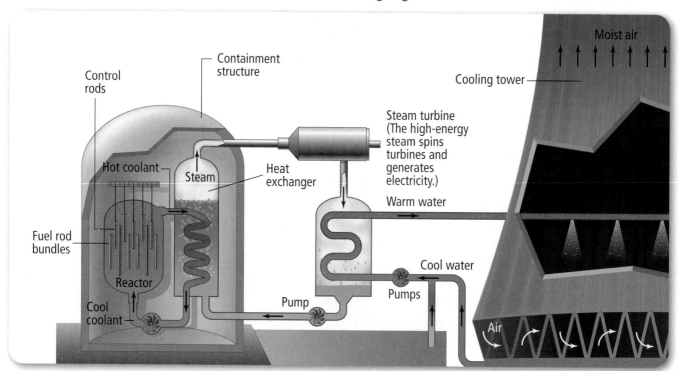

During a safety check on April 26, 1986, a nuclear reactor at the Chernobyl Nuclear Power Plant near Pripyat, Ukraine, exploded causing the worst nuclear disaster of all time.

Pripyat was abandoned after the explosion. Though it remains uninhabited, it is possible to obtain permits to visit the area affected by the explosion.

■ Figure 15 A steam explosion in the nuclear reactor at Chernobyl Nuclear Power Plant melted fuel rods and ignited the reactor's graphite cover setting the facility on fire.

The release of radioactivity Nuclear power plants around the world operate safely every day. However, one of the serious risks of nuclear power is the release of harmful radiation from power plants. The fuel rods contain radioactive elements. Some of these radioactive elements could harm living organisms if they are released from the reactor core of a nuclear power plant. To prevent accidents, nuclear reactors have elaborate systems of safeguards, strict safety precautions, and highly trained workers. In spite of this, accidents have still occurred.

For example, an accident occurred when a reactor core at the Chernobyl Nuclear Power Plant near Pripyat, Ukraine, overheated during a routine safety test on April 26, 1986. Materials in the core caught fire and caused a chemical explosion that blew a hole in the reactor, as shown in **Figure 15**. This resulted in the release of radioactive materials that were carried by winds and deposited over a large area. As a result of the accident, 50 people died from acute radiation sickness and about 4,000 cancer-related cases have been attributed to the release of radioactivity from the explosion.

The World Health Organization estimates that approximately 600,000 people were exposed to levels of radiation that continue to pose a risk to their health. Newer nuclear power plants are designed to prevent accidents like the one that occurred at Chernobyl. But there is always a possibility that an accident might occur.

VOCABULARY

SCIENCE V. COMMON USE

System

Science usage
 the particular reaction or process being studied
 The universe consists of a system and its surroundings.

Common usage
 an organized or established procedure
 She designed a system in which each person would have an equal opportunity to earn a raise.

The Disposal of Nuclear Waste

After about three years of use, the amount of U-235 in the fuel pellets in the reactor core is too small for the chain reaction to continue. The fuel pellets left are now referred to as spent fuel. The spent fuel includes radioactive fission products in addition to some leftover U-235. Spent fuel is a form of nuclear waste. **Nuclear waste** is any radioactive material that results when radioactive materials are used.

 **Reading Check Describe** the formation of spent fuel.

While many people support the idea of using nuclear energy as an alternative to fossil fuels, they do not necessarily support the idea of nuclear waste disposal in their state. Many people refer to this antinuclear attitude as the "Not in My Backyard" syndrome. Nuclear waste disposal has been a controversial subject and continues to fuel debate about nuclear energy use.

Low-level waste Low-level nuclear waste usually contains a small amount of radioactive material. Additionally, low-level waste usually contains radioactive materials with short half-lives. Low-level waste is a by-product of electricity generation, medical research and treatments, the pharmaceutical industry, and food preparation. Low-level wastes also include used water and air filters from nuclear power plants and discarded smoke detectors. Low-level waste is kept isolated from people and the environment. It is treated as hazardous material and is stored in spill-safe containers underground.

 **Apply Science**

Can a contaminated radioactive site be reclaimed?

With the discovery of radium in the early 1900s, extensive mining for the element began in the Denver, Colorado, area. Radium is a radioactive element that was used to make watch dials and instrument panels that glowed in the dark. After World War I, the radium industry collapsed. The area was left contaminated with 97,000 tons of radioactive soil and debris containing heavy metals and radium, which is now known to cause cancer. The soil was used as fill or foundation material, left in place, or mishandled.

Identify the Problem

In the 1980s, one area became known as the Denver Radium Superfund Site and was cleaned up by the Environmental Protection Agency. The land then was reclaimed by a local commercial establishment.

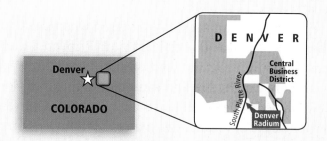

Solve the Problem

1. The contaminated soil was placed in one area and a protective cap was placed over it. This area was also restricted from being used for residential homes. Explain why it is important for the protective cap to be maintained and why homes could not be built in this area.

2. The advantages of cleaning this site are economical, environmental, and social. Give an example of each.

High-level waste High-level nuclear waste is generated in nuclear power plants and by nuclear weapons programs. After spent fuel is removed from a reactor, it is stored in steel-lined concrete pools filled with water, as shown in **Figure 16,** or in airtight steel or concrete and steel canisters.

Many of the radioactive materials in high-level nuclear waste become nonradioactive after a relatively short amount of time. However, the spent fuel also contains materials that will remain radioactive for tens of thousands of years. For this reason, high-level waste must be disposed of in extremely durable, safe, and stable containers.

Reading Check Describe What are the differences between low-level and high-level nuclear wastes?

One method proposed for the disposal of high-level waste is to seal the waste in ceramic glass, which is placed in protective metal containers. The containers then are buried hundreds of meters belowground in stable rock formations or in salt deposits.

■ **Figure 16** Spent nuclear fuel is stored in spill-safe containers at nuclear power plants and often submerged within specially designed pools.

Section 2 Review

Section Summary

▶ Nuclear power plants produce about 8 percent of the energy used each year in the United States.

▶ Nuclear reactors use the energy released in the fission of U-235 to produce electricity.

▶ The energy released in the fission reaction is used to make steam. The steam spins a turbine that powers an electric generator.

▶ Nuclear power generation produces high-level nuclear waste.

8. **MAIN Idea Compare and contrast** the advantages and disadvantages of nuclear power plants and those that burn fossil fuels.

9. **Describe** nuclear fission and how the chain reaction in a nuclear reactor is controlled.

10. **Describe** nuclear fusion and the problems associated with using nuclear fusion reactions as an energy source.

11. **Explain** why a chain reaction occurs when uranium-235 undergoes fission.

12. **Think Critically** A research project produced 10 g of nuclear waste with a short half-life. How would you classify this waste, and how would it be disposed?

Apply Math

13. **Calculate** Naturally occurring uranium contains 0.72 percent of the isotope uranium-235. What is the mass of uranium-235 in 2,000 kg of naturally occurring uranium?

Reading Preview

Essential Questions

▶ What are renewable energy resources?

▶ What are some methods for converting different types of renewable resources into electrical energy?

Review Vocabulary

radiant energy: the energy carried by electromagnetic waves

New Vocabulary

renewable resource
photovoltaic cell
hydroelectricity
geothermal energy
biomass

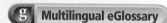 **Multilingual eGlossary**

Renewable Energy Resources

MAIN Idea Renewable energy resources help lessen human dependence on fossil fuels.

Real-World Reading Link What types of activities did you engage in this morning that required energy resources? Whether it involved turning on lights, preparing breakfast on the stove, or riding a bus to school, many daily activities use Earth's energy resources.

Energy Options

The demand for energy is increasing every day as Earth's population increases. As demand increases, our supply of non-renewable energy resources decreases. Use of nuclear energy produces high-level waste that has to be disposed of safely. As a result, alternative energy sources are being developed to meet increasing energy demands. Some alternative energy sources are renewable resources. A **renewable resource** is an energy source that is replaced by natural processes faster than humans can consume the resource.

Energy from the Sun The average amount of solar energy that shines down on the United States in one year is 1,000 times more energy than the total energy used in one year. Because the Sun is expected to continue producing energy for billions of years, solar energy is inexhaustible in our lifetime. Solar energy is a renewable resource.

Despite being renewable, only 1 percent of the energy in the United States is produced using solar power. There are several ways to produce solar power. One way is to use a photovoltaic cell, as shown in **Figure 17**. A **photovoltaic cell** converts radiant energy directly into electrical energy. Photovoltaic cells are also called solar cells..

 Inquiry Video Lab

■ **Figure 17** Photovoltaic cells convert radiant energy into electrical energy. Some vehicles have optional photovoltaic panels made of solar cells that are used to cool the car without the use of the engine.

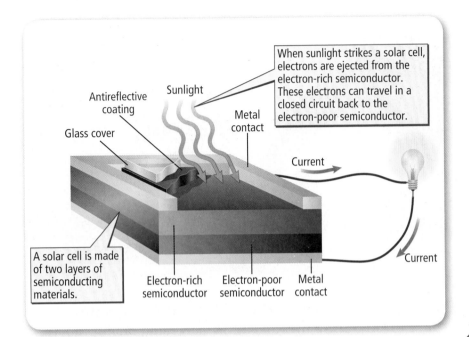

When sunlight strikes a solar cell, electrons are ejected from the electron-rich semiconductor. These electrons can travel in a closed circuit back to the electron-poor semiconductor.

Antireflective coating

Sunlight

Glass cover

Metal contact

Current

A solar cell is made of two layers of semiconducting materials.

Electron-rich semiconductor

Electron-poor semiconductor

Metal contact

Current

Figure 18 Radiant energy from sunlight strikes the surface of a solar cell, exciting electrons and causing them to flow through an electric circuit.

Identify *two devices that use solar cells for power.*

How solar cells work Solar cells are made of two layers of semiconducting material sandwiched between two layers of conducting metal, as shown in **Figure 18.** One layer of the semiconducting material is rich in electrons, and the other layer is electron poor. When sunlight strikes the surface of the solar cell, electrons flow through an electric circuit from the electron-rich material to the electron-poor material. This process of converting radiant energy from the Sun directly to electrical energy is only about 7–11 percent efficient.

The transformation of radiant energy into electrical energy using solar cells is more expensive than the transformation of thermal energy into electrical energy by combustion. However, in remote areas where power lines are not available, solar cells are a practical energy source.

Parabolic troughs Other promising solar technologies concentrate solar power into a receiver. One such system is called the parabolic trough. The trough focuses the sunlight on a tube that contains a heat-absorbing fluid, such as synthetic oil or liquid salt. Sunlight heats the fluid, which circulates through a boiler, where it turns water to steam that spins a turbine to generate an electric current.

One of the world's largest concentrating solar power plants is located in the Mojave Desert in California. This facility consists of nine units that generate more than 350 megawatts of power. These nine units can generate enough electricity to meet the demands of approximately 500,000 people. The units also use natural gas as a backup power source for generating an electric current at night and on cloudy days when solar energy is unavailable.

MiniLab

? Inquiry MiniLab

Use Solar Power at Home

Procedure

1. Read the procedure and safety information, and complete the lab form.
2. Use **scissors** to cut a piece of **cloth** into four pieces of equal size.
3. Wet the pieces with **water** and wring them out so they are the same dampness.
4. Spread out the pieces to dry—two pieces inside and two pieces outside. One piece of each set should be in direct sunlight, and one piece should be in the shade.
5. Record the time that it takes for each cloth piece to dry.

Analysis

1. **Evaluate** How long did it take for each cloth piece to dry?
2. **Analyze** What conditions determined how quickly the cloth dried?
3. **Infer** how you can use solar energy in your home to conserve electricity.

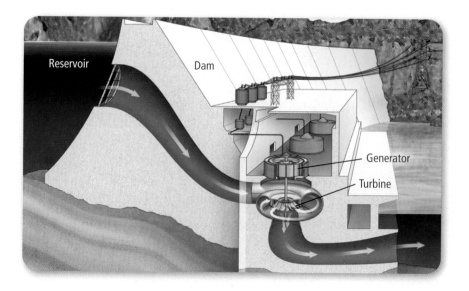

■ **Figure 19** The gravitational potential energy of water behind the dam is converted to electrical energy in a hydroelectric power plant.

Explain *the energy conversions that occur as a hydroelectric dam produces electrical energy.*

■ **Figure 20** A fish ladder is like a series of steps that allows fish to swim past a dam.

Energy from water Just as the expansion of steam can spin a turbine and power an electric generator, rapidly moving water can do the same. The gravitational potential energy of water is great when the water is retained by a dam. This energy is released when the water flows through tunnels near the base of the dam. **Figure 19** shows how the rushing water spins a turbine, converting gravitational potential energy to mechanical energy and then to electrical energy. Dams built for this purpose are called hydroelectric dams.

Hydroelectricity Electric current produced from the energy of moving water is called **hydroelectricity.** About nine percent of the electrical energy used in the United States comes from hydroelectric power plants. Hydroelectric power plants convert mechanical energy into electrical energy with almost no pollution. They are almost twice as efficient as fossil fuel–burning or nuclear power plants.

   In addition to efficiency, another advantage of hydroelectric power is that the bodies of water held back by dams can form lakes that can provide water for drinking and crop irrigation. These lakes can also be used for boating and swimming. After the initial cost of dam construction, hydroelectric power plants are more cost effective than other energy resources.

   However, dams and hydroelectric power plants can disturb the balance of natural ecosystems. Some species of fish that live in the ocean migrate back to the rivers in which they were hatched to breed. This migration can be blocked by dams, causing a decline in the fish population. Fish ladders, such as those shown in **Figure 20,** have been designed to enable fish to migrate upstream past some dams. Also, operation of a hydroelectric power plant can change the temperature of the water, which affects plant and animal habitats. Finally, river sediments can build up behind the dam and affect life downstream.

Energy from the oceans The gravitational pull of the Moon and the Sun on Earth's oceans causes tides. Hydroelectric power can be produced by these tides. As the tide rises, water spins a turbine, which transforms mechanical energy into electrical energy. The water is then trapped behind a dam. At low tide, the water behind the dam is released to flow back out to sea, converting even more energy to electricity.

Hydroelectric power can also be produced by waves. There are several new technologies that harness wave energy. One type focuses wave energy into a channel. As waves enter a channel, they spin turbines, converting mechanical energy into electrical energy. Plans are also in place to harness mechanical energy from ocean currents, as shown in **Figure 21.**

Energy from the ocean is nearly pollution free, and the efficiency of tidal and wave power plants is similar to that of hydroelectric power plant. However, only a few places on Earth have large enough differences between high and low tide for oceans to be a useful energy source.

Wind energy Windmills can convert wind energy into electrical energy. As the wind blows, it spins a propeller that is connected to an electric generator. The greater the wind speed and the longer the wind blows, the greater the amount of wind energy that is converted into electrical energy. Windmill farms, like the one shown in **Figure 22,** can contain several hundred windmills.

One disadvantage of wind energy is that only a few places on Earth have enough wind to meet our energy needs. Even then, wind energy cannot be stored without the use of batteries. Windmills can be noisy and change the appearance of a landscape. They also can disrupt the migration patterns of some birds.

The advantages of using wind energy are that wind generators do not consume nonrenewable resources and they do not pollute air or water. Research is underway to improve the design of wind generators and to increase their efficiency. By 2030, the U.S. Department of Energy wants to increase our use of wind energy so it provides 20 percent of our total electrical power.

■ **Figure 21** In Florida, researchers estimate that underwater turbines spun by the Gulf Stream currents could generate the equivalent of ten nuclear power plants' worth of electricity.

■ **Figure 22** Wind energy is converted to electrical energy as the spinning propeller turns a generator.

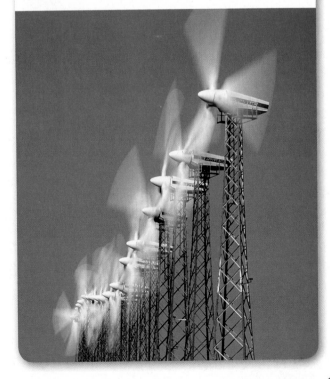

Energy from inside Earth Unstable radio-active elements in Earth's core transform nuclear energy into thermal energy. As these unstable elements decay, thermal energy is transferred from the core into Earth's mantle and crust. This is called geothermal heat. Geothermal heat can cause the rock beneath Earth's crust to melt. Molten rock beneath Earth's surface is called magma. The thermal energy that is contained in and around magma is called **geothermal energy.**

✓ **Reading Check Identify** the process that transforms energy inside Earth into thermal energy.

In some areas, Earth's crust has cracks or areas of weakness that allow magma to rise toward the surface. Active volcanoes, for example, permit hot gases and magma from deep within Earth to escape. Perhaps you have seen a geyser, such as Old Faithful in Yellowstone National Park, spewing hot groundwater and steam. The groundwater erupting from the geyser is heated by magma close to Earth's surface. In some areas, hot groundwater is pumped directly into homes to provide warmth.

Geothermal power plants Geothermal energy can be converted into electrical energy, as shown in **Figure 23.** Where magma is close to the surface, the surrounding rocks are hot. Water is pumped into the ground through a well, where it comes into contact with hot rock and changes into steam. The steam then returns to the surface, where it spins a turbine that powers an electric generator.

The efficiency of geothermal power plants is about 16 percent. Although geothermal power plants can produce sulfur-based compounds, pumping the water that condenses from steam back into the ground reduces this pollution. This makes geothermal power plants a source of clean energy. However, one disadvantage is that the use of geothermal energy is limited to volcanically active areas where magma is close to the surface. Five states in the United States have geothermal power plants—California, Nevada, Hawaii, Montana, and Utah.

FOLDABLES
Incorporate information from this section into your Foldable.

Concepts in Motion Animation

■ **Figure 23** Geothermal power plants convert geothermal energy to electrical energy. Water is changed to steam by the hot rock. The steam is pumped to the surface where it drives a turbine attached to an electric generator.

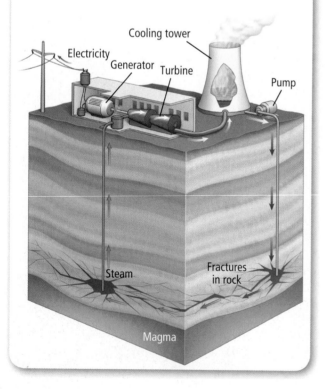

Alternative Fuels

The use of fossil fuels would be greatly reduced if cars could run on alternative energy resources alone. For example, cars have been developed that use electrical energy supplied by batteries as their primary power source. Hybrid cars use electric motors and gasoline engines.

Hydrogen Hydrogen fuel cells are another possible alternative energy resource. A fuel cell behaves like a battery. It combines hydrogen with oxygen in air to generate electrical energy, water, and heat. There are several problems with using hydrogen fuel as an alternative energy resource, however. First, obtaining hydrogen requires more energy than the energy that is released by the fuel-cell reaction. Second, hydrogen fuel cells are built from expensive platinum parts. And third, there is a lack of hydrogen fueling stations, as storing hydrogen is considered to be dangerous and difficult.

Biomass Are there any other materials that can be used to heat water and produce electricity other than fossil fuels, nuclear fission, or hydrogen? Biomass is one of the oldest energy sources. **Biomass** is renewable organic matter, such as wood, soy, corn, sugarcane fibers, rice hulls, and animal manure. It can be burned in the presence of oxygen, which converts the stored chemical potential energy to thermal energy. **Figure 24** shows a bus powered by recycled cooking oil derived from biomass.

■ **Figure 24** Soybean oil and recycled cooking oils can be used as alternative fuels for transportation.

Section 3 Review

Section Summary

▶ Solar cells convert radiant energy into electrical energy.

▶ Hydroelectric power plants convert gravitational potential energy into electrical energy.

▶ Wind energy is converted into electrical energy using a propeller attached to an electric generator.

▶ Alternative energy sources, such as the Sun, water, wind, and Earth's internal heat, can help reduce human dependence on fossil fuels.

14. **MAIN Idea Explain** the need to develop and use alternative energy sources.

15. **Describe** three ways that solar energy can be used.

16. **Explain** how the generation of electricity by hydroelectric, tidal, and wind sources are similar to each other.

17. **Infer** why geothermal energy is unlikely to become a major energy source.

18. **Think Critically** On what single energy source do most energy alternatives depend, either directly or indirectly?

Apply Math

19. **Use Percentages** A house uses solar cells that generate 1.5 kW of electrical power to supply some of its energy needs. If the solar panels supply the house with 40 percent of the power it needs, how much power does the house use?

✓ **Assessment** Online Quiz

LAB Solar Heating

Objectives

■ **Demonstrate** solar heating.

■ **Compare and contrast** the effectiveness of heating items of different colors.

Background: Radiant energy from the Sun is absorbed by Earth's atmosphere, land, and water. In a similar way, radiant energy is also absorbed by solar collectors to heat water and buildings.

Question: *Does the rate at which an object absorbs radiant energy depend on the color of the object?*

Preparation

Materials

small cardboard boxes
black, white, and colored paper
scissors
tape or glue
thermometer
watch with a second hand

Safety Precautions

Procedure

1. Read the procedure and safety information, and complete the lab form.

2. Cover at least three small boxes with colored paper. The colors should include black and white as well as at least one other color.

3. Create a data table to show the temperature change of different colored boxes over time.

4. Place the three objects on a windowsill or outside in a sunny spot, and note the starting time.

5. Measure and record the temperature inside each box at 2-min intervals for at least 20 min.

Conclude and Apply

1. **Graph** your data using a line graph.

2. **Describe** the shapes of the lines on your graph. Which color heated the fastest? Which heated the slowest?

3. **Explain** why the colored boxes heated at different rates.

4. **Infer** Suppose you wanted to heat a tub of water using radiant energy. Based on the results of this activity, what color would you want the tub to be? Explain.

5. **Explain** why you might want to wear a white or light-colored shirt on a hot, sunny, summer day.

Compare your results with those of other students in your class. Discuss any differences found in your graphs, particularly if different colors were used by different groups.

? **Inquiry** Lab

Reading Preview

Essential Questions

▶ How does human population affect Earth's carrying capacity?

▶ What are the causes and effects of pollution on land, water, and air?

▶ What are some methods used to control the types and sources of pollution?

▶ How can you help protect and preserve Earth's natural resources?

Review Vocabulary

temperature: measure of the average kinetic energy of all the particles in an object

New Vocabulary

population
carrying capacity
pollutant
hazardous waste
photochemical smog
acid precipitation

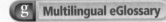 **Multilingual eGlossary**

Environmental Impacts

MAIN ◀Idea **Human impact on land, water, and air affects the natural resources available for use.**

Real-World Reading Link By 2050, Earth's population could increase to more than 9 billion people—almost one and a half times more than what it is today. Try to picture the impact that such a large human population will have on our natural resources and the environment.

Population and Carrying Capacity

A **population** includes all the individuals of one species living in a particular area. You can see in **Figure 25** that it took thousands of years for the human population to reach 1 billion people. In the mid-1800s, human population began to increase at a rapid rate because of advances in modern medicine and the availability of clean water and better nutrition. People began to live longer. In addition, the number of births increased because more people survived to a child-bearing age.

Carrying capacity Every person alive today uses and is dependent on Earth's natural resources. However, Earth has a carrying capacity. **Carrying capacity** is the largest number of individuals of a particular species that the environment can support, given the natural resources available. If natural resources are consumed too quickly or the environment becomes threatened, then populations suffer. Unless Earth's natural resources are treated with care, the human population could reach its carrying capacity.

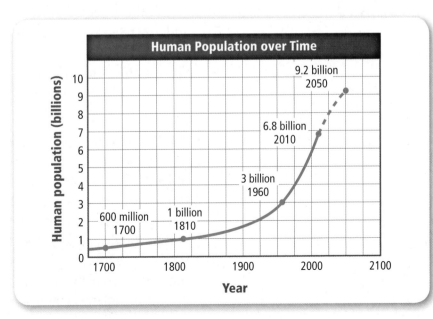

■ **Figure 25** Human population growth rates remained fairly steady until the middle of the nineteenth century. The growth rate then began to increase rapidly.

Estimate *What will the human population be in the year 2025?*

Figure 26 At a water treatment plant, pollutants are removed from wastewater to help keep our waterways clean.

People and the Environment

You have an impact on the environment every day. The electrical energy that you use most likely comes from burning fossil fuels. The cars and buses you use for transportation burn fossil fuel. Fossil fuels are mined from Earth and have an impact on the air that you breathe. The water that you use must be treated, as shown in **Figure 26,** to remove as many pollutants as possible before it is recycled back into waterways. **Pollutants** include any substance that contaminates the environment.

You also use plastics and paper every day. Plastics are petroleum-based products. When petroleum is refined, it produces pollutants. In the process of harvesting trees to make paper, trees are cut down. They are transported using fossil fuels, and water and air can be polluted in the paper-making process.

Impact on Land

Land is affected when resources such as fossil fuels, water, soil, or trees, are extracted from Earth. You might not think of land as a natural resource, but it is as important as fossil fuels, clean water, and clean air. We use land for agriculture, forests, urban development, and even waste management. These uses impact the land and the natural resources it provides.

Agriculture The pears and apples that you purchase at the grocery store were grown on farms, which cover 16 million km² of Earth's total land area. To feed the world's growing population, some farmers are planting higher-yielding seeds and using stronger nitrate- and phosphate-based fertilizers. Herbicides and pesticides are also used for weed and pest control. These methods increase the amount of food grown, but if they not managed properly, they can have a negative impact by possibly polluting soil and water and endangering animals.

Organic farms Organic farming methods, as shown in **Figure 27,** use natural fertilizers, crop rotation, and biological pest controls. These methods help reduce pollution and other negative impacts on land. However, organic farming methods cannot currently produce the food that is necessary to feed the world's growing population.

Deforestation

Approximately 25 percent of Earth's total land area is covered by forests. Whether you are writing on paper with a pencil, sitting in a wooden chair, or wiping your face with a napkin, you are using products derived from wood. This wood comes from forests worldwide.

Deforestation is the clearing of forest land for agriculture, grazing, urban development, or logging. It is estimated that the amount of forested land decreases by 94,000 km^2 each year. Many of these forests are home to diverse populations of plants and animals. Cutting down trees could lead to the extinction of some of these organisms. In addition, plants remove carbon dioxide from the atmosphere. Deforestation increases the concentration of carbon dioxide in the atmosphere. Scientists believe that an increase in carbon dioxide has contributed to an increase in atmospheric temperatures worldwide.

Urban development

With a growing population, the percentage of land area devoted to urban development has increased. Highways, office buildings, stores, housing developments, and parking lots are under construction every day. This development can lead to negative impacts on land. For example, paving land prevents water from soaking into the soil. Instead, water runs off into sewers or streams, increasing stream discharge and the threat of flooding. Because water is unable to seep through pavement, this also decreases the amount of water that seeps into the ground.

Some communities, businesses, and private organizations preserve areas rather than pave them. As population grows, more urban areas have been set aside for recreation and preservation for future generations to enjoy. Some urban areas have been designated as historic sites, parks, and monuments by the U.S., state, and local governments, such as Central Park in New York City shown in **Figure 28.**

Waste

Whether or not you realize it, you impact land when you throw garbage into your trash can. About 55 percent of our garbage is disposed of in sanitary landfills. The rest is recycled or burned. Some of the wastes release substances, such as lead from batteries, that are harmful to humans and animals. Wastes that are poisonous, that cause cancer, or that can catch fire are classified as **hazardous wastes.**

■ **Figure 27** Organic farms can reduce the environmental impact of fertilizers, pesticides, and herbicides on land.

■ **Figure 28** Some land in urban areas, such as New York City's Central Park, is preserved for recreation.

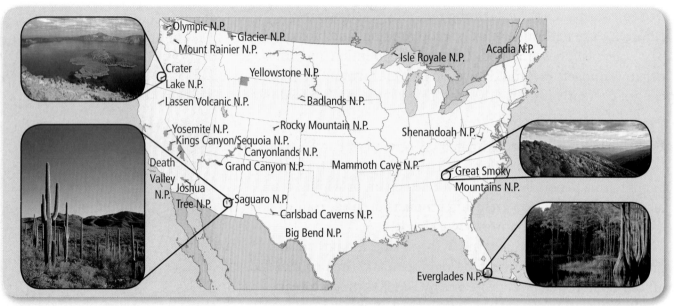

Figure 29 The U.S. National Park Service is committed to preserve and protect nearly 400 natural, cultural, and recreational areas across the country.

Identify *any national parks located close to where you live.*

National and state parks National and state parks are areas of land, like those shown in **Figure 29,** that are preserved and protected by the U.S. government. These forests, wetlands, grasslands, and parks in the United States are safe from urban development, waste disposal, and extensive deforestation. Parks are home to plants, animals, and waterways. Millions of people visit parks, such as Grand Canyon National Park each year.

Many countries throughout the world also set aside land for protection and preservation. As the world population grows, the impact on land may worsen. Preserving this land in its natural state will benefit generations to come.

Impact on Water

Life on Earth would not be possible without water. Plants need water to convert radiant energy into food energy. Some animals, such as fish, frogs, and whales, make bodies of water their homes. Approximately 60 percent of the human body is composed of water. How are living things affected when water becomes polluted?

Sources of water pollution Many streams and lakes in the United States are polluted. Polluted water contains harmful chemicals and sometimes organisms that cause disease. Water can also be polluted with sediments, such as silt and clay. Sediment from runoff makes water cloudy and can limit the sunlight and oxygen supply, which then affects fish and wildlife.

Industry Mining can release metals into water. Metals such as mercury, lead, nickel, and cadmium are poisonous. However, environmental laws limit the amount of these harmful chemicals that can be released into the environment, and they protect natural resources and the people that depend upon them.

Prince William Sound, Alaska

Gulf of Mexico, USA

Oil and gas Oil and gas can run off roads and parking lots into lakes and rivers when it rains. It can also leak from oil tankers or pipelines associated with offshore drilling, as shown in **Figure 30**. Oil and gas are pollutants, which can cause cancer. Today, environmental laws require that all new gasoline storage tanks have a double layer of steel or fiberglass to prevent spills. These laws help protect soil and water from oil spills.

Human waste When you flush a toilet or take a shower, you create wastewater. Wastewater, also called sewage, contains human waste, household detergents, and soaps. Sewage contains harmful organisms that can make people ill.

In most cities in the United States, underground pipes route water from homes, schools, and businesses to sewage treatment plants. Sewage treatment plants remove pollutants in a series of steps. These steps purify the water by removing solid materials from the sewage, killing harmful microorganisms, and reducing the amount of nitrogen and phosphorous in the water. The water is then recycled back into the environment.

Impact on Air

Like water, air is essential for all life on Earth. Air pollution can affect human health and threaten plants and animals. Air pollution comes from natural and manufactured sources. For example, cars, buses, and trucks burn fuel for energy and release exhaust into the atmosphere. Factories and power plants emit pollutants during production, as shown in **Figure 31**. Dust from farms and construction sites also contributes to air pollution. Natural sources of pollution include particles and gases emitted into air from erupting volcanoes and forest fires.

✓ **Reading Check List** the sources of land, water, and air pollution.

■ **Figure 30** An oil tanker accident near Prince William Sound, Alaska on March 24, 1989 resulted in nearly two decades of environmental clean-up. A deadly explosion in the Gulf of Mexico on April 20, 2010 resulted in the release of millions of gallons of oil. Recovery efforts to try and save and rehabilitate marine habitats, fish, and wildlife will likely continue here for decades.

■ **Figure 31** Two significant sources of air pollution include cars and factories.

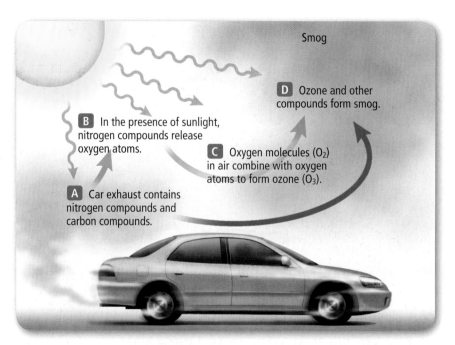

Figure 32 Exhaust from cars contributes to smog formation. Sunlight helps fuel reactions that form smog compounds. These compounds include nitrogen-based compounds and ozone.

Smog

B In the presence of sunlight, nitrogen compounds release oxygen atoms.

C Oxygen molecules (O₂) in air combine with oxygen atoms to form ozone (O₃).

D Ozone and other compounds form smog.

A Car exhaust contains nitrogen compounds and carbon compounds.

Types of air pollution Have you ever observed a thick, brown haze on the horizon? The brown haze that you see forms from vehicle exhaust and factory and power plant pollution. This haze is often referred to as photochemical smog. **Photochemical smog** is a term used to describe the pollution that results from the reaction between sunlight and vehicle or factory exhaust.

Smog Major sources of photochemical smog include cars, factories, and power plants. Pollutants are released into the air when fossil fuels, such as gasoline are burned, as shown in **Figure 32,** emitting sulfur-, nitrogen-, and carbon-based compounds. These compounds react with oxygen in the presence of sunlight. One of the products of this reaction is ozone (O_3). Ozone that forms high in the atmosphere protects you from ultraviolet (UV) radiation from the Sun. Ozone near Earth's surface, however, can cause breathing problems.

CFCs The protective ozone high in the atmosphere is concentrated in a layer roughly 20 km above Earth's surface. This layer is called the ozone layer, and it is at risk of being destroyed. Chlorofluorocarbons (CFCs) are compounds that leak from old air conditioners and refrigerators and react with ozone. This reaction destroys ozone molecules. Even though the use of CFCs has been declining due to environmental laws, these compounds can remain in the atmosphere for decades.

Acid precipitation When sulfur-, nitrogen-, and carbon-based compounds from vehicles and factories react with moisture in the air, they form acids. When acidic moisture falls from the sky as precipitation, it is called **acid precipitation.** Acid precipitation can corrode metals and cause harm to plants and animals.

Reducing Pollution

Pollution is often difficult to contain. Airborne pollutants travel wherever the wind carries them. Even if one state or country reduces air pollution, pollutants from another state or country can blow across borders. For example, burning coal in the midwestern United States might cause acid precipitation in Canada. Water pollution can enter a river or stream and travel several kilometers downstream, into groundwater supplies, and across state borders.

How can you help? The United States uses more natural resources per person than most countries in the world. There are ways that you can help conserve resources. You can reduce the amount of consumable materials you use. You can compost some yard and kitchen waste rather than throwing it in the trash. You can also reuse and recycle many different materials, as shown in **Figure 33.**

Energy-efficient appliances can help your family reduce its energy dependence. Low-flush toilets, leak-free faucets, and dishwashers and washing machines that run on less water will help you reduce your water use. Driving fuel-efficient vehicles or using alternate modes of transportation, such as a bicycle or a bus, will help lessen your impact on air.

■ **Figure 33** Many communities offer recycling programs where paper, plastic, and glass are reused rather than going into landfills.

Section 4 Review

Section Summary

▶ Land resources are threatened by agriculture, deforestation, industry, and waste.

▶ All life on Earth requires healthy air and water.

▶ Water can be contaminated with sediment, industrial pollutants, and human waste.

▶ Smog, acid precipitation, and CFCs cause air pollution.

20. **MAIN Idea Discuss** what you can do to lessen your environmental impact on natural resources such as land, water, and air.

21. **Describe** how urban development can increase flooding.

22. **Infer** the effect of deforestation on the carrying capacity of the Amazon rain forest.

23. **Identify** three pollutants released into the air when fossil fuels are burned.

24. **Think Critically** Southern Florida is home to many dairy and sugarcane farms. Everglades National Park, including its shallow river system, is also located there. What kinds of pollutants might affect plants and animals in the Everglades?

Apply Math

25. **Calculate pH** A decrease of one unit on the pH scale means a solution is ten times more acidic. A decrease of two units means the solution is 100 times more acidic. How much more acidic is acid precipitation (pH = 4.0) than pure water (pH = 7.0)?

Energy-Efficient Buildings

USE THE INTERNET

Objectives

- **Research** new technologies that are used to construct energy-efficient buildings.
- **Compare and contrast** energy-efficient building materials.
- **Research** how you could implement these technologies into the design of an energy-efficient home.

Background: Buildings can be designed to be energy efficient. For example, engineers and architects choose materials that store thermal energy, such as solar cells. They design buildings that incorporate these materials to help consumers decrease their monthly gas or electric bills and to conserve natural resources.

Question: *What types of new technologies are available to build energy-efficient homes?*

Preparation

Data Source
Access to the Internet or reference materials

Safety Precautions

Make a Plan

1. Read the procedure and safety information, and complete the lab form.
2. Research energy-efficient buildings, and choose three new technologies that have been designed to conserve energy.
3. Working in groups of three or four, identify these new technologies.
4. As a group, decide how you might determine which technology is the most energy efficient.

5. Prepare a summary of your research. Describe all three energy-efficient technologies, their design and application, and how cost effective they are.
6. Design a building that incorporates the most energy-efficient technology that you researched.
7. Decide which materials you will use to construct your building using this new technology. Possible materials that you might use include glass or clear plastic squares, sturdy cardboard boxes, scissors, tape, glue, thermometers, white and black paint, paper, aluminum foil , polystyrene, stone, mirrors, fabric, and a light source.

Follow Your Plan

1. Make sure your teacher approves your plan before you begin.
2. Construct an energy-efficient building that incorporates the results of your research. Construct a control building as a comparison for your energy-efficient building. Use the same building materials in both designs. Leave the new energy-efficient technology out of the control-building design.

Data Table	Temperature Variations of Buildings	
Time (min)	Energy-Efficient Building (°C)	Control Building (°C)
5		
10		
15		
20		
25		

3. Test your building's energy efficiency. For example, you might heat the energy-efficient building and the control building and evaluate how well each building is insulated. **WARNING:** *Make sure the heat source is far enough away from the building material so the material does not burn or melt.*

4. Record your temperature data in a table like the one shown above.

5. Make modifications to the design to improve your building's energy efficiency.

6. In your summary, include an analysis about whether your energy-efficient building design was successful.

Analyze Your Data

1. **Analyze** Of the new technologies that you investigated, which one is the most energy efficient?

2. **Analyze** What problems did you encounter in your building design, and how did you solve them?

3. **Compare and contrast** the design of your energy-efficient building to the control building.

Conclude and Apply

1. **Conclude** Was the building that you designed more energy efficient than the control building?

2. **Predict** Would your design work in a home in your community? In a community with a different climate? Why or why not?

3. **Research** other new building designs that help conserve natural resources.

4. **Propose** how your design could be improved.

Investigate How did your design compare to the energy-efficient buildings designed and built by your classmates? Prepare a brochure to highlight the advantages of your energy-efficient design.

SCIENCE & HISTORY

EARTH DAY, 1970

Environmental crises captured national attention in the late 1960s. In 1966, more than 160 deaths were caused by chemical smog that settled in New York City for three days. In 1969, an oil well exploded and spread crude oil across 55 kilometers of coastline near Santa Barbara, California. During that same year, the contaminated Cuyahoga River in Ohio caught fire. These disasters inspired an environmental movement that would eventually lead to new laws that protect and preserve our natural resources.

Change in the air An environmental movement swept across the United States in the 1970s. Influential books like *Silent Spring*, written by biologist Rachel Carson, painted a bleak picture of a polluted world. Many think, however, that the oil spill in Santa Barbara, California, shown in **Figure 1,** and nationally broadcast images of volunteers rescuing oil-covered seals and seabirds was the catalyst for change. In 1970, one year after the catastrophe, 20 million Americans participated in the first Earth Day.

Safeguarding the environment Earth Day led the U.S. government to respond to calls for environmental reform. In 1970, President Nixon created the Environmental Protection Agency (EPA). The EPA was responsible for conducting research and proposing new laws to protect the environment. The agency also had the ability to enforce these laws and to hold individuals and corporations accountable for meeting new environmental standards.

Figure 1 Volunteers clean the Santa Barbara coastline after the devastating 1969 oil spill.

The Clean Air and Water Acts One of the EPA's first laws was the Clean Air Act of 1970. The law created air pollution standards, limiting the amount of carbon monoxide, nitrogen dioxide, and ozone released into the environment. It also required the eventual phase-out of lead as an ingredient in gasoline. The Clean Water Act followed in 1972, limiting the amount of pollutants that could be released into rivers, lakes, and streams.

Looking forward Protecting natural resources has become a worldwide effort. More than 200 countries signed the Montreal Protocol to end the production of ozone-destroying chemicals. Eighty world leaders convened at the UN Climate Change Conference in Copenhagen, Denmark, to commit to worldwide reductions of greenhouse gases. Many hope a global effort will create a healthier planet for future generations.

WebQuest **Analyze and Design a Logo** The EPA logo contains symbols that represent a clean environment. Research symbols used for this logo. Write about what each symbol represents to you. Use these ideas to design a logo that symbolizes the Montreal Protocol or UN Climate Change Conference.

THEME FOCUS Energy

The energy resources that we use can be renewable or nonrenewable. Our use of these resources impacts our daily lives and the environment in which we live.

BIG (Idea Energy can be transformed from one form to another for human use.

Section 1 Fossil Fuels

fossil fuel (p. 235)
nonrenewable resource (p. 240)
petroleum (p. 236)

MAIN (Idea The burning of fossil fuels converts chemical potential energy to thermal energy, which is then converted into other useful forms.

- Energy cannot be created or destroyed but can only be transformed from one form to another.
- Petroleum, natural gas, and coal are fossil fuels.
- Petroleum is a mixture of hydrocarbons.
- Power plants burn fossil fuels to extract chemical potential energy that spins turbines and powers electric generators.
- Fossil fuels are nonrenewable resources.

Section 2 Nuclear Energy

fission (p. 241)
fusion (p. 241)
nuclear reactor (p. 242)
nuclear waste (p. 246)

MAIN (Idea Nuclear power plants transform nuclear energy into electrical energy.

- Nuclear power plants produce about 8 percent of the energy used each year in the United States.
- Nuclear reactors use the energy released in the fission of U-235 to produce electricity.
- The energy released in the fission reaction is used to make steam. The steam spins a turbine that powers an electric generator.
- Nuclear power generation produces high-level nuclear waste.

Section 3 Renewable Energy Resources

biomass (p. 253)
geothermal energy (p. 252)
hydroelectricity (p. 250)
photovoltaic cell (p. 248)
renewable resource (p. 248)

MAIN (Idea Renewable energy resources help lessen human dependence on fossil fuels.

- Solar cells convert radiant energy into electrical energy.
- Hydroelectric power plants convert gravitational potential energy into electrical energy.
- Wind energy is converted into electrical energy using a propeller attached to an electric generator.
- Alternative energy sources, such as the Sun, water, wind, and Earth's internal heat, can help reduce human dependence on fossil fuels.

Section 4 Environmental Impacts

acid precipitation (p. 260)
carrying capacity (p. 255)
hazardous waste (p. 257)
photochemical smog (p. 260)
pollutant (p. 256)
population (p. 255)

MAIN (Idea Human impact on land, water, and air affects the natural resources available for use.

- Land resources are threatened by agriculture, deforestation, industry, and waste.
- All life on Earth requires healthy air and water.
- Water can be contaminated with sediment, industrial pollutants, and human waste.
- Smog, acid precipitation, and CFCs cause air pollution.

Use Vocabulary

Complete each sentence with the correct term from the Study Guide.

26. A(n) _____ converts radiant energy to electrical energy.

27. _____ makes use of thermal energy inside Earth.

28. _____ is precipitation that can be harmful to plants and animals.

29. _____ are nonrenewable resources such as oil, natural gas, and coal.

30. The Sun, wind, water, and Earth's internal heat are _____ because they are replaced faster than they are used.

31. The maximum number of individuals of a particular species that the environment can support is called the _____.

32. Special caution should be taken in disposing of radioactive materials, such as _____.

Check Concepts

33. Why are fossil fuels considered to be nonrenewable resources?
   A) They are no longer being produced.
   B) They are being produced as fast as they are being used.
   C) They are not being produced as fast as they are being used.
   D) They contain hydrocarbons.

34. What combines with moisture in the air to form acid precipitation?
   A) ozone   **C)** lead
   B) sulfur dioxide   **D)** oxygen

35. To generate electric current, nuclear power plants produce which of the following?
   A) steam   **C)** plutonium
   B) carbon dioxide   **D)** water

36. Which of the following is the source of almost all of Earth's energy resources?
   A) plants   **C)** magma
   B) the Sun   **D)** fossil fuels

37. How are spent nuclear fuel rods disposed?
   A) burying them in a community landfill
   B) storing them in a deep pool of water
   C) burying them at the reactor site
   D) releasing them into the air

Use the figure below to answer question 38.

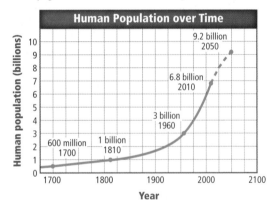

38. Between 1960 and 2010, the world population increased by how many billions of people?
   A) 1.0   **C)** 4.2
   B) 3.8   **D)** 5.9

39. Solar cells would be more practical to use if they were which of the following?
   A) pollution-free   **C)** less expensive
   B) nonrenewable   **D)** larger

40. What do hydrocarbons react with when fossil fuels are burned?
   A) carbon dioxide   **C)** oxygen
   B) carbon monoxide **D)** water

Interpret Graphics

41. **BIG Idea** Copy and complete the table below describing renewable energy sources and the energy transformations that occur.

Conservation of Energy	
Renewable Energy Source	Energy Transformation
Hydroelectricity	**a.**
b.	radiant energy to electrical energy
Wind	mechanical energy to electrical energy
c.	thermal energy to electrical energy

42. Copy and complete this concept map.

((○) Concepts in Motion Interactive Concept Map

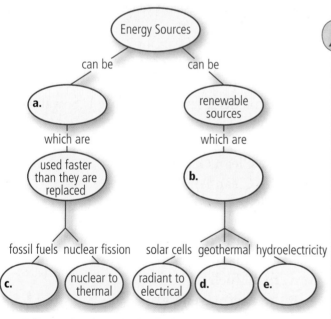

Think Critically

43. **Infer** why alternative energy resources are not more widely used.

44. **Explain** why it is important to place stricter regulations on pollutants from cars, power plants, and factories as the U.S. population grows.

45. **Infer** whether fossil fuels should be conserved if renewable energy sources are being developed.

46. **THEME FOCUS** **Explain** why coal is considered a nonrenewable energy source, but biomass such as wood is considered a renewable energy source.

47. **Predict** Forests in Germany are dying due to acid precipitation. What effects might this loss of trees have on the environment?

48. **Create** a table identifying two advantages and two disadvantages for each of the following energy sources: fossil fuels, hydroelectricity, wind turbines, nuclear fission, solar cells, geothermal energy.

Apply Math

49. **Convert Units** Crude oil is sold on the world market in units called barrels. A barrel of crude oil contains 42 gallons. If 1 gallon is equal to 3.8 L, how many liters are in a barrel of crude oil?

Use the table below to answer question 50.

High-Production Coal Mines	
Coal Mine	Metric Tons/Year
North Antelope Rochelle	6.78×10^7
Black Thunder	6.13×10^7

50. **Use Percentages** Nine of the top coal-producing mines are located in Wyoming. Production information on two of the mines is shown in the table above. A total of about 1.02×10^9 metric tons is produced per year in the United States. What percentage do these two coal mines contribute to the total yearly coal production in the United States?

Standardized Test Practice

Multiple Choice

Record your answers on the answer sheet provided by your teacher or on a sheet of paper.

Use the figure below to answer questions 1 and 2.

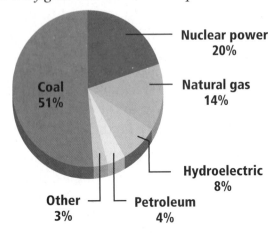

Nuclear power
20%

Natural gas
14%

Coal
51%

Hydroelectric
8%

Other
3%

Petroleum
4%

1. The graph above shows the percentage of electrical power generated in the United States that came from various energy resources. According to the graph, about what percentage came from fossil fuels?
   A. 51 percent
   B. 55 percent
   C. 69 percent
   D. 84 percent

2. The graph shows that approximately what percentage of electrical energy came from renewable energy resources?
   A. 11 percent
   B. 51 percent
   C. 65 percent
   D. 93 percent

3. Which of the following is a substance that contaminates the environment?
   A. compost
   B. development
   C. pollutant
   D. groundwater

4. Which of the following best describes windmills that are used for the production of electric current?
   A. They are quiet.
   B. They can be used anywhere.
   C. They are 90 percent efficient.
   D. They are nonpolluting.

5. Which term describes all the individuals of one species that occupy an area?
   A. population explosion
   B. carrying capacity
   C. population
   D. community

6. Which of the following is NOT a source of nuclear waste?
   A. products of fission reactors
   B. U-235
   C. some medical and industrial products
   D. products of coal-burning power plants

7. Which federal law protects the U.S. water supply?
   A. the Clean Water Act
   B. the Clean Ocean Act
   C. the Clean Air Act
   D. the Clean Hydrosphere Act

8. Which form of energy comes from the presence of magma in Earth's crust?
   A. fossil fuels
   B. geothermal energy
   C. wind energy
   D. biomass

9. Which of the following sources contributes to the formation of acid precipitation?
   A. coal-fired power plants
   B. geothermal plants
   C. wind power plants
   D. nuclear power plants

✓ Assessment   Standardized Test Practice

Short Response

Record your answers on the answer sheet provided by your teacher or on a sheet of paper.

10. Explain why hydroelectric power plants are almost twice as efficient as fossil fuel and nuclear power plants.

11. How is 70 percent of all the electrical energy used in the United States produced?

12. Why is it better for the environment for countries to work together to reduce pollution?

13. Describe the five stages in the production of electrical energy at a fossil fuel–burning power plant.

14. Describe the typical disposal method for high-level nuclear waste.

15. The core of a nuclear reactor might contain hundreds of fuel rods. Describe the composition of a fuel rod.

16. Fusion is one of the most concentrated energy sources known. Why is it not used at nuclear plants as a source for electrical energy?

17. What device can lower sulfur dioxide emissions from coal-burning power plants?

18. Should household hazardous waste be included with normal trash? Explain.

Extended Response

Record your answers on a sheet of paper.

Use the image below to answer questions 19 and 20.

19. Explain how the nuclear power plant shown above can transform nuclear energy into electrical energy.

20. What is the purpose of the large cement tower in the image above?

21. Explain how the steam that is used to run turbines is produced at a geothermal power plant.

22. Describe the processes that form petroleum, natural gas, and coal.

NEED EXTRA HELP?																						
If You Missed Question . . .	1	2	3	4	5	6	7	8	9	10	11	12	13	14	15	16	17	18	19	20	21	22
Review Section . . .	1	1	4	3	4	2	4	3	4	3	1	4	1	2	2	2	4	4	2	2	3	1

UNIT 3
Waves

THEMES

Energy All waves carry energy.

Motion and Forces The properties of waves depend on the source of the vibration and the medium in which the waves travel.

Scientific Inquiry Technologies, such as seismographs, microscopes, and sonar, allow us to collect data and ask questions that we would otherwise be unable to test.

Technology Technologies enable us to detect, learn about, and harness energy from waves.

WebQuest **STEM UNIT PROJECT**

Model Research how musical instruments create sound. Design and build your own musical instrument.

CHAPTER 9

Introduction to Waves

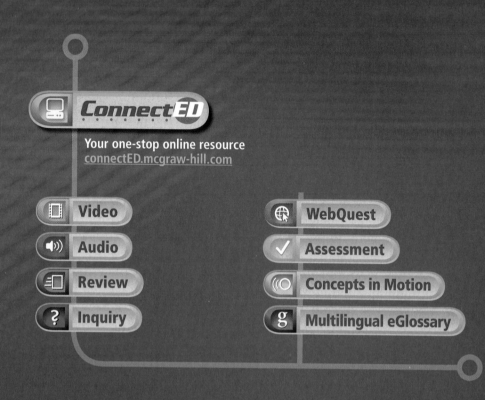

ConnectED

Your one-stop online resource
connectED.mcgraw-hill.com

- Video
- Audio
- Review
- Inquiry
- WebQuest
- Assessment
- Concepts in Motion
- g Multilingual eGlossary

Launch Lab
Energy of Waves

Light enters your eyes, and sound strikes your ears, enabling you to sense the world around you. Light waves and sound waves transfer energy from one place to another. Do waves transfer anything else along with their energy? Do waves transfer matter, too? In this activity, you will observe a model of how waves transfer energy.

For a lab worksheet, use your StudentWorks™ Plus Online.

 Inquiry Launch Lab

FOLDABLES

Make a Venn diagram foldable. Label it as shown. Use it to organize your notes on transverse and longitudinal waves.

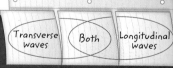

Transverse waves | Both | Longitudinal waves

THEME FOCUS Energy
Waves are one of many ways to transfer energy.

BIG((Idea)) Waves transfer energy from place to place without transferring matter.

Reading Preview

Essential Questions

▶ How do waves transfer energy?

▶ What are mechanical waves?

▶ How do transverse waves differ from longitudinal waves?

Review Vocabulary

matter: anything that has mass and takes up space

New Vocabulary

wave
medium
mechanical wave
transverse wave
longitudinal wave

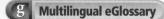
Multilingual eGlossary

The Nature of Waves

MAIN Idea **Waves travel through matter as energy is transferred from particle to particle.**

Real-World Reading Link When you catch a fastball from a baseball pitcher, you can feel the slight sting on your hand. The baseball carries both matter and energy as it travels to your hand. On the other hand, a wave carries only energy as it travels.

Waves Defined

Figure 1 shows a disturbance from a pebble splashing into a pond. This disturbance travels outward from the spot where the pebble splashed. The pebble produced a small wave in the pond. A **wave** is a repeating disturbance that transfers energy through matter or space. Other examples of waves include ocean waves, sound waves, seismic waves, and light waves. Do these and other types of waves have anything in common with one another?

Waves and Energy

Because a falling pebble is moving, the falling pebble has kinetic energy. As the falling pebble splashes into the pool in **Figure 1,** the pebble transfers some of its energy to nearby water molecules, causing them to move. Those molecules then pass the energy along to neighboring water molecules, which, in turn, transfer it to their neighbors. The energy travels farther and farther from the source of the disturbance. However, the water itself does not move outward. What you see is energy traveling in the form of a wave on the surface of the water.

■ **Figure 1** Falling pebbles transfer their kinetic energy to the molecules of water in a pond, forming waves. The waves travel outward in all directions from the source of the splash.

Waves and matter Imagine that you are in a boat on a lake. Approaching waves bump against your boat, but they do not carry it along with them as they pass. The boat does move up and down and maybe even a short distance back and forth because the waves transfer some of their energy to the boat.

But after the waves have passed, the boat is still in nearly the same place. The waves do not even carry the water along with them. The water that is around the boat after the wave passes is the same water as was there before the wave passed. Only the energy travels along with the waves. All waves, whether they are water waves, sound waves, light waves or earthquake waves, have this property: they carry energy without transporting matter from place to place.

Reading Check **Identify** what waves carry.

Making waves A wave will travel only as long as it has energy to carry. When you drop a pebble into a pond, small waves form, as shown in **Figure 1.** These waves carry energy. However, this energy is gradually transferred to the surrounding water and air. At the same time, the remaining energy in the waves spreads out as the waves spread out. As the energy spreads out and is transferred away from the waves, those waves shrink and disappear.

Suppose you are holding a rope at one end, and you give it a shake. You would create a pulse that would travel along the rope to the other end, and then the rope would be still again, as shown in **Figure 2.**

Now suppose you shake your end of the rope up and down for a while. You would make waves that travel along the rope. When you stop shaking your hand up and down, the rope will be still again. It is the up-and-down motion of your hand that creates the wave.

Anything that moves up and down or back and forth in a rhythmic way is vibrating. The vibrating movement of your hand at the end of the rope created the wave. In fact, all waves are produced by something that vibrates.

Reading Check **Identify** what produces waves.

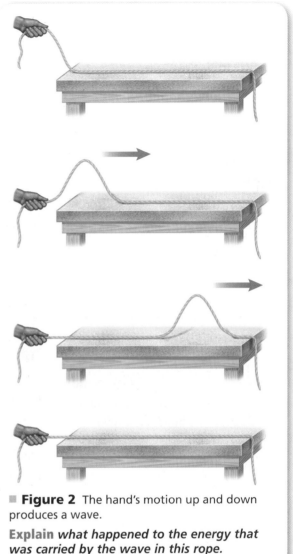

■ **Figure 2** The hand's motion up and down produces a wave.

Explain *what happened to the energy that was carried by the wave in this rope.*

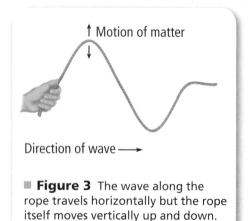

↑ Motion of matter
↓

Direction of wave ⟶

■ **Figure 3** The wave along the rope travels horizontally but the rope itself moves vertically up and down.

Concepts in Motion

Animation

■ **Figure 4** A water wave causes water to move back and forth, as well as up and down. Water is pushed back and forth to form the high and low points on the wave.

Mechanical Waves

Sound waves travel through the air to reach your ears. Ocean waves travel through water to reach the shore. A **medium** is matter through which a wave travels. The medium can be a solid, a liquid, a gas, or a combination of these. Not all waves need a medium. Some waves, such as light and radio waves, can travel through a vacuum. **Mechanical waves,** such as sound waves, are waves that can travel only through matter. Mechanical waves can be either transverse waves or longitudinal waves.

Transverse waves

In a **transverse wave,** particles in the medium moves back and forth at right angles to the direction that the wave travels. For example, **Figure 3** shows how a wave along a rope travels horizontally, but the portions of the rope that the wave passes through move up and down. When you shake one end of a rope while your friend holds the other end, you are making transverse waves.

✓ **Reading Check** **Compare** the direction that a transverse wave travels with the direction that matter in that wave vibrates.

Water waves

When the wind blows across the surface of the ocean, water waves form. Water waves are often thought of as transverse waves, but this is not entirely correct. The water in water waves does move up and down as the waves go by. But the water also moves short distances back and forth along the direction that the wave is traveling.

This movement happens because the low part of the wave can be formed only by pushing water forward or backward toward the high part of the wave, as shown on the left in **Figure 4.** This is much like a child pushing sand into a pile. Sand must be pushed in from the sides to make the pile. As the wave passes, the water that was pushed aside moves back to its initial position, as shown on the right in **Figure 4.**

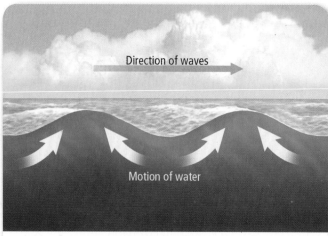

The low point of a water wave is formed when water is pushed aside and up to the high point of the wave.

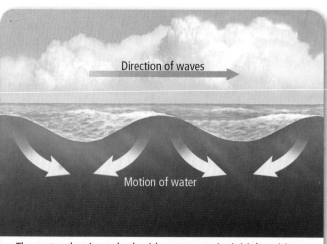

The water that is pushed aside returns to its initial position.

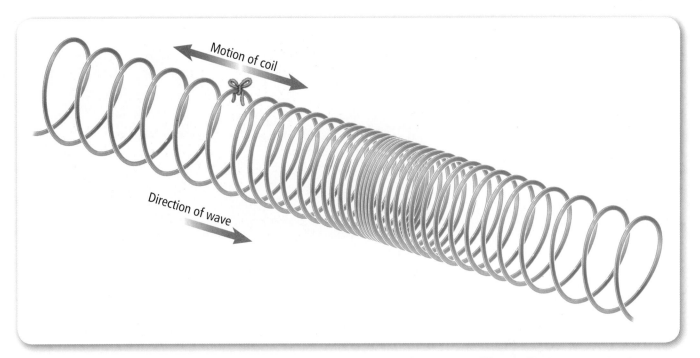

Motion of coil

Direction of wave

Longitudinal waves

Longitudinal waves In a **longitudinal wave,** matter in the medium moves back and forth along the same direction that the wave travels. You can model longitudinal waves with a coiled spring toy, as shown in **Figure 5.** Squeeze several coils together at one end of the spring. Then let go of the coils, still holding onto coils at both ends of the spring. A wave will travel along the spring. As the wave moves, it looks as if the whole spring is moving toward one end.

Suppose you watched the coil with yarn tied to it, as in **Figure 5.** You would see that the yarn moves back and forth as the wave passes and then stops moving after the wave has passed. The wave carries energy, but not matter, forward along the spring. Longitudinal waves are sometimes called compressional waves.

Sound waves Sound waves are longitudinal waves. When a noise is made, such as when a locker door slams shut, nearby molecules in the air are pushed together by the vibrations caused by the slamming door. The molecules in the air are squeezed together similar to the coils in the coiled spring toy in **Figure 5.** These compressions travel through the air to make a sound wave. The sizes of a sound wave's compressions, as well as the distances between those compressions, determines the nature of that sound.

Sound waves in liquids and solids Sound waves can also travel through liquids and solids, such as water and wood. Particles in these mediums are pushed together and move apart as the sound waves travel through them.

✔ **Reading Check Describe** how sound waves travel through solids.

■ **Figure 5** The longitudinal wave through this spring toy moves in only one direction, but the coils in this spring toy move back and forth.

Describe *another example of a longitudinal wave.*

VOCABULARY ....
SCIENCE USAGE V. COMMON USAGE
Medium
Science usage
the matter through which waves travel
Air is often the medium through which a sound wave travels.

Common usage
something in a middle position
She set the video game to a medium difficulty level. ....

Figure 6 A fault is a break in Earth's crust. The red arrows show the direction that Earth's crust is moving at a fault. When Earth's crust shifts or breaks, the energy that is released is transmitted outward, causing an earthquake. The point where the earthquake originates is called the focus.

Explain *how earthquake waves and sound waves are similar.*

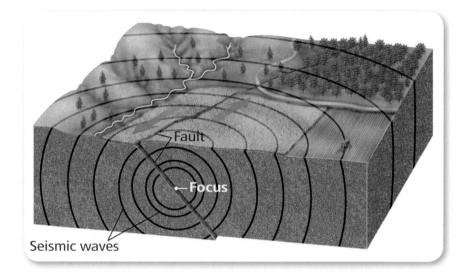

Fault

←Focus

Seismic waves

Seismic waves Forces within Earth's interior can cause regions of Earth's crust to move, bend, or even break. Movement in the crust, which occurs along faults, can result in a rapid release of energy. This energy travels away from the fault in the form of seismic (SIZE mihk) waves, as shown in **Figure 6.** Seismic waves can be longitudinal waves or transverse waves. Scientists have found out much about Earth's interior by studying these seismic waves.

Seismic waves can travel through Earth, as well as along Earth's surface. When the energy from seismic waves is transferred to objects on Earth's surface, those objects move and shake.

Section 1 Review

Section Summary

▶ A wave is a repeating disturbance that transfers energy through matter or space.

▶ Waves carry energy without transporting matter.

▶ In a transverse wave, matter in the medium moves at right angles to the direction that the wave travels.

▶ In a longitudinal wave, matter in the medium moves back and forth along the direction that the wave travels.

1. **MAIN Idea Describe** the motion of an unanchored rowboat when a water wave passes. Does the wave move the boat forward?

2. **Contrast** how you would move a spring to make a transverse wave with how you would move a spring to make a longitudinal wave.

3. **Identify** evidence that seismic waves transfer energy without transferring matter.

4. **Identify** a mechanical wave that is also a longitudinal wave.

5. **Think Critically** Describe how the world would be different if all waves were mechanical waves.

Apply Math

6. **Calculate Time** The average speed of sound in water is 1,500 m/s. How long would it take a sound wave to travel 9,000 m in water?

✓ Assessment   Online Quiz

Reading Preview

Essential Questions

▶ How are wavelength and period related?

▶ What is the relationship between frequency and wavelength?

▶ How do you calculate the speed of a wave?

Review Vocabulary

vibration: a back and forth movement

New Vocabulary

crest
trough
compression
rarefaction
wavelength
frequency
period
amplitude

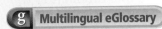
g Multilingual eGlossary

Video BrainPOP

■ **Figure 7** Different types of waves are described in different ways. Transverse waves have crests and troughs, but longitudinal waves have compressions and rarefactions.

Wave Properties

MAIN ◀Idea Wave properties depend on the vibrations of the wave source and the material in which the wave travels.

Real-World Reading Link When you adjust the volume on a stereo or adjust the brightness on a computer monitor, you are manipulating wave properties. Wave properties affect how we see objects and hear sounds. When we adjust color, brightness, pitch, or loudness, we are changing certain characteristics of the waves produced by the computer monitor or stereo system.

The Parts of a Wave

What makes sound waves, water waves, and seismic waves different from each other? Waves can differ in how much energy they carry and in how fast they travel. Waves also have other characteristics that make them different from each other.

Suppose you shake the end of a rope and make a transverse wave. The transverse wave in **Figure 7** has alternating high points and low points. **Crests** are the high points of a transverse wave. **Troughs** are the low points of a transverse wave. The imaginary line that is half the vertical distance between a crest and a trough is called the rest position.

On the other hand, a longitudinal wave has no crests and troughs. When a longitudinal wave passes through a medium, it creates regions where the medium becomes crowded together and more dense, as in **Figure 7.** These regions are compressions. A **compression** is the more dense region of a longitudinal wave. **Figure 7** also shows that the coils in the region next to a compression are spread apart, or less dense. The less-dense region of a longitudinal wave is called a **rarefaction.**

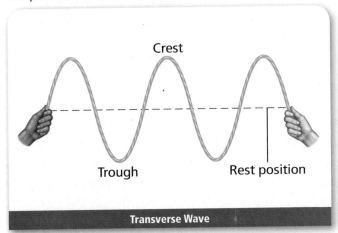

Crest

Trough

Rest position

Transverse Wave

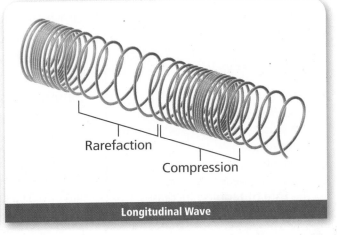

Rarefaction

Compression

Longitudinal Wave

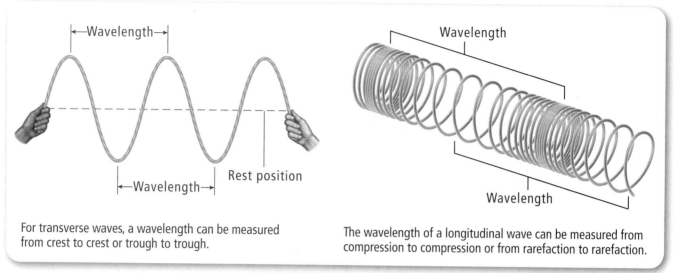

For transverse waves, a wavelength can be measured from crest to crest or trough to trough.

The wavelength of a longitudinal wave can be measured from compression to compression or from rarefaction to rarefaction.

■ **Figure 8** One wavelength is the distance from one point on a wave to the nearest point just like it.

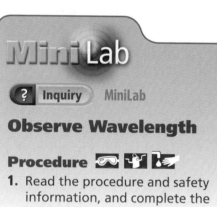

MiniLab

? Inquiry   MiniLab

Observe Wavelength

Procedure 🥽 💧 🧤

1. Read the procedure and safety information, and complete the lab form.
2. Fill a **pie plate** or other wide pan about 2 cm deep with **water**.
3. Lightly tap your finger once per second on the surface of the water, and observe the spacing of the water waves.
4. Increase the rate of your tapping, and observe the spacing of the water waves.

Analysis

1. **Describe** how the spacing of the water waves relates to their wavelength.
2. **Describe** how the spacing of the water waves changes when the rate of tapping increases.

Wavelength

All waves have wavelength. **Wavelength** is the distance between one point on a wave and the nearest point just like it. **Figure 8** shows that for transverse waves, the wavelength is the distance from crest to crest or trough to trough. These two distances are equal on a transverse wave.

A wavelength in a longitudinal wave is the distance from the middle of one compression to the middle of the next compression, as shown in **Figure 8.** The wavelength in a longitudinal wave is also the distance from the middle of one rarefaction to the middle of the next rarefaction. These two distances are equal on a longitudinal wave.

 Reading Check Describe how wavelength is defined for both transverse waves and longitudinal waves.

Frequency and Period

When you tune your radio to a station, you are choosing radio waves of a certain frequency. The **frequency** of a wave is the number of wavelengths that pass a fixed point each second. Frequency is also the number of times that a point on a wave moves up and down or back and forth each second. You can find the frequency of a transverse wave by counting the number of crests that pass a point each second. You can find the frequency of a longitudinal wave by counting the number of compressions that pass a point each second.

Frequency is expressed in hertz (Hz). A frequency of 1 Hz means that one wavelength passes by in 1 s. In SI units, 1 Hz is the same as 1/s. The **period** of a wave is the amount of time it takes one wavelength to pass a point. As the frequency of a wave increases, the period decreases. In SI units, period has units of seconds.

Wavelength is related to frequency.

If you make transverse waves with a rope, you increase the frequency by moving the rope up and down faster. Moving the rope faster also makes the wavelength shorter. This relationship is always true—as frequency increases, wavelength decreases. If you double the frequency of a wave, you halve the wavelength of that wave. If you double the wavelength of a wave, you halve that wave's frequency. **Figure 9** compares the wavelengths and frequencies of two different waves.

The frequency of a wave is always equal to the rate of vibration of the source that creates it. If you move the rope up, down, and back up in 1 s, the frequency of the wave that you generate is 1 Hz. If you move the rope up, down, and back up five times in 1 s, the resulting wave has a frequency of 5 Hz.

Reading Check **Describe** how the wavelength and the frequency of a wave are related.

Wave Speed

Suppose you are at a large stadium watching a baseball game, but you are up high in the bleachers, far from the action. The batter swings and, you see the ball rise. An instant later, you hear the crack of the bat hitting the ball. You see the impact before you hear it because light waves travel much faster than sound waves. Therefore, the light waves reflected from the flying ball reach your eyes before the sound waves created by the crack of the bat reach your ears.

The speed of a wave depends on the medium through which it is traveling. Sound waves usually travel faster in liquids and solids than they do in gases. However, light waves travel more slowly in liquids and solids than they do in gases or in a vacuum. Also, sound waves usually travel faster in a material if the temperature of the material is increased. For example, sound waves travel faster in air at 20°C than in air at 0°C.

■ Figure 9 When frequency increases, the number of wavelengths that pass in one second increases.

Identify *which of the waves in this figure has the higher frequency.*

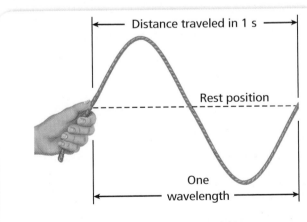

Frequency = 1 Hz
The rope is moved down, up, and down again one time in 1 s. The wave has a frequency of 1 Hz.

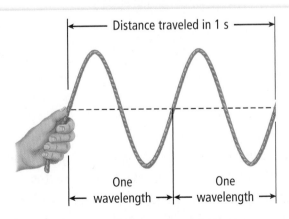

Frequency = 2 Hz
The rope is shaken down, up, and down again twice in 1 s. The wave has a frequency of 2 Hz.

Calculating wave speed The speed of a wave depends on the medium in which the wave travels. However, the wave speed, wave frequency, and the wavelength are related. The speed of a wave can be calculated from the following equation.

> **Wave Speed Equation**
>
> speed (in m/s) = frequency (in Hz) × wavelength (in m)
> $$v = f\lambda$$

In this equation, v represents the wave speed, f is the frequency, and the Greek letter λ (lambda) represents the wavelength. Why does multiplying the frequency unit Hz by the distance unit m give the unit for speed m/s? Recall that the SI unit Hz is the same as 1/s. Therefore, m × Hz is the same as m × 1/s. Both are equivalent to m/s.

EXAMPLE Problem 1

Solve for Wave Speed What is the speed of a sound wave that has a wavelength of 2.00 m and a frequency of 170.5 Hz?

Identify the Unknown: wave speed: v

List the Knowns: wavelength: $\lambda = 2.00$ m
frequency: $f = 170.5$ Hz

Set Up the Problem: $v = f\lambda$

Solve the Problem: $v = (170.5$ Hz$)(2.00$ m$)$
$= (170.5$ waves/s$)(2.00$ m$)$
$= 341$ m/s
The speed of the sound wave is 341 m/s.

Check the Answer: The speed of sound through air varies but is typically close to 340 m/s. Therefore, this answer seems reasonable.

PRACTICE Problems

Find **Additional Practice Problems** in the back of your book.

7. A wave traveling in water has a frequency of 250 Hz and a wavelength of 6.0 m. What is the speed of the wave?

8. The lowest-pitched sounds humans can generally hear have a frequency of roughly 20 Hz. What is the approximate wavelength of these sound waves if their wave speed is 340 m/s?

9. A particular radio station broadcasts radio waves at 100 MHz (100 million Hz). If radio waves travel at the speed of light (300 million m/s), then what is the wavelength of the radio waves that the station is broadcasting?

10. **Challenge** A sound wave with a frequency of 100.0 Hz travels in water with a speed of 1,500 m/s and then travels in air with a speed of 340 m/s. Approximately how many times larger is the wavelength in water than in air?

Review
Additional Practice Problems

Amplitude

Why do some earthquakes cause terrible damage, while others are hardly felt? This is because the disturbance from a wave can vary. **Amplitude** is a measure of the size of the disturbance from a wave. If the wave's amplitude is greater, then the disturbance from that wave is also greater. Amplitude is measured differently for longitudinal waves and transverse waves.

Amplitude of longitudinal waves The amplitude of a longitudinal wave is related to how tightly the medium is pushed together at the compressions and how much the medium is pulled apart at the rarefactions. The more tightly pushed together the medium is at the compressions, the denser the medium. The denser the medium is at the compressions, the larger the wave's amplitude is and the greater the disturbance from the wave. The denser the medium is at the compressions, the less dense the medium is in the rarefactions. Therefore, another indication of high amplitude is whether the medium is stretched out more in the rarefactions. **Figure 10** shows longitudinal waves with different amplitudes.

VOCABULARY . . . .

WORD ORIGIN

Amplitude
  comes from the Latin word *amplitudinem*, which means "wide extent or width"
  Waves with high amplitudes are usually very noticeable. . . . .

■ **Figure 10** The coils in the high-amplitude wave's compression are closer together than the coils in the low-amplitude wave's compression.

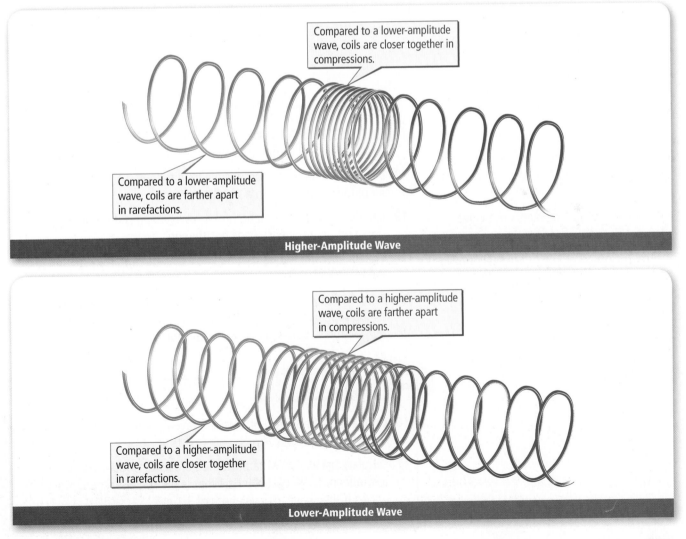

Compared to a lower-amplitude wave, coils are closer together in compressions.

Compared to a lower-amplitude wave, coils are farther apart in rarefactions.

Higher-Amplitude Wave

Compared to a higher-amplitude wave, coils are farther apart in compressions.

Compared to a higher-amplitude wave, coils are closer together in rarefactions.

Lower-Amplitude Wave

■ Figure 11 The amplitude of a transverse wave is half the vertical distance between the crests and the troughs.

Describe *how you could create waves with different amplitudes in a piece of rope.*

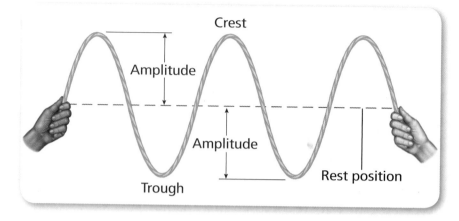

Crest

Amplitude

Amplitude

Trough

Rest position

Amplitude of transverse waves If you have ever been knocked over by an ocean wave, you know that the higher the wave, the greater the disturbance from that wave. Remember that the amplitude of a wave increases as the disturbance from that wave increases. So, a tall ocean wave has a greater amplitude than a short ocean wave does.

The amplitude of any transverse wave is the vertical distance from the crest or trough of the wave to the rest position of the medium, as shown in **Figure 11.** Tall waves have large amplitude, and short waves have small amplitude. The amplitude of any transverse wave is also half the vertical distance from crest to trough.

Section **2**   **Review**

Section Summary

▶ Wavelength is the distance between a point on a wave and the nearest point just like it.

▶ Wave frequency is the number of wavelengths passing a fixed point each second.

▶ Wave period is the amount of time it takes one wavelength to pass a fixed point.

▶ The speed of a wave is the product of its frequency and its wavelength.

▶ As the amplitude of a wave increases, the disturbance from that wave increases.

11. **MAIN Idea Identify** a wave that speeds up when it passes from air to water as well as one that slows down.

12. **Describe** the difference between a longitudinal wave with a large amplitude and one with a small amplitude.

13. **Describe** how the wavelength of a wave changes if the wave slows down but its frequency does not change.

14. **Explain** how the frequency of a wave changes when the period of the wave increases.

15. **Think Critically** You make a transverse wave by shaking the end of a long rope up and down. Explain how you would shake the end of the rope to make the wavelength shorter.

Apply Math

16. **Calculate** the frequency of a water wave that has a wavelength of 0.5 m and a speed of 4 m/s.

17. **Calculate Speed** An FM radio station broadcasts radio waves with a frequency of 96,000,000 Hz. What is the speed of these radio waves if they have a wavelength of 3.1 m?

✓ **Assessment**   Online Quiz

LAB

Wave Speed and Tension

Objectives

■ **Determine** the relationship between tension and wave speed.

Background: Before playing her violin, the musician must adjust the tension, or the amount of force pulling on each string, to tune the violin.

Question: *How does the tension in a material affect the waves traveling through that material?*

Preparation

Materials

coiled-spring toy
meterstick
stopwatch

Safety Precautions

Procedure

1. Read the procedure and safety information, and complete the lab form.
2. Create a data table similar to the one shown.
3. Attach one end of the spring to a chair leg so that the spring rests on a smooth floor.
4. Stretch the spring to a length of 1.0 m.
5. Make a longitudinal wave by squeezing several coils together, and then releasing them.
6. Have your partner time how long the wave takes to travel two or three lengths of the spring. Record the time in your data table. Record the distance the wave traveled in your data table.
7. Repeat steps 4 and 5 two more times for waves 2 and 3.
8. Stretch the spring to a length of 1.5 m. Repeat steps 4 and 5 for waves 4, 5, and 6.

Data Table

	Distance (m)	Wave Time (s)	Wave Speed (m/s)
Wave 1			
Wave 2			
Wave 3			
Wave 4			
Wave 5			
Wave 6			

Conclude and Apply

1. **Calculate** the speed of each wave. Use the formula speed = distance/time.
2. **Calculate** the average speed of the waves on the spring when the spring has a length of 1.0 m.
3. **Calculate** the average speed of the waves on the spring when the spring has a length of 1.5 m.
4. **Describe** how the tension in the spring changes as the length of the spring is increased.
5. **Describe** how the wave speed depends on the tension. How could you make the waves travel even faster? Test your prediction.
6. **Predict** how you could increase the speed of waves along a violin string.

COMMUNICATE YOUR DATA

Compare your results with those of other students in your class. Form a hypothesis about what you might observe if the coiled-spring toy were made of another material. How would you test your hypothesis?

Essential Questions

▶ What is the law of reflection?

▶ Why do waves change direction when they travel from one material to another?

▶ How are refraction and diffraction similar? How are they different?

▶ What happens when waves interfere with each other?

Review Vocabulary

perpendicular: a line that forms a 90-degree angle with another line

New Vocabulary

refraction
diffraction
interference
standing wave
node
resonance

g Multilingual eGlossary

The Behavior of Waves

MAIN Idea Waves interact with matter and with each other.

Real-World Reading Link Think about what you might see when you look at the surface of a calm lake. You might see your reflection, or you might see distorted images of what is below the surface of the lake. Both of these examples are results of the behavior of light waves.

Reflection

If you are one of the last people to leave your school building at the end of the day, you will probably find that the hallways are quiet and empty. When you close your locker door, the sound echoes down the empty hall. Your footsteps also make a hollow sound. Thinking you are all alone, you might be startled by your own reflection in a classroom window. The echoes and your image looking back at you from the window are caused by wave reflection. Without wave reflection, you could not even see the lockers in your school's hallway.

Reflection occurs when a wave strikes an object and bounces off it. All types of waves—including sound, water, and light waves—can be reflected. How does the reflection of light allow the girl in **Figure 12** to see herself in the mirror? It happens in two steps. First, light strikes her face and bounces off her face. Then, the light reflected off her face strikes the mirror and is reflected into her eyes.

■ **Figure 12** Some of the light that strikes this girl's face is reflected into the mirror. Some of that light then reflects off the mirror into her eyes.

Describe *how the path of the light reflected from the girl's face would be different if the mirror was not present.*

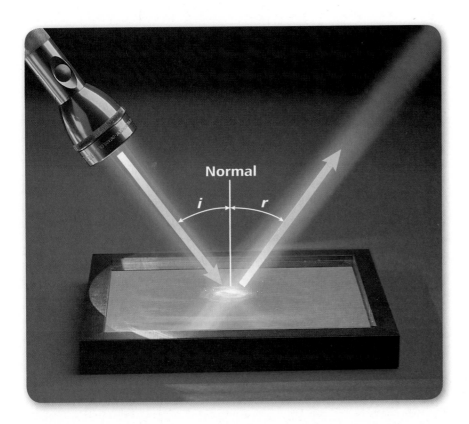

■ **Figure 13** The angle between the incident light beam and the normal is equal to the angle between the reflected light beam and the normal.

Identify *If the angle of incidence is 40°, what is the angle of reflection?*

Echoes Surfaces often reflect sound waves as well as light waves. Echoes are a result of reflecting sound waves. Sound waves form when your foot hits the floor, and the waves travel through the air to both your ears and other objects. When sound waves reach another object, such as a row of lockers, they reflect off that object. Sometimes, those reflected waves travel back to your ears.

Some animals use echoes to learn about their surroundings. For example, dolphins make clicking sounds and listen to the echoes. These echoes enable the dolphin to locate objects.

The law of reflection Look at the two light beams in **Figure 13.** The beam striking the mirror is called the incident beam. The beam that bounces off the mirror is called the reflected beam. The line drawn perpendicular to the surface of the mirror is called the normal. The angle formed by the incident beam and the normal is the angle of incidence, labeled *i*. The angle formed by the reflected beam and the normal is the angle of reflection, labeled *r*.

According to the law of reflection, the angle of incidence is always equal to the angle of reflection. All reflected waves, including light waves, sound waves, and water waves, obey this law. Objects that bounce from a surface sometimes behave like waves that are reflected from a surface. For example, suppose you throw a bounce pass while playing basketball. The angle between the ball's direction and the normal to the floor is the same before and after it bounces.

■ **Figure 14** Light waves refract as they enter the water. This results in the illusion of the broken straw.

Refraction

Do you notice anything unusual in **Figure 14**? The straw looks as if it is broken into two pieces. But if you pulled the straw out of the water, you would see that it is unbroken. This illusion is caused by refraction. How does refraction work?

Remember that a wave's speed depends on the medium through which it is traveling. When a wave passes from one medium to another, such as when a light wave passes from air to water, it changes speed. If the wave is traveling at an angle when it passes from one medium to another, it changes direction, or bends, as it changes speed.

Refraction is the bending of a wave caused by a change in its speed as it travels from one medium to another. The greater the change in speed, the more the wave bends. The top panel in **Figure 15** shows how a wave refracts when it passes into a material in which that wave slows down. The wave is refracted (bent) toward the normal. The bottom panel in **Figure 15** shows how a wave refracts when it passes into a medium in which it speeds up. Then, the wave is refracted away from the normal.

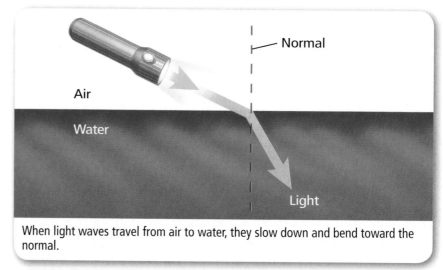

When light waves travel from air to water, they slow down and bend toward the normal.

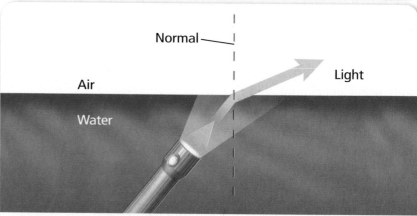

When light waves travel from water to air, they speed up and bend away from the normal.

■ **Figure 15** Light waves travel more slowly in water than in air. This causes light waves to refract when they move from water to air or air to water.

Predict *how the beam would bend if the speed of light were the same in both air and water.*

■ **Figure 16** The fish in this photo are farther from the surface of the water than they appear. Refraction causes this pond to appear to be much shallower than it actually is.

Refraction of light in water Have you ever gazed at fish in a pond, such as those in **Figure 16?** You may have noticed that objects that are underwater seem closer to the surface than they really are. **Figure 17** shows how refraction causes this illusion.

In **Figure 17,** the light waves reflected from the swimmer's foot are refracted away from the normal and enter the girl's eyes. However, the brain assumes that all light waves have traveled in a straight line. The light waves that enter the girl's eyes seem to have come from a foot that was higher in the water.

This is also why the straw in **Figure 14** seems to be broken. The light waves coming from the part of the straw that is underwater are refracted, but your brain interprets them as if they have traveled in a straight line. However, the light waves coming from the part of the straw above the water are not refracted. So, the part of the straw that is underwater looks as if it has shifted.

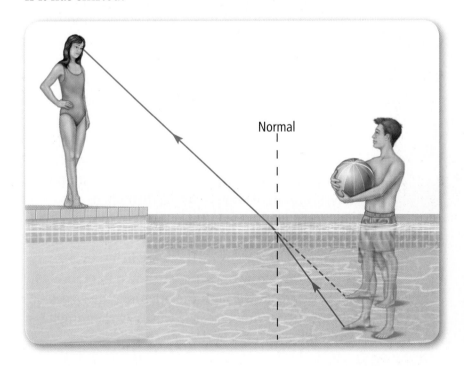

■ **Figure 17** From the girl's perspective, the boy appears to be shorter than he actually is due to refraction. Light rays from the boy's foot refract downward at the water's surface.

Figure 18 Diffraction can cause ocean waves to change direction as they pass around an island.

Figure 19 Waves diffract and spread out when they pass through a small opening or around a small obstacle.

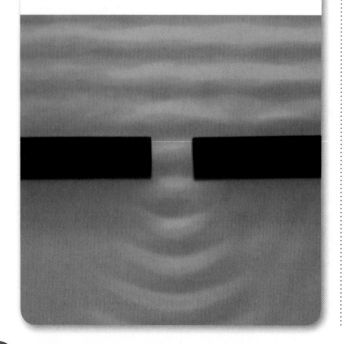

Diffraction

When waves strike an object, several things can happen. The waves can be reflected. If the object is transparent, light waves can be refracted as they pass through it. Often, some waves are reflected and some waves are refracted. If you look into a glass window, sometimes you can see your reflection in the window, as well as objects behind it. Light is passing through the window and is also being reflected at its surface.

Waves can also behave another way when they strike an object. The waves can bend around the object. **Figure 18** shows ocean waves changing direction and bending after they strike an island. **Diffraction** is the bending of a wave around an object. Diffraction and refraction both cause waves to bend. The difference is that refraction occurs when waves pass through an object, while diffraction occurs when waves pass around an object. All waves, including water waves, sound waves, and light waves, can be diffracted.

✓ **Reading Check** **Contrast** refraction with diffraction.

Waves also can be diffracted when they pass through a narrow opening, as shown in **Figure 19.** After they pass through the opening, the waves spread out. In this case, the waves are bending around the corners of the opening.

Less diffraction occurs if the wavelength is smaller than the obstacle.

More diffraction occurs if the wavelength is the same size as the obstacle.

Diffraction and wavelength How much does a wave bend when it strikes an object or an opening? The amount of diffraction that occurs depends on how big the obstacle or opening is compared to the wavelength, as shown in **Figure 20.**

When an obstacle is roughly the same size as or smaller than the wavelength of a wave, the wave bends around it. But when the obstacle is much larger than the wavelength, the waves do not diffract as much. If the obstacle is much larger than the wavelength, almost no diffraction occurs. Instead, the obstacle casts a shadow.

Hearing around corners Suppose you are walking down the hallway, and you hear sounds coming from a classroom on the left before you reach the open classroom door. However, you cannot see into the room until you reach the doorway.

Why can you hear the sound waves but not see the light waves while you are still in the hallway? The wavelengths of sound waves are similar in size to a door opening. Sound waves diffract around the door and spread out down the hallway. Light waves have a much shorter wavelength. They are hardly diffracted at all by the door. So, you cannot see into the room until you get to the door.

Diffraction of radio waves Diffraction also affects your radio's reception. AM radio waves have longer wavelengths than FM radio waves do. Because of their longer wavelengths, AM radio waves diffract around obstacles, such as buildings and mountains.

The FM waves with their short wavelengths do not diffract as much. As a result, AM radio reception is often better than FM reception around tall buildings and natural barriers, such as hills.

■ **Figure 20** The diffraction of waves around an obstacle depends on how the wavelength compares with the size of the obstacle.

Use Vocabulary

Answer each question using the correct term from the Study Guide.

24. Which type of wave has points, called nodes, that do not move?

25. Which part of a longitudinal wave has the lowest density?

26. What can occur when a wave passes from one medium into another?

27. What occurs when waves overlap?

28. What is half the vertical distance from the crest to the trough of a transverse wave?

29. What can be measured in Hz?

Check Concepts

30. THEME FOCUS Which do waves transfer?
  A) matter
  B) energy
  C) matter and energy
  D) the medium

31. What is the formula for calculating wave speed?
  A) $v = f\lambda$
  B) $v = f - \lambda$
  C) $v = \dfrac{f}{\lambda}$
  D) $v = \lambda + f$

32. When a longitudinal wave travels through a medium, in which direction does the matter in the medium move?
  A) in circles
  B) perpendicular to the rest position
  C) along the same direction the wave travels
  D) in all directions

Use the figure below to answer question 33.

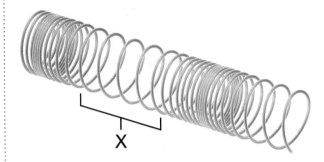

33. What is region *X* on the above wave called?
  A) crest
  B) compression
  C) rarefaction
  D) trough

34. If the frequency of a vibrating object decreases, how does the wavelength of the resulting wave change?
  A) It stays the same.
  B) It decreases.
  C) It vibrates.
  D) It increases.

35. If the amplitude of a wave changes, which also changes?
  A) disturbance from the wave
  B) frequency
  C) wave speed
  D) refraction

36. Which term describes the bending of a wave around an object?
  A) refraction
  B) interference
  C) diffraction
  D) reflection

37. Which is equal to the angle of reflection?
  A) angle of refraction
  B) angle of diffraction
  C) angle of bouncing
  D) angle of incidence

✓ Assessment   Online Test Practice

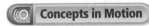

Interpret Graphics

38. Copy and complete the following concept map.

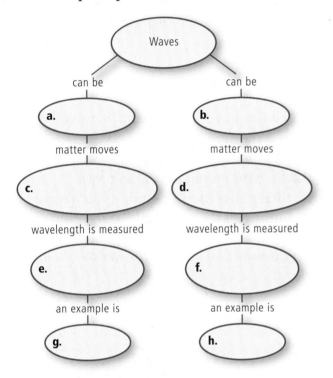

Think Critically

39. **BIG Idea** **Explain** An earthquake on the ocean floor produces a tsunami that reaches a remote island. Is the water that reaches the island the same water that was above the earthquake on the ocean floor? Explain.

40. **Diagram** a transverse wave. Label the crest, trough, wavelength, and amplitude. Identify the period and frequency for the wave.

41. **Describe** What happens to the frequency of a wave if the period of that wave is increased?

42. **Explain** why you can hear a fire engine coming around a street corner before you can see it.

43. **Describe** the objects or materials that vibrated to produce three of the sounds that you have heard today.

44. **Explain** why it is easier to hear low-frequency sounds than high-frequency sounds when a large tree is between you and the sound source.

45. **Form a Hypothesis** In 1831, soldiers marching over the Broughton Suspension Bridge in England caused that bridge to collapse. Use what you have learned about wave behavior to form a hypothesis that explains why this happened.

46. **Make and Use Tables** Find information in newspaper articles or magazines describing five recent earthquakes. Construct a table for each earthquake that shows the date, location, magnitude, and whether the damage caused by the earthquake was light, moderate, or heavy.

Apply Math

47. **Calculate Wavelength** Calculate the wavelength of a wave traveling on a spring when the wave travels at 0.2 m/s and has a period of 0.5 s.

48. **Calculate Wavespeed** The microwaves produced inside a microwave oven have a wavelength of 12 cm and a frequency of 2,500,000,000 Hz. At what speed (in m/s) do the microwaves travel?

49. **Calculate Frequency** Water waves on a lake travel toward a dock with a speed of 2.0 m/s and a wavelength of 0.5 m. How many wave crests strike the dock each second?

Standardized Test Practice

Multiple Choice

Record your answers on the answer sheet provided by your teacher or on a sheet of paper.

1. When a transverse wave travels through a medium, which way does matter in the medium move?
   A. in circles
   B. in all directions
   C. at right angles to the direction that the wave travels
   D. parallel to the direction the wave travels

Use the illustration below to answer questions 2 and 3.

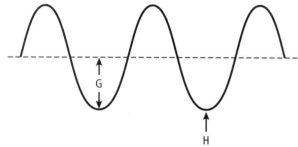

2. What wave property is shown by G?
   A. amplitude   C. crest
   B. wavelength   D. trough

3. What part of the wave is shown at H?
   A. amplitude   C. crest
   B. wavelength   D. trough

4. What is the number of waves that passes a point in a certain time called?
   A. wavelength
   B. amplitude
   C. period
   D. frequency

5. The period of a wave can be directly calculated from which?
   A. angle of incidence
   B. amplitude
   C. frequency
   D. wavelength

6. To what is the size of the disturbance from a wave related?
   A. frequency   C. amplitude
   B. wave speed   D. refraction

7. What happens when two waves pass through each other?
   A. refraction   C. reflection
   B. diffraction   D. interference

8. When the crests of two identical waves meet, what is the amplitude of the resulting wave?
   A. half the amplitude of each wave
   B. twice the amplitude of each wave
   C. three times the amplitude of each wave
   D. four times the amplitude of each wave

Use the illustration below to answer questions 9 and 10.

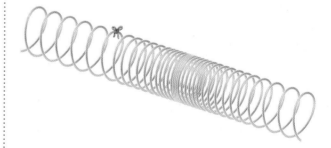

9. What kind of wave is shown?
   A. electromagnetic
   B. longitudinal
   C. transverse
   D. sound

10. What happens to the yarn that is tied to the coil?
   A. It moves back and forth as the wave passes.
   B. It moves up and down as the wave passes.
   C. It does not move as the wave passes.
   D. It moves forward along the coil along with the wave.

✓ Assessment   Standardized Test Practice

Short Response

Record your answers on the answer sheet provided by your teacher or on a sheet of paper.

11. Explain why water waves traveling toward a swimmer on a float do not move the float forward along with the wave.

Use the illustration below to answer questions 12 and 13.

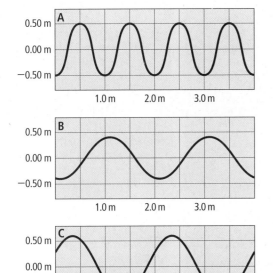

12. Determine the amplitudes and the wavelengths of each of the three waves.

13. If the length of the *x*-axis on each diagram represents 2 s of time, what is the frequency of each wave?

14. A tuning fork vibrates at a frequency of 440 Hz. The wavelength of the sound produced by the tuning fork is 0.75 m. What is the speed of the wave?

Extended Response

Record your answers on a sheet of paper.

15. Describe how a standing wave forms and why it has nodes.

16. Explain why objects that are underwater seem to be closer to the surface than they actually are.

17. Compare and contrast refraction and diffraction of waves.

18. In a science-fiction movie, a huge explosion occurs on the surface of a planet. People in a spaceship heading toward the planet see and hear the explosion. Is this realistic? Explain.

Use the illustration below to answer questions 19 and 20.

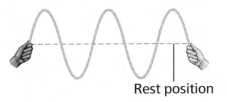

Rest position

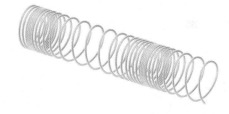

19. Describe how amplitude is related to the density of the coils in the bottom drawing.

20. Describe how you would change both drawings to show waves that cause a greater disturbance.

NEED EXTRA HELP?																				
If You Missed Question . . .	1	2	3	4	5	6	7	8	9	10	11	12	13	14	15	16	17	18	19	20
Review Section . . .	1	2	2	2	2	2	3	3	1	1	1	2	2	2	3	3	3	1	2	2

CHAPTER 10
Sound

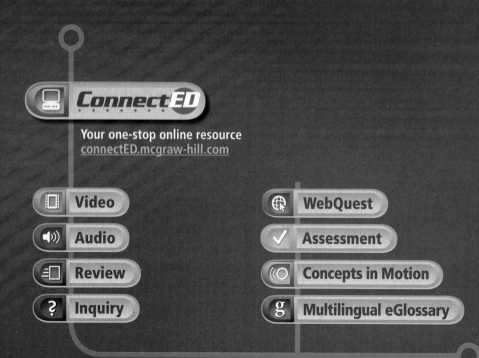

ConnectED

Your one-stop online resource
connectED.mcgraw-hill.com

- 🎞 Video
- 🔊 Audio
- ▤ Review
- ❓ Inquiry
- 🌐 WebQuest
- ✓ Assessment
- ◉ Concepts in Motion
- g Multilingual eGlossary

Launch Lab
Sound Energy

The sounds that you hear all around you every day are caused by sound waves that reach your ears. Some sounds that you hear are loud, and others are soft. What is the difference between loud and soft sounds?

For a lab worksheet, use your StudentWorks™ Plus Online.

❓ **Inquiry** Launch Lab

FOLDABLES

Make a three-tab book. Label it as shown. Use it to organize your notes on how sound travels through solids, liquids, and gases.

Sound

through solids | through liquids | through gases

THEME FOCUS Energy
Sound waves carry energy that can be detected by the ear.

BIG Idea Sound waves are longitudinal waves produced by vibrations.

Reading Preview

Essential Questions

▶ How does sound travel through different mediums?

▶ What affects the speed of sound?

▶ How does your ear enable you to hear?

Review Vocabulary

amplify: to make louder or greater

New Vocabulary

eardrum
cochlea

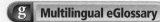 **g** | Multilingual eGlossary

The Nature of Sound

MAIN ‹Idea Sound waves are longitudinal waves that travel only through matter.

Real-World Reading Link Notice the wide variety of sounds all around you—footsteps, voices, wind, and falling water. These sounds and all others result from vibrations. Your ears are extremely sensitive vibration detectors.

Vibrations and Sound

An amusement park can be a noisy place. The sounds from the rides and games can make it hard to hear what your friends say. All of these sounds have something in common: a vibrating object produces each one. For example, your friends' vibrating vocal cords produce their voices. Vibrating speakers produce the music from a carousel.

Sound Waves

Recall that air is composed of matter. When an object such as a radio speaker vibrates, it collides with some of the particles that make up the nearby air, transferring some energy to those particles. These particles then collide with other particles, passing the energy on farther. The energy originally transferred by the vibrating object continues to travel through the air in this way. This process of energy transfer forms a sound wave.

Sound waves are longitudinal waves. Remember that a longitudinal wave is composed of two types of regions called compressions and rarefactions. If you look at **Figure 1,** you will see that when a radio speaker vibrates outward, the particles near the speaker are pushed together, forming a compression. When the speaker moves inward, the particles near the speaker are spread apart, and a rarefaction forms. As long as the speaker continues to vibrate back and forth, sound waves are produced.

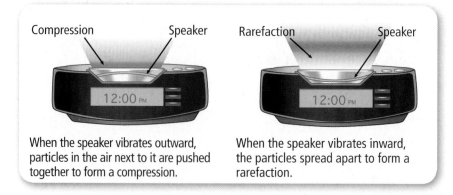

When the speaker vibrates outward, particles in the air next to it are pushed together to form a compression.

When the speaker vibrates inward, the particles spread apart to form a rarefaction.

■ **Figure 1** A vibrating speaker cone produces longitudinal waves.

■ **Figure 2** A vacuum is a region where no matter is present. Outer space is not a perfect vacuum, but there is too little matter for sound waves to travel through outer space. Therefore, these astronauts must use radios to communicate with each other.

Explain *why the astronauts need radios in order to talk to each other.*

Sound in liquids and solids If you have been underwater and heard garbled voices, you know that sound travels through water as well as air. If you have put your ear to your desk and heard sounds, you know that sound travels through solids. Sound waves travel through any type of matter—solid, liquid, or gas. The matter through which a wave travels is called the medium. A sound wave produces compressions and rarefactions in a medium as it travels through that medium.

What happens when there is no matter through which sound can travel? Could sound be transmitted without matter to compress, expand, and collide? In order for astronauts, such as the ones in **Figure 2,** to talk to each other, they must use electronic communication equipment because there is no atmosphere to transmit sound waves. Sound waves cannot travel through the vacuum of outer space.

The speed of sound The speed of a sound wave through a medium depends on that medium's composition and whether that medium is solid, liquid, or gas. **Table 1** shows the speed of sound through some common mediums. The temperature, density, and elasticity of the medium also affect the speed of sound.

Temperature and sound speed The speed of sound through a fluid depends on the temperature of that fluid. This relationship is particularly pronounced for gases but also exists for liquids, such as water. As the temperature of a fluid increases, the particles that make up that fluid move faster. This makes them more likely to collide with each other. If the particles that make up a medium collide more often, more energy can be transferred in a shorter amount of time. As a result, the sound waves travel faster.

FOLDABLES

Incorporate information from this section into your Foldable.

Table 1	Speed of Sound in Different Mediums
Medium	**Speed of Sound (m/s)**
Air (0°C)	330
Air (20°C)	340
Cork	500
Water (0°C)	1,400
Water (20°C)	1,500
Copper	3,600
Bone	4,000
Steel	5,800

When the people are far away from each other, like the particles that make up a gas, it takes longer to transfer the bucket of water from person to person.

When the people are close together, like the particles that make up a solid or a liquid, the bucket can be transferred quickly from person to person.

■ **Figure 3** A line of people passing a bucket is a model for particles transferring the energy of a sound wave.

Explain *why sound would travel more slowly in cork than in water.*

VOCABULARY ·······················

ACADEMIC VOCABULARY

Section
  a part set off as distinct
  The outer ear is a section of the ear. ····

Density and sound speed Sound usually travels fastest in solids and slowest in gases. One reason for this is that the individual particles that make up liquids and solids are usually closer together than the particles that make up gases.

You can understand why solids and liquids transmit sound well by picturing a large group of people standing in a line. Imagine that they are passing a bucket from person to person. If everyone stands far apart, each person has to walk a long distance to transfer the bucket, as in the left photo of **Figure 3.** However, if everyone stands close together, as in the right photo of **Figure 3,** the bucket quickly moves down the line.

The people standing close to each other are like the particles that make up solids and liquids. Those standing far apart are like the particles that make up gases. When the particles that make up a medium are farther apart, sound travels more slowly through that medium.

Elasticity and sound speed Elasticity is the tendency of an object to rebound to its original state when it is deformed. A rubber ball is elastic. It tends to spring back to its original shape after someone squeezes it. A ball of clay is much less elastic.

Elastic objects also rebound more quickly when sound waves travel through them. Therefore, sound waves travel more quickly through elastic objects. Usually, solids are more elastic than liquids and liquids are more elastic than gases. This is another reason why sound waves typically travel fastest through solids, slower through liquids, and slowest through gases.

✓ **Reading Check Identify** two reasons why sounds usually travel faster through solids than through gases.

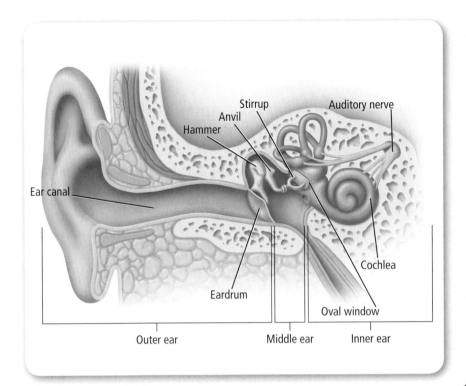

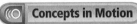

Figure 4 The three sections of the ear gather sound, amplify sound, and convert sound into an electric signal.

Identify *the three sections of the human ear.*

Animation

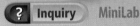

The Ear

Your ears and brain work together to interpret sound waves. When you think of your ear, you probably picture just the fleshy, visible, outer part. But, as shown in **Figure 4,** the human ear has three sections called the outer ear, the middle ear, and the inner ear. Each section of the ear has a different function.

The outer ear The visible part of your ear, the ear canal, and the eardrum make up the outer ear. The outer ear gathers sound waves. The gathering process starts with the outer part of your ear, which is shaped to help capture and direct sound waves into the ear canal. The ear canal is a passageway that is 2-cm to 3-cm long and is a little narrower than your index finger. The sound waves travel along this passageway, which leads to the eardrum. The **eardrum** is a tough membrane about 0.1 mm thick that transmits sound from the outer ear to the middle ear.

Reading Check Identify what makes the eardrum vibrate.

The middle ear When the eardrum vibrates, it passes the sound waves into the middle ear, where three tiny bones start to vibrate. These bones are called the hammer, the anvil, and the stirrup. They make a lever system that increases the force and pressure exerted by the sound waves. The bones amplify the sound wave. The stirrup is connected to a membrane on a structure called the oval window, which vibrates as the stirrup vibrates.

MiniLab

? Inquiry MiniLab

Compare Sounds

Procedure

1. Read the procedure and safety information, and complete the lab form.
2. Hold a **wood block** next to your ear, and have a partner hold a ringing **cellular telephone** next to the block. Note the sound of the telephone through the wood.
3. Hold an **empty water bottle** next to your ear. Hold the cellular telephone against the other side of the water bottle, and note the sound of the telephone through the air in the bottle.
4. Repeat step 3 with the water bottle filled with **water.**

Analysis

1. **Compare and contrast** the sound of the cellular telephone when it traveled through the three different mediums.

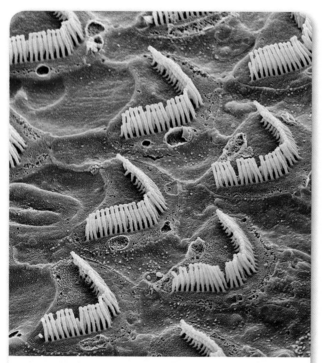

Figure 5 Hair cells in the human ear send nerve impulses to the brain when sound waves cause them to vibrate. In this photo, the hair cells are magnified 5,500 times.

The inner ear When the membrane in the oval window vibrates, the sound vibrations are transmitted into the inner ear. The inner ear contains the cochlea (KOH klee uh). The **cochlea** is a spiral-shaped structure that is filled with liquid and contains tiny hair cells. These hair cells are shown in **Figure 5**. When the tiny hair cells in the cochlea begin to vibrate, nerve impulses are sent through the auditory nerve to the brain. It is the cochlea that converts sound waves to nerve impulses.

Hearing loss When a person's hearing is damaged, it is usually because the tiny hair cells in the cochlea are damaged or destroyed, often by loud sounds. This damage is permanent. The hair cells in the cochlea of humans and other mammals do not grow back when damaged or destroyed.

However, current research suggests that doctors may be able to repair damaged or even destroyed hair cells in the future. Much of this research centers on birds. Unlike in mammals, the hair cells in birds do grow back when damaged or destroyed.

Section 1 Review

Section Summary

▶ Sound waves are longitudinal waves produced by vibrating objects.

▶ The compressions and rarefactions of sound waves transfer energy.

▶ Sound cannot travel through a vacuum.

▶ The ear detects sound waves and converts them to electrical impulses.

1. **MAIN Idea Explain** how sound travels from your vocal cords to your friend's ears when you talk.

2. **Summarize** the physical reasons that sound waves travel at different speeds through different mediums.

3. **Explain** why sound speeds up when temperature increases.

4. **Describe** each section of the human ear and its role in hearing.

5. **Think Critically** Some people hear ringing in their ears, called tinnitus, even in the absence of sound. Form a hypothesis to explain why this occurs.

Apply Math

6. **Calculate Time** Using **Table 1**, calculate how long it takes a sound wave to travel 1.0 km through air when the temperature is 0.0°C.

7. **Calculate Time** Using **Table 1**, calculate how long the same wave takes to travel 1.0 km in air when the temperature is 20.0°C.

Assessment   Online Quiz

Reading Preview

Essential Questions

▶ How are amplitude, intensity, and loudness related?

▶ How is sound intensity measured?

▶ What is the relationship between frequency and pitch?

▶ What is the Doppler effect?

Review Vocabulary

frequency: the number of wavelengths that pass a fixed point each second

New Vocabulary

intensity
loudness
decibel
pitch
Doppler effect

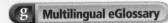

g Multilingual eGlossary

Properties of Sound

MAIN ‹ Idea The loudness of a sound depends on its intensity, and the pitch of a sound depends on its frequency.

Real-World Reading Link If the phone rings while you are listening to music, you might have to turn down the volume on the stereo to be able to hear the person on the telephone. When you turn down the volume, you are lowering the intensity of the sound coming from that stereo.

Intensity and Loudness

Recall that the degree of disturbance from a wave corresponds to its amplitude. For a longitudinal wave, amplitude is related to how close the particles of the medium are together in the compressions. **Figure 6** compares longitudinal waves of low amplitude and high amplitude. Increasing the amplitude of a longitudinal wave pushes the particles in that wave's compressions closer together.

To produce a sound wave with greater amplitude, more energy must be transferred from the vibrating object to the medium. This greater energy is then transferred through the medium as the sound wave is transmitted.

What happens to the sound waves from your stereo when you adjust the volume? The notes sound the same, but you decreased the amplitude of the sound waves. For sound waves, the amplitude of the wave is related to its intensity.

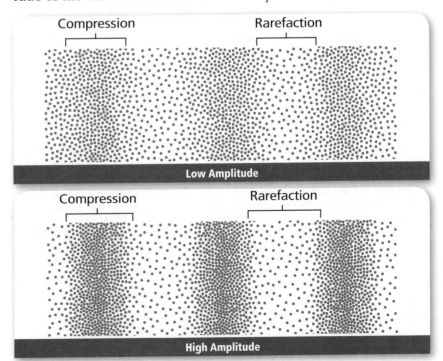

■ **Figure 6** The amplitude of a sound wave depends on the density of the medium in the compressions and rarefactions.

Identify *the areas of highest and lowest density for each wave.*

Figure 7 The intensity of the sound waves from the computer speakers is related to the rate at which energy passes through the imaginary rectangle and how far the listener is from the computer speaker.

Describe *how the intensity would be different for a person standing 10 m away from the computer speakers.*

Intensity Picture sound waves traveling through the air from a computer's speakers to your ears. If you also picture a rectangle between you and the computer speakers, as in **Figure 7,** and could measure how much energy passed through the loop in one second, you would measure intensity. **Intensity** is the amount of energy that passes through a certain area in a specific amount of time. When you turn down the volume on your computer, you reduce the energy carried by the sound waves, so you also reduce their intensity.

Distance and intensity Intensity influences how far away a sound can be heard. If you and a friend whisper a conversation, the sound waves you produce have low intensity and are not heard at a far distance. However, when you shout, you can hear each other from much farther apart. The sound waves made by your shouts have high intensity and can be heard farther away.

Sound intensity decreases with distance for two reasons. First, the energy that a sound wave carries spreads out as the sound wave spreads out. Second, some of a sound wave's energy converts to other forms of energy, usually thermal energy, as the sound travels through matter. As the sound wave travels farther, more of its energy converts into other forms. Some materials, such as soft, thick curtains, are very effective at converting sound energy to other forms of energy.

Loudness Some sounds are so loud that they can be painful to hear. **Loudness** is the human perception of sound volume and primarily depends on sound intensity. When sound waves of high intensity reach your ear, they cause your eardrum to move back and forth a greater distance than when sound waves of low intensity reach your ear. As a result, you hear a loud sound.

✓ **Reading Check Relate** intensity and loudness.

The decibel scale It is hard to say how loud is too loud. Two people are unlikely to agree on what is too loud because people vary in their perceptions of loudness. A sound that seems fine to you may seem earsplitting to your teacher. Even so, the intensity of sound can be described using a measurement scale.

A **decibel** (DE suh bel), abbreviated dB, is a unit of sound intensity. The loudest sounds that you hear are probably more than 10 billion times more intense than the softest sounds that you can hear. In order to handle this wide-range of intensities, the decibel measurement scale is set up in a special way.

Every increase in 10 dB on the decibel scale represents a tenfold increase in intensity. This means that a 50-dB sound is 10 times more intense than a 40-dB sound. You might think that this means a 60-dB sound is 20 times more intense than a 40-dB sound. However, a 60-dB sound is 10×10, or 100, times more intense than a 40-dB sound. A 100-dB sound is 10^7, or 10 million, times more intense than a 30-dB sound. A sound can also have an intensity of less than 0 dB. However, people generally cannot hear these sounds.

Sustained sounds above about 90 dB can cause permanent hearing loss. Even short, sudden sounds with intensity levels above 120 dB may cause pain and permanent hearing loss. During some rock concerts, sounds reach this damaging intensity level. Wearing ear protection, such as earplugs, around loud sounds can help protect against hearing loss. **Figure 8** shows some sounds and their intensity levels in decibels.

■ **Figure 8** The volumes of different sounds are often measured in decibels.

Identify *where a normal speaking voice would fall on the decibel scale.*

Loudness in Decibels

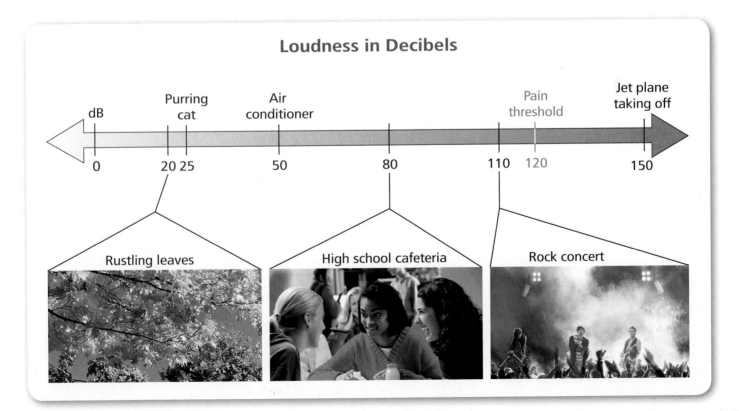

Pitch

If you have ever studied music, you are probably familiar with the musical scale *do, re, mi, fa, so, la, ti, do*. If you were to sing this scale, your voice would start low and become higher with each note. As you sang, you would have heard a change in pitch. **Pitch** is how high or low a sound seems to be. The notes on the right side of a piano have high pitches. The pitch of a sound is primarily related to the frequency of the sound waves.

Frequency and pitch Frequency is a measure of how many wavelengths pass a particular point each second. For a longitudinal wave, such as sound, the frequency is the number of compressions or the number of rarefactions that pass by each second. Frequency is measured in hertz (Hz). One Hz means that one wavelength passes by in one second.

When sound waves with high frequency reach your ears, many compressions reach your eardrums each second. The waves cause your eardrums and all the other parts of your ears to vibrate more quickly than a sound wave with a low frequency. Your brain interprets these fast vibrations caused by high-frequency waves as a sound with a high pitch.

As the frequency of sound waves decreases, the pitch becomes lower. **Figure 9** shows different notes and their frequencies. Notice that when you sing *do, re, mi, fa, so, la, ti, do*, the second *do* has twice the frequency of the first *do*.

Animals differ in their sensitivities to different ranges of sound frequencies. A human teenager with typical hearing can hear sounds with frequencies from about 20 Hz to 20,000 Hz. However, the highest frequency that a typical human can hear decreases with age. The human ear is most sensitive to sounds in the range of 440 Hz to about 7,000 Hz. This roughly corresponds to the notes on the upper half of a piano. Dogs can hear sounds with frequencies up to about 35,000 Hz, and bats can detect frequencies higher than 100,000 Hz.

VOCABULARY ....
SCIENCE USAGE V. COMMON USAGE
Pitch
Science Usage:
   how high or low a sound seems to be
   The highest note on a violin has a higher pitch than the highest note on a cello.

Common Usage:
   a throw, usually toward a particular person or point
   The batter struck out when he swung at a fast pitch. ....

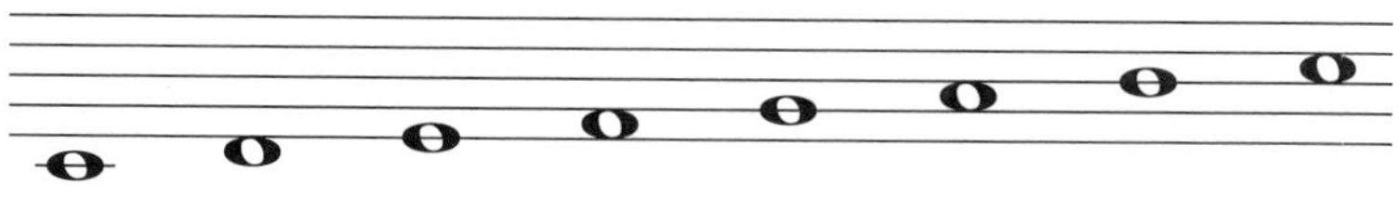

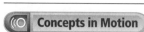

Concepts in Motion

Animation

■ **Figure 9** Every musical note has a distinct frequency, which gives that note a distinct pitch.

Describe *how pitch changes when frequency increases.*

C	D	E	F	G	A	B	C
do	re	mi	fa	sol	la	ti	do
262 Hz	294 Hz	330 Hz	349 Hz	392 Hz	440 Hz	494 Hz	523 Hz

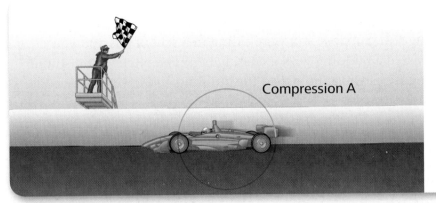

The race car sends out a sound wave as it moves, producing compression A. Compression A continues to move outward, and the car continues to move forward.

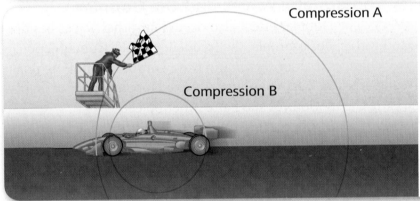

The car is closer to the flagger when it creates compression B. Compressions A and B are closer together in front of the car, so the flagger hears a higher-pitched sound.

The Doppler Effect

Imagine that you are standing at the side of a racetrack with race cars zooming past. When the cars are moving toward you, the pitches of their engines are higher. When the cars are moving away from you, the pitches are lower. The **Doppler effect** is the change in wave frequency due to a wave source moving relative to an observer or an observer moving relative to a wave source. **Figure 10** shows how the Doppler effect occurs.

 **Reading Check Describe** the Doppler effect.

Moving sound sources As a race car moves, it sends out sound waves in the form of compressions and rarefactions. In the top panel of **Figure 10,** the race car produces a compression labeled A as that race car speeds toward the flagger. Compression A travels through the air toward the flagger.

By the time compression B leaves the race car in the bottom panel of **Figure 10,** the car has moved forward. Because the car has moved since the time it created compression A, compressions A and B are closer together in front of the car. Because the compressions are closer together, the frequency is higher and the flagger hears a higher pitch. The compressions behind the moving car are farther apart, resulting in the flagger observing a lower pitch after the car passes.

■ **Figure 10** The Doppler effect occurs when the source of a sound wave is moving relative to a listener.

Explain *why the flagger will hear a lower pitch sound once the car passes him.*

Concepts in Motion
Animation

Figure 11 Police use radar guns to measure the speeds of motorists on highways. Radar guns function based on the Doppler effect.

Moving observers You can also observe the Doppler effect when you are moving past a sound source that is standing still. Suppose you were riding in a school bus and you passed a building with a ringing bell. The pitch would sound higher as you approached the building and lower as you rode away from it.

The Doppler effect happens any time a sound source is moving relative to an observer. It occurs whether the sound source or the observer is moving. The faster the change in position, the greater the change in frequency and pitch.

Electromagnetic waves and the Doppler effect

The Doppler effect also occurs for other waves besides sound waves. For example, the frequency of electromagnetic waves changes if an observer and wave source are moving relative to each other. Astronomers use the Doppler effect to help measure the motions of stars and other objects.

In addition, police radar guns, such as the one shown in **Figure 11,** use the Doppler effect to measure the speeds of cars. The radar gun sends radar waves toward a moving car. The waves are reflected from the car and their frequency is shifted, depending on the speed and direction of the car. From the Doppler shift of the reflected waves, the radar gun determines the car's speed.

Section 2 Review

Section Summary

▶ Tight, dense compressions in a sound wave mean higher intensity, loudness, and more energy.

▶ Sound intensity is measured in decibels.

▶ Pitch is most strongly related to frequency.

▶ The Doppler effect is the change in wave frequency due to a wave source moving relative to an observer or an observer moving relative to a wave source.

8. **MAIN Idea Determine** which will change if you turn up a radio's volume: *wave velocity, intensity, pitch, frequency, wavelength, loudness.* Explain.

9. **Identify** the range of human hearing in decibels and the level at which sound can damage human ears.

10. **Compare and contrast** frequency and pitch.

11. **Draw and label** a diagram that explains the Doppler effect.

12. **Think Critically** Why would a passing car exhibit a greater sound frequency change when it moves at 30 m/s than when it moves at 12 m/s?

Apply Math

13. **Make a Table** Using the musical scale in **Figure 9,** make a table that shows how many wavelengths will pass you in one minute for each musical note. What is the relationship between frequency and the number of wavelengths that pass you in one minute?

Assessment   Online Quiz

Reading Preview

Essential Questions

▶ What is the difference between noise and music?

▶ Why does a guitar sound different from a horn, even when both play the same note?

▶ How do string, wind, and percussion instruments produce music?

▶ What are beats, and why do they occur?

Review Vocabulary

resonance: the process by which an object is made to vibrate by absorbing energy at its natural frequencies

New Vocabulary

music
sound quality
overtone
resonator

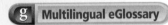 **Multilingual eGlossary**

? Inquiry   Video Lab

Music

MAIN ⟨Idea A musical instrument produces combinations of frequencies that determine how the instrument sounds.

Real-World Reading Link From ancient rhythms and harmonies to modern rock and hip-hop, music has probably been a part of human culture for more than fifty thousand years. The sounds that musical instruments make depend on their sizes, their shapes, and the materials from which they are made.

Making Music

To someone else, your favorite music might sound like a jumble of noise. A sound wave from noise and a sound wave from music are both shown in **Figure 12**. Noise has random patterns and pitches. **Music** is any collection of sounds that are deliberately used in a regular pattern.

Natural frequencies Every material or object has a particular set of frequencies at which it vibrates. These frequencies are called natural frequencies. Musical instruments employ the natural frequencies of various objects, such as strings, to control pitch. When you pluck a guitar string, the pitch you hear depends on the string's natural frequencies. Each string on a guitar has a different set of natural frequencies.

Resonance The sound produced by musical instruments is amplified by resonance. Recall that resonance occurs when a material or an object is made to vibrate at its natural frequencies by absorbing energy from something that is also vibrating at those frequencies. The vibrations of a person's lips in a brass instrument or the reed in a wind instrument cause the air inside the instrument to absorb energy and vibrate at its natural frequencies. The vibrating air makes the instrument sound louder.

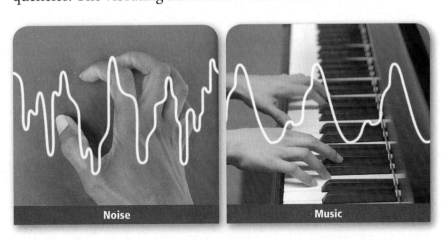

■ **Figure 12** The sound waves that make up noise have a random pattern. The sound waves that make up music have deliberate patterns.

Noise

Music

Sound Quality

Suppose your friend played a note on a flute and then a note of the same pitch and loudness on a piano. Even if you closed your eyes, you could tell the difference between the two instruments. Their sounds would not be the same. Each of these instruments has a unique sound quality. **Sound quality** describes the differences between sounds of the same pitch and loudness. Sound quality results from overtones.

✔ **Reading Check** **Compare** sound quality and pitch.

Overtones You might think that a musical instrument vibrates at only one frequency when you play only one note on that instrument. In fact, a musical instrument vibrates at many different frequencies when you play it, even when you play just one note.

The lowest of these frequencies is called the fundamental frequency. On a guitar, for example, the fundamental frequency is produced by the entire string vibrating back and forth, as shown in **Figure 13.** The fundamental frequency primarily determines the pitch of the note that you hear.

The other frequencies are called overtones. An **overtone** is a vibration whose frequency is a multiple of the fundamental frequency. Overtones determine the sound quality of the note that you hear. The first two guitar-string overtones are also shown in **Figure 13.**

Consider the example of your friend playing the same note on a flute and on a piano again. Your friend produces the same fundamental frequency on both instruments. However, the loudness of each overtone differs between the two instruments.

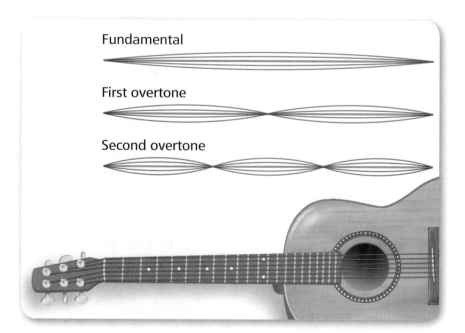

■ **Figure 13** The vibration patterns on a plucked guitar string for the fundamental frequency, first overtone, and second overtone are shown. A plucked guitar string usually vibrates at multiple frequencies simultaneously.

Infer *how the string would vibrate to produce the third overtone.*

Musical Instruments

A musical instrument is any device used to produce a musical sound. Violins, cellos, oboes, bassoons, horns, flutes, and kettledrums are musical instruments that you might have seen and heard in your school orchestra. The members of a jazz band might play clarinets, trumpets, and saxophones. You have probably also heard guitars, pianos and keyboards, and banjos.

These familiar examples form just a small sample of the diverse assortment of instruments that people play throughout the world. For example, Australian Aborigines accompany their songs with a woodwind instrument called the didgeridoo (DIH juh ree dew). Caribbean musicians use rubber-tipped mallets to play steel drums, and a flutelike instrument called the nay is played throughout the Arab world.

Strings Compared to the overall history of musical instruments, string instruments are quite modern. The first manufactured string instruments probably did not appear until about 6,000 years ago. Today, you might hear soft violins, screaming electric guitars, and elegant harps. In string instruments, sound is produced by plucking, striking, or drawing a bow across tightly stretched strings.

Because the sound of a vibrating string is soft, almost all string instruments include a way to amplify sound. Many string instruments, such as the violin in **Figure 14,** have a resonator. A **resonator** (RE zuh nay tur) is a hollow chamber filled with air that amplifies sound when the air inside of it vibrates. Other string instruments, such as electric guitars, have pick-ups. Pick-ups convert the sound from strings into electric signals. These signals are then amplified before being converted back into sound.

VOCABULARY ....

ACADEMIC VOCABULARY

Diverse
differing from one another
They had many diverse interests. ....

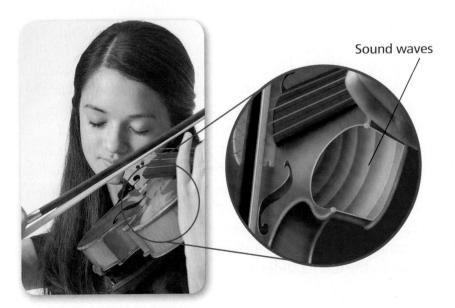

Sound waves

■ **Figure 14** Violins, acoustic guitars, and grand pianos use resonators to amplify sound. A violin's resonator causes air to vibrate much more than that air would from the violin string alone.

Identify *what causes the air to vibrate. What amplifies the vibrations?*

■ **Figure 15** The Wanamaker Organ in Philadelphia, Pennsylvania, is a wind instrument. It has more than 28,000 pipes built into a seven-story structure.

Wind and brass instruments The vibrations of air inside of wind and brass instruments determine the frequencies that those instruments produce. These instruments have been around for much longer than string instruments. Humans created their first wind instruments at least 30,000 years ago. Some scientists think that the first wind instruments may have been created more than 45,000 years ago.

The largest wind instrument today is the pipe organ. The Wanamaker Organ, a small part of which is shown in **Figure 15,** has more than 28,000 pipes. Other modern brass and wind instruments include tubas, horns, oboes, and flutes. All brass and wind instruments use various methods to make air vibrate inside the instrument.

For example, a flute player blows a stream of air over the edge of the flute's mouth hole. This causes the air inside the flute to vibrate. For an instrument such as a saxophone, the player blows across a reed, causing it to vibrate. The vibrating reed then causes the air in the instrument to vibrate. For a brass instrument, such as a trumpet, the musician vibrates his or her lips to make the air inside the instrument vibrate.

In brass and wind instruments, the length of the vibrating tube of air determines the pitch of the sound produced. In flutes and trumpets, the musician changes the length of the resonator by opening and closing finger holes or valves. In a trombone, the tubing slides in and out to become shorter or longer.

■ **Figure 16** The air inside the resonators of these drums amplify the sounds produced.

Describe *how the natural frequencies of the air in the drums affect the sounds those drums produce.*

Percussion Does the sound of a bass drum make your heart start to pound? Percussion instruments are probably the oldest of all instruments other than the human voice. Since ancient times, people have used drums and other percussion instruments to send signals, accompany important rituals, and entertain one another.

Percussion instruments are struck, shaken, rubbed, or brushed to produce sound. Some, such as the marching band drums shown in **Figure 16,** have a membrane stretched over a resonator. When the drummer strikes the membrane, the membrane vibrates and causes the air inside the resonator to vibrate. The resonator amplifies the sound made when the membrane is struck. Some drums have a fixed pitch, but others have a pitch that can be changed by tightening or loosening the membrane.

Caribbean steel drums and xylophones are both examples of percussion instruments where you can control the pitch of the instrument. Caribbean steel drums were developed in the 1940s in Trinidad. As many as 32 different striking surfaces hammered from the ends of 55-gallon oil barrels create different pitches of sound. The side of a drum acts as the resonator.

The xylophone shown in **Figure 17** is another type of percussion instrument. It has a series of wooden bars, each with its own tube-shaped resonator. The musician strikes the bars with mallets that affect the sound quality. Hard mallets make crisp sounds, while softer rubber mallets produce sounds that are more like the drops of water from a leaky faucet. Other types of percussion instruments include cymbals, rattles, and even old-fashioned washboards.

■ **Figure 17** Xylophones are made with many wooden bars, each of which has its own resonator tube. The size of each wooden bar helps to determine the pitch that the musician produces when he strikes that wooden bar.

Explain *why the bars on a xylophone are different sizes.*

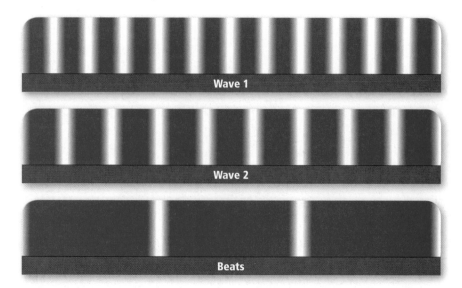

Figure 18 Beats occur when sound waves with slightly different frequencies combine. These sound waves interfere with each other, producing a low frequency, pulsing variation in loudness called beats. The top two panels illustrate two slightly different sound waves. The bottom panel illustrates the beats that occur when those two sound waves interfere.

Wave 1

Wave 2

Beats

Beats

Have you ever heard two flutes play the same note when they were not properly tuned? You might have heard a pulsing variation in loudness. This variation in loudness is called beats and can be unpleasant to the listener. As shown in **Figure 18,** when compressions and rarefactions overlap each other, loudness decreases. When compressions overlap compressions, loudness increases.

If two waves of different frequencies interfere, a new wave is produced that has a different frequency. The frequency of this wave is the difference between the frequencies of the two component waves. The frequency of the beats that you hear decreases as the two waves become closer in frequency. If two flutes play a note at the same frequency, no beats are heard.

Section 3 Review

Section Summary

▶ Quality of sound describes the differences between sounds of the same pitch and loudness.

▶ Sound quality results from specific combinations of frequencies produced in various musical instruments.

▶ The interference of two waves with different frequencies produces beats.

14. **MAIN Idea** **Compare and contrast** music and noise.

15. **Explain** how two instruments could be used to produce a pulsing sound, and identify the name for this pulsing sound.

16. **Explain** how a flute, a violin, and a kettledrum each produce sound.

17. **Think Critically** Two musical notes have the same pitch and volume. However, they sound very different from each other. How is this possible?

Apply Math

18. **Calculate Frequencies** A string on a guitar vibrates with a frequency of 440 Hz. Two beats per second are heard when this string and a string on another guitar are played at the same time. What are the possible frequencies of vibration of the second string?

Assessment Online Quiz

Objectives

- **Demonstrate** how to make music using water and test tubes.
- **Predict** how the tones will change when there is more or less water in a test tube.

Background: There are many different types of musical instruments. Early instruments were made from materials that were easily obtained, such as clay, shells, skins, wood, and reeds. These materials were fashioned into various instruments that produced pleasing sounds. In this lab, you are going to create a musical instrument using materials that are available to you, just as your ancestors did.

Question: *How can you control pitch in an instrument?*

Preparation

Materials

test tubes
test-tube rack
water

Safety Precautions

Procedure

1. Read the procedure and safety information, and complete the lab form.
2. Put different amounts of water into each of the test tubes.
3. Predict any differences that you expect in how the tones from the different test tubes will sound.
4. Blow across the top of each test tube.
5. Record any differences that you notice in the tones that you hear from each test tube.

Conclude and Apply

1. **Describe** how the tones change depending on the amount of water in the test tube.
2. **Explain** why the pitch depends on the height of the water.
3. **Summarize** why each test tube produces a different tone.
4. **Explain** how resonance amplifies the sound from a test tube.
5. **Explain** how the natural frequencies of the air columns in each of the tubes differ.
6. **Infer** how you might control the pitch in a flute.

COMMUNICATE YOUR DATA

Discuss When you are listening to music with family or friends, describe to them what you have learned about how musical instruments produce sound. Write a brief summary of your discussion.

Reading Preview

Essential Questions

▶ What are some of the factors that affect the design of concert halls and movie theaters?

▶ How do some animals use sound waves to hunt and navigate?

▶ How does sonar work?

▶ How is ultrasound used in medicine?

Review Vocabulary

echo: the reflection of a sound from a surface

New Vocabulary

acoustics
echolocation
sonar
ultrasound

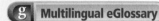
g Multilingual eGlossary

Using Sound

MAIN ⟨Idea Sound waves are used to locate objects, to form images, and to treat medical problems.

Real-World Reading Link Have you ever seen an ultrasound of a human fetus? Doctors use ultrasound, which is high frequency sound, to monitor human fetuses, to diagnose a variety of medical problems, and even to treat some ailments.

Acoustics

When an orchestra stops playing, does it seem as if the sound of its music lingers for a couple of seconds? The sounds and their reflections reach your ears at different times, so you hear echoes. This echoing effect produced by many reflections of sound is called reverberation (re vur bu RAY shun).

During an orchestra performance, reverberation can ruin the sound of the music. To prevent this problem, the people who design concert halls must understand how the size, shape, and furnishings of the room affect the reflection of sound waves.

These scientists and engineers specialize in acoustics (uh KEWS tihks). **Acoustics** is the study of sound. People who study acoustics know that soft, porous materials, such as curtains, can reduce excess reverberation. **Figure 19** shows a concert hall that has been designed to produce a good listening environment.

Echolocation

At night, bats swoop around in darkness without bumping into anything. They even manage to find insects and other prey in the dark. Most species of bats depend on echolocation. **Echolocation** is the process of locating objects by emitting sounds and then interpreting the sound waves that are reflected from those objects. **Figure 20** explains more about echolocation.

■ **Figure 19** Cloth drapes, cushioned seats, and carpeted floors help reduce reverberations in this concert hall. The amount of reverberation can be controlled slightly by opening and closing the drapes.

Explain *whether the drapes are more likely to absorb or to reflect sound energy.*

FIGURE 20

Visualizing Bat Echolocation

Bats and dolphins navigate and locate prey by using echolocation. The diagrams below show how a bat uses echolocation in hunting for prey. Recent research suggests that some species of moths may be able to jam a bat's echolocation by emitting their own sounds.

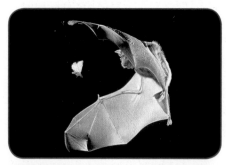

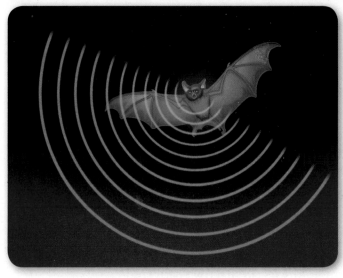

▲ A. The bat emits sound waves.

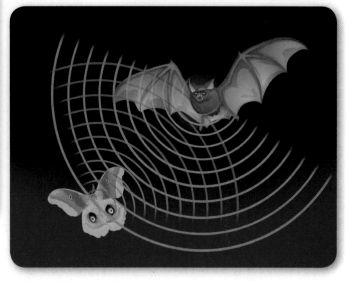

▲ B. Some of the sound waves reflect off a moth.

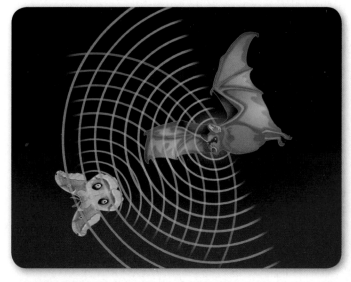

▲ C. The bat determines the moth's location and velocity from the reflected waves.

▲ D. The bat continues to emit sound waves and locate the moth until the moth is captured.

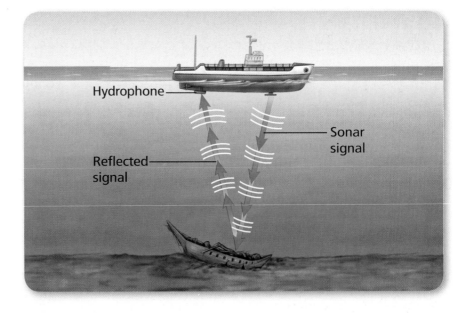

■ **Figure 21** People use sonar to find objects that are underwater. Here, scientists use sonar to locate a sunken ship.

 Concepts in Motion
Animation

Hydrophone

Reflected signal

Sonar signal

Sonar

More than 140 years ago, a ship named the *SS Central America* disappeared in a hurricane off the coast of South Carolina. In its hold lay ten tons of newly minted gold coins and bars. When the shipwreck occurred, there was no way to search for the ship in the deep water where it sank. The *SS Central America* and its treasures laid at the bottom of the ocean until 1988, when crews used sonar to locate the wreck under 2,400 m of water. Over $100 million in gold was eventually recovered.

Sonar is a system that uses the reflection of underwater sound waves to detect objects. First, a sound pulse is emitted toward the bottom of the ocean. The sound travels through the water and is reflected when it hits something solid, as shown in **Figure 21.** A sensitive underwater microphone, called a hydrophone, picks up the reflected signal. Because the speed of sound in water is known, the distance to the object can be calculated by measuring how much time passes between emitting the sound pulse and receiving the reflected signal.

 Reading Check **Describe** how sonar detects underwater objects.

The idea of using sonar to detect underwater objects was first suggested as a way to avoid icebergs, but many other uses have been developed for it. Navy ships use sonar for detecting, identifying, and locating submarines. Fishing crews also use sonar to find schools of fish, and scientists use it to map the ocean floor.

More detail can be revealed by using sound waves of high frequency. As a result, most sonar systems are ultrasonic. **Ultrasound** is sound with frequency above 20,000 Hz and cannot be heard by humans.

Ultrasound in Medicine

Ultrasonic waves are commonly used in medicine. Medical professionals use ultrasound to examine many parts of the body, including the heart, liver, gallbladder, pancreas, spleen, kidneys, breasts, and eyes. Medical professionals can also use ultrasonic imaging, which is much safer than X-ray imaging, to monitor a human fetus.

When ultrasound is used for medical imaging, an ultrasound technician directs the ultrasound waves toward a target area of a patient's body. The sound waves reflect off the targeted area, and the reflected waves are used to produce electronic signals. A computer program converts these signals into video images called sonograms. A sonogram of a human fetus is shown in **Figure 22**.

Kidney stones and ultrasound
Medical professionals can also use ultrasound to treat certain medical problems. For example, ultrasound can be used to break up kidney stones. Bursts of ultrasound create vibrations that cause the stones to break into small pieces. These fragments then pass out of the body with the urine. Without ultrasound, surgery would be necessary to remove the kidney stones.

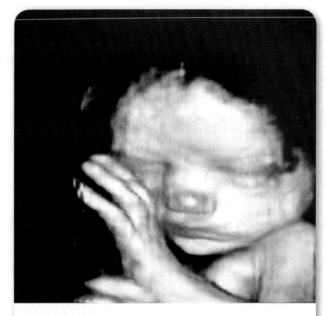

■ **Figure 22** Ultrasonic waves are directed into a pregnant woman's uterus to form images of her fetus. This allows doctors to safely monitor the fetus's growth.

Section 4 Review

Section Summary

▶ Acoustics is the study of sound.

▶ Bats locate objects by emitting sounds and then interpreting their reflected sound waves.

▶ Humans using sonar can interpret reflected sound to locate objects underwater.

▶ High-frequency sound waves are useful for detecting and monitoring certain medical conditions.

19. **MAIN Idea** Describe at least three different ways that people use sound.

20. **Describe** some differences between a gym and a concert hall that might affect the amount of reverberation in each.

21. **Compare and contrast** echolocation and sonar.

22. **Explain** how ultrasonic imaging works.

23. **Think Critically** How might sonar technology be useful in locating deposits of oil and minerals?

Apply Math

24. **Calculate Distance** Sound travels at about 1,500 m/s in seawater. How far will a sonar pulse travel in 46 s?

25. **Calculate Time** How long will it take for an undersea sonar pulse to travel 3 km?

Chapter 10 Review

Use Vocabulary

Complete each sentence with the correct term from the Study Guide.

26. The _____ is filled with fluid and contains tiny hair cells that vibrate.

27. _____ is the study of sound.

28. A change in pitch or wave frequency due to a moving wave source is an instance of the _____.

29. _____ is a combination of sounds and pitches that follows a specified pattern.

30. Bats use _____ to locate moths and other prey.

31. _____ primarily depends on the intensity of sound.

32. A(n) _____ is a unit of sound intensity.

Check Concepts

33. For a sound with a low pitch, what else is always low?
    A) amplitude   C) wavelength
    B) frequency   D) wave velocity

34. **THEME FOCUS** The amount of energy that a sound carries decreases when which decreases?
    A) beats   C) quality
    B) wavelength   D) amplitude

35. To which is sound quality most closely related?
    A) pitch   C) loudness
    B) overtones   D) resonance

36. Sound can travel through all but which?
    A) solids   C) gases
    B) liquids   D) outer space

37. Sounds with the same pitch and loudness traveling in the same medium may differ in which property?
    A) frequency   C) quality
    B) amplitude   D) wavelength

38. What part of a musical instrument amplifies sound waves?
    A) resonator   C) mallet
    B) string   D) reed

39. What is the name of the method used to find objects that are underwater?
    A) sonogram   C) sonar
    B) ultrasonic bath   D) percussion

Use the figure below to answer question 40.

40. Which will occur in this situation from the flagger's perspective?
    A) The sound's speed will decrease as the car passes.
    B) The sound's speed will increase as the car passes.
    C) The sound's frequency will decrease as the car passes.
    D) The sound's frequency will increase as the car passes.

✓ **Assessment** **Online Test Practice**

Interpret Graphics

Use the table below to answer question 41.

Federally Recommended Noise Exposure Limits	
Sound Level (dB)	Time Permitted (hours per day)
90	8
95	4
100	2
105	1
110	0.5

41. You use a lawn mower with a sound level of 100 dB. Using the table above, determine the maximum number of hours a week that you can safely mow lawns without ear protection.

42. **BIG Idea** Copy and fill in the answers for **a, b,** and **c** for the following table on musical instruments.

Characteristics of Musical Instruments			
	Guitar	Flute	Bongo Drum
How played	plucked	blown into	**a.**
Role of resonator	amplifies sound	**b.**	amplifies sound
Type of instrument	**c.**	wind	percussion

Think Critically

43. **Infer** A car comes to a railroad crossing. The driver hears a train's whistle, and its pitch becomes lower. What can be assumed about how the train is moving?

44. **Explain** Sound travels slower in air at high altitudes than at low altitudes. Explain why this is so.

45. **Apply** Acoustic scientists sometimes conduct research in rooms that absorb all sound waves. How could such a room be used to study how bats find their food?

46. **Communicate** Some people enjoy using snowmobiles. Others object to the noise that they make. Write a proposal for a policy that seems fair to both groups for the use of snowmobiles in a state park.

Apply Math

Use the wave speed equation, $v = f\lambda$, to answer questions 47–49.

47. **Calculate Frequency** A sonar pulse has a wavelength of 3.0 cm. The pulse has a speed in water of 1,500 m/s. What is the pulse's frequency?

48. **Estimate Wavelength** What is the approximate wavelength of a sound wave in air with a frequency of 440 Hz if the speed of sound through the air is 340 m/s?

49. **Calculate Wave Speed** An earthquake produces a seismic wave that has a wavelength of 650 m and a frequency of 10.0 Hz. How fast does the wave travel?

50. **Calculate Distance** The sound of thunder travels at a speed of 340 m/s and reaches you 2.5 s after you see a lightning flash. How far away is the lightning?

51. **Calculate Time** The shipwrecked *SS Central America* was discovered lying beneath 2,400 m of water. If the speed of sound in seawater is 1,500 m/s, how long will it take a sonar pulse to travel to the shipwreck and return?

Standardized Test Practice

Multiple Choice

Record your answers on the answer sheet provided by your teacher or on a sheet of paper.

1. What does a sound's frequency most influence?
   A. pitch
   B. amplitude
   C. intensity
   D. energy

2. Through which medium does sound travel fastest?
   A. empty space C. gases
   B. liquids   D. solids

Use the table below to answer questions 3 and 4.

Sound	Loudness (dB)
Jet taking off	150
Pain threshold	120
Chain saw	115
Power mower	110
Vacuum cleaner	75
Average home	50
Purring cat	25

3. Which would sound the loudest?
   A. vacuum cleaner
   B. chain saw
   C. power mower
   D. purring cat

4. Which statement is true about a sound of 65 decibels?
   A. It causes intense pain.
   B. It can cause permanent hearing loss.
   C. It cannot be heard by anyone.
   D. It can be heard without discomfort or damage.

5. Which instrument does not have a resonator?
   A. kettledrum
   B. flute
   C. classical violin
   D. electric guitar

6. Which sound wave property is most related to loudness?
   A. wavelength   C. frequency
   B. amplitude   D. wave speed

Use the illustration below to answer questions 7 and 8.

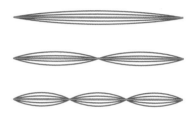

7. What is shown for the top guitar string in the figure above?
   A. first overtone
   B. fundamental frequency
   C. second overtone
   D. third overtone

8. What is shown for the bottom guitar string in the figure above?
   A. first overtone
   B. fundamental frequency
   C. second overtone
   D. third overtone

9. Two instruments produce 1 Hz beats when played simultaneously. Which might be the frequencies of the notes from the instruments?
   A. 1 Hz and 10 Hz
   B. 260 Hz and 262 Hz
   C. 330 Hz and 330 Hz
   D. 440 Hz and 441 Hz

✓ Assessment   Standardized Test Practice

Record your answers on the answer sheet provided by your teacher or on a sheet of paper.

Use the table below to answer questions 10–12.

Medium	Speed of Sound (m/s)
Air	340
Water	1,500
Iron	5,100

10. A "fishfinder" sends out a pulse of ultrasound and measures the time needed for the sound to travel to a school of fish and back to the boat. If the fish are 15 m below the surface, how long would it take sound to make the round trip in the water?

11. Suppose you are sitting in the bleachers at a baseball game 170 m from home plate. How long after the batter hits the ball do you hear the "crack" of the ball and bat?

12. Suppose your friend is 510 m away along a long iron railing while you have your ear to that railing. She taps the railing with a stone. How long will the sound take to reach your ear?

13. Explain how placing your hand on a bell that has just been rung stops the sound.

14. What produces the different sound qualities in different musical instruments?

Record your answers on a sheet of paper.

15. Explain why people who work on the ground near jet runways wear big ear muffs filled with a sound insulator.

16. Describe the path of sound from the time it enters the ear until a message reaches the brain.

17. Would sound waves traveling through the outer ear travel faster or slower than those traveling through the inner ear? Explain.

18. Compare the way a violin, a flute, and a drum produce sound waves. What acts as a resonator in each instrument?

Use the figure below to answer question 19.

19. Explain how the ship's crew in the figure is able to measure the depth of the water.

20. Suppose you have been hired to reduce the reverberation that occurs in classrooms. What recommendations would you make?

NEED EXTRA HELP?																				
If You Missed Question . . .	1	2	3	4	5	6	7	8	9	10	11	12	13	14	15	16	17	18	19	20
Review Section . . .	2	1	2	2	3	2	3	3	3	4	1	1	1	3	2	1	1	3	4	4

CHAPTER 11

Electromagnetic Waves

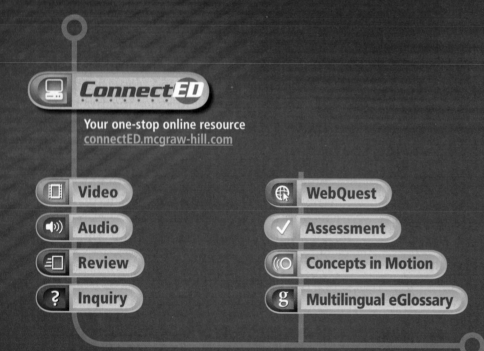

X-Ray

ConnectED

Your one-stop online resource
connectED.mcgraw-hill.com

- Video
- Audio
- Review
- Inquiry
- WebQuest
- Assessment
- Concepts in Motion
- Multilingual eGlossary

Launch Lab

Electromagnetic Waves and Matter

You often hear about the danger of the Sun's ultraviolet rays, which can damage the cells of your skin. When the exposure is not too great, your cells repair themselves. But too much at one time causes a painful sunburn. Exposure to the Sun can also cause skin cancer. In this lab, observe how energy carried by ultraviolet waves causes changes in other materials.

For a lab worksheet, use your StudentWorks™ Plus Online.

 Inquiry Launch Lab

FOLDABLES®

Make a two-tab book. Label it as shown. Use it to organize your notes on electromagnetic waves.

How do electromagnetic waves travel through space?

How do electromagnetic waves transfer energy to matter?

Optical

Infrared

Ultraviolet

Composite

THEME FOCUS Technology
Many modern devices operate by using electromagnetic waves.

BIG Idea Electromagnetic waves transfer energy through matter and through space.

Reading Preview

Essential Questions

▶ How does a vibrating electric charge produce an electromagnetic wave?
▶ What properties describe electromagnetic waves?
▶ How do electromagnetic waves transfer energy?

Review Vocabulary

magnetic field: surrounds a magnet and exerts a force on magnetic materials

New Vocabulary

electromagnetic wave
photon

g Multilingual eGlossary

What are electromagnetic waves?

MAIN Idea Vibrating electric charges produce electromagnetic waves.

Real-World Reading Link No matter where you are, you are surrounded by electromagnetic waves. They enable you to see, and they make your skin feel warm. They carry cell phone, radio, and television signals, and they allow popcorn to be prepared in a microwave oven.

Waves in Matter

Waves are produced by something that vibrates, and they carry energy from one place to another. Look at the water wave and the sound wave in **Figure 1.** Both waves are moving through matter. The water wave is moving through water, and the sound wave is moving through air. These waves travel because energy is transferred from particle to particle. Without matter to transfer the energy, these waves cannot move. However, there is another type of wave that does not require matter to transfer energy.

Electromagnetic Waves

Electromagnetic waves are made by vibrating electric charges. Electromagnetic waves are composed of changing electric fields and magnetic fields. Instead of transferring energy from particle to particle, electromagnetic waves travel by transferring energy between the electric and magnetic fields. Electromagnetic waves do not require matter to travel because electric fields and magnetic fields can exist where matter is not present.

✓ **Reading Check Identify** what produces an electromagnetic wave.

■ **Figure 1** Some waves, such as water waves and sound waves, require matter to move. As the wave travels through the matter, energy is transferred from one particle to the next.

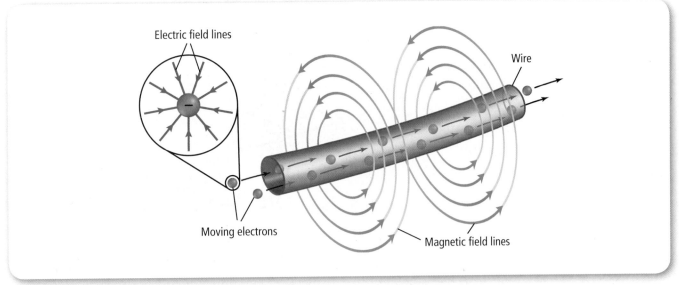

Electric field lines

Wire

Moving electrons

Magnetic field lines

Electric and magnetic fields
Recall that electric charges are surrounded by electric fields and that magnets are surrounded by magnetic fields. These fields exert a force even when the charge or magnet is not in contact with an object. Fields exist around an electric charge or a magnet even in a vacuum. A vacuum is a volume of space that contains little or no matter.

You might also recall that a moving electric charge, such as the current in the wire shown in **Figure 2,** is surrounded by a magnetic field. Similarly, a moving magnet is surrounded by an electric field. A changing electric field creates a magnetic field, and a changing magnetic field creates an electric field.

Making electromagnetic waves
When an electric charge vibrates, the electric field around it changes. Because the electric charge is in motion, it also has a magnetic field around it. This magnetic field also changes as the charge vibrates. As a result, the vibrating electric charge is surrounded by a changing electric field and a changing magnetic field.

How do the vibrating electric field and magnetic field around the charge become a wave that travels through space? The changing electric field around the charge creates a changing magnetic field. This changing magnetic field then creates a changing electric field. This process continues, with the magnetic field and electric field continually creating each other.

■ **Figure 2** All moving electric charges, such as the electrons in this wire, are surrounded by an electric field and a magnetic field.

FOLDABLES
Incorporate information from this section into your Foldable.

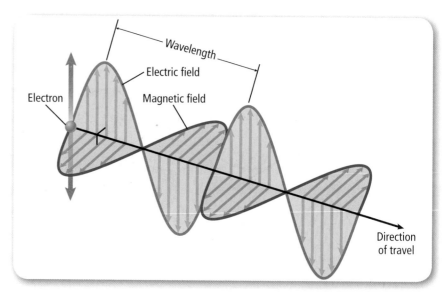

Figure 3 A vibrating electric charge creates an electromagnetic wave. The wave travels outward in all directions from the charge. Here, the wave is shown in only one direction.

State *whether an electromagnetic wave is a transverse wave or a longitudinal wave.*

Properties of Electromagnetic Waves

The vibrating electric and magnetic fields of an electromagnetic wave are perpendicular to each other. That is, they are at right angles (90°) to each other. They travel outward from the moving charge, as shown in **Figure 3.** Because the electric and magnetic fields vibrate at right angles to the direction the wave travels, an electromagnetic wave is a transverse wave.

Speed In a vacuum, all electromagnetic waves travel at 300,000 km/s. Because light is a type of electromagnetic wave, the speed of electromagnetic waves in a vacuum is usually called the "speed of light." The speed of light is nature's speed limit—nothing travels faster than the speed of light. The speed of an electromagnetic wave in matter depends on the material through which the wave travels. However, it is always slower than the speed of light in a vacuum. In matter, electromagnetic waves are usually the slowest in solids and faster in gases.

Table 1 in **Figure 4** lists the speed of electromagnetic waves in a vacuum and several common materials. **Figure 4** illustrates that light travels slower and refracts when it enters glass.

(Concepts in Motion logo) **Concepts in Motion**

Interactive Table

Figure 4 Electromagnetic waves travel slower in glass than in air. This difference in speed leads to refraction when the wave passes from air into glass.

Table 1	Speed of Electromagnetic Waves	
Material	**Speed (km/s)**	
None (Vacuum)	300,000	
Air	299,000	
Water	226,000	
Glass	200,000	
Diamond	124,000	

How do scientists work with large numbers?

The speed of light in a vacuum is 300,000,000 m/s. That is a lot of zeros to keep track of when doing a calculation! To make large numbers easier to work with, scientists use a system called scientific notation.

A number written in scientific notation has the form $M \times 10^N$. N is the number of places the decimal point in the number has to be moved so that the number M that results has only one digit to the left of the decimal point.

For example, the speed of light in a vacuum is 300,000,000 m/s = 3.00000000×10^8 m/s. We moved the decimal 8 spaces to the left, so $N = 8$ and $M = 3.00000000$. In this case, the zeros are just place holders, so we can drop them and write the speed of light in a vacuum as 3×10^8 m/s.

Very small numbers can also be written in scientific notation. For example, $0.00000045 = 4.5 \times 10^{-7}$. The minus sign in the exponent indicates that the decimal was moved to the right instead of to the left.

Identify the Problem

In order to make calculations neat and efficient, place the values for the speed of light in different materials into scientific notation.

Speed of Light in Various Materials		
Material	Speed (m/s)	Speed (m/s) in Scientific Notation
None (Vacuum)	300,000,000	3×10^8
Air	299,000,000	
Water	226,000,000	
Glass	200,000,000	
Diamond	124,000,000	

Solve the Problem

1. Copy the table, and write each speed in scientific notation.
2. Would it be helpful to write your mass (about 50 kg) in scientific notation? Explain.
3. Write the following measurements in scientific notation:
   a) 0.00005291 s
   b) 0.000000246 m
   c) 0.030042 kg

Frequency and wavelength Like all waves, electromagnetic waves can be described by their wavelengths and frequencies. The wavelength of an electromagnetic wave is the distance from one crest to another, as shown in **Figure 3**. The frequency is the number of wavelengths that pass a point in one second. The units for frequency are hertz. The frequency of an electromagnetic wave equals the frequency of the vibrating charge that produces the wave. This frequency is the number of vibrations of the charge in one second. Electromagnetic waves follow the wave speed equation, $v = f\lambda$. As the frequency (f) increases, the wavelength (λ) becomes smaller.

Matter and Electromagnetic Waves

All matter contains charged particles that are always in motion. As a result, all objects emit electromagnetic waves. Objects can emit electromagnetic waves at many wavelengths. However, the dominant wavelength emitted becomes shorter as the temperature of the material increases.

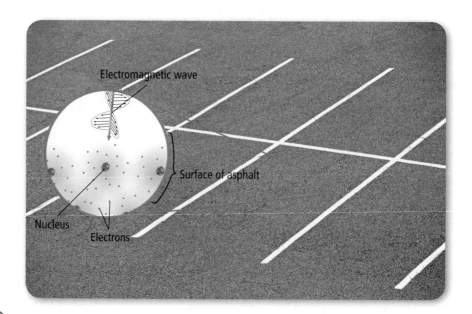

Figure 5 As an electromagnetic wave from the Sun strikes asphalt, charged particles in the asphalt gain energy from the vibrating electric and magnetic fields. This transfer of energy is why asphalt gets hot on sunny, summer days.

Electromagnetic wave

Surface of asphalt

Nucleus

Electrons

MiniLab

? Inquiry   MiniLab

Investigate Electromagnetic Waves

Procedure

1. Read the procedure and safety information, and complete the lab form.
2. Point your **television remote control,** which uses electro-magnetic waves, in different directions and observe whether it will still control the **television.**
3. Place various materials in front of the television's infrared receiver, and observe whether the remote still controls the television. Some materials you might try are **glass,** a **book, your hand, paper,** or a **metal pan.**

Analysis

1. **Explain** Was it necessary for the remote control to be point-ing exactly toward the receiver to control the television? Explain.
2. **Analyze** Did the remote con-tinue to work when the various materials were placed between it and the receiver? Explain.

Electromagnetic waves interact with matter

As an electromagnetic wave moves, it encounters objects. The vibrating electric and magnetic fields of the wave exert forces on the charged particles and magnetic materials that make up the object. This interaction causes the particles in the object to gain energy. For example, electromagnetic waves from the Sun cause electrons in asphalt to vibrate and gain energy, as shown in **Figure 5.** This energy can make the asphalt hot. The energy carried by an electromagnetic wave is called radiant energy. Radiant energy makes fire feel warm and enables you to see.

Waves and Particles

The difference between a wave and a particle might seem obvious—a wave is a disturbance that carries energy, and a particle is a piece of matter. However, in reality, the difference is not so clear.

Waves as particles In 1887, Heinrich Hertz found that he could create a spark by shining light on a metal. (Today, we know that this spark means that electrons were ejected from the metal.) Hertz found that whether sparks occurred depended on the frequency of the light and not the amplitude. Because the energy carried by a sound wave or water wave depends on its amplitude and not its frequency, this result was mysterious.

In 1905, Albert Einstein provided an explanation. An elec-tromagnetic wave can behave as a particle called a photon. A **photon** is a massless bundle of energy that behaves like a particle. The photon's energy depends on the frequency of the wave. The photon's energy increases as the wave's frequency increases.

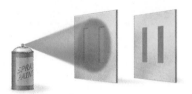

Paint particles sprayed through two slits coat only the area behind the slits.

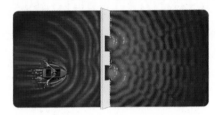

Water waves produce an interference pattern after passing through two slits.

Electrons fired at two slits form a wavelike interference pattern.

Particles as waves Because electromagnetic waves could behave as particles, other scientists wondered whether particles, such as electrons, could behave as waves. If a beam of electrons was sprayed at two tiny slits, you might expect that the electrons would strike only the area behind the slits, like the spray paint at the left of **Figure 6.**

But scientists found that the electrons formed an interference pattern typical of waves, as seen in the right of **Figure 6.** When waves pass through narrow slits, they interfere with each other. This experiment showed that electrons can behave like waves. It is now known that all particles can behave like waves. However, this does not mean that particles travel in wavy lines. Rather, this means that they display behavior, such as interference, that was once associated only with waves.

■ **Figure 6** When sent through two narrow slits, electrons behave as a wave, not as particles.

((O)) **Concepts in Motion**
Animation

Section 1 Review

Section Summary

▶ An electromagnetic wave consists of a vibrating electric field and a vibrating magnetic field.

▶ Electromagnetic waves carry radiant energy.

▶ In empty space, electromagnetic waves travel at 300,000 km/s, the speed of light.

▶ Electromagnetic waves travel more slowly in matter, with a speed that depends on the material.

▶ Electromagnetic waves can behave as particles that are called photons.

1. **MAIN Idea Infer** Would a vibrating proton produce an electromagnetic wave? Would a vibrating neutron? Explain.

2. **Compare** the frequency of an electromagnetic wave with the frequency of the vibrating charge that produces the wave.

3. **Describe** how electromagnetic waves transfer energy to matter.

4. **Explain** how an electromagnetic wave can travel through space that contains no matter.

5. **Think Critically** Would a stationary electron produce an electromagnetic wave? Would a stationary magnet? Explain.

Apply Math

6. **Calculate** How many minutes does it take an electromagnetic wave to travel from the Sun to Earth (150,000,000 km)?

LAB   The Speed of Light

Objectives

- **Recognize** that microwaves are a type of electromagnetic wave.
- **Determine** the speed of light using a microwave oven and the wave-speed equation, $v = f\lambda$.

Background: The speed of light in air is 299,000 km/s.

Question: *How can we measure the speed of light?*

Preparation

Materials

microwave oven with turntable removed
microwave-safe plate
push-pin size disks
mini marshmallows

waxed paper
metric ruler
oven mitts

Safety Precautions

WARNING: *Plate and melted marshmallows will be hot. Do not eat any items used in the lab. Do not use metal objects in the microwave.*

Procedure

1. Read the procedure and the safety information, and complete the lab form.
2. Prepare a data table like the one shown.
3. Locate the manufacturer's label on the back or inside of your microwave, and record the microwave's frequency.
4. Cover the plate with wax paper. Spread a single layer of marshmallows onto the plate. Make sure the marshmallows are evenly spread and cover the plate.
5. Microwave the plate for about 45 s. Use oven mitts to remove the plate, and allow the plate to cool.

Data Table

Pair	Distance Between Hot Spots (cm)	Wavelength (km)	Speed of Light (km/s)
1			
2			

6. Use the small disks to mark the hot spots where the marshmallow melted the most.
7. For five pairs of nearby hot spots, measure from the center of one disk to the center of a nearby disk. Record your data in your table.

Conclude and Apply

1. **Determine** the wavelength. The distance between hot spots is half a wavelength, $d = \lambda / 2$.
2. **Convert Units** Wavelengths should have units of kilometers, and the frequency should have units of hertz. *Hint: 1 MHz = 1,000,000 Hz.*
3. **Calculate** the speed of light using the equation $v = f\lambda$. Do this for each pair of hot spots.
4. **Determine** the largest, smallest, and average values for the speed of light.
5. **Compare** these speeds to the accepted speed of light in air (299,000 km/s).
6. **Think Critically** Describe another way that scientists could measure the speed of light.

COMMUNICATE YOUR DATA

Compare your average speed of light with other groups in your class. Identify some reasons why some calculations were closer to the accepted speed of light than others.

Reading Preview

Essential Questions

▶ What are the main divisions of the electromagnetic spectrum?

▶ What are the properties of each type of electromagnetic wave?

▶ What are some common uses of each type of electromagnetic wave?

Review Vocabulary

radiation: the transfer of thermal energy by electromagnetic waves

New Vocabulary

radio waves
microwaves
infrared waves
visible light
ultraviolet waves
X-rays
gamma rays

g Multilingual eGlossary

? Inquiry   Virtual Lab

■ **Figure 7** Each region of the electromagnetic spectrum spans a range of frequencies.

The Electromagnetic Spectrum

MAIN ⟨Idea The electromagnetic spectrum is divided into several sections, each with a certain range of frequencies and specific properties.

Real-World Reading Link Cell phones and X-ray machines both use electromagnetic waves. But you cannot take a picture of a broken bone with your cell phone. Cell phones and X-ray machines use two different types of electromagnetic waves.

A Range of Frequencies

Electromagnetic waves have a wide variety of frequencies. They might vibrate once each second or trillions of times each second. The entire range of electromagnetic wave frequencies is called the electromagnetic spectrum. A spectrum is a continuous sequence arranged by a particular property.

Each region of the electromagnetic spectrum has a specific name, as shown in **Figure 7.** Each region interacts with matter differently. The human eye detects only a small portion of the electromagnetic spectrum called visible light. Various devices have been developed to detect other frequencies. For example, the antenna of your radio detects radio waves.

Radio Waves

Even though you cannot see them, radio waves are all around you. **Radio waves** are electromagnetic waves with wavelengths longer than 10 cm. Radio waves have long wavelengths and low frequencies, and their photons have low energies. Radio waves have many uses, including communications and medical imaging.

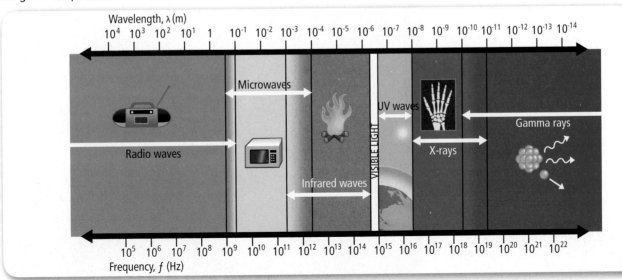

Figure 8 Radar uses radio waves to track airplanes and ships.

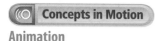

Concepts in Motion

Animation

Vocabulary ....

Academic Vocabulary

Transmit

to send from one place to another
My cell phone transmitted the message to my sister. ....

Audio transmission Some radio waves carry an audio signal from a radio station to a radio. However, even though these radio waves carry information that a radio uses to create sound, you cannot hear radio waves. You hear sound when your radio changes the radio wave into a sound wave.

Radar Another use for radio waves is to find the position and movement of objects by a method called radar. Radar stands for **RA**dio **D**etecting **A**nd **R**anging. With radar, radio waves are transmitted toward an object. By measuring the time required for the waves to bounce off the object and return to a receiving antenna, the location of the object can be found. Radar is used for tracking the movement of aircraft, watercraft, and space-craft, as shown in **Figure 8.** Law enforcement officers also use radar to measure how fast a vehicle is moving.

Magnetic resonance imaging (MRI) In the early 1980s, medical researchers developed a technique called magnetic resonance imaging, which uses radio waves to help diagnose ill-ness. The patient lies inside a large cylinder, like the one shown in **Figure 9.** The cylinder contains a powerful magnet, a radio wave emitter, and a radio wave detector.

Protons in hydrogen atoms in bones and soft tissue behave like magnets and align with the strong magnetic field created by the machine's magnet. Some of the protons absorb energy from the radio waves and flip their alignments. The amount of energy a proton absorbs and then re-emits depends on the type of tissue it is part of. A radio receiver detects this released energy. This information is then used to create a map of the different tissues. A picture of the inside of the patient's body is produced painlessly.

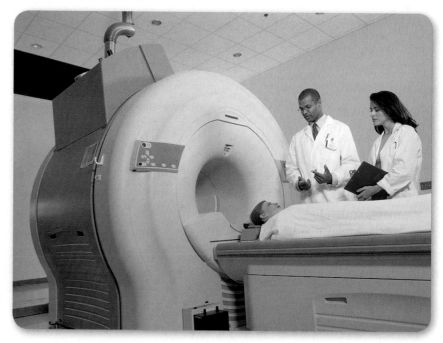

Figure 9 Magnetic resonance imaging is used to produce images of soft tissues.

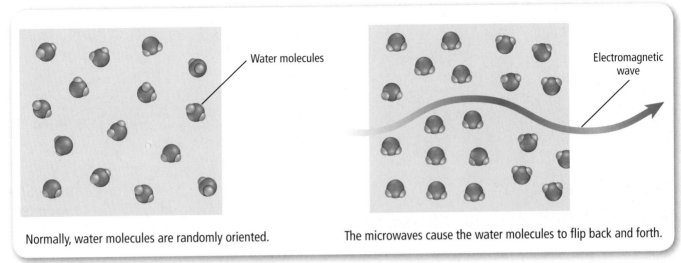

Water molecules

Normally, water molecules are randomly oriented.

Electromagnetic wave

The microwaves cause the water molecules to flip back and forth.

Microwaves

Electromagnetic waves with wavelengths between 0.1 mm and 30 cm are called **microwaves.** Microwaves with wavelengths of about 1 cm to 20 cm are widely used for communication, such as for cellular telephones and satellite signals. However, you are probably most familiar with microwaves because of their use in microwave ovens.

Reading Check **Describe** the differences between microwaves and radio waves.

Microwave ovens In a microwave oven, microwaves interact with the water molecules in food, as shown in **Figure 10.** Each water molecule has a slight positive charge on one side and a slight negative charge on the other side, so it will align in an electric field. The vibrating electric field inside a microwave oven causes water molecules in food to rotate back and forth billions of times each second.

This rotation causes a type of friction between water molecules that generates thermal energy. The thermal energy produced by the water molecules' interactions causes your food to cook.

Foods with plenty of water cook well in the microwave. Frozen water, however, cannot be warmed using microwaves because the water molecules are bound in a crystallized structure and cannot rotate. On many microwave ovens, there is a special defrost setting. This setting heats the partly melted water on the surface of the food. The inside of the food is then warmed by conduction until all the water is liquid again.

■ **Figure 10** Microwave ovens use electromagnetic waves to transfer energy to water particles in food.

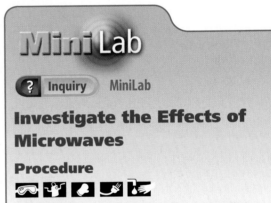

MiniLab

? Inquiry   MiniLab

Investigate the Effects of Microwaves

Procedure

WARNING: *Use caution when handling hot beakers. Do not use metal objects in the microwave.*

1. Read the procedure and safety information, and complete the lab form.
2. Obtain two **small beakers**. Place 50 mL of **dry sand** into each.
3. To one of the beakers, add 20 mL of **room-temperature water** and stir well.
4. With a **thermometer,** record the temperature of the sand in each beaker.
5. Together, **microwave** both beakers of sand for 10 s and immediately record the temperature again.

Analysis

1. **Compare** the initial and final temperatures of the wet and dry sand.
2. **Infer** why there was a difference.

■ **Figure 11** Infrared images of homes and other buildings can provide information about the structure's energy efficiency.

Identify *where energy is escaping from the house.*

Infrared Waves

Most of the warm air in a fireplace moves up the chimney, yet you feel the warmth of the blazing fire when you stand in front of a fireplace. Why do you feel the warmth? The warmth you feel is thermal energy transmitted to you by infrared waves. **Infrared waves** are electromagnetic waves with wavelengths between about one-thousandth of a meter and about 700-billionths of a meter.

Using infrared waves Every object emits infrared waves. Hotter objects emit more infrared waves than cooler objects. Infrared detectors can form images of objects from the infrared waves they emit. These images, like the one in **Figure 11,** can help determine how energy-efficient a structure is. Other devices that use infrared waves include television remote controls and CD-ROM drives.

Visible Light

Visible light is the range of electromagnetic waves that you detect with your eyes. Visible light differs from radio waves, microwaves, and infrared waves only by its frequency and wavelength. Visible light has wavelengths around 700-billionths to 400-billionths of a meter. Color is the brain's interpretation of the wavelengths of the light absorbed by substances in the eye. These colors range from short-wavelength violet to long-wavelength red, as illustrated in **Figure 12.** If all colors of light are present in the same place, you see the light as white.

■ **Figure 12** This spectrum of visible light shows all colors of light from long-wavelength red to short-wavelength violet.

Red (7.00×10^{-7} m) — Violet (4.00×10^{-7} m)

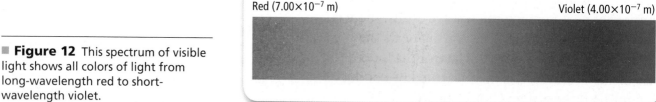

Ultraviolet Waves

Ultraviolet waves are electromagnetic waves with wavelengths from about 400-billionths to 10-billionths of a meter. Ultraviolet waves (UV waves) can enter cells, making ultraviolet waves both useful and harmful.

Useful UVs You probably know that UV waves cause sunburn. But some exposure to ultraviolet waves is healthy. Ultraviolet waves striking the skin enable your body to make vitamin D, which is needed for healthy bones and teeth.

Ultraviolet waves are also used to disinfect food, water, and medical supplies, as shown in **Figure 13.** When ultraviolet light enters a cell, it damages protein and DNA. For some single-celled organisms, such as bacteria, this damage can mean death.

Ultraviolet waves make some materials fluoresce (floo RES). Materials that fluoresce absorb ultraviolet waves and reemit the energy as visible light. Police detectives sometimes use fluorescent powder to reveal fingerprints.

Harmful UVs When you spend time in the Sun, you might wear sunscreen to prevent sunburn. Most of the UV waves that reach Earth's surface are longer-wavelength UVA rays. The shorter-wavelength UVB rays are the primary cause of sunburn and skin cancers, but UVA rays contribute to skin cancers and skin damage, such as wrinkling.

The ozone layer About 20 to 50 km above Earth's surface is a region called the ozone layer. Ozone is a molecule composed of three oxygen atoms. The ozone layer is vital to life on Earth because it absorbs most of the Sun's harmful ultraviolet waves and prevents them from reaching Earth's surface, as shown in **Figure 14.**

■ **Figure 13** Ultraviolet light is used to kill bacteria in the water supply to make it safe for drinking. This purifier also gives off visible blue light to let the user know that the device is working.

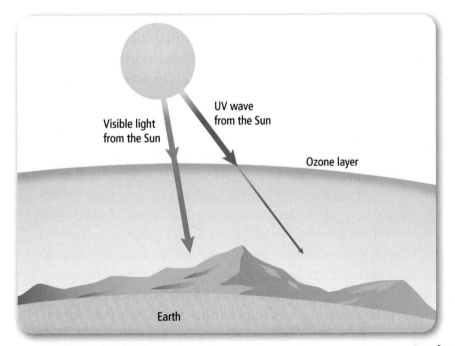

Visible light from the Sun

UV wave from the Sun

Ozone layer

Earth

■ **Figure 14** The ozone layer absorbs most of the Sun's UV waves. As a result, few UV waves reach Earth's surface where they are harmful to organisms.

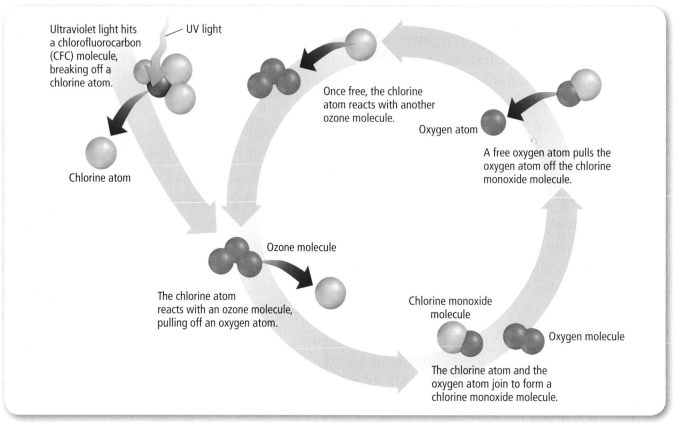

Ultraviolet light hits a chlorofluorocarbon (CFC) molecule, breaking off a chlorine atom.

UV light

Chlorine atom

Once free, the chlorine atom reacts with another ozone molecule.

Oxygen atom

A free oxygen atom pulls the oxygen atom off the chlorine monoxide molecule.

Ozone molecule

The chlorine atom reacts with an ozone molecule, pulling off an oxygen atom.

Chlorine monoxide molecule

Oxygen molecule

The chlorine atom and the oxygen atom join to form a chlorine monoxide molecule.

■ **Figure 15** Ultraviolet radiation breaks apart CFCs, which produces single chlorine atoms. The highly-reactive chlorine atoms destroy ozone molecules.

Concepts in Motion   Animation

■ **Figure 16** X-rays pass through soft tissue, such as skin and muscle, but they are absorbed by the denser bones. The image of a bone on an X-ray is the shadow cast by the bone.

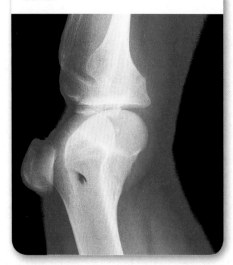

Damage to the ozone layer   The ozone layer changes naturally with the seasons. But in the 1980s and early 1990s, scientists noticed an overall decrease in the amount of ozone in the ozone layer. Averaged globally, the decrease is about four percent, but it is greater at higher latitudes. Many scientists think that certain chemicals, such as chlorofluorocarbons (CFCs), caused the reduction of ozone in the ozone layer.

CFCs were widely used in air conditioners, refrigerators, and cleaning fluids, and some were released into the air. CFCs can rise over time into the upper regions of Earth's atmosphere. When CFCs reach the ozone layer, they might react with ozone molecules, as illustrated in **Figure 15**. One chlorine atom from a CFC molecule can break apart thousands of ozone molecules. To prevent the loss of more ozone, many countries worked together to reduce the use of CFCs and other ozone-depleting substances.

X-rays

Electromagnetic waves with wavelengths between about ten-billionths of a meter and ten-trillionths of a meter are called **X-rays.** X-rays have shorter wavelengths than UV waves and their photons have larger energies. X-rays penetrate skin and soft tissue but not denser materials, such as teeth and bones. Doctors and dentists use low doses of X-rays to form images, like the one in **Figure 16,** of bones and teeth. X-rays are also used in airport screening devices to examine the contents of luggage.

Gamma Rays

Electromagnetic waves with wavelengths shorter than about 100-trillionths of a meter are called **gamma rays.** Gamma rays have high frequencies and have the highest-energy photons. They have enough energy to penetrate several centimeters of lead. Gamma rays are produced by processes that occur in the nuclei of atoms.

Both X-rays and gamma rays are used in a technique called radiation therapy to kill diseased cells in the human body. A beam of X-rays or gamma rays can damage the biological molecules in living cells, causing both healthy and diseased cells to die. By carefully controlling the amount of X-ray or gamma ray radiation and focusing it on the diseased area, the damage to healthy cells can be reduced during treatment. **Figure 17** shows a patient receiving radiation to treat cancer. The gamma rays are focused on the tumor and kill the cancer cells, while doing little damage to the surrounding healthy cells.

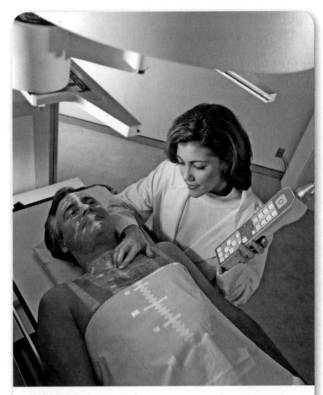

■ **Figure 17** This patient is undergoing radiation therapy. Cancer cells can be killed with carefully controlled beams of X-rays and gamma rays.

Section 2 Review

Section Summary

▶ The entire range of frequencies of electromagnetic waves is called the electromagnetic spectrum.

▶ Radio waves and microwaves have the longest wavelengths.

▶ All objects emit infrared waves.

▶ The human eye can detect visible light.

▶ Ultraviolet waves, X-rays, and gamma rays are both helpful and harmful to humans.

7. **MAIN Idea Compare and contrast** the properties and uses of radio waves, infrared waves, and ultraviolet rays.

8. **Explain** A mug of tea is heated in a microwave oven. Explain why the tea gets hotter than the mug.

9. **Identify** the beneficial effects and the harmful effects of human exposure to ultraviolet rays.

10. **Name** three objects in a home that produce electromagnetic waves, and describe how the electromagnetic waves are used.

11. **Think Critically** How could infrared imaging be used to find a lost hiker?

Apply Math

12. **Use Scientific Notation** Express the range of wavelengths corresponding to visible light, ultraviolet waves, and X-rays in scientific notation.

13. **Convert Units** A nanometer, abbreviated nm, equals one-billionth of a meter, or 10^{-9} m. Express the range of wavelengths corresponding to visible light, ultraviolet waves, and X-rays in nanometers.

Reading Preview

Essential Questions

▶ How are carrier waves modulated to transmit information?

▶ What is the difference between amplitude modulation and frequency modulation?

▶ What technologies use radio waves and microwaves for communication?

Review Vocabulary

amplitude: a measure of the size of the disturbance a wave produces

New Vocabulary

carrier wave
modulation
analog signal
digital signal
transceiver
Global Positioning System (GPS)

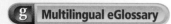 **Multilingual eGlossary**

Radio Communication

MAIN ⟨Idea⟩ Radio waves and microwaves can be used to transmit signals and information.

Real-World Reading Link When listening to the radio, you may have noticed that each radio station is labeled with a number. This number is not just randomly assigned. It is the frequency of the radio wave that the station broadcasts.

Radio Transmissions

Each radio station uses an assigned frequency to avoid interfering with other radio broadcasts. Television stations and cell phone companies are also assigned specific frequencies. These frequency ranges are shown in **Figure 18.** The remaining radio frequencies are assigned for other purposes, such as navigation and radio astronomy.

Changing the channel on your radio or television allows you to select a particular frequency carrying the information you want to listen to or watch. An electromagnetic wave with the specific frequency that a station is assigned is called a **carrier wave.**

Modulation The station must do more than simply transmit a carrier wave. It must also send information about the sounds that you are to receive. The sounds produced at the radio station are converted into electric signals. This electric signal is called the signal wave and is used to modify the carrier wave. The process of adding the signal wave to the carrier wave is called **modulation.** There are two ways to modulate carrier waves: amplitude modulation (AM) and frequency modulation (FM).

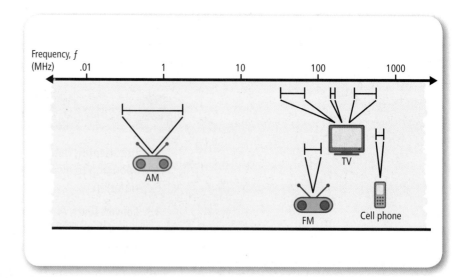

■ **Figure 18** To avoid interference, cell phones, TVs, and radios broadcast at assigned frequencies between 500,000 Hz and 1 billion Hz.

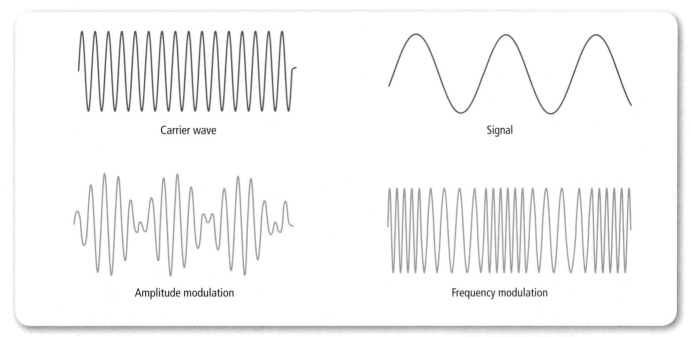

Carrier wave

Signal

Amplitude modulation

Frequency modulation

AM radio An AM radio station broadcasts information by varying the amplitude of the carrier wave, as shown at the left in **Figure 19**. AM carrier wave frequencies range from 540,000 to 1,600,000 Hz.

FM radio In FM radio signals, the signal wave is used to vary the frequency of the carrier wave, as shown at the right in **Figure 19**. Because the strength of the FM waves is kept fixed, FM signals tend to be more clear than AM signals. FM carrier frequencies range from 88 million to 108 million Hz. These frequencies are much higher than AM frequencies.

 **Reading Check** **Compare and contrast** AM and FM radio signals.

Broadcasting radio waves The modified carrier wave is converted from an electric signal to a radio wave by using an antenna, like the one in **Figure 20**. The electric signal causes electrons in the antenna to vibrate. These vibrating electrons create electromagnetic waves that travel outward from the antenna in all directions.

The signal from the radio station is strongest closer to the broadcasting antenna and becomes weaker as you move away. Eventually, the signal will be too weak to be detected by your radio. This is why radios in New York City do not pick up FM radio stations broadcast in Los Angeles. Bad weather, surrounding mountains, and artificial structures can also interfere with radio transmissions.

Reading Check **Describe** how a radio signal's strength changes as you move away from the tower.

■ **Figure 19** The carrier wave is a wave with the frequency assigned to the station. The signal wave contains information about what you will hear. Modulation is the process of adding the signal wave to the carrier wave. A carrier wave can be modulated with amplitude modulation (AM) or frequency modulation (FM).

■ **Figure 20** In a broadcast antenna, vibrating electrons produce an electromagnetic wave that travels outward in all directions.

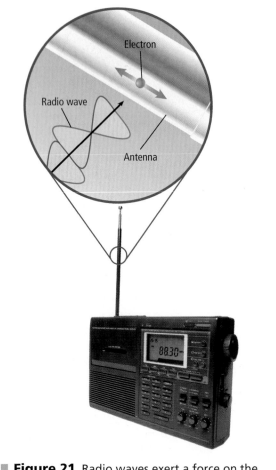

Figure 21 Radio waves exert a force on the electrons in a receiving antenna, causing the electrons to vibrate. The radio then filters out the carrier wave and converts the signal to a sound wave for you to hear.

Receiving radio waves As electromagnetic waves pass by your radio's antenna, the electrons in the metal vibrate, as illustrated in **Figure 21.** These vibrating electrons produce a changing electric current that contains the information about the music and words. This current is used to make the speakers vibrate, creating the sound waves that you hear.

The Digital Revolution

Until the early 21st century, information was sent to TV sets in the same way it was sent to radios. The audio information was sent using FM, and the visual information was sent using AM. The information signals were analog signals. **Analog signals** are electric signals whose values change smoothly over time.

In 2009, full-power television stations in the United States began broadcasting only digital signals. A **digital signal** is an electric signal where there are only two possible values: ON and OFF. This is similar to a light switch where the light can be on or off, but it cannot be half-on or half-off.

There are many ways to modulate radio waves using this on-and-off information. The simplest methods, however, resemble traditional AM and FM and are called Amplitude-Shift Keying (ASK) and Frequency-Shift Keying (FSK). These types of digital modulation are shown in **Figure 22.** More complex ways of digital modulation allow more information to be carried by a single wave. In the United States, television stations use multiple amplitude modulations to encode data on the carrier wave.

Figure 22 A digital signal can be used to modulate the amplitude or frequency of the carrier wave.

Compare and contrast *amplitude-shift keying with amplitude modulation.*

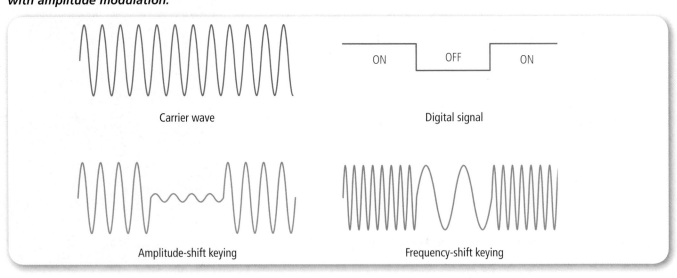

Carrier wave

Digital signal

Amplitude-shift keying

Frequency-shift keying

Telephones

Just a few decades ago, telephones had to be connected with wires. Today, cell phones are seen everywhere. When you speak into a telephone, a microphone converts the sound waves into an electric signal. In a cell phone, this signal is transmitted to and from microwave towers using microwaves or radio waves. The towers, like the ones in **Figure 23,** are several kilometers apart and each covers an area called a cell. If you move from one cell to another, an automated control station transfers the signal to the new cell and its tower.

Transceivers A cell phone is a transceiver. A **transceiver** transmits one radio signal and receives another radio signal. Using two signals with different frequencies allows you to talk and listen at the same time without interference.

Cordless telephones are also transceivers. However, you must remain close to the base unit when using a cordless phone. Another drawback is that if someone nearby is using a cordless telephone at the same frequency, you could hear that conversation on your phone. For this reason, many cordless phones have a channel button that allows you to switch to another frequency.

Pagers Some hospitals ban cell phone use because there are concerns that transceivers might interfere with medical equipment. So, many doctors carry small, portable radio receivers called pagers. To contact the doctor, a caller leaves a callback number or a text message at a central terminal. The message is changed into an electronic signal and transmitted by radio waves along with the identification number of the desired pager. The pager receives all messages transmitted at its assigned frequency, but it only responds to messages with its identification number. Restaurants also use pagers, like the one in **Figure 24,** to notify customers that their tables are ready.

■ **Figure 23** The antennae on these microwave towers are transceivers that send signals to and receive signals from nearby cell phones.

■ **Figure 24** Each pager at a restaurant responds to its own assigned frequency to alert customers that a table is ready.

Figure 25 Communications satellites are transceivers. They receive signals at one frequency and send signals at a different frequency. The solar panels on either side of the satellite allow the satellite to obtain energy from the Sun.

VOCABULARY ....

WORD ORIGIN

Satellite
comes from the Latin word *satelles* which means "attendant"
The Moon is a satellite that travels around Earth. ....

? Inquiry   Video Lab

Communications Satellites

Since satellites were first developed, thousands have been launched into Earth's orbit. Many of these, like the one in **Figure 25,** are used for communication. The sender broadcasts a microwave signal to the satellite. The satellite receives the signal, amplifies it, and transmits it to a particular region on Earth. Like cell phones, satellites are transceivers. To avoid interference, the satellite receives signals at one frequency and broadcasts signals at a different frequency.

Satellite telephone systems Some mobile telephones can be used when sailing across the ocean, even though there are no nearby cell phone towers. The telephone transmits the signal directly to a satellite. The satellite relays the signal to a ground station, and the call is passed on to the telephone network. Satellite links work well for one-way transmissions, but two-way communications can have a delay caused by the large distance the signals travel to and from the satellite.

Television satellites The satellite-reception dishes that you sometimes see in yards or attached to houses are receivers for television satellite signals. Satellite television is used as an alternative to ground-based transmission. Communications satellites use microwaves rather than the radio waves used for normal television broadcasts. Microwaves have shorter wavelengths and travel more easily through the atmosphere. The ground receivers are dish-shaped to help focus the microwaves onto an antenna.

The Global Positioning System

Getting lost while hiking is not uncommon; but if you are carrying a Global Positioning System receiver, it is much less likely to happen. The **Global Positioning System (GPS)** is a system of satellites, ground monitoring stations, and receivers that determine your exact location at or above Earth's surface. The 24 satellites necessary for 24-hour, around-the-world coverage became fully operational in 1995. **Figure 26** illustrates how these satellites are arranged in orbit. Signals from four satellites are used to determine the location of an object using a GPS receiver.

GPS satellites are owned and operated by the United States Department of Defense, but the microwave signals they send out can be used by anyone. Several other countries are working to develop similar systems. Airplanes, ships, cars, and cell phones can use GPS for navigation. Some pet collars also contain GPS receivers. If the pet runs away or is lost, the GPS receiver in the collar can be used to locate the animal.

■ **Figure 26** A GPS receiver uses signals from four of the 24 orbiting satellites to determine the receiver's location.

((○)) **Concepts in Motion** Animation

Section 3 Review

Section Summary

▶ Radio stations transmit radio waves that receivers convert to sound waves.

▶ Modulation is the process of adding the information signal to the carrier wave.

▶ Telephones contain transceivers and convert sound waves into electric signals and electric signals into sound waves.

▶ Microwave towers and satellites are used to transmit telephone signals.

▶ The Global Positioning System uses a system of satellites to determine your exact position.

14. **MAIN Idea Identify and describe** the steps that a radio station uses to broadcast sounds to your radio receiver.

15. **Explain** the difference between AM and FM radio. Make a sketch of how a carrier wave is modulated in AM and FM radio signals.

16. **Describe** what happens to your signal when you are talking on a cell phone and you travel from one cell to another cell.

17. **Explain** some of the uses of the Global Positioning System. Why might emergency vehicles be equipped with GPS receivers?

18. **Think Critically** Why do cordless telephones stop working when you move too far from the base unit?

Apply Math

19. **Calculate a Ratio** A TV screen is composed of many points of light called pixels. A standard TV has 460 pixels horizontally and 360 pixels vertically. A high-definition TV has 1,920 horizontal and 1,080 vertical pixels. What is the ratio of the number of pixels in a high-definition TV to the number in a standard TV?

USE THE INTERNET

Objectives

- **Research** which frequencies are used by different radio stations.
- **Observe** the relationship between broadcasting power and the range that the signal reaches.
- **Make a chart** of your findings and communicate them to other students.

Background: The signals from many radio stations broadcasting at different frequencies are hitting your radio's antenna at the same time. When you tune to your favorite station, the electronics inside your radio amplify the signal at the frequency broadcast by the station.

While you are listening to a radio station in your car, you might have noticed sometimes the station gets fuzzy and you will hear another station at the same time. Sometimes you lose the station completely.

Questions: *How far away from the radio station can you travel before that happens? Does the distance vary depending on the station you listen to? Which type of signal, AM or FM, has a greater range?*

Preparation

Data Source

Use approved Internet sites to find information about radio stations in your area.

The Federal Communications Commission's Web site provides official technical data.

Make a Plan

1. Read the procedure and safety information, and complete the lab form.
2. Explore the frequencies used by local AM and FM radio stations by tuning a radio. Select two AM and two FM stations to research for more information.
3. Copy the table below, and record the frequency for each station. Write down any other information that you know about the station, such as its call letters or its broadcast location.
4. Evaluate how well the signal can be detected. Is it strong and clear, or does it come and go?
5. Hypothesize how distance between the broadcast antenna and your radio affects the strength of the signal.

Data Table

	1st FM Station	2nd FM Station	1st AM Station	2nd AM Station
Frequency				
Call letters				
How well is the signal received?				
Broadcast location				
Broadcast power				
Distance between broadcast antenna and your radio				
Coverage area of station				

6. Form a hypothesis about how the broadcast power affects the size of the radio station's broadcast area. The broadcast power is the power of the signal that the radio station broadcasts.

7. Determine what information you will need to find to check your two hypotheses.

Follow Your Plan

1. Make sure your teacher approves your plan before you start.

2. Research each of your radio stations using approved Web sites.

3. Visit the Federal Communications Commission's Web site to find official data on each station. You will need to go to the Media Bureau Home Page Audio Division.

4. Use the FCC's FM Query option to find information about each of your FM stations. On the information page, locate the location, signal strength, and call letters of each station.

5. Locate the Service Contour Map option, and use the map there to find the radius of each station's coverage area.

6. Use the FCC's AM Query option to find information about each of your AM stations. Find each station's location, signal strength, and call letters.

7. Record your data in your data table.

Analyze Your Data

1. **Make** a map of the radio stations that you researched. Mark the location of each broadcast antenna and your location.

2. **Draw** the approximate coverage areas for each FM station.

3. **Describe** how the broadcast power affects the coverage area of the station.

4. **Describe** how the distance between the broadcast antenna and your radio affects the signal strength.

Conclude and Apply

1. **Compare** your findings to those of your classmates. Did they find the same relationship between broadcast power and coverage area that you did?

2. **Identify** other possible reasons for how well a station can be received.

3. **Infer** If you wanted to build a radio station that uses the same frequency and power as one of your FM stations, how close could you place it? Mark a possible location on your map.

COMMUNICATE YOUR DATA

Write a proposal for your new FM radio station. Include the location, frequency, and power of the station. Present your proposal to your class. As a class, evaluate each proposal.

How SCIENCE Works

A New Kind of Rays

Have you ever broken a bone? Visited the dentist? If so, you were exposed to a powerful, invisible form of electromagnetic radiation. Well-understood and widely applied today, this phenomenon was once so mysterious that it was simply called "X"—a mathematical symbol representing something unknown. The discovery of X-rays shocked and inspired scientists around the world.

A ghostly, green glow On a November afternoon in 1895, German physicist William Roentgen was studying the effects of passing electric current through gas in a device called a cathode ray tube. To eliminate light that would interfere with his experiment, Roentgen covered the tube with thick black paper. Roentgen observed something he could not explain: a screen, which was coated with chemicals, was shimmering with a green glow.

Why did the chemical-coated screen glow? Roentgen knew it was impossible for light produced by the cathode ray tube to penetrate the heavy paper and travel the distance to the screen. He concluded that some unknown, invisible ray—quickly dubbed "X"—had penetrated the glass walls of the gas tube and the heavy black paper, crossed the air, and contacted the screen, causing it to glow.

Famous fingers Roentgen conducted weeks of experiments and was amazed to discover that the powerful X-rays could pass through wood, skin, most metal objects, and even the thick walls of his laboratory. In late December, Roentgen recorded an X-ray image of his wife Bertha's hand. The now famous photo is shown in **Figure 1.**

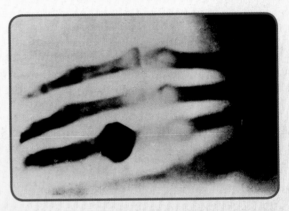

Figure 1 The image of Bertha Roentgen's hand clearly shows the shadows cast by her bones and rings as the X-rays traveled through her flesh.

In December 1895, Roentgen published his findings in a paper entitled "On a New Kind of Rays." Scientists worldwide quickly and enthusiastically duplicated, expanded upon, and applied Roentgen's findings.

Immediate application A few months later, 14-year-old Eddie McCarthy fell while ice skating in New Hampshire and injured his left wrist. Physicians recognized the potential of X-rays to study internal injuries and used X-rays to photograph the fracture. The medical application of X-ray technology was born. In a relatively brief time, researchers unveiled the once-hidden properties of X-ray radiation. Their discoveries created a powerful tool that revolutionized medicine and scientific research.

WebQuest **Investigate and Analyze** Research ways that X-ray technology is applied in the medical field. Create a spreadsheet showing several specific uses of the technology, the branch or branches of medicine to which each applies, and a description of the benefits of using X-ray technology in each setting.

THEME FOCUS Technology

Electromagnetic waves are essential for modern medicine and communications. X-rays and MRI allow us to see inside the human body. Radio waves and microwaves carry information and allow us to communicate over long distances.

BIG Idea Electromagnetic waves transfer energy through matter and through space.

Section 1 What are electromagnetic waves?

electromagnetic wave (p. 338)
photon (p. 342)

MAIN Idea Vibrating electric charges produce electromagnetic waves.

- An electromagnetic wave consists of a vibrating electric field and a vibrating magnetic field.
- Electromagnetic waves carry radiant energy.
- In empty space, electromagnetic waves travel at 300,000 km/s, the speed of light.
- Electromagnetic waves travel more slowly in matter, with a speed that depends on the material.
- Electromagnetic waves can behave as particles called photons.

Section 2 The Electromagnetic Spectrum

gamma ray (p. 351)
infrared wave (p. 348)
microwave (p. 347)
radio wave (p. 345)
ultraviolet wave (p. 349)
visible light (p. 348)
X-ray (p. 350)

MAIN Idea The electromagnetic spectrum is divided into several sections, each with a certain range of frequencies and specific properties.

- The entire range of frequencies of electromagnetic waves is called the electromagnetic spectrum.
- Radio waves and microwaves have the longest wavelengths.
- All objects emit infrared waves.
- The human eye can detect visible light.
- Ultraviolet waves, X-rays, and gamma rays are both helpful and harmful to humans.

Section 3 Radio Communication

analog signal (p. 354)
carrier wave (p. 352)
digital signal (p. 354)
Global Positioning System (GPS) (p. 357)
modulation (p. 352)
transceiver (p. 355)

MAIN Idea Radio waves and microwaves can be used to transmit signals and information.

- Radio stations transmit radio waves that receivers convert to sound waves.
- Modulation is the process of adding the information signal to the carrier wave.
- Telephones contain transceivers and convert sound waves into electric signals and electric signals into sound waves.
- Microwave towers and satellites are used to transmit telephone signals.
- The Global Positioning System uses a system of satellites to determine your exact position.

Review Vocabulary eGames

Use Vocabulary

Complete each sentence with the correct term from the Study Guide.

20. _____ are the type of electromagnetic waves that are used in radar and MRI.

21. A remote control uses _____ to communicate with a television set.

22. If you stay outdoors too long, your skin might be burned by exposure to _____ from the Sun.

23. A radio station broadcasts a radio wave called a(n) _____ that has the specific frequency assigned to the station.

24. A(n) _____ is a device that transmits at one frequency and receives at another.

25. Transverse waves that are produced by vibrating electric charges and consist of vibrating electric and magnetic fields are _____.

Check Concepts

26. Which of the following describes X-rays?
A) short wavelength, high frequency
B) short wavelength, low frequency
C) long wavelength, high frequency
D) long wavelength, low frequency

27. Which type of electromagnetic wave has wavelengths greater than about 10 cm?
A) ultraviolet waves
B) radio waves
C) gamma rays
D) X-rays

28. Electromagnetic waves can behave like particles called _____.
A) electrons   C) photons
B) molecules   D) atoms

29. Which type of electromagnetic wave enables skin cells to produce vitamin D?
A) infrared waves   C) visible light
B) ultraviolet waves D) X-rays

Use the figure below to answer questions 30 and 31.

The Electromagnetic Spectrum						
← Low frequency					High frequency →	
(A)	Microwaves	(B)	Visible light	(C)	X-rays	(D)

30. Which type of electromagnetic wave is the correct label for section B?
A) ultraviolet waves C) radio waves
B) gamma rays   D) infrared waves

31. Which type of electromagnetic wave is the correct label for section D?
A) ultraviolet waves C) radio waves
B) gamma rays   D) infrared waves

32. Which of the following is modified in an AM radio wave?
A) speed   C) amplitude
B) frequency   D) wavelength

33. What is the name of the ability of some materials to absorb ultraviolet light and re-emit it as visible light?
A) modulation   C) transmission
B) interference   D) fluorescence

34. Which of these colors of visible light has the shortest wavelength?
A) violet   C) red
B) yellow   D) green

✓ Assessment   **Online Test Practice**

Interpret Graphics

35. Copy and complete the following table about the electromagnetic spectrum.

Uses of Electromagnetic Waves	
Type of Electromagnetic Waves	Examples of How Electromagnetic Waves Are Used
a.	radio, TV transmission
Microwaves	b.
Infrared waves	c.
Visible light	vision
d.	fluorescent materials
X-rays	e.
f.	destroying harmful cells

36. Copy and complete the following events chain about the destruction of ozone molecules in the ozone layer by CFC molecules.

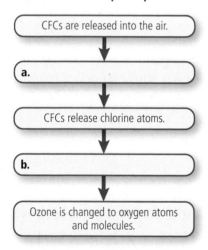

CFCs are released into the air.

↓

a.

↓

CFCs release chlorine atoms.

↓

b.

↓

Ozone is changed to oxygen atoms and molecules.

Think Critically

37. **BIG Idea Predict** whether an electromagnetic wave would travel through space if its electric and magnetic fields were not changing with time. Explain your reasoning.

38. **Explain** why X-rays are used in medical imaging.

39. **THEME FOCUS Classify** Look around your home, school, and community. Make a list of the devices that use electromagnetic waves. Beside each device, write the type of electromagnetic wave that the device uses.

40. **Form a hypothesis** to explain why communications satellites do not use ultraviolet waves to receive information and transmit signals to Earth's surface.

41. **Compare** the energy of photons corresponding to infrared waves with the energy of photons corresponding to ultraviolet waves.

42. **Determine** Under what conditions will an electromagnetic wave travel slower than the speed of light? At the speed of light? Faster than the speed of light? Explain.

43. **Compare and contrast** sound waves and electromagnetic waves.

Apply Math

44. **Use Fractions** When visible light waves travel in ethyl alcohol, their speed is three-fourths the speed of light in air. What is the speed of light in ethyl alcohol?

45. **Use Scientific Notation** The speed of light in a vacuum has been determined to be 299,792,458 m/s. Express this number using scientific notation and round to four total digits.

46. **Calculate** A radio wave has a frequency of 600,000 Hz and travels at a speed of 300,000 km/s. Use the wave speed equation to calculate the wavelength of the radio wave. Express your answer in meters.

Standardized Test Practice

Multiple Choice

Record your answers on the answer sheet provided by your teacher or on a sheet of paper.

1. Which of the following produces electromagnetic waves?
   A. vibrating electric charge
   B. direct current
   C. static electric charge
   D. constant magnetic field

Use the figure below to answer questions 2 and 3.

2. What type of signal is shown above?
   A. analog signal
   B. carrier signal
   C. digital signal
   D. modulated signal

3. Which type of broadcast would be most likely to use this signal?
   A. AM radio
   B. FM radio
   C. newspapers
   D. television

4. Which term refers to the energy carried by an electromagnetic wave?
   A. kinetic energy
   B. potential energy
   C. radiant energy
   D. nuclear energy

5. The frequency of an electromagnetic wave has which unit?
   A. newtons
   B. hertz
   C. nanometers
   D. meters/second

6. Which explains how interference is avoided between the signals that cordless phones receive and the signals that they broadcast?
   A. The signals travel at different speeds.
   B. The signals have different amplitudes.
   C. The signals have different frequencies.
   D. The signals are only magnetic.

7. Which person explained how light can behave as a particle, called a photon, whose energy depends on the frequency of light?
   A. Einstein   C. Newton
   B. Hertz   D. Galileo

Use the table below to answer questions 8 and 9.

Regions of the Electromagnetic Spectrum		
Infrared waves	Radio waves	Gamma rays
X-rays	Visible light	Ultraviolet waves

8. If you arranged the list of electromagnetic waves shown above in order from shortest to longest wavelength, which would be first on the list?
   A. radio waves   C. gamma rays
   B. X-rays   D. visible light

9. Which region of the electromagnetic spectrum is not listed in the table above?
   A. alpha rays   C. water waves
   B. sound waves   D. microwaves

10. The warmth that you feel from a fire is thermal energy transmitted to you by which type of electromagnetic waves?
    A. X-rays
    B. microwaves
    C. ultraviolet waves
    D. infrared waves

✓ Assessment   Standardized Test Practice

Short Response

Record your answers on the answer sheet provided by your teacher or on a sheet of paper.

Use the figure below to answer questions 11 and 12.

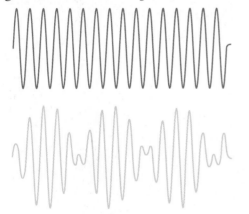

11. The illustration above shows two radio waves broadcast by a radio station. What is the upper, unmodulated wave called?

12. The lower figure shows the same wave after it was modulated to carry sound information. What type of modulation does it show?

13. What distance would an electromagnetic wave in a vacuum travel in one year?

14. Even on a cloudy day, you can get sunburned outside. What does this tell you about ultraviolet waves?

15. Which type of radio station transmits radio waves at higher frequencies: AM stations or FM stations?

Extended Response

Record your answers on a sheet of paper.

16. A CD player converts the musical information on a CD to a varying electric current. Describe how the varying electric current produced by a CD player in a radio station is converted into radio waves.

17. Explain how an electromagnetic wave that strikes a material transfers radiant energy to the atoms in the material.

18. How would a change in the amount of ozone in the ozone layer affect the amount of the ultraviolet and visible light waves emitted by the Sun that reach Earth's surface?

19. Explain why all objects emit electromagnetic waves.

Use the figure below to answer question 20.

20. The illustration above shows microwaves interacting with water molecules in food. How does the electric field in microwaves affect water molecules?

21. Describe how thermal energy inside food is produced by microwaves interacting with water molecules.

NEED EXTRA HELP?																					
If You Missed Question . . .	1	2	3	4	5	6	7	8	9	10	11	12	13	14	15	16	17	18	19	20	21
Review Section . . .	1	3	3	1	1	3	1	2	2	2	3	3	1	2	3	3	1	2	1	2	2

CHAPTER 12
Light

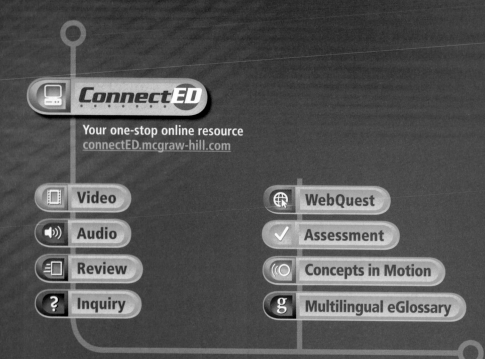

ConnectED

Your one-stop online resource
connectED.mcgraw-hill.com

- Video
- Audio
- Review
- Inquiry
- WebQuest
- Assessment
- Concepts in Motion
- Multilingual eGlossary

Launch Lab
Rainbows of Light

Light from the Sun looks white on a sunny day. Yet, you know that if the Sun shines at the proper angle into a rain shower, a rainbow can form. Is white light really white?

For a lab worksheet, use your StudentWorks™ Plus Online.

? Inquiry Launch Lab

FOLDABLES®

Make a two-tab book. Label it as shown. Use it to organize your notes on producing and using light.

Producing Light | Using Light

THEME FOCUS Energy
Light waves carry radiant energy.

BIG Idea Visible light waves are electromagnetic waves that enable us to see.

Reading Preview

Essential Questions

▶ How are transparent, translucent, and opaque materials different?

▶ What is the difference between regular and diffuse reflection?

▶ What is the index of refraction of a material?

▶ Why does a prism separate white light into different colors?

Review Vocabulary

visible light: electromagnetic waves with wavelengths between 700- and 400- billionths of a meter that can be detected by the human eye

New Vocabulary

opaque
translucent
transparent
index of refraction
mirage

 **g** **Multilingual eGlossary**

■ **Figure 1** Different materials interact differently with light. Materials can absorb, reflect, scatter, and transmit light.

The Behavior of Light

MAIN Idea Materials can absorb, transmit, or reflect light waves.

Real-World Reading Link Have you ever been in a cave or a room where there was no light at all? What did you see? Nothing! Without visible light, we cannot see anything.

Light and Matter

Look around your darkened room at night. After your eyes adjust to the darkness, you can see some familiar objects. Brightly colored objects look gray or black in the dim light. If you turn on the light, however, you might see all the objects in the room, including their colors. What you see depends on the amount of light in the room and the color of the objects. To see an object, it must reflect some light into to your eyes.

Opaque, translucent, and transparent Objects can absorb, reflect, and transmit light. Objects that transmit light allow light to pass through them. An object's material determines the amount of light it absorbs, reflects, and transmits. The material in the first candleholder in **Figure 1** is opaque (oh PAYK). **Opaque** materials only absorb and reflect light; no light passes through them. As a result, you cannot see the candle.

Some materials, such as the second candleholder in **Figure 1,** are translucent (trans LEW sunt). **Translucent** materials transmit light but also scatter it. You cannot see clearly through translucent materials, and objects appear blurry.

The third candleholder shown in **Figure 1** is transparent. **Transparent** materials transmit light without scattering it, so you can see objects clearly through them.

Opaque

Translucent

Transparent

Reflection of Light

Just before you left for school this morning, did you glance in a mirror to check your appearance? For you to see your reflection in the mirror, light had to reflect off you, hit the mirror, and reflect off the mirror into your eye. Reflection occurs when a light wave strikes an object and bounces off.

The law of reflection Like all waves, light obeys the law of reflection. According to the law of reflection, the angle at which a light wave strikes a surface is the same as the angle at which it is reflected. This law is illustrated in **Figure 2.** Light reflected from any surface—a mirror or a sheet of paper—behaves this way.

Regular and diffuse reflection If light always obeys the law of reflection, why can you see your reflection in a store window but not in a brick wall? The answer involves the smoothness of the surfaces. A smooth, even surface like a pane of glass produces a sharp image by reflecting parallel light waves in only one direction. Reflection of light waves from a smooth surface is regular reflection. A brick wall has an uneven surface that causes incoming parallel light waves to be reflected in many directions, as shown in **Figure 3.** The reflection of light from a rough surface is diffuse reflection. Diffuse reflection does not produce an image.

Reading Check **Identify** some objects that produce regular reflections and some objects that produce diffuse reflections.

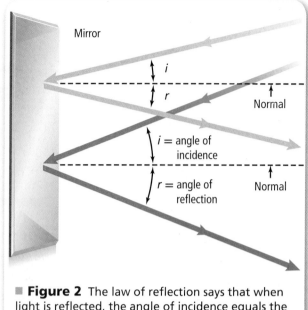

■ **Figure 2** The law of reflection says that when light is reflected, the angle of incidence equals the angle of reflection.

Concepts in Motion Animation

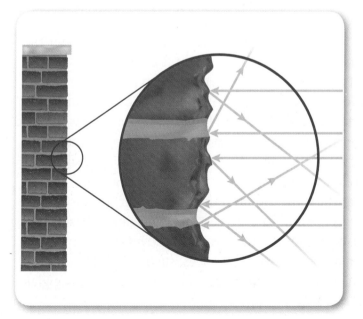

■ **Figure 3** The uneven surface of this brick wall produces a diffuse reflection.

Explain *Use the law of reflection to explain why a rough surface causes parallel light waves to be reflected in many directions.*

Figure 4 Surfaces that appear smooth to the human eye are seen to be rough at high magnification. These surfaces produce diffuse reflections.

Microscopic roughness Even a surface that appears to be smooth can be rough enough to cause diffuse reflection. For example, a metal pot might seem to be smooth, but the surface shows rough spots at high magnification, as shown in **Figure 4**. To cause a regular reflection, the sizes of the surface irregularities must be less than the wavelengths of the light that the surface reflects.

Refraction of Light

What occurs when a light wave passes from one material to another—from air to water, for example? Light rays refract, or bend. Refraction is caused by a change in the speed of a wave when it passes from one material to another. If the light wave is traveling at an angle other than 90° to the boundary between the materials and the speed that light travels is different in the two materials, then the wave will bend. If light hits a boundary at 90°, it will change speed but not bend.

✓ **Reading Check Identify** when refraction occurs.

The index of refraction The amount of bending depends on the speed of light in each material. The greater the difference in speeds, the more the light is bent as it crosses the boundary at an angle. Every material has an index of refraction. The **index of refraction** is a property of a material that indicates how much the speed of light in the material is reduced compared to the speed of light in a vacuum. In most materials, the index of refraction also depends on the light's wavelength. Longer wavelengths have smaller indices of refraction.

The larger the index of refraction, the slower the speed of light will be in the material. For example, because glass has a larger index of refraction than air, light moves more slowly in glass than in air. Many useful devices, such as eyeglasses, binoculars, cameras, and microscopes, form images by using glass lenses to refract light.

MiniLab

? Inquiry MiniLab

Observe Refraction in Water

Procedure 🥽 🧤 🧹

1. Read the procedure and safety information, and complete the lab form.
2. Place a **penny** at the bottom of a **short, opaque cup.** Set the cup on a table in front of you.
3. Have a partner slowly slide the cup away from you until you cannot see the penny.
4. Without disturbing the penny or the cup and without moving your position, have your partner slowly pour **water** into the cup until you can see the penny.
5. Reverse roles and repeat the experiment.

Analysis

1. **Describe** what you observed. Explain how this is possible.
2. **Sketch** the light path from the penny to your eye after the water was added.

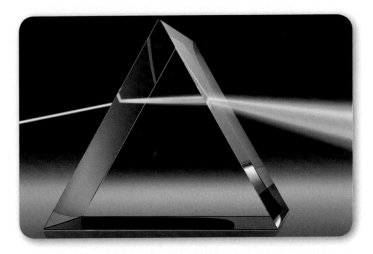

■ **Figure 5** When white light enters a prism, each color of light is bent a different amount. As a result, the white light is split into the different colors of light.

Prisms A sparkling glass prism hangs in a sunny window, refracting the sunlight and projecting a colorful pattern onto the walls of the room. How does the bending of light create these colors? It occurs because the amount of bending usually depends on the wavelength of the light. Wavelengths of visible light range from the longer red waves to the shorter violet waves. White light, such as sunlight, is composed of this whole range of wavelengths.

Figure 5 shows what occurs when white light passes through a prism. The triangular prism refracts the light twice—once when it enters the prism and again when it leaves the prism and reenters the air. Because a longer wavelength of light has a smaller index of refraction, it is refracted less than a shorter wavelength. As a result of these different amounts of bending, the different colors are separated when they emerge from the prism. Red light is bent the least.

✅ **Reading Check** **Predict** which color of light you would expect to bend the most.

Rainbows Does the light leaving the prism in **Figure 5** remind you of a rainbow? Like prisms, rain droplets also refract light. The refraction of the different wavelengths can cause white light from the Sun to separate into the individual colors of visible light, as shown in **Figure 6.** In a rainbow, the human eye can usually distinguish only about seven colors clearly. In order of decreasing wavelength, these colors are red, orange, yellow, green, blue, indigo, and violet.

■ **Figure 6** Water droplets can act as prisms. As white light passes through the water droplet, different wavelengths are refracted by different amounts. This produces the separate colors seen in a rainbow.

Identify *Which color of light is refracted the most as it leaves the water droplet? Which color is refracted the least?*

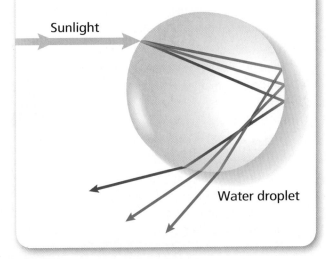

Sunlight

Water droplet

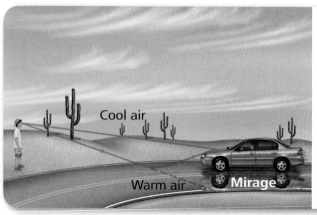

Cool air

Warm air   Mirage

■ **Figure 7** Light waves reflected from an object are refracted when air near the ground is much warmer or cooler than the air above. These refracted waves create one or more additional images.

Mirages You might have seen what looks like the reflection of an oncoming car in a pool of water on the road ahead. As you get closer, the water seems to disappear. You saw a **mirage,** an image of a distant object produced by the refraction of light through air layers of different densities.

Mirages, such as those shown in **Figure 7,** occur when the air at ground level is much warmer or cooler than the air above it. The density of air increases as air cools. Light waves travel slower as the density of air increases, so light travels slower in cooler air. As a result, light waves refract as they pass through air layers with different temperatures. These refracted light waves form additional images of objects.

Section 1 Review

Section Summary

▶ Objects can absorb, reflect, or transmit light.

▶ Light waves always follow the law of reflection.

▶ Reflection can be regular or diffuse.

▶ Refraction occurs when light changes speed in moving from one material to another at an angle to the normal.

▶ Different wavelengths of light are refracted by different amounts.

1. MAIN Idea **Describe** two ways that you could direct a light wave around a corner.

2. **Predict** how rubbing a mirror with sandpaper will affect how the mirror reflects light.

3. **Identify** what an object's index of refraction indicates.

4. **Explain** what happens to white light when it passes through a prism.

5. **Think Critically** Decide whether the lens of your eye, your fingernails, your skin, and your tooth are opaque, translucent, or transparent. Explain.

Apply Math

6. **Find an Angle** A light ray strikes a mirror at an angle of 42° from the surface of the mirror. What angle does the reflected ray make with the normal?

7. **Find an Angle** A ray of light hits a mirror at 27° from the normal. What is the angle between the reflected ray and the normal?

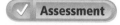
✓ Assessment   Online Quiz

Reading Preview

Essential Questions

▶ How do you see color?

▶ What is the difference between light color and pigment color?

▶ What happens when different colors are mixed?

Review Vocabulary

wavelength: distance between one point on a wave and the nearest point just like it

New Vocabulary

filter
pigment

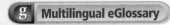
g Multilingual eGlossary

Video BrainPOP

Light and Color

MAIN ⟨Idea⟩ Light waves of different wavelengths or combinations of wavelengths cause the human eye to detect different colors.

Real-World Reading Link You have probably seen an old movie in black and white. That is what our world might look like if our eyes could only distinguish the amount of light present and could not distinguish between different wavelengths of light.

Colors

Why do some apples appear to be red, and others look green or yellow? An object's color depends on the wavelengths of light that it reflects and that our eyes detect. You know that white light is a blend of all colors of visible light. When a red apple is struck by white light, it reflects more red light than green or blue light. **Figure 8** shows white light striking a green leaf. The leaf reflects more green light than other colors and appears green.

Although some objects appear to be black, black is not a color that is present in visible light. Objects that are black absorb all colors of light and reflect little or no light back to your eye. White objects are white because they reflect all colors of visible light.

✓ **Reading Check** **Explain** why a white object is white.

Seeing Color

As you approach a busy intersection, the color of the traffic light changes from green to yellow to red. On the cross street, the color changes from red to green. How do your eyes detect the differences between red, yellow, and green light?

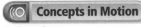
Concepts in Motion
Animation

■ **Figure 8** When white light hits this green leaf, more green light is reflected than red or blue light. The leaf absorbs more red and blue light than green light, so the leaf appears green.

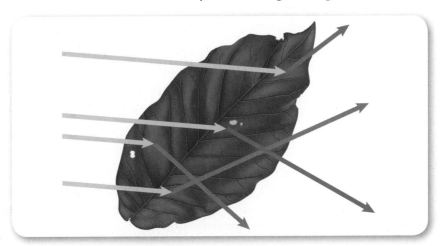

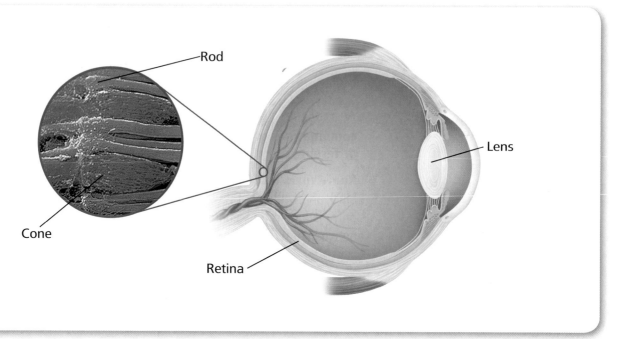

Rod

Lens

Cone

Retina

■ **Figure 9** Light enters the eye and hits the retina. The retina is made up of two types of light-detecting cells. Rod cells are sensitive to dim light. Cone cells detect different wavelengths of light. The cone cells send nerve impulses to the brain, which interprets the different combinations of wavelengths as different colors.

Light and the eye As shown in **Figure 9,** light enters a healthy eye through the lens and is focused on the retina, an area on the inside of your eyeball. The retina is composed of two types of cells that absorb light. When these cells absorb light, chemical reactions convert light's radiant energy into nerve impulses that are transmitted to the brain. One type of cell in the retina, called a cone, allows you to distinguish colors and detailed shapes of objects. Cones are most effective in daytime vision. The second type of cell, called a rod, is sensitive to dim light and is useful for night vision.

Your eyes have three types of cones, each of which responds to a different range of wavelengths. Red cones respond mostly to red and yellow, green cones respond mostly to yellow and green, and blue cones respond mostly to blue and violet.

✔ **Reading Check Identify** the colors of light detected by each type of cone cell.

Interpreting color Why does a banana appear to be yellow? The light reflected by the banana causes the cone cells that are sensitive to red and green light to send signals to your brain. Your brain would get the same signal if a mixture of red light and green light reached your eye. Again, your red and green cones would respond, and you would see yellow light because your brain cannot perceive the difference between incoming yellow light and yellow light produced by combining red and green light. The next time that you are at a play or a concert, look at the lighting above the stage. Observe how the colored lights combine to produce effects onstage.

Color blindness If one or more of your sets of cones did not function properly, you would not be able to distinguish between certain colors. About eight percent of men and one-half percent of women have a form of color blindness. Most people who are said to be color-blind are not truly blind to color, but they have difficulty distinguishing between a few colors, most commonly red and green. **Figure 10** shows an example of a red-green color blindness test. Because these two colors are used in traffic signals, severely color-blind drivers and pedestrians must use the position of the light, instead of color, to know when to stop and go.

Filtering Colors

Wearing tinted glasses changes the color of almost everything that you see. If the lenses are yellow, the world takes on a golden glow. If they are rose-colored, everything looks rosy. Something similar would occur if you placed a colored, transparent plastic sheet over this white page. The paper would appear to be the same color as the plastic. The plastic sheet and the tinted lenses are filters. A **filter** is a transparent material that selectively transmits light. For example, color filters transmit one or more colors of light but absorb all others. The color of a filter is the color of the light that it transmits.

Figure 11 shows what happens when you look at a colored object through various colored filters. On the left of **Figure 11,** a blue bowl looks blue because it primarily reflects blue light and absorbs more of the other colors of light. If you look at the bowl through a blue filter as in the center of **Figure 11,** the bowl still looks blue because the filter transmits the reflected blue light. The right image of **Figure 11** shows how the bowl looks when you examine it through a red filter.

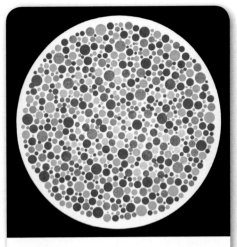

■ **Figure 10** This circle of red and green dots is a common, simple test for red-green color blindness.

Describe *what you see in the dots. What might this tell you about your ability to distinguish between red and green?*

■ **Figure 11** The color of the bowl seems to change when it is viewed through different colored filters.

Explain *Why does the blue bowl appear to be black when viewed through a red filter?*

The bowl appears to be blue in white light.

The bowl appears to be blue when viewed through a blue filter.

The bowl appears to be black when viewed through a red filter.

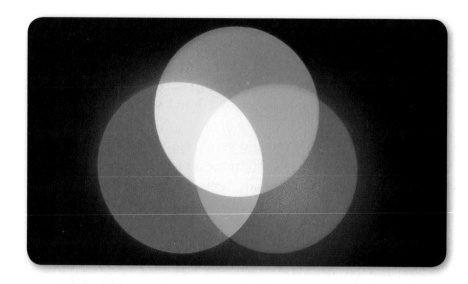

■ Figure 12 The three primary colors of light are red, green, and blue. White light is produced when the three primary colors of light are mixed in equal amounts.

② Inquiry Virtual Lab

Mixing Colors

If you have ever browsed through a paint store, you have probably seen displays where customers can select paint samples of almost every imaginable color. For example, you might have mixed blue and yellow paint to produce green paint. What would happen if you mixed blue and yellow light? Would you get green light?

Mixing colored lights From the glowing orange of a sunset to the deep blue of a mountain lake, all the colors that you see can be made by mixing three colors of light. These three colors—red, green, and blue—are the primary colors of light. They correspond to the three different types of cones in the retina of your eye. When they are mixed together in equal amounts, they produce white light, as **Figure 12** shows. Mixing the primary colors in different proportions produces all the colors that you see.

 Reading Check Identify the primary colors of light.

Paint pigments If you mixed equal amounts of red, green, and blue paint, would you get white paint? If mixing colors of paint were like mixing colors of light, you would. But mixing paint is different. The variety of colors of paint is a result of mixtures of pigments. A **pigment** is a colored material that is used to change the color of other substances.

The color of a pigment results from the different wavelengths of light that the pigment reflects. Paint pigments are usually made of chemical compounds. For example, titanium oxide is a bright white pigment that reflects all colors of light. Another example is lead chromate, which is used to make paint for yellow lines on highways. Other colors can be obtained by mixing various pigments together.

VOCABULARY

WORD ORIGIN
Pigment
from the Latin *pigmentum,* meaning "paint"
The painter made green paint by adding blue and yellow pigments to white paint.

Mixing pigments You can make any pigment color by mixing different amounts of the three primary pigments—magenta (bluish red), cyan (greenish blue), and yellow. In fact, color printers use those pigments to make full-color prints like the pages in this book. However, color printers also use black ink to produce a true black color. A primary pigment's color depends on the color of light that it reflects.

Pigments absorb and reflect a range of colors in sending a single color message to your eye. For example, a yellow pigment viewed in white light appears to be yellow because it reflects yellow, red, orange, and green light but absorbs blue and violet light. The color of a mixture of two primary pigments is determined by the primary colors of light that both pigments reflect.

In **Figure 13,** the area in the center where the colors overlap appears to be black because the three blended primary pigments absorb all the primary colors of light. Recall that the primary colors of light combine to produce white light. They are called additive colors. However, the primary pigment colors combine to produce black. So, the primary pigments are called subtractive colors.

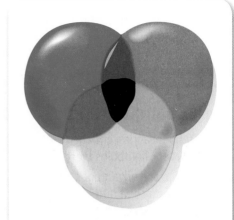

■ **Figure 13** The three primary pigment colors are magenta, yellow, and cyan. When the three primary pigment colors are mixed in equal amounts, they appear to be black.

Section 2 Review

Section Summary

▶ The color of an object depends on the wavelengths of light it reflects.

▶ Rod and cone cells are light-sensitive cells found in the human eye.

▶ The color of a filter is the color of the light that the filter transmits.

▶ All light colors can be created by mixing the primary light colors—red, green, and blue.

▶ All pigment colors can be formed by mixing the primary pigment colors—magenta, cyan, and yellow.

8. **MAIN Idea Explain** why a white fence appears to be white. In your answer, include the colors of light that your eye detects and tell how your brain interprets those colors.

9. **Identify** what color would be seen if equal amounts of red light and green light were mixed.

10. **Compare and contrast** the primary colors of light and the primary pigment colors.

11. **Describe** how your eyes detect color.

12. **Think Critically** Light reflected from an object passes through a green filter, then a red filter, and finally a blue filter. What color will the object appear to be?

Apply Math

13. **Use Percentages** In the human eye, there are about 120,000,000 rods. If 90,000,000 rods trigger at once, what percent of the total number of rods are triggered?

14. **Convert Units** The wavelengths of a color are measured in nanometers (nm), which is 0.000000001 meters (one-billionth of a meter). Find the wavelength in meters of a light wave that has a wavelength of 690 nm.

Reading Preview

Essential Questions

▶ How do incandescent and fluorescent lightbulbs work?

▶ What are the advantages and disadvantages of different lighting devices?

▶ How does a laser produce coherent light?

▶ What are some uses for lasers?

Review Vocabulary

thermal energy: the sum of the kinetic energy and potential energy of a system of particles

New Vocabulary

incandescent light
fluorescent light
coherent light
incoherent light

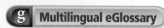
g **Multilingual eGlossary**

? **Inquiry** Video Lab

Figure 14 Fluorescent lightbulbs use phosphors to convert ultraviolet light to visible light. They are commonly used in houses, schools, and offices.

Producing Light

MAIN ⟨Idea⟩ Heating a filament and passing a current through a gas are common ways of producing light.

Real-World Reading Link You might have a lamp beside your bed. How is the lightbulb in that lamp different from the lightbulbs that light your classroom? Do they just look different, or do they produce light in different ways?

Incandescent Lights

Many of the lightbulbs in your house probably produce incandescent light. **Incandescent light** is light generated by heating a piece of metal until it glows. Inside an incandescent lightbulb is a small wire coil, called a filament, that usually is made of tungsten metal. When there is an electric current in the filament, the electric resistance of the metal causes the filament to become hot enough to give off light. However, about 90 percent of the energy given off by an incandescent bulb is in the form of thermal energy.

Fluorescent Lights

Your school might have fluorescent (floo REH sunt) lights, like the ones shown in **Figure 14. Fluorescent light** is light generated by using phosphors to convert ultraviolet radiation to visible light. A phosphor is a substance that absorbs ultraviolet radiation and then emits visible light. A fluorescent bulb is filled with a gas at low pressure and is coated on the inside with phosphors. An electrode is at each end of the tube. Electrons are given off when the electrodes are connected in a circuit. These electrons collide with the gas atoms, which then emit ultraviolet radiation. The wavelength of light emitted depends on the type of gas used. The phosphors absorb this radiation and convert it to visible light.

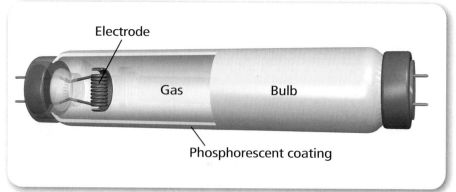
Electrode
Gas
Bulb
Phosphorescent coating

Efficient lighting Fluorescent lights use as little as one-fifth of the electrical energy to produce the same amount of light as incandescent bulbs. Fluorescent bulbs also last much longer than incandescent bulbs. This higher efficiency can mean lower energy costs over the life of the bulb. Reduced energy usage could reduce the amount of fossil fuels that are burned to generate electricity. This would also decrease the amount of carbon dioxide and pollutants released into Earth's atmosphere.

Hospitals, schools, office buildings, and factories have been using long, tube-shaped fluorescent bulbs for many years. Many homes are now using compact fluorescent bulbs, which can be screwed into traditional lightbulb sockets. This gives consumers one way to save energy without having to replace the entire light fixture.

Neon Lights

The vivid, glowing colors of neon lights, such as the one shown in **Figure 15,** make them a popular choice for signs and eye-catching decorations on buildings. These lighting devices are glass tubes filled with gas—often neon—and work similarly to fluorescent lights. When there is an electric current through the tube, electrons collide with the gas molecules. In this case, however, the gas molecules emit visible light. If the tube contains only neon, the light is bright red-orange. Other gases and phosphor coatings are used to make other colors.

✔️ **Reading Check Identify** what causes the color in a neon light.

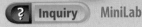

MiniLab

Discover Energy Waste in Lightbulbs

Procedure 🥽 👐 🧤 🧼 ✋

WARNING: *Lightbulbs will be hot. Do not touch the lightbulbs. Keep supplies from touching the lightbulbs.*

1. Read the procedure and safety information, and complete the lab form.
2. Obtain an **incandescent bulb** and a **fluorescent bulb** of identical wattage. Connect each bulb to a **power source** and turn it on.
3. Make a heat collector by covering the top of a **foam cup** with a piece of **plastic food wrap** to make a window. Carefully make a small hole (diameter less than the thermometer's) in the side of the cup. Carefully push a **thermometer** through the hole.
4. Measure the temperature of the air inside the cup. Then hold the window of the tester 1 cm from one of the lights for 2 min and measure the temperature. Use a **stopwatch** to keep time.
5. Cool the heat collector and thermometer. Repeat step 4 using the second bulb.

Analysis

1. **State** the temperature for each bulb.
2. **Evaluate** Which bulb appears to give off more thermal energy? Explain why this occurs.

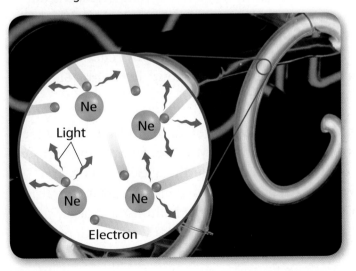

■ **Figure 15** The bright color of this neon light is produced when a current is passed through the neon tube. Electrons collide with neon atoms, causing them to emit a red-orange light.

■ **Figure 16** Tungsten-halogen lightbulbs are often used in car headlights because they are brighter than incandescent lightbulbs.

Identify *some advantages and disadvantages of tungsten-halogen headlights.*

Sodium-Vapor Lights

Sodium-vapor lights are often used for streetlights and other outdoor lighting. Inside a sodium-vapor lamp is a tube that contains a mixture of neon gas, a small amount of argon gas, and a small amount of sodium metal. When the lamp is turned on, the gas mixture becomes hot. The hot gases cause the sodium metal to turn to vapor. The hot sodium vapor emits a yellow-orange glow.

Tungsten-Halogen Lights

Tungsten-halogen lights are sometimes used to create intensely bright light. These lights have a tungsten filament inside a quartz bulb or tube. The tube is filled with a gas that contains one of the halogen elements, such as fluorine or chlorine. The presence of this gas enables the filament to become much hotter than the filament in an ordinary incandescent bulb. As a result, the light is much brighter.

Another advantage of tungsten-halogen bulbs is that they last longer than incandescent bulbs. Their long lifetime is due to the chemical interactions between the halogen gas and the tungsten filament. Tungsten-halogen lights are sometimes used on movie sets and in underwater photography. They are also used in many headlights for cars, as shown in **Figure 16.**

Lasers

From laser surgery to a laser light show, lasers have become a large part of the world in which you live. Lasers can be made with many different materials, including gases, liquids, and solids. The wavelength of the laser depends on the materials used. One of the most common is the helium-neon laser, which produces a beam of red light with a wavelength of 632 billionths of a meter.

Amplifying light In a helium-neon laser, a mixture of helium and neon gases is sealed in a tube with mirrors at both ends. These atoms absorb radiant energy from a flashtube, which is a device that produces a short, bright flash of light. The atoms then release their excess energy by emitting light waves of a certain wavelength. The light waves are emitted in all directions. Most of the waves escape through the sides of the tube, but those traveling along the tube are reflected between the two mirrors. As they travel, these light waves stimulate other atoms to emit more light waves. The result is an increase in the amount of light traveling back and forth along the tube.

To allow a small amount of the light out of the tube, one of the mirrors is coated only partially with reflective material. A narrow, intense laser beam is emitted from the partially reflective end of the tube. **Figure 17** illustrates how a laser creates a beam of light.

■ **Figure 17** Lasers produce narrow, intense beams of light.

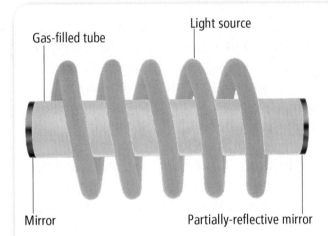

Gas-filled tube
Light source

Mirror
Partially-reflective mirror

A helium-neon laser consists of a glass tube filled with neon and helium. Mirrors are located on each end of the tube. The mirror on the right transmits a small amount of light. The tube is surrounded by a source of intense light.

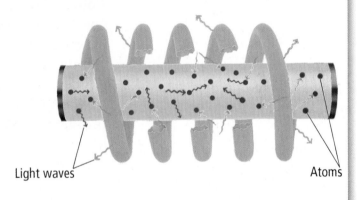

Light waves
Atoms

The light source produces a bright flash of light. This radiant energy is absorbed by the gas atoms inside the tube. The atoms then emit the energy as a single wavelength of light.

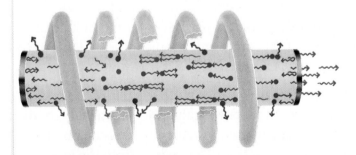

Most of the emitted light waves escape through the sides of the tube and are wasted. However, some waves are reflected back and forth between the mirrors. About 1 percent of light waves that hit the mirror on the right are transmitted.

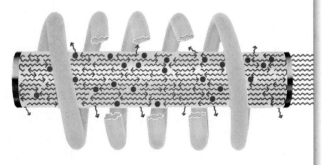

The reflected light can cause other atoms to emit light waves. This results in an increase in the amount of light traveling back and forth in the tube. The laser now emits a narrow, intense beam of light in a single direction.

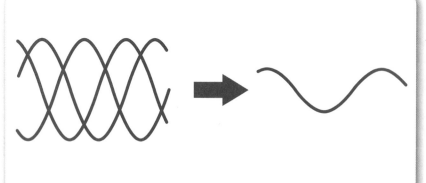

These waves are coherent because they have the same wavelength and they travel in one direction with a constant distance between their corresponding crests. They combine to form a single wave.

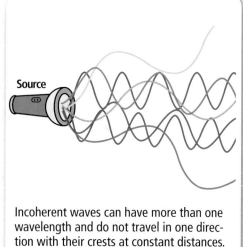

Source

Incoherent waves can have more than one wavelength and do not travel in one direction with their crests at constant distances.

■ **Figure 18** Light waves can be coherent or incoherent.

Coherent and incoherent light The beams from a laser light do not spread out because laser light is coherent. **Coherent light** is light of only one wavelength that travels in one direction with a constant distance between the corresponding crests of the waves. This is illustrated in **Figure 18.** Notice that the coherent waves combine to form a wave with constant wavelength and frequency.

The light from an ordinary lightbulb is incoherent. **Incoherent light** can have more than one wavelength, can travel in more than one direction, and does not travel with a constant distance between the corresponding crests of the waves. This is also illustrated in **Figure 18.** The waves do not travel in the same direction, so the beam spreads out and the energy carried by the light waves is spread over a large area.

Using lasers A laser beam does not spread out as it travels over long distances. Therefore, it can apply large amounts of energy to small areas. Videodisc players, surgical tools, and many other useful devices take advantage of this property. In industry, powerful lasers are used for cutting and welding materials. Surveyors and builders use lasers for measuring distances and for leveling. In communications, information can be coded in pulses of light from lasers.

Lasers in medicine In the eye and in other parts of the body, surgeons can use lasers in place of scalpels to cut through body tissues. Lasers are routinely used to remove cataracts, reshape the cornea, and repair the retina. The energy from the laser seals blood vessels in the incision and reduces bleeding. Most lasers do not penetrate deeply through the skin, so they can be used to remove small tumors or birthmarks on the surface without damaging deeper tissues.

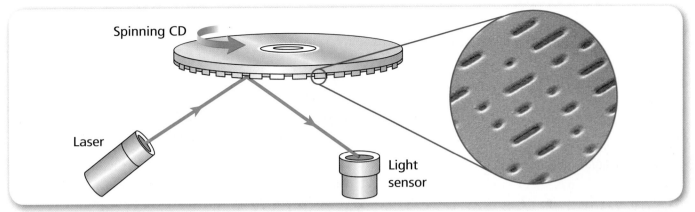

CDs and videodiscs CDs and videodiscs are plastic discs with reflective surfaces that are used to store sound, images, and text in digital form. When a disc is produced, the information is burned into the surface of the disc with a laser. The laser creates millions of tiny pits in a spiral pattern that start at the center of the disc and move out to the edge. A videodisc or CD player also uses a laser to read the disc. As illustrated in **Figure 19,** as the laser beam strikes a pit or flat spot, different amounts of light are reflected to a light sensor. The reflected light is converted to an electric signal that can be converted into sound or images.

■ **Figure 19** The pits on the bottom of a videodisc or CD store information. A videodisc player or CD player uses a laser to detect the pits on the disc. The information is then converted to an electric signal.

Section 3   Review

Section Summary

▷ Incandescent and fluorescent lightbulbs are often used in homes, schools, and offices.

▷ When electrons collide with neon gas, red light is emitted.

▷ A tungsten-halogen bulb is brighter and hotter than an ordinary incandescent bulb.

▷ Lasers emit narrow beams of coherent light.

15. **MAIN ⟨Idea⟩ Compare and contrast** the two main types of bulbs found in your home. Explain how they produce light.

16. **Discuss** the advantages of using a fluorescent bulb instead of an incandescent bulb.

17. **Describe** the difference between coherent and incoherent light.

18. **Describe** the processes used to produce light in a laser.

19. **Identify** several uses of lasers.

20. **Think Critically** Which type of lighting device would you use for each of the following needs: an economical light source in a manufacturing plant, an eye-catching sign that will be visible at night, and a baseball stadium? Explain.

Apply Math

21. **Calculate Efficiency** A 25-W fluorescent light emits 5.0 J of thermal energy each second. What is the efficiency of the fluorescent light?

22. **Use Percentages** If 90 percent of the energy emitted by an incandescent bulb is thermal energy, how much thermal energy is emitted by a 60-W bulb each second?

Reading Preview

Essential Questions

▶ What is the difference between polarized light and unpolarized light?

▶ How is a hologram made?

▶ When does total internal reflection occur?

▶ How are optical fibers used?

Review Vocabulary

interference: occurs when two or more waves overlap and form a new wave

New Vocabulary

linearly polarized light
holography
total internal reflection
optical scanner

g **Multilingual eGlossary**

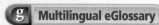

■ **Figure 20** A polarizing filter transmits only light waves with magnetic fields vibrating parallel to the lines of crystals in the filter. When multiple filters are used, the amount of light transmitted depends on how the two filters are oriented.

Using Light

MAIN ◁Idea Light can be used to form three-dimensional images and to transmit information in optical fibers.

Real-World Reading Link You may have a pair of sunglasses with a sticker that reads "polarized." Unlike regular sunglasses, polarized sunglasses can reduce glare. How do polarized lenses do this?

Polarized Light

You can make transverse waves in a rope vibrate in any direction—horizontal, vertical, or anywhere in between. Light also is a transverse wave and can vibrate in any direction. The direction of vibration for a light wave refers to the direction that its electric or magnetic field vibrates. **Linearly polarized light** is light with a magnetic field that vibrates in only one direction. As always, the electric field is perpendicular to the magnetic field.

Polarizing filters Light can be polarized by using a special filter with lines of crystals that act like a group of parallel slits. Only light waves with magnetic fields that vibrate in the same direction as the lines of crystals can pass through. This is similar to the rope wave traveling through the fence in **Figure 20.**

Polarized lenses When light is reflected from a horizontal surface, such as a lake, some of the light is polarized with its magnetic field vibrating vertically. The lenses of polarizing sunglasses are polarizing filters that block the reflected light with vertically vibrating magnetic fields, thus reducing glare.

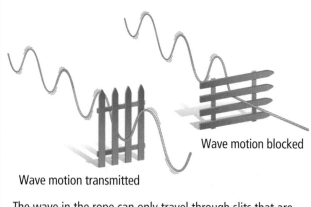

Wave motion blocked

Wave motion transmitted

The wave in the rope can only travel through slits that are positioned in the same direction as its vibrations.

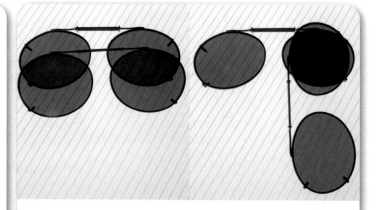

When polarizing filters are stacked with their lines of crystals in the same direction, light can pass through them. When the lines are positioned at right angles, the light is blocked.

■ **Figure 21** Holograms can be displayed as three-dimensional art. An embossed copy of a hologram can be placed over a reflective background to create the security features on important documents.

Holography

Science museums often have exhibits where a three-dimensional image seems to float in space, like the one shown in **Figure 21.** You can see the image from different angles, just as you would if you viewed the real object. These images are produced by holography. **Holography** is a technique that produces a hologram, a complete three-dimensional photographic image of an object. The three-dimensional images on some credit cards are also produced by holography.

Making holograms Lasers are needed to produce holograms. The laser beam is split into two parts. One part illuminates the object and reflects onto photographic film. At the same time, the second part of the beam is also directed at the film. The light from the two beams creates an interference pattern on the film. The pattern looks nothing like the original object, but when laser light shines on the pattern on the film, a holographic image is produced.

Reading Check **Describe** how holographic images are produced.

Information in light An ordinary photographic image captures only the brightness or intensity of light reflected from an object's surface, but a hologram records the direction as well as the intensity. As a result, it conveys more information to your eye than a conventional two-dimensional photograph does. It is also more difficult to copy. In addition to credit cards, holographic images are used on identification cards and on the labels of some products to help prevent counterfeiting. Using X-ray lasers, scientists can produce holographic images of microscopic objects. It may be possible to create three-dimensional views of biological cells.

FOLDABLES

Incorporate information from this section into your Foldable.

Optical Fibers

When laser light must travel long distances or be sent into hard-to-reach places, optical fibers often are used. These transparent glass fibers transmit light from one place to another by a process called total internal reflection.

Partial Reflection Remember that refraction can happen when light changes speed as it travels from one medium to another. When light travels from water to air, it speeds up and the direction of the light ray is bent away from the normal, as shown in **Figure 22.** However, not all the light passes through the surface. Some light is reflected back into the water.

As the underwater light ray makes larger angles with the normal, the refracted light rays in the air bend closer to the surface of the water. Additionally, more light is reflected back into the water. At a certain angle, called the critical angle, the refracted ray has been bent so that it is traveling along the surface of the water, as shown in **Figure 22.** For a light ray traveling from water into air, the critical angle is about 49°.

Total Internal Reflection **Figure 22** also shows what happens if the underwater light ray strikes the boundary between the air and water at an angle larger than the critical angle. There is no longer any refraction, and the light ray does not travel in the air. Instead, the light ray is reflected at the boundary, just as if a mirror were there. This complete reflection of light at the boundary between two different materials is called **total internal reflection.** The light wave follows the law of reflection. For total internal reflection to occur, light must travel slower in the first medium and must strike the boundary at an angle that is greater than the critical angle.

✔ **Reading Check Identify** when total internal reflection occurs.

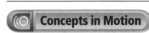
Concepts in Motion
Animation

■ **Figure 22** At the surface between water and air, some light is refracted and some is reflected. The greater the angle of incidence is, the greater both the angle of refraction and the amount of reflected light will be. At the critical angle, the refracted wave is traveling along the water surface. At angles greater than the critical angle, total internal reflection occurs.

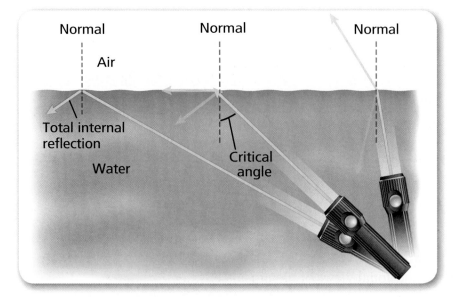

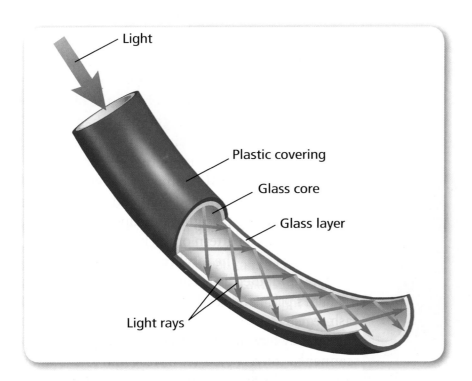

Light

Plastic covering

Glass core

Glass layer

Light rays

Using total internal reflection Total internal reflection makes light transmission in optical fibers possible. As shown in **Figure 23,** light entering one end of the fiber is reflected continuously from the sides of the fiber until it emerges from the other end. The light moves like water through a pipe—almost no light is lost or absorbed in optical fibers. The core of the fiber and the surrounding layer are made from glasses with two different indexes of refraction. Light moves more slowly in the core than in the surrounding layer. Therefore, total internal reflection can occur at the surface between the two layers.

Using optical fibers Optical fibers are most often used in communications. Telephone conversations, television programs, and computer data can be coded in light beams. Signals cannot leak from one fiber to another and interfere with other messages, so the signal is transmitted clearly. To send telephone conversations through an optical fiber, sound is converted into digital signals consisting of pulses of light from a light-emitting diode (LED) or from a laser. Some systems use multiple lasers, each with its own wavelength, to fit multiple signals into the same fiber. You could send a million copies of your favorite book in one second on a single fiber. **Figure 24** shows the size of typical optical fibers.

Optical fibers also have medical uses. Doctors use them to explore the inside of the human body. One bundle of fibers transmits light while another carries the reflected light back to the doctor. Physicians can also treat blocked arteries by sending laser light into the body through an optical fiber.

■ **Figure 24** Despite its small size, just one of these optical fibers can carry thousands of phone conversations at the same time.

Use Vocabulary

Answer each question using the correct term from the Study Guide.

29. What is a colored material that is used to change the color of other substances?

30. What process would produce a complete three-dimensional image of an object?

31. How would you describe an object that you can see through clearly?

32. What process enables optical fibers to transmit signals over long distances?

33. What is an image of a distant object produced by the refraction of light through air layers of different densities?

34. What type of light has magnetic fields that vibrate in only one direction?

Check Concepts

35. Which type of cells in your eyes allows you to see the color violet?
A) red cones   C) blue cones
B) green cones   D) rods

36. **BIG Idea** What do you see when you note the color of an object?
A) the light it reflects
B) the light it absorbs
C) the light it refracts
D) the light it diffracts

37. Which term best describes the light from a laser?
A) incoherent   C) incandescent
B) coherent   D) opaque

38. Which property of a material indicates how much light slows down when traveling in the material?
A) pigment   C) filter
B) index of refraction   D) mirage

Use the figure below to answer question 39.

39. How would you describe the candleholder?
A) translucent   C) transparent
B) opaque   D) diffuse

40. Which event explains why a prism separates white light into the colors of the rainbow?
A) interference   C) diffraction
B) fluorescence   D) refraction

Use the figure below to answer question 41.

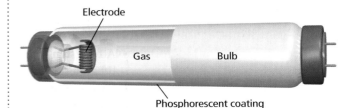

Electrode

Gas   Bulb

Phosphorescent coating

41. How does the lightbulb shown produce light?
A) A filament is heated until it glows.
B) Phosphors convert ultraviolet waves into visible light.
C) Stimulated electrons emit coherent light.
D) Electrons collide with gas molecules to produce red light.

✓ **Assessment**   **Online Test Practice**

Interpret Graphics

42. Copy and complete this concept map to show the steps in the production of fluorescent light.

Concepts in Motion **Interactive Concept Map**

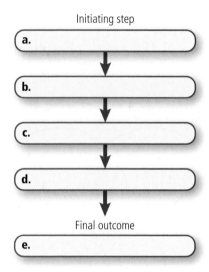

Initiating step

a.

b.

c.

d.

Final outcome

e.

43. Copy and complete the following concept map about using light.

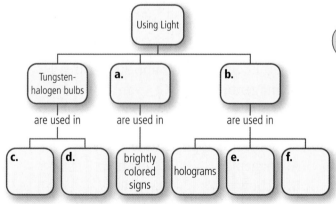

Using Light

Tungsten-halogen bulbs | a. | b.

are used in | are used in | are used in

c. | d. | brightly colored signs | holograms | e. | f.

Think Critically

44. **Summarize** the function of the electric current in fluorescent lights and neon lights.

---

45. **Explain** The speed of light is greater in air than in glass. Explain whether total internal reflection could occur when a light wave traveling in air strikes the glass.

46. **Compare and contrast** the reflection of light from a white wall that has a rough surface with the reflection of light from a mirror.

47. **THEME FOCUS** **Explain** how light is produced by an incandescent bulb. What is a disadvantage of these bulbs?

48. **Predict** what color a white shirt would appear to be if the light reflected from the shirt passed through a red filter and then through a green filter. Explain.

49. **Infer** Some mammals cannot see colors. Infer how a color-blind mammal's retina might be different from the human retina.

50. **Compare** White light passes through a green pane of glass and shines on a white shirt. Compare the colors of light that are absorbed, reflected, and transmitted by the glass and by the shirt.

Apply Math

51. **Calculate Angle of Incidence** A light ray is reflected from a mirror. If the angle between the incident ray and the reflected ray is 136°, what is the angle of incidence?

52. **Calculate Speed** A material's index of refraction equals the speed of light in a vacuum (300,000 km/s) divided by the speed of light in the material. What is the speed of light in water if water's index of refraction is 1.33?

53. **Calculate Efficiency** A 60-W incandescent lightbulb gives of 50 J of thermal energy per second. What is its efficiency?

Standardized Test Practice

Multiple Choice

Record your answers on the answer sheet provided by your teacher or on a sheet of paper.

1. Which word describes materials that transmit light so objects behind them are seen clearly?
   A. translucent   C. opaque
   B. transparent   D. diffuse

Use the figure below to answer questions 2 and 3.

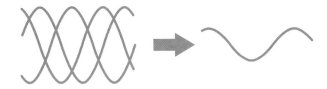

2. Which word best describes the light waves shown in the figure above?
   A. coherent   C. opaque
   B. incoherent   D. transparent

3. Which device produces these light waves?
   A. fluorescent lightbulb
   B. sodium-vapor lightbulb
   C. laser
   D. incandescent lightbulb

4. In which situation is light refracted?
   A. Light hits a wall.
   B. Light travels in a vacuum.
   C. Light hits a mirror.
   D. Light travels from air into water.

5. Which part of the eye detects colors?
   A. rods   C. cones
   B. retina   D. lens

6. Why does an apple appear to be red?
   A. It reflects red light.
   B. It absorbs red light.
   C. It reflects all colors of light.
   D. It absorbs all colors of light.

7. Which process is used to transmit light in optical fibers?
   A. diffuse reflection
   B. total internal reflection
   C. polarization
   D. incandescence

8. Which substance inside a fluorescent bulb absorbs ultraviolet radiation and emits visible light?
   A. neon   C. phosphors
   B. tungsten   D. sodium vapor

Use the figure below to answer questions 9 and 10.

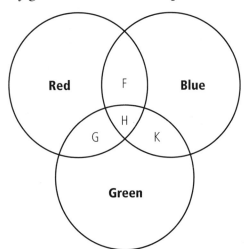

9. When the primary colors of light are added together, which color appears in area H?
   A. magenta   C. cyan
   B. yellow   D. white

10. Which color appears in area G?
    A. magenta   C. cyan
    B. yellow   D. white

11. Which light source produces light by making a piece of metal hot enough to glow?
    A. fluorescent bulb
    B. incandescent bulb
    C. neon bulb
    D. laser

Record your answers on the answer sheet provided by your teacher or on a sheet of paper.

12. If green light shines on a black-and-white striped shirt, what colors will the stripes appear to be?

Use the figure below to answer questions 13 and 14.

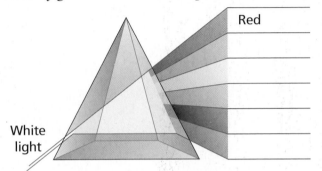

13. Identify the object shown and explain how it affects white light that passes through it.

14. Identify the colors that would be seen on the right side of the object, in order of decreasing wavelength.

15. Describe the difference between an object that appears to be black and an object that appears to be white.

16. For regular reflection to occur, how must the roughness of a surface compare to the wavelengths of light that it reflects?

Record your answers on a sheet of paper.

17. Explain why a laser beam can deliver a large amount of energy to a small area.

18. Infer whether a large bottle made of green glass would be a suitable container in which to grow small plants.

Use the figure below to answer questions 19–21.

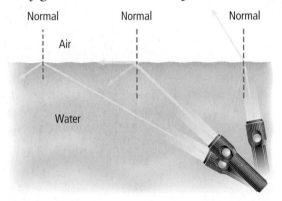

19. When a light wave traveling in water reaches the water's surface, under what circumstances will total internal reflection occur?

20. Suppose the flashlight were in the air and the direction of the light beams shown in the figure were reversed. Under what circumstances would total internal reflection occur? Explain.

21. If the air were replaced by glass, under what circumstances would total internal reflection occur? *Hint: Glass has a higher index of refraction than water.*

NEED EXTRA HELP?																				
If You Missed Question . . .	1	2	3	4	5	6	7	8	9	10	11	12	13	14	15	16	17	18	19	20	21
Review Section . . .	1	3	3	1	2	2	4	3	2	2	3	2	1	1	2	1	3	2	4	4	4

CHAPTER 13

Mirrors and Lenses

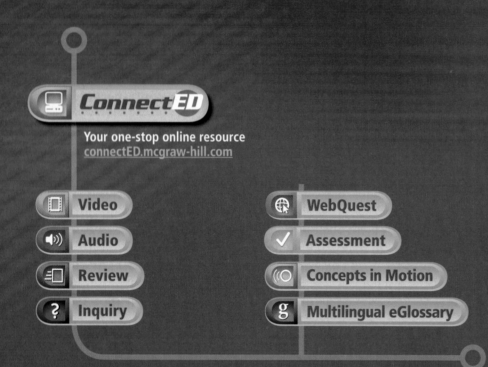

ConnectED

Your one-stop online resource
connectED.mcgraw-hill.com

- Video
- Audio
- Review
- Inquiry
- WebQuest
- Assessment
- Concepts in Motion
- g Multilingual eGlossary

Launch Lab
Water Lenses

Have you ever used a magnifying glass, a camera, a microscope, or a telescope? If so, you were using a lens to create an image. A lens is a transparent material that bends rays of light and forms an image. In this activity, you will use water to create a lens.

For a lab worksheet, use your StudentWorks™ Plus Online.

 **Inquiry** Launch Lab

FOLDABLES

Make a three-tab book. Label it as shown. Use it to organize your notes about mirrors.

Plane Mirror | Concave Mirror | Convex Mirror

THEMES FOCUS Technology
Mirrors and lenses are integral parts of telescopes, microscopes, cameras, and other devices.

BIG⟮ Idea⟯ Mirrors and lenses form images by causing light rays to change direction.

Reading Preview

Essential Questions

▶ How do different types of mirrors form images?

▶ What are real images and virtual images?

▶ What are some examples of plane, convex, and concave mirrors?

Review Vocabulary

reflection: the return of waves or particles from a surface

New Vocabulary

plane mirror
virtual image
concave mirror
optical axis
focal point
focal length
real image
convex mirror

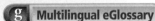 **g** Multilingual eGlossary

Mirrors

MAIN ◀Idea Mirrors form images by reflecting light rays.

Real-World Reading Link You might use a mirror to check your appearance before going to school. Astronomers use very large concave mirrors to form images of distant stars and galaxies.

Light and Vision

Have you ever tried to find an address on a house or an apartment at night on a poorly lit street? It is harder to do those activities in the dark than it is when there is plenty of light. Your eyes see by detecting light, so when you can see something, it is because light came from that object to your eyes. Light is emitted from a light source, such as the Sun or a lightbulb, and then reflects off an object, such as the page of a book, as shown in **Figure 1**.

When light travels from an object to your eye, you see the object. Light can reflect more than once. For example, light can reflect off an object into a mirror and then reflect into your eyes. When no light is available to reflect off objects and into your eyes, you cannot see anything. This is why it is hard to see an address in the dark.

Light rays Light sources send out light waves that travel in all directions. These waves spread out from the light source, just as ripples on the surface of water spread out from the point of impact of a pebble.

You could also think of the light coming from the source as traveling in narrow beams. Each narrow beam travels in a straight line and is called a light ray. Even though light rays can change direction when they are reflected or refracted, your brain interprets images as if light rays travel in a straight line.

■ **Figure 1** Light from the lamp reflects off the book and into the person's eyes. People see objects when their eyes detect light emitted or reflected from those objects.

■ **Figure 2** Light reflects off of the girl's forehead and then off of the mirror before entering the girl's eyes.

Plane Mirrors

Greek mythology tells the story of a handsome young man named Narcissus who noticed his image in a pond and fell in love with himself. Like pools of water, mirrors are smooth surfaces that reflect light to form images. Just as Narcissus did, you can see yourself as you glance into a quiet pool of water or walk past a shop window. Most of the time, however, you probably look for your image in a flat, smooth mirror. A flat, smooth mirror is a **plane mirror.**

✓ **Reading Check** **Define** What is a plane mirror?

Reflections from plane mirrors What do you see when you look into a plane mirror? Your reflection is upright. If you were one meter in front of the mirror, your image would appear to be one meter behind the mirror, or two meters from you. You might notice that the reflection of any writing in a plane mirror appears backward.

Figure 2 shows how you see yourself in a plane mirror. First, light rays from a light source strike you. Every point that is struck by the light rays reflects these rays so they travel outward in all directions. If your friend were looking at you, these reflected light rays coming from you would enter her eyes so she could see you. However, if a mirror is placed between you and your friend, the light rays are reflected from the mirror into your eyes.

VOCABULARY ....
SCIENCE USAGE V. COMMON USAGE
Plane
Science usage
  having no elevations or depressions
  A plane mirror is a flat mirror.

Common usage
  short for *airplane*; a powered, heavier-than-air aircraft with fixed wings
  We took a plane from Houston to Orlando. ....

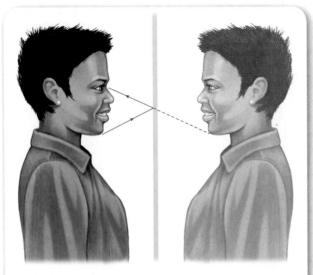

Figure 3 The light rays that reflect off of a plane mirror appear to originate behind that mirror. This gives the illusion that objects exist behind the plane mirror.

Virtual images You can understand your brain's interpretation of your reflection in a mirror by looking at **Figure 3.** The light waves that are reflected off you travel in all directions. Light rays reflected from your chin strike the mirror at different places. Then, they reflect off the mirror in different directions. A few of these light rays reflect off the mirror in just the right way to enter your eyes.

Recall that your brain always interprets light rays as if they have traveled in a straight line. It does not realize that the light rays have been reflected and that they changed direction. Your reflected image appears to be behind the mirror.

An image that your brain perceives even though no light rays pass through the location of that image is a **virtual image.** The imaginary light rays that appear to come from virtual images are called virtual rays. The dashed line in **Figure 3** is a virtual ray. Plane mirrors always form upright, virtual images.

Concave Mirrors

Not all mirrors are flat like plane mirrors. A **concave mirror** is a mirror whose surface curves inward. Concave mirrors, like plane mirrors, reflect light waves to form images. However, a concave mirror's curved surface produces different images from a plane mirror's flat surface.

Features of concave mirrors A concave mirror has an optical axis. The **optical axis** is an imaginary straight line drawn perpendicular to the surface of the mirror at the mirror's center. Concave mirrors are made so that every light ray traveling toward the mirror parallel to the optical axis is reflected through a point on the optical axis called the focal point.

The **focal point** for a concave mirror is the point on the optical axis on which light rays that are initially parallel to the optical axis converge after they reflect off the mirror. The distance from the center of the mirror to the focal point is the **focal length.** Using the focal point and the optical axis, you can diagram how some of the light rays that travel to a concave mirror are reflected, as shown in **Figure 4.**

Figure 4 A concave mirror has an optical axis and a focal point. When light rays travel toward the mirror parallel to the optical axis, they reflect through the focal point. Light rays that travel through the focal point before hitting the mirror are reflected parallel to the optical axis.

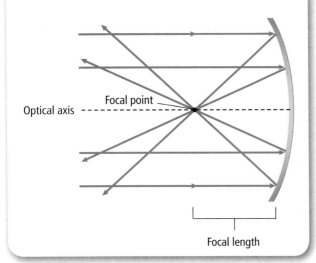

Optical axis · · · · · · Focal point

Focal length

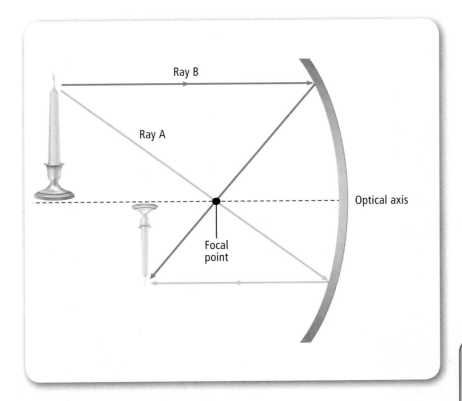

Ray B

Ray A

Optical axis

Focal point

■ **Figure 5** Ray A first passes through the focal point and then reflects parallel to the optical axis. Ray B is first parallel to the optical axis and then reflects through the focal point. An image of the candle forms where the two rays converge.

Diagram *how other points on the image of the candle are formed.*

Ray tracing for concave mirrors

You can diagram how concave mirrors form images by tracing some of the light rays involved. Suppose that the distance between an object, such as the candle in **Figure 5,** and the mirror is greater than the focal length. Light rays bounce off the candle in all directions. One light ray, labeled Ray A, starts from a point on the flame of the candle and passes through the focal point on its way to the mirror. Ray A is then reflected parallel to the optical axis.

Another ray, Ray B, starts from the same point on the candle's flame, but it travels parallel to the optical axis as it moves toward the mirror. The mirror then reflects Ray B through the focal point. The place where Ray A and Ray B meet after they are reflected is a point on the reflected image of the flame.

More points on the reflected image can be located in this way. From each point on the candle, one ray can be drawn that passes through the focal point and is reflected parallel to the optical axis. Another ray can be drawn that travels parallel to the optical axis and then reflects through the focal point. The point where the two rays meet is on the reflected image.

Real images

The image that is diagrammed in **Figure 5** is not virtual. Rays of light pass through the location of the image. A **real image** is an image that is formed when light rays converge to form the image. You could hold a sheet of paper at the location of a real image and see the image projected on the paper.

MiniLab

? **Inquiry** MiniLab

Observe Images in a Spoon

Procedure

1. Read the procedure and safety information, and complete the lab form.
2. Look at a **concave mirror.** Move it close to your face and then far away. The place where your image blurs is the focal point.
3. Hold the inside of the mirror facing a **bright light** a little farther away than the focal length of the mirror.
4. Place a piece of **poster board** between the light and the mirror without blocking all of the light.
5. Move the poster board between the mirror and the light until you see the reflected light on the poster board.

Analysis

1. **Identify** which of the images that you observed were real and which were virtual.

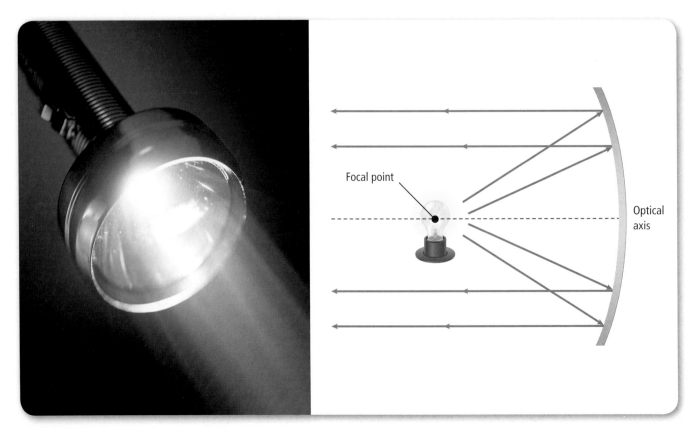

Figure 6 A light beam forms whenever someone places a light source at a concave mirror's focal point.

Explain *why the reflected rays of light in the diagram are parallel to each other.*

Figure 7 An enlarged and virtual image forms where the virtual rays converge when an object is placed between a concave mirror and that mirror's focal point.

Infer *why this image could not be projected on a screen.*

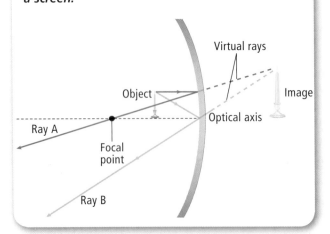

Spotlights What happens when you place an object exactly at the focal point of a concave mirror? **Figure 6** shows that when the object is at the focal point, the mirror reflects all light rays parallel to the optical axis. The rays never meet, and no image forms. Even the virtual rays that extend behind the mirror do not meet. Therefore, a light placed at the focal point is reflected in a beam. Car headlights, flashlights, spotlights, and other devices use concave mirrors in this way to produce light beams with nearly parallel rays.

Mirrors that magnify A concave mirror magnifies an object when you place that object between the concave mirror and that mirror's focal point. **Figure 7** shows that the reflected rays diverge and a virtual image forms.

Just as it does with a plane mirror, your brain interprets the diverging rays as if they came from one point behind the mirror. You can find this point by imagining virtual rays that extend behind the mirror. The resulting image is magnified. Shaving mirrors and makeup mirrors are concave mirrors that are used for magnification. They form enlarged, upright images of a person's face so that it is easier to see small details.

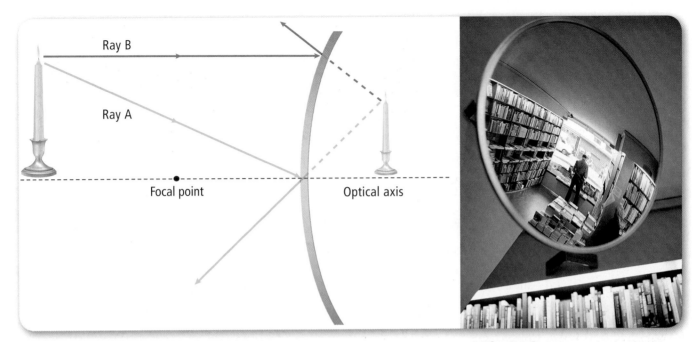

Ray B

Ray A

Focal point

Optical axis

Convex Mirrors

Why do you think the security mirrors in banks and stores are shaped the way that they are? The next time that you are in a store, look at one of the back corners or at the end of an aisle to see if a large, rounded mirror is mounted there. You can see a large area of the store in the mirror. A **convex mirror** is a mirror that curves outward, like the back of a spoon.

Light rays that hit a convex mirror spread apart after they are reflected. Look at **Figure 8** to see how the rays from an object are reflected off a convex mirror to form an image. The reflected rays diverge and never meet, so the image formed by a convex mirror is a virtual image. The image is also always upright and is smaller than the actual object.

✓ **Reading Check** **Describe** the image formed by a convex mirror.

Uses of convex mirrors Because convex mirrors cause light rays to diverge, they allow large areas to be viewed. As a result, a convex mirror is said to have a wide field of view. In addition to increasing the field of view in places like grocery stores and factories, convex mirrors can widen the view of traffic that can be seen in rear-view or side-view mirrors of automobiles.

However, because the image that a convex mirror forms is smaller than the object, your perception of distance can be distorted. Objects look farther away than they truly are in a convex mirror. Distances and sizes seen in a convex mirror are not realistic, so most convex side mirrors on cars carry a printed warning that states "Objects in mirror are closer than they appear."

FOLDABLES
Incorporate information from this section into your Foldable.

Table 1	Images Formed by Mirrors	Concepts in Motion	Interactive Table	
Mirror Shape	Distance of Object from Mirror	Virtual/ Real	Image Created Upright/ Upside Down	Size
Plane	any distance	virtual	upright	same as object
Concave	object more than two focal lengths from mirror	real	upside down	smaller than object
	object between one and two focal lengths	real	upside down	larger than object
	object at focal point	none	none	none
	object within focal length	virtual	upright	larger than object
Convex	any distance	virtual	upright	smaller than object

Mirror images The different shapes of plane, concave, and convex mirrors cause them to reflect light in distinct ways. For example, concave mirrors are the only mirrors that magnify images. Convex mirrors always make objects appear to be smaller and farther away than they actually are. Each type of mirror has different uses. Most wall mirrors are plane mirrors. Most makeup and shaving mirrors are concave mirrors. Most store security mirrors are convex mirrors. **Table 1** summarizes the characteristics of plane mirrors, concave mirrors, and convex mirrors.

Section 1 Review

Section Summary

▶ You see an object because your eyes detect the light reflected from that object.

▶ Plane mirrors are smooth and flat.

▶ No light rays pass through the location of a virtual image.

▶ A concave mirror curves inward.

▶ A convex mirror curves outward.

1. **MAIN Idea Diagram** how both concave mirrors and convex mirrors form images.

2. **Identify** at least one example of a plane mirror, one example of a concave mirror, and one example of a convex mirror.

3. **Describe** the image of an object that is 38 cm from a concave mirror that has a focal length of 10 cm.

4. **Infer** whether a virtual image can be photographed.

5. **Think Critically** An object is less than one focal length from a concave mirror. How does the size of the image change as the object gets closer to the mirror?

Apply Math

6. **Calculate Distance** If you stand 2 m away from a plane mirror, how far away does your reflection appear to be from you?

Assessment   Online Quiz

Objectives

■ **Infer** how the number of reflections depends on the angle between mirrors.

Background: How can you see the back of your head? If you have ever been to a barbershop or a hair salon, you probably know the answer. You can use two mirrors to view a reflection of a reflection of the back of your head.

Question: *How many reflections can you see with two mirrors?*

Preparation

Materials

plane mirrors (2)
masking tape
protractor
paper clip

Safety Precautions

WARNING: *Handle glass mirrors and paper clips carefully.*

Procedure

1. Read the procedure and safety information, and complete the lab form.
2. Lay one mirror on top of the other with the mirror surfaces inward. Tape them together so they will open and close. Use tape to label them "L" and "R."
3. Stand up the mirrors on a sheet of paper. Using the protractor, close the mirrors to an angle of 72°.
4. Bend one leg of a paper clip up 90°, and place it close to the front of the R mirror.

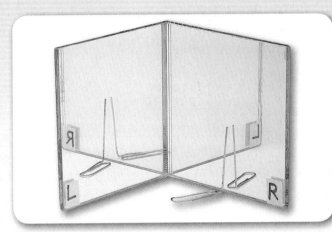

5. Count the number of images of the clip that you see in the L and R mirrors. Record these numbers in a data table.
6. The mirror arrangement creates an image of a circle divided into wedges by the mirrors. Record the number of wedges in your data table.
7. Hold the R mirror still, and slowly open the L mirror to 90°. Count and record the images of the clip and the wedges in the circle. Repeat, this time opening the mirrors to 120°.

Conclude and Apply

1. **Infer** the relationship between the number of wedges and paper clip images that you can see.
2. **Determine** the angle that would divide a circle into six wedges. Hypothesize how many images would be produced.
3. **Infer** how many images would be produced if two mirrors were directly opposite one another, as in a barbershop or a hair salon.

COMMUNICATE YOUR DATA

Demonstrate Prepare a short video for middle school students that demonstrates the relationship between the angle of the mirrors and the number of reflections.

Essential Questions

▶ In what ways do convex lenses and concave lenses bend light rays?
▶ What types of images do convex lenses and concave lenses form?
▶ How are lenses used to correct vision problems?

Review Vocabulary

transparent: material that transmits light without scattering so that objects are clearly visible through it

New Vocabulary

convex lens
concave lens
cornea
retina

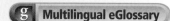 **Multilingual eGlossary**

Lenses

MAIN Idea Lenses form images by refracting light rays.

Real-World Reading Link Anyone who wears glasses uses lenses to improve their vision. Without lenses, even people who do not wear glasses or contacts could not see. Each human eye contains a pair of lenses to help bring images into focus.

What is a lens?

What do your eyes have in common with cameras and eyeglasses? Each of these things contains at least one lens. A lens is a transparent material with at least one curved surface that causes light rays to bend, or refract, as those rays pass through the lens. The image that a lens forms depends on the shape of the lens. Like curved mirrors, a lens can be convex or concave.

Convex Lenses

A **convex lens** is a lens that is thicker in the middle than at the edges. Its optical axis is an imaginary straight line that is perpendicular to the surface of the lens at its thickest point. When light rays approach a convex lens traveling parallel to its optical axis, the rays are refracted toward the center of the lens, as shown in **Figure 9.**

All light rays traveling parallel to the optical axis in **Figure 9** are refracted so they pass through a single point, which is the focal point of the lens. The focal length of the lens depends on the shape of the lens. If the sides of a convex lens are less curved, light rays are bent less. As a result, lenses with flatter sides have longer focal lengths. **Figure 9** also shows that light rays traveling along the optical axis are not bent at all.

■ **Figure 9** A convex lens bends light inward, toward the optical axis. A light ray that passes straight through the center of the lens does not refract.

Identify *the focal point of the lens.*

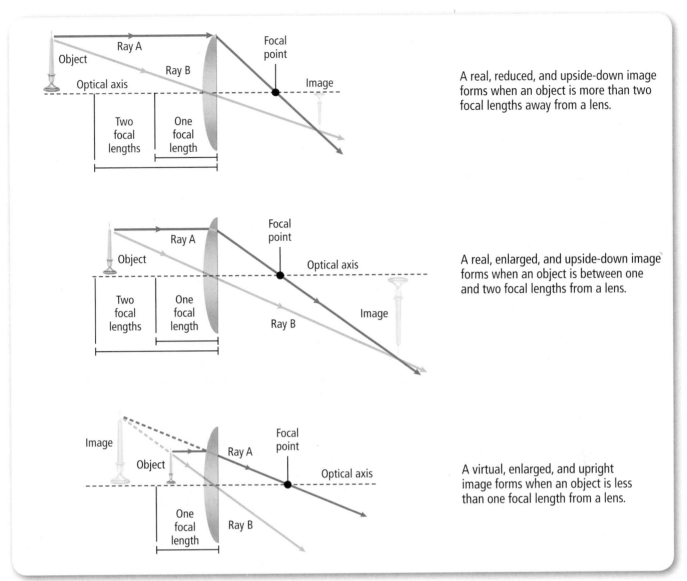

A real, reduced, and upside-down image forms when an object is more than two focal lengths away from a lens.

A real, enlarged, and upside-down image forms when an object is between one and two focal lengths from a lens.

A virtual, enlarged, and upright image forms when an object is less than one focal length from a lens.

■ **Figure 10** The image that a convex lens forms depends on the relative positions of the lens and the object.

Identify *the type of mirror that produces images that are similar to the images produced by a convex lens.*

Forming images with convex lenses The type of image that a convex lens forms depends on where the object is relative to the focal point of the lens. If an object is more than two focal lengths from the lens, as in the top panel of **Figure 10,** the image is real, reduced, inverted, and on the opposite side of the lens from the object.

As the object moves closer to the lens, the image gets larger. The middle panel of **Figure 10** shows the image formed when the object is between one and two focal lengths from the lens. Now the image is larger than the object but is still inverted.

When an object is less than one focal length from the lens, as shown in the bottom panel of **Figure 10,** the image becomes an enlarged, virtual image. The image is virtual because the light rays from the object are not converging after they have passed through the lens. When you use a magnifying glass, you move a convex lens so that it is less than one focal length from an object. This causes the image of the object to be magnified.

■ **Figure 11** A concave lens causes light rays to diverge.

Classify *Does a concave lens behave more like a concave mirror or a convex mirror?*

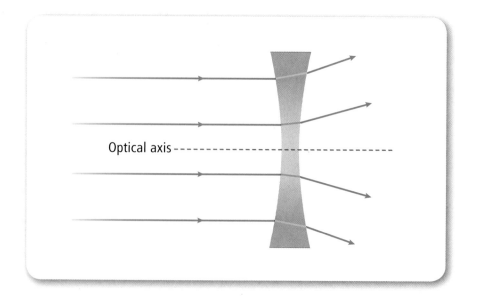

Optical axis

Concave Lenses

A **concave lens** is a lens that is thinner in the middle and thicker at the edges. As shown in **Figure 11,** light rays that pass through a concave lens bend outward, away from the optical axis. The rays spread out and never meet at a focal point, so they never form a real image. However, a concave lens can form virtual images. These virtual images are always upright and smaller than the actual object. Notice that concave lenses and convex mirrors both produce the same types of images.

Concave lenses are used in some types of eyeglasses and in some microscopes. Concave lenses are usually placed in combination with other lenses. A summary of the images formed by concave and convex lenses is shown in **Table 2.**

Table 2	**Images Formed by Lenses** ((◎ Concepts in Motion) Interactive Table			
Lens Shape	**Location of Object**	**Vitual/Real**	**Type of Image Upright/Inverted**	**Size**
Convex	object beyond 2 focal lengths from lens	real	inverted	smaller than object
	object between 1 and 2 focal lengths	real	inverted	larger than object
	object within 1 focal length	virtual	upright	larger than object
Concave	object at any position	virtual	upright	smaller than object

How do object distance and image distance compare?

The size and orientation of an image formed by a lens depends on the location of the object and on the nature of the lens. Convex lenses form both real images and virtual images. Concave lenses can form only virtual images. What happens to the location of the image formed by a lens as the object moves closer to or farther from the lens? The distance from the lens to the object is the object distance, and the distance from the lens to the image is the image distance. How are the focal length, object distance, and image distance related to each other?

Identify the Problem

A 5-cm-tall object is placed at different lengths from a convex lens with a focal length of 15 cm. The table at the right shows the different object and image distances. How are these two measurements related?

Object and Image Distances		
Focal Length (cm)	Object Distance (cm)	Image Distance (cm)
15.0	45.0	22.5
15.0	30.0	30.0
15.0	20.0	60.0

Solve the Problem

1. Describe the relationship between the object distance and the image distance.
2. The lens equation describes the relationship between the focal length and the image and object distances.

$$\frac{1}{focal\ length} = \frac{1}{object\ distance} + \frac{1}{image\ distance}$$

Using this equation, calculate the image distance when the object is placed at a distance of 60.0 cm from the lens.

Eyesight and Lenses

What determines how well you can see the words on this page? If you do not need eyeglasses, the structure of your eye gives you the ability to focus on these words and on other objects around you. Look at **Figure 12.** Light enters your eye through a transparent covering on your eyeball called the **cornea** (KOR nee uh). The cornea causes light rays to bend so that they converge.

Reading Check **Describe** the function of the cornea.

After passing through the cornea, the light then passes through an opening called the pupil. Behind the pupil is a flexible, convex lens, called the eye lens. The eye lens helps focus light rays so that a sharp image is formed on your retina. Your **retina** is the inner lining of your eye, which has cells that convert the light image into electrical signals. These electric signals are then carried along the optic nerve to your brain where they can be interpreted.

Reading Check **Describe** the function of the retina.

■ **Figure 12** The cornea is a convex lens that causes light rays from distant objects to converge on the retina. The eye lens helps bring closer objects into focus.

Concepts in Motion Animation

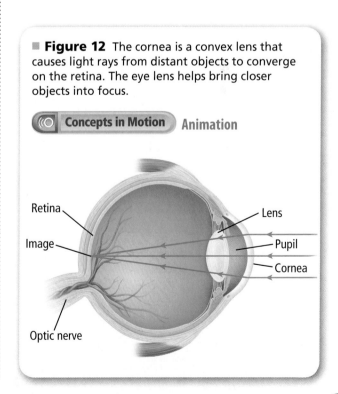

Focusing on near and far How can your eyes focus both on close objects, such as the watch on your wrist, and distant objects, such as a clock across the room? For you to see an object clearly, its image must be focused sharply on your retina. However, the retina is always a fixed distance from the lens. Remember that the location of an image formed by a convex lens depends on the focal length of the lens and the location of the object.

For an image to be formed on the retina, the focal length of the lens needs to be able to change as the distance to the object changes. The lens in your eye is flexible, and muscles attached to it change its shape and its focal length. This is why you can see objects that are near and far away.

Look at **Figure 13.** When you focus on an object far from your eye, the muscles around the lens relax. This pulls the lens into a less convex shape. When you focus on a nearby object, these muscles make the lens more curved, causing the focal length to decrease.

✓ **Reading Check Describe** how the shape of the lens in your eye changes when you focus on a nearby object.

■ **Figure 13** The eye lens changes shape so that you can focus on objects at different distances.

Infer why you are more likely to get eyestrain by looking at a nearby object than by looking at a faraway object.

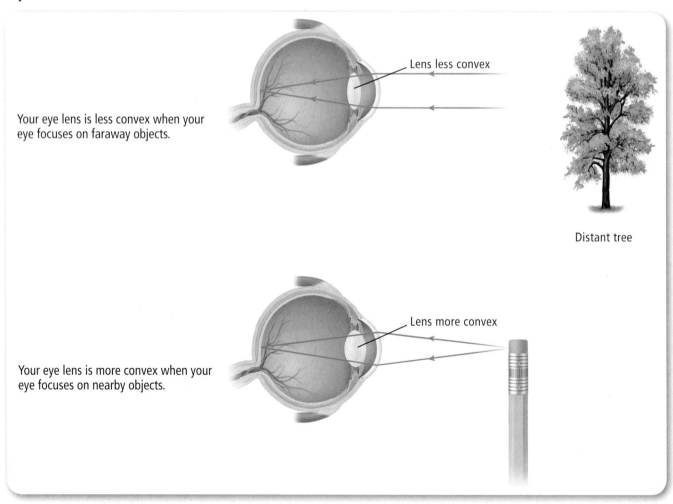

Your eye lens is less convex when your eye focuses on faraway objects.

Lens less convex

Distant tree

Your eye lens is more convex when your eye focuses on nearby objects.

Lens more convex

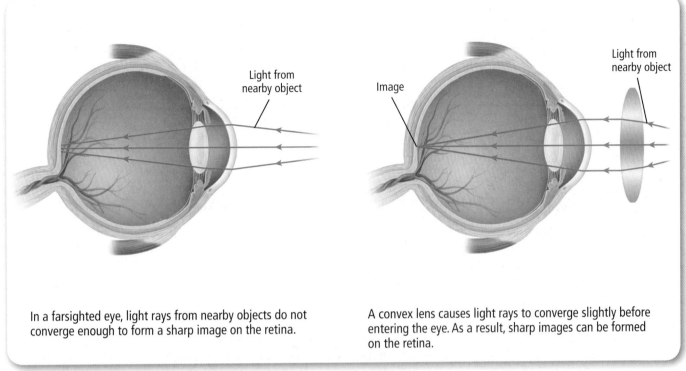

In a farsighted eye, light rays from nearby objects do not converge enough to form a sharp image on the retina.

A convex lens causes light rays to converge slightly before entering the eye. As a result, sharp images can be formed on the retina.

Vision Problems
People with good vision can see objects clearly that are about 25 cm or farther away from their eyes. However, people with the most common vision problems see objects clearly only at some distances, or they see all objects as being blurry.

Astigmatism
One vision problem, called astigmatism, occurs when the surface of the cornea is unevenly curved. When people have astigmatism, their corneas are more oval than round in shape. Astigmatism causes blurry vision at all distances. Corrective lenses for astigmatism also have an uneven curvature, canceling out the effect of an uneven cornea.

Farsightedness
Another vision problem is farsightedness. A farsighted person can see distant objects clearly, but cannot bring nearby objects into focus. Light rays from nearby objects do not converge enough after passing through the cornea and the lens to form a sharp image on the retina, as shown in **Figure 14.** The problem can be corrected with a convex lens that bends light rays so they are less spread out before they enter the eye, also shown in **Figure 14.**

Farsightedness is often related to age. As many people age, the lenses in their eyes become less flexible. The muscles around the lenses still contract as they try to change the shape of the lens. However, the lenses have become more rigid and cannot be made curved enough to focus on close objects. People who are more than 40 years old might not be able to focus on objects closer than 1 m from their eyes.

■ **Figure 14** A farsighted person can see faraway objects clearly, but he or she has trouble focusing on nearby objects. For example, a farsighted person could watch a football game from the stands without glasses. However, a farsighted person would have more difficulty reading without proper glasses. Farsightedness can be corrected by convex lenses.

? **Inquiry** Virtual Lab

▣ **Video** BrainPOP

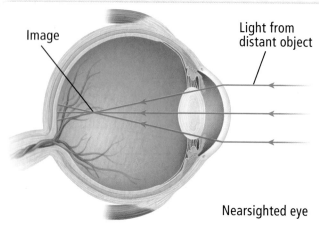

Image · Light from distant object

Nearsighted eye

In a nearsighted eye, light rays from distant objects converge too much and form a sharp image in front of the retina.

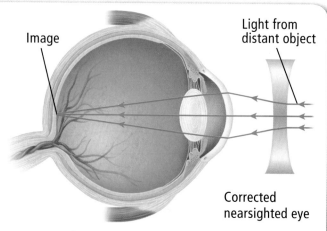

Image · Light from distant object

Corrected nearsighted eye

A concave lens makes the light rays spread out before entering the eye, enabling a sharp image to be formed on the retina.

■ **Figure 15** People use concave lenses to correct nearsightedness.

Nearsightedness A person who is nearsighted can see objects clearly only when those objects are nearby. Objects that are far away appear blurred. In a nearsighted eye, the cornea and the lens form a sharp image of a distant object before the light reaches the retina, as shown in **Figure 15.**

To correct this problem, a nearsighted person can wear concave lenses. **Figure 15** shows how a concave lens causes incoming light rays to diverge before they enter the eye. Then the light rays from distant objects can be focused by the eye to form a sharp image on the retina and not in front of it.

Section 2 Review

Section Summary

▶ A convex lens is thicker in the middle than at the edges. Light rays are refracted toward the optical axis.

▶ The image formed by a convex lens depends on the distance of the object from the lens.

▶ A concave lens is thinner in the middle and thicker at the edges. Light rays are refracted away from the optical axis.

▶ The cornea and the lens focus light onto the retina.

7. **MAIN Idea Sketch** light rays as they pass through a convex lens and then through a concave lens.

8. **Compare** the image of an object less than one focal length from a convex lens with the image of an object more than two focal lengths from the lens.

9. **Describe** the image formed by a concave lens.

10. **Explain** how lenses are used to correct vision problems.

11. **Think Critically** If image formation by a convex lens is similar to image formation by a concave mirror, describe the image formed by a light source placed at the focal point of a convex lens.

Apply Math

12. **Calculate Object Distance** If you looked through a convex lens with a focal length of 15 cm and saw a real, inverted, enlarged image, what is the maximum distance between the lens and the object?

✓ Assessment   Online Quiz

Reading Preview

Essential Questions

▶ What is the difference between a refracting telescope and a reflecting telescope?
▶ How does a microscope magnify images?
▶ How does a camera work?

Review Vocabulary

refraction: the change in direction of a wave when it changes speed as it moves from one medium to another

New Vocabulary

refracting telescope
reflecting telescope
microscope

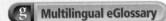
Multilingual eGlossary

Optical Instruments

MAIN◁Idea Lenses and mirrors are used to make objects easier to see.

Real-World Reading Link With a good digital camera, you can zoom in on distant objects, bring them into focus, and record images to view later. Optical instruments, such as cameras, telescopes, and microscopes allow us to see things that we could not see without them.

Telescopes

You know from your experience that it is difficult to see far-away objects clearly. When you look at an object, only some of the light reflected from its surface enters your eyes. As you move farther away from the object, the amount of light entering your eyes decreases, as shown in **Figure 16.** As a result, the object appears dimmer and less detailed.

A telescope uses a lens or a concave mirror that is much larger than your eye to gather more of the light from distant objects. The largest telescopes can gather more than a million times more light than the human eye. As a result, objects such as distant galaxies appear much brighter. Because the image formed by a telescope is so much brighter, more detail can be seen when the image is magnified.

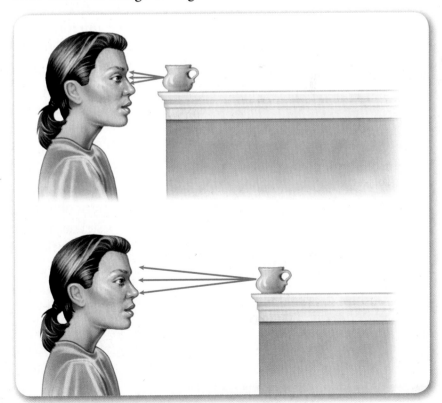

■ **Figure 16** All three of the light rays reflected off the mug enter the girl's eye in the top panel, but only one of those light rays enters the girl's eye in the bottom panel. As the girl gets farther away, fewer light rays reflected from the mug enter the girl's eyes.

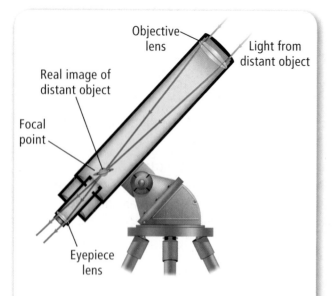

■ **Figure 17** A refracting telescope uses an objective lens and an eyepiece lens to gather light from distant objects so that scientists can observe and study those objects.

Refracting telescopes One common type of telescope is the refracting telescope. A telescope that uses lenses to gather light from distant objects is called a **refracting telescope.** A simple refracting telescope, shown in **Figure 17,** uses two convex lenses to gather and focus light from distant objects.

Incoming light from distant objects passes through the first lens, called the objective lens. Light rays from distant objects are nearly parallel to the optical axis of the lens. As a result, the objective lens forms a real image at the focal point of the lens, within the body of the telescope.

The second convex lens, called the eyepiece lens, magnifies this real image. When you look through the eyepiece lens, you see an enlarged, inverted, virtual image of the real image formed by the objective lens.

In order to form detailed images of distant objects, the objective lens of a refracting telescope must be as large as possible. A telescope lens can be supported only around its edge. A large lens can sag or flex due to its own weight, distorting the image that it forms. Another class of telescopes, called reflecting telescopes, do not have this problem.

Reflecting telescopes A telescope that uses mirrors and lenses to collect and focus light from distant objects is a **reflecting telescope.** Mirrors, unlike lenses, can be supported from behind. This additional support for mirrors prevents mirrors from sagging inside reflecting telescopes. As a result, reflecting telescopes can be much larger than refracting telescopes. **Figure 18** shows a reflecting telescope.

For this reflecting telescope, light from a distant object enters one end of the telescope and strikes a concave mirror at the opposite end. The light reflects off this mirror and converges. Before it converges at a focal point, the light hits a plane mirror inside the telescope tube. The light is then reflected from the plane mirror toward the telescope's eyepiece. The light rays converge at the focal point, creating a real image of the distant object. Just like a refracting telescope, a convex lens in the eyepiece then magnifies this image.

■ **Figure 18** A reflecting telescope uses two mirrors and an eyepiece lens to gather light from distant objects.

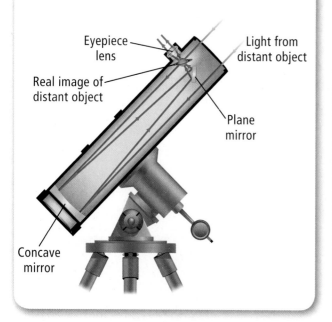

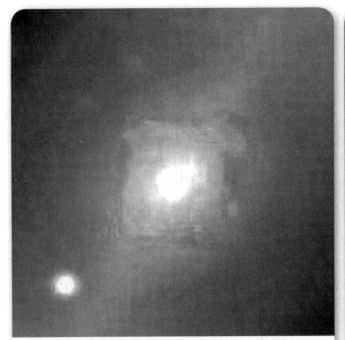

The distorting effects of Earth's atmosphere can cause telescopes on Earth to form blurry images.

The *Hubble Space Telescope* is above Earth's atmosphere and forms clearer images of objects in space.

Space telescopes Imagine being at the bottom of a swimming pool and trying to read a sign by the pool's edge. The motion of the water in the pool would distort your view of any object beyond the water's surface. In a similar way, Earth's atmosphere distorts the view of objects in space.

To overcome the blurriness of humans' view into space, the National Aeronautics and Space Administration (NASA) built a telescope called the *Hubble Space Telescope* and launched it into space, high above Earth's atmosphere. Because *Hubble* is above Earth's atmosphere, it has produced incredibly sharp and detailed images. **Figure 19** shows the difference in the images produced by telescopes on Earth and the *Hubble* telescope.

With the *Hubble Space Telescope*, scientists can detect light from planets, stars, and galaxies that would otherwise be scattered by Earth's atmosphere. *Hubble* is not the only space telescope. Other space telescopes, such as the *Chandra X-Ray Observatory* and the *Spitzer Space Telescope*, help scientists study the universe through X-ray and infrared radiation.

✓ **Reading Check** **Explain** why a space telescope is able to produce clearer images than telescopes on Earth.

The *Hubble* telescope is a type of reflecting telescope that uses two mirrors to collect and focus light to form an image. The primary mirror in the telescope is 2.4 m across. A next-generation space telescope, called the *James Webb Space Telescope*, is due to be launched in 2014. The primary mirror on the *James Webb Space Telescope* will be 6.5 m across.

■ **Figure 19** The view from telescopes on Earth is different from the view from telescopes in space. Both of these images show the core of the same galaxy, and both were taken through telescopes of similar size. However, the photograph on the left was taken with the McDonald Observatory on Earth, but the photograph on the right was taken with the *Hubble Space Telescope.*

VOCABULARY ....
ACADEMIC VOCABULARY
Distort
  to twist out of the true meaning or proportion
  People who lie distort the truth. ....

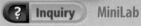
Experiment with Focal Lengths

Procedure 🖐️ 🧤 🧫

1. Read the procedure and safety information, and complete the lab form.
2. Fill a **glass test tube** with **water,** and seal it with a **lid** or a **stopper.**
3. Type or print the compound name *SULFUR DIOXIDE* in capital letters on a **piece of paper** or a **note card.**
4. Set the test tube horizontally over the words, and observe them. What do you notice?
5. Hold the tube 1 cm over the words, and observe them again. Record your observations. Repeat, holding the tube at several other heights above the words.

Analysis

1. **Describe** your observations of the words at the different distances.
2. **Identify** whether the image that you see at each height is real or virtual.

■ **Figure 20** A microscope uses two convex lenses to magnify small objects. The lens closest to the object that is being studied is called the objective lens. In a microscope, unlike in a refracting telescope, more than one lens magnifies the object.

Explain *why this microscope's light source is placed below the bug instead of above the bug.*

Microscopes

A telescope would be useless if you were trying to study the cells in a butterfly wing, a sample of pond scum, or the differences between a human hair and a horse hair. You would need a microscope to look at such small objects.

A **microscope** is a device that uses two convex lenses with relatively short focal lengths to magnify small, close objects. A microscope, like a telescope, has an objective lens and an eyepiece lens. However, it is designed differently because the objects viewed are close to the objective lens.

Figure 20 shows a simple microscope. The object to be viewed is placed on a transparent slide and is illuminated from below. The light passes by or through the object on the slide and then travels through the objective lens. The objective lens is a convex lens.

It forms a real, enlarged image of the object because the distance from the object to the lens is between one and two focal lengths. The real image is then magnified again by the eyepiece lens (another convex lens) to create a virtual, enlarged image. This final image can be hundreds of times larger than the actual object, depending on the focal lengths of the two lenses. The total magnification is the magnification of the objective times the magnification of the eyepiece.

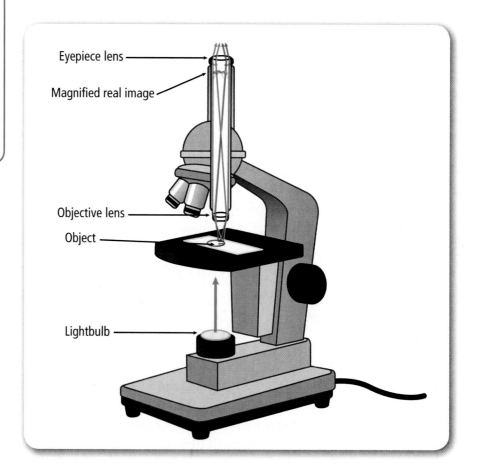

Eyepiece lens

Magnified real image

Objective lens

Object

Lightbulb

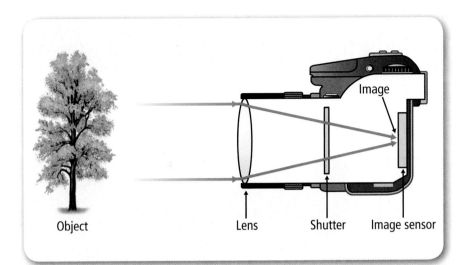

Object · Lens · Shutter · Image sensor

Image

■ **Figure 21** A camera's lens focuses an image onto the image sensor. An image sensor converts the light from an image into a set of electric signals.

Compare *a digital camera with the human eye.*

Cameras

With the click of a button, you can capture a beautiful scene in a photo. How does a digital camera make a reduced image of a life-sized scene? **Figure 21** shows the path that light follows as it enters a camera from a distant object. The light rays from distant objects are almost parallel to each other. When you take a picture with a camera, a shutter opens to allow light to enter the camera for a specific length of time.

The light reflected off the object enters the camera through an opening called the aperture. The camera lens focuses the image onto an image sensor, which converts light into electric signals. A computer then processes these signals into an image that can be displayed on a screen or printed.

Section 3 Review

Section Summary

▶ Refracting telescopes use two convex lenses to gather and focus light.

▶ Reflecting telescopes use a concave mirror, a plane mirror, and a convex lens to collect, reflect, and focus light.

▶ Placing a telescope in orbit avoids the distorting effects of Earth's atmosphere.

▶ A microscope uses two convex lenses with short focal lengths to magnify small, close objects.

▶ A camera lens focuses light onto an image sensor.

13. **MAIN Idea Identify** the advantage to making the objective lens larger in a refracting telescope.

14. **Describe** the image formed by the objective lens in a microscope.

15. **Explain** why the largest telescopes are reflecting telescopes instead of refracting telescopes.

16. **Think Critically** Which optical instrument—a telescope, a microscope, or a camera—forms images in a way most like your eye? Explain.

Apply Math

17. **Calculate Magnification** Suppose the objective lens in a microscope forms an image that is 100 times the size of an object. The eyepiece lens magnifies this image 10 times. What is the total magnification?

Use Vocabulary

Complete each sentence with the correct term from the Study Guide.

18. A flat, smooth surface that reflects light and forms an image is a(n) _____.

19. A(n) _____ uses two convex lenses to magnify small, close objects.

20. Every light ray that travels parallel to the optical axis before hitting a concave mirror is reflected such that it passes through the _____.

21. A(n) _____ is thicker in the middle than at the edges.

22. The inner lining of the eye that converts light images into electric signals is called the _____.

Check Concepts

23. Which best describes image formation by a plane mirror?
  A) A real image is formed in front of the mirror.
  B) A real image is formed behind the mirror.
  C) A virtual image is formed in front of the mirror.
  D) A virtual image is formed behind the mirror.

24. Which can form an enlarged image?
  A) convex mirror   **C)** convex lens
  B) plane mirror   **D)** concave lens

25. Which is NOT part of a reflecting telescope?
  A) plane mirror   **C)** convex lens
  B) concave mirror   **D)** concave lens

Use the figure below to answer question 26.

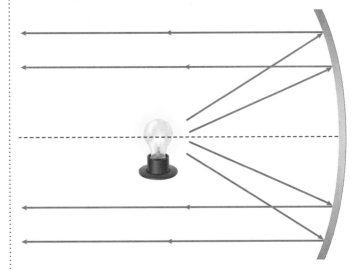

26. Which is being used in the figure above?
  A) concave lens   **C)** concave mirror
  B) convex lens   **D)** convex mirror

27. What do lenses do?
  A) reflect light   **C)** diffract light
  B) refract light   **D)** interfere with light

Use the figure below to answer question 28.

28. **BIG** **Idea** Which way does the lens shown bend light that is parallel to the optical axis?
  A) toward its optical axis
  B) toward its focal point
  C) away from its optical axis
  D) away from its edges

29. What type of lens is used to correct farsightedness?
  A) flat lens
  B) convex lens
  C) concave lens
  D) plane lens

Interpret Graphics

Use the figure below to answer question 30.

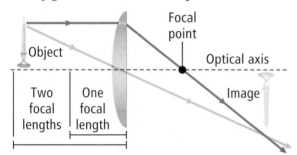

30. Suppose the image of the candle moves away from the focal point. How did the position of the candle change?

31. Copy and complete the following table about image formation by lenses and mirrors.

Image Formation by Lenses and Mirrors		
Type of Lens or Mirror	**Position of Object**	**Type of Image**
Concave lens	all positions of object	virtual, upright, reduced
Convex lens	closer than one focal length	a.
	between one and two focal lengths	b.
	farther than two focal lengths	real, inverted, reduced
Concave mirror	closer than one focal length	c.
	object placed at focal point	d.
	farther than two focal lengths	e.
Convex mirror	all positions of object	f.

Think Critically

32. Infer Could a person who is nearsighted use his or her glasses to focus light and start a fire?

33. THEME FOCUS Compare and contrast a refracting telescope and a microscope.

34. Infer why a convex mirror and a concave lens can never produce a real image.

35. Explain The top half of a bifocal lens helps a person to focus on distant objects. The bottom half of a bifocal lens helps a person to focus on nearby objects. Why might a person need glasses with bifocal lenses?

36. Infer why it would be easier to make a concave mirror for a reflecting telescope than an objective lens of the same size for a refracting telescope.

37. Compare A concave lens made of plastic is placed in a liquid. Light rays traveling in the liquid are not refracted when they pass through the lens. Compare the speed of light in the plastic and in the liquid.

Apply Math

38. Calculate Magnification The magnification of a refracting telescope can be calculated by dividing the focal length of the objective lens by the focal length of the eyepiece lens. If an objective lens has a focal length of 1 m and the eyepiece has a focal length of 1 cm, what is the magnification of the telescope?

39. Infer Object Distance You hold an object in front of a concave mirror with a 30-cm focal length. You don't see a reflected image. How far from the mirror is the object?

Standardized Test Practice

Multiple Choice

Record your answers on the answer sheet provided by your teacher or on a sheet of paper.

1. How far is an object from a concave mirror if the image formed is upright?
   A. one focal length
   B. less than one focal length
   C. more than two focal lengths
   D. two focal lengths

Use the figure below to answer questions 2 and 3.

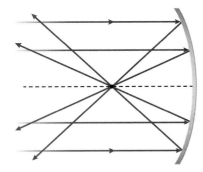

2. Which describes a light ray that passes through the focal point and then is reflected by the mirror?
   A. It travels parallel to the optical axis.
   B. It forms a real image.
   C. It is reflected back through the focal point.
   D. It forms a virtual image.

3. If the mirror becomes flatter and the focal point moves farther from the mirror, which best describes the reflection of the parallel rays shown in the figure?
   A. They pass through the old focal point.
   B. They do not pass through either the old or the new focal point.
   C. They pass through the new focal point.
   D. They reverse direction.

4. Which describes the image formed by a convex mirror?
   A. real   C. inverted
   B. enlarged   D. virtual

5. What is an advantage to increasing the diameter of the concave mirror in a reflecting telescope?
   A. The mirror forms brighter images.
   B. The mirror forms larger images.
   C. The mirror forms more magnified images.
   D. The focal length increases.

Use the table below to answer questions 6–8.

Image Magnification by a Convex Lens		
Object Distance (cm)	Image Distance (cm)	Magnification
250.0	62.5	0.25
200.0	66.7	0.33
150.0	75.0	0.50
100.0	100.0	1.00
75.0	150.0	2.00

6. How does the image change as the object gets closer to the lens?
   A. It gets larger.
   B. It gets smaller.
   C. It gets closer.
   D. It becomes real.

7. Which is the best estimate of the magnification if the object is 225 cm from the lens?
   A. 0.20
   B. 0.30
   C. 64
   D. 68

8. What should the object distance be if the lens is to be used as a magnifying glass?
   A. 150 cm
   B. 100 cm
   C. greater than 250 cm
   D. less than 100 cm

Short Response

Record your answers on the answer sheet provided by your teacher or on a sheet of paper.

9. Describe how you could determine whether the image formed by a lens or a mirror is a real image or a virtual image.

10. The objective lens in a microscope has a magnification of 30. What is the magnification of the microscope if the eyepiece lens has a magnification of 20?

11. Describe how the focal length of a convex lens changes as the lens becomes more curved.

Use the figure below to answer questions 12 and 13.

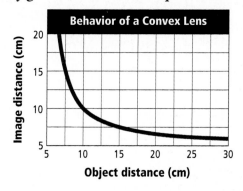

12. Determine how far the image is from the lens when the object is 15 cm from the lens.

13. At what object distance are the image distance and the object distance equal?

Extended Response

Record your answers on a sheet of paper.

Use the figure below to answer questions 14 and 15.

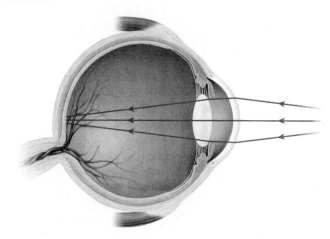

14. Describe the vision problem shown in the figure. Why does this vision problem become more serious as people age?

15. Explain how the vision problem shown in the figure can be corrected.

16. Predict whether a camera that uses a concave lens to focus light onto the image sensor would work.

17. Describe the change in the lenses in your eyes when you look at this book and then look out the window at a distant object.

18. Explain why objects become dimmer and less detailed as they move farther away.

NEED EXTRA HELP?																		
If You Missed Question . . .	1	2	3	4	5	6	7	8	9	10	11	12	13	14	15	16	17	18
Review Section . . .	1	1	1	1	3	2	2	2	2	3	2	2	2	2	2	3	2	3

UNIT 4

Matter

THEMES

Changes of Matter Substances can change state and be physically combined into mixtures without changing their chemical identities.

Energy All physical and chemical changes absorb or release energy.

Motion and Forces Forces of attraction and repulsion are balanced within the stable atom.

Scientific Inquiry The current model of the atom and atomic theory are the culmination of observations and investigations that began in ancient Greece.

Structure and Properties of Matter The atomic structure of an element determines its properties.

WebQuest **STEM UNIT PROJECT**

Technology Research robots used in extreme environments. Present the materials that would be used to construct your own extreme robot.

CHAPTER 14

Solids, Liquids, and Gases

ConnectED

Your one-stop online resource
connectED.mcgraw-hill.com

- Video
- Audio
- Review
- ? Inquiry
- WebQuest
- ✓ Assessment
- Concepts in Motion
- g Multilingual eGlossary

Launch Lab
Floating and Density

Why do ships float? Do they actually weigh less or is there some other force pushing up on them? What about objects that sink, such as rocks? Is there a force that pushes up on them too?

For a lab worksheet, use your StudentWorks™ Plus Online.

? Inquiry Launch Lab

FOLDABLES®

Create a concept map book. Label the tabs as shown. Use it to organize your notes about solids, liquids, and gases.

States of Matter

Solid | Liquid | Gas

THEME FOCUS Changes of Matter
Changes in the arrangement and motion of particles occur as matter changes state.

BIG **Idea** Each state of matter—solid, liquid, or gas—has unique properties, defined by the motion of its particles.

Section 1 • Matter and Thermal Energy

Section 2 • Properties of Fluids

Section 3 • Behavior of Gases

Reading Preview

Essential Questions

▶ What is the kinetic theory of matter?

▶ How do particles move in the different states of matter?

▶ How do particles behave at the boiling and melting points?

Review Vocabulary

kinetic energy: energy of motion

New Vocabulary

kinetic theory
melting point
heat of fusion
boiling point
heat of vaporization
sublimation
plasma
thermal expansion

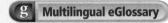 **g** Multilingual eGlossary

FOLDABLES
Incorporate information from this section into your Foldable.

Matter and Thermal Energy

MAIN ◀Idea Matter can exist as a solid, a liquid, a gas, or a plasma.

Real-World Reading Link Water can be a cool, refreshing drink, a hard surface for sliding on, or a dangerously hot gas. How water behaves depends on its state of matter.

Kinetic Theory

You encounter solids, liquids, and gases every day. Look at **Figure 1.** Can you identify the states of matter present? The tea is in the liquid state. The ice cubes dropped into the tea to cool it are in the solid state. Surrounding the glass, as part of the air, is water in the gas state. How do these states compare?

Gas state To understand the states of matter, we must think about the particles that make up matter. Consider the air around you: it is composed of nitrogen, oxygen, and water, along with other gases. These atoms and molecules—the particles that make up the air—are constantly moving. The **kinetic theory** is an explanation of how the particles in gases behave. To explain the behavior of particles, it is necessary to make some basic assumptions. The assumptions of the kinetic theory are as follows:

1. All matter is composed of tiny particles (atoms, molecules, and ions).
2. These particles are in constant, random motion.
3. The particles collide with each other and with the walls of any container in which they are held.
4. The amount of energy that the particles lose from these collisions is negligible.

■ **Figure 1** Water is a substance that can exist in all three common states of matter at the same time.

Identify *the solid and liquid states of water in this photo.*

Figure 2 demonstrates the kinetic theory, showing the particles that make up a substance in the gas state. Because their particles are in constant motion, colliding with each other and with the walls of their container, gases do not have a fixed volume or shape. Instead, the particles that make up a gas spread out so that they fill whatever container they are in.

Liquid state Although the kinetic theory explains the behaviors of gas particles, some of the assumptions of the theory apply to liquids and solids as well. The particles of a substance in the liquid state, shown in **Figure 2,** are also constantly moving, although they are not moving as quickly as they would be if the substance were in the gas state. Therefore, the particles that make up a substance have less kinetic energy when in its liquid state than when in its gas state.

Because they have less energy, the particles are less able to overcome their attractions to each other. They can slide past each other, allowing a liquid to flow and take the shape of its container. However, because the particles that make up a liquid have not completely overcome the attractive forces between them, the particles cling together, giving the liquid a definite volume.

Solid state Unlike a gas or a liquid, a solid has a definite shape and volume. The particles that make up a solid are closely packed together, as shown in **Figure 2.** They are still in motion, but they have so little kinetic energy that the particles are unable to overcome their attractions to each other.

Many solids are crystalline, which means their particles have specific geometric arrangements. **Figure 3** shows the geometric arrangement of ice. Notice that the hydrogen and oxygen atoms alternate in the arrangement.

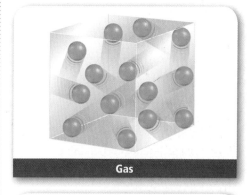
Gas

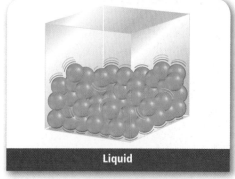
Liquid

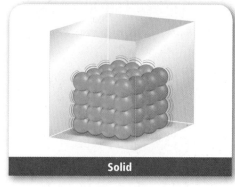
Solid

■ **Figure 2** Solids, liquids, and gases differ in how their particles move. These differences account for their physical properties.

Compare and contrast *the shape and volume of each state of matter.*

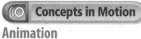 **Concepts in Motion**
Animation

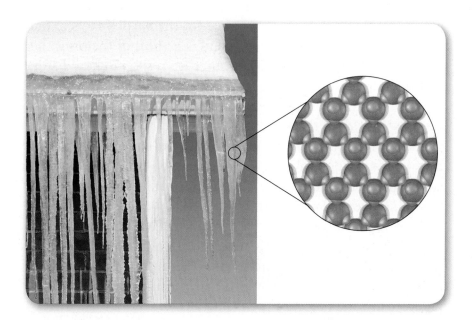

■ **Figure 3** Ice is a crystalline solid—its particles have a specific geometric arrangement. Although ice does not look like it is moving, its molecules are vibrating in place.

BrainPOP

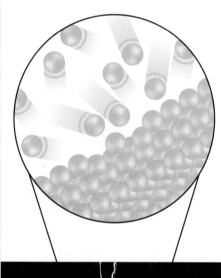

■ **Figure 4** When ice is placed in water, energy from the particles of the liquid water is transferred to the particles of solid ice, melting the ice and cooling the water.

Thermal energy Think about the ice in **Figure 3.** How can frozen, solid ice have motion? The particles that make up solids are held tightly in place by the attractions between the particles. Those attractions give solids a definite shape and volume. However, the particles that make up a solid are still in constant motion. The particles' thermal energy causes them to vibrate.

Thermal energy is the total energy of a material's particles. This includes both the kinetic energy of the particles as well as their potential energy. Energy from the motions of individual particles and energy from forces that act within or between particles are both forms of thermal energy. Energy from the motion of an object as a whole and energy from its interactions with its surroundings are not thermal energy.

Temperature *Temperature* is the term used to explain how hot or cold an object is. Temperature represents the average kinetic energy of the particles that make up a substance. On average, molecules of water at 0°C have less kinetic energy than molecules of water at 100°C.

Changes of State

What happens to a solid when thermal energy is added to it? Think about the iced water in **Figure 4.** The particles that make up the water are moving fast and colliding with the particles that make up the ice cube. Those collisions transfer energy from the water to the ice. The particles at the surface of the ice cube vibrate faster, transferring energy to other particles in the ice cube.

Melting and freezing Soon, the particles that make up the ice have enough kinetic energy to overcome the attractive forces holding them in their crystalline structure. The ice melts. The **melting point** is the temperature at which a solid becomes a liquid. Energy is required for the particles to slip out of the ordered arrangement of a solid. The **heat of fusion** is the energy required to change a substance from solid to liquid at its melting point.

The transfer of energy between particles of liquid and particles of solid causes the ice to melt, but what happens to the particles of liquid after they collide with the solid? They slow down because they have less kinetic energy. As more of these collisions occur, the average kinetic energy of the particles of the liquid decreases, and the liquid cools.

Freezing is the reverse of melting. When a liquid's temperature is lowered, the average kinetic energy of the molecules decreases. When enough energy has been removed, the molecules become fixed into position. The freezing point is the temperature at which a liquid turns into a solid.

Vaporization and condensation How does a liquid become a gas? Remember that the particles that make up a liquid are constantly moving. When particles move fast enough to escape the attractive forces of other particles, they enter the gas state. This process is called vaporization. Vaporization can occur in two ways: evaporation and boiling. The process in which a gas becomes a liquid is called condensation. Condensation is the reverse of vaporization.

Evaporation Evaporation occurs at the surface of a liquid and can happen at nearly any temperature. To evaporate, particles must be at the liquid's surface and have enough kinetic energy to escape the attractive forces of the liquid.

Boiling Shown in **Figure 5,** boiling is the second way that a liquid can vaporize. Unlike evaporation, boiling occurs throughout a liquid at a specific temperature, depending on the pressure on the surface of the liquid.

The **boiling point** of a liquid is the temperature at which the pressure of the vapor in the liquid is equal to the external pressure acting on the surface of the liquid. This external pressure pushes down on the liquid, keeping particles from escaping. Particles require energy to overcome this pressure. The **heat of vaporization** is the amount of energy required for the liquid at its boiling point to become a gas.

Sublimation At certain pressures, some substances can change directly from solids into gases without going through the liquid phase. **Sublimation** is the process of a solid changing directly to a gas without forming a liquid. **Figure 6** shows frozen carbon dioxide, also known as dry ice, which is a common substance that undergoes sublimation.

■ **Figure 5** As temperature increases, the particles that make up a substance in its liquid state move faster. When their energy creates sufficient pressure to surpass the air pressure above the liquid, the liquid boils.

Infer *What is inside the bubbles of the boiling liquid?*

■ **Figure 6** Carbon dioxide (CO_2) turns from a solid directly to a gas. Because this gas is very cold, it causes water in the air to condense, forming a white fog.

Figure 7 Although thermal energy is added at a constant rate, the temperature of the water only increases at *A, C,* and *E.* At *B* and *D,* the added energy is used to overcome the attractions between the particles.

Infer *how this graph would be different if 2.0 kg of water were being heated instead of 1.0 kg of water. How would this graph be different if 0.5 kg of water were being heated?*

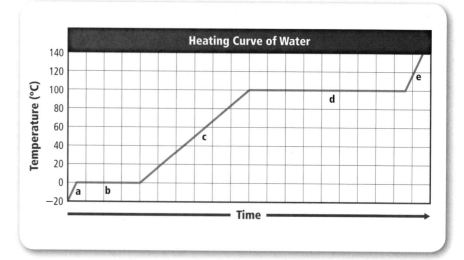

Heating Curve of Water

Temperature (°C) / Time

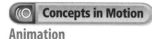

Concepts in Motion

Animation

Heating curves A graph of temperature v. time for heating of 1.0 kg of water is shown in **Figure 7.** This type of graph is called a heating curve. It shows how temperature changes over time as thermal energy is continuously added. Notice the two areas on the graph where the temperature does not change. At 0°C, ice is melting. All of the energy put into the ice at this temperature is used to overcome the attractive forces between the particles. The flat line on the graph indicates that temperature remains constant during melting.

After the attractive forces are overcome, the particles move more freely and their temperature increases. At 100°C, water is boiling, the temperature remains constant again, and the graph is flat. All of the energy that is put into the water goes to overcoming the remaining attractive forces between the particles. When all of the attractive forces between the particles are overcome, the energy goes into increasing the temperature again.

Plasma State

So far, you have learned about the three familiar states of matter—solids, liquids, and gases. However, there is a state of matter beyond the gas state. **Plasma** is matter that has enough energy to overcome not just the attractive forces between its particles but also the attractive forces within its atoms. The atoms that make up a plasma collide with such force that the electrons are completely stripped off the atoms.

You may be surprised to learn that most of the ordinary matter in the universe is in the plasma state. Every star that you can see in the sky, including the Sun, is composed of matter in the plasma state. Most of the matter between the stars and galaxies is also in the plasma state. The familiar states of matter—solid, liquid, and gas—are extremely rare in the universe.

Thermal Expansion

Have you ever wondered why a concrete sidewalk has seams? When thermal energy is transferred to a concrete sidewalk, the concrete expands. Without the seams, a concrete sidewalk would crack in hot weather. The kinetic theory can help to explain this behavior.

Recall that particles move faster and farther apart as the temperature rises. This separation of particles results in an expansion of the entire object, known as thermal expansion. **Thermal expansion** is an increase in the size of a substance when the temperature is increased. Substances also contract when they cool.

Thermometers A common example of liquids undergoing thermal expansion occurs in thermometers, like the one shown in **Figure 8.** The addition of energy causes the particles that make up the liquid in the thermometer to move faster. As their motion increases, the particles that make up the liquid in the narrow thermometer tube start to move farther apart. This causes the liquid in a thermometer to expand and rise as the temperature increases.

Hot-air balloons An application of gases undergoing thermal expansion is shown in **Figure 9.** Hot-air balloons are able to rise due to the thermal expansion of air. The air in the balloon is heated, causing the distance between the particles that make up the air to increase. As the hot-air balloon expands, the number of particles per cubic centimeter decreases. This expansion results in a decreased density of hot air. Because the density of the air in the hot-air balloon is lower than the density of the cooler air outside, the balloon will rise.

■ **Figure 8** As the air temperature goes up, the liquid in the thermometer is heated and expands. As a result, the level of the liquid rises. The liquid in the thermometer contracts with a decrease in temperature.

■ **Figure 9** When the air inside a hot-air balloon is heated, its particles move farther apart. The balloon rises because the air inside it is less dense than the surrounding air.

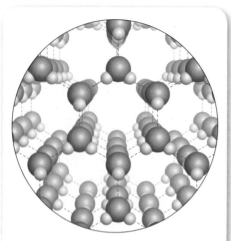

Figure 10 When water freezes, the positively and negatively charged ends of the water molecules interact, creating empty spaces in the crystal lattice.

Explain *why ice floats in water.*

Water's strange behavior Ordinarily, substances contract as their temperatures decrease. However, an exception to this rule is water. Over a small range of temperatures, water expands as the temperature decreases. At first, it behaves like other substances. As the temperature begins to drop, the particles that make up the water move closer together. This continues until the water reaches 4°C.

Water molecules are unusual in that they have highly positive and highly negative areas. These charged regions affect the behavior of water. As the temperature of water continues to drop under 4°C, the molecules line up so that only positive and negative areas are near each other, as shown in **Figure 10.** As a result, empty spaces occur in the structure. Water expands as it cools from about 4°C to 0°C and becomes less dense than liquid water. That is why ice floats in liquid water.

Solid or Liquid?

Other substances also show unusual behaviors when changing states. Amorphous solids and liquid crystals are two classes of materials that do not react as you would expect when they are changing states.

Amorphous solids Ice melts at 0°C, and lead melts at 327°C. But not all solids have a specific temperature at which they melt. Consider a stick of butter. Instead of having a specific melting point, butter softens and melts over a range of temperatures.

Some solids are like butter. Instead of having a specific melting point, they soften and gradually turn into a liquid over a temperature range. These solids lack a crystalline structure and are called amorphous solids. One common amorphous solid is glass, shown in **Figure 11.**

Figure 11 Glass lacks the repeating crystalline structure of solids like ice. Rather than melting at a specific temperature, glass becomes increasingly soft and malleable as temperature increases.

Liquid crystals Liquid crystals form another group of materials that do not change states in the usual manner. Normally, the ordered geometric arrangement of a solid is lost when the substance goes from the solid state to the liquid state. Liquid crystals start to flow during the melting phase, similar to a liquid. But they do not lose their ordered arrangement completely, as most substances do. Liquid crystals will retain their geometric order in specific directions.

Liquid crystals are placed in classes, depending upon the type of order they maintain when they liquify. They are highly responsive to temperature changes and electric fields. Scientists use the unique properties of liquid crystals to make liquid crystal displays (LCD) for cell phones, calculators, and netbooks, shown in **Figure 12.** LCD screens are composed of individual crystal picture elements, or "pixels" for short. Varying the amount of electricity that passes through the pixel determines how the crystals are aligned and whether light is able to pass through them.

■ **Figure 12** Many small computing devices and electronics, such as MP3 players, cell phones, TVs, and netbooks, utilize liquid crystal displays (LCDs).

Section **1** Review

Section Summary

▶ There are four main states of matter: solid, liquid, gas, and plasma.

▶ The kinetic theory is an explanation of how the particles that make up gases behave.

▶ Thermal energy is the total energy of the particles that make up a material, including kinetic and potential energy.

▶ Temperature is the average kinetic energy of a substance.

1. **MAIN Idea Describe** the movement of the particles in solids, liquids, and gases.

2. **State** the basic assumptions of the kinetic theory.

3. **Describe** how the particles of a substance behave at its melting point.

4. **Describe** how the particles of a substance behave at its boiling point.

5. **Think Critically** How would the heating curve for glass be different from the heating curve for water?

Apply Math

6. **Interpret Data** Using the graph in **Figure 7,** describe the energy changes that are occurring when water goes from −15°C to 120°C.

7. **Make and Use Graphs** The melting point of acetic acid is 17°C, and the boiling point is 118°C. Draw a graph similar to the graph in **Figure 7** showing the phase changes for acetic acid. Clearly mark the three phases, the boiling point, and the melting point on the graph.

Objective
- **Observe** the thermal energy changes that occur as matter goes from the solid to the gas state.

Background: Matter can be changed from one state to another. Commonly observed changes of state include ice melting and water boiling.

Question: *How much energy is involved in the phase changes that we commonly witness?*

Preparation

Materials

500-mL beaker
ice
thermometer
hot plate

Safety Precautions

Procedure

1. Read the procedure and safety information, and complete the lab form.

2. Set up the equipment as pictured. Ice should be placed in the beaker. Prepare a data table for tracking the temperature of the water in the beaker over time.

3. Gently heat the ice in the beaker. Every 3 minutes, record your observations and the temperature of the water in the beaker. Do not touch the thermometer to the bottom or sides of the container.

4. After the ice in the beaker melts and the water begins to boil, observe the system for several more minutes and record your observations.

5. Turn off the heat, and let your system completely cool before you clean up.

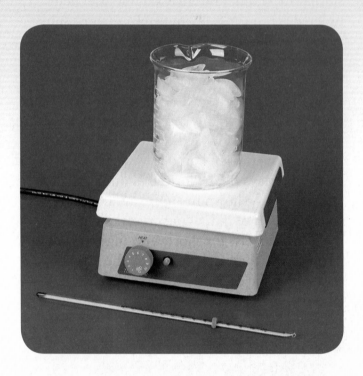

Conclude and Apply

1. **Draw** a picture of the system used in this lab.

2. **Label** the state in which the water started in the beaker, the state into which it changed in the beaker, and the state above the beaker.

3. **Make and Use Graphs** Draw a temperature vs. time graph using your data from this lab.

4. **Apply** Explain the shape of your graph in terms of energy. Why are there flat lines in temperature despite the constant addition of heat?

COMMUNICATE YOUR DATA

Compare your graph to those of your classmates. Decide what your curve would look like if you started with water vapor and cooled it.

? Inquiry Lab

Reading Preview

Essential Questions

▶ What is Archimedes' principle?
▶ What is Pascal's principle?
▶ What is Bernoulli's principle?
▶ What are some applications of Archimedes', Pascal's, and Bernoulli's principles?

Review Vocabulary

density: mass per unit volume of a material

New Vocabulary

buoyancy
pressure
viscosity

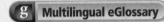

g Multilingual eGlossary

Properties of Fluids

MAIN ◄Idea Fluids flow and exert forces on objects.

Real-World Reading Link Have you ever seen an aircraft carrier? Although these ships are composed of metal and sometimes weigh more than 100,000 tons, they still float. How is this possible?

Archimedes' Principle and Buoyancy

Some ships are like floating cities. For example, aircraft carriers are large enough to allow airplanes to take off and land on their decks. Despite their weights, these ships float. There is a force pushing up on the ship that opposes the gravitational force, pulling the ship down.

What is the force pushing up on the ship? It is called the buoyant force. If the buoyant force is equal to the object's weight, the object will float. If the buoyant force is less than the object's weight, the object will sink. **Buoyancy** is the ability of a fluid—a liquid or a gas—to exert an upward force on an object immersed in it.

Archimedes' principle In the third century B.C., a Greek mathematician named Archimedes made a discovery about buoyancy. Archimedes found that the buoyant force on an object is equal to the weight of the fluid displaced by the object. For example, if you place a block of wood in water, it will push water out of the way as it begins to sink—but only until the weight of the water displaced equals the block's weight.

When the weight of water displaced—the buoyant force—becomes equal to the weight of the block, it floats. If the weight of the water displaced is less than that of the block, the object sinks. **Figure 13** shows the forces that affect objects in fluids.

✓ **Reading Check Infer** why rocks sink and rubber balls float in water.

■ **Figure 13** The steel block sinks because the buoyant force from the fluid is less than the gravitational force on the object. When the buoyant force is the same as or greater than the gravitational force—as with the wooden block—the object floats.

Compare *the volumes of the wood block and the steel block. How do the masses of the wood block and the steel block compare?*

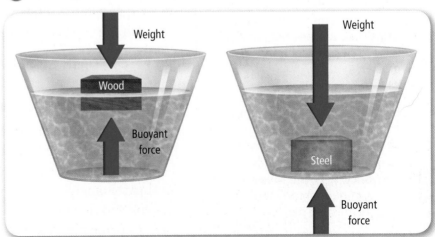

■ **Figure 14** The overall density of a large ship is lower than the density of water because its empty hull contains mostly air.

Infer *why a boat cannot be made of solid steel.*

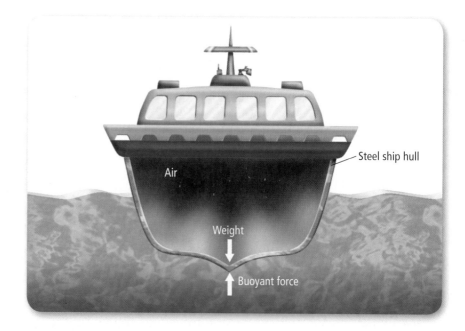

Steel ship hull

Air

Weight

Buoyant force

(((O))) **Concepts in Motion**   Animation

MiniLab

(?) **Inquiry**   MiniLab

Relate Density and Buoyancy

Procedure 🥽 🧤 👐
1. Read the procedure and safety information, and complete the lab form.
2. Pour 10 mL of **corn syrup** into a **100-mL beaker**. In **another beaker,** add 3 to 4 drops of **food coloring** to 10 mL of **water.** Pour the dyed water into the 100-mL beaker containing corn syrup. Add 10 mL of **vegetable oil** to the beaker.
3. Drop a 0.5-cm² piece of **aluminum foil**, a **steel nut**, and a **whole peppercorn** into the beaker.

Analysis
1. **Explain,** using the concept of density, why the contents of the beaker separated into layers.
2. **Explain,** using the concept of buoyancy, why the foil, steel nut, and peppercorn settle in their respective places.

Comparing buoyancy and weight Look again at the wood and steel blocks in **Figure 13.** They both displace the same volume and weight of water when submerged. Therefore, the buoyant forces on the blocks are equal. Yet, the steel block sinks and the wood block floats. What is different?

The steel block weighs much more than the wood block. The gravitational force on the steel block is enough to make the steel block sink. The gravitational force on the wood block is not enough to make the wood block sink.

Density and buoyancy One way to know whether an object will float or sink is to compare its density to the density of the fluid in which it is placed. An object floats if its density is less than that of the fluid. Remember that density is mass per unit volume. The density of the steel block is greater than the density of water. The wood block's density is less than that of water.

Suppose you formed the steel block into the shape of the hull of a ship filled with air, as in **Figure 14.** Now the same mass takes up a larger volume. The overall density of the steel boat and air is less than the density of water. The boat will now float.

✓ **Reading Check** **Explain** why a steel block sinks but a steel ship floats.

Pascal's Principle and Pressure

Blaise Pascal (1623–1662), a French scientist, discovered that pressure applied to a fluid is transmitted throughout the fluid. This is why when you squeeze one end of a toothpaste tube, toothpaste emerges from the other end. The pressure has been transmitted through the toothpaste. In order to understand Pascal's principle, you must first understand pressure.

Pressure Right now, pressure from the air is pushing on you from all sides like the pressure you feel underwater in a swimming pool. **Pressure** is force exerted per unit area.

Pressure Equation

$$\text{Pressure (Pa)} = \frac{\text{force (N)}}{\text{area (m}^2\text{)}}$$

$$P = \frac{F}{A}$$

VOCABULARY . . . .
SCIENCE USAGE V. COMMON USAGE
Pressure
Science usage
    force per unit area
    Increasing the pressure on a gas decreases its volume.

Common usage
    the burden of physical or mental distress
    Teachers often feel a lot of pressure to help their students do well in school. . . . .

The SI unit of pressure is the pascal (Pa). Because pressure is the amount of force divided by area, one pascal is one newton per square meter (N/m²). Most pressures are given in kilopascals (kPa) because 1 Pa is a very small amount of pressure.

EXAMPLE Problem 1

Calculate Force Atmospheric pressure at sea level is about 101 kPa. With how much total force does Earth's atmosphere push on an average human being at sea level? Assume that the surface area of an average human is 1.80 m².

List the Unknown: force: **F**

List the Knowns: pressure: **P = 101 kPa = 101,000 Pa**
                            area: **A = 1.80 m²**

Set Up the Problem: $P = \frac{F}{A}$

Solve the Problem: $101{,}000 \text{ Pa} = \dfrac{F}{1.80 \text{ m}^2}$

                            $F = 101{,}000 \text{ Pa} \times 1.80 \text{ m}^2$

                            $= 182{,}000 \text{ Pa} \cdot \text{m}^2 = 182{,}000 \,\frac{\text{N}}{\cancel{\text{m}^2}} \cdot \cancel{\text{m}^2} = \textbf{182,000 N}$

Check the Answer: You've set up your equation correctly if the units are the same on both sides: units of pressure = Pa = N/m² and (units of force)/(units of area) = N/m².
                            The units on both sides of the equation match. Now, just double-check the calculation.

PRACTICE Problems

Find **Additional Practice Problems** in the back of your book.

8. A diver who is 10.0 m underwater experiences a pressure of 202 kPa. If the diver's surface area is 1.50 m², with how much total force does the water push on the diver?

9. A car weighs 15,000 N, and its tires are inflated to a pressure of 190 kPa. How large is the area of the car's tires that are in contact with the road?

10. **Challenge** The atmospheric pressure at the surface of Venus is 91 times the pressure at sea level on Earth. With approximately how much total force would Venus's atmosphere push on an average human being at sea level? Assume that the surface area of an average human being is 1.8 m².

Review

Additional Practice Problems

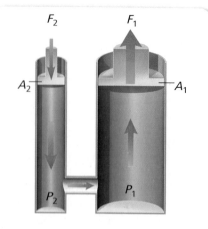

F_2 F_1

A_2 A_1

P_2 P_1

$P_2 = P_1$

Figure 15 The pressure of the fluid on one side of a hydraulic lift is equal to the pressure on the other side.

Concepts in Motion
Animation

Pascal's principle The idea that pressure is transferred through a fluid can be written as an equation: pressure in = pressure out. Since pressure is force over area, Pascal's principle can be written another way.

Pascal's Principle

$$\frac{\text{input force (N)}}{\text{input area (m}^2)} = \frac{\text{output force (N)}}{\text{output area (m}^2)}$$

$$\frac{F_{in}}{A_{in}} = \frac{F_{out}}{A_{out}}$$

Hydraulic lifts Auto repair shops often make use of hydraulic lifts, which move heavy loads in accordance with Pascal's principle. A pipe that is filled with fluid connects small and large cylinders, as shown in **Figure 15.** Pressure applied to the small cylinder is transferred through the fluid to the large cylinder. With a hydraulic lift, you could use your weight to lift something much heavier than you.

EXAMPLE Problem 2

Calculate Forces A hydraulic lift is used to lift a heavy machine that is pushing down on a 2.8-m² platform with a force of 3,700 N. What force must be exerted on a 0.072-m² piston to lift the heavy machine?

List the Unknowns: force on piston: F_{in}

List the Knowns: force on platform: F_{out} = **3,700 N**
area of platform: A_{out} = **2.8 m²**
area of piston: A_{in} = **0.072 m²**

Set Up the Problem: $\dfrac{F_{in}}{A_{in}} = \dfrac{F_{out}}{A_{out}}$

Solve the Problem: $F_{in} = \left(\dfrac{F_{out}}{A_{out}}\right) A_{in} = \left(\dfrac{3{,}700 \text{ N}}{2.8 \text{ m}^2}\right) 0.072 \text{ m}^2 = \textbf{95 N}$

Check the Answer: The ratio of the forces should be the same as the ratio of the areas. The area of the platform is about 40 times the area of the piston. Therefore, the force on the platform should be about 40 times the force on the piston. 3,700 N is about 40 times greater than 95 N, so the answer is reasonable.

PRACTICE Problems

Find **Additional Practice Problems** in the back of your book.

11. A car weighing 15,000 N is on a hydraulic lift platform measuring 10 m². What is the area of the smaller piston if a force of 1,100 N is used to lift the car?

Review
Additional Practice Problems

12. Challenge A heavy crate applies a force of 1,500 N on a 25-m² piston. The smaller piston is 1/30 the size of the larger one. What force is needed to lift the crate?

Bernoulli's Principle

Daniel Bernoulli (1700–1782) was a Swiss scientist who studied the properties of moving fluids, such as water and air. Bernoulli found that fluid velocity increases when the flow of the fluid is restricted. Placing your thumb over a running garden hose, as shown in **Figure 16,** demonstrates this effect. When the opening of the hose is decreased in size, the water flows out more quickly.

Bernoulli examined the relationship between fluid flow and pressure. You might think that increasing the velocity of a fluid's flow would increase its pressure, but Bernoulli found the opposite to be true. According to Bernoulli's principle, as the velocity of a fluid increases, the pressure exerted by that fluid decreases. He published this discovery in 1738.

One application of Bernoulli's principle is the hose-end sprayer. This sprayer is used to apply fertilizers and pesticides to yards and gardens. To use this sprayer, a concentrated solution of the chemical that is to be applied is placed in the sprayer. The sprayer is attached to a garden hose, as shown in **Figure 17.** A straw-like tube is attached to the lid of the unit. The end of the tube is submerged into the concentrated chemical. The water to the garden hose is turned to a high flow rate.

When you are ready to apply the chemicals to the lawn or plant area, you must push a trigger on the sprayer attachment. This allows the water in the hose to flow at a high rate of speed, creating a low pressure area above the straw-like tube. The concentrated chemical solution is sucked up through the straw and into the stream of water. The concentrated solution mixes with water, reducing the concentration to the appropriate level and creating a spray that is easy to apply.

Reading Check **Describe** how pressure changes as the velocity of a fluid increases.

■ **Figure 16** Bernoulli's principle explains why covering the end of a hose causes the water to flow faster. When the flow of a fluid is restricted, its velocity increases.

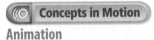
Concepts in Motion
Animation

■ **Figure 17** A hose-end sprayer makes use of Bernoulli's principle.

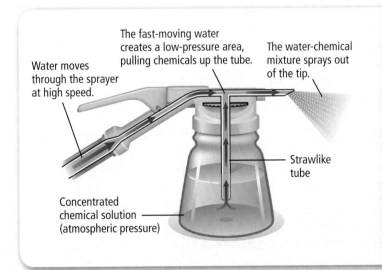

Water moves through the sprayer at high speed.

The fast-moving water creates a low-pressure area, pulling chemicals up the tube.

The water-chemical mixture sprays out of the tip.

Strawlike tube

Concentrated chemical solution (atmospheric pressure)

Figure 18 This maple syrup flows slowly because it has high viscosity.

Identify *other examples of liquids with high viscosities.*

Viscosity

Another property exhibited by a fluid is its tendency to flow. While all fluids flow, they vary in the rates at which they flow. **Viscosity** is the resistance of a fluid to flowing. For example, when you take syrup out of the refrigerator and pour it, as shown in **Figure 18,** the flow of syrup is slow. But if this syrup were heated, it would flow much faster. Water flows easily because it has low viscosity. Cold syrup flows slowly because it has high viscosity.

What causes viscosity? When a container of liquid is tilted to allow flow to begin, the flowing portion of the liquid transfers energy to the portion of the liquid that is stationary. In effect, the flowing portion of the liquid is pulling the stationary portion of the liquid, causing it to flow, too.

If the flowing portion of the liquid does not effectively pull the other portions of the liquid into motion, then the liquid has a high viscosity, which is a high resistance to flow. If the flowing portion of the liquid pulls the other portions of the liquid into motion easily, then the liquid has low viscosity, or a low resistance to flow.

 Inquiry Video Lab

Section 2 Review

Section Summary

▶ If the buoyant force on an object is equal to or greater than the gravitational force on that object, the object will float. If the buoyant force on an object is less than the gravitational force on that object, the object will sink.

▶ Pascal's principle states that pressure applied to a fluid is transmitted throughout the fluid.

▶ Bernoulli's principle states that as the velocity of a fluid increases, the pressure exerted by the fluid decreases.

▶ The resistance to flow by a fluid is called viscosity.

13. **MAIN Idea Describe** how fluids exert forces on objects.

14. **Explain** why a steel boat floats on water but a steel block does not.

15. **Explain** why squeezing a plastic mustard bottle forces mustard out the top.

16. **Describe,** using Bernoulli's principle, how roofs are lifted off buildings in tornados.

17. **Think Critically** If you blow up a balloon, tie it off, and release it, it will fall to the floor. Why does it fall instead of float? Explain what would happen if the balloon contained helium instead of air.

Apply Math

18. **Calculate Force** The density of water is 1.0 g/cm³. How many kilograms of water does a submerged 120-cm³ block displace? Recall that 1.0 kg weighs 9.8 N on Earth. What is the buoyant force on the block?

19. **Solve an Equation** To lift an object weighing 21,000 N, how much force is needed on a piston with an area of 0.060 m² if the platform being lifted has an area of 3.0 m²?

✓ Assessment   Online Quiz

Reading Preview

Essential Questions

▶ How does a gas exert pressure on its container?

▶ How is a gas affected when pressure, temperature, or volume change?

Review Vocabulary

temperature: a measure of the average kinetic energy of all the particles in an object

New Vocabulary

Boyle's law
Charles's law

 Multilingual eGlossary

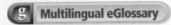 Inquiry Virtual Lab

Behavior of Gases

MAIN Idea Gases respond to changes in pressure, temperature, and volume in predictable ways.

Real-World Reading Link If you've ever climbed high up on a mountain, you might have had the experience of panting as it becomes more difficult to breathe. Modern jet airplanes fly much higher, at altitudes where breathing is close to impossible. The cabins of these jets must be specially pressurized for the people on board. In order to do this effectively, an understanding of the behavior of gases is necessary.

Boyle's Law–Volume and Pressure

Have you ever seen a weather balloon, like the one shown in **Figure 19?** They carry sensing instruments to very high altitudes to detect weather information. A weather balloon is inflated near Earth's surface with a low-density gas.

Recall that a gas completely fills its container. The balloon remains inflated because of collisions that the gas particles inside the balloon have with the balloon itself. In other words, these collisions between gas particles and the container wall cause the gas to exert pressure on the container. As the balloon rises, the atmospheric pressure outside the balloon decreases. This decrease in pressure allows the balloon to expand, eventually reaching a volume between 30 and 200 times its original size. Boyle's law describes the relationship between gas pressure and volume that explains the behavior of weather balloons.

✓ **Reading Check Describe** what happens to weather balloons as they rise.

■ **Figure 19** A weather balloon expands as it rises due to decreased external pressure. Eventually, the balloon ruptures, and the instruments fall back to the ground.

Volume and pressure Because a balloon is flexible, its volume can change. In the case of the weather balloon, the volume increases as the external pressure decreases. The volume of the gas inside the weather balloon continues to increase until the balloon can no longer contain it. At this point, the balloon ruptures and the sensing instruments it was carrying fall to the ground.

From the weather balloon, we know what happens to volume when you decrease pressure. What happens to the pressure from a gas if you decrease its volume—for example, by decreasing the size of the container in which the gas is held? Think about the kinetic theory of matter. The pressure from a gas depends on how often its particles strike the walls of the container. If you squeeze gas into a smaller space, its particles will strike the walls more often, causing increased pressure. The opposite is also true, too. If you give the particles that make up the gas more space, increasing the volume, they will hit the walls less often and the pressure from the gas will be reduced.

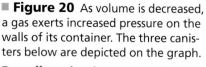 **Reading Check** **Explain** the relationship between pressure and volume.

Robert Boyle (1627–1691), a British scientist, described this property of gases. According to **Boyle's law,** if you decrease the volume of a container of gas and hold the temperature constant, the pressure from the gas will increase. An increase in the volume of the container causes the pressure to drop, if the temperature remains constant. **Figure 20** shows this relationship as the volume of a gas is decreased from 10 L to 5 L to 2.5 L. Note the points on the graph that correspond to each of these volumes.

■ **Figure 20** As volume is decreased, a gas exerts increased pressure on the walls of its container. The three canisters below are depicted on the graph.

Describe *what happens to the volume of a gas if the pressure on that gas is doubled.*

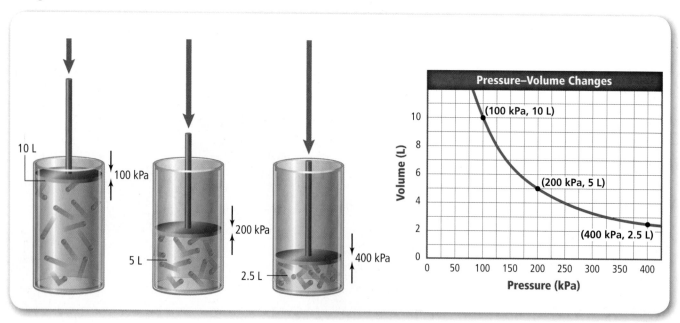

An equation for Boyle's law Boyle's law can be expressed with a mathematical equation. When the temperature of a gas is constant, then the product of the pressure and volume of that gas does not change.

> **Boyle's Law Equation**
>
> initial pressure × initial volume =
> final pressure × final volume
>
> $$P_iV_i = P_fV_f$$

The product of the initial pressure and volume—designated with the subscript *i*—is equal to the product of the final pressure and volume—designated with the subscript *f*. You can use this equation to find one unknown value when you have the other three.

The equation will work with any units for either volume or pressure, as long as you use the same pressure units for P_i and P_f and the same volume units for V_i and V_f.

EXAMPLE Problem 3

Boyle's Law A weather balloon has a volume of 100.0 L when it is released from sea level, where the pressure is 101 kPa. What will be the balloon's volume when it reaches an altitude where the pressure is 43.0 kPa?

Identify the Unknown:	final volume: V_f
List the Knowns:	initial pressure: $P_i = $ **101 kPa**
	initial volume: $V_i = $ **100.0 L**
	final pressure: $P_f = $ 43.0 kPa
Set Up the Problem:	$P_iV_i = P_fV_f$
	$V_f = V_i \left(\dfrac{P_i}{P_f} \right)$
Solve the Problem:	$V_f = 100.0 \text{ L} \left(\dfrac{101 \text{ kPa}}{43.0 \text{ kPa}} \right)$
	$= $ **235 L**
Check the Answer:	You can do a quick estimate to check your answer. The pressure was slightly more than halved. Therefore, the volume should slightly more than double. The final volume of 235 L is slightly more than twice the initial volume of 100.0 L. Therefore, the answer seems reasonable.

PRACTICE Problems

Find **Additional Practice Problems** in the back of your book.

20. A volume of helium occupies 11.0 L at a pressure of 98.0 kPa. What is the new volume if the pressure drops to 86.2 kPa?

Review

Additional Practice Problems

21. **Challenge** A weather balloon has a volume of 90.0 L when it is released from sea level. What is the atmospheric pressure on the balloon when it has grown to a size of 175.0 L?

■ **Figure 21** As the temperature of a sample of gas at constant pressure increases, the volume also increases. The dotted lines represent extrapolations of experimental data. Notice that all the extrapolated lines converge at 0 K.

Identify *which gas had the greatest volume change.*

Concepts in Motion
Animation

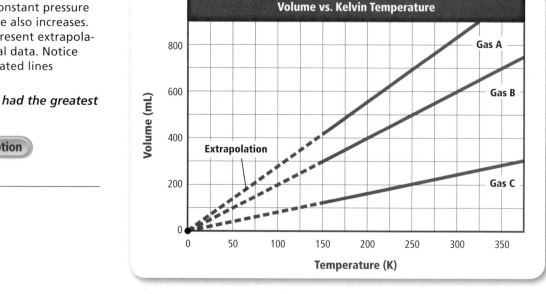

MiniLab

? Inquiry MiniLab

Observe Pressure
Procedure
1. Read the procedure and safety information, and complete the lab form.
2. Blow up a **balloon** to about half its maximum size.
3. Place the balloon on a **beaker** filled with **ice water** for five minutes, and observe the results.
4. Heat **another beaker** filled with **water** on a **hot plate** until it almost boils. Remove the beaker from the hot plate.
5. Place the balloon on the beaker filled with hot water for five minutes, and observe the results. Be careful not to allow the balloon to touch the hot plate.

Analysis
1. **Explain** what happened to the balloon when you placed it on the beaker filled with ice water.
2. **Explain** what happened when you placed the balloon on the beaker with hot water.

Charles's Law—Temperature and Volume

If you've watched a hot-air balloon being inflated, you know that gases expand when they are heated. Jacques Charles (1746–1823), a French scientist, also noticed this. According to **Charles's law,** the volume of a gas increases with increasing temperature, as long as the pressure on the gas does not change. As with Boyle's law, the reverse is also true. The volume of a gas shrinks with decreasing temperature, as shown in **Figure 21**.

The kinetic theory and Charles's law Charles's law can be explained using the kinetic theory of matter. As a gas is heated, the particles that make up that gas move faster and faster. Because the particles that make up the gas move faster, they strike the walls of their container more often and with more force. In the hot-air balloon, the walls have room to expand. So instead of increased pressure, the volume increases.

An equation for Charles's law Like Boyle's law, Charles's law can be expressed mathematically. When the pressure on a gas is constant, then the ratio of the volume to the absolute temperature does not change. The absolute temperature is the temperature measured in kelvins.

> ### Charles's Law Equation
> $$\frac{\text{initial volume}}{\text{initial temperature (K)}} = \frac{\text{final volume}}{\text{final temperature (K)}}$$
> $$\frac{V_i}{T_i} = \frac{V_f}{T_f}$$

This shows that the ratio of the initial volume to the initial temperature is equal to the ratio of the final volume to the final temperature. Remember that temperature must be in kelvins.

Use Charles's Law A 2.0-L balloon at room temperature (20.0°C) is placed in a refrigerator at 3.0°C. What is the volume of the balloon after it cools in the refrigerator?

Identify the Unknown:	final volume: V_f
List the Knowns:	initial volume: $V_i = $ **2.0 L**
	initial temperature: $T_i = $ **20°C** $= 20.0°C + 273 = $ **293 K**
	final temperature: $T_f = $ **3.0°C** $= 3.0°C + 273 = $ **276 K**

Set Up the Problem:

$$\frac{V_i}{T_i} = \frac{V_f}{T_f}$$

$$V_f = V_i\left(\frac{T_f}{T_i}\right)$$

Solve the Problem:

$$V_f = 2.0\text{ L}\left(\frac{276\text{ K}}{239\text{ K}}\right)$$
$$= 2.3\text{ L}$$

Check the Answer: A good way to check your answer here is through experiment! If you place a balloon in a refrigerator, you will notice that the balloon shrinks, but not very much. This is consistent with our answer above.

PRACTICE Problems

Find **Additional Practice Problems** in the back of your book.

22. What would be the final size of the balloon in the example problem above if it were placed in a −18°C freezer?

▤ **Review**

Additional Practice Problems

23. **Challenge** A gas is heated so that it expands from a volume of 1.0 L to a volume of 1.5 L. If the initial temperature of the gas was 5.0°C, then what is the final temperature of the gas?

Section 3 Review

Section Summary

▶ Boyle's law states that if the temperature is constant, as the volume of a gas decreases, the pressure increases.

▶ Charles's law states that at constant pressure, the volume of a gas increases with increasing temperature.

▶ Both Boyle's law and Charles's law can be expressed as mathematical equations.

24. **MAIN ⟨Idea Describe** what would happen to the volume of a gas if the pressure on it were decreased and then the gas's temperature were increased.

25. **Predict,** using Boyle's law, what will happen to a balloon that an ocean diver takes to a pressure of 202 kPa.

26. **Think Critically** Predict what would happen to the volume of a gas if the pressure on that gas were doubled and then the absolute temperature of the gas were doubled.

Apply Math

27. **Calculate Volume** A helium balloon has a volume of 2.00 L at 101 kPa. As the balloon rises the pressure drops to 97.0 kPa. What is the new volume?

28. **Solve One-Step Equations** If a 5-L balloon at 25°C were gently heated to 30°C, what new volume would the balloon have?

How SCIENCE Works

Detecting Dark Matter

The Andromeda Galaxy, shown in **Figure 1,** is a group of approximately one trillion stars held together by the force of gravity. Scientists estimate that the matter we can see, which makes up the stars, planets, and gas clouds in the galaxy, is about 15 percent of the galaxy's mass. The other 85 percent is in a form we do not currently understand.

"Missing" mass Fritz Zwicky, working in the 1930s on gravity's effects on the motion of galaxies, proposed that much of the matter in the universe does not emit or absorb light. While Zwicky's data hinted at mysterious "dark matter," his idea was largely ignored because his measurements were not accurate enough to be conclusive.

New data, renewed interest In the 1970s, astronomer Vera Rubin used newly developed technology to study light coming from the Andromeda Galaxy. Her extremely accurate data supported the existence of dark matter and renewed interest in investigating Zwicky's proposal. Today, scientists are investigating dark matter in an attempt to understand the 85 percent of the universe that is not explained by current scientific knowledge.

Figure 1 Much of the visible matter is concentrated at the center of the Andromeda Galaxy. Dark matter is spread throughout.

Figure 2 These detectors attempt to measure rare interactions between dark matter and normal matter.

Undiscovered particles In an effort to explain dark matter, scientists are trying to detect massive particles that do not usually interact with other particles. Detecting these particles requires sensitive equipment, like that shown in **Figure 2.** All attempts at detecting dark matter particles have been unsuccessful so far.

A new theory of gravity Some scientists are trying to develop a gravitational theory that can explain the motion of distant galaxies without relying on mysterious undiscovered particles. To date, however, none of these proposals are able to explain the observed phenomena. Until a new particle is discovered or a new theory is written, dark matter will be a scientific mystery.

WebQuest **Describing Dark Matter**
The undiscovered particles of dark matter are commonly known by the descriptive acronym WIMP. What does WIMP mean? Why is it an accurate description of the particles under consideration?

THEME FOCUS Changes of Matter

Changes in the arrangement and motion of particles occur as matter changes state. The energy of the particles defines how they interact and whether they can overcome the attractive forces between them.

BIG Idea Each state of matter—solid, liquid, or gas—has unique properties, defined by the motion of its particles.

Section 1 Matter and Thermal Energy

boiling point (p. 435)
heat of fusion (p. 434)
heat of vaporization (p. 435)
kinetic theory (p. 432)
melting point (p. 434)
plasma (p. 436)
sublimation (p. 435)
thermal expansion (p. 437)

MAIN Idea Matter can exist as a solid, a liquid, a gas, or a plasma.

- The kinetic theory is an explanation of how the particles that make up gases behave.
- Thermal energy is the total energy of the particles that make up a material, including kinetic and potential energy.
- Temperature is the average kinetic energy of a substance.

Section 2 Properties of Fluids

buoyancy (p. 441)
pressure (p. 443)
viscosity (p. 446)

MAIN Idea Fluids flow and exert forces on objects.

- If the buoyant force on an object is equal to or greater than the gravitational force on that object, the object will float. If the buoyant force on an object is less than the gravitational force on that object, the object will sink.
- Pascal's principle states that pressure applied to a fluid is transmitted throughout the fluid.
- Bernoulli's principle states that as the velocity of a fluid increases, the pressure exerted by the fluid decreases.
- The resistance to flow by a fluid is called viscosity.

Section 3 Behavior of Gases

Boyle's law (p. 448)
Charles's law (p. 450)

MAIN Idea Gases respond to changes in pressure, temperature, and volume in predictable ways.

- Boyle's law states that if the temperature is constant, as the volume of a gas decreases, the pressure increases.
- Charles's law states that at constant pressure, the volume of a gas increases with increasing temperature.
- Both Boyle's law and Charles's law can be expressed as mathematical equations.

Use Vocabulary

Answer each question using the correct term from the Study Guide.

29. Which law states that volume increases with temperature (at constant pressure)?

30. What is the SI unit of pressure?

31. What term is used to describe the amount of force exerted per unit of area?

32. What is the temperature when a solid begins to liquefy?

33. What theory is used to explain the behavior of particles in gases?

34. What is the ability of a fluid to exert an upward force on an object?

Check Concepts

35. At what temperature is the pressure of the vapor in a liquid equal to the external pressure on that liquid?
A) absolute zero   **C)** melting point
B) boiling point   **D)** heat of fusion

36. What is the most common state of matter in the universe?
A) solid   **C)** liquid
B) gas   **D)** plasma

37. What is the amount of energy needed to change a solid to a liquid at its melting point?
A) heat of fusion
B) temperature
C) heat of vaporization
D) absolute zero

38. Which is a unit of pressure?
A) gram   **C)** newtons
B) kilopascals   **D)** kilograms

Use the figure below to answer question 39.

39. What force does the upward arrow in the diagram above represent?
A) pressure   **C)** buoyant force
B) density   **D)** gravitational force

40. Which uses Pascal's principle?
A) aerodynamics   **C)** hydraulic lift
B) buoyancy   **D)** changes of state

41. Which uses Bernoulli's principle?
A) hose-end sprayer   **C)** piston
B) skateboard   **D)** snowboard

Interpret Graphics

Use the figure below to answer question 42.

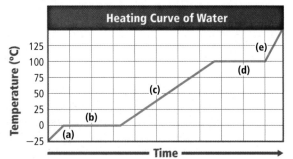

42. **BIG** **Idea** A student continuously heated ice until it turned to steam and graphed the change in temperature over time. Explain what is happening at each letter (A, B, C, D, and E) in the graph, shown above.

✓ **Assessment**   **Online Test Practice**

Concepts in Motion Interactive Concept Map

43. Copy and complete this concept map.

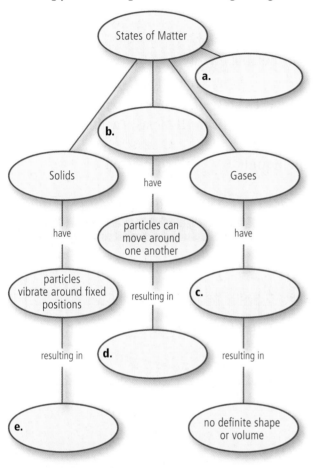

Think Critically

44. THEME FOCUS **Describe** the changes that occur inside a helium balloon as it rises from sea level.

45. Explain The Dead Sea is a solution that is so dense that you easily float on it. Explain why you are able to float easily, using the terms *density* and *buoyant force*.

46. Describe how you would shape a piece of aluminum foil so that it floats on water.

47. Predict what would happen to a plasma if it were placed in a chamber with positively charged walls.

Apply Math

48. Calculate A sample of gas has a volume of 25 L at a pressure of 200 kPa and a temperature of 25°C. What would be the volume if the pressure were increased to 250 kPa and the temperature were decreased to 0°C?

49. Calculate The water pressure at the bottom of the Marianas Trench is approximately 1,100 kPa. With how much force would the water pressure at the bottom of the Marianas Trench push on a fish with a surface area of 0.50 m²?

Use the figure below to answer question 50.

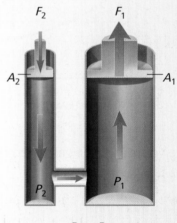

$$P_2 = P_1$$

50. Solve One-Step Equations A hydraulic lift is used to lift a heavy box that is pushing down on a 3.0-m² piston with a force of 1,500 N. What force needs to be exerted on a 0.08-m² piston to lift the box?

51. Calculate A balloon has a volume of 1.5-L at 25.0°C. What would be the volume of the balloon if it were placed in a container of hot water at 90.0°C?

52. Use Ratios A balloon has a volume of 25.0 L at a pressure of 98.7 kPa. What will be the new volume when the pressure is 51.2 kPa?

Standardized Test Practice

Multiple Choice

*Record your answers on the answer sheet
provided by your teacher or on a sheet of paper.*

1. In which state of matter would you expect to
   find water on Earth's surface if the temperature is –25°C?
   A. solid   C. gas
   B. liquid   D. plasma

Use the graph below to answer questions 2 and 3.

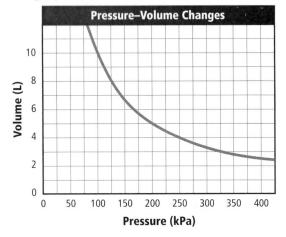

2. What is the volume of the gas when the
   pressure on the gas is 325 kPa?
   A. 1 L
   B. 2 L
   C. 3 L
   D. 4 L

3. Which scientific concept does this graph best
   represent?
   A. Boyle's law
   B. Charles's law
   C. Pascal's principle
   D. Bernoulli's principle

4. Which is unlikely to contain plasma?
   A. a star
   B. a lightning bolt
   C. a neon light
   D. a glass of water

5. Which describes the energy required for a
   liquid at its boiling point to become a gas?
   A. heat of vaporization
   B. diffusion
   C. heat of fusion
   D. thermal energy

6. In which state of matter do particles stay close
   together, yet can slide past each other?
   A. solid   C. gas
   B. liquid   D. plasma

Use the graph below to answer questions 7 and 8.

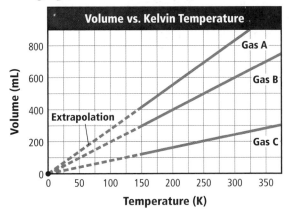

7. Which statement is true?
   A. Gas A had the greatest increase in volume
      with an increase in temperature.
   B. Gas B had the greatest increase in volume
      with an increase in temperature.
   C. Gas C had the greatest increase in volume
      with an increase in temperature.
   D. The gases had the same increase in
      volume.

8. At approximately what temperature is the
   volume of Gas B about 600 mL?
   A. 100 K
   B. 200 K
   C. 300 K
   D. 400 K

Short Response

Record your answers on the answer sheet provided by your teacher or on a sheet of paper.

9. If you place two wood blocks in water and one sinks while the other floats, what do you know about the densities of the blocks?

Use the figure below to answer question 10.

10. A weather balloon is inflated near Earth's surface with a low-density gas. Explain why the balloon rises when it is released.

11. If volume is held constant, what happens to the pressure of a gas as temperature is decreased? Explain.

12. A car on a 25-m² hydraulic lift platform weighs 15,000 N. If the force on the smaller piston required to lift it is 1/100 its weight, what is the area of the smaller piston?

Extended Response

Record your answers on a sheet of paper.

13. Explain how a hot-air balloon can still float in the air when it is carrying people.

14. Explain why a hot-air balloon descends when the burner is turned off while the balloon is in the air.

Use the figure below to answer question 15.

15. Describe how the above device works.

16. Scuba divers often wear buoyancy vests to maintain neutral buoyancy, so they neither sink to the bottom nor float to the surface. If the diver takes a breath, the diver will rise slightly. When the diver exhales, the diver will sink slightly. Explain why this happens.

17. Explain why a penny sinks in a beaker of water but floats in a beaker of mercury.

18. Compare and contrast evaporation and boiling.

NEED EXTRA HELP?																		
If You Missed Question . . .	1	2	3	4	5	6	7	8	9	10	11	12	13	14	15	16	17	18
Review Section . . .	1	3	3	1	1	1	3	3	2	3	3	3	1	1	2	2	2	1

CHAPTER 15

Classification of Matter

ConnectED

Your one-stop online resource
connectED.mcgraw-hill.com

- Video
- Audio
- Review
- Inquiry
- WebQuest
- Assessment
- Concepts in Motion
- g Multilingual eGlossary

Launch Lab
Distillation of Water

Various types of matter have different properties, such as different boiling points and freezing points. These properties could be very important to you if you were stranded on a desert island and needed a drink of water. For example, the process of distillation produces purified water by taking advantage of some of its properties.

For a lab worksheet, use your StudentWorks™ Plus Online.

 **Inquiry** Launch Lab

FOLDABLES

Make a two tab book. Label it as shown. Use it to organize your notes on the classification of matter.

Pure Substances | Mixtures

THEME FOCUS Structure and Properties of Matter

Matter can be divided into pure substances and mixtures.

BIG Idea Matter can be classified by its composition, by its physical properties, and by its chemical properties.

Section 1 • Composition of Matter

Section 2 • Properties of Matter

Reading Preview

Essential Questions

▶ What are the differences between substances and mixtures?

▶ How are elements and compounds identified?

▶ How are suspensions, solutions, and colloids related?

Review Vocabulary

property: characteristic or essential quality

New Vocabulary

substance
element
compound
heterogeneous mixture
suspension
colloid
Tyndall effect
homogeneous mixture
solution

g Multilingual eGlossary

Figure 1 All the atoms of the same element are alike.

Composition of Matter

MAIN Idea Matter exists as either a pure substance or a mixture.

Real-World Reading Link The bottom of a pail holding ocean water will have sediment that has settled over time. The water itself, however, will still be salty. Why doesn't the salt settle with the sediment?

Substances

Recall that matter is anything that takes up space and has mass. Everything around us is matter, including things that we cannot see—like the salt in a pail of ocean water. Matter is either a pure substance or a mixture of substances. A pure substance, or simply a **substance,** is a type of matter with a fixed composition. A substance can be an element or a compound. Some substances you might recognize are helium, aluminum, water, and salt.

Elements All substances are built from atoms. An **element** is a substance made up of atoms that are all alike. The graphite in your pencil point and the copper used in electrical wiring are examples of elements. In graphite, all of the atoms are carbon atoms. In a copper sample, all of the atoms are copper atoms. Some elements that you might recognize are shown in **Figure 1.** Examples of less common elements and their uses are shown in **Figure 2.** About ninety elements are found naturally on Earth. More than 20 others have been made in laboratories, but most of these are unstable and exist only for short periods of time.

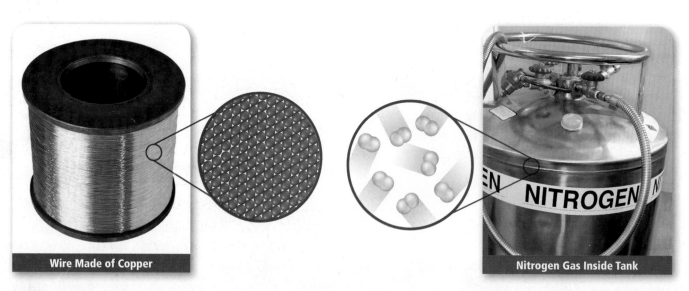

Wire Made of Copper

Nitrogen Gas Inside Tank

FIGURE 2

Visualizing Elements

When you think of elements, you probably think about those that you see every day, such as silver used to make jewelry or aluminum used to make lightweight bicycles and baseball bats. Many other elements are not as commonly known, but you might see them in everyday objects.

Titanium Titanium is strong and lightweight and is used for bone or joint replacements and aircraft construction. In rare instances, titanium panels are used in building construction.

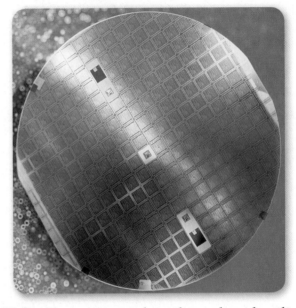

Silicon Present in sand as silicon dioxide, silicon is used to make window glass as well as the silicon chips that run computers.

Americium The synthetic, radioactive element americium is used in smoke detectors.

Magnesium Chlorophyll—the substance that makes plants green—contains magnesium. Used in metal mixtures, magnesium is lightweight, strong, and resistant to corrosion. Because of these physical properties, it is used in jet engines.

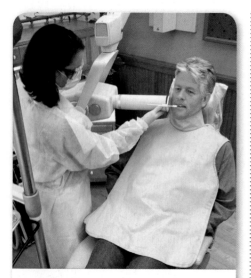

Figure 3 Lead is used as a barrier to protect dental patients from X-ray exposure.

Other applications of elements Lead is an important element due to its high density. Lead blocks harmful radiation because its high density makes it difficult for radiation to pass through. Lead aprons are used while dental X-rays are being taken in order to protect the rest of the body from X-ray exposure, as shown in **Figure 3.**

Aluminum is bendable and resistant to corrosion. It is becoming one of the world's most widely used elements. Because it is strong and lightweight, aluminum is used in automobile parts, airplanes, bicycles, and pots and pans.

Compounds Two or more elements can combine to form substances called compounds. A **compound** is a substance in which the atoms of two or more elements are chemically combined in a fixed proportion. For example, water is a compound in which two atoms of the element hydrogen combine with one atom of the element oxygen. Chalk contains one atom each of calcium and carbon and three atoms of oxygen.

Have you put something made from a silvery metal and a greenish-yellow, poisonous gas on your food? Table salt is a chemical compound that fits this description. Even though it is composed of white crystals and adds flavor to food, its components in their elemental forms—sodium and chlorine—are neither white nor salty, as illustrated in **Figure 4.** Like salt, the properties of compounds differ from the properties of the elements that combine to make the compounds. The elements that make up a compound cannot be separated by physical means.

Reading Check **Compare** How are elements and compounds related?

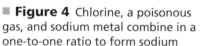

Figure 4 Chlorine, a poisonous gas, and sodium metal combine in a one-to-one ratio to form sodium chloride, table salt.

Chlorine Gas

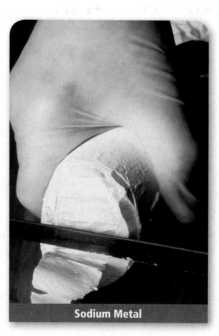

Sodium Metal

Sodium Chloride Crystals

■ **Figure 5** Salad dressings are mixtures of oil, vinegar, and seasoning. Notice the visible herbs and spices floating in the salad dressings.

Mixtures

Salad dressings, such as the examples shown in **Figure 5,** are mixtures. A mixture is matter composed of two or more substances that can be separated by physical means.

Heterogeneous mixtures In the salad dressing, all of the items in the dressing are in contact, but they do not react with one another. If the dressing is allowed to sit undisturbed long enough, the oil and vinegar will separate. Because the different components remain distinct, this salad dressing is considered an example of a heterogeneous mixture. A mixture in which different materials remain distinct is called a **heterogeneous** (he tuh ruh JEE nee us) **mixture.** Like the dressing, salad is a heterogeneous mixture. The vegetables in a salad remain distinct. You can remove the vegetables if you do not care to eat them.

Some components of heterogeneous mixtures are easy to see, like the components of the salad, but others are not. For example, the shirt shown in **Figure 6** is also a heterogeneous mixture, but you cannot see its individual components. However, with the help of a microscope, you can see the distinct cotton and polyester threads.

MiniLab

? **Inquiry** MiniLab

Separate Mixtures

Procedure 🥽 👕 ✋

1. Read the procedure and safety information, and complete the lab form.
2. Place equal amounts of **soil, clay, sand, gravel,** and **pebbles** in a **clear, plastic container.** Add **water** until the container is almost full.
3. Stir or shake the mixture thoroughly. Predict the order in which the materials will settle.
4. Observe what happens, and compare your observations to your predictions.

Analysis

1. **Describe** In what order did the materials settle?
2. **Explain** why the materials settled in the order that they did.

? **Inquiry** Virtual Lab

■ **Figure 6** Even though you cannot see the individual components that make up this shirt with your naked eye, the different colors and fibers types are clearly visible under the microscope. The shirt is a heterogeneous mixture.

Figure 7 River water is a suspension that carries soil and sediment. If river water slows or sits undisturbed, the suspended particles settle out.

Explain *How can you tell that river water is a suspension?*

VOCABULARY ....
SCIENCE USAGE V. COMMON USAGE
Suspend
Science usage
  to keep from settling
  The seasonings were suspended in the oil and vinegar mixture.

Common usage
  to cause to stop temporarily
  The game was suspended due to the bad weather. ....

FOLDABLES
Incorporate information from this section into your Foldable.

Suspensions A **suspension** is a heterogeneous mixture made of a liquid and solid particles that settle. Recall the oil and vinegar salad dressing in **Figure 5**. The dressing also has seasoning particles. The seasoning particles in the liquid will settle to the bottom of the container if allowed to sit undisturbed.

River deltas are large scale examples of how particles in a suspension settle. Rivers flow swiftly through narrow channels, picking up soil and sediment along the way. As the river widens, it flows more slowly. Suspended particles settle and form deltas at the mouth of the river, as shown in **Figure 7**.

Colloids Milk is an example of another kind of heterogeneous mixture called a colloid. It contains water, fats, and proteins in varying proportions. Unlike a suspension, however, its components will not settle if left standing. A **colloid** (KAH loyd) is a heterogeneous mixture with particles that never settle.

Paint is a liquid colloid with suspended particles. Gases and solids can contain colloidal particles, too. Fog, like that shown in **Figure 8**, consists of particles of liquid water suspended in air. Smoke contains solids suspended in air.

Figure 8 Fog is a colloid composed of water droplets suspended in air.

Fog—a colloid containing suspended water droplets— scatters the light produced by the vehicle's headlights.

Water droplets suspended in air allow you to see the sunlight as it streams through the forest fog.

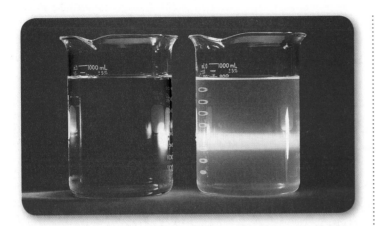

Identifying colloids One way to identify a colloid is by its appearance. Fog appears white because its particles are large enough to scatter light, as shown in **Figure 8.** Sometimes, it is not so obvious that a liquid is a colloid. These colloids can look very much like solutions, which are also mixtures where the particles cannot be seen.

You can identify whether a liquid is a colloid by passing a beam of light through it. A light beam is invisible as it passes through a solution but can be seen as it passes through a colloid, as shown in **Figure 9.** This occurs because the particles in the colloid are large enough to scatter light, but those in the solution are not. The scattering of a light beam as it passes through a colloid is called the **Tyndall effect.**

Homogeneous mixtures Soft drinks contain water, sugar, flavoring, coloring, and carbon dioxide gas. **Figure 10** will help you visualize some of these particles in a liquid soft drink. A soft drink in a sealed bottle is an example of a homogeneous mixture. A **homogeneous** (hoh muh JEE nee us) **mixture** is a mixture that remains constantly and uniformly mixed and has particles that are so small that they cannot be seen with a microscope. Due to the interactions between particles, particles in a homogeneous mixture will never settle to the bottom of their container.

A **solution** is the same thing as a homogeneous mixture. The most familiar solutions might be solids dissolved in liquids, but solutions can also be mixtures of a solid and a gas, a solid and a solid, a gas and a liquid, and so on. Tea, vinegar, steel alloys, and the compressed gas used by divers are all examples of solutions.

■ **Figure 10** A soft drink can be either heterogeneous or homogeneous. As carbon dioxide fizzes out it is a heterogeneous mixture. The resulting flat soft drink is a homogeneous mixture of water, sugar, flavor, color, and some remaining carbon dioxide.

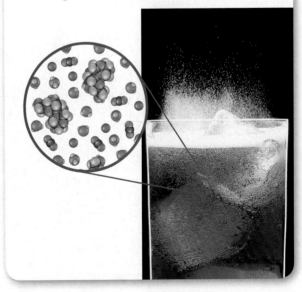

Figure 11 The concept map shows that mixtures can be either heterogeneous or homogeneous. Pure substances can be elements or compounds.

Examine *Where on this chart would you classify pizza?*

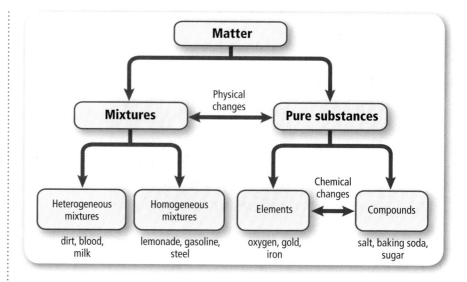

Comparing mixtures and substances Mixtures, unlike compounds, do not always contain the same proportions of the substances of which they are made. Additionally, mixtures can be physically separated unlike pure substances. Recall that a substance is matter that has a fixed composition. In contrast, mixtures can have widely different compositions. For example, you could dissolve a small amount of salt in a tank of water, or a large amount of salt in the same tank. Mixtures are not substances, but they are composed of two or more substances. The differences between mixtures and substances are summarized in **Figure 11**.

Section 1 Review

Section Summary

▶ An element is a substance with the same kind of atoms.

▶ There are approximately 90 naturally occurring elements found on Earth and over 25 that have been created in laboratories.

▶ A compound is a substance that has two or more elements chemically combined in a fixed proportion.

▶ Mixtures can be heterogeneous or homogeneous and can be separated by physical means.

1. **MAIN Idea Distinguish** a substance from a mixture. Give two examples of each.

2. **Compare and Contrast** How is a compound similar to a homogeneous mixture? How is it different?

3. **Identify** three elements and three compounds. How are they similar? How are they different?

4. **Summarize** Make a table that compares the properties of suspensions, colloids, and solutions.

5. **Think Critically** Why do the words "Shake well before using" indicate that the fruit juice in a carton is a suspension? Why are these words not used on a milk container?

Apply Math

6. **Conversion** The weather report this morning stated there is a thick fog in your town. Visibility is less than 500 feet. How many kilometers in front of your vehicle can you see?

✓ **Assessment** Online Quiz

Reading Preview

Essential Questions

▶ What are physical and chemical properties?

▶ What are the differences and similarities of physical and chemical changes?

▶ How does the law of conservation of mass apply to chemical changes?

Review Vocabulary

boiling point: the temperature at which the vapor pressure of the liquid is equal to the external pressure acting on the surface of the liquid

New Vocabulary

physical property
physical change
distillation
chemical property
chemical change
law of conservation of mass

 Multilingual eGlossary

Properties of Matter

MAIN ‹Idea A physical property can be observed without changing the identity of the material; a chemical property can be observed when one or more new substances are formed.

Real-World Reading Link When you make a chocolate cake, a mixture of ingredients such as cocoa, flour, baking powder, butter, sugar, and eggs is placed in the oven. The delicious baked dessert results from chemical interactions called chemical reactions among the ingredients while in the oven.

Physical Properties

You can stretch a rubber band, but you cannot stretch a piece of string much, if at all. You can bend a piece of wire, but you cannot easily bend a matchstick. The abilities to stretch and to bend substances are physical properties. The identity of the substances—rubber, string, wire, wood—does not change. Any characteristic of a material that you can observe without changing the identity of the substance is a **physical property.** Some examples of physical properties are color, shape, size, density, melting point, and boiling point.

Appearance The appearance of substances, such as the ones shown in **Figure 12,** is a physical property. How would you describe a tennis ball? You could begin by describing its shape, color, and state of matter. You might describe the tennis ball as a brightly colored, hollow sphere. You can measure some physical properties—for example, the diameter of the ball. What physical property of the ball is measured with a balance?

To describe a soft drink in a cup, you could start by calling it a liquid with a brown color. You could measure its volume and temperature. Each of these characteristics is a physical property of that soft drink.

■ **Figure 12** Appearance is a physical property. Appearance includes color, shape, size, texture, and volume.

Compare *the physical properties of the rubber bands and the tennis ball.*

VOCABULARY ....

ACADEMIC VOCABULARY

Specific
characterized by precise formulation or accurate restriction
Some diseases have specific symptoms. ....

■ **Figure 13** The best way to separate mixtures depends on their physical properties.

Size is the property used to separate sesame seeds from sunflower seeds.

Magnetism easily separates iron from sand.

Behavior Some physical properties describe the behavior of a material or a substance. As you might know, objects that contain iron, such as a safety pin, are attracted by a magnet. Attraction to a magnet is a physical property of iron.

Every substance has a specific combination of physical properties that make it useful for certain tasks. Some metals, such as copper, can be drawn out into wires. Others, such as gold, can be pounded into sheets as thin as 0.1 micrometers (μm), about four-millionths of an inch. This property of gold makes it useful for decorating picture frames and other objects. Gold that has been beaten or flattened in this way is called gold leaf.

Think again about a soft drink. If you knock over the cup, the drink will spread over the table or floor. If you knock over a jar of molasses, however, it does not flow as easily. Viscosity, the resistance to flow, is a physical property of liquids.

Using physical properties to separate mixtures Removing the seeds from a watermelon can be done easily based on the physical properties of the seeds compared to the rest of the fruit. **Figure 13** shows a mixture of sesame seeds and sunflower seeds. You can identify the two kinds of seeds by differences in color, shape, and size. By sifting the mixture, you can quickly separate the sesame seeds from the sunflower seeds because their sizes differ.

Now look at the mixture of iron filings and sand shown in **Figure 13.** You probably will not be able to sift out the iron filings because they are similar in size to the sand particles. What you can do is pass a magnet through the mixture. The magnet attracts only the iron filings and pulls them from the sand. This is an example of how a physical property, such as magnetic attraction, can be used to separate substances in a mixture. A similar method is used to separate iron from aluminum and other refuse for recycling. Strong magnets are used in scrap yards and landfills to remove iron for recycling and reuse in an effort to conserve natural resources.

✓ **Reading Check** **Describe** how you could use physical properties to separate sand from sugar.

Physical Change

Physical properties can change while composition remains fixed. If you tear a piece of chewing gum, you change some of its physical properties—its size and shape. However, you have not changed the identity of the materials that make up the gum.

The identity remains the same When a substance, such as water, freezes, boils, evaporates, or condenses, it undergoes a physical change. A change in size, shape, or state of matter in which the identity of the substance remains the same is called a **physical change.** These changes might involve energy changes, but the kind of substance—the identity of the element or compound—does not change. Because all substances have distinct properties such as density, specific heat, and melting and boiling points, these properties can often be used to help identify a substance when a particular mixture contains more than one unknown material.

A substance can change states if it absorbs or releases enough energy. Iron, for example, will melt at high temperatures. Yet, whether in solid or liquid state, iron has physical properties that identify it as iron. Color changes are physical changes, too. For example, when iron is first heated, it glows red. Then, if it is heated to a higher temperature, it turns white, as shown in **Figure 14.**

✓ **Reading Check** **Infer** Does a change in state mean that a new substance has formed? Explain.

Using physical changes A cool drink of water is something most people take for granted; but in some parts of the world, drinkable water is scarce. Not enough drinkable water can be obtained from wells. Many such areas that lie close to the sea obtain drinking water by using physical properties of water to separate it from the salt. One method, which uses the property of boiling point, is a type of distillation.

 Video   BrainPOP

 Inquiry   Virtual Lab

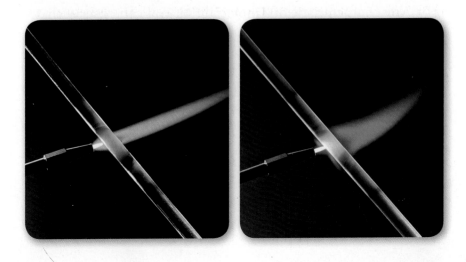

■ **Figure 14** Heating iron raises its temperature and changes its color. These changes are physical changes because it is still iron.

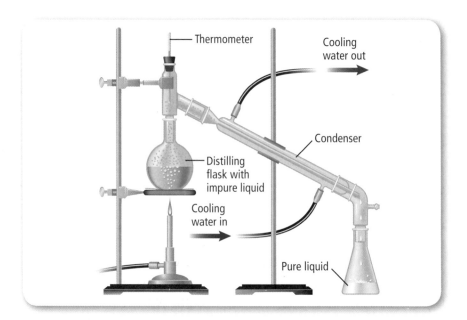

Figure 15 Distillation can separate liquids from solids dissolved in them. The liquid is heated until it evaporates and moves up the column. Then, as it touches the water-cooled surface of the condenser, it becomes liquid again.

Identify *where the solids would be found after distillation is complete.*

Thermometer

Cooling water out

Condenser

Distilling flask with impure liquid

Cooling water in

Pure liquid

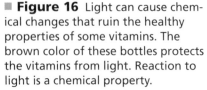

? Inquiry Video Lab

Distillation The process of separating substances, such as salt and water, in a mixture by evaporating a liquid and recondensing its vapor is **distillation.** Distillation is done in the laboratory using an apparatus similar to the one shown in **Figure 15.**

Two liquids with different boiling points can be separated in this way. The mixture is heated slowly until it begins to boil. The liquid with the lowest boiling point vaporizes first and is condensed and collected. Then, as temperature increases the second liquid boils, vaporizes, condenses, and is collected. Distillation is often used in industry. For instance, crude oil obtained from drilling is distilled to separate many different compounds in order to make products, such as the gasoline used to fuel automobiles.

Chemical Properties and Chemical Changes

You have probably seen warnings on cans of paint thinner and lighter fluid for charcoal grills that state these liquids are flammable (FLA muh buhl). The tendency of a substance to burn, called its flammability, is an example of a chemical property. Any characteristic of a material that you can observe that produces one or more new substances is a **chemical property.** Flammability is a chemical property because burning produces new substances. As a result, a chemical change, also called a chemical reaction, has occurred. Many other substances used around the home are flammable. Knowing which ones are flammable helps you to use them safely.

A less dramatic chemical change can affect some medicines. Look at **Figure 16.** You have probably seen bottles like this in a pharmacy. Many medicines are stored in dark bottles because the medicines contain compounds that can chemically change if they are exposed to light.

Figure 16 Light can cause chemical changes that ruin the healthy properties of some vitamins. The brown color of these bottles protects the vitamins from light. Reaction to light is a chemical property.

Detecting Chemical Change

If you leave a pan of chili cooking unattended on the stove for too long, your nose soon tells you that something is wrong. Instead of a spicy aroma, you detect an unpleasant smell that alerts you that something is burning. This burnt odor is a clue that a new substance has formed.

The identity changes The smell of rotten eggs and the formation of rust on bikes and car fenders are also signs that a chemical change has taken place. A change of one substance to another is a **chemical change.** Bubble formation produced by the foaming of an antacid tablet in a glass of water is a sign of new substances being produced. In some chemical changes, a rapid release of energy—detected as heat, light, and sound—is a clue that changes are occurring. A display of fireworks in the night sky is an example. **Figure 17** illustrates another visual clue—the formation of a solid precipitate. What is another example of a chemical change forming a solid?

 **Reading Check Define** What is a chemical change?

Clues such as heat, cooling, or the formation of bubbles or solids in a liquid are indicators that a reaction is taking place. However, the only sure proof is that a new substance is produced. Consider the following examples. The heat, light, and sound produced when hydrogen gas combines with oxygen in a rocket engine are clear evidence that a chemical reaction has taken place. But no clues announce the onset of the reaction that combines iron with oxygen to form rust. The only clue that iron has changed into a new substance is the visible presence of rust. Burning and rusting are chemical changes because new substances form.

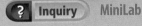

MiniLab

? Inquiry MiniLab

Identify Changes

Procedure
1. Read the procedure and safety information, and complete the lab form.
2. Add 250 mL of **water** to a **400-mL beaker**.
3. Add 5 g of **baking soda** to the water, stir, and observe what happens.
4. Add 5 mL of **vinegar** to the solution, and observe.

Analysis
1. **Explain** Is dissolving a chemical or a physical change?
2. **Summarize** What evidence of a chemical change did you see?
3. **Infer** When baking soda and vinegar are mixed, does a chemical change occur or does the baking soda dissolve?

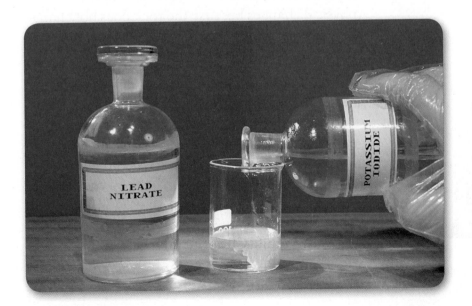

■ **Figure 17** When clear solutions of lead(II) nitrate and potassium iodide mix, a reaction takes place and a yellow solid, lead(II) iodide, appears. The yellow solid that is produced in the chemical reaction is called a precipitate.

Using chemical changes One case where you might separate substances using a chemical change is in cleaning tarnished silver, such as jewelry. Tarnishing, a chemical reaction between silver metal and sulfur compounds in the air, results in silver sulfide. A chemical reaction in a warm water bath with baking soda and aluminum can change silver sulfide back into silver.

Separating substances using chemical changes is rarely done in the home, but it is commonly done in industrial and laboratory settings. For example, many metals are separated from their ores and then purified using chemical changes.

Weathering

The forces of nature continuously shape Earth's surface. Rocks split, deep canyons are carved, sand dunes shift, and limestone formations decorate caves. Do you think these changes, referred to as weathering, are physical or chemical? The answer is both. Geologists, who use the same criteria that you have learned in this chapter, say that some weathering changes are physical and some are chemical.

Reading Check **Determine** Is weathering a physical change or a chemical change?

Physical weathering Large rocks can split when water seeps into small cracks, freezes, and expands. Streams can smooth and sculpt hard rock, as shown at left in **Figure 18.** These are physical changes because the rock does not change into another substance.

Chemical weathering In other cases, the change is chemical. For example, solid calcium carbonate, a compound found in limestone, reacts with water if it is slightly acidic, such as when it contains some dissolved carbon dioxide. The calcium carbonate reacts to form calcium bicarbonate. This change in limestone is a chemical change because the identity of the substances changes. This chemical change contributes to the weathering of the White Cliffs of Dover, shown in **Figure 18,** and also produces the icicle-shaped rock formations that are found in caves.

■ **Figure 18** Weathering can involve physical change and chemical change.

Flowing water shaped and smoothed these rocks in a physical process.

Both chemical and physical changes shaped the famous White Cliffs of Dover, which line the English Channel.

The Conservation of Mass

Wood burns, which means it undergoes combustion. Combustion is a chemical change. Suppose you burn a large log in a fireplace until nothing is left but a small pile of ashes. Smoke, heat, and light are given off, and the changes in the composition of the log confirm that a chemical change took place.

At first, you might think that matter was lost as the log burned because the pile of ashes looks much smaller than the log looked. In fact, the mass of the ashes is less than that of the log. However, suppose that you could collect all of the oxygen in the air that was combined with the log during the burning and all of the smoke and gases that escaped from the burning log and measure their masses too. You would find that no mass was lost after all.

Mass is not gained or lost during any chemical change. In fact, matter is neither created nor destroyed during a chemical change. According to the **law of conservation of mass,** the mass of all substances that are present before a chemical change, known as the reactants, equals the mass of all of the substances that remain after the change, which are called the products.

The Law of Conservation of Mass
total mass of the reactants = total mass of the products

Figure 19 illustrates the law of conservation of mass. Solid sodium bicarbonate in the balloon reacts with liquid hydrochloric acid in the flask. The gas, carbon dioxide, is released, which expands the balloon. Without the balloon in place, the gas would escape and you might think that mass was not conserved. With the balloon to collect the gas, the mass on the scale remains the same. The mass of the reactants is the same as the mass of the products.

Concepts in Motion
Animation
■ **Figure 19** The reaction between sodium bicarbonate and hydrochloric acid produces carbon dioxide gas, which is collected in the balloon.

Describe *How can you tell that matter was not created or destroyed in this reaction?*

Calculate Total Mass of Product When hydrogen reacts with chlorine, the only product is hydrochloric acid. If 18 g of hydrogen react completely with 633 g of chlorine, how many grams of hydrochloric acid are formed?

Identify the Unknown:	mass of hydrochloric acid
List the Knowns:	mass of hydrogen = 18 g mass of chlorine = 633 g
Set Up the Problem:	total mass of the product = total mass of the reactants mass of hydrochloric acid = mass of hydrogen + mass of chlorine
Solve the Problem:	mass of hydrochloric acid = 18 g + 633 g The mass of hydrochloric acid is 651 g.
Check the Answer:	The mass of reactants and products are equal because the equation was set up according to the law of conservation of mass.

PRACTICE Problems

Find **Additional Practice Problems** in the back of your book.

7. When methane reacts with oxygen, the products are carbon dioxide and water. How many grams of water are formed if 24 g of methane react completely with 96 g of oxygen to form 66 g of carbon dioxide?

Review

Additional Practice Problems

8. **Challenge** Sulfur dioxide reacts with bromine and water to produce hydrogen bromide and sulfuric acid. If 64.1 g of sulfur dioxide react completely with 159.9 g of bromine and an unknown amount of water to form 161.9 g of hydrogen bromide and 98.1 g of sulfuric acid, then how many grams of water react?

Section 2 Review

Section Summary

▶ Physical properties can be used to distinguish and separate substances.

▶ A chemical change is sometimes indicated by the cooling, heating, or formation of solids or bubbles.

▶ The law of conservation of mass states that matter is neither created nor destroyed in a chemical reaction.

9. MAIN Idea **Explain** why evaporation of water is a physical change and not a chemical change.

10. **Identify** four physical properties that describe a liquid. Identify a chemical property.

11. **Explain** how the law of conservation of mass applies to chemical changes.

12. **Think Critically** Does the law of conservation of mass apply to physical changes? How could you test this for melting ice? For the distillation of water?

 A pply Math

13. **Calculate** Bismuth and fluorine react to form bismuth fluoride. If 417.96 g of bismuth reacts completely with 113.99 g of fluorine, how many grams of bismuth fluoride are formed?

✓ **Assessment** **Online Quiz**

LAB

Pure Substances and Mixtures

Objectives

■ **Classify** matter.

■ **Compare** the pure substances and mixtures.

Background: Everything that you see is made of matter and is either a pure substance or a mixture. Some things, such as iron nails and aluminum foil, are elements; others, such as water and salt, are compounds. Steel, lemonade, concrete, and a bowl of fruit are all examples of mixtures.

Question: *How do physical properties help to identify pure substances and mixtures?*

Preparation

Materials

salt and iron filings mixture
pepper water
sugar water
graduated cylinder
250-mL beakers (5)
magnet
balance
funnel
filter paper
stirring rod
magnifying lens
hot plate
watch glass

Safety Precautions

WARNING: *To evaporate water from a mixture, use a low setting on the hot plate. Use caution when touching the hot plate or watch glass—they will not look hot.*

Procedure

1. Read the procedure and the safety information, and complete the lab form.
2. Create a data table using the heads *Substance, Color, Magnetic Attraction,* and *Matter Classification.* Record all data in your table.
3. Determine the color of each sample.
4. Determine whether each sample is magnetic.
5. Try to separate the salt and iron filings mixture.
6. Try to separate the particles of the pepper-water mixture and the sugar-water mixture.
7. Classify each material as an element, a compound, or a mixture.

Conclude and Apply

1. **Describe** the separation methods of the mixture samples.
2. **Conclude** if the different particles of each sample can be separated using physical properties.
3. **Infer** how you can identify a substance versus a mixture.
4. **State** whether any of the materials used are compounds.

COMMUNICATE YOUR DATA

Compare your results with classmates. Identify possible sources of error that might cause any difference in your data.

Objectives

- **Measure** the total mass of water and antacid tablets before and after the tablets are added to the water.

- **Compare** the total mass of water and tablets before and after the tablets are dissolved in the water.

- **Infer** whether the law of conservation of mass applies to antacid tablets dissolving in water.

Background: Have you ever watched burning logs in a fireplace? If you have, you might have noticed many large logs being burned in the hearth during an evening. At the end of the night, nothing more than a pile of ash remains. The other substances produced were gases that went up the chimney. In this lab, your group will design an investigation to verify the law of conservation of mass.

Question: *Is the mass of antacid tablets conserved after they are dissolved?*

Preparation

Possible Materials

antacid tablets
empty plastic drink bottle
balloon
beaker
water
spatula
balance
mortar
pestle
funnel
weighing paper

Safety Precautions

WARNING: *Do not eat the antacid tablet.*

Form a Hypothesis

Based on your understanding of the law of conservation of mass, form a hypothesis about the total mass of antacid tablets and water before and after the tablets are dissolved.

Make a Plan

1. Read the procedure and safety information, and complete the lab form.

2. As a group, agree upon and write the hypothesis.

3. Plan an investigation to test your hypothesis. List the steps of your procedure.

4. List the materials that you need to test your hypothesis.

5. Decide upon any needed safety equipment or safety procedures to ensure the safety of your group during the experiment.

6. Have one group member reread your entire procedure aloud to the group to make certain that you have all of the necessary materials and that your procedure can be easily followed.

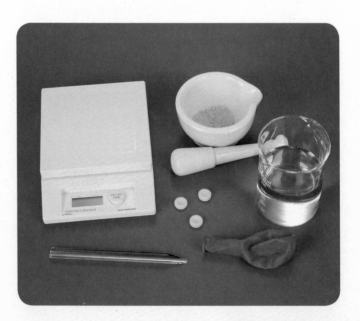

 Inquiry Lab

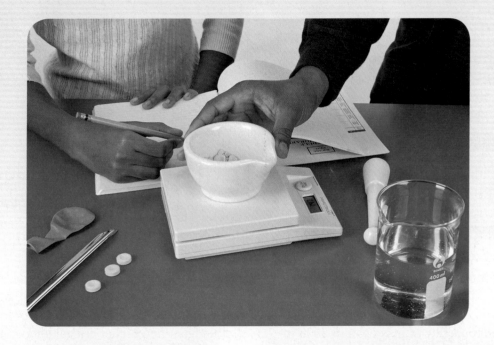

Follow Your Plan

1. Make sure your teacher approves your plan.
2. Copy the data table to record measurements.
3. While doing the investigation, record your observations and complete the data table.

Analyze Your Data

1. **Describe** the effects of mixing the antacid powder and the water.
2. **Compare** the total mass of the substances before mixing to the total mass after the reaction.
3. **Calculate** the percentage of error in the investigation. Use the following equation:

   Δmass = mass of reactants − mass of products

   $$\frac{\Delta \text{mass}}{\text{mass of reactants}} \times 100 = \% \text{ error}$$

4. **Graph** the mass of the substances before and after the reaction using a bar graph.
5. **Explain** whether your data support your hypothesis.

Data Table

Trial Number	Mass Before Reaction (g)	Mass After Reaction (g)	Percent Error

Conclude and Apply

1. **Infer** how mass might have been lost or added between the initial and final weighing of the substances.
2. **Summarize** How does this experiment support the law of conservation of mass?

YOUR DATA

Compare the data your group collected with the data collected by the other groups, and discuss possible reasons why percent error might not be zero.

SCIENCE & TECHNOLOGY

Room Temperature Superconductors

Maglev trains have been used as people–movers since 1984. *Maglev* stands for *magnetic levitation.* The first commercial use of high–speed maglev trains began in 2003 in Shanghai, China. While floating on a cushion of air, high–speed maglev trains can travel twice as fast as traditional railroad commuter trains.

Figure 1 Like two magnets whose north poles push apart, magnetic fields hold the maglev train above the rails and propel the train forward.

Maglev trains do not need the type of engine typical trains use. They rely on repulsion between magnetic fields produced by powerful electromagnets to propel them, as shown in **Figure 1.** These trains hover over the rails rather than riding directly on them. The next generation of maglev trains rely on electromagnets made with superconductors.

Superconductor electromagnets A superconductor is a metallic element, compound, or mixture with zero resistance to the flow of electricity. This produces large current in smaller–sized electromagnets and intense magnetic fields. The material must be cooled to near absolute zero, –273°C (0K) to reach superconductivity. Although energy must be used to cool them, operating costs are lower.

The temperature at which a material becomes a superconductor is called the critical temperature. In recent years, scientists have discovered materials with higher and higher critical temperatures. The recognized world record is –135°C (138K), although there have been reports of critical temperatures over –73°C (200K). But scientists are aiming even higher in their pursuit of room temperature superconductors, which would not need to be cooled.

Benefits There are many obstacles to overcome in the application of superconducting materials, but the benefits may outweigh the difficulties. Maglev trains are expensive to construct because they are not compatible with traditional rail lines. But maglev trains are very efficient, do not create pollution, operate quietly, are safer and compared to traditional modes of transportation.

Another application of this technology involves the wires that currently carry electricity to homes and businesses. They lose energy while conducting electricity. Replacing them with superconducting wires and cables creates many obstacles, including the expense of cooling the wires and the brittleness of the superconducting materials used to make the wires. However, room temperature superconductor wires would make the process much more efficient.

> **WebQuest**
>
> **Investigate** the potential application of room temperature superconductors used in electrical energy storage devices. These devices are called Superconducting Magnetic Energy Storage (SMES). Report your findings to your class.

THEME FOCUS Structure and Properties of Matter

Matter can be classified as substances and mixtures. Substances can be either an element, such as oxygen, or a compound, such as salt. Mixtures are made of two or more substances.

BIG Idea Matter can be classified by its composition, by its physical properties, and by its chemical properties.

Section 1 Composition of Matter

colloid (p. 466)
compound (p. 464)
element (p. 462)
heterogeneous mixture (p. 465)
homogeneous mixture (p. 467)
solution (p. 467)
substance (p. 462)
suspension (p. 466)
Tyndall effect (p. 467)

MAIN Idea Matter exists as either a pure substance or a mixture.

- An element is a substance with the same kind of atoms.
- There are approximately 90 naturally occurring elements found on Earth and over 25 that have been created in laboratories.
- A compound is a substance that has two or more elements combined in a fixed proportion.
- Mixtures can be heterogeneous or homogeneous and can be separated by physical means.

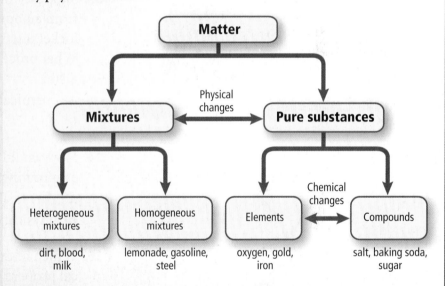

Section 2 Properties of Matter

chemical change (p. 473)
chemical property (p. 472)
distillation (p. 472)
law of conservation of mass (p. 475)
physical change (p. 471)
physical property (p. 469)

MAIN Idea A physical property can be observed without changing the identity of the material; a chemical property can be observed when one or more new substances are formed.

- Physical properties can be used to distinguish and separate substances.
- A chemical change is sometimes indicated by the cooling, heating, or formation of solids or bubbles.
- The law of conservation of mass states that matter is neither created nor destroyed in a chemical reaction.

Use Vocabulary

Complete each sentence with the correct term from the Study Guide.

14. Substances formed from atoms of two or more elements are called _____.

15. A(n) _____ is a heterogeneous mixture in which solid particles settle.

16. Freezing, boiling, and evaporation are all examples of _____.

17. Compounds are made from the atoms of two or more _____.

18. _____ is the process that can separate two liquids using physical change.

Check Concepts

Use the figure below to answer question 19.

19. What type of property is represented by the illustration?
    A) chemical
    B) physical
    C) conservation
    D) element

20. Which is an example of a chemical change?
    A) boiling   C) evaporation
    B) burning   D) melting

21. Which type of substance is milk?
    A) colloid   C) substance
    B) compound   D) suspension

22. A visible sunbeam is an example of which of the following?
    A) a chemical change
    B) a physical property
    C) a suspension
    D) the Tyndall effect

23. Suppose you start to eat some potato chips from an open bag that you found in your locker and notice that they taste unpleasant. What process resulted in this unpleasant taste?
    A) chemical change   C) chemical property
    B) physical change   D) physical property

24. How would you describe the process of evaporating fresh water from seawater?
    A) chemical change   C) chemical property
    B) physical change   D) physical property

25. Which of these warnings refers to a chemical property of the material?
    A) Fragile
    B) Sharp Object
    C) Flammable
    D) Shake Well

26. Which of the following is evidence that a physical change has occurred?
    A) broken glass
    B) formation of bubbles
    C) rust
    D) formation of a solid precipitate

✓ Assessment   Online Test Practice

Interpret Graphics

27. Copy and complete the concept map about matter.

Concepts in Motion
Interactive Concept Map

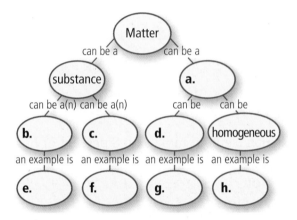

Use the table below to answer question 28.

Common Colloids	
Colloid	**Example**
Solid in a liquid	gelatin
Solid in a gas	a.
Gas in a solid	b.
Solid in a liquid	c.
Liquid in a gas	d.

28. Different colloids can involve different states of matter. For example, gelatin is formed from solid particles in a liquid. Complete the table using these colloids: smoke, marshmallow, fog, and paint.

Think Critically

29. **THEME FOCUS** **Describe** the contents of a carton of milk using at least four physical properties.

30. **Explain** Carbon and the gases hydrogen and oxygen combine to form sugar. How do you know sugar is a compound?

31. **Explain** How does a nail rusting in air follow the law of conservation of mass?

32. **BIG Idea** **Explain** Mai says that ocean water is a solution. Tom says that it is a suspension. Can they both be correct?

33. **Determine** Marcos took a 100-cm³ sample of a suspension, shook it well, and poured equal amounts into four different test tubes. He placed one test tube in a rack, one in hot water, one in warm water, and the fourth test tube in ice water. He then observed the time it took for each suspension to settle. What was the independent variable in the experiment? What was the control? What was one constant?

Apply Math

Use the graph below to answer question 34.

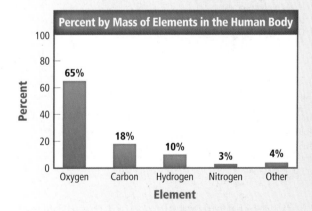

34. **Use Proportions** The average human body has a mass of 75 kg. Determine the mass of each element in an average human body based on the percent shown in the graph.

Standardized Test Practice

Record your answers on the answer sheet provided by your teacher or on a sheet of paper.

1. Which statement about elements is FALSE?
   A. The same kind of atoms exist in an element.
   B. There are about 1,000 elements found in nature.
   C. Some elements have been made in laboratories.
   D. Zinc, copper, and iron are elements.

2. $CaCO_3$ is an example of which type of material?
   A. element   C. compound
   B. mixture   D. colloid

Use the graph below to answer question 3.

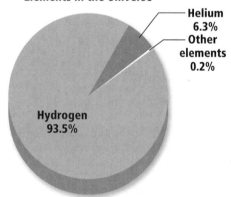

Elements in the Universe

Helium 6.3%
Other elements 0.2%
Hydrogen 93.5%

3. What percentage do the elements hydrogen and helium account for in the universe?
   A. 100%   C. 98%
   B. 99.9%   D. 99.8%

4. The most plentiful element in the universe readily burns in air. What is this chemical property called?
   A. flammability
   B. ductility
   C. density
   D. boiling point

Use the graph below to answer questions 5 and 6.

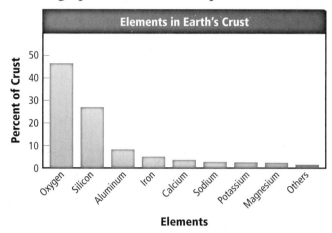

5. Which element makes up 8 percent of Earth's crust?
   A. iron
   B. aluminum
   C. silicon
   D. oxygen

6. Which elements each make up approximately the same percentage of Earth's crust?
   A. iron, calcium, and sodium
   B. sodium, potassium, and magnesium
   C. oxygen, silicon, and aluminum
   D. aluminum, iron, and calcium

7. Which statement best describes the law of conservation of mass?
   A. The mass of the products is always greater than the mass of the materials that react in a chemical change.
   B. The mass of the products is always less than the mass of the materials that react in a chemical change.
   C. A certain mass of material must be present for a reaction to occur.
   D. Matter is neither lost nor gained during a chemical change.

 Assessment Standardized Test Practice

Short Response

*Record your answers on the answer sheet
provided by your teacher or on a sheet of paper.*

Use the images below to answer question 8.

8. Compare and contrast the physical properties of these minerals.

9. Why are some medicines stored in dark bottles?

10. Explain how two dangerous elements can form a compound that is edible.

11. What physical and chemical properties of titanium make it useful for airplane manufacturing?

12. Compare and contrast the properties of heterogeneous and homogeneous mixtures. What is another name for a homogeneous mixture?

13. You are given a mixture of iron filings, sand, and salt. Describe how to separate this mixture.

Extended Response

Record your answers on a sheet of paper.

Use the figure below to answer question 14.

14. Describe what type of changes are taking place in the photo.

15. Design an experiment that shows that this type of chemical change is governed by the law of conservation of mass.

16. Describe the distillation of seawater using a multi-step process. On what physical property is this process based?

17. Give two examples of a compound. Describe why the ratios of elements in these compounds do not change.

NEED EXTRA HELP?																
If You Missed Question . . .	1	2	3	4	5	6	7	8	9	10	11	12	13	14	15	16	17
Review Section . . .	1	1	2	2	1	1	2	2	2	1, 2	1	1	1, 2	2	2	2	1

CHAPTER 16

Properties of Atoms and the Periodic Table

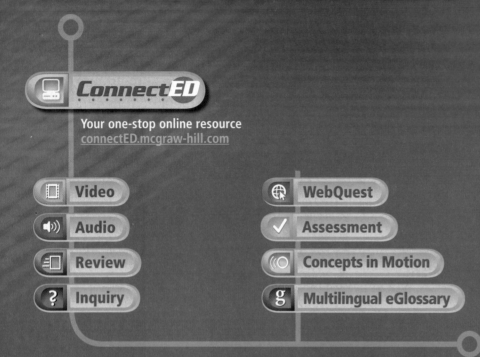

ConnectED

Your one-stop online resource
connectED.mcgraw-hill.com

- Video
- Audio
- Review
- Inquiry
- WebQuest
- Assessment
- Concepts in Motion
- g Multilingual eGlossary

Launch Lab
Hidden Information

Much of the fun of receiving a wrapped gift is trying to figure out what is inside before you open it. Chemists have had similar experiences trying to determine the structure of the atom. How good are your skills of observation and inference?

For a lab worksheet, use your StudentWorks™ Plus Online.

? Inquiry   Launch Lab

FOLDABLES

Make a three-tab book. Label it as shown. Use it to organize your notes on atoms.

Know? | Like to know? | Learned?

THEME FOCUS Structures and Properties of Matter

Atoms are classified based on their physical and chemical properties.

BIG Idea The properties of an element are determined by the structure of its atoms.

Reading Preview

Essential Questions

▶ What are the names and symbols of common elements?

▶ What is the structure of the atom?

▶ How do scientists study quarks?

▶ What is the electron cloud model of the atom?

Review Vocabulary

element: substance with atoms that are all alike

New Vocabulary

atom
nucleus
proton
neutron
electron
quark
electron cloud

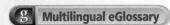

g Multilingual eGlossary

Structure of the Atom

MAIN ◀Idea Protons and neutrons form the nucleus of an atom, and electrons occupy a space surrounding the nucleus.

Real-World Reading Link Try to picture an outer belt freeway system of a large city that measures thirty kilometers in diameter. This freeway outer belt represents the size of an atom. The nucleus of that same atom is represented by a small fish tank in the center of the city.

Scientific Shorthand

Do you use abbreviations for long words, street addresses, or the names of states? Scientists also use abbreviations. In fact, scientists have developed their own shorthand for naming the elements.

Do the letters C, Al, Ne, and Au mean anything to you? Each letter or pair of letters is a chemical symbol, which is a short or abbreviated name of an element. Chemical symbols, such as those in **Table 1,** consist of one capital letter or a capital letter plus one or two lowercase letters. For some elements, the symbol is the first letter of the element's name. For other elements, the symbol is the first letter of the name plus another letter from its name. Some symbols are derived from Latin. For instance, *argentum* is Latin for silver. The chemical symbol for silver is Ag.

Elements are named in a variety of ways. Some elements are named to honor scientists, for places, or for their properties. For example, the element curium was named to honor Pierre and Marie Curie, scientists who researched radioactivity. Other elements, like germanium, were named after a country. Regardless of the origin of the name, scientists derived the international system of symbols for convenience. It is much easier to write H for hydrogen, O for oxygen, and H_2O for dihydrogen monoxide (water). Because scientists worldwide use this system, everyone recognizes what these symbols represent.

((O)) Concepts in Motion

Interactive Table

Table 1	Symbols of Common Elements		
Element	Symbol	Element	Symbol
Aluminum	Al	Iron	Fe
Calcium	Ca	Mercury	Hg
Carbon	C	Nitrogen	N
Chlorine	Cl	Oxygen	O
Gold	Au	Potassium	K
Hydrogen	H	Sodium	Na

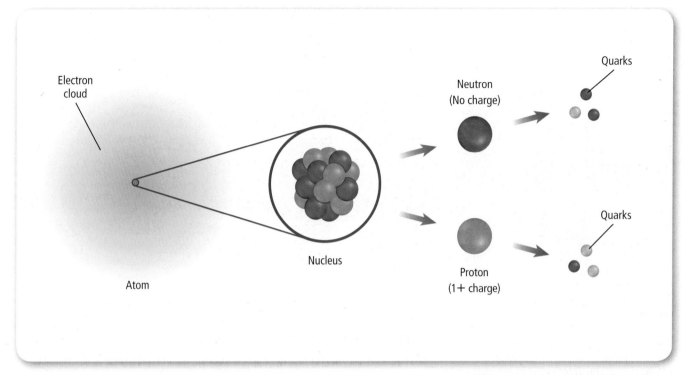

Subatomic Particles

An element is matter that is composed of only one type of atom. An **atom** is the smallest particle of an element that retains the element's properties. For example, the element iron is composed of only iron atoms and the element hydrogen is composed of only hydrogen atoms. Atoms are composed of even smaller particles—subatomic particles—called protons, neutrons, and electrons, as shown in **Figure 1.** The small, positively charged center of the atom is called the **nucleus.** The nucleus contains protons and neutrons. **Protons** are particles in the nucleus with an electric charge of 1+. The number of protons in the nucleus is unique for each element. **Neutrons** are electrically neutral particles in the nucleus; they do not have a charge. **Electrons** are particles with an electric charge of 1−. They occupy the space surrounding the nucleus of an atom.

Reading Check **Identify** the three types of subatomic particles.

Quarks—even smaller particles

Are the protons, electrons, and neutrons that make up atoms the smallest particles that exist? Scientists have inferred that protons and neutrons are composed of smaller particles called **quarks.** Electrons, however, are not made of smaller particles. So far, scientists have confirmed the existence of six uniquely different quarks. A particular arrangement of three of these quarks produces a proton. Another arrangement of three quarks produces a neutron. The search for the composition of protons and neutrons is a continuing effort.

■ **Figure 1** The nucleus of the atom contains protons and neutrons. The proton has a positive charge, and the neutron has no charge. Protons and neutrons are themselves composed of quarks. Electrons occupy a space surrounding the nucleus, which is called the electron cloud.

Compare and contrast *protons, neutrons, and electrons.*

VOCABULARY ....

WORD ORIGIN

Atom
comes from the Greek word *atomos,* meaning *indivisible* or *uncuttable The basic building block of all matter is the atom.* ....

FOLDABLES
Incorporate information from this section into your Foldable.

Tevatron

Large Hadron Collider

■ **Figure 2** At 6.4km in circumference, the Tevatron in Batavia, Illinois was the world's most powerful particle accelerator when it was completed in 1985. When the Tevatron stopped colliding particles in 2011, the Large Hadron Collider (LHC) had become the largest accelerator. The LHC, which is 27 km in circumference, is designed to use 9,300 superconducting magnets.

■ **Figure 3** Wire chambers can be used by scientists to study the tracks left by subatomic particles. The chamber is a complex structure. This image is a view of the inside of a wire chamber.

The search for quarks To study quarks, scientists accelerate charged particles to tremendous speeds and then force them to collide with—or smash into—protons. These collisions cause the protons to break apart. **Figure 2** shows two particle accelerators. These giant machines use electric and magnetic fields to accelerate, focus, and collide fast-moving particles. The Fermi National Accelerator Laboratory in Batavia, Illinois, housed a machine called the Tevatron that could generate the forces that are required to break apart protons. The Large Hadron Collider (LHC), also shown in **Figure 2,** is a particle accelerator in Geneva, Switzerland. The LHC is capable of even greater forces, hopefully leading to a deeper understanding of the structure of the universe. The ultimate goal is to discover new particles.

Scientists use a variety of collection devices to obtain detailed information about the particles created in a collision. Just as police investigators can reconstruct traffic accidents from tire marks and other evidence at the scene, scientists are able to examine data collectors for evidence of the tiniest of particles. Scientists use inference to identify subatomic particles and to reveal information about each particle's structure. For example, the wire chambers in **Figure 3** help scientists examine the varying tracks made by different types of particles formed in high-speed collisions.

The sixth quark Finding evidence for the existence of quarks was not an easy task. Scientists discovered five quarks and hypothesized that a sixth quark existed. However, it took several years for a team of nearly 450 scientists from around the world to find the sixth quark. The tracks of the sixth quark were hard to detect because only about one-billionth of a percent of the proton collisions performed showed the presence of a sixth quark, typically referred to as the top quark.

Models—Tools for Scientists

Scientists and engineers use models to represent objects or ideas that are difficult to visualize or to picture in your mind. You might have seen models or blueprints of buildings, planetary models of the solar system, or even a model airplane. These are scaled-down models. Scaled-down models allow you to see something that is too large to visualize all at once or something that has not yet been built.

Scaled-up models are often used to visualize things that are too small to see. Models of atoms are examples of scaled-up models. To give you an idea of how small the atom is, it would take about 50,000 aluminum atoms stacked one on top of the other to equal the thickness of a sheet of aluminum foil. To study the atom, scientists have developed scaled-up models that they can use to visualize an atom. For a model of the atom to be useful, it must support the accepted ideas about atomic structure and behavior. As new discoveries about atoms are made, scientists must include these new details in the model.

The atomic model You now know that all matter is composed of atoms, but this was not always accepted as truth. Around 400 B.C., the Greek philosopher Democritus proposed the idea that atoms are tiny particles that make up all matter. Another philosopher, Aristotle, disputed Democritus's idea and proposed that matter was uniform throughout and was not composed of such small particles. Aristotle's incorrect idea was accepted for about 2,000 years. In the 1800s, the English scientist John Dalton was able to present evidence to suggest that atoms exist.

Dalton's atomic theory, highlighted in **Table 2,** led to his model of the atom. This model has changed somewhat over time with further investigations by other scientists, as shown on the next page in **Figure 4.** Dalton's modernization of Democritus' idea of the atom provided a physical explanation for chemical reactions. Due to this discovery, scientists could finally express these reactions in quantitative terms using chemical symbols and equations.

Inquiry   Video Lab

Video   BrainPOP

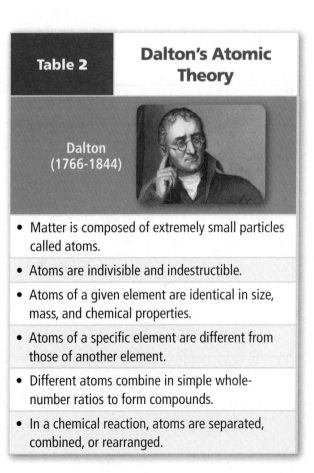

Table 2	Dalton's Atomic Theory

Dalton (1766-1844)

- Matter is composed of extremely small particles called atoms.
- Atoms are indivisible and indestructible.
- Atoms of a given element are identical in size, mass, and chemical properties.
- Atoms of a specific element are different from those of another element.
- Different atoms combine in simple whole-number ratios to form compounds.
- In a chemical reaction, atoms are separated, combined, or rearranged.

FIGURE 4

Visualizing the Early Atomic Models

The currently accepted model of the atom evolved from the ideas and
the work of many scientists.

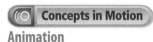
Concepts in Motion
Animation

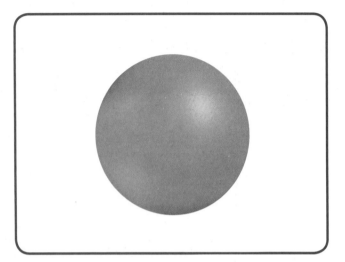

400 B.C. Democritus Model Democritus first proposed that elements consisted of tiny, solid particles that
could not be subdivided. He called these particles *atomos,*
meaning "uncuttable." Democritus's ideas were criticized
by Aristotle, who believed that empty space could not exist.
Because Aristotle was one of the most influential philoso-
phers of his time, Democritus's atomic theory was rejected.

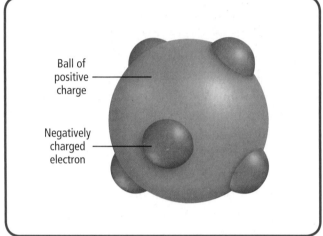

1904 Thomson Model English physicist Joseph
John Thomson proposed a model that consisted of a
spherical atom containing small, negatively charged particles. He thought these "electrons" (in red) were evenly
embedded throughout a positively charged sphere, much
like chocolate chips in a ball of cookie dough.

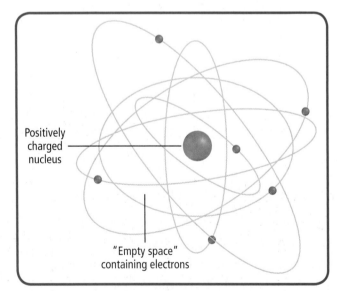

1911 Rutherford Model English physicist Ernest
Rutherford proposed that all the positive charge of an
atom was concentrated in a central atomic nucleus that
was surrounded by electrons.

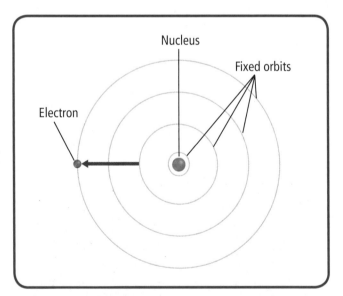

1913 Bohr Model Danish physicist Niels Bohr
hypothesized that electrons traveled in fixed orbits. The
electrons could jump between orbits as they absorbed
or released specific amounts of energy. The Bohr model
worked very well for hydrogen, but did not work as well
for atoms with many electrons.

The electron cloud model By 1926, scientists developed the electron cloud model of the atom, which is the model that is accepted today. An **electron cloud** is the area around the nucleus of an atom where electrons are most likely to be found. The electron cloud is 100,000 times larger in diameter than the diameter of the nucleus of an atom. In contrast, each electron in the cloud is significantly smaller in mass than a single proton or single neutron.

✓ **Reading Check Explain** the difference between the Bohr model and the electron cloud model.

Because an electron's mass is negligible compared to the nucleus and the electron is moving so quickly around the nucleus, it is impossible to describe its exact location in an atom at any moment. Picture the spokes on a moving bicycle wheel. The spokes are moving so quickly that you cannot pinpoint any single spoke in the wheel. All that you see is a blur that contains all the spokes somewhere with-in it. In a similar way, an electron cloud is a blur of activity containing all of an atom's electrons somewhere within it. **Figure 5** illustrates the location of the nucleus and the electron cloud in the electron cloud model of the atom.

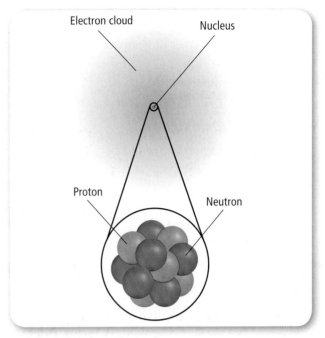

■ **Figure 5** Most of an atom is empty space. The electron cloud represents the area in which the electrons are moving.

Section 1 Review

Section Summary

▶ Scientists use chemical symbols to abbreviate element names.

▶ Atoms are composed of protons, neutrons, and electrons.

▶ Scientists have confirmed the existence of six different quarks.

▶ The electron cloud model is the current atomic model.

1. **MAIN Idea Identify** the names, charges, and locations of three types of subatomic particles that make up an atom.

2. **Identify** the chemical symbols for the elements carbon, aluminum, hydrogen, oxygen, and sodium.

3. **Describe** how quarks were discovered.

4. **Think Critically** Explain how a rotating electric fan might be used to model the atom. Explain how the rotating fan is unlike an atom.

Apply Math

5. **Estimate** A proton's mass is estimated to be 1.6726×10^{-24} g, and the mass of an electron is estimated to be 9.1093×10^{-28} g. How many times larger is the mass of a proton compared to the mass of an electron?

Reading Preview

Essential Questions

▶ How do you determine the atomic mass and mass number of an atom?

▶ What are isotopes?

▶ How do you determine the average atomic mass of an element?

Review Vocabulary

mass: amount of matter in an object

New Vocabulary

atomic number
mass number
isotope
average atomic mass

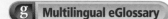

ⓖ **Multilingual eGlossary**

Masses of Atoms

MAIN ‹Idea› All atoms of the same element have the same number of protons but can have different numbers of neutrons.

Real-World Reading Link You are probably aware that numbers are often used for identification. For example, a computer can be identified by its IP address and a cell phone by its phone number. Similarly, elements are identified by a number—the number of protons in the nucleus.

Atomic Mass

The nucleus contains almost all of the atom's mass because protons and neutrons are far more massive than electrons. The mass of a proton is roughly the same as that of a neutron—about 1.67×10^{-24} g, as shown in **Table 3.** The mass of each is more than 1,800 times greater than the mass of the electron. The electron's mass is so small that it can be ignored when evaluating the mass of an atom.

If you were asked to estimate the height of your school building, you would probably not give an answer in kilometers. Considering the scale of the building, you would more likely give the height in meters. When thinking about the mass of an atom, scientists discovered that even grams were not small enough to use for measurement. Scientists needed a more manageable unit.

The unit of measurement used to quantify an atom's mass is the atomic mass unit (amu). The mass of a proton or a neutron is almost equal to 1 amu. This is not coincidence—the unit was defined that way. The atomic mass unit is defined as one-twelfth of the mass of a carbon atom containing six protons and six neutrons, as shown in **Figure 6.** Remember that the mass of an atom is contained almost entirely in the mass of the protons and neutrons in the nucleus. Therefore, each of the 12 particles in the carbon nucleus must have a mass nearly equal to 1 amu.

■ **Figure 6** The masses in **Table 3** show how similar the mass of a proton and a neutron are. An electron's mass is negligible compared to an atom's mass. The atomic mass unit is equal to one-twelfth of the mass of a carbon atom, shown to the right.

Identify *Where is most of the mass of an atom located?*

Table 3	Subatomic Particle Masses	
Particle	**Mass (g)**	
Proton	1.6726×10^{-24}	
Neutron	1.6749×10^{-24}	
Electron	9.1093×10^{-28}	

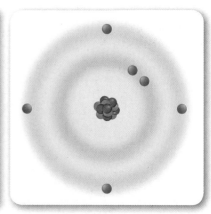

Table 4	Mass Numbers of Atoms	Concepts in Motion	Interactive Table		
Element	Symbol	Atomic Number	Protons	Neutrons	Mass Number
Boron	B	5	5	6	11
Carbon	C	6	6	6	12
Oxygen	O	8	8	8	16
Sodium	Na	11	11	12	23
Copper	Cu	29	29	34	63

Atomic number Recall that an element is made of one type of atom. What determines the type of atom? In fact, the number of protons identifies the type of atom. For example, every carbon atom has six protons. Also, any atom with six protons is a carbon atom. Atoms of different elements have different numbers of protons. For example, atoms with eight protons are oxygen atoms.

The number of protons in an atom's nucleus is equal to its **atomic number.** The atomic number of carbon is six. Oxygen has an atomic number equal to eight, as shown in **Table 4.** Therefore, if you are given any one of the following—the name of the element, the number of protons for the element, or the atomic number of the element—you can identify the other two. For example, if your teacher asked you to identify an atom with an atomic number of 11, you would know that the atom has eleven protons and it is sodium, as indicated in **Table 4.**

 Reading Check Interpret Table 4 to identify the element and the atomic number of the element with 29 protons.

Mass Number The **mass number** of an atom is the sum of the number of protons and the number of neutrons in the nucleus of an atom.

Mass number = number of protons + number of neutrons

For example, you can calculate the mass number of the copper atom listed in **Table 4**: 29 protons plus 34 neutrons equals a mass number of 63.

Also, if you know the mass number and the atomic number of an atom, you can calculate the number of neutrons in the nucleus. The number of neutrons is equal to the mass number minus the atomic number. In fact, if you know two of the three numbers—mass number, atomic number, number of neutrons—you can always calculate the third.

Isotopes

Atoms of the same element can have different mass numbers. For example, carbon atoms can have a mass number of 12 and also a mass number of 14. The number of protons for each element never changes. So, for an atom's mass number to differ, the number of neutrons must change. Atoms of the same element that have different numbers of neutrons are called **isotopes.**

To identify isotopes, scientists write the name of the element followed by the element's mass number. Carbon with a mass number of 12 is written as carbon-12. Carbon-12 has six protons and six neutrons. Carbon-14 has six protons and eight neutrons. Carbon-12 and carbon-14 are isotopes of the element carbon. Some properties of carbon-12 and carbon-14 are unique due to differences in the number of neutrons that each element contains. For example, carbon-14 is radioactive, but carbon-12 is not.

Suppose you have a sample of the element boron. Naturally occurring isotopes of boron have mass numbers of 10 or 11. How many neutrons does each isotope contain? Locate boron in **Table 4** on the previous page and determine the number of protons in an atom of boron. You can then calculate that boron-10 has 5 neutrons and boron-11 has 6 neutrons.

Apply Science

How can radioactive isotopes help tell time?

Some isotopes are radioactive, which means they decay over time. The time that it takes for half of the radioactive isotope to decay into another isotope is called its half-life. Scientists use the half-lives of radioactive isotopes to measure geologic time.

Identify the Problem

The table to the right shows the half-lives of some radioactive isotopes (parent isotopes) and the isotopes into which they decay (daughter isotopes). For example, it would take 5,730 years for half of the carbon-14 atoms in a sample to change into atoms of nitrogen-14. After another 5,730 years, half of the remaining carbon-14 atoms will change, and so on. Because the number of carbon-14 atoms change while the number of carbon-12 atoms do not, the ratio of the number of carbon-14 atoms to carbon-12 atoms can be used to determine the length of time that has passed.

Half-Lives of Radioactive Isotopes		
Parent Isotope	Daughter Isotope	Half-Life
Uranium-238	Lead-206	4,470 billion years
Potassium-40	Argon-40, calcium-40	1,260 billion years
Rubidium-87	Strontium-87	48,800 billion years
Carbon-14	Nitrogen-14	5,730 years

Solve the Problem

1. How many years would it take half of the rubidium-87 atoms in a piece of rock to change into strontium-87? How many years would it take for three-quarters of the atoms to change?
2. After a long period, only one-quarter of the parent uranium-238 atoms in a sample of rock remain. How many years old would you predict the rock to be?

Average atomic mass How do scientists account for and represent the different atomic masses of isotopes? As an example, models of two naturally occurring isotopes of boron are shown in **Figure 7.** Because most elements, including boron, naturally occur as more than one isotope, each element can be described by an average atomic mass of the isotopes. The **average atomic mass** of an element is the weighted average mass of all naturally occurring isotopes of an element, measured in atomic mass units (amu), according to their natural abundances.

For example, eighty percent or four out of five atoms of boron are boron-11 and twenty percent or one out of five is boron-10. The following calculation gives the weighted average of these two masses.

$$\frac{4}{5}(11 \text{ amu}) + \frac{1}{5}(10 \text{ amu}) = 10.8 \text{ amu}$$

The average atomic mass of the element boron is 10.8 amu. Note that the average atomic mass of boron is closest to the mass of its most abundant isotope, boron-11.

✓ **Reading Check Define** average atomic mass and explain how it is calculated.

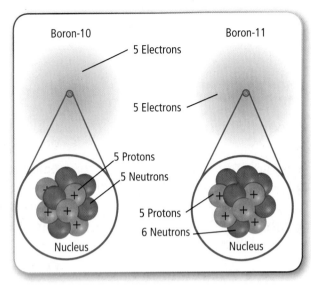

■ **Figure 7** Boron-10 and boron-11 are two isotopes of boron. These two isotopes differ by one neutron. Like boron, most naturally occurring elements have more than one naturally occurring isotope.

Explain *why these atoms are isotopes.*
_____

Section 2 Review

Section Summary

▶ Protons and neutrons make up most of an atom's mass.

▶ Each element has a unique number of protons.

▶ Atoms of the same element with different numbers of neutrons are called isotopes.

▶ The average atomic mass of an element is the weighted average mass of all naturally occurring isotopes of that element.

6. **MAIN Idea Determine** the mass number and the atomic number of a chlorine atom that has 17 protons and 18 neutrons.

7. **Explain** how the isotopes of an element are alike and how are they different.

8. **Explain** why the atomic mass of an element is a weighted-average mass.

9. **Calculate** the number of neutrons in potassium-40.

10. **Think Critically** Chlorine has an average atomic mass of 35.45 amu. The two naturally occurring isotopes of chlorine are chlorine-35 and chlorine-37. Do most chlorine atoms contain 18 neutrons or 20 neutrons? Why?

Apply Math

11. **Determine** Use the information in **Table 3** to determine the mass in kilograms of each subatomic particle.

Reading Preview

Essential Questions

▶ How is the periodic table organized?

▶ What are the trends on the periodic table?

▶ What are the properties of metals, nonmetals, and metalloids?

Review Vocabulary

chemical property: any characteristic of a substance that indicates whether it can undergo a certain chemical change

New Vocabulary

periodic table
period
group
electron dot diagram

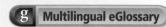 **Multilingual eGlossary**

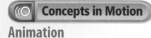 **Concepts in Motion**
Animation

? Inquiry Virtual Labs

The Periodic Table

MAIN ◀Idea Atoms of elements that are in the same group on the periodic table have similar physical and chemical properties.

Real-World Reading Link It is easier to find matching pairs of shoes or clean shirts when your closet is neat and tidy. Like a tidy closet, elements are organized in rows and columns on the periodic table to make it easy for chemists to infer their physical and chemical properties.

Organizing the Elements

On a clear night, you can see one of the various phases of the Moon. Each month, the Moon appears to grow larger and then smaller in a predictable pattern. This type of change is periodic. Periodic means "repeated in a pattern." For example, a calendar is a periodic table of the days and months of the year. The days of the week are also periodic because they repeat themselves every seven days.

In the late 1800s, a Russian chemist named Dmitri Mendeleev presented a way to organize all the known elements. While studying the physical and chemical properties of the elements, Mendeleev found that these properties repeated in predictable patterns based on an element's atomic mass. Because the pattern repeated, it was considered to be periodic.

Figure 8 shows one of Mendeleev's early periodic charts. Mendeleev arranged elements in rows based on increasing atomic mass and in columns based on elements that shared similar physical and chemical properties. Today, this arrangement is called the periodic table of elements. In the modern **periodic table,** the elements are arranged by increasing atomic number—not atomic mass—and by periodic changes in physical and chemical properties.

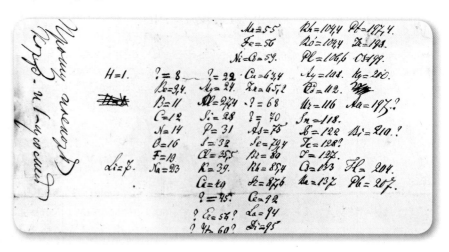

■ **Figure 8** When Mendeleev organized the known elements in order of increasing atomic mass, he discovered that the elements had a periodic pattern in their chemical properties.

Infer *What do the question marks in Mendeleev's chart represent?*

Table 5   Mendeleev's Predictions

Predicted Properties of Ekasilicon (Es)	Actual Properties of Germanium (Ge)
Existence Predicted: 1871	*Actual Discovery: 1886*
Atomic mass = 72	Atomic mass = 72.61
High melting point	Melting point = 938°C
Density = 5.5 g/cm³	Density = 5.323 g/cm³
Dark-gray metal	Gray metal
Density of EsO_2 = 4.7 g/cm³	Density of GeO_2 = 4.23 g/cm³

Mendeleev's predictions Mendeleev had to leave blank spaces in his periodic table. He studied the physical and chemical properties and the atomic masses of the elements surrounding all the blank spaces. From this information, he was able to predict the probable properties and the atomic masses for the missing elements that had not yet been discovered. **Table 5** shows a few of Mendeleev's predicted physical and chemical properties for germanium, which he called ekasilicon. His predictions proved to be accurate when compared to the actual properties of germanium. Scientists later confirmed the identities of missing elements and found that their properties were similar to what Mendeleev had suggested.

Changes in the periodic table Although Mendeleev's arrangement of the elements was a success, it required some changes. The atomic mass gradually increased from left to right on Mendeleev's table. If you look at the modern periodic table, in **Figure 9,** on the next page, you can locate instances where atomic mass decreases from left to right, such as nickel and cobalt.

You might also observe that the atomic number always increases from left to right. In 1913, a young English scientist named Henry G.J. Moseley arranged all the known elements based on increasing atomic number instead of atomic mass. This new arrangement seemed to solve the problem of fluctuating mass. The modern periodic table uses Moseley's arrangement of the elements.

✓ **Reading Check Explain** how Mendeleev organized his periodic table.

MiniLab

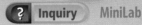 **Inquiry**   MiniLab

Organize Elements
Procedure
1. Read the procedure and safety information, and complete the lab form.
2. Make a set of **element cards** based on fictitious elements provided in the table below.

Symbol	Mass (g)	State	Color
Ad	52.9	solid/liquid	orange
Ax	108.7	ductile solid	light blue
Bp	69.3	gas	red
Cx	112.0	brittle solid	light green
Lq	98.7	ductile solid	blue
Pd	83.4	brittle solid	green
Qa	68.2	ductile solid	dark blue
Rx	106.9	liquid	yellow
Tu	64.1	brittle solid	hunter green
Xn	45.0	gas	crisom red

3. Organize the cards by increasing atomic mass, and start placing them in a **3 × 4 grid.**
4. Place each card based on its properties and leave gaps where necessary. Make a table reflecting your arrangement of cards.

Analysis
1. **Describe** the trends for color in your new table across rows and down columns.
2. **Describe** similar trends for mass in your new table. Explain your placement of any elements that do not fit the trend.
3. **Predict** the placement of a newly found element, Ph, which is a deep-pink gas. What would be an expected range for the mass of Ph?

Figure 9　The Periodic Table of the Elements

PERIODIC TABLE OF THE ELEMENTS

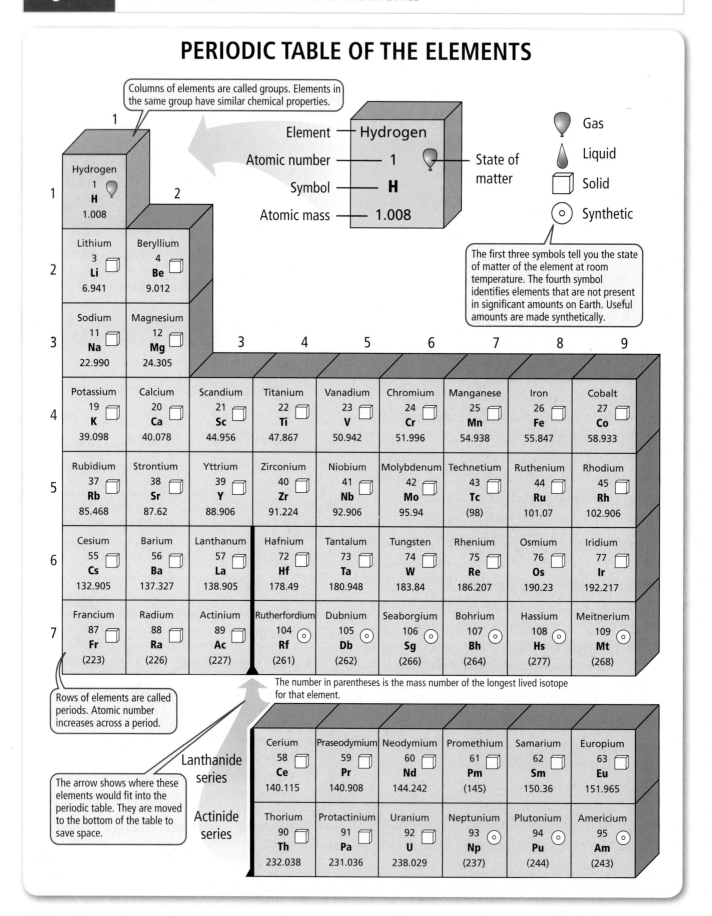

Columns of elements are called groups. Elements in the same group have similar chemical properties.

Element — Hydrogen
Atomic number — 1
Symbol — **H**
Atomic mass — 1.008
State of matter

Gas
Liquid
Solid
Synthetic

The first three symbols tell you the state of matter of the element at room temperature. The fourth symbol identifies elements that are not present in significant amounts on Earth. Useful amounts are made synthetically.

	1		2		3	4	5	6	7	8	9
1	Hydrogen 1 **H** 1.008										
2	Lithium 3 **Li** 6.941		Beryllium 4 **Be** 9.012								
3	Sodium 11 **Na** 22.990		Magnesium 12 **Mg** 24.305								
4	Potassium 19 **K** 39.098		Calcium 20 **Ca** 40.078		Scandium 21 **Sc** 44.956	Titanium 22 **Ti** 47.867	Vanadium 23 **V** 50.942	Chromium 24 **Cr** 51.996	Manganese 25 **Mn** 54.938	Iron 26 **Fe** 55.847	Cobalt 27 **Co** 58.933
5	Rubidium 37 **Rb** 85.468		Strontium 38 **Sr** 87.62		Yttrium 39 **Y** 88.906	Zirconium 40 **Zr** 91.224	Niobium 41 **Nb** 92.906	Molybdenum 42 **Mo** 95.94	Technetium 43 **Tc** (98)	Ruthenium 44 **Ru** 101.07	Rhodium 45 **Rh** 102.906
6	Cesium 55 **Cs** 132.905		Barium 56 **Ba** 137.327		Lanthanum 57 **La** 138.905	Hafnium 72 **Hf** 178.49	Tantalum 73 **Ta** 180.948	Tungsten 74 **W** 183.84	Rhenium 75 **Re** 186.207	Osmium 76 **Os** 190.23	Iridium 77 **Ir** 192.217
7	Francium 87 **Fr** (223)		Radium 88 **Ra** (226)		Actinium 89 **Ac** (227)	Rutherfordium 104 **Rf** (261)	Dubnium 105 **Db** (262)	Seaborgium 106 **Sg** (266)	Bohrium 107 **Bh** (264)	Hassium 108 **Hs** (277)	Meitnerium 109 **Mt** (268)

Rows of elements are called periods. Atomic number increases across a period.

The number in parentheses is the mass number of the longest lived isotope for that element.

The arrow shows where these elements would fit into the periodic table. They are moved to the bottom of the table to save space.

Lanthanide series

Cerium 58 **Ce** 140.115	Praseodymium 59 **Pr** 140.908	Neodymium 60 **Nd** 144.242	Promethium 61 **Pm** (145)	Samarium 62 **Sm** 150.36	Europium 63 **Eu** 151.965

Actinide series

Thorium 90 **Th** 232.038	Protactinium 91 **Pa** 231.036	Uranium 92 **U** 238.029	Neptunium 93 **Np** (237)	Plutonium 94 **Pu** (244)	Americium 95 **Am** (243)

Metal
Metalloid
Nonmetal
Recently observed

The color of an element's block tells you if the element is a metal, nonmetal, or metalloid.

	18
	Helium 2 He 4.003

13	14	15	16	17	
Boron 5 B 10.811	Carbon 6 C 12.011	Nitrogen 7 N 14.007	Oxygen 8 O 15.999	Fluorine 9 F 18.998	Neon 10 Ne 20.180
Aluminum 13 Al 26.982	Silicon 14 Si 28.086	Phosphorus 15 P 30.974	Sulfur 16 S 32.066	Chlorine 17 Cl 35.453	Argon 18 Ar 39.948

10	11	12	13	14	15	16	17	18
Nickel 28 Ni 58.693	Copper 29 Cu 63.546	Zinc 30 Zn 65.39	Gallium 31 Ga 69.723	Germanium 32 Ge 72.61	Arsenic 33 As 74.922	Selenium 34 Se 78.96	Bromine 35 Br 79.904	Krypton 36 Kr 83.80
Palladium 46 Pd 106.42	Silver 47 Ag 107.868	Cadmium 48 Cd 112.411	Indium 49 In 114.82	Tin 50 Sn 118.710	Antimony 51 Sb 121.757	Tellurium 52 Te 127.60	Iodine 53 I 126.904	Xenon 54 Xe 131.290
Platinum 78 Pt 195.08	Gold 79 Au 196.967	Mercury 80 Hg 200.59	Thallium 81 Tl 204.383	Lead 82 Pb 207.2	Bismuth 83 Bi 208.980	Polonium 84 Po 208.982	Astatine 85 At 209.987	Radon 86 Rn 222.018
Darmstadtium 110 Ds (281)	Roentgenium 111 Rg (272)	Copernicium 112 Cn (285)	Ununtrium ★ 113 Uut (284)	Ununquadium ★ 114 Uuq (289)	Ununpentium ★ 115 Uup (288)	Ununhexium ★ 116 Uuh (291)		Ununoctium ★ 118 Uuo (294)

★ The names and symbols for elements 113, 114, 115, 116, and 118 are temporary. Final names will be selected when the elements' discoveries are verified.

Gadolinium 64 Gd 157.25	Terbium 65 Tb 158.925	Dysprosium 66 Dy 162.50	Holmium 67 Ho 164.930	Erbium 68 Er 167.259	Thulium 69 Tm 168.934	Ytterbium 70 Yb 173.04	Lutetium 71 Lu 174.967
Curium 96 Cm (247)	Berkelium 97 Bk (247)	Californium 98 Cf (251)	Einsteinium 99 Es (252)	Fermium 100 Fm (257)	Mendelevium 101 Md (258)	Nobelium 102 No (259)	Lawrencium 103 Lr (262)

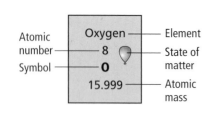

	Oxygen	
Atomic number	8	Element
		State of matter
Symbol	**O**	
	15.999	Atomic mass

■ **Figure 10** Each box on the periodic table contains information such as the element's name, its atomic number, its chemical symbol, its atomic mass, and its state of matter.

Evaluate *What is the atomic mass of oxygen?*

Concepts in Motion

Animation

■ **Figure 11** Energy levels in atoms can be represented by a flight of stairs. Each stair step away from the nucleus represents an increase in the amount of energy within the electrons. The higher energy levels can hold more electrons.

The Atom and the Periodic Table

The modern periodic table consists of boxes, each containing information such as element name, symbol, atomic number, and atomic mass. A typical box is shown in **Figure 10.** As you have learned, elements on the periodic table are organized based on similarities in their physical and chemical properties. The horizontal rows of elements in the periodic table are called **periods** and are numbered 1 through 7. The vertical columns in the periodic table are called **groups** (also called families), and they are numbered 1 through 18. Elements in each group share similar properties. For example, the elements in group 11—including copper, silver, and gold—are all similar. Each element is a shiny metal and a good conductor of heat and electricity. Why are these elements so similar?

Electron cloud structure You have learned about the nucleus of an atom and the fact that protons and neutrons are located there. But where are the electrons? How many are there? Because an atom does not have an overall charge, the number of electrons is equal to the number of protons. Therefore, a carbon atom has six protons and six electrons. An oxygen atom has eight protons and eight electrons. Electrons are located in an area surrounding the nucleus called the electron cloud.

Energy Levels Scientists have discovered that electrons within the electron cloud have different amounts of energy. Scientists model the energy differences between electrons by placing electrons in energy levels, as shown in **Figure 11.** Electrons located in energy levels close to the nucleus have less energy than electrons in energy levels farther away. Electrons occupy energy levels in a predictable pattern from the inner to the outer levels.

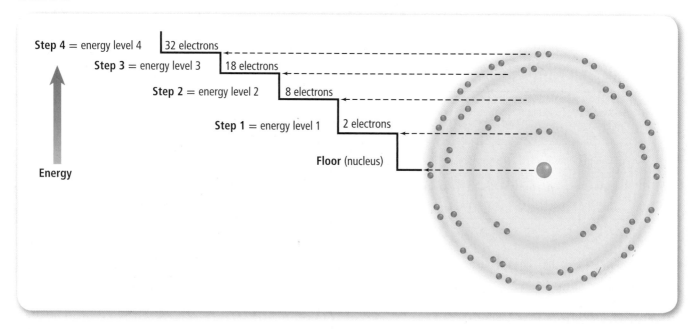

Step 4 = energy level 4   32 electrons
Step 3 = energy level 3   18 electrons
Step 2 = energy level 2   8 electrons
Step 1 = energy level 1   2 electrons
Floor (nucleus)

Energy

■ **Figure 12** One proton and one electron are added to each element as you travel across a row on the periodic table. When a certain level is complete—as with helium and neon—the next electron added starts a new row.

Explain *how the elements in each group are similar.*

Elements in the same group have the same number of electrons in their outermost energy levels. These electrons are called valence electrons. It is the number of valence electrons that determines the chemical properties of each individual element. It is important to understand the relationship between the location of an element in the periodic table, the element's chemical properties, and the element's atomic structure.

Rows on the periodic table Energy levels coincide with the number of rows on the periodic table. These energy levels are named using numbers one to seven. The maximum number of electrons that can be placed in each of the first four levels is shown in **Figure 11.** For example, energy level one can hold a maximum of two electrons. Energy level two can hold a maximum of eight electrons. For energy levels two and higher, the outer energy level is stable when it holds eight electrons. Notice, however, that energy levels three and four can contain more than eight electrons. The way in which energy levels split into sublevels allows for energy levels three and higher to contain more than eight electrons. These additional electrons are added to inner sublevels; the outer energy level is still stable when it contains eight electrons.

Filling the first row Remember that the atomic number found on the periodic table is equal to the number of electrons in a neutral atom. Look at the elements in **Figure 12.** The first row has hydrogen with one electron and helium with two electrons, both in energy level one. Because energy level one is the outermost energy level containing an electron, hydrogen has one outer electron. Helium has two outer electrons. Recall from **Figure 11** that energy level one can hold a maximum of two electrons. Therefore, helium has a full outer energy level and is chemically stable.

MiniLab

? Inquiry MiniLab

Model an Aluminum Atom

Procedure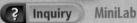

1. Read the procedure and safety information, and complete the lab form.
2. Arrange **thirteen 3-cm diameter circles** cut from **orange paper** and **fourteen 3-cm diameter circles** cut from **blue paper** on a flat surface to represent the nucleus of an atom. Each orange circle represents one proton, and each blue circle represents one neutron.
3. Position **two holes punched** from **red paper** about 20 cm from your nucleus.
4. Position **eight punched holes** in pairs about 40 cm from your nucleus.
5. Position **three punched holes** individually about 60 cm from your nucleus.

Analysis

1. **Explain** how your circles model an aluminum atom.
2. **Explain** why your model does not accurately represent the true size and distance between particles in an aluminum atom.

H·

Li·

Na·

K·

Rb·

Cs·

Fr·

Figure 13 The elements in group 1 have one electron in their outermost energy levels.

Figure 14 Electron dot diagrams show the electrons in an element's outermost energy level.

Filling higher rows The second row starts with lithium, which has three electrons—two in the first energy level and one in the second energy level. Lithium is followed by beryllium with two outer electrons, boron with three, and so on. Neon has a complete outermost energy level with eight outer electrons. Electrons begin filling energy level three for elements in the third row. The row ends with argon, which has eight outer electrons.

Electron dot diagrams Elements in the same group have the same number of electrons in their outermost energy levels. These electrons determine the chemical properties of an element. They are so significant that American chemist G.N. Lewis created a diagram to represent an element's outermost electrons while teaching a college chemistry class. An **electron dot diagram** uses the chemical symbol of an element surrounded by dots to represent the number of electrons in the outermost energy level. **Figure 13** shows the electron dot diagrams for the group 1 elements.

Same group—similar properties The electron dot diagrams for the elements in group 1 show that all members of a group have the same number of outermost electrons. Remember that the number of outermost electrons determines the chemical properties for each element.

A common chemical property of group 1 metals is the tendency to react with nonmetals in group 17. The nonmetals in group 17 have electron dot diagrams similar to chlorine, as shown in **Figure 14**. For example, the group 1 element sodium reacts easily with the group 17 element chlorine. The result is the formation of the compound sodium chloride (NaCl)—ordinary table salt.

Group 18 Not all elements will combine easily with other elements. The elements in group 18 have complete outermost energy levels, meaning that they cannot hold any more electrons. This special configuration makes many of the group 18 elements unreactive. **Figure 14** shows the electron dot diagram for neon, a member of group 18.

The electron dot diagram for group 17 consists of three sets of paired dots and one single dot.

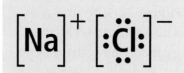

Sodium combines with chlorine to give each element a complete outer energy level in the resulting compound.

Neon, a member of group 18, has a full outer energy level. Neon has eight electrons in its outer energy level, making it unreactive.

Regions of the Periodic Table

The periodic table has areas with specific names. Recall that the horizontal rows of elements are called periods. The elements increase by one proton and one electron as you move from left to right across a period.

All the elements in the blue squares in **Figure 15** are metals. Iron, zinc, and copper are examples of a few common metals. Most metals occur as solids at room temperature. They are shiny, can be drawn into wires, can be pounded into sheets, and are good conductors of heat and electricity.

The elements on the right side of the periodic table that appear in the yellow squares are classified as nonmetals. Oxygen, bromine, and carbon are examples of nonmetals. Most nonmetals are gases at room temperature or brittle solids. They are poor conductors of heat and electricity. The elements in the green squares are metalloids. They exhibit properties of metals and nonmetals. Boron and silicon are examples of metalloids.

New elements Scientists around the world continue their research into the synthesis of elements. In 1994, scientists at the Heavy Ion Research Laboratory in Darmstadt, Germany, discovered element 111. The International Union of Pure and Applied Chemistry (IUPAC) confirmed the discovery in 2003. The name Roentgenium (Rg) was officially approved in 2004. Element number 112 was discovered at the same laboratory. Synthesis of the element was reported in 1996. IUPAC confirmed the discovery in 2009, and the element was officially named Copernicium (Cn) in 2010. These elements are produced in the laboratory by joining smaller atoms into a single, larger atom. The search for elements with higher atomic numbers continues. Scientists think that they have synthesized elements 113, 114, 115, 116, and 118.

VOCABULARY ....
ACADEMIC VOCABULARY
Occur
to be found; to come into existence
Many tornadoes occur in the central plains of the United States. ....

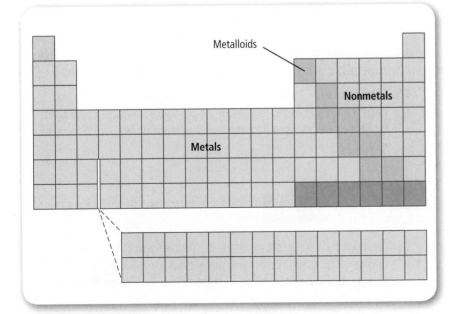

■ **Figure 15** Metalloids are located along the green stair-step line in the periodic table. Metals are located to the left of the metalloids and are shown in blue. Nonmetals are mainly located to the right of the metalloids (except for hydrogen) and are shown in yellow.

Elements in the Universe

With the development of new technologies, scientists have been able to study the chemistry of the universe. Because the universe is so vast, they have been able to study only a small section of the universe. However, scientists have learned that many of the same elements are found throughout the universe. These include lightweight elements, such as hydrogen and helium, and heavier elements, such as silicon, oxygen, and iron.

Many scientists think that hydrogen and helium are the building blocks of all other elements. Atoms fuse together within stars to produce heavier elements with atomic numbers greater than the atomic numbers of hydrogen and helium. Exploding stars, called supernovas, like the one shown in **Figure 16,** provide evidence to support this theory.

When stars explode, a mixture of elements, including heavy elements like iron, are expelled into the galaxy. Many scientists think that supernovas have scattered heavy, naturally occurring elements throughout the universe. Promethium, technetium, and elements with atomic numbers greater than 92 are rare or are not found on Earth. Some of these elements, such as neptunium and plutonium, are found only in trace amounts in Earth's crust as a result of uranium decay. Others have been found only in stars.

■ **Figure 16** The Crab Nebula is a remnant of a supernova that occurred in A.D. 1054. This is now an area of expanding gas and elements that will be incorporated into newly forming stars.

Section 3 Review

Section Summary

▶ Mendeleev organized the elements based on atomic mass and chemical and physical properties.

▶ Moseley built upon Mendeleev's periodic table by further organizing elements by increasing atomic number.

▶ Elements in the same vertical column on the periodic table are known as a group. They share similar physical and chemical properties.

▶ Elements in the same horizontal row on the periodic table are known as a period. They have the same number of energy levels.

▶ Elements on the periodic table are classified as metals, nonmetals, or metalloids.

12. **MAIN ‹Idea Identify** Use the periodic table to find the name, atomic number, atomic mass, and the number of outermost electrons for each of the following elements: N, P, As, and Sb.

13. **Provide** the symbol, the group number, and the period of each of the following elements: nitrogen, sodium, iodine, and mercury.

14. **Classify** each of the following elements as a metal, a nonmetal, or a metalloid and give the full name of each element: K, Si, and S.

15. **Think Critically** The Mendeleev and Moseley periodic charts had gaps for undiscovered elements. Why do you think the chart used by Moseley was more accurate at predicting where new elements would be placed?

Apply Math

16. **Construct** a circle graph showing the percentage of elements classified as metals, metalloids, and nonmetals. Use markers or colored pencils to distinguish clearly between each section on the graph.

✓ Assessment   Online Quiz

LAB

A Periodic Table of Foods

Objectives

- **Organize** 20 of your favorite foods into a periodic table of foods.

- **Analyze** your periodic table for similar characteristics among groups or family members.

- **Infer** where new foods added to your table would be placed.

Background: Mendeleev's task of organizing a collection of loosely related items probably seemed daunting at first. But as he searched for patterns and similarities, the task became more manageable and organized. Look for similarities and patterns as you create a periodic table.

Question: *How does organizing your favorite foods to create your own periodic table resemble the task that Mendeleev took on?*

Preparation

Materials

11 × 17 paper
metric ruler
colored pencils or markers

Procedure

1. Read the procedure and safety information, and complete the lab form.

2. Create a list of 20 of your favorite food and drink items.

3. Describe basic characteristics of each of these items. For example, you might describe the primary ingredient, nutritional value, taste, and color of each item. You could also identify the food group to which each item belongs, such as fruits, vegetables, grains, dairy products, meat, and sweets.

4. Create a data table to organize the information that you collect.

5. Construct a periodic table of foods on an 11 × 17 sheet of paper. Determine which characteristics you will use to organize your items. Create groups (columns) of food and drink items that share similar characteristics on your table. For example, potato chips, pretzels, and cheese-flavored crackers could be combined as a group of salty-tasting foods. Create as many groups as you need. You do not need to have the same number of items in every group.

Conclude and Apply

1. **Evaluate** the characteristics you used to create groups on your periodic table. Do the characteristics of each group adequately describe each of its members? Do the characteristics of each group distinguish its members from the members of another group?

2. **Explain** the way your rows of food are organized.

3. **Analyze** the reasons why some items did not necessarily fit into a group.

4. **Infer** why chemists have not created a periodic table of compounds.

Construct a bulletin board of the periodic tables of foods created by the class. Are the tables similar? Why or why not?

Use Vocabulary

Complete each sentence with the correct term from the Study Guide.

17. Two elements with the same number of protons but a different number of neutrons are called _____.

18. _____ is the weighted-average mass of all the known isotopes for an element.

19. The positively charged center of an atom is called the _____.

20. The particles that make up protons and neutrons are called _____.

21. A(n) _____ is a horizontal row in the periodic table.

22. The _____ is the sum of the number of protons and neutrons in an atom.

23. In the current model of the atom, the electrons are located in the _____.

Check Concepts

24. Most of the elements to the left of the stair-step line in the periodic table exist as _____ at room temperature.
   A) gases   C) plasmas
   B) liquids   D) solids

25. What is the term for a repeating pattern?
   A) isotopic   C) periodic
   B) metallic   D) transition

26. Which element has properties that are similar to neon?
   A) aluminum
   B) argon
   C) arsenic
   D) silver

27. Which term describes boron?
   A) metal   C) noble gas
   B) metalloid   D) nonmetal

28. How many outermost electrons do lithium and potassium have?
   A) 1   C) 3
   B) 2   D) 4

29. Which of the following is NOT found in the nucleus of an atom?
   A) proton   C) electron
   B) neutron   D) quark

Use the figure below to answer question 30.

30. Which atomic model is represented in the figure above?
   A) Democritus Model
   B) Dalton Model
   C) Rutherford Model
   D) Thomson Model

31. In which of the following states is nitrogen found at room temperature?
   A) gas   C) metal
   B) plastic   D) liquid

32. Which of the elements below is a shiny element that conducts heat and electricity?
   A) chlorine   C) hydrogen
   B) sulfur   D) magnesium

33. The atomic number of Re is 75. The atomic mass of one of its isotopes is 186. How many neutrons are in an atom of this isotope?
   A) 75   C) 186
   B) 111   D) 261

✓ Assessment   Online Test Practice

Interpret Graphics

Concepts in Motion   Interactive Concept Map

34. Copy and complete the concept map below.

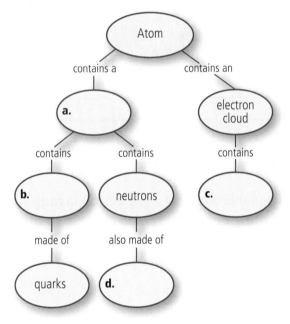

35. Construct As a star dies, it becomes denser. Its temperature rises to a point where helium (He) nuclei fuse with other nuclei. When this happens, the atomic numbers of the other nuclei are increased by 2 because each gains the two protons contained in the He nucleus. For example, chromium (Cr) fuses with He to become iron (Fe). Copy and complete the concept map showing the first four steps in He fusion.

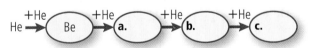

Think Critically

36. Infer Lead and mercury are two environmental pollutants. Why are they called heavy metals?

37. Explain why it is necessary to change models as new information becomes available.

38. THEME FOCUS Infer If you discovered a new element, what steps would you follow to classify it?

39. Infer Germanium and silicon are used in making semiconductors. Locate each element on the periodic table and explain why they are not efficient conductors.

40. BIG Idea Infer Calcium-40 and strontium-90 are isotopes. Calcium-40 is used by the body to make bones and teeth. Strontium-90 is radioactive. Calcium-40 is safe for people, and strontium-90 is hazardous. Why is strontium-90 hazardous to people?

Apply Math

41. Solve Equations The atomic number of yttrium is 39. The atomic mass of one of its isotopes is 89. How many neutrons are in an atom of this isotope?

Use the table below for question 42.

Electrons per Energy Level	
Energy Level	Maximum Number of Electrons
1	2
2	a.
3	b.
4	c.

42. Use Tables Use the information in **Figure 11** to determine how many electrons should be in the second, third, and fourth energy levels for argon, atomic number 18. Copy and complete the table above with the number of electrons for each energy level.

Standardized Test Practice

Multiple Choice

Record your answers on the answer sheet provided by your teacher or on a sheet of paper.

1. What particle identifies each particular element?
   - A. electrons
   - C. protons
   - B. photons
   - D. quarks

2. Which group of elements on the periodic table is unreactive?
   - A. group 1
   - C. group 17
   - B. group 2
   - D. group 18

Use the figure below to answer questions 3 and 4.

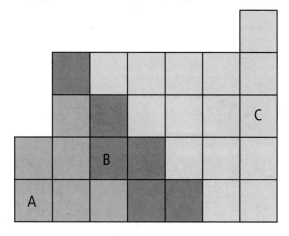

3. What is the name for all the elements in region B on the periodic table above?
   - A. metals
   - B. metalloids
   - C. nonmetals
   - D. transition metals

4. Which region shown on the periodic table contains elements that are unreactive gases at room temperature?
   - A. region A
   - B. region B
   - C. region C
   - D. none of the regions

5. The periodic table is organized into columns called _____.
   - A. groups
   - C. periods
   - B. categories
   - D. rows

Use the table below to answer questions 6 and 7.

Element	Electrons in a Neutral Atom	Electrons in the Outer Energy Level
Carbon	6	4
Oxygen	8	6
Neon	10	8
Sodium	11	1
Chlorine	17	7

6. The table above shows electron arrangements of some elements. Which element in the table has a complete outermost energy level?
   - A. carbon
   - B. oxygen
   - C. neon
   - D. sodium

7. Which element would you expect to be located in group 1 of the periodic table?
   - A. oxygen
   - B. neon
   - C. sodium
   - D. chlorine

8. How many quarks have been found?
   - A. 6
   - C. 10
   - B. 8
   - D. 12

9. The element nickel has five naturally occurring isotopes. Which of the following describes the relationship of these isotopes?
   - A. same mass, same atomic number
   - B. same mass, different atomic number
   - C. different mass, same atomic number
   - D. different mass, different atomic number

✓ Assessment   Standardized Test Practice

Short Response

Record your answers on the answer sheet provided by your teacher or on a sheet of paper.

10. According to the periodic table, an atom of lead has an atomic number of 82. How many neutrons does lead-207 have?

11. About three out of four chlorine atoms are chlorine-35, and about one out of four are chlorine-37. What is the average atomic mass?

Use the figure below to answer question 12.

$$\left[\text{Na}\right]^{+}\left[:\overset{\cdot\cdot}{\underset{\cdot\cdot}{\text{Cl}}}:\right]^{-}$$

12. What does the above diagram represent? What do the dots represent?

13. What characteristic determines the chemical properties of an element?

14. Silicon's atomic mass is listed as 28.09 amu on the periodic table. A student claims that no silicon atom has this atomic mass. Is this true? Explain why or why not.

15. How can you use the periodic table to determine the average number of neutrons an element has, even though the number of neutrons is not listed?

Extended Response

Record your answers on a sheet of paper.

Use the figure below to answer questions 16 and 17.

16. The illustration above shows the currently accepted model of atomic structure. Describe this model.

17. Compare and contrast the model shown above with Bohr's model of an atom.

18. Describe the concept of energy levels and how they relate to the organization of elements on the periodic table.

19. Describe how Dalton's modernization of the ancient Greeks' ideas of elements, atoms, and compounds provided a basis for understanding chemical reactions. Give an example.

NEED EXTRA HELP?																			
If You Missed Question . . .	1	2	3	4	5	6	7	8	9	10	11	12	13	14	15	16	17	18	19
Review Section . . .	2	3	3	3	3	3	3	1	2	2	2	3	2	2	2	1	1	3	1

CHAPTER 17

Elements and Their Properties

ConnectED

Your one-stop online resource
connectED.mcgraw-hill.com

- Video
- Audio
- Review
- Inquiry
- WebQuest
- Assessment
- Concepts in Motion
- g Multilingual eGlossary

Launch Lab
Colorful Clues

It is the distinct properties of each element that identify one element from another. In this lab, you will observe how the heated atoms of some elements absorb energy and then release the absorbed energy, which you see as colored light.

For a lab worksheet, use your StudentWorks™ Plus Online.

- ? Inquiry Launch Lab
- Concepts in Motion Animation

FOLDABLES®

Make a concept map book. Label it as shown. Use it to organize your notes on the properties of the elements.

Metals Metalloids Nonmetals

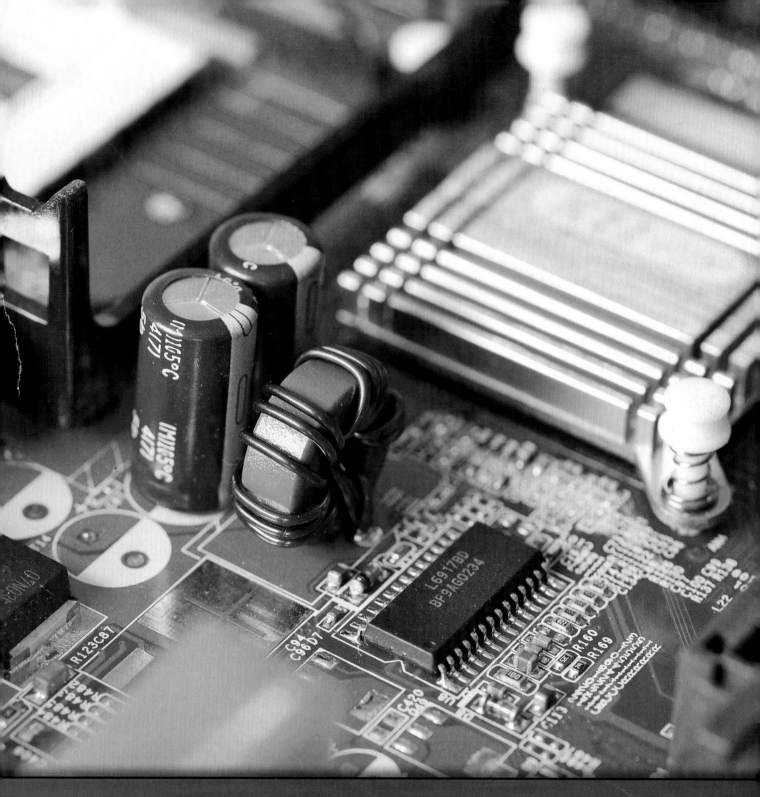

THEME FOCUS Structure and Properties
of Matter
**Elements in the same group on the periodic table
have similar properties.**

BIG ((Idea) Elements are classified into three main
types—metals, nonmetals, and metalloids.

Reading Preview

Essential Questions

▶ What are the properties of a typical metal?

▶ How do atoms bond in metallic bonding?

▶ Which elements are alkali metals and which are alkaline earth metals?

▶ What are of some common uses of the transition elements?

Review Vocabulary

group: vertical column in the periodic table

New Vocabulary

metal
malleable
ductile
metallic bonding
radioactive element
transition element

g Multilingual eGlossary

? Inquiry Virtual Lab

■ **Figure 1** The various properties of metals make them useful.

Describe *some uses for metal sheets and wires.*

Metals

MAIN ◀Idea Metals are located on the left side of the periodic table and are generally shiny, malleable, ductile, and good conductors.

Real-World Reading Link Grocery stores set aside different areas for different types of food, such as vegetables, meat, and dairy products. In a similar way, scientists classify the elements into different groups.

Properties of Metals

Coins, paper clips, and some baseball bats are made of metals. **Metals** are elements that are shiny, malleable, ductile, and good conductors of heat and electricity. Except for mercury, metals are solids at room temperature. The shiny property of metals is called metallic luster. Metals are **malleable** (MA lee uh bul), which means they can be hammered or rolled into sheets. Metals are also **ductile,** which means they can be drawn into wires. **Figure 1** shows the malleability and ductility of metals.

These properties make metals suitable for use in objects ranging from eyeglass frames to computers to buildings. The specific properties of a metal depend on its unique configuration of electrons, protons, and neutrons. However, elements in the same group of the periodic table have similar properties.

Metals and the periodic table In the periodic table, metals are found to the left of the stair-step line. In the periodic tables in this book, the metal element blocks are colored blue. Notice that most elements are metals. Except for hydrogen, all the elements in groups 1 through 12 are metals, as are the elements under the stair-step line in groups 13 through 15.

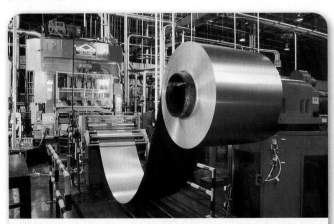

Metals are malleable: they can be hammered into thin sheets.

Metals are ductile: they can be drawn into wires.

Bonding in metals The atoms of metals generally have one to three electrons in their outer energy levels. In chemical reactions, metals tend to give up electrons easily because they are not strongly held by the protons in the nucleus.

Bonding with nonmetals When metals combine with non-metals, the atoms of the metals tend to lose electrons to the atoms of nonmetals. The metal atoms become positive ions, and the nonmetal atoms become negative ions. Ions are charged particles with more or fewer electrons than the neutral atom. Both metals and nonmetals become more chemically stable when they form ions.

A positively charged metal ion and a negatively charged nonmetal ion are attracted because of the electric force between them. They form a bond called an ionic bond. For example, a sodium atom can lose an electron to a chlorine atom. As shown in **Figure 2,** the sodium ion and chloride ion bond to form the compound sodium chloride (NaCl), also known as table salt.

Metallic bonding A different type of bonding occurs between the atoms of metals. In **metallic bonding,** positively charged metallic ions are surrounded by a sea of electrons. Outer-level electrons are not held tightly to the nucleus of an atom but move freely among many positively charged ions, as shown in **Figure 3.**

Metallic bonding explains many of the properties of metals. For example, when a metal is hammered into a sheet or drawn into a wire, it does not break because the ions are in layers that slide past one another without losing their attraction to the electron sea. Metals are also good conductors of heat and electricity because the outer-level electrons are weakly held and travel relatively freely.

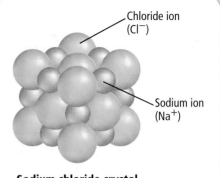

Sodium chloride crystal

- **Figure 2** Metals can form ionic bonds with nonmetals. In a crystal of table salt (NaCl), the positive ions come from the metal sodium and the negative ions come from the nonmetal chlorine.

FOLDABLES

Incorporate information from this section into your Foldable.

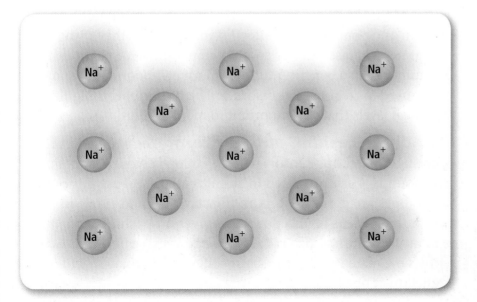

- **Figure 3** In metallic bonding, the electrons in the shared electron sea are not attached to any one metal ion. This allows the electrons to move.

Explain *Why do metals conduct electricity?*

Figure 4 Alkali metals are very reactive. For example, the vigorous reaction between potassium and water releases enough thermal energy to ignite the hydrogen gas that forms.

The Alkali Metals

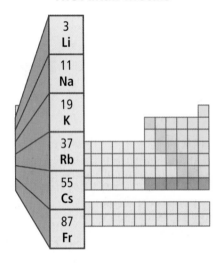

The Alkali Metals

The elements in group 1 of the periodic table are the alkali (AL kuh li) metals. Like other metals, alkali metals are shiny, malleable, ductile, and good conductors of heat and electricity. However, they are softer than most other metals and are the most reactive metals. They react rapidly and sometimes violently with oxygen and water, as shown in **Figure 4.** Because they are so reactive, alkali metals do not occur naturally in their elemental forms, and pure samples must be stored in oil to prevent reaction with oxygen and water in the air.

Atomic structure explains the reactive nature of alkali metals. Each atom of an alkali metal has one electron in its outer energy level. This electron is easily given up when an alkali metal combines with a nonmetal. As a result, the alkali metal atom becomes a positively charged ion in a compound such as sodium chloride (NaCl) or potassium bromide (KBr).

Lithium, sodium, and potassium Look carefully at the nutritional information on a cereal box. You will notice that sodium and potassium are often listed. You and other living things need potassium and sodium compounds to stay healthy. Lithium can also benefit health. Lithium compounds are sometimes used to treat bipolar disorder. The lithium helps regulate chemical levels that are important to mental health.

Rubidium, cesium, and francium The operation of some light-detecting sensors depends upon rubidium or cesium compounds. Cesium is used in atomic clocks because some of its isotopes are radioactive. A **radioactive element** is one in which the nucleus breaks down and gives off particles and energy. Francium is also radioactive and is extremely rare. Scientists estimate that Earth's crust contains less than 30 g of francium at one time.

The Alkaline Earth Metals

The alkaline earth metals make up group 2 of the periodic table. Like most metals, these metals are shiny, malleable, and ductile. Like the alkali metals, they combine readily with other elements and are not found as free elements in nature.

Each atom of an alkaline earth metal has two electrons in its outer energy level. These electrons are given up when an alkaline earth metal combines with a nonmetal. The alkaline earth metal atom becomes the positively charged ion in a compound, such as calcium fluoride (CaF_2). Some compounds of the alkaline earth metals are used to color fireworks, like those in **Figure 5**.

 Reading Check **Compare and contrast** the alkali metals and the alkaline earth metals.

Magnesium Magnesium's lightness and strength make it a good material for cars, planes, spacecraft, household ladders, and baseball and softball bats. Most life on Earth depends upon chlorophyll, a magnesium-containing compound that enables plants to absorb light and make food.

Calcium Calcium is seldom used as a free metal, but its compounds are useful and essential for life. Marble statues and some countertops are made of calcium carbonate ($CaCO_3$). Calcium carbonate is the major component of the mineral limestone, which is found in many caves. You might take a vitamin with calcium. Calcium phosphate ($CaPO_4$) in your bones helps make them strong.

Other alkaline earth metals The compound barium sulfate ($BaSO_4$) is used to diagnose some digestive disorders because it absorbs X-ray radiation well. First, the patient swallows a barium compound. Next, an X-ray is taken while the barium compound is going through the digestive tract. A doctor can then see where the barium is in the body and can use this information to diagnose internal abnormalities.

Radium, the last element in group 2, is radioactive and is found associated with uranium. It was once used to treat cancers but is being replaced with more readily available radioactive elements.

The Alkaline Earth Metals

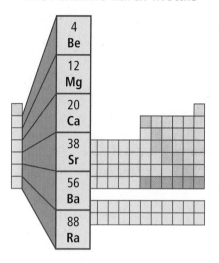

Figure 5 Alkaline earth metals make spectacular fireworks. The labels on each burst indicate the compound that is primarily responsible for the color of the firework.

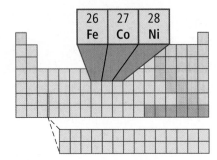

MiniLab

Discover What's in Cereal

Procedure 🥽 👕 🧤

WARNING: *Never eat any food in the lab.*

1. Read the procedure and safety information, and complete the lab form.
2. **Tape** a small, strong **magnet** to a **pencil** at the eraser end.
3. Place some dry **fortified cold cereal** in a **plastic bag**.
4. Thoroughly crush the cereal.
5. Pour the crushed cereal into a deep **bowl** and cover it with **water**.
6. Stir the mixture for about 10 min with your pencil and magnet. Stir slowly for the last minute.
7. Remove the magnet and examine it carefully. Record your observations.

Analysis

1. **Identify** the element that was attracted to your magnet.
2. **Explain** Why is this element added to cereal?

The Iron Triad

26	27	28
Fe	Co	Ni

■ **Figure 6** Chromium compounds give rubies their red color and emeralds their green color. Cobalt is used to make blue glass.

The Transition Elements

Many transition elements, such as iron and gold, are familiar because they are less reactive than the metals in groups 1 and 2 and often occur in nature as uncombined elements. **Transition elements** are the elements in groups 3 through 12 in the periodic table. They are called transition elements because they are considered to be in transition between the main group elements. The main group elements are groups 1 and 2 and groups 13 through 18. Main group elements are sometimes called the representative elements.

A glowing lightbulb filament is made from the transition element tungsten. Titanium, another transition element, is used in bike frames, ships, and golf clubs because of its lightness and strength. Chromium is used in making steel and in chrome plating. Platinum is used to make jewelry because it is rare, resistant to corrosion, and more valuable than gold.

Transition elements often form colored compounds, as shown in **Figure 6.** The gems' colors come from chromium compounds. The blue glass gets its color from cobalt. Cadmium yellow and cobalt blue paints are made from compounds of transition elements. However, their uses are limited because they are so toxic.

Iron, cobalt, and nickel Iron, cobalt, and nickel form a unique cluster of transition elements sometimes called the iron triad. They are the most common magnetic elements and are used in steel and other metal mixtures.

Iron is the second most abundant metal in Earth's crust and the most widely used of all metals. It is the main component of steel. Other metals, such as nickel and cobalt, are added to steel to give it various characteristics. Nickel is also used to give a shiny, protective coating to other metals.

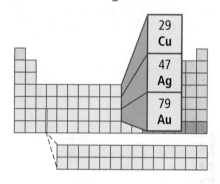

■ **Figure 7** Copper is often mixed with zinc or nickel to make modern coins. Gold, silver, and bronze—an alloy of copper and tin—are used to make athletic medals.

Concepts in Motion
Animation

Copper, silver, and gold You are probably familiar with copper, silver, and gold—three of the elements in group 11. Because they are so stable and malleable and can be found as free elements in nature, these metals were once used widely to make coins. For this reason, they are known as the coinage metals. Because they are so expensive, silver and gold are rarely used in coins anymore. The United States stopped making everyday coins with gold in 1933 and coins with silver in 1964. Most coins now are mixtures of nickel, zinc, and copper, as shown in **Figure 7.**

The coinage metals have a variety of other uses, such as in the athletic medals in **Figure 7.** Copper is often used in electrical wiring because of its superior ability to conduct electricity and its relatively low cost. The compounds silver iodide (AgI) and silver bromide (AgBr) are used to make photographic film and paper because they break down when they are exposed to light. Silver and gold are used in jewelry because of their attractive colors, relative softness, resistance to corrosion, and rarity.

☑ **Reading Check** **Explain** why gold's relative softness makes it a good choice for jewelry.

Zinc, cadmium, and mercury Zinc, cadmium, and mercury are found in group 12 of the periodic table. Zinc and oxygen in the air combine to form a thin, protective coating of zinc oxide on its surface. Zinc as well as cadmium, which also forms a protective coating, are often used to coat metals such as iron. Cadmium also is used in rechargeable batteries.

Mercury is the only metal that is a liquid at room temperature. It is used in thermostats, switches, and batteries. Mercury is toxic and can accumulate in the body, so it is rarely used in modern thermometers. People have died from mercury that built up in their bodies from repeatedly eating fish that lived in mercury-contaminated water.

The Coinage Metals

			29 Cu
			47 Ag
			79 Au

Zinc, Cadmium, and Mercury

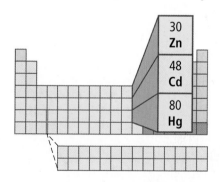

			30 Zn
			48 Cd
			80 Hg

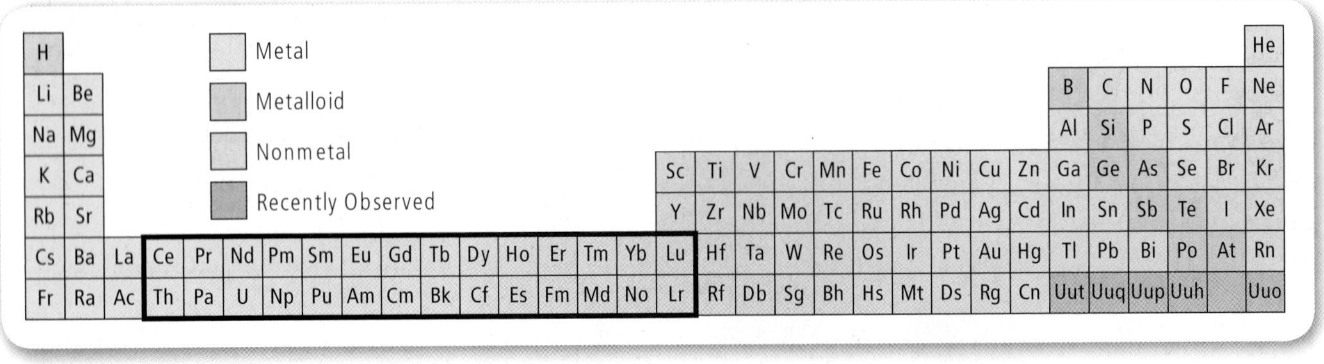

																																He
H																																
Li	Be																								B	C	N	O	F	Ne
Na	Mg																								Al	Si	P	S	Cl	Ar
K	Ca									Sc	Ti	V	Cr	Mn	Fe	Co	Ni	Cu	Zn	Ga	Ge	As	Se	Br	Kr
Rb	Sr									Y	Zr	Nb	Mo	Tc	Ru	Rh	Pd	Ag	Cd	In	Sn	Sb	Te	I	Xe
Cs	Ba	La	Ce	Pr	Nd	Pm	Sm	Eu	Gd	Tb	Dy	Ho	Er	Tm	Yb	Lu	Hf	Ta	W	Re	Os	Ir	Pt	Au	Hg	Tl	Pb	Bi	Po	At	Rn
Fr	Ra	Ac	Th	Pa	U	Np	Pu	Am	Cm	Bk	Cf	Es	Fm	Md	No	Lr	Rf	Db	Sg	Bh	Hs	Mt	Ds	Rg	Cn	Uut	Uuq	Uup	Uuh		Uuo

Legend: Metal — Metalloid — Nonmetal — Recently Observed

Figure 8 This is what the periodic table would look like if the inner transition elements were positioned where they should be. To save space, they are usually placed below the periodic table.

Identify *an advantage and a disadvantage to the above arrangement of the periodic table.*

VOCABULARY ....

ACADEMIC VOCABULARY

Transition
a movement or evolution from one stage to another
The transition from middle school to high school is difficult for many students. ....

The Inner Transition Elements

The two rows of elements that seem to be disconnected from the rest of the periodic table are called the inner transition elements. They are called this because they fit within the transition metals on the periodic table. The inner transition elements are located between groups 3 and 4 in periods 6 and 7. To save room, they are usually listed below the table. **Figure 8** shows what the periodic table would look like if the inner transition elements were not written below the table.

The lanthanides The first row of the inner transition elements includes elements with atomic numbers of 58 to 71. These elements are called the lanthanide series because they follow the element lanthanum. Lanthanum, cerium, praseodymium, and samarium are used with carbon to make a compound that is used extensively for movie lighting. Compounds of europium, gadolinium, and terbium are used as colored phosphors. Recall that phosphors change ultraviolet light into visible light. Televisions once used these compounds to produce the colors that you see.

 Reading Check **Explain** why the first row of inner transition elements is called the lanthanide series.

The actinides The second row of inner transition elements includes elements with atomic numbers 90 to 103. These elements are called the actinide series because they follow the element actinium. All of the actinides are radioactive and unstable. Because they are unstable, the actinides are rare or nonexistent in nature. Their instability also makes them difficult to research.

Thorium and uranium are the only actinides found in the Earth's crust in usable quantities. Thorium is used in making the glass for high-quality camera lenses because it bends light without much distortion. Uranium is best known for its use in nuclear reactors and in weapons, but one of its compounds has been used as photographic toner.

Metals in Earth's Crust

Earth's hardened outer layer, called the crust, contains many compounds and a few uncombined metals such as gold and copper. Metals that are found in Earth's crust are minerals. Minerals are often found in ores. Ores are mixtures of minerals, clay, and rock that occur naturally in Earth's crust.

Most metals must be mined and separated from their ores. After an ore is mined, the mineral is separated from the rock and clay. Then the mineral often is converted to another physical form. This step usually involves heat and is called roasting. Finally, the metal is refined into a pure form. Later, it can be alloyed with other metals. Removing the waste rock can be expensive. If the cost of removing the waste rock becomes greater than the value of the desired material, the mineral mixture is no longer classified as an ore.

Mines can be found throughout the world. The western United States has several copper mines, like the one in **Figure 9.** Most of the world's platinum is found in South Africa. Chromium is important because it is used to harden steel, to manufacture stainless steel, and to form other alloys. The United States imports most of its chromium from South Africa, the Philippines, and Turkey.

■ **Figure 9** Copper mines, such as this one near Kennecott, Utah, are found in the western United States.

Section 1 Review

Section Summary

▶ Metals tend to form ionic and metallic bonds.

▶ Group 1 elements are called alkali metals and are the most reactive metals.

▶ Group 2 elements are called alkaline earth metals and are very reactive.

▶ Transition elements are elements in groups 3–12 in the periodic table. They have a wide variety of properties and uses.

▶ Inner transition elements fit in the periodic table between groups 3 and 4 in periods 6 and 7.

1. **MAIN Idea Describe** how to test a sample of an element to see if it is a metal.

2. **Compare and contrast** the uses of the iron triad and the use of the coinage metals.

3. **Classify** the following as alkali metals, alkaline earth metals, transition elements, or inner transitional elements: calcium, gold, iron, magnesium, plutonium, potassium, sodium, and uranium.

4. **Discuss** how metallic bonding accounts for the common properties of metals.

5. **Think Critically** Suppose you discovered a new element with 120 protons and 2 electrons in its outer level. In what group does this new element belong? What properties would you expect it to have?

Apply Math

6. **Use Percentages** Pennies used to be made of copper and zinc and had a mass of 3.1 g. Today, pennies are made of copper-plated zinc and have a mass of 2.5 g. A new penny's mass is what percent of an old penny's mass?

✓ Assessment Online Quiz

Essential Questions

▶ How do nonmetals bond?

▶ What properties of hydrogen make it a nonmetal?

▶ What are the properties and uses of the halogens?

▶ Why are noble gases unreactive?

Review Vocabulary

sublimation: the process of a solid changing directly to a vapor without forming a liquid

New Vocabulary

nonmetal
diatomic molecule

 **Multilingual eGlossary**

 **Concepts in Motion** Animation

■ **Figure 10** Humans are composed of mostly nonmetals.

Nonmetals

MAIN Idea Nonmetals are located on the right side of the periodic table and are generally dull, brittle, and poor conductors.

Real-World Reading Link The air around you is not a metal. It is not a shiny solid, you cannot roll it into sheets or pull it into wires, and it is a poor conductor. The elements that make up air are part of a different group of elements called nonmetals.

Properties of Nonmetals

Most of your body's mass is made of oxygen, carbon, hydrogen, and nitrogen, as shown in **Figure 10.** Calcium, phosphorus, sulfur, and chlorine are among the other elements found in your body. Except for the metal calcium, these elements are nonmetals. **Nonmetals** are elements that are usually gases or solids at room temperature. Solid nonmetals are not malleable or ductile but are brittle or powdery. Nonmetals are poor conductors of heat and electricity because the electrons in nonmetals are not free to move as they do in metals.

Nonmetals and the periodic table In the periodic table, all nonmetals except hydrogen are found to the right of the stair-step line. On the table in back of your book, the non-metal element blocks are colored yellow. The noble gases, group 18, make up the only group of elements that are all nonmetals. Group 17 elements are called the halogens. The halogens, except astatine, are also nonmetals. The other nonmetals are found in groups 13 through 16.

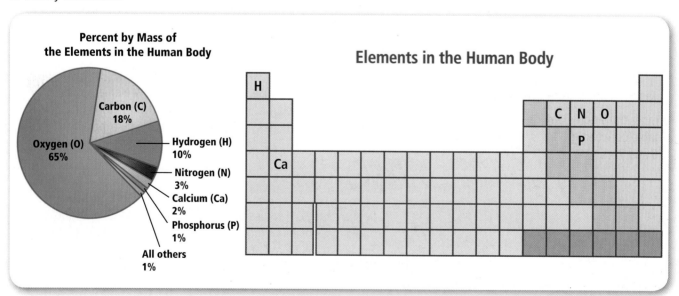

One calcium atom bonds ionically with two fluorine atoms to form calcium fluoride (CaF_2).

Carbon and oxygen bond covalently to form carbon dioxide (CO_2).

■ **Figure 11** Nonmetals form ionic bonds with metals, and covalent bonds with other nonmetals.

Bonding in nonmetals

Nonmetals become negative ions when they gain electrons from metals. An example of an ionic compound is calcium fluoride (CaF_2), which is shown in **Figure 11.** Calcium fluoride forms from the nonmetal fluoride and the metal calcium.

When bonded with other nonmetals, atoms of nonmetals usually share electrons to form covalent bonds. Compounds made of atoms that are covalently bonded are called covalent compounds. For example, the covalent compound carbon dioxide (CO_2) is shown in **Figure 11.** Carbon dioxide is a gas that you exhale and that plants need to survive.

Hydrogen

If you could count all the atoms in the universe, you would find that about 90 percent of them are hydrogen atoms. Most hydrogen on Earth is found in the compound water. In fact, the word *hydrogen* comes from the Greek word *hydro,* which means "water." When water is broken down into its elements, hydrogen becomes a gas composed of diatomic molecules. A **diatomic molecule** consists of two atoms of the same element in a covalent bond.

✓ **Reading Check Describe** what a diatomic molecule is.

Hydrogen is highly reactive. A hydrogen atom has a single electron, which the atom shares when it combines with other nonmetals. For example, hydrogen burns in oxygen to form water (H_2O), in which hydrogen shares electrons with oxygen. Hydrogen can gain an electron when it combines with alkali and alkaline earth metals. The compounds formed are hydrides, such as sodium hydride (NaH).

FOLDABLES
Incorporate information from this section into your Foldable.

Hydrogen

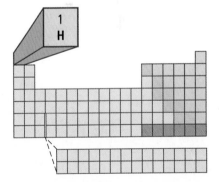

The Halogens

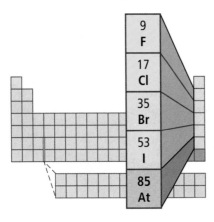

VOCABULARY ·

WORD ORIGIN

Halogen
from the Greek words *hals,* meaning "salt" and *–gen,* meaning "making."
Chlorine is a halogen found in table salt. · · · · · · · · · · · · · ·

The Halogens

Fluorine, chlorine, bromine, iodine, and astatine are called halogens and make up group 17. They are very reactive in their elemental forms, and their compounds have many uses. For example, halogen lightbulbs contain small amounts of bromine or iodine vapor.

Because an atom of a halogen has seven electrons in its outer energy level, only one electron is needed to complete this energy level. If a halogen gains an electron from a metal, an ionic compound called a salt is formed. An example of this is sodium chloride (NaCl). You know this compound as table salt. In the gaseous state, the halogens form reactive diatomic molecules and can be identified by their distinctive colors. Chlorine is greenish-yellow, bromine is reddish- orange, and iodine is violet.

Fluorine Fluorine is the most chemically active of the nonmetal elements. As you can see in **Figure 12,** fluorine compounds have many uses. Fluorine compounds, called fluorides, are added to toothpastes and to city water systems to help prevent tooth decay. Hydrofluoric acid, a mixture of hydrogen fluoride and water, is used to etch glass and to frost the inner surfaces of lightbulbs. It is also used in the fabrication of semiconductors.

■ **Figure 12** Fluorine's compounds have many uses.

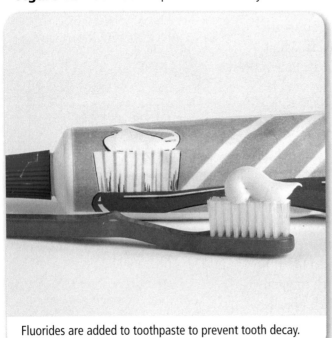

Fluorides are added to toothpaste to prevent tooth decay.

Hydrofluoric acid is used to etch glass.

Chlorine compounds are used to disinfect water in swimming pools.

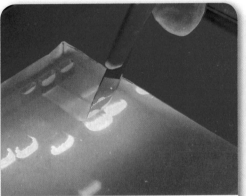
Scientists use a bromine compound to stain DNA samples.

Iodine will sublime at room temperature.

■ **Figure 13** Halogens have a wide variety of uses and properties.

Chlorine and bromine The odor that you sometimes smell near a swimming pool is chlorine. Chlorine compounds, like the one shown in **Figure 13,** disinfect water. Chlorine, the most abundant halogen, is obtained from seawater at ocean-salt recovery sites. Household and industrial bleaches that are used to whiten flour, clothing, and paper also contain chlorine compounds.

Bromine, the only nonmetal that is a liquid at room temperature, also is extracted from compounds in seawater. Some hot tubs use bromine compounds instead of chlorine compounds to disinfect water. Bromine compounds were once used in cosmetics and as flame retardants. But, due to health concerns, these compounds are being used less frequently.

Another bromine compound is used to study genetic material, such as DNA. The bromine compound binds to the DNA in certain areas and acts as a sort of tag. Under fluorescent light, the bromine compound absorbs the fluorescent light and emits a reddish visible light, as shown in **Figure 13.**

✔ **Reading Check** **Name** some uses of chlorine and bromine compounds.

Iodine and astatine Iodine, a shiny purple-gray solid at room temperature, is obtained from seawater. When heated, iodine sublimes to a purple vapor, as seen in **Figure 13.** Iodine is essential in your diet for the production of the hormone thyroxin and to prevent goiter, an enlarging of the thyroid gland in the neck. Iodine compounds are also used as disinfectants.

Astatine is the last member of group 17. It is radioactive and rare but has many properties similar to those of the other halogens. Because it is so rare in nature, scientists usually make astatine for research purposes. Medical researchers are investigating the possibility of using astatine's radioactive properties to treat cancer.

MiniLab

? **Inquiry** MiniLab

Identify Chlorine Compounds in Your Water

Procedure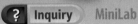

WARNING: *Avoid contact with the silver nitrate solution. Silver nitrate is corrosive and toxic. It can stain skin and clothes.*

1. Read the procedure and safety information, and complete the lab form.
2. In three labeled **test tubes,** obtain 2 mL of **chlorine standard solution, distilled water,** and **drinking water.**
3. Carefully add five drops of **silver nitrate solution** to each and stir.

Analysis
1. **Identify** which solutions showed a presence of chlorine.
2. **Infer** Which result most resembled your drinking water, the chlorine standard solution or the distilled water? What does this tell you about your drinking water?

Helium

Neon

Argon

Krypton

Xenon

■ **Figure 14** Each noble gas glows a different color when a current is passed through it.

The Noble Gases

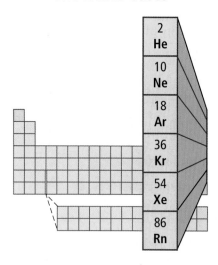

The Noble Gases

The noble gases exist as isolated atoms. They are stable because their outermost energy levels are full. No naturally occurring noble gas compounds are known, but several compounds of argon, krypton, and xenon, primarily with fluorine, have been created in a laboratory.

The stability of noble gases makes them useful. For example, helium is less dense than air but does not burn in oxygen. This makes it safer than hydrogen to use in blimps and balloons. An electric current will cause the noble gases to glow, as shown in **Figure 14.** For this reason, some noble gases, such as neon and argon, are used for brightly colored signs. The noble gases are also used in many lasers. One common type of laser is a helium-neon laser. Helium neon lasers produce beams of intense, red light.

Section 2 Review

Summary

▶ Nonmetals are usually gases or brittle solids that are not shiny and do not conduct heat or electricity.

▶ Hydrogen is the most common element in the universe and is highly reactive.

▶ Halogens are in group 17 and are highly reactive.

▶ Noble gases exist as isolated atoms in nature.

7. **MAIN Idea Explain** how solid nonmetals are different from solid metals.

8. **Describe** two ways in which nonmetals combine with other elements.

9. **Identify** the nonmetal in these compounds: MgO, NaH, $AlBr_3$, and FeS.

10. **Identify** some common uses for the halogen compounds.

11. **Explain** why the noble gases are unreactive.

12. **Think Critically** How you can tell that a gas is a halogen?

Apply Math

13. **Calculate Mass** If a chlorine atom has a mass of 35.5 atomic mass units and a sodium atom has a mass of 23.0 atomic mass units, what is the mass of one NaCl unit?

✓ Assessment **Online Quiz**

Short Response

Record your answers on the answer sheet provided by your teacher or on a sheet of paper.

9. Define the general properties of metals that make them useful and versatile materials.

10. Use the electron configurations of sodium and potassium to explain why they do not naturally occur in elemental form.

Use the figure below to answer questions 11 and 12.

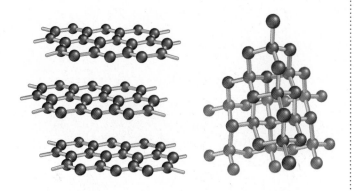

11. Define the term *allotrope* and identify these allotropes of carbon.

12. Compare the structures of these carbon allotropes, and relate the structures to the properties of these materials.

13. Describe some unique properties of hydrogen.

14. What halogens are commonly obtained from seawater? What are their uses?

Extended Response

Record your answers on a sheet of paper.

15. Recent Federal Drug Administration statements advise limiting consumption of tuna and salmon. Which transition metal is the source of the problem? Explain why this element poses a potential risk.

Use the figure below to answer question 16.

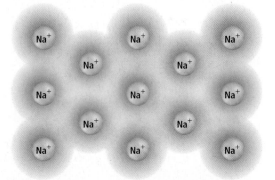

16. Identify the type of bond shown. Describe the properties of a material with these bonds.

17. Compare the two types of bonds that nonmetals can form.

18. Explain why fluorine is very reactive and why neon is very unreactive.

19. Why is helium instead of hydrogen used in blimps?

20. Explain the importance of organisms that convert nitrogen from its diatomic form into other compounds.

NEED EXTRA HELP?																				
If You Missed Question . . .	1	2	3	4	5	6	7	8	9	10	11	12	13	14	15	16	17	18	19	20
Review Section . . .	1	1	1	1	2	1	3	3	1	1	3	3	2	2	1	1	2	2	2	3

UNIT 5

Reactions

THEMES

Changes of Matter Elements can combine or radioactively decay to form new substances.

Energy Chemical reactions can release energy in forms of thermal energy, light, sound, and electricity.

Scientific Inquiry By studying chemical reactions, we have learned about important processes, such as nuclear reactions, photosynthesis, and combustion.

Structure and Properties of Matter The properties of compounds are generally different from the properties of the elements they contain.

(⊕ WebQuest) **STEM UNIT PROJECT**

Technology Research the Large Hadron Collider (LCH). Draw a diagram of the LHC that compares the size of the LHC with other colliders, such as the Tevatron.

CHAPTER 18
Chemical Bonds

ConnectED

Your one-stop online resource
connectED.mcgraw-hill.com

- Video
- Audio
- Review
- Inquiry
- WebQuest
- Assessment
- Concepts in Motion
- g Multilingual eGlossary

Launch Lab
Molecules and Mixing

You may have noticed that some liquids will mix easily, but others quickly separate no matter how much you stir or shake them. Why does this happen? The answer lies in understanding how molecules can be alike and different.

For a lab worksheet, use your StudentWorks™ Plus Online.

? Inquiry Launch Lab

FOLDABLES®

Make a vocabulary book with eight tabs. Label the tabs with groups 1, 2, and 13–18. Write the electron-dot structure and your notes for that group underneath.

THEME FOCUS Changes of Matter
Based on their positions in the periodic table, atoms combine in predictable ways to form compounds with different properties.

BIG Idea Elements link together with chemical bonds.

Section 1 • Stability in Bonding

Section 2 • Types of Bonds

Section 3 • Writing Formulas and Naming Compounds

Reading Preview

Essential Questions

▶ How does a compound differ from its component elements?

▶ What does a chemical formula represent?

▶ How do electron dot diagrams help predict chemical bonding?

▶ Why does chemical bonding occur?

Review Vocabulary

compound: substance formed from two or more elements in which the exact combination and proportion of elements is always the same

New Vocabulary

chemical formula
chemical bond

g Multilingual eGlossary

■ **Figure 1** Copper sulfate isn't shiny and copper-colored like elemental copper. Nor is it a pale, yellow solid like sulfur or a colorless, odorless gas like oxygen.

Stability in Bonding

MAIN Idea When atoms form compounds, each atom is more stable in the compound than it was by itself.

Real-World Reading Link A bridge must be built with the proper materials in the correct arrangement in order to remain stable. Similarly, chemical bonds form only when stability is achieved by a particular combination and arrangement of atoms.

Combined Elements

Have you ever noticed the color of the Statue of Liberty? Why is it green? Was it painted that way? No, the Statue of Liberty was not painted. It is made of the metal copper, which is an element. Pennies have a coating made of copper. But copper isn't green. Uncombined, elemental copper is a bright, shiny copper color. So, again, we ask: Why is the Statue of Liberty green?

Compounds The matter around you includes pure elements such as copper, sulfur, and oxygen. But, like many other elements, these elements combine chemically to form compounds when the conditions are right. For example, the copper in the Statue of Liberty combines with oxygen and sulfur in the air to form a number of different compounds, including copper sulfate. Copper sulfate is one of the compounds responsible for the blue-green patina seen on the Statue of Liberty. **Figure 1** compares copper sulfate, in the patina on the Statue of Liberty, with elemental copper and elemental sulfur. Although it is formed from these elements, copper sulfate looks very different from either copper or sulfur and has its own unique properties.

Copper

Sulfur

Copper Sulfate

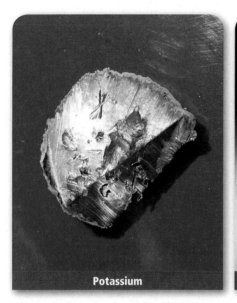

Potassium

Iodine

Potassium Iodide

New properties A compound formed when elements combine often has properties that aren't anything like those of the individual elements. For example, potassium iodide, shown in **Figure 2,** is a compound made from the elements potassium and iodine. Potassium is a shiny, soft, silvery metal that reacts violently with water. Iodine is a blue-black solid that turns to a purple gas at room temperature. Would you have guessed that these elements combine to make a white, crystalline salt?

Formulas

Every element has a chemical symbol. For example, the chemical symbol Na represents the element sodium, and the symbol Cl represents the element chlorine. When written as NaCl, these symbols make up the formula for the compound sodium chloride. A **chemical formula** shows what elements a compound contains and the exact number of the atoms of each element in a unit of that compound.

Another compound with which you are familiar is H_2O, which is more commonly known as water. This formula contains the symbols H for the element hydrogen and O for the element oxygen. Notice the subscript number 2 written after the H for hydrogen. *Subscript* means "written below." A subscript written after a symbol tells how many atoms of that element are in a single unit of the compound. If a symbol has no subscript, the unit contains only one atom of that element. A unit of H_2O contains two hydrogen atoms and one oxygen atom.

Look at the formulas for each compound listed in **Table 1.** What elements combine to form each compound? How many atoms of each element are required to form each of the compounds?

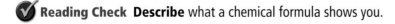 **Reading Check** **Describe** what a chemical formula shows you.

■ **Figure 2** The properties of a compound can be very different from those of its component elements.

Describe *how the properties of potassium iodide are different from those of potassium and iodine.*

 **Concepts in Motion**
Interactive Table

Table 1	Some Familiar Compounds	
Common Name	Chemical Name	Chemical Formula
Sand	silicon dioxide	SiO_2
Milk of magnesia	magnesium hydroxide	$Mg(OH)_2$
Cane sugar	sucrose	$C_{12}H_{22}O_{11}$
Lime	calcium oxide	CaO
Vinegar	acetic acid	CH_3COOH
Laughing gas	dinitrogen oxide	N_2O
Grain alcohol	ethanol	C_2H_5OH
Battery acid	sulfuric acid	H_2SO_4
Stomach acid	hydrochloric acid	HCl

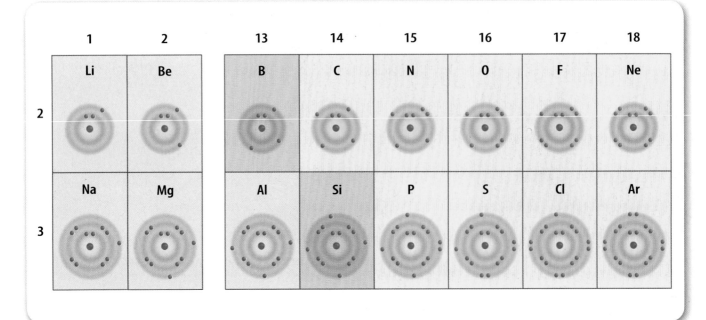

	1	2	13	14	15	16	17	18
2	Li	Be	B	C	N	O	F	Ne
3	Na	Mg	Al	Si	P	S	Cl	Ar

■ **Figure 3** Simplified representations of electron distribution can help you understand why atoms bond. Note that the number of electrons in each group's outer level increases across the table. The noble gases in group 18 have filled outer levels.

 **Inquiry** Virtual Lab

■ **Figure 4** Electron dot diagrams of noble gases show that they all have a stable, filled outer energy level.

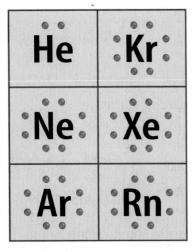

Chemical Bond Formation

Why do atoms form compounds? The answer to that question is in the last column on the periodic table. The six noble gases in group 18 seldom form compounds. In fact, the first compound containing a noble gas was not made until 1962. Why? Atoms of noble gases are unusually stable. The reason for this stability lies in the arrangement of the electrons.

Electron dot diagrams To understand the stability of atoms, it is helpful to show the electrons in the outer energy level of an atom—its valence electrons—in electron dot diagrams. Electron dot diagrams contain the chemical symbol for an element surrounded by dots representing its valence electrons.

How do you know how many dots to make in electron dot diagrams? For groups 1, 2, and 13 through 18, you can use a periodic table or the portion of it shown in **Figure 3**. Look at the ring depicting the outer energy level of each of the elements. Group 1 elements each have one outer electron. The elements in group 2 have two. Group 13 elements have three, group 14 have four, and so on to group 18, the noble gases, which each have eight.

The unique noble gases An atom is chemically stable when its outer energy level is complete. The outer energy levels of helium and hydrogen are stable with two electrons. The outer energy levels of all other elements are stable with eight.

The noble gases are stable because they each have a full outer energy level. **Figure 4** shows electron dot diagrams of some of the noble gases. Notice that eight dots surround Kr, Ne, Xe, Ar, and Rn, and two dots surround He.

Unfilled and filled energy levels How do the electron dot diagrams represent elements, and how does that relate to their abilities to make compounds? Hydrogen and helium, the elements in period 1 of the periodic table, can hold a maximum of two electrons in their outer energy levels. Hydrogen contains one electron in its lone energy level. The dot diagram for hydrogen has a single dot next to its symbol. This means that hydrogen's outer energy level is not full. It is more stable when it is part of a compound.

In contrast, helium's outer energy level contains two electrons as represented by its electron dot diagram with two dots—a pair of electrons—next to its symbol. Helium already has a full outer energy level by itself and is chemically stable. As a result, helium rarely forms compounds. Compare the electron dot diagrams of helium and hydrogen below.

Ḣ   Ḧe

Outer levels—getting their fill How does hydrogen, or any other element, change the number of outer electrons to become stable? Atoms with unstable outer energy levels can lose, gain, or share electrons to obtain a stable outer energy level. They do this by combining with other atoms that also have partially complete outer energy levels. As a result, each achieves stability.

Gaining and losing electrons **Figure 5** shows electron dot diagrams for sodium (Na) and chlorine (Cl). When they combine, sodium loses one electron and chlorine gains one electron. You can see from the electron dot diagram that chlorine now has a stable outer energy level similar to a noble gas.

Concepts in Motion
Animation

FOLDABLES
Incorporate information from this section into your Foldable.

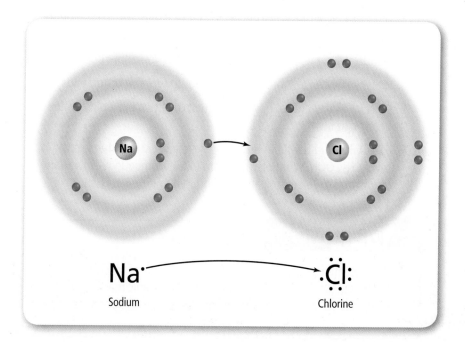

Na· → ·C̈l:
Sodium   Chlorine

■ **Figure 5** Chlorine has seven electrons in its outer energy level; sodium has only one. When sodium loses an electron and chlorine gains one, they each have a stable outer energy level.

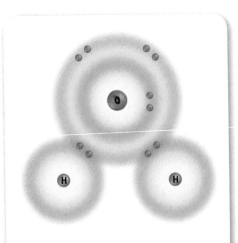

■ **Figure 6** In water, hydrogen contributes one electron and oxygen contributes the other in each hydrogen-oxygen bond. The atoms share the electrons, instead of giving them up, to achieve a complete outer energy level for each atom in the compound.

As shown in **Figure 5** on the previous page, sodium had only one electron in its outer energy level, which it lost to combine with chlorine in sodium chloride. As a result, the next-to-the-last energy level becomes the new outermost energy level. Sodium now has an outer energy level that is stable with eight electrons. When the outer electron of sodium is removed, a complete inner energy level is revealed and becomes the outer energy level. Sodium and chlorine are now stable because of the exchange of an electron.

Sharing electrons A hydrogen atom has one electron in its outer energy level. So, it needs one electron to fill its outer energy level. An oxygen atom has six electrons in its outer energy level. It needs two electrons for its outer level to be stable with eight electrons. Hydrogen and oxygen become stable and form bonds in a different way than sodium and chlorine. Instead of gaining or losing electrons, they share them. **Figure 6** shows how hydrogen and oxygen share electrons to achieve a more stable arrangement of electrons to form water.

Chemical bond formation When atoms gain, lose, or share electrons, an attraction forms between the atoms, pulling them together to form a compound. This attraction is called a chemical bond. A **chemical bond** is the force that holds atoms together in a compound.

Section 1 Review

Section Summary

▶ A chemical formula describes the numbers and types of atoms in a compound.

▶ The elements of group 18, the noble gases, rarely combine with other elements.

▶ Most atoms need eight electrons to complete their outer energy levels.

▶ Electron dot diagrams show the electrons in the outer energy level of an atom.

▶ A chemical bond is the force that holds atoms together in a compound.

1. **MAIN Idea** Explain why some elements are stable on their own while others are more stable in compounds.

2. **Compare and contrast** the properties of potassium (K) and iodine (I) with the compound KI.

3. **Identify** what the electron dot diagram tells you about bonding.

4. **Explain** why electric forces are essential to forming compounds.

5. **Describe** why chemical bonding occurs. Give two examples of how bonds can form.

6. **Think Critically** The label on a box of cleanser states that it contains CH_3COOH. What elements are in this compound? How many atoms of each element can be found in a unit of CH_3COOH?

Apply Math

7. **Use Percentages** Given that the molecular mass of magnesium hydroxide ($Mg(OH)_2$) is 58.32 amu and the atomic mass of an atom of oxygen is 15.999 amu, what percentage of this compound is oxygen?

✓ Assessment   Online Quiz

LAB

Atomic Trading Cards

Objectives
- **Display** the electrons of elements according to their energy levels.
- **Classify** elements according to their outer energy levels.

Background: Perhaps you have seen or collected trading cards for famous athletes. Usually, each card has a picture of the athlete on one side with important statistics related to the sport on the back. Atoms can also be identified by their properties and statistics.

Question: *How can a model show how energy levels fill when atoms combine?*

Preparation

Materials
4-in. × 6-in. index cards
pencil
periodic table

Procedure

1. Read the procedure and safety information, and complete the lab form.
2. Get an assigned element from the teacher. Write the following information for your element on your index card: name, symbol, group number, atomic number, atomic mass, metal/nonmetal/metalloid.
3. On the other side of your index card, show the number of protons and neutrons in the nucleus (for example, 6*p* for six protons and 6*n* for six neutrons for carbon).

4. Draw circles around the nucleus to represent the energy levels of your element. The number of circles you will need is the same as the row the element is in on the periodic table.
5. Draw dots on each circle to represent the electrons in each energy level. Remember, elements in period 1 become stable with two outer electrons while all others become stable with eight electrons.
6. Look at the picture side only of five of your classmates' cards. Determine which element they have and to which group it belongs.

Conclude and Apply

1. **Analyze** As you classify the elements according to their group numbers, what pattern do you see in the number of electrons in the outer energy level?
2. **Apply** Atoms that give up electrons combine with atoms that gain electrons in order to form compounds. How can the model help you to predict a pair of elements that would combine in this way? Identify several combinations of elements that would form stable compounds.

COMMUNICATE YOUR DATA

Make a graph that relates the groups to the number of electrons in their outer energy levels. Compare your graph to those of your classmates.

Reading Preview

Essential Questions

▶ What are ionic bonds and covalent bonds?

▶ Which particles are produced by different types of bonding?

▶ How do nonpolar and polar covalent bonds compare?

Review Vocabulary

atom: the smallest piece of matter that still retains the properties of the element

New Vocabulary

ion
ionic bond
covalent bond
molecule
nonpolar bond
polar bond
polar molecule
nonpolar molecule

 **Multilingual eGlossary**

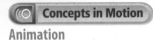 **Concepts in Motion**
Animation

Types of Bonds

MAIN ‹Idea› Atoms form ionic bonds by transferring electrons and form covalent bonds by sharing electrons.

Real-World Reading Link Some toy models need to be glued together, and others have interlocking pieces that snap together. While either type might be used to build a car, they're different in how the individual pieces are connected to each other. Compounds are the same way.

Ions

When you participate in a sport, you might talk about gaining or losing an advantage. To gain an advantage, you want to have a better time or score than your opponent. It is important that you keep practicing because you don't want to lose that advantage. Gaining or losing an advantage happens as you try to meet a standard for your sport.

Atoms, too, lose or gain to meet a standard—stability. They do not lose or gain an advantage. Instead, atoms lose or gain electrons. An atom that has gained or lost an electron is called an ion. An **ion** is a charged particle that has either more or fewer electrons than protons. When an atom loses electrons, it becomes a positively charged ion known as a *cation*. When an atom gains electrons, it becomes a negatively charged ion known as an *anion*. The electric forces between oppositely charged particles can hold ions together.

Some of the most common compounds are made by the loss and gain of just one electron. These compounds contain an element from group 1 and an element from group 17 on the periodic table. Some examples are sodium chloride (NaCl), commonly known as table salt, and potassium iodide (KI), an ingredient in iodized salt. **Figure 7** shows the importance of iodine in nutrition.

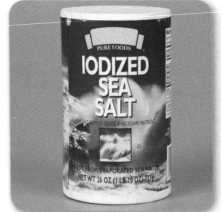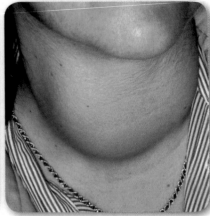

■ **Figure 7** Iodized salt is an important dietary source of iodine. A lack of iodine causes health problems. For example, a goiter, an enlargement of the thyroid gland in the neck, can be caused by an iodine deficiency.

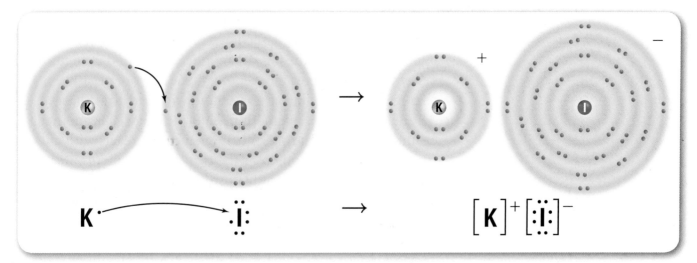

Transfer of electrons What happens when potassium and iodine atoms come together? A neutral atom of potassium has one electron in its outer energy level. This is not a stable outer energy level. Recall that for most elements a stable outer energy level contains eight electrons. When it forms a compound with iodine, potassium loses the one electron from its fourth level. With the fourth level gone, the third level is a complete outer energy level.

Although the complete outer energy level means the atom is now stable, because it has lost an electron, it is no longer neutral. The potassium atom has become an ion. When a potassium atom loses an electron, the atom becomes a positively charged ion because there is one electron fewer in the atom than there are protons in the nucleus. The 1+ charge of the potassium cation is shown as a superscript written after the element's symbol, K^+, to indicate its charge. Superscript means "written above."

✓ **Reading Check** **Explain** What part of an ion's symbol indicates its charge?

An iodine atom has seven electrons in its outer energy level. It needs one more electron in order to have a stable outer energy level. During the reaction with potassium, the iodine atom gains an electron, giving its outer energy level eight electrons. This atom is no longer neutral because it has gained an extra negative particle. It now has a charge of 1– and is called an iodide anion, written as I^-.

The compound formed between potassium and iodine is called potassium iodide. The electron dot diagrams for the process are shown in **Figure 8.** As they lose or gain electrons, the two atoms become stable ions. Notice that the resulting compound has a neutral charge because the 1+ positive charge of K^+ and the 1– negative charge of I^- cancel each other.

The ionic bond When ions combine in this way, a bond is formed. An **ionic bond** is the force of attraction between the opposite charges of the ions in an ionic compound. The number of positive charges must equal the number of negative charges in order to form a compound with a neutral charge.

The formation of magnesium chloride ($MgCl_2$) is another example of ionic bonding. When magnesium reacts with chlorine, a magnesium atom loses the two electrons in its outer energy level and becomes the positively charged ion Mg^{2+}. At the same time, two chlorine atoms gain one electron each to become negatively charged chloride ions, Cl^-. In this case, a magnesium atom has two electrons to donate, but a single chlorine atom needs to accept only one electron. Therefore, it takes two chlorine atoms, as shown in **Figure 9,** to accept the two electrons that the magnesium ion donates.

✓ **Reading Check** **Explain** What is the charge of an ionic compound?

Zero net charge As with potassium iodide, the result of these ionic bonds is a neutral compound. The compound as a whole is neutral because the sum of the charges on the ions is zero. The positive charge of the magnesium ion is exactly equal to the negative charge of the two chloride ions. In other words, when atoms form an ionic compound, their electrons are shifted between the individual atoms, but the overall number of protons and electrons of the combined atoms remains equal and unchanged. Therefore, the compound is neutral.

Ionic bonds are usually formed between metals and non-metals. Ionic compounds are often formed by elements across the periodic table from each other. They are typically crystalline solids with high melting points.

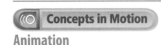

Animation

■ **Figure 9** A magnesium atom gives an electron to each of two chlorine atoms to form $MgCl_2$.

Compare and contrast *formation of KI with the formation of* $MgCl_2$.

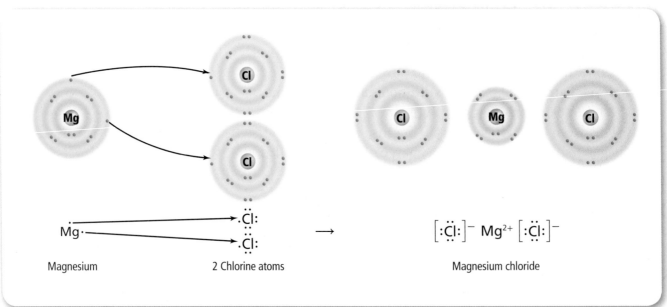

Magnesium   2 Chlorine atoms   Magnesium chloride

Molecules

Some atoms of nonmetals are unlikely to lose or gain electrons. For example, group 14 elements would have to either gain or lose four electrons to achieve a stable outer energy level. The loss of this many electrons takes a great deal of energy. Each time an electron is removed, the nucleus holds the remaining electrons even more tightly. Therefore, these atoms become more chemically stable by sharing electrons, rather than by becoming ions.

The attraction that forms between atoms when they share electrons is known as a **covalent bond.** The neutral particle that forms as a result of electron sharing is called a **molecule.**

The covalent bond A single covalent bond is composed of two shared electrons. Usually, one shared electron comes from each atom in the bond. A water molecule contains two single bonds, as shown in **Figure 10.** In each bond, a hydrogen atom contributes one electron to the bond, and the oxygen atom contributes the other. The two electrons are shared, forming a single bond. The result of this type of bonding is a stable outer energy level for each atom in the molecule.

Multiple covalent bonds A covalent bond can also contain more than one pair of electrons. In the diatomic molecule of oxygen (O_2), each oxygen atom has six electrons in its outer energy level and needs to gain two electrons to become stable. It can do this by sharing two of its electrons with another oxygen atom. When each atom contributes two electrons to the bond, the bond contains four electrons, or two pairs of electrons. Each pair of electrons represents a bond. Therefore, two pairs of electrons represent two bonds, called a double bond. Each oxygen atom is stable with eight electrons in its outer energy level. Similarly, a bond that contains three shared pairs of electrons is a triple bond. The diatomic molecule N_2 is an example shown in **Figure 11.**

Covalent bonds form between nonmetallic elements. Many covalent compounds are solids or gases at room temperature.

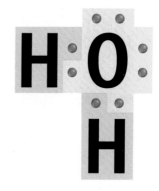

■ **Figure 10** Hydrogen and oxygen each contribute one electron to the hydrogen-oxygen bonds in a water molecule.

VOCABULARY . . . .
WORD ORIGIN
Molecule
comes from the Latin word *moles,* which means "mass"
A water molecule is composed of two hydrogen atoms and one oxygen atom.

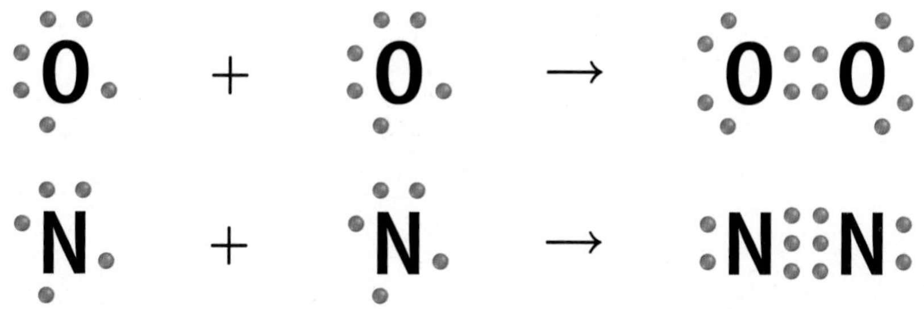

■ **Figure 11** Electron dot diagrams for O_2 and N_2 show that they each share multiple pairs of electrons.

Equal sharing When electrons are shared in covalent bonds by similar or identical atoms, such as in the N_2 or O_2 molecules just discussed, the electron charge is shared equally across the bond. A **nonpolar bond** is a covalent bond in which electrons are shared equally by both atoms.

✓ **Reading Check** **Describe** the atoms involved in a nonpolar bond.

Unequal sharing In some molecules, however, electrons are not shared equally, and the electron charge is concentrated more on one end of the molecule than the other. Why aren't electrons always shared equally? Different types of atoms exert different levels of attraction for the electrons in a covalent bond. The strength of the attraction of each atom to its electrons is related to the size of the atom, the charge of the nucleus, and the total number of electrons the atom contains. Recall how a magnet has a stronger pull when it is right next to a piece of metal. In a similar way, a nucleus has a stronger attraction to electrons nearby. In addition, you know that a strong magnet holds metal more firmly than a weak magnet. Similarly, more positive charge in a nucleus attracts electrons more strongly.

Partial charges One example of unequal sharing in a covalent bond is found in a molecule of hydrogen chloride (HCl), shown in **Figure 12.** Chlorine atoms have a stronger attraction for electrons than hydrogen atoms have. As a result, the shared electrons will spend more time near the chlorine atom. The chlorine atom has a partial negative charge, which is represented by a lower-case Greek letter delta with a negative superscript (δ^-). Because the electrons spend less time near the hydrogen atom, it has a partial positive charge, represented by a δ^+.

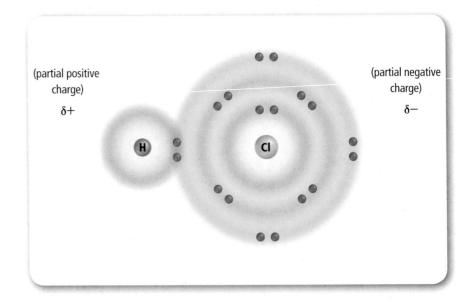

■ **Figure 12** The electrons shared by hydrogen and chlorine spend more time near the chlorine atom, giving the hydrogen chloride molecule partial charges on its ends.

■ **Figure 13** In a tug-of-war, the stronger dog can pull harder and has the advantage.

Compare and Contrast *How is a tug-of-war similar to unequal sharing of electrons?*

Tug-of-war When electrons are shared unequally, the bond is said to be polar. The term *polar* means "having opposite ends." A **polar bond** is a bond in which electrons are shared unequally, resulting in a slightly positive end and a slightly negative end. It might help you to visualize a polar bond as the rope in a tug-of-war, as shown in **Figure 13.** Think of the shared electrons as being in the space in the center. As the two dogs in the tug-of-war pull, the middle of the rope ends up closer to the stronger dog. Similarly, each atom in a bond attracts the electrons that they share, but the electrons will be held more closely to the atom with the stronger pull.

Polar and nonpolar molecules Look again at the molecule of hydrogen chloride (HCl) in **Figure 12.** The atom holding the electrons more closely will always have a slightly negative charge. This polar bond results in the molecule being polar. A **polar molecule** is one in which the unequal sharing of electrons results in a slightly positive end and a slightly negative end, although the overall molecule is neutral. On the other hand, a **nonpolar molecule** is a molecule that does not have oppositely charged ends.

Polarity and geometry Just because a molecule is nonpolar, that doesn't mean its electrons are all shared equally. When a molecule contains just two atoms, such as HCl, it is easy to see how the polar bond creates a polar molecule. When a molecule contains three or more atoms, however, like the molecule of H_2O shown in **Figure 14,** you need to consider both the polarity of the bonds and the shape of the molecule to determine whether the molecule is polar or nonpolar. **Figure 15,** on the next page, shows how determining polarity is dependent on multiple factors. Polar bonds can be found in nonpolar molecules.

■ **Figure 14** Water is a polar molecule. Water's polarity accounts for many of its unique properties.

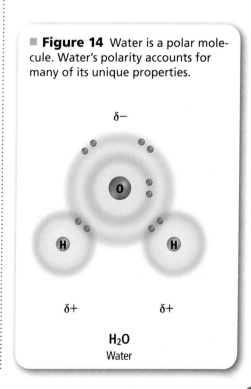

$\delta-$

$\delta+$ $\delta+$

H_2O
Water

FIGURE 15

Polar and Nonpolar Molecules

In molecules with a single bond, polarity is determined by the bond. However, in molecules with multiple bonds, polarity is determined by the shape of the molecule as well as the polarity of each of the bonds. It is possible to have a nonpolar molecule with polar bonds.

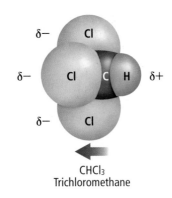

CHCl₃
Trichloromethane

In a molecule of trichloromethane ($CHCl_3$), each carbon-chlorine bond is polar with a partial negative charge on the chlorine atom. Because of the shape of the molecule, there is an overall negative charge on the chlorine end and an overall positive charge on the hydrogen end.

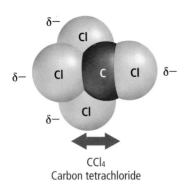

CCl₄
Carbon tetrachloride

Carbon tetrachloride (CCl_4) is very similar to trichloromethane. As with $CHCl_3$, each of the carbon-chlorine bonds is polar. But because the chlorines are evenly distributed around the carbon, the charges are also evenly distributed. As a result, the molecule as a whole is nonpolar.

Section 2 Review

Section Summary

▶ An ion is a charged particle that has either fewer or more electrons than protons, resulting in a net positive or negative charge.

▶ An ionic bond is the force of attraction between oppositely charged ions.

▶ Some atoms share electrons instead of losing or gaining them. These atoms form covalent bonds with other atoms.

▶ When the electrons in a covalent bond are shared unequally, the result is a polar bond.

8. **MAIN Idea Compare and contrast** ionic and covalent bonds.

9. **Determine** the type of bonding in CaO and in SO_2.

10. **Name** the type of particle formed by covalent bonds.

11. **Identify** the following compounds as polar covalent or nonpolar covalent: HBr, Cl_2, and H_2O.

12. **Think Critically** From the given symbols, choose two elements that are likely to form an ionic bond: O, Ne, S, Ca, and K. Next, select two elements that would likely form a covalent bond. Explain.

Apply Math

13. **Solve One-Step Equations** Aluminum oxide (Al_2O_3) can be produced during rocket launches. Show that the sum of the positive and negative charges in a unit of Al_2O_3 equals zero.

✓ Assessment   Online Quiz

Reading Preview

Essential Questions

▶ How are oxidation numbers determined?

▶ How are formulas for ionic and covalent compounds written?

▶ How are ionic and covalent compounds named?

Review Vocabulary

ion: an atom that has gained or lost electrons

New Vocabulary

oxidation number
binary compound
polyatomic ion
hydrate

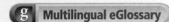

g **Multilingual eGlossary**

Writing Formulas and Naming Compounds

MAIN ‹Idea **For both ionic and covalent compounds, you can write a name from the chemical formula or a chemical formula from the name.**

Real-World Reading Link How do you refer to specific individuals on a basketball team? On their jerseys, you'll find both their names and their numbers. In the same way, compounds may be identified by either their names or their chemical formulas.

Writing Chemical Formulas

Does the table in **Figure 16** look like it has anything to do with chemistry? It is an early table of the elements made for alchemy, a practice from the Middle Ages that laid the foundations for modern chemistry. Alchemists used symbols like these to write the formulas of substances created when individual elements combined. Before you can write a chemical formula, you must have all the needed information at your fingertips. What will you need to know?

Oxidation numbers To write a chemical formula, you need to know the elements involved and the number of electrons they gain, lose, or share to become stable. This last piece of information is what chemists refer to as an element's oxidation number. An **oxidation number** is a positive or negative number that indicates how many electrons an atom has gained, lost, or shared to become stable.

For ionic compounds, the oxidation number is the same as the charge on the ion. For example, a sodium ion has a charge of 1+ and an oxidation number of 1+. A chloride ion has a charge of 1− and an oxidation number of 1−.

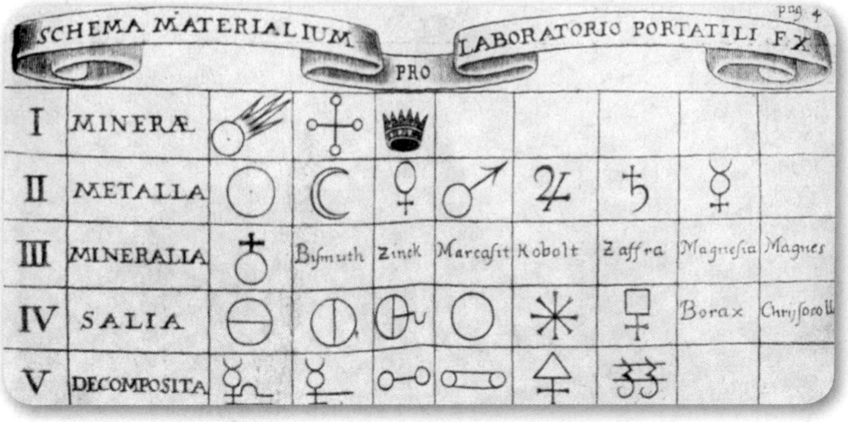

■ **Figure 16** Early representations of elements used pictorial symbols to identify known substances.

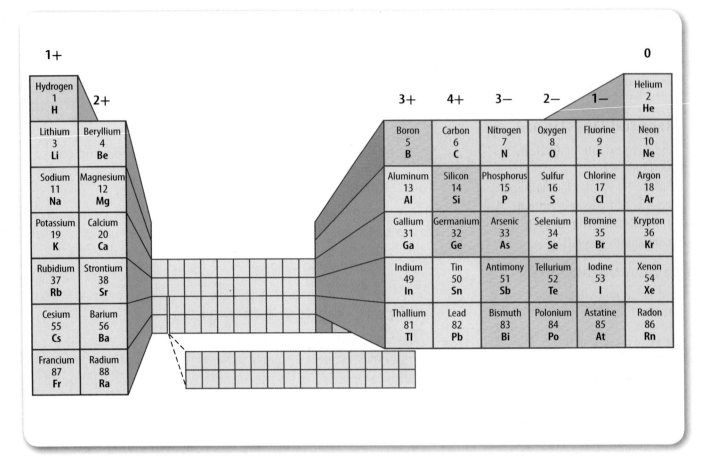

■ **Figure 17** The number at the top of each column is the most common oxidation number of the elements in that group.

Oxidation numbers and the periodic table Many elements have one common oxidation number, as shown in **Figure 17.** Notice how they fit with the periodic table groupings.

Many of the transition elements can have more than one oxidation number, as shown in **Table 2.** When naming these compounds, the oxidation number is expressed in the name with a roman numeral. For example, the oxidation number of iron in iron(III) oxide is 3+.

Binary ionic compounds The easiest compounds to write formulas for are **binary compounds,** which are composed of two elements. When writing formulas, remember that although the individual ions that make up ionic compounds carry charges, the compound itself is neutral. For example, lithium fluoride is composed of a lithium ion with a 1+ charge and a fluoride ion with a 1– charge. One of each ion put together makes a neutral compound with the formula LiF.

Some compounds require more figuring. Aluminum oxide contains an aluminum ion with a 3+ charge and an oxide ion with a 2– charge. To determine the overall positive and negative charge, you must find the least common multiple of 3 and 2, which is 6. In order to have a 6+ charge, you need two aluminum ions. In order to have a 6– charge, you need three oxygen ions. This gives the neutral compound Al_2O_3.

Table 2	Special Ions
Ion Name	**Oxidation Number**
Copper(I)	1+
Copper(II)	2+
Iron(II)	2+
Iron(III)	3+
Chromium(II)	2+
Chromium(III)	3+
Lead(II)	2+
Lead(IV)	4+

Writing formulas Once you've found the oxidation numbers and their least common multiple, you can write formulas for binary ionic compounds by using the rules below.

1. Write the symbol of the element that has the positive oxidation number or charge.
   - All metals have positive oxidation numbers. Hydrogen often does.
2. Write the symbol of the element with the negative oxidation number.
   - Nonmetals have negative oxidation numbers. Hydrogen does occasionally, when bonded to a metal.
3. Find the least common multiple of the charges of each ion.
   - The charge (without the sign) of one ion becomes the subscript of the other ion. Reduce the subscripts to the smallest whole numbers that retain the ratio of the ions.

EXAMPLE Problem 1

Determine a Chemical Formula What is the formula for lithium nitride?

List the Unknown: formula for lithium nitride

List the Knowns: symbol and oxidation number of the positive ion:
    lithium Li^{1+}
symbol and oxidation number of the negative ion:
    nitrogen N^{3-}

Set Up the Problem: Write the symbol of the element with the positive charge, followed by the symbol of the element with the negative charge:
    $Li^{1+}N^{3-}$

Solve the Problem: The charge (without the sign) of one ion becomes the subscript of the other:
    Li_3N_1 or Li_3N

In this case, the subscripts already reflect the smallest whole numbers that retain the ratios of the ions.

Check the Answer: Check the answer by determining whether your compound is neutral. The charge on three lithium ions is 3+, and the charge on one nitrogen is 3–. The compound is neutral, so the formula is correct.

PRACTICE Problems

Find **Additional Practice Problems** in the back of your book.

14. What is the formula for potassium sulfide?

15. What is the formula for calcium chloride?

16. **Challenge** What is the formula for lead(IV) phosphide?

Review
Additional Practice Problems

Table 3	Elements in Binary Compounds	
Element	**–ide Name**	
Oxygen	oxide	
Phosphorus	phosphide	
Nitrogen	nitride	
Sulfur	sulfide	

Naming Name binary ionic compounds with these rules.

1. Write the name of the positive ion.
2. Using **Table 2,** check to see whether the positive ion forms more than one oxidation number. If so, determine the oxidation number of the ion from the formula of the compound. Remember, the overall charge of the compound is zero, and the negative ion has only one possible charge. Write the charge of the positive ion using roman numerals in parentheses after the ion's name.
3. Write the root name of the anion. The root is the first part of the element's name. Chlorine is *chlor–*; oxygen is *ox–*.
4. Add the ending *–ide* to the root, as shown in **Table 3.** Chlorine becomes chlor*ide* and oxygen becomes ox*ide.*

Subscripts are not part of the name for ionic compounds.

EXAMPLE Problem 2

Name a Binary Ionic Compound What would a chemist name the compound CuCl?

List the Unknown: compound name for CuCl

List the Knowns: the names of the atoms in the compound:
copper and chlorine

Set Up the Problem:
1. Name the positive ion in the compound.
2. Check **Table 2** to determine whether the positive ion can have more than one oxidation number. If it can, determine which one to use, name the positive ion, and write the charge using roman numerals in parentheses.
3. Write the root name of the negative ion.
4. Add the ending *–ide* to the root of the negative ion.

Solve the Problem:
1. The positive ion is copper.
2. Copper has two oxidation numbers. The oxidation number for chlorine is 1–, so copper has to be 1+. The positive ion is *copper(I)*.
3. The negative atom is chlorine. The root is *chlor–*.
4. Adding *–ide* to the negative root name is *chloride*.

The compound name is copper(I) chloride.

Check the Answer: In an ionic compound, the positive ion is a metal and the negative ion is a nonmetal—the name should be in this order. The name *copper(I) chloride* has the metal first and the nonmetal second.

PRACTICE Problems

Find **Additional Practice Problems** in the back of your book.

17. What is the name of $AlCl_3$?

18. What is the name of Li_2S?

19. **Challenge** What is the name of Cr_2O_3?

Review

Additional Practice Problems

Compounds With Complex Ions

Not all compounds are binary. Many common compounds contain more than two atoms. Baking soda—used in cooking, as a medicine, and for brushing your teeth—has the formula $NaHCO_3$. This is an example of an ionic compound that is not binary. Some compounds, including baking soda, contain polyatomic ions. The prefix *poly-* means "many," so the term *polyatomic* means "having many atoms." A **polyatomic ion** is a positively or negatively charged, covalently bonded group of atoms. Polyatomic ions as a whole contain two or more elements. Even though polyatomic ions contain more than one element, they act like individual ions in forming compounds. The polyatomic ion in baking soda is the hydrogen carbonate ion (HCO_3^-), which is commonly called bicarbonate.

Writing formulas **Table 4** lists several common polyatomic ions. To write formulas for compounds containing these ions, follow the rules for binary compounds, with one addition. When more than one polyatomic ion is needed to balance the charges of the ions, write parentheses around the polyatomic ion before adding the subscript.

How would you write the formula of barium chlorate? First, identify the symbol of the positive ion. Barium has a symbol of Ba and forms a 2+ ion, Ba^{2+}. Next, identify the negative chlorate ion. **Table 4** shows that it is ClO_3^-. Finally, you need to balance the charges of the ions to make the compound neutral. It will take two chlorate ions with a 1– charge to balance the 2+ charge of the barium ion. Because the chlorate ion is polyatomic, you use parentheses before adding the subscript. Therefore, the formula is $Ba(ClO_3)_2$. Another example of naming complex compounds is shown in **Table 5**.

Naming First, write the name of the positive ion. Then write the name of the negative ion. What is the name of $Sr(OH)_2$? Begin by writing the name of the positive ion, strontium. Then find the name of the polyatomic ion OH^-. **Table 4** lists it as hydroxide. Thus, the name is strontium hydroxide.

Table 4	Polyatomic Ions	
Charge	Name	Formula
1+	ammonium	NH_4^+
1–	acetate	$C_2H_3O_2^-$
	chlorate	ClO_3^-
	hydroxide	OH^-
	nitrate	NO_3^-
2–	carbonate	CO_3^{2-}
	sulfate	SO_4^{2-}
3–	phosphate	PO_4^{3-}

Table 5	Naming Complex Compounds
To write the chemical formula for ammonium phosphate, answer these questions:	
1. What is the positive ion and its charge? The positive ion is NH_4^+, and its charge is 1+.	
2. What is the negative ion, and what is its charge? The negative ion is PO_4^{3-}, and its charge is 3–.	
3. How can the charges be balanced in order to make the compound neutral? Three NH_4^+ ions (3+) balances one PO_4^{3-} (3–) ion. Add parentheses for subscripts greater than one. The chemical formula for ammonium nitrate is $(NH_4)_3PO_4$.	

Table 6	Prefixes for Covalent Compounds	
Number of Atoms	Prefix	Example
1	mono–	carbon monoxide
2	di–	sulfur dioxide
3	tri–	phosphorous trichloride
4	tetra–	carbon tetrachloride
5	penta–	dinitrogen pentoxide
6	hexa–	uranium hexafluoride
7	hepta–	dichlorine heptoxide
8	octa–	xenon octaflouride

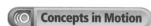 Concepts in Motion   Interactive Table

Inquiry   MiniLab

Make a Hydrate

Procedure

1. Read the procedure and safety information, and complete the lab form.
2. Mix 150 mL of **plaster of paris** with 75 mL of **water** in a **small bowl.**
3. Let the plaster dry overnight, and take the hardened plaster out of the bowl.
4. Lightly tap the plaster with a **rubber hammer.**
5. Heat the plaster with a **hair dryer** on the hottest setting and observe.
6. Place a **towel** over the sample and lightly tap it with the hammer again.

Analysis

1. **Compare** what happened to the plaster when you tapped it before heating to what happened when you tapped it after heating.
2. **Explain** what happened to the plaster as you heated it.

Naming Binary Covalent Compounds

Covalent compounds are those formed between elements that are nonmetals. Some pairs of nonmetals can form more than one compound with each other. For example, nitrogen and oxygen can combine to form a number of different compounds, including N_2O, NO, NO_2, and N_2O_5. In the system of naming using oxidation numbers that you have just learned, each of these compounds would be called nitrogen oxide. From that name, you would not know the composition of the compound. So, a different system of naming must be used for covalent compounds.

Using prefixes Scientists use the Greek prefixes in **Table 6** to indicate how many atoms of each element are in a binary covalent compound. The nitrogen and oxygen compounds N_2O, NO, NO_2, and N_2O_5 would be named dinitrogen monoxide, nitrogen monoxide, nitrogen dioxide, and dinitrogen pentoxide, respectively. Notice that the last vowel of the prefix is dropped when the second element begins with a vowel, as in monoxide and pentoxide. The prefix *mono–* is always omitted from the name of the first element of the compound. For example, CO is carbon monoxide as opposed to monocarbon monoxide.

Compounds with Added Water

Some compounds have water molecules as part of their structures. These compounds are called hydrates. A **hydrate** is a compound that has water chemically attached to its atoms and written into its chemical formula.

Common hydrates The term *hydrate* comes from a word that means "water." When a solution of cobalt chloride evaporates, pink crystals that contain six water molecules for each unit of cobalt chloride are formed. The formula for this compound is $CoCl_2 \cdot 6H_2O$ and is called cobalt chloride hexahydrate. You can remove water from these crystals by heating them. The resulting blue compound is called anhydrous cobalt chloride. The word *anhydrous* means "without water." When anhydrous (blue) $CoCl_2$ is exposed to water, even from the air, it will revert back to its hydrated state.

■ **Figure 18** Gypsum is used in building materials. The fact that it is a hydrate means that it can be heated for some time before it has released all of its water.

Infer *How does gypsum's ability to hold water make it a useful material for building construction?*

 Video Lab

VOCABULARY ....
SCIENTIFIC USAGE V. COMMON USAGE
Hydrate
Science usage
   a compound that has water chemically attached to its atoms
   Gypsum is a hydrate.

Common usage
   to supply water to a person in order to maintain proper fluid balance
   It's important to hydrate before and after physical activity. ....

The plaster of paris shown in **Figure 18** also forms a hydrate when water is added. It is made from calcium sulfate dihydrate, which is also known as gypsum. When writing a formula for a hydrate, the water is shown after a " · ". Following the dot, write the number of water molecules attached to the compound. For example, calcium sulfate dihydrate (gypsum) is written $CaSO_4 \cdot 2H_2O$. Notice that when naming hydrates, you use the same prefixes listed in **Table 6** to indicate the number of water molecules.

Section 3 Review

Section Summary

▶ A binary compound is one composed of two elements.

▶ The oxidation number tells how many electrons an atom has gained, lost, or shared to become stable.

▶ An ionic compound is made of ions, but the compound as a whole is neutral.

▶ A polyatomic ion is a positively or negatively charged, covalently bonded group of atoms.

▶ Greek prefixes are used to indicate how many atoms of each element are in a binary covalent compound.

▶ A hydrate is a compound that has water chemically attached to its ions.

20. **MAIN Idea** Write formulas for the following compounds: potassium iodide, phosphorus pentachloride, magnesium hydroxide, aluminum sulfate, dichlorine heptoxide, and calcium nitrate trihydrate.

21. **Identify** Write the names of these compounds: Al_2O_3, $Ba(ClO_3)_2$, SO_2, NH_4Cl, PCl_3, and $Mg_3(PO_4)_2 \cdot 4H_2O$.

22. **Determine** the oxidation number of each atom in the following compounds: sodium chloride and iron(II) oxide.

23. **Think Critically** Explain whether sodium and potassium will react to form a bond with each other.

Apply Math

24. **Solve One-Step Equations** The overall charge on the polyatomic sulfate ion is 2–. Its formula is $SO_4{}^{2-}$. If the oxygen ion has a 2– oxidation number, determine the oxidation number of sulfur in this polyatomic ion.

Nonstick Surfaces

In your kitchen, there is probably a pan coated with the same nonstick substance used in artificial hearts and in NASA spacecraft. The substance, known as polytetrafluoroethylene (PTFE), has been used in hundreds of applications.

Losing the ability to bond PTFE, represented in **Figure 1,** is a long string of carbon atoms surrounded by fluorine atoms. The bonds between carbon and fluorine result in a compound that is very stable. PTFE is so stable that it will not bond with most other chemicals. PTFE can be used as a protective barrier between materials that would otherwise bond, such as use in laboratory equipment that comes in contact with corrosive chemicals.

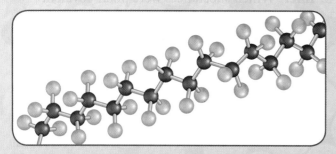

Figure 1 PTFE can contain thousands of carbon atoms, each bonded to two fluorine atoms.

Nonstick surface The long chain structure of PTFE forms a low-friction surface. Materials coated with PTFE become smooth and slippery because the long molecules fill in the rough crevices of the underlying surface, as shown in **Figure 2.**

These two properties of PTFE—resistance to bonding and low friction—make the nonstick pan in your kitchen so useful. Not only will food not bond to the surface of the pan, but it also slides easily across the pan's smooth surface.

Figure 2 Long PTFE molecules create a smooth, low-friction surface on which food can slide easily.

Medical uses PTFE has many medical uses because it will not deteriorate over time. It is used in attaching artificial heart valves to existing tissue and can also be woven into a mesh that can be surgically implanted into arteries. These strong, flexible PTFE implants reinforce and repair damaged blood vessels.

PTFE in space Reactive ions in space bombard the *Hubble Space Telescope* and can bond to and degrade the telescope's sensitive equipment. A PTFE coating helps prevent the damaging ions from bonding to the telescope. Research continues into the next generation of nonstick coatings that will help protect future space missions.

WebQuest

Environmental Impact Some scientific advances have had significant environmental impacts. Work with a partner to research potential negative impacts caused by PTFE. Create a presentation that weighs the benefits of PTFE against its drawbacks.

Theme Focus Changes of Matter

Based on their positions in the periodic table, atoms combine in predictable ways to form compounds with different properties. By forming ions or sharing electrons, atoms join together into ionic compounds or covalent compounds held together with chemical bonds.

BIG (Idea) Elements link together with chemical bonds.

Section 1 Stability in Bonding

chemical bond (p. 556)
chemical formula (p. 553)

MAIN (Idea) When atoms form compounds, each atom is more stable in the compound than it was by itself.

- A chemical formula describes the numbers and types of atoms in a compound.
- The elements of group 18, the noble gases, rarely combine with other elements.
- Most atoms need eight electrons to complete their outer energy level.
- Electron dot diagrams show the electrons in the outer energy level of an atom.
- A chemical bond is the force that holds atoms together in a compound.

Section 2 Types of Bonds

covalent bond (p. 561)
ion (p. 558)
ionic bond (p. 560)
molecule (p. 561)
nonpolar bond (p. 562)
nonpolar molecule (p. 563)
polar bond (p. 563)
polar molecule (p. 563)

MAIN (Idea) Atoms form ionic bonds by transferring electrons and form covalent bonds by sharing electrons.

- An ion is a charged particle that has either fewer or more electrons than protons, resulting in a net positive or negative charge.
- An ionic bond is the force of attraction between oppositely charged ions.
- Some atoms share electrons instead of losing or gaining them. These atoms form covalent bonds with other atoms.
- When the electrons in a covalent bond are shared unequally, the result is a polar bond.

Section 3 Writing Formulas and Naming Compounds

binary compound (p. 566)
hydrate (p. 570)
oxidation number (p. 565)
polyatomic ion (p. 569)

MAIN (Idea) For both ionic and covalent compounds, you can write a name from the chemical formula or a chemical formula from the name.

- A binary compound is one composed of two elements.
- The oxidation number tells how many electrons an atom has gained, lost, or shared to become stable.
- An ionic compound is made of ions, but the compound as a whole is neutral.
- A polyatomic ion is a positively or negatively charged, covalently bonded group of atoms.
- Greek prefixes are used to indicate how many atoms of each element are in a binary covalent compound.
- A hydrate is a compound that has water chemically attached to its ions.

Use Vocabulary

Match each phrase with the correct term from the Study Guide.

25. a charged group of atoms

26. a compound composed of two elements

27. a molecule with partially charged areas

28. a positively or negatively charged particle

29. a chemical bond between oppositely charged ions

30. a bond formed from shared electrons

31. a crystalline substance that contains water

32. a particle made of covalently bonded atoms

33. indicates how many electrons must be lost, gained, or shared to achieve stability

34. shows which elements are in a compound and their ratios

Check Concepts

Choose the word or phrase that best answers the question.

35. Which elements are least likely to react with other elements?
  A) metals
  B) nonmetals
  C) noble gases
  D) transition elements

36. What is the name of CuO?
  A) copper oxide
  B) copper(I) oxide
  C) copper(II) oxide
  D) copper(III) oxide

37. Which of the following formulas represents a nonpolar molecule?
  A) N_2 **C)** NaCl
  B) H_2O **D)** HCl

Use the figure below to answer question 38.

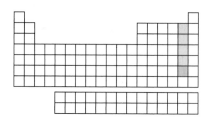

38. How many electrons are in the outer energy levels of the highlighted group of elements?
  A) 1   **C)** 7
  B) 2   **D)** 17

39. Which is a binary ionic compound?
  A) O_2 **C)** H_2SO_4
  B) NaF   **D)** $Cu(NO_3)_2$

40. Which of these is an example of an anhydrous compound?
  A) H_2O **C)** $CuSO_4 \cdot 5H_2O$
  B) $CaSO_4$ **D)** $CaSO_4 \cdot 2H_2O$

41. Which of the following is an atom that has gained an electron?
  A) negative ion
  B) positive ion
  C) polar molecule
  D) nonpolar molecule

42. Which of these is an example of a covalent compound?
  A) sodium chloride
  B) calcium fluoride
  C) calcium chloride
  D) sulfur dioxide

Interpret Graphics

Concepts in Motion   Interactive Concept Map

43. Copy and complete the concept map using the terms *stable, different, covalent bonds, ionic bonds, shared, lost or gained.*

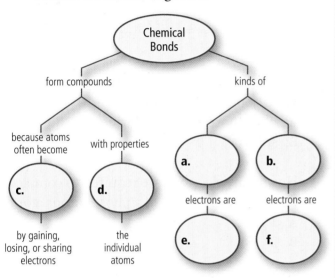

Use the table below to answer question 44.

Which compounds exist?	
Formula	Possible Compounds
SF_6	AlF_6 or TeF_6
K_2SO_4	Na_2SO_4 or Ba_2SO_4
CO_2	CCl_2 or CS_2
$CaCO_3$	OCO_3 or $BaCO_3$

44. **THEME FOCUS** Elements from one group of the periodic table generally combine with elements from another group and polyatomic ions in the same ratio. For example, Ca combines with two Cl atoms to form $CaCl_2$ as it does with F to form CaF_2. The table above shows the correct formula of a compound in the left column. Use the periodic table to predict which of the two possible compounds in the right column of the table above is correct.

Think Critically

45. **Draw** Anhydrous magnesium chloride is used to make wood fireproof. Draw a dot diagram of magnesium chloride.

46. **Explain** Artificial diamonds are made using thallium carbonate. If thallium has an oxidation number of 1+, what is the formula for the compound?

47. **Apply** Baking soda, which is sodium hydrogen carbonate, and vinegar, which contains hydrogen acetate, can both be used as household cleaners. Write the chemical formulas for these two compounds.

48. **Explain** If the linear molecule beryllium dichloride ($BeCl_2$) has polar bonds because Cl attracts electrons more strongly than Be, why is the molecule nonpolar?

49. **BIG Idea Explain** what electric forces between oppositely charged electrons and protons have to do with chemical bonds.

50. **Apply** Write the chemical names for Fe_2S_3, $Cu(ClO_3)_2$, $Ca_3(PO_4)_2$, and $(NH_4)_2SO_4$.

51. **Model** Draw an electron dot diagram for $BaCl_2$. Explain how your diagram shows that all atoms in the compound are stable.

Apply Math

52. **Calculate** What is the oxidation number of iron (Fe) in the compound Fe_2S_3?

53. **Use Percentages** If the mass of a single unit of ammonium sulfate ((NH_4)$_2SO_4$) is 132.14 amu and the mass of one nitrogen atom is 14.007 amu, what percentage of the compound's mass does nitrogen account for?

Standardized Test Practice

Multiple Choice

Record your answers on the answer sheet provided by your teacher or on a sheet of paper.

1. Which element is NOT part of the compound NH_4NO_3?
   A. nitrogen
   B. nickel
   C. oxygen
   D. hydrogen

2. Aside from hydrogen and helium, when an atom is chemically stable, how many electrons are in its outer energy level?
   A. 0
   B. 4
   C. 7
   D. 8

Use the figure below to answer questions 3 and 4.

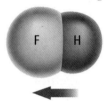

3. What type of bond holds the atoms of this molecule together?
   A. covalent
   B. ionic
   C. triple
   D. double

4. Which statement about this molecule is true?
   A. This is a nonpolar molecule.
   B. The electrons are shared equally in the bonds of this molecule.
   C. This molecule does not have oppositely charged ends.
   D. This is a polar molecule.

5. What is the oxidation number of sodium in the compound Na_3PO_4?
   A. 3–
   B. 1–
   C. 1+
   D. 3+

Use the figure below to answer questions 6–8.

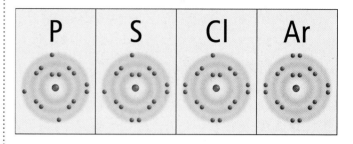

6. Which element is least likely to form an ionic bond with sodium?
   A. phosphorous
   B. sulfur
   C. chlorine
   D. argon

7. How many electrons are required to complete the outer energy level of a phosphorous atom?
   A. 1
   B. 2
   C. 3
   D. 4

8. How many electrons are in an argon atom?
   A. 8
   B. 10
   C. 18
   D. 26

9. What do the group 17 elements become when they react with group 1 elements?
   A. negative ions
   B. positive ions
   C. neutral
   D. polyatomic ions

10. What is the name of $KC_2H_3O_2$?
    A. potassium carbide
    B. potassium acetate
    C. potassium hydroxide
    D. potassium oxide

11. How many electrons are in the outer energy level of a helium atom?
    A. 1
    B. 2
    C. 3
    D. 4

✓ Assessment Standardized Test Practice

Short Response

Record your answers on the answer sheet provided by your teacher or on a sheet of paper.

Use the figure below to answer questions 12 and 13.

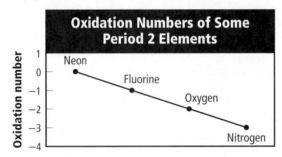

Oxidation Numbers of Some Period 2 Elements

12. Describe the trend in the oxidation numbers of these period 2 elements.

13. Why do the oxidation numbers of nitrogen and fluorine differ?

14. Draw electron dot diagrams for carbon and hydrogen. Draw a dot diagram for methane (CH$_4$) one of many compounds formed by these two elements.

15. Give several examples of ionic compounds. What are two properties often shared by these compounds?

16. A compound has the formula MgSO$_4 \cdot$ 7H$_2$O. Identify and define this type of compound. Using the appropriate prefix, write its name.

17. What information is given in a chemical formula?

Extended Response

Record your answers on a sheet of paper.

Use the figure below to answer question 18.

$$ \cdot \ddot{N} \cdot \ + \ \cdot \ddot{N} \cdot \ \rightarrow \ \ddot{\underset{..}{N}} \vdots \vdots \ddot{\underset{..}{N}} \vdots $$

18. Describe the bond holding the atoms together in this molecule.

19. Nitrogen occurs naturally as a diatomic molecule because N$_2$ molecules are more stable than nitrogen atoms. H$_2$, O$_2$, F$_2$, Cl$_2$, Br$_2$, and I$_2$ are other diatomic molecules. Draw dot diagrams for three of these molecules.

20. Explain why elements in group 4, which have four electrons in the outer energy level, are unlikely to lose all of the electrons in the outer energy level.

21. What factors affect how strongly an atom is attracted to its electrons?

22. Create a chart that compares the properties of polar and nonpolar molecules. Your chart should include several examples of each type of molecule.

23. What is the difference between nitrogen monoxide and dinitrogen pentoxide? Why are prefixes used in this situation?

NEED EXTRA HELP?																						
If You Missed Question . . .	1	2	3	4	5	6	7	8	9	10	11	12	13	14	15	16	17	18	19	20	21	22	23
Review Section . . .	1	1	2	2	3	1	1	1	2	3	1	3	3	1	2	3	1	2	1	1	2	2	3

CHAPTER 19

Chemical Reactions

ConnectED

Your one-stop online resource
connectED.mcgraw-hill.com

- Video
- Audio
- Review
- Inquiry
- WebQuest
- Assessment
- Concepts in Motion
- g Multilingual eGlossary

Launch Lab

Rusting—A Chemical Reaction

Rusting is a chemical reaction in which iron metal combines with oxygen. Other metals combine with oxygen, too—some more readily than others. In this lab, you will compare how iron and aluminum react with oxygen.

For a lab worksheet, use your StudentWorks™ Plus Online.

? **Inquiry** Launch Lab

FOLDABLES®

Create a vocabulary book using the titles shown. Use it to organize your notes on the types of chemical reactions.

Chemical Reactions
- Double-Displacement
- Single-Displacement
- Decomposition
- Synthesis
- Combustion

THEME FOCUS Changes of Matter
Chemical changes occur in a predictable manner at a definite rate.

BIG Idea A chemical reaction changes one or more substances into a different substance or substances.

g **Multilingual eGlossary**

Chemical Changes

MAIN Idea A balanced chemical equation describes the rearrangement of atoms in a chemical reaction.

Real-World Reading Link When you eat an apple, what happens to it? Chemical reactions occur inside your body that break down the apple and provide you with energy and nutrients.

Lavoisier and the Conservation of Mass

Chemical reactions take place all around you, and even within you. A **chemical reaction** is a change in which one or more substances are converted into new substances. The starting substances that react are called **reactants.** The new substances produced are called **products.** This relationship between reactants and products can be written as follows: reactants → products.

By the 1770s, the pseudoscience of alchemy was starting to be replaced by chemistry. While alchemy imitated science, alchemists did not provide science-based explanations about the natural world. However scientists, such as the French chemist Antoine Lavoisier, studied chemical reactions using scientific methods. As a result of such study, Lavoisier established that the total mass of the products always equals the total mass of the reactants. This principle is demonstrated in **Figure 1.**

The mystery of exactly what happens when substances change made Lavoisier curious. In one experiment, Lavoisier placed a carefully measured mass of solid mercury(II) oxide into a sealed container. When he heated this container, he noted a dramatic change. The red powder transformed into a silvery liquid that he recognized as mercury metal, and a gas was produced. When he determined the mass of the liquid mercury and the gas, their combined masses were exactly the same as the mass of the red powder he started with.

■ **Figure 1** The mass of the candle and oxygen before burning equals the mass of the candle and gaseous products after burning.

Before burning   After burning

Lavoisier's experiment also established that the gas produced by heating mercury(II) oxide was a component of air. He did this by heating mercury metal with air and saw that a portion of the air combined with the mercury metal to produce red mercury(II) oxide.

mercury(II) oxide		oxygen	plus	mercury
10.0 g	=	0.7 g	+	9.3 g

Notice that the mass of the mercury(II) oxide reactant is equal to the combined masses of the mercury metal and the oxygen gas products. Hundreds of experiments carried out in Lavoisier's laboratory confirmed that matter is not created or destroyed, but conserved in a chemical reaction. This important principle became known as the law of conservation of mass. This means that the total starting mass of all reactants is equals the total final mass of all products.

✓ **Reading Check** **Explain** the law of conservation of mass.

Writing Equations

If you wanted to describe the chemical reaction shown in **Figure 2,** you might write something like this:

Nickel(II) chloride, dissolved in water, plus sodium hydroxide, dissolved in water, produces solid nickel(II) hydroxide plus sodium chloride, dissolved in water.

This series of words is rather cumbersome, but all of the information is important. The same is true of descriptions of most chemical reactions.

VOCABULARY ....
ACADEMIC VOCABULARY
Establish
to show to be valid or true; to prove
The lawyer established the facts of the case. ....

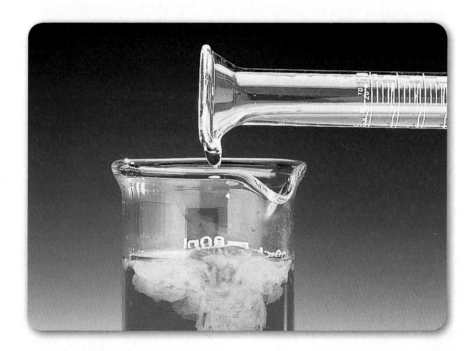

■ **Figure 2** When solutions of nickel(II) chloride and sodium hydroxide mix, a chemical reaction occurs that produces nickel(II) hydroxide as a white solid.

Table 1	Symbols Used in Chemical Equations
Symbol	**Meaning**
$\rightarrow$	produces or yields
+	plus
(s)	solid
(l)	liquid
(g)	gas
(aq)	aqueous—a substance is dissolved in water
$\xrightarrow{heat}$	The reactants are heated.
$\xrightarrow{light}$	The reactants are exposed to light.
$\xrightarrow{elec.}$	An electric current is applied to the reactants.

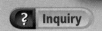

MiniLab

(?) **Inquiry**   **MiniLab**

Design a Team Equation

Procedure

1. Read the procedure and safety information, and complete the lab form.
2. Obtain **15 index cards** and mark each as follows: five with *Guard,* five with *Forward,* and five with *Center.*
3. Group the cards to form as many complete basketball teams as possible. Each team needs two guards, two forwards, and one center.

Analysis

1. **Write** the formula for a team. Write the formation of a team as an equation. Use coefficients in front of each type of player needed for a team.
2. **Explain** how your team equation is like a chemical equation. Why can't you use the remaining cards?
3. **Infer** How do the remaining cards illustrate the law of conservation of matter in this example?

Many words are needed to state all the important information about reactions. As a result, scientists developed a shorthand method to describe chemical reactions. A **chemical equation** is a way to describe a chemical reaction using chemical formulas and other symbols. Some of the symbols used in chemical equations are listed in **Table 1.**

The chemical equation for the reaction shown in **Figure 2** looks like this:

$$NiCl_2(aq) + 2NaOH(aq) \rightarrow$$
$$Ni(OH)_2(s) + 2NaCl(aq)$$

It is much easier to quickly and clearly identify what is happening by writing the information in this form. Chemical equations quickly convey information such as the states of matter of the reactants and products. Later, you will learn how chemical equations make it easier to calculate the quantities of reactants that are needed and of products that are formed.

Coefficients

A chemical equation lists more than just the chemical formulas of the reactants and products. Notice the numbers next to the formulas for sodium hydroxide (NaOH) and sodium chloride (NaCl) above. What do they mean? These numbers, called **coefficients,** represent the number of units of each substance taking part in a reaction.

When no coefficient appears before a substance in a balanced equation, a coefficient of one is assumed. In the equation above, for example, one unit of nickel(II) chloride ($NiCl_2$) combined with two units of sodium hydroxide (NaOH). The products were one unit of nickel(II) hydroxide ($Ni(OH)_2$) and two units of sodium chloride (NaCl). Remember that according to the law of conservation of mass, atoms are rearranged but never destroyed in a chemical reaction. Notice in the equation above that there are equal numbers of each type of atom on each side of the arrow.

✓ **Reading Check** **Summarize** Describe the purpose of coefficients in a chemical equation.

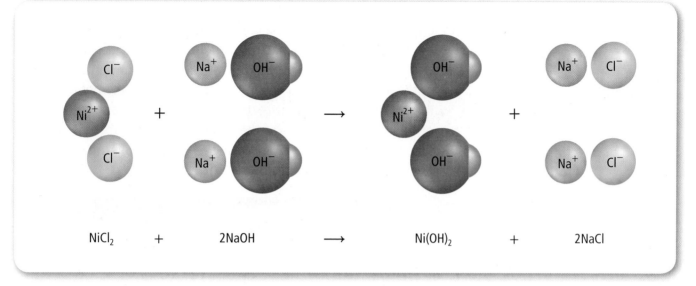

$$NiCl_2 + 2NaOH \longrightarrow Ni(OH)_2 + 2NaCl$$

Knowing the number of units of reactants enables chemists to add the correct amounts of reactants to a reaction. Also, these coefficients tell them exactly how much product will form. For example, you would react one unit of $NiCl_2$ with two units of NaOH to produce one unit of $Ni(OH)_2$ and two units of NaCl. You can see this reaction in **Figure 3,** above.

Balancing Equations

Lavoisier's mercury(II) oxide reaction, shown in **Figure 4,** can be written as:

$$HgO(s) \xrightarrow{\text{heat}} Hg(l) + O_2(g)$$

Notice the number of mercury atoms is the same on both sides of the equation but that the number of oxygen atoms is not:

Numbers and Kinds of Atoms				
	HgO(s) $\longrightarrow$	$O_2(g)$	+	Hg(l)
Hg	1			1
O	1	2		

In this equation, one oxygen atom appears on the reactant side of the equation and two appear on the product side. But according to the law of conservation of mass, one oxygen atom cannot just become two. Nor can you simply add the subscript 2 and write HgO_2 instead of HgO. The formulas HgO_2 and HgO do not represent the same compound. In fact, HgO_2 does not exist. The formulas in a chemical equation must accurately represent the compounds that react. Fixing this equation requires a process called balancing. Balancing an equation doesn't change what happens in a reaction—it simply changes the way the reaction is represented.

Figure 3 There are equal numbers of each type of atom on both sides of the arrow.

Inquiry Virtual Lab

Figure 4 Mercury metal and oxygen form when mercury oxide is heated.

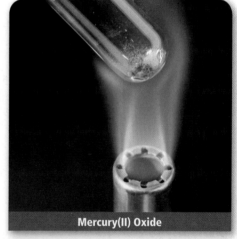
Mercury(II) Oxide

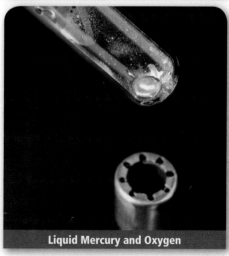
Liquid Mercury and Oxygen

Choosing coefficients The balancing process involves changing coefficients in a reaction in order to achieve a **balanced chemical equation,** which is a chemical equation with the same number of atoms of each element on both sides of the arrow. In the equation for Lavoisier's experiment, the number of mercury atoms is balanced, but one oxygen atom is on the left and two are on the right. Oxygen can be balanced by placing a 2 before the HgO on the left. Then, placing a 2 before Hg on the right balances mercury.

Numbers and Kinds of Atoms				
	2HgO(s) $\longrightarrow$	O$_2$(g)	+	2Hg(l)
Hg	2			2
O	2	2		

Try your balancing act Magnesium (Mg) burns with such a brilliant white light that it is often used in fireworks, as shown in **Figure 5.** Burning leaves a white powder called magnesium oxide (MgO). To write a balanced chemical equation, follow these steps.

Step 1 Write a chemical equation for the reaction of magnesium with oxygen. Recall that oxygen is a diatomic molecule.

$$Mg(s) + O_2(g) \longrightarrow MgO(s)$$

Step 2 Count the atoms in reactants and products. The magnesium atoms are balanced, but the oxygen atoms are not.

Numbers and Kinds of Atoms				
	Mg(s)	+ O$_2$(g)	$\longrightarrow$	MgO(s)
Mg	1			1
O		2		1

Step 3 Choose coefficients that balance the equation. Remember, never change subscripts of a correct formula to balance an equation. Try putting a coefficient of 2 before MgO.

$$Mg(s) + O_2(g) \longrightarrow 2MgO(s)$$

Step 4 Recheck the numbers of each atom on each side of the equation and adjust coefficients again if necessary. Now two Mg atoms are on the right side and only one is on the left side. So a coefficient of 2 is needed for Mg to balance the equation.

$$2Mg(s) + O_2(g) \longrightarrow 2MgO(s)$$

■ **Figure 5** When magnesium combines with oxygen, it produces a bright white flame, making it ideal for applications such as sparklers, fireworks, and flares.

Balance Equations A sample of barium sulfate ($BaSO_4$) is placed on a piece of paper, which is then ignited. Barium sulfate reacts with the carbon (C) from the burned paper producing barium sulfide (BaS) and carbon monoxide (CO). Write a balanced equation for this reaction.

List the Knowns: We know the substances that are involved in the reaction: barium sulfate ($BaSO_4$), carbon (C), barium sulfide (BaS), and carbon monoxide (CO).

Set Up the Problem: From this, we can write a chemical equation showing reactants and products:

$$BaSO_4(s) + C(s) \longrightarrow BaS(s) + CO(g)$$

Then, count and list the atoms found on both sides of the equation.

	Numbers and Kinds of Atoms			
	$BaSO_4(s)$ +	C(s) $\longrightarrow$	BaS(s) +	CO(g)
Ba	1		1	
S	1		1	
O	4			1
C		1		1

Solve the Problem: The oxygen atoms are unbalanced. Try putting a 4 in front of CO. Now you have 4 oxygen atoms on the right, which balances oxygen, but the carbon atoms become unbalanced. To fix this, add a 4 in front of the C in the reactants. The balanced equation looks like this:

$$BaSO_4(s) + 4C(s) \longrightarrow BaS(s) + 4CO(g)$$

Check the Answer: Count the number of atoms on each side of the equation and verify that they are equal.

PRACTICE Problems

Find **Additional Practice Problems** in the back of your book.

1. Balance this equation: $MgCl_2(aq) + AgNO_3(aq) \longrightarrow Mg(NO_3)_2(aq) + AgCl(s)$.

2. Balance this equation: $NaOH(aq) + CaBr_2(s) \longrightarrow Ca(OH)_2(s) + NaBr(aq)$.

3. HCl is slowly added to aqueous Na_2CO_3 forming NaCl, H_2O, and CO_2. Write a balanced equation for this reaction.

4. **Challenge** Write the balanced equation for the reaction of hydrogen gas and oxygen gas that forms water.

 Review

Additional Practice Problems

Figure 6 When charcoal burns, carbon (C) reacts with oxygen (O_2). The product of the reaction is carbon dioxide (CO_2).

Understanding Chemical Equations

Have you watched someone cook food outdoors on a charcoal grill? When charcoal burns, as shown in **Figure 6,** heat is liberated by the chemical reaction between carbon in the charcoal and oxygen in the air. Molecules of carbon dioxide gas are produced. Note in the balanced equation for this reaction that one oxygen molecule (O_2) is required for each carbon atom and that one carbon dioxide molecule (CO_2) is produced. Also, we know from the periodic table that the average mass of a carbon atom is 12.01 amu and the average mass of an oxygen atom is 16.00 amu. Therefore, the average mass of an O_2 molecule is 32.00 amu (2 × 16.00 amu) and the average mass of a CO_2 molecule is 44.01 amu (12.01 amu + (2 × 16.00 amu)).

C(s)	+	O_2(g)	→	CO_2(g)
1 atom		1 molecule		1 molecule
12.01 amu		32.00 amu		44.01 amu

In the laboratory, selecting a single carbon atom (mass 12.01 amu) and reacting it with a single oxygen molecule (mass 32.00 amu) is virtually impossible. Instead, chemists use masses in *grams,* rather than *amu.* For example, 12.01 *grams* of carbon reacts with 32.00 *grams* of oxygen, the same ratio of masses as in the balanced equation. That's because the number of carbon atoms in 12.01 grams of carbon must be very nearly equal to the number of oxygen molecules in 32.00 grams of oxygen. In fact, 12.01 grams of carbon contains 6.02×10^{23} carbon atoms and 32.00 grams of oxygen contains 6.02×10^{23} oxygen molecules.

Reading Check **Explain** why chemists use masses in grams instead of amu.

Moles Because the number of particles involved in most chemical reactions is so large, chemists use a counting unit called the mole (mol). One **mole** is the amount of a substance that contains 6.02×10^{23} particles of that substance. The reaction between one mole of carbon and one mole of oxygen, yielding one mole of carbon dioxide is summarized in **Table 2.**

Table 2	Moles, Mass, and Particles		
Equation	C(s) +	O_2(g) →	CO_2(g)
Number of moles	1	1	1
Mass	12.01 g	32.00 g	44.01 g
Number of particles	6.02×10^{23} atoms	6.02×10^{23} molecules	6.02×10^{23} molecules

Molar mass The mass in grams of one mole of a substance is called its **molar mass.** Just as the mass of a dozen eggs is different from the mass of a dozen watermelons, different substances have different molar masses. The atomic mass of titanium (Ti), for example, is 47.87 amu, and the molar mass is 47.87 g/mol. By comparison, the atomic mass of sodium (Na) is 22.99 amu, and its molar mass is 22.99 g/mol. For a compound such as nitrogen dioxide (NO_2), the molar mass is the sum of the masses of its component atoms. The nitrogen dioxide (NO_2) molecule contains one nitrogen atom (1×14.01 amu) and two oxygen atoms (2×16.00 amu = 32.00 amu). So NO_2 has a molar mass of 46.01 g/mol (14.01 g/mol + 32.00 g/mol = 46.01 g/mol).

Mole-mass conversions Given the mass of a substance, you can use the molar mass as a conversion factor to calculate the number of moles. The following example uses this method to calculate the number of moles in 50.00 g of NO_2.

Review
Additional Practice Problems

$$50.00 \text{ g } NO_2 \times \frac{1 \text{ mol } NO_2}{46.01 \text{ g } NO_2} = 1.087 \text{ mol } NO_2$$

Similarly, given the number of moles of a substance, you can use the molar mass as a conversion factor to calculate the mass. What is the mass of 0.2020 mol of NO_2?

$$0.2020 \text{ mol } NO_2 \times \frac{46.01 \text{ g } NO_2}{1 \text{ mol } NO_2} = 9.294 \text{ g } NO_2$$

Section 1 Review

Section Summary

▶ A chemical reaction is a process that involves one or more reactants changing into one or more products.

▶ A balanced chemical equation indicates relative amounts of reactants and products.

▶ A mole (mol) is the amount of a substance that contains 6.02×10^{23} particles of that substance.

5. MAIN Idea **Identify** the reactants and the products in the following chemical equation.
$$Cd(NO_3)_2(aq) + H_2S(g) \rightarrow CdS(s) + 2HNO_3(aq)$$

6. **Explain** the importance of the law of conservation of mass.

7. **Explain** why oxygen gas must be written as O_2 in a chemical equation.

8. **Think Critically** Explain why the sum of the coefficients on the reactant side of a balanced equation does not have to equal the sum of the coefficients on the product side of the equation.

Apply Math

9. **Use Numbers** Balance the following equation:
$$Fe(s) + Cl_2(g) \rightarrow FeCl_3(s).$$

10. **Calculate** How many moles are in 125 g of water?

11. **Calculate** What is the mass of 3.000 mol of calcium?

✓ Assessment   Online Quiz

Reading Preview

Essential Questions

▶ What are the five general types of chemical reactions?

▶ How can you predict if a metal will replace another in a compound?

▶ What do the terms *oxidation* and *reduction* mean?

▶ How are redox reactions identified?

Review Vocabulary

states of matter: the physical forms in which all matter exists, most commonly solid, liquid, and gas

New Vocabulary

combustion reaction
synthesis reaction
decomposition reaction
single-displacement reaction
double-displacement reaction
precipitate
oxidation
reduction

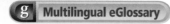

g **Multilingual eGlossary**

FOLDABLES
Incorporate information from this section into your Foldable.

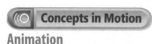
Concepts in Motion

Animation

■ **Figure 7** *Burning* is another word for a combustion reaction. The reactants in this reaction are carbon and hydrogen (in the wood) and oxygen (from the air). The products are carbon dioxide (CO_2) and water (H_2O).

Classifying Chemical Reactions

MAIN Idea Reactions can be classified based on how atoms are rearranged.

Real-World Reading Link You may have noticed that books in a library are arranged according to type or subject. Mystery novels are in one section; home-improvement books are in another. You can tell a lot about a book based on what section it comes from.

Types of Reactions

You might have noticed that there are all sorts of chemical reactions. In fact, there are literally millions of chemical reactions that occur every day, and scientists have described many of them and continue to describe more. In order to organize these reactions into manageable areas of study, chemists have defined five main categories of chemical reactions: combustion, synthesis, decomposition, single displacement, and double displacement.

Combustion reactions If you have ever observed something burning, you have observed a combustion reaction, such as the one shown in **Figure 7.** As mentioned previously, Lavoisier was one of the first scientists to accurately describe combustion. He deduced that the process of burning (combustion) involves the combination of a substance with oxygen. Our definition states that a **combustion reaction** occurs when a substance reacts with oxygen to produce energy in the form of heat and light. The combustion reaction shown in **Figure 7** creates flames of heat and light as carbon in the wood reacts with oxygen in the air to form carbon dioxide (CO_2). Many combustion reactions also will fit into other categories of reactions. For example, the reaction between carbon and oxygen also is a synthesis reaction.

Synthesis reactions One of the easiest reaction types to recognize is a synthesis reaction. In a **synthesis reaction,** two or more substances combine to form another substance. The generalized formula for this reaction type is $A + B \longrightarrow AB$. The reaction in which hydrogen gas (H_2) combines with chlorine gas (Cl_2) to form hydrogen chloride (HCl) is an example of a synthesis reaction.

$$H_2(g) + Cl_2(g) \xrightarrow{\text{light}} 2HCl(g)$$

This synthesis reaction requires the addition of light. It can be explosive in direct sunlight. Another synthesis reaction with which you may be familiar is the combination of oxygen (O_2) with iron (Fe) in the presence of water to form hydrated iron(III) oxide (Fe_2O_3), which is known as rust.

Decomposition reactions A decomposition reaction is just the reverse of a synthesis. Instead of two substances coming together to form a third, a **decomposition reaction** occurs when one substance breaks down, or decomposes, into two or more substances. The general formula for this type of reaction can be expressed as $AB \longrightarrow A + B$.

Most decomposition reactions require the input of heat, light, or electricity. For example, hydrogen peroxide (H_2O_2), shown in **Figure 8,** will slowly decompose in the presence of light, producing oxygen gas (O_2) and water (H_2O).

$$2H_2O_2(l) \xrightarrow{\text{light}} O_2(g) + 2H_2O(l)$$

Single displacement The chemical reaction in which one element replaces another element in a compound is called a **single-displacement reaction.** Single-displacement reactions—sometimes called *single-replacement reactions*—are described by the general equation $A + BC \longrightarrow AC + B$. Here you can see that atom A displaces atom B to produce a new molecule, AC. A single displacement reaction is illustrated in **Figure 9,** where a copper wire is put into a solution of silver nitrate. Because copper is a more active metal than silver, it replaces the silver, forming a blue copper(II) nitrate solution. The silver, which is not soluble, forms crystals on the wire.

$$Cu(s) + 2AgNO_3(aq) \longrightarrow Cu(NO_3)_2(aq) + 2Ag(s)$$

✓ **Reading Check Summarize** Describe what happens in a single-displacement reaction.

■ **Figure 8** In the presence of light, hydrogen peroxide (H_2O_2) will decompose, producing oxygen gas (O_2) and water (H_2O).

Infer *Why is the hydrogen peroxide bottle brown?*

■ **Figure 9** Copper from the wire replaces silver in silver nitrate, forming a blue-tinted solution of copper(II) nitrate.

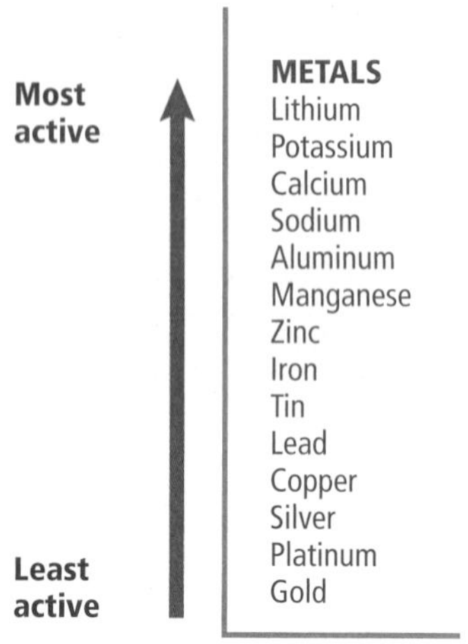

METALS
Lithium
Potassium
Calcium
Sodium
Aluminum
Manganese
Zinc
Iron
Tin
Lead
Copper
Silver
Platinum
Gold

■ **Figure 10** An activity series is a useful tool for determining whether a chemical reaction will occur and for determining the result of a single-replacement reaction.

Sometimes single-displacement reactions can cause problems. For example, if iron-containing vegetables such as spinach are cooked in aluminum pans, aluminum can displace iron from the vegetable. This causes a black deposit of iron to form on the sides of the pan. For this reason, it is better to use stainless steel or enamel cookware when cooking spinach.

We can predict which metal will replace another using the diagram shown in **Figure 10,** which lists metals according to how reactive they are. A metal will replace any less active metal. Notice that silver and gold are two of the least active metals on the list. That is why these elements often occur as deposits of the pure element. More reactive metals often occur as compounds.

Double displacement The positive ion of one compound replaces the positive ion of the other to form two new compounds in a **double-displacement reaction**—sometimes called a *double-replacement reaction.* You know that a double-displacement reaction is taking place if a precipitate, water, or a gas forms when two ionic compounds in solution are combined. A **precipitate** is an insoluble compound that comes out of solution during this type of reaction. The generalized formula for this reaction is $AB + CD \rightarrow AD + CB$.

✓ **Reading Check Classify** Which reaction produces a precipitate?

The reaction of copper(II) chloride with sodium hydroxide is an example of this type of reaction. A precipitate—copper(II) hydroxide—forms, as shown in **Figure 11.**

$$2NaOH(aq) + CuCl_2(aq) \rightarrow Cu(OH)_2(s) + 2NaCl(aq)$$

■ **Figure 11** When solutions of copper(II) chloride ($CuCl_2$) and sodium hydroxide (NaOH) are mixed, a solid—copper(II) hydroxide ($Cu(OH)_2$)—is formed. Formation of a precipitate is a sign of a double-displacement reaction.

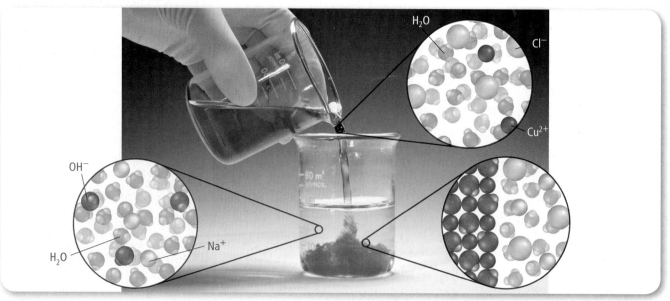

Oxidation-Reduction Reactions

One characteristic that is common to many chemical reactions is the tendency of the substances involved to lose or gain electrons. Chemists use the term **oxidation** to describe the loss of electrons and **reduction** to describe the gain of electrons. Chemical reactions involving electron transfer of this sort often involve oxygen, which is very reactive, pulling electrons from metallic elements. Corrosion of metal is a visible result, as shown in **Figure 12.**

The cause and effect of oxidation and reduction can be taken one step further by describing the substances after the electron transfer. The substance that gains an electron or electrons becomes more negative, and we say it is reduced. On the other hand, the substance that loses an electron or electrons then becomes more positive, and we say it is oxidized. Reduction is the partner to oxidation; the two always work as a pair, which is commonly referred to as redox.

■ **Figure 12** Rust is a common form of corrosion. Oxygen (O_2) from the air pulls electrons from iron (Fe) to form iron(III) oxide (Fe_2O_3).

Infer *Which element is oxidized and which is reduced?*

Concepts in Motion
Animation

Section 2 Review

Section Summary

▶ Chemical reactions are organized into five basic classes: combustion, synthesis, decomposition, single displacement, and double displacement.

▶ An activity series predicts which metal will replace another in a single-displacement reaction.

▶ Some reactions produce a solid called a precipitate when two ionic substances are combined.

▶ Oxidation is the loss of electrons and reduction is the corresponding gain of electrons.

12. **MAIN Idea Classify** each of the following reactions:
   a. $CaO(s) + H_2O(l) \longrightarrow Ca(OH)_2(aq)$
   b. $Fe(s) + CuSO_4(aq) \longrightarrow FeSO_4(aq) + Cu(s)$
   c. $C_{10}H_8(l) + 12O_2(g) \longrightarrow 10CO_2(g) + 4H_2O(g)$
   d. $NaCl(aq) + AgNO_3(aq) \longrightarrow NaNO_3(aq) + AgCl(s)$
   e. $NH_4NO_3(s) \longrightarrow N_2O(g) + 2H_2O(g)$

13. **Describe** what happens in a combustion reaction.

14. **Compare and contrast** synthesis reactions and decomposition reactions.

15. **Determine,** using **Figure 10,** if zinc will displace gold in a chemical reaction and explain why or why not.

16. **Think Critically** Describe one possible economic impact of redox reactions. How might that impact be lessened?

 pply Math

17. **Use Proportions** The following chemical reaction is balanced, but the coefficients used are larger than necessary. Rewrite this balanced equation using the smallest coefficients possible.
   $$9Fe(s) + 12H_2O(l) \longrightarrow 3Fe_3O_4(s) + 12H_2(g)$$

18. **Use Coefficients** Sulfur trioxide (SO_3), a pollutant released by coal-burning plants, can react with water (H_2O) in the atmosphere to produce sulfuric acid (H_2SO_4). Write the balanced equation for this reaction.

Essential Questions

▶ How can the source of energy changes in chemical reactions be identified?

▶ How do exergonic and endergonic reactions compare?

▶ How do exothermic and endothermic reactions compare?

▶ Is energy conserved during a chemical reaction?

Review Vocabulary

chemical bond: the force that holds two atoms together

New Vocabulary

exergonic reaction
exothermic reaction
endergonic reaction
endothermic reaction

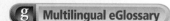 **Multilingual eGlossary**

Chemical Reactions and Energy

MAIN ◀Idea Exergonic reactions release energy and endergonic reactions absorb energy.

Real-World Reading Link How does a burning match compare to an egg being cooked on the stove? Both are chemical reactions, but they are very different: One gives off heat and the other requires heat to continue.

Chemical Reactions—Energy Exchanges

Crowds often gather to watch a rocket launch. Hundreds of kilograms of solid and liquid rocket fuel are converted into a gas, pushing the enormous rocket skyward. The combustion of rocket fuel is an example of a rapid chemical reaction.

Most chemical reactions proceed more slowly, but all chemical reactions release or absorb energy. This energy can take many forms, such as thermal energy, light, sound, or electricity. The thermal energy produced by a wood fire and the light emitted by a glow stick are examples of energy released by chemical reactions.

Chemical bonds are the source of this energy. When most chemical reactions take place, some chemical bonds in the reactants are broken, which requires energy called activation energy. In order for products to be produced, new bonds must form. Bond formation releases energy. Reactions such as dynamite combustion, shown in **Figure 13,** require much less energy to break chemical bonds than the energy released when new bonds are formed. The result is a release of energy and an explosion.

■ **Figure 13** When new bonds are formed, energy is released. In a reaction like exploding dynamite, the energy that is released from the products' formation is much greater than the energy required to break bonds in the reactants.

More Energy Out

Many of the reactions with which you are most familiar involve the release of energy. Chemical reactions that release energy are called **exergonic reactions.** In these reactions, the activation energy required to break the original bonds is less than the energy that is released when new bonds form. As a result, some form of energy, such as light or thermal energy, is given off by the reaction. The abdomen of a firefly, as shown in **Figure 14,** glows as a result of an exergonic reaction that produces visible light.

Thermal energy released In many reactions, the energy given off is thermal energy. This is the case with some heat packs that are used to treat muscle aches and other problems. When the energy given off is primarily in the form of thermal energy, the reaction is called an **exothermic reaction.** Wood burning and the explosion of dynamite are exothermic reactions. Iron rusting is also exothermic, but, under typical conditions, the reaction proceeds so slowly that it's difficult to detect any temperature change.

■ **Figure 14** The chemical reactions happening inside the abdomen of a firefly produce light.

Infer *How do you know these are exergonic reactions?*

Reading Check Infer Why is a log fire considered to be an exothermic reaction?

Exothermic reactions provide most of the power used in homes and industries, as shown in **Figure 15.** Fossil fuels contain carbon. These fuels, such as coal, petroleum, and natural gas, combine with oxygen to yield carbon dioxide gas and energy. Unfortunately impurities in these fuels, such as sulfur, burn as well, producing pollutants such as sulfur dioxide. Sulfur dioxide combines with water in the atmosphere, producing acid rain.

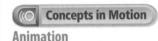

Concepts in Motion

Animation

■ **Figure 15** Most of the electricity in the United States is produced by burning coal. The energy released from burning coal heats water, which turns to steam and turns an electricity-generating turbine. The graph below shows the release of energy as the reaction progresses.

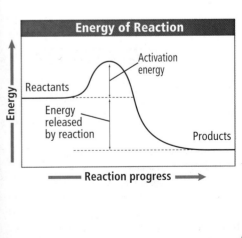

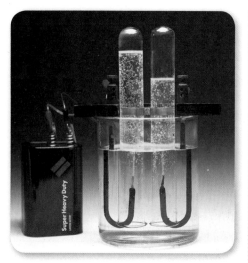

Figure 16 Water is stable, but the addition of an electrical current will cause it to decompose into hydrogen and oxygen.

More Energy In

Sometimes a chemical reaction requires more energy to break bonds than is released when new ones are formed. These reactions are called **endergonic reactions.** The energy absorbed can be in the form of light, thermal energy, or electricity.

Electricity is often used to supply energy to endergonic reactions. For example, an electric current passed through water produces hydrogen and oxygen, shown in **Figure 16.** Also, aluminum metal is obtained from its ore using the following endergonic reaction:

$$2Al_2O_3(l) \xrightarrow{\text{elec.}} 4Al(l) + 3O_2(g)$$

In these cases, electricity provides the energy necessary for the reactions.

Thermal energy absorbed When the energy needed to keep a reaction going is in the form of thermal energy, the reaction is called an **endothermic reaction.** The terms exothermic and endothermic are not just related to chemical reactions. They can also describe physical changes. If you ever had to soak a swollen ankle in an Epsom salt solution, you probably noticed that when you mixed the Epsom salt in water, the solution became cold. The dissolving of Epsom salt absorbs thermal energy. Thus, it is a physical change that is endothermic.

Cooking requires the addition of thermal energy to bring about chemical changes in the food. When baking cookies, you might add baking soda ($NaHCO_3$) to the dough mixture. Through an endothermic reaction, the baking soda breaks down into sodium carbonate (Na_2CO_3), carbon dioxide gas (CO_2), and water vapor (H_2O). As these gases are released, tiny pockets form in the dough and the cookies puff up, as shown in **Figure 17.**

Figure 17 Baking involves endothermic reactions such as the decomposition of baking soda. The graph below shows how energy is absorbed during these chemical reactions.

Compare *How did the cookies change when they were baked?*

Energy of Reaction

Activation energy

Products

Energy absorbed by reaction

Reactants

Energy

Reaction progress

Conservation of Energy in Chemical Reactions

You learned in an earlier chapter that energy can change from one form to another, but the total amount of energy never changes. This principle is usually stated as the law of conservation of energy. Does the total amount of energy remain constant in chemical reactions, too? Consider the exergonic burning of methane (CH_4), the major component of natural gas, as described by the following equation. Note that energy is included as a product.

$$CH_4(g) + 2O_2(g) \rightarrow CO_2(g) + 2H_2O(g) + energy$$

During this process, some of the chemical energy of the reactants is released as thermal energy and light. However, the sum of the energy released and the chemical energy of the products is exactly equal to the chemical energy of the reactants in an exergonic chemical reaction.

chemical energy = chemical energy + energy released
  of reactants   of products

So the total amount of energy before and after the reaction remains the same. Similarly, the total amount of energy remains the same in endergonic chemical reactions. Summarizing, the law of conservation of energy applies to chemical reactions, as well as to other types of energy transformations.

VOCABULARY ....
SCIENCE USAGE V. COMMON USAGE
Conserve
Science usage
  to hold a physical quantity, such as energy or mass, constant during a physical or chemical change
  Matter is conserved in chemical reactions.

Common usage
  to manage resources wisely; to save resources
  It is important to conserve wetlands, a valuable natural resource. ....

Section 3 Review

Section Summary

▶ Breaking chemical bonds absorbs energy.

▶ Forming chemical bonds releases energy.

▶ Exergonic chemical reactions release energy; endergonic chemical reactions absorb energy.

▶ Exothermic reactions give off thermal energy. Endothermic reactions absorb thermal energy.

19. **MAIN Idea** Classify the chemical reaction photosynthesis as endergonic or exergonic. Explain.

20. **Explain** why the total amount of energy does not decrease in an exergonic chemical reaction.

21. **Explain** How can an endergonic reaction not be endothermic?

22. **Classify** the reaction that makes a firefly glow in terms of energy input and output.

23. **Think Critically** To develop a product that warms people's hands, would you use an exothermic or endothermic reaction? Why?

Apply Math

24. **Calculate** If an endothermic reaction begins at 26°C and decreases by 2°C per minute, how long will it take to reach 0°C?

25. **Use Graphs** Create a graph of the data in question 24. After 5 minutes, what is the temperature of the reaction?

Essential Questions

▶ How do chemists express the rates of chemical reactions?

▶ How do catalysts and inhibitors affect reaction rates?

▶ What is equilibrium?

▶ How does Le Châtelier's principle explain shifts in equilibria?

Review Vocabulary

pressure: the force acting on a unit area of surface

New Vocabulary

reaction rate
collision model
catalyst
inhibitor
reversible reaction
equilibrium
Le Châtelier's principle

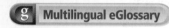
g Multilingual eGlossary

■ **Figure 18** The main engine for the space shuttle combines hydrogen and oxygen to produce water. The shuttle is able to overcome Earth's gravity because this reaction occurs very rapidly.

Reaction Rates and Equilibrium

MAIN ‹Idea Every chemical reaction proceeds at a definite rate, which can be speeded up or slowed down by changing the conditions of the reaction.

Real-World Reading Link Which is the fastest: an ice cream truck, a school bus, or a race car? They're all vehicles, but some move faster than others. Chemical reactions are the same way.

Reaction Rates

Some chemical reactions, such as the combustion of rocket fuel, take place rapidly and release tremendous amounts of energy in a matter of seconds. Other reactions, such as the rusting of steel, proceed so slowly that you hardly notice any change from one week to the next. How can you express the rate of a chemical reaction? The **reaction rate** is the rate at which reactants change into products.

Consider the synthesis reaction, shown in **Figure 18,** described by the following chemical equation.

$$2H_2(l) + O_2(l) \rightarrow 2H_2O(g) + energy$$

A chemist might choose one of several ways to state the rate of this reaction. Three examples are the rate at which one of the two reactants is used up, the rate at which water is produced, and the rate at which energy is released. Because many factors influence the rates of chemical reactions, the chemist would also state the conditions under which the reaction occurred.

Factors Affecting Reaction Rates

You already know that sugar dissolves faster in hot water than it does in cold water. Sugar dissolving in water is not a chemical reaction. However, the rates of most chemical reactions, too, vary with temperature. Chemists use a commonsense idea to explain why reaction rates depend upon temperature and other factors, such as concentration and surface area. This idea is called the collision model. The **collision model** states that atoms, ions, and molecules must collide in order to react. Understanding the collision model will help to explain why changing the conditions of a chemical reaction can have an effect on the reaction rate.

Temperature You normally store perishable foods such as milk, eggs, and vegetables in a refrigerator. That's because lowering the temperature decreases the rates of the chemical reactions that cause spoilage. Conversely, increasing the temperature of chemical reactions generally increases their reaction rates.

Why does temperature affect reaction rate? Recall that the temperature of a substance is a measure of the average kinetic energy of all of its particles. At higher temperatures, therefore, reacting particles move faster and collide more frequently. A higher collision frequency alone, however, does not completely explain the increase in reaction rate. Because the particles are moving faster at higher temperatures, they collide with greater energy. As a result, a greater percentage of collisions result in a reaction between colliding particles.

Reading Check **Explain** the effect of increased temperature in terms of the collision model.

Concentration Another way you can change the rate of a chemical reaction is by changing the concentration of one or more of the reactants. Concentration describes the number of particles of a substance per unit volume. Chemists usually express concentration as moles of a substance per liter (mol/L).

Consider the two test tubes shown in **Figure 19.** Each tube has a magnesium (Mg) ribbon immersed in a solution of hydrochloric acid (HCl). The difference between the two tubes is the concentration of the acid solution. As the magnesium and hydrochloric acid react, hydrogen gas (H_2) is released as a product, so you can compare the rates of the two reactions by how rapidly bubbles are formed. Why does magnesium ribbon react faster with the more concentrated hydrochloric acid? The more concentrated acid solution contains more reacting particles per unit volume, resulting in more opportunities for collisions between reacting particles. As a result, the reaction rate is greater.

■ **Figure 19** The rate of the reaction between magnesium and hydrochloric acid increases when the concentration of the acid is increased.

Mg in Dilute HCl

Mg in Concentrated HCl

Volume and pressure For chemical reactions involving gases, volume and pressure are important considerations because they relate to the concentrations of the reacting gases. For example, decreasing the volume of a flask containing gases while the temperature remains constant increases the concentrations of the gases. Just as with liquid solutions, increasing the concentrations of gases increases the rate at which the particles collide with each other and with the walls of the container. The pressure inside the flask increases. More importantly, the reaction rate increases as well because the reacting gas particles collide with each other more frequently. The effect of increased pressure and decreased volume on gas particles is demonstrated in **Figure 20.**

✅ **Reading Check Compare and contrast** the effects of increased concentration of liquid reactants and decreased volume of gaseous reactants.

Surface area Which dissolves more quickly: granulated sugar or a sugar cube? As you probably guessed, the answer is granulated sugar because the individual grains of sugar have much greater total surface area compared to the sugar cube. Dissolving sugar is a physical change, but increased surface area also increases the rate of chemical reactions.

Operators of grain elevators must take measures to ensure that grain dust and oxygen in the air do not combine in a combustion reaction. Even on a scorching-hot day, there is little danger that whole grains of wheat or kernels of corn will react rapidly with oxygen in the air. However, the fine particles that make up grain dust can react explosively on a hot day, as shown in **Figure 21.** The larger total surface area of the grain dust greatly increases the rate at which reacting particles collide. With more collisions per unit time, the rate of the combustion reaction increases dramatically.

■ **Figure 21** Grain dust can be explosive because of its increased surface area.

Catalysts and inhibitors Some reactions proceed too slowly to be useful. To speed up such a reaction, a catalyst can be added. A **catalyst** is a substance that speeds up a chemical reaction without being permanently changed itself. When you add a catalyst to a reaction, the mass of the product that is formed remains the same, but it will form more rapidly. The catalyst remains unchanged and often is recovered and reused. Catalysts are used to speed many reactions in industry, such as the process of polymerization to make plastics and fibers. In order to break down food, your body utilizes special catalysts, called enzymes.

At times, it is worthwhile to prevent certain reactions from occurring. For example, foods often spoil because they react with oxygen from the air. Substances called **inhibitors** are used to slow down the rates of chemical reactions or prevent a reaction from happening at all. Food preservatives are inhibitors that prevent the reactions that lead to the spoilage of certain foods.

One thing to remember when thinking about catalysts and inhibitors is that they do not change the amount of product produced. They only change the rate of production.

✔️ **Reading Check Compare and contrast** catalysts and inhibitors in how they affect reaction rates.

Equilibrium

Think of a one-way street, a street on which vehicles may travel in only one direction. Now examine the general formula for a chemical reaction (reactants → products). Do you notice a similarity?

The single reaction arrow may lead you to suppose that every chemical reaction proceeds in only one direction, from reactants to products. Under certain conditions, some reactions do exactly that. And when such a reaction continues until at least one reactant is completely consumed, the reaction is said to "go to completion." The decomposition of potassium chlorate ($KClO_3$) into potassium chloride (KCl) and oxygen (O_2) is just such a reaction.

$$2KClO_3(s) \rightarrow 2KCl(s) + 3O_2(g)$$

Unlike reactions that go to completion, many reactions under certain conditions can occur in both directions. These reactions are said to be "reversible." A **reversible reaction** is one that can occur in both the forward and reverse directions. Think of a reversible reaction as a two-way street, shown in **Figure 22.** Vehicles may proceed in both directions at the same time.

■ **Figure 22** Some reactions move in only one direction—from reactants to products—like a one-way street. Many reactions are more like a two-way street, capable of moving in both directions.

VOCABULARY ....
WORD ORIGIN
Equilibrium
comes from the Latin word *aequus*, meaning *equal*
When reactants and products are created at equal rates, the reaction is said to be in a state of equilibrium. ....

When a reversible reaction's forward and reverse reactions take place at exactly the same rate, a state of balance, or equilibrium, exists. **Equilibrium** (plural *equilibria*) is a state in which forward and reverse reactions or processes proceed at equal rates. An equilibrium state is indicated with double reaction arrows, as shown below. Chemists generally call the left-to-right reaction the forward reaction and the right-to-left reaction the reverse reaction.

$$reactants \rightleftharpoons products$$

✓ **Reading Check** **Compare** the forward and reverse reactions.

Types of equilibria Some equilibria involve physical changes, rather than chemical reactions. When opposing physical changes take place at equal rates, a state of physical equilibrium exists. In a sealed bottle of soda, for example, CO_2 molecules are continually escaping from the solution. At the same time—and at an identical rate—CO_2 molecules are reentering the solution. This state of physical equilibrium is shown in **Figure 23.**

$$CO_2(aq) \rightleftharpoons CO_2(g)$$

Likewise, when opposing chemical reactions take place at equal rates, a state of chemical equilibrium exists. An example is the chemical equilibrium established in the Haber process, used to manufacture ammonia (NH_3) by reacting nitrogen (N_2) with hydrogen (H_2). The Haber-process equilibrium reactions are described by the following equation.

$$N_2(g) + 3H_2(g) \rightleftharpoons 2NH_3(g) + energy$$

When this reaction is at equilibrium, ammonia is constantly being formed. At the same time and at the same rate, nitrogen and hydrogen molecules are being reformed.

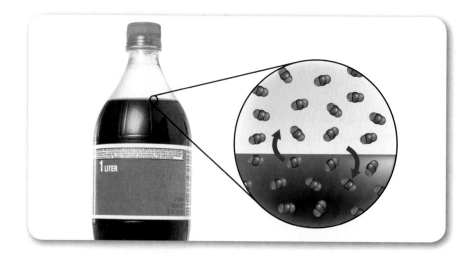

■ **Figure 23** In a sealed bottle of soda, dissolved CO_2 molecules are continually coming out of solution and reentering solution.

Infer *What happens to the equilibrium when the bottle is opened?*

Factors affecting equilibria When a state of equilibrium exists, the forward and reverse reactions are taking place at equal rates. The net amounts of both reactants and products remain constant. In the previous example of the sealed bottle of soda, what would happen if the bottle were opened? As you can assume, the system would no longer be at equilibrium and, for a time, physical changes would proceed toward the right in the following equation:

$$CO_2(aq) \rightleftharpoons CO_2(g)$$

Chemical reactions at equilibrium can likewise change. An equilibrium system may be subjected to stresses that speed up or slow down one of the opposing reactions. Instead of remaining constant, the net amounts of reactants and products favor one of the directions of the reaction. The equilibrium becomes temporarily unbalanced. But in time, the forward and reverse reactions again reach a state of balance. A new equilibrium state is established, now with changed amounts of reactants and products.

When a stress is imposed on an equilibrium system, the equilibrium responds to the stress according to a general rule known as Le Châtelier's (luh SHAHT uhl yays) principle. **Le Châtelier's principle** states that if a stress is applied to a system at equilibrium, the equilibrium shifts in the direction that opposes the stress. A stress is any kind of change that disturbs the equilibrium. Common stresses include the following: changing concentration by adding or removing a reactant or product; changing temperature by adding or removing heat, as shown in **Figure 24**; and changing volume and pressure. When a forward or reverse reaction rate increases or decreases in response to a stress, the equilibrium is said to "shift."

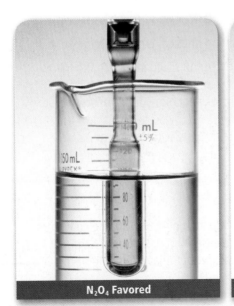

N₂O₄ Favored

2NO₂ Favored

MiniLab

(?) Inquiry MiniLab

Model Equilibrium

Procedure 🥽 👕 🧤 🗑️

1. Read the procedure and safety information, and complete the lab form.
2. Measure 20 mL of water in a **graduated cylinder** and pour it into a **100-mL beaker.** Fill the graduated cylinder to the 20-mL mark with water. Add two drops of **food coloring** to the water in each container.
3. Obtain **two glass tubes of equal diameter.** Place one tube in the graduated cylinder and the other in the beaker.
4. With the ends of the tubes at the bottoms of their containers, cover the open ends of the glass tubes with your index fingers so that water becomes trapped in the tubes. Simultaneously, move each tube to the other container and release your fingers to release the water.
5. Repeat the transfer process about 25 times. Record your observations.

Analysis

1. **Describe** your observations during the transfers.
2. **Explain** Would the final result be different if you had continued the transfer process longer?
3. **Infer** How does this process model equilibrium?

■ **Figure 24** In the reaction between dinitrogen tetroxide and nitrogen dioxide ($N_2O_4 \rightleftharpoons 2NO_2$), the equilibrium shifts toward the reddish-brown NO_2 when placed in boiling water (right) and toward the colorless N_2O_4 when placed in the cold water bath (left).

How could a chemical engineer apply Le Châtelier's principle to maximize the production of ammonia (NH_3) in the following equilibrium system?

$$N_2(g) + 3H_2(g) \rightleftharpoons 2NH_3(g) + energy$$

Changing concentration Suppose that the manufacturing process is engineered to remove ammonia (NH_3) as it is formed. The concentration of ammonia decreases, which causes the rate of the reverse reaction to decrease. As a result, the forward reaction is temporarily faster than the reverse reaction, described as a shift to the right. More ammonia is formed.

Changing temperature Suppose that the engineer lowers the temperature by removing energy. The equilibrium responds by reacting to release energy and raise the temperature. A shift to the right occurs and more ammonia is formed.

Changing volume and pressure Because the reaction vessel contains gases, decreasing the volume increases the pressure. If possible, the equilibrium responds to reduce the pressure. The pressure can be reduced by decreasing the number of gas molecules. Because the product (NH_3) side of the equation has fewer gas molecules (2) than the reactant side (4), the equilibrium shifts to the right. More ammonia is formed.

Section 4 Review

Section Summary

▶ The rates of chemical reactions can be manipulated by changing conditions under which the reaction takes place.

▶ A state of equilibrium exists when forward and reverse reactions or processes take place at equal rates.

▶ Le Châtelier's principle describes how an equilibrium responds to a stress.

26. **MAIN ‹Idea** List four ways to change the rate of a chemical reaction.

27. **Describe** two ways in which you might state the rate of a chemical reaction.

28. **Explain** what must happen in order for two molecules to react.

29. **Compare and contrast** chemical and physical equilibrium.

30. **Think Critically** Describe two ways you could influence the following equilibrium to produce more ethanal (CH_3CHO). Use Le Châtelier's principle to explain why each of your methods would produce the desired result.

$$C_2H_2(g) + H_2O(g) \rightleftharpoons CH_3CHO(g) + energy$$

Apply Math

31. **Calculate** For the reaction described in question 30, the concentration of CH_3CHO is found to increase from 0.0300 mol/L to 0.0500 mol/L in 42.5 seconds. Express the average rate of the reaction in mol CH_3CHO produced/L·s.

Assessment Online Quiz

LAB

To Glow or Not to Glow

Objectives

- **Observe** the effect of temperature on a light stick.

- **Explain** how temperature affects the rate of a chemical reaction.

Background: Many chemical reactions release energy as light. The light seen from a light stick is the result of a chemical reaction.

Question: *How does changing the water temperature affect the amount of light produced from a light stick?*

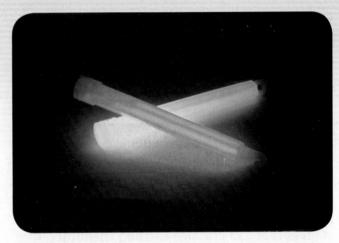

Preparation

Materials

light sticks (3)
400-mL beakers (4)
water
ice
hot plate
thermometer
graduated cylinder

Safety Precautions

Procedure

1. Read the procedure and safety information, and complete the lab form.

2. Write a hypothesis about how the light stick will be affected by temperature.

3. Prepare a hot water bath by heating 200 mL of water in a beaker on a hot plate. Record the temperature of the hot water.

4. Prepare a 200-mL ice water bath and a 200-mL bath of room-temperature water. Record the temperature of the water in each bath.

5. Bend a light stick until the inner capsule snaps. Shake the light stick for 10 seconds. Place the light stick in the hot water bath.

6. Repeat step 5 with the remaining light sticks using the ice water bath and the room-temperature bath. Record your observations.

Conclude and Apply

1. **Summarize** your observations.

2. **Explain** why the amount of light released is different above and below the water level.

3. **Evaluate** your hypothesis.

4. **Infer** how temperature affects the light intensity. How does the intensity of the light relate to the rate of the chemical reaction?

5. **Explain** the importance of using room-temperature water in your investigation. What purpose does the room-temperature water serve?

Compare your results with those of your classmates. Write a report that explains any differences among the results.

Objectives

■ **Evaluate** the effect of concentration on the rate of a chemical reaction.

■ **Examine** the effect of temperature on the rate of a chemical reaction.

Background: Many people believe that you cannot perform chemical reactions without expensive equipment or costly chemicals. But this isn't true; chemical reactions happen everywhere. All you need is a food store to find many substances that can produce exciting chemical reactions.

Question: *What factors determine how much product is produced in a chemical reaction or how fast a reaction occurs?*

Preparation

Materials

water	vinegar solution
baking soda	balloons (3)
plastic 0.5-liter	test tube
soft-drink bottles (3)	150-mL beakers (2)
marker	500-mL beaker
stopwatch, or clock	100-mL graduated
with second hand	cylinder
tape	

Safety Precautions

Procedure

1. Read the procedure and safety information, and complete the lab form.

2. Make data tables similar to those shown on the next page.

3. Prepare a 50 percent vinegar solution by mixing 30 mL of vinegar with 30 mL of water. This is solution *A*.

4. Prepare a 30 percent vinegar solution by mixing 30 mL of vinegar with 70 mL of water. This is solution *B*.

5. Prepare a 10 percent vinegar solution by mixing 30 mL of vinegar with 270 mL of water. This is solution *C*.

6. Pour the vinegar solutions into their associated 0.5-L plastic bottles labeled *A*, *B*, and *C*.

7. Mark a small test tube about 1–2 cm from its bottom. Fill the test tube to the line with baking soda. Pour the baking soda into one balloon.

8. Repeat step 7 with two more balloons. Be sure the amount of baking soda in each balloon is the same.

9. Place the mouth of one balloon over the mouth of one 0.5-L bottle. Do not let any of baking soda fall into the vinegar solution.

10. Repeat step 9 with the other two balloons and the remaining bottles.

? Inquiry Lab

Concentration Data Table

Vinegar Concentration	Solution A (50%)	Solution B (30%)	Solution C (10%)
Observations			

Temperature Data Table

Solution Temperature	Cold	Room Temperature	Hot
Observations			

11. Lift each balloon to allow the baking soda to fall into each vinegar solution. Time how long it takes for the reaction to finish. Measure how much each balloon inflates. Record your observations in your Concentration data table.

12. Carefully remove the balloons from the bottles.

13. Rinse the plastic bottles with water.

14. Prepare a 30 percent vinegar solution by mixing 30 mL of vinegar with 70 mL of cold water.

15. Prepare two more similar solutions with room-temperature water and hot water.

16. Place the three solutions in the three 0.5-L plastic bottles.

17. Repeat steps 7 and 8 to refill the balloons with baking soda.

18. Place the balloons back on the bottles, repeating steps 9 and 10.

19. Repeat step 11. Time how long it takes for the reaction to finish.

20. Measure how much each balloon inflates. Record your observations in your Temperature data table.

Analyze Your Data

1. **Describe** how increasing the concentration of a solution affects the rate of a chemical reaction.

2. **Summarize** how temperature affects the rate of a chemical reaction.

3. **Explain** why the balloons become inflated.

Conclude and Apply

4. **Infer** why the vinegar solutions in steps 3, 4, and 5 were different volumes. Why couldn't the volumes be the same?

5. **Predict** what factors might affect the amount of product that is produced. What factors affect the rate at which products are produced?

COMMUNICATE
YOUR DATA

Draw a diagram that visually represents your observations. Compare your diagram to those of your classmates.

CHAPTER 20

Radioactivity and Nuclear Reactions

ConnectED

Your one-stop online resource
connectED.mcgraw-hill.com

- Video
- Audio
- Review
- Inquiry
- WebQuest
- Assessment
- Concepts in Motion
- Multilingual eGlossary

Launch Lab
The Size of a Nucleus

Do you realize you are made up mostly of empty space? Your body is made of atoms, and atoms are made of electrons speeding around a small nucleus of protons and neutrons. The size of an atom is the size of the space in which the electrons move around the nucleus. In this lab, you will find out how the size of an atom compares with the size of a nucleus.

For a lab worksheet, use your StudentWorks™ Plus Online.

 Inquiry Launch Lab

FOLDABLES

Create a half-book. Label it as shown. Use it to organize your notes on radioactivity and nuclear reactions.

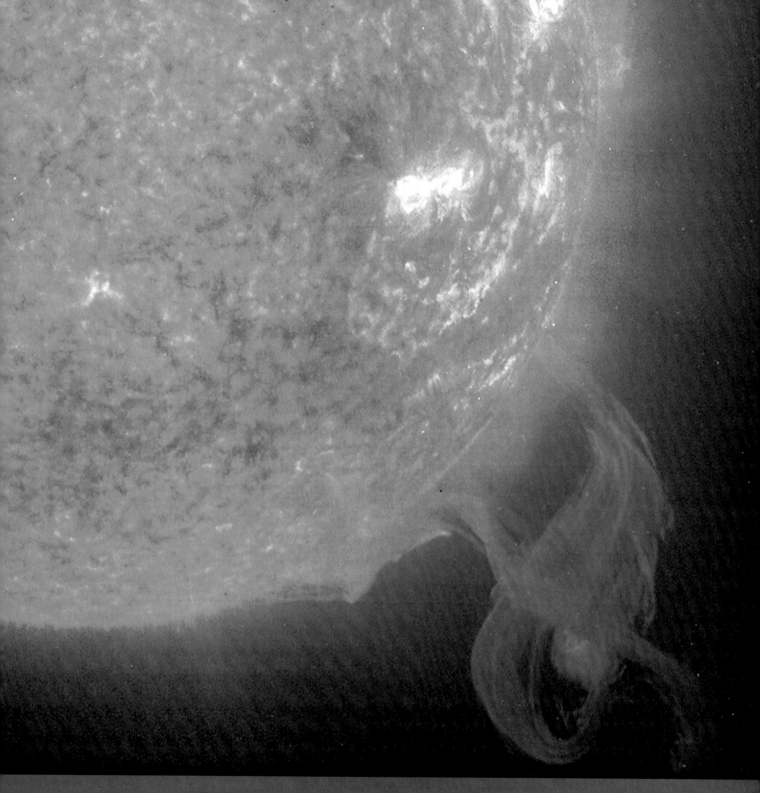

THEME FOCUS Changes of Matter
One element can be changed into another
element through radioactivity and nuclear
reactions.

BIG ((**Idea** (Forces inside the nucleus can cause changes
that release particles and energy.

Section 1 • The Nucleus

Section 2 • Nuclear Decays and
Reactions

Section 3 • Radiation
Technologies and
Applications

Reading Preview

Essential Questions

▶ What force holds the atomic nucleus together?

▶ How do the strong force and the electric force compare to each other?

▶ How are radioactive atomic nuclei different from stable nuclei?

Review Vocabulary

electric force: force between electric charges such as protons and electrons

New Vocabulary

strong force
radioactivity

 Multilingual eGlossary

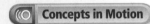 **Concepts in Motion**

Animation

■ **Figure 1** The size of a nucleus in an atom can be compared to a marble sitting in the middle of an empty soccer stadium. The diameter of an atom is approximately 10,000 times greater than the diameter of an atomic nucleus.

The Nucleus

MAIN ◀**Idea** The strong force holds an atomic nucleus together.

Real-World Reading Link With every breath you take, your body takes in a very small amount of carbon-14, a radioactive material. Radioactive materials, such as carbon-14 and radon gas, are all around us. Studying the fundamentals of radioactivity can help us to understand its potential uses and dangers.

Describing the Nucleus

Recall that atoms are composed of protons, neutrons, and electrons. The nucleus of an atom is composed of protons and neutrons. Protons have a positive electric charge. Neutrons have no electric charge. So, the number of protons in a nucleus determines that nucleus's total charge.

Negatively charged electrons are attracted to the positively charged nucleus and swarm around it. An electron has a charge that is equal but opposite to a proton's charge. Atoms contain the same number of protons as electrons.

Size of the nucleus The protons and neutrons that make up a nucleus are packed together tightly. The region outside the nucleus where the electrons are located is large compared to the size of the nucleus. As **Figure 1** helps show, the nucleus occupies only a tiny fraction of the space in an atom.

If an atom were enlarged so that it was 1 km in diameter, its nucleus would have a diameter of only a few centimeters. But the nucleus has almost all of the atom's mass. Neutrons are slightly more massive than protons. However, both protons and neutrons are almost 2,000 times as massive as electrons.

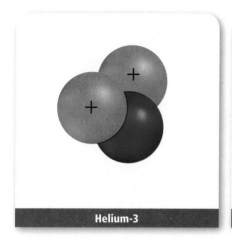

Helium-3

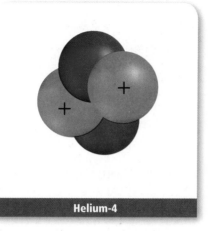

Helium-4

Isotopes

The atoms of an element all have the same number of protons in their nuclei. For example, the nuclei of all carbon atoms have six protons. However, naturally occurring carbon nuclei can have six, seven, or eight neutrons.

Isotopes are nuclei that have the same number of protons but different numbers of neutrons. Each element has many different isotopes. For example, the element carbon has three isotopes that occur naturally. The atoms of all isotopes of an element have the same chemical properties. However, each isotope has its own nuclear properties. For example, some carbon isotopes are radioactive, but other carbon isotopes are not radioactive. **Figure 2** shows two isotopes of helium.

Nucleus numbers

You can describe a nucleus with its numbers of protons and neutrons. The atomic number is the number of protons that are a part of a nucleus. The total mass of all the protons and neutrons that make up a nucleus is nearly the same as the mass of the atom. As a result, the number of protons plus neutrons is called the mass number.

✓ **Reading Check** **Define** atomic number.

You can represent a nucleus with its atomic number, mass number, and the symbol of the element to which it belongs. The representation for the nucleus of the most common isotope of carbon is shown below as an example.

$$\text{mass number} \rightarrow {}^{12}_{6}\text{C} \leftarrow \text{element symbol}$$
$$\text{atomic number} \rightarrow$$

This isotope is called carbon-12. The number of neutrons equals the mass number minus the atomic number. The number of neutrons that are a part of carbon-12 is $12 - 6 = 6$. The nucleus of carbon-12 is composed of six protons and six neutrons. Look at **Figure 2** again. How many protons does helium-4 have? How many neutrons does helium-4 have? What is the total number of protons plus neutrons?

VOCABULARY
WORD ORIGIN
Isotope
   comes from the Greek word *isos*, which means *equal*, and the Greek word *topos*, which means *place* Isotopes have the same atomic numbers but different mass numbers.

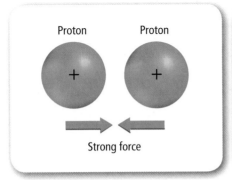

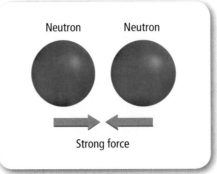

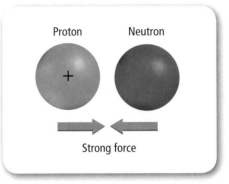

Figure 3 The particles in the nucleus are attracted to each other by the strong force. The strong force is one of four fundamental forces in nature. The other three include the weak force, the electromagnetic force, and gravity.

Forces in the Nucleus

How are the protons and neutrons that make up a nucleus held together so tightly? Positive electric charges repel each other, so why do the protons that are part of a nucleus not push each other away? The **strong force** is the force that causes protons and neutrons to be attracted to each other. **Figure 3** illustrates the strong force between protons and neutrons.

The strong force is one of the four basic forces in nature. It is about 100 times stronger than the electromagnetic force. The attractive forces between all of the protons and neutrons that make up a nucleus keep the nucleus together.

However, protons and neutrons have to be extremely close together to be attracted by the strong force. The strong force is a short-range force that quickly becomes extremely weak as protons and neutrons get farther apart. The electromagnetic force is a long-range force, so protons that are farther apart are still repelled by the electric force, as shown in **Figure 4.**

✓ **Reading Check Identify** the force that produces the attraction between protons and neutrons.

Figure 4 The total force between two protons depends on how far apart those protons are. The strong force is very strong but also very short-ranged. The electromagnetic force is much weaker but can act over much greater distances.

Infer *whether the net force between two protons could ever be zero.*

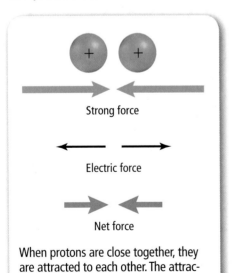

When protons are close together, they are attracted to each other. The attraction due to the short-range strong force is much stronger than the repulsion due to the long-range electric force.

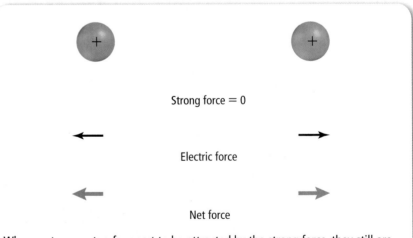

When protons are too far apart to be attracted by the strong force, they still are repelled by the electric force between them. Then the net force between them is repulsive.

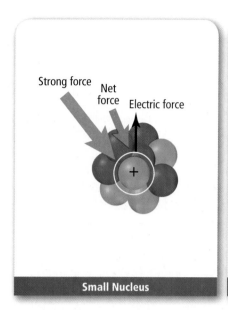

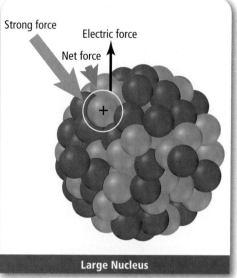

Small Nucleus

Large Nucleus

■ **Figure 5** For nuclei with few protons, the repulsive force on a proton due to the other protons is small. For nuclei with many protons, the attractive strong force is exerted only by the nearest neighbors, but all of the protons exert repulsive forces. The total repulsive force is then large.

Compare *the net force holding the smaller nucleus together with the net force holding the larger nucleus together.*

Nuclei with few protons The left panel of **Figure 5** shows the forces in a small nucleus, one with relatively few protons. If a nucleus has only a few protons and neutrons, they are all close enough together to be attracted to each other by the strong force. Because only a few protons are present, the total electric force repelling the protons from each other is small. As a result, the net forces between the protons and the neutrons hold the nucleus together tightly.

Nuclei with many protons Some nuclei, such as uranium nuclei, are composed of many protons and neutrons. In these cases, each proton or neutron is attracted to only a few neighbors by the strong force, as shown in the right panel of **Figure 5.** The other protons and neutrons are too far away. Therefore, the strong force holding a proton or neutron in place in a large nucleus is about the same as for a small nucleus.

However, all of the protons in a large nucleus exert a repulsive electric force on each other. Thus, the electric repulsive force on a proton in a nucleus with many protons is large. As a result, a nucleus with many protons is held together less tightly than a nucleus with fewer protons.

Neutron to proton ratios There are no repulsive electric forces between neutrons. Why are there no atomic nuclei composed completely of neutrons? Such an atomic nucleus would not have a stable ratio of neutrons to protons. In less massive elements, an isotope is stable when the ratio is about 1:1. In more massive elements, an isotope is stable when the ratio of neutrons to protons is about 3:2. Nuclei with too many neutrons compared to the number of protons are unstable. Larger nuclei have a higher neutron to proton ratio because neutrons contribute to the attractive strong force but not to the repulsive electric force within the nucleus.

? Inquiry MiniLab

Model Forces Between Protons

Procedure 🥽 👔 🧤

1. Read the procedure and safety information, and complete the lab form.
2. Cover the outer edges of **two disk magnets** with **double-sided tape.**
3. Arrange both magnets so that their north poles are facing up. The magnets should repel each other.
4. Slowly bring the magnets together until they are touching. Observe the total force between the magnets as you do this.

Analysis

1. **Identify** the fundamental force that the repulsive force between the magnets represents. Explain.
2. **Identify** the fundamental force that the double-sided tape represents. Explain.

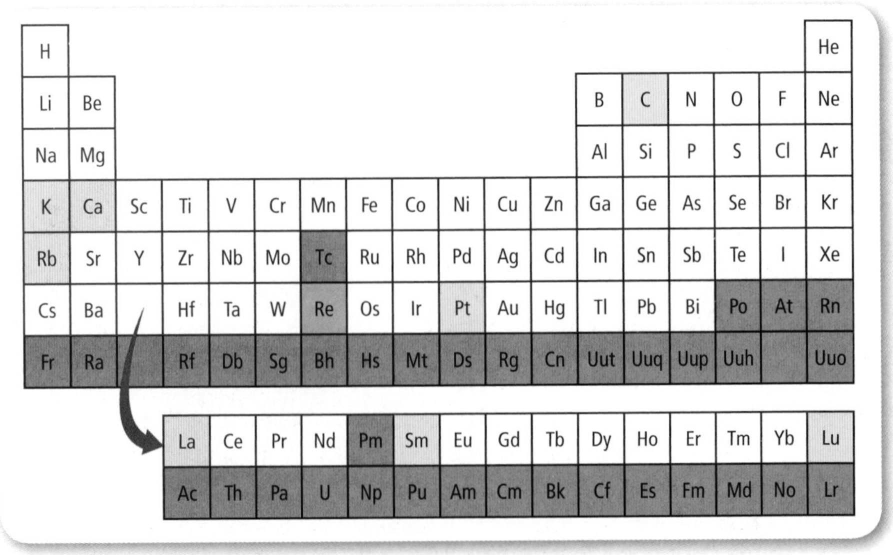

Figure 6 In this diagram of the periodic table, the redder the element box, the more likely it is that an isotope of that element will be radioactive. The white boxes indicate elements that do not have measurable percentages of radioactive isotopes.

Identify *where radioactive isotopes tend to be on the periodic table.*

Radioactivity

In most nuclei, the strong force is able to keep the nucleus permanently together. These nuclei are stable. When the strong force is not large enough to hold a nucleus together tightly, the nucleus can decay. When a nucleus decays, it emits matter and energy. **Radioactivity** is the process of nuclei decaying and emitting matter and energy.

All nuclei that contain more than 83 protons are radioactive. However, some nuclei that contain fewer than 83 protons are also radioactive, such as carbon-14. In addition, no nuclei with more than 92 protons is stable enough to occur naturally. Instead, people must synthesize these elements, usually in a lab. **Figure 6** shows which elements have measurable percentages of radioactive isotopes.

Section 1 Review

Section Summary

▶ Isotopes of an element have the same number of protons but different numbers of neutrons.

▶ The atomic number is the number of protons in a nucleus. The mass number is the number of protons and neutrons in a nucleus.

▶ A nucleus might decay due to its neutron to proton ratio.

▶ Radioactivity is the process of nuclear decay.

1. **MAIN Idea Compare** the properties of the strong force to the properties of the electromagnetic force.

2. **Compare** the forces in a small nucleus to the forces in a large nucleus.

3. **Explain** why large nuclei tend to be radioactive.

4. **Think Critically** Explain whether you would expect helium-6 to be radioactive or stable.

Apply Math

5. **Calculate a Ratio** What is the approximate ratio of neutrons to protons in a nucleus of radon-222?

Assessment   Online Quiz

Reading Preview

Essential Questions

▶ What are alpha particles, beta particles, and gamma rays?
▶ How does nuclear fission differ from nuclear fusion?
▶ How are mass and energy related?

Review Vocabulary

gamma ray: electromagnetic wave with no mass and no charge that travels at the speed of light

New Vocabulary

alpha particle
beta particle
transmutation
chain reaction

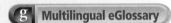 **Multilingual eGlossary**

Nuclear Decays and Reactions

MAIN Idea Incredibly high-energy particles are released during nuclear decays and reactions.

Real-World Reading Link Have you ever heard of medieval alchemists attempting to change common metals into gold? We can actually achieve this feat today through nuclear reactions. Unfortunately, the process is too expensive to make transforming common metals into gold profitable.

Nuclear Decays

When a nucleus decays, particles and energy are emitted from it. We call this emission nuclear radiation. Three types of nuclear radiation are alpha, beta, and gamma radiation. Alpha and beta radiation are composed of particles. Gamma radiation is composed of electromagnetic waves.

Alpha particles When the strong force is not strong enough to hold a nucleus together, that nucleus emits alpha particles. An **alpha particle** is a particle that is composed of two protons and two neutrons. An alpha particle is the same as a helium-4 nucleus.

✔ **Reading Check Identify** the components of an alpha particle.

Alpha particles are extremely massive when compared with other nuclear radiation. For example, an alpha particle is about 7,000 times more massive than a beta particle. Alpha particles also have twice as much charge as beta particles. Because of their high mass and charge, alpha particles interact with other matter frequently.

As a result, alpha particles transfer energy to their surroundings very quickly as they travel through solids, liquids, and gases. Alpha particles are the least penetrating form of nuclear radiation. A sheet of paper will stop most alpha particles. **Table 1** summarizes the properties of alpha particles.

Table 1	Alpha Particle	
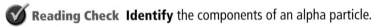	Description	high-energy helium-4 nucleus
	Symbol	^4_2He
	Mass	approximately 4 hydrogen atoms
	Charge	+2
	Can be stopped by...	a sheet of paper

Table 2	Beta Particle	
	Description	high-energy electron
	Symbol	$_{-1}^{0}e$
	Mass	approximately 1/7000th the mass of an alpha particle
	Charge	−1
	Can be stopped by...	a sheet of aluminum that is 3 mm thick

Beta particles and the weak force So far, you have learned about three fundamental forces in nature. They are the gravitational force, the electromagnetic force, and the strong force. The fourth and final fundamental force is called the weak force. The weak force causes beta decay.

Like the strong force, the weak force is very short-ranged. The weak force is also weaker than all the other fundamental forces except for gravity. However, the weak force can cause neutrons to decay into protons when the neutron to proton ratio in a nucleus is too high. When this decay happens, the nucleus emits an electron. A **beta particle** is a high-energy electron that is emitted when a neutron decays into a proton.

Beta particles are much faster and more penetrating than alpha particles. It takes a sheet of aluminum 3-mm thick to absorb most beta radiation. **Table 2** summarizes the properties of beta particles.

Gamma rays Gamma rays are extremely high-energy electromagnetic waves. Lead bricks or other heavy materials are necessary to stop gamma rays. They are usually emitted from a nucleus during alpha decay and beta decay. Gamma rays have no mass and no charge and travel at the speed of light. A nucleus loses energy, but no particles, during gamma decay. **Table 3** summarizes the properties of gamma rays.

VOCABULARY ·

ACADEMIC VOCABULARY

Fundamental
of or relating to essential structure, function, or facts
Some scientists study and try to understand the most fundamental laws of nature. ·

Table 3	Gamma Ray	
	Description	high-energy, high-frequency electromagnetic wave
	Symbol	γ
	Mass	0
	Charge	0
	Can be stopped by...	thick blocks of lead

Damage from nuclear decay Alpha, beta, and gamma radiation can all be dangerous to human tissue. Biological molecules inside your body are large and easily damaged. A single alpha or beta particle or gamma ray can damage many fragile biological molecules. Damage from radiation can cause cells to function improperly, leading to illness and disease.

Transmutation Another result of nuclear decay is transmutation. Transmutation occurs during alpha and beta decay. **Transmutation** is the process of changing one element into a different element.

During alpha decay, a nucleus loses two protons and two neutrons. As a result, the resulting nucleus has two fewer protons and two less neutrons. The nucleus transmutes into an entirely different element. The new element has an atomic number two less than that of the original element. The mass number of the new element is four less than that of the original element. The top panel of **Figure 7** shows a transmutation caused by alpha decay.

During beta decay, a neutron emits an electron and becomes a proton. The resulting nucleus has one more proton. It becomes the element with an atomic number one greater than that of the original element. However, the total number of protons and neutrons does not change during beta decay. Therefore, the mass number of the new element is the same as that of the original element. The bottom panel of **Figure 7** shows a transmutation caused by beta decay.

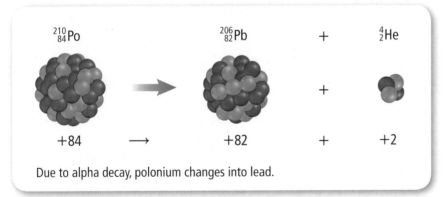

Due to alpha decay, polonium changes into lead.

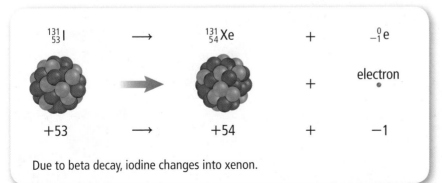

Due to beta decay, iodine changes into xenon.

■ **Figure 7** Elements transmute into new elements whenever they emit alpha or beta particles. Gamma rays do not have mass or charge. Therefore, a nucleus will not transmute into another nucleus if it emits only gamma rays. However, gamma rays are often emitted along with alpha or beta particles.

Compare *the total charge before a radioactive decay with the total charge after a radioactive decay.*

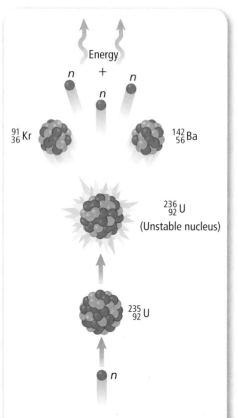

Figure 8 Some large nuclei split apart when they absorb a neutron. Here, a neutron splits a uranium-235 nucleus into two smaller nuclei and three more neutrons.

Predict *whether carbon-12 would be likely to undergo nuclear fission.*

Video   BrainPOP

Nuclear Fission

Energy is released whenever a nucleus emits nuclear radiation. However, the amount of energy released during a nuclear decay is very small compared to the amount of energy that can be released during nuclear fission.

Recall that nuclear fission is the process of splitting a nucleus into two or more smaller nuclei. Scientists can do this by bombarding the larger nucleus with neutrons. **Figure 8** shows a nuclear fission reaction for uranium-235. Other isotopes, including plutonium-239 and uranium-233 also undergo nuclear fusion. The nuclear bomb that was dropped on Hiroshima during World War II was powered by the fission of uranium-235.

Chain reactions The products of a fission reaction usually include several neutrons in addition to the smaller nuclei. These neutrons can then strike other nuclei in the sample, causing them to split as well. These reactions then release more neutrons, causing additional nuclei to split. A **chain reaction** is a series of repeated fission reactions caused by the release of additional neutrons in every fission. A chain reaction is shown in **Figure 9.**

If a chain reaction is uncontrolled, an enormous amount of energy is released in an instant. This is what happens when a nuclear fission bomb is detonated. However, a chain reaction can be controlled by adding materials that absorb neutrons. If enough neutrons are absorbed, the reaction will be controlled. This is how a nuclear power plant operates.

Figure 9 A chain reaction occurs when neutrons emitted from a split nucleus cause other nuclei to split and emit additional neutrons. An uncontrolled chain reaction occurs when a nuclear fission bomb is detonated. In a nuclear power plant, the chain reaction is controlled with rods of material that absorb some of the neutrons.

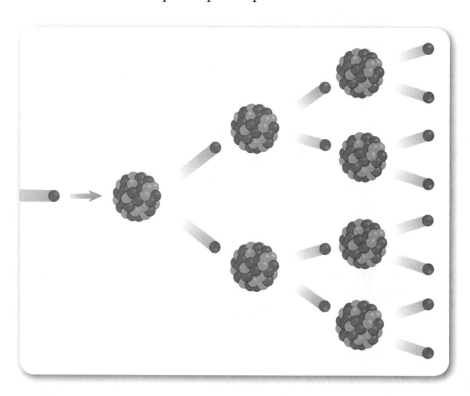

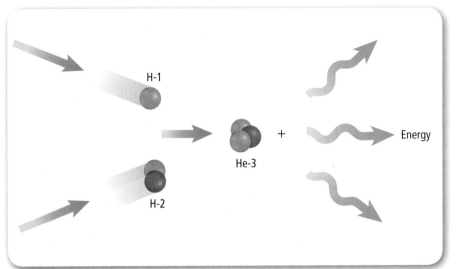

Figure 10 One form of fusion is shown here. Helium-3 is produced when a hydrogen-1 isotope combines with a hydrogen-2 . This reaction is common inside the Sun.

Nuclear Fusion

Nuclear fusion reactions can release even more energy than nuclear fission reactions can. Recall that nuclear fusion is the process of two or more nuclei combining to form a nucleus of larger mass. **Figure 10** shows an example of nuclear fusion that occurs inside the Sun. Fission splits nuclei apart. Fusion fuses nuclei together.

Temperature and fusion For nuclear fusion to occur, nuclei must get close to each other. However, all nuclei have positive electric charge. Therefore, they repel each other. In order to fuse, the nuclei need to have enough kinetic energy to overcome this repulsion.

Remember that the kinetic energy of atoms increases as their temperature increases. Only at temperatures of millions of degrees Celsius are nuclei moving so fast that they can get close enough for fusion to occur. These extremely high temperatures are found in the centers of stars, including the Sun.

Every atom that exists, other than hydrogen-1, was originally constructed through nuclear fusion. This nuclear fusion occurs in the cores of stars and during the explosions of those stars. Nuclear fusion also occurred in the first moments after the Big Bang. Thermal energy and temperature are extremely high in each of these situations.

The Sun and fusion The Sun emits more than 3.8×10^{26} J of energy every second. This radiant energy is first extracted from hydrogen nuclei in the Sun's core through a series of nuclear fusion reactions. The most important series of fusion reactions that occurs in the Sun is shown in **Figure 11**. This series is also common to other stars that have masses similar to the mass of the Sun. In more massive stars, another series of reactions is more common.

FOLDABLES

Incorporate information from this section into your Foldable.

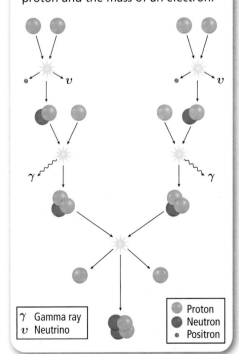

Figure 11 A series of fusion reactions inside the Sun's core provide its power. A neutrino is a tiny, nearly massless particle. A positron is a particle with the charge of a proton and the mass of an electron.

| γ | Gamma ray |
| υ | Neutrino |

	Proton
	Neutron
	Positron

■ **Figure 12** A small amount of mass is the same thing as a very large amount of energy. Albert Einstein first explained mass-energy equivalence in 1905, although others had suggested this connection between mass and energy before then.

Identify *the amount of energy that is equivalent to 2 g of mass.*

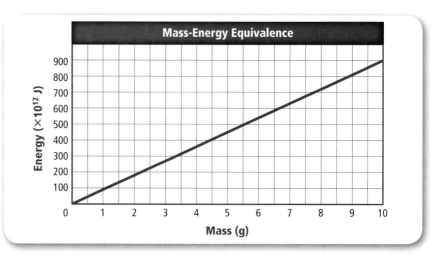

Mass and Energy

During the fusion reactions that take place inside the Sun, no matter is ejected from the Sun. However, the total mass of all the particles in the reaction is greater before the reaction than after the reaction. How can there be less mass after a reaction if no matter leaves the Sun? Isn't mass conserved?

According to the theory of special relativity, a tremendous amount of energy is the same thing as a small amount of mass. Mass is energy, and energy is mass. **Figure 12** shows this relationship. One gram of mass is enough energy to launch the Washington Monument into orbit.

However, unless incredibly large energies are involved, the mass-energy relationship is almost impossible to observe. With nuclear fission and fusion reactions, such tremendous energies are involved that the total amount of matter after the reaction can be noticeably different from the total amount of matter before the reaction. This change in the total amount of matter occurs even when the total number of protons plus neutrons does not change.

Converting between mass and energy is just a conversion of units. It is similar to converting from miles per hour to meters per second. To convert from units of mass to units of energy, multiply by the speed of light in a vacuum squared (c^2).

Mass-Energy Equation

units of energy (joules) =
[units of mass (kg)] × [speed of light in a vacuum (m/s)]^2

$$E = mc^2$$

The speed of light in a vacuum is 300,000,000 m/s. The example problem and practice problems on the next page will help you to further explore mass-energy equivalence.

Convert Units of Energy to Units of Mass The Sun emits approximately 3.8×10^{26} J of radiant energy every second. How much mass does the Sun lose every second due to this energy emission?

Identify the Unknown:	mass: m
List the Knowns:	Energy: $E = 3.8 \times 10^{26}$ **J**
	speed of light in a vacuum: $c = 3.0 \times 10^8$ **m/s**
Set Up the Problem:	$E = mc^2$
Solve the Problem:	3.8×10^{26} **J** $= m(3.0 \times 10^8$ **m/s**$)^2$
	$m = \dfrac{3.8 \times 10^{26} \text{ J}}{(3.0 \times 10^8 \text{ m/s})^2}$
	$m = 4.2 \times 10^9$ **kg** (roughly the mass of 12 Empire State Buildings)
Check the Answer:	Put the answer back into the equation $E = mc^2$. Then, $E = 4.2 \times 10^9$ kg $\times (3.0 \times 10^8$ m/s$)^2 = 3.8 \times 10^{26}$ J. This is the same as the energy given in the problem, so the math in this problem is correct.

PRACTICE Problems

Find **Additional Practice Problems** in the back of your book.

6. How much energy is equal to 1 kg of mass?

7. **Challenge** The population of the United States releases approximately 17×10^{18} J of energy every year by burning gasoline for trucks and automobiles. Approximately how much mass is this?

Review

Additional Practice Problems

Section 2 Review

Section Summary

▶ Radioactivity can result in alpha particles, beta particles, or gamma rays.

▶ Nuclear fission occurs when a neutron strikes a nucleus, causing it to split into smaller nuclei.

▶ Nuclear fusion occurs when two nuclei combine to form another nucleus.

▶ A small amount of mass is the same as a tremendous amount of energy.

8. MAIN Idea **Contrast** the energy that can be released during a nuclear fission reaction with the energy that can be released during a nuclear fusion reaction.

9. **Contrast** alpha particles, beta particles, and gamma rays.

10. **Explain** why mass-energy equivalence is not apparent for chemical reactions.

11. **Think Critically** Explain why high temperatures are needed for fusion reactions to occur but not for fission reactions to occur.

Apply Math

12. **Calculate Number of Nuclei** In a chain reaction, two additional fissions occur for each nucleus that is split. If one nucleus is split in the first step of the reaction, how many nuclei will have been split after the fifth step?

LAB Chain Reactions

Objectives
- **Model** a controlled chain reaction.
- **Model** an uncontrolled chain reaction.
- **Compare** the two types of chain reactions.

Background: In an uncontrolled nuclear chain reaction, the number of reactions increases as additional neutrons split more nuclei. In a controlled nuclear reaction, neutrons are absorbed, so the reaction continues at a constant rate.

Question: *How could you model a controlled and an uncontrolled nuclear reaction in the classroom?*

Preparation

Materials
dominoes
stopwatch

Procedure

1. Read the procedure and safety information, and complete the lab form.
2. Set up a single line of dominoes standing on end. When the first domino is pushed over, it should knock over the second, and each domino should knock over the one following it.
3. Using the stopwatch, time how long it takes from the moment that the first domino is pushed over until the last domino falls over. Record the time.
4. Line up the same number of dominoes in the shape of a Y, as shown above. Be sure that both dominoes at the split in the Y will get knocked down by the falling dominoes.
5. Repeat step 3.

 Inquiry   Video Lab

Conclude and Apply

1. **Compare** the amount of time that it took for all of the dominoes to fall in each of your two arrangements.
2. **Determine** the average number of dominoes that fell per second in both domino arrangements.
3. **Explain** how your models represented controlled and uncontrolled chain reactions.
4. **Explain** how adding more branches to your model would affect the average number of dominoes that fall per second. What would adding more domino branches to your model represent in terms of the number of neutrons released in each fission reaction?
5. **Predict** Assuming that they had equal amounts of materials, which would finish faster: a controlled or an uncontrolled nuclear chain reaction?

COMMUNICATE YOUR DATA

Create a Web site that explains how a controlled nuclear chain reaction can be used in nuclear power plants to generate electricity.

 Inquiry   Lab

Essential Questions

▶ How can radioactivity be detected?

▶ What are some common sources of background radiation?

▶ What is the half-life of a radioactive material?

▶ How can radioactivity be used to help find the age of an object?

Review Vocabulary

electric current: the net movement of electric charges in a single direction

New Vocabulary

tracer
half-life

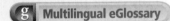 **Multilingual eGlossary**

Radiation Technologies and Applications

MAIN ⟨Idea⟩ Nuclear radiation can be dangerous, but it also has beneficial applications.

Real-World Reading Link Have you ever heard of doctors using radiation to diagnose illnesses? Medical doctors sometimes administer radioactive dyes to their patients. The doctors can then track the progress of the dye through the patient's body.

Detecting Nuclear Radiation

Special equipment is necessary to detect and study nuclear radiation. One such piece of equipment is a Geiger counter. A Geiger counter, as shown in **Figure 13,** has a tube with a positively charged wire running through the center of a negatively charged metal tube. This tube is filled with a low-density gas.

When radiation enters the tube at one end, it knocks electrons from the gas molecules. These electrons then knock more electrons off other gas molecules, producing an electron avalanche. The positive wire attracts these electrons, resulting in a current in that wire. This current is amplified and produces a clicking sound. The number of clicks each second indicates the intensity of radiation.

Often, scientists want to do more than just detect nuclear radiation. Unlike a Geiger counter, a wire chamber can track the trajectories of subatomic particles as well as detect those particles' presence. A wire chamber works much like a Geiger counter. However, a wire chamber contains an array of positively charged wires instead of just one wire.

Concepts in Motion

Animation

■ **Figure 13** Subatomic particles, including alpha particles and beta particles, can be detected by a Geiger counter. The larger the amount of radiation, the larger the electron avalanche and the larger the number of clicks. Many Geiger counters also include a gauge with a needle that measures the incoming radiation.

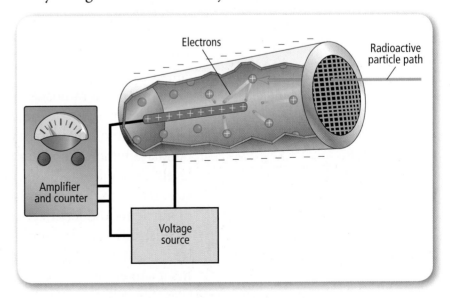

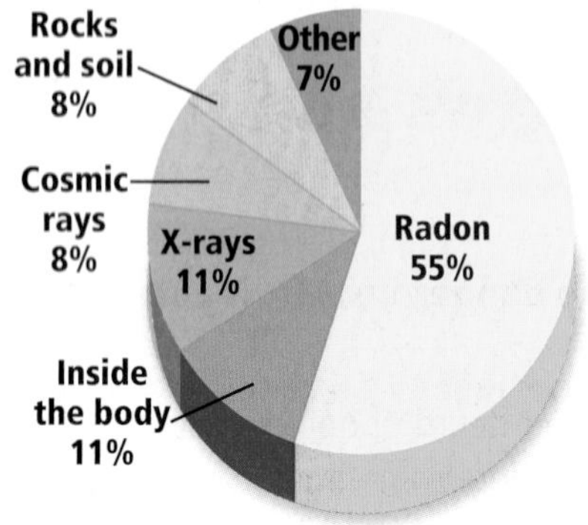

Sources of Background Radiation

Rocks and soil 8%
Other 7%
Cosmic rays 8%
X-rays 11%
Radon 55%
Inside the body 11%

■ **Figure 14** This circle graph shows the sources of background radiation received on average by a person living in the United States. Most of the radiation that originates inside our bodies is from potassium-40, which we consume in our food. Most X-ray exposure is from medical X-rays.

Video What's PHYSICAL and EARTH SCIENCE Got to Do With It?

Background Radiation

You might be surprised to learn that humans have been bathed in radiation for millions of years. This radiation, called background radiation, is not produced by humans. Instead, it is emitted mainly by radioactive isotopes found in Earth's rocks, soils, and atmosphere. This background radiation is low-level but is still detectable.

Building materials, such as bricks, wood, and stones, contain traces of radioactive materials. Traces of naturally occurring radioactive isotopes are also in our food, water, and air. Background radiation is even emitted from inside our own bodies. For example, our bodies contain the isotopes carbon-14 and potassium-40. Both of these isotopes are radioactive.

Sources of background radiation Background radiation comes from several sources, as shown in **Figure 14.** The most common source of background radiation, radon gas, can seep into houses and basements from surrounding soil and rocks. In addition, some background radiation comes from high-speed particles that strike Earth's atmosphere from outer space. These high-speed particles are called cosmic rays.

The amount of background radiation that a person receives can vary greatly. The amount depends on the types of rocks underground, types of materials used to construct the person's home, and the elevation at which the person lives, among other things. However, some amount of background radiation is present for everyone and has been present throughout history and prehistory.

Using Nuclear Radiation in Medicine

It would be easier to find a friend in a crowded area if she told you that she would be wearing a red hat. In a similar way, scientists can find one molecule in a large group of molecules if it is "wearing" something unique. A molecule cannot wear a red hat; however, it can be found easily if it has a radioactive atom in it. A radioactive atom emits radiation that doctors can detect.

A **tracer** is a radioactive isotope that doctors use to locate molecules in an organism. Doctors use tracers to follow where particular molecules go in a human body and to study how organs function. This might seem harmful, but the radiation levels are too low to be harmful or dangerous. Agricultural scientists also use tracers in agriculture to monitor the uptake of nutrients and fertilizers. Common tracers include technetium-99m and iodine-131. These are useful tracers because they emit gamma rays that medical imaging equipment can easily detect.

Iodine tracers in the thyroid Tracers can be used to detect problems in your thyroid, which helps regulate several body processes, including growth. Iodine accumulates in the thyroid. Iodine-131, a radioactive isotope of iodine, emits gamma rays. A patient can ingest a capsule that contains iodine-131, which can be easily absorbed by the patient's thyroid.

Gamma rays from the iodine-131 penetrate the skin. Doctors can detect these gamma rays, producing an image like the one shown in **Figure 15**. If the detected radiation is not intense, then the thyroid has not properly absorbed the iodine-131. This could be due to the presence of a tumor.

Reading Check **Describe** a medical use for iodine-131.

Cancer treatments When a person has cancer, a group of cells in that person's body grows out of control. Cancer is a harmful and often fatal disease. The left panel of **Figure 16** shows two cancerous cells. The right panel of **Figure 16** shows a cancer patient undergoing radiation therapy. Doctors can use radiation to stop some types of cancerous cells from growing and dividing.

Remember that radiation can ionize nearby atoms. If a source of radiation is placed near cancer cells, atoms in those cells can be ionized. If the ionized atoms are in a critical molecule, such as DNA or RNA, then the molecule may no longer function properly. The cell then may stop growing or may even die.

Noncancerous cells can also be damaged during radiation therapy. For this reason, doctors must be careful to focus the radiation on the cancer cells as much as possible. However, radiation therapy still often harms healthy cells. Cancer patients often experience severe side effects when they receive radiation therapy.

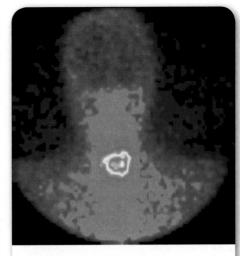

■ **Figure 15** Radioactive iodine-131 builds up in the thyroid gland and emits gamma rays, which can be detected to form an image of this cancer patient's thyroid.

■ **Figure 16** During radiation therapy, a radioactive source is placed close to the tumor. The goal in such a procedure is to damage the tumor as much as possible while minimizing damage to the surrounding tissue.

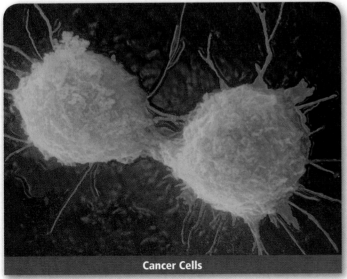

Cancer Cells

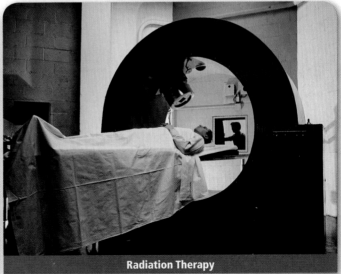

Radiation Therapy

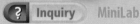

MiniLab

Inquiry MiniLab

Model Radioactive Decay

Procedure

1. Read the procedure and safety information, and complete the lab form.
2. Place **200 pennies** in a **shoe box**.
3. Close the shoe box. Shake the shoe box for 3 s while holding the lid tightly.
4. Open the shoe box, and remove all pennies that are now tails up. Count and record the number of pennies remaining.
5. Repeat steps 3 and 4 until you have removed all pennies from the box.

Analysis

1. **Graph** your data as number of pennies left v. time. Record each shake-and-remove step as a 3-s interval.
2. **Describe** how this experiment models radioactive decay.
3. **Predict** how many more shakes would be necessary for all your pennies to decay if you started with 400 pennies instead of 200.

Inquiry Virtual lab

■ **Figure 17** One half of a radioactive sample decays every half-life.

Identify *how many half-lives would be necessary for three quarters of a radioactive sample to decay.*

Half-Life

How can you tell when a radioactive isotope is going to decay? Suppose you shake a box filled with hundreds of pennies. You then remove every penny that comes up tails. You would remove about half the pennies every time you did this.

You could not predict exactly which pennies would come up tails each time. However, you could predict approximately how many pennies would come up tails after each shake. You could also predict how many times you can repeat this process before you have removed all of the pennies from the box.

Radioactive decay works in a similar way. You cannot know when a specific radioactive nucleus will decay. However, you can accurately predict approximately how many radioactive nuclei will decay in a given amount of time.

If a large number of radioactive nuclei are present in a sample, it is possible to consider the half-life of that sample. **Half-life** is the amount of time it takes for half the nuclei in a sample of an isotope to decay. Half-life is similar to the amount of time between shakes for the pennies in the shoe box.

The radioactive nucleus is called the parent nucleus. The nucleus left after the isotope decays is called the daughter nucleus. **Figure 17** shows the proportion of decaying nuclei left after each half-life. Notice that one half of the original parent nuclei remain after one half-life. One quarter of the original parent nuclei remain after two half-lives.

Half-lives vary widely. The half-life of polonium-214 is less than a thousandth of a second. Carbon-14 has a half-life of slightly less than 6,000 years. Uranium-238 has a half-life of 4.5 billion years. Scientists can use their knowledge of the half-life of an isotope to calculate the ages of rocks, fossils, and artifacts.

✓ **Reading Check** **Define** *daughter nucleus.*

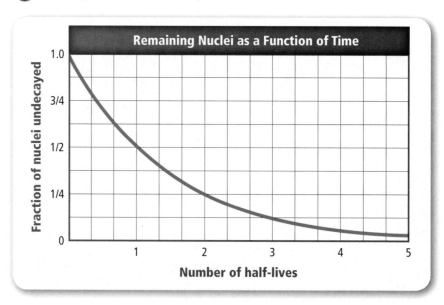

Remaining Nuclei as a Function of Time

Fraction of nuclei undecayed

1.0
3/4
1/2
1/4
0

Number of half-lives
1 2 3 4 5

Radiometric dating The rock shown in **Figure 18** is from the Moon and is more than 4 billion years old. How can scientists learn the age of something that is thousands, millions, or even billions of years old? Scientists use many different methods to determine the ages of samples. One of their most effective methods of dating samples involves an understanding of radioactivity and half-life. Scientists call this method radiometric dating.

First, scientists measure the amounts of radioactive isotope and daughter isotope in a sample. Then, they calculate the number of half-lives that need to pass to give the measured amounts. The number of half-lives can then be multiplied by the length of each half-life. This gives the amount of time that has passed since the isotope began to decay. This is usually close to the total amount of time that has passed since the object was formed.

Different isotopes are useful in dating various types of materials. Carbon-14 can be used to date the fossils of once-living organisms that are tens of thousands of years old. However, carbon-14 cannot be used to date materials that were not once a part of a living organism or materials that are more than about 60,000 years old. Uranium-235, which has a much longer half-life, can be used to date rocks and minerals that are billions of years old. Other isotopes that are used in radiometric dating are potassium-40, rubidium-87, and samarium-147.

■ **Figure 18** This rock is commonly known as the Genesis Rock. It was returned from the Moon during the Apollo 15 mission. Radiometric dating techniques showed this rock to be more than 4 billion years old.

Section 3 Review

Section Summary

▶ Alpha and beta particles can be detected by Geiger counters and in wire chambers.

▶ Background radiation is low-level radiation emitted mainly by radioactive isotopes in Earth's rocks, soils, and atmosphere.

▶ Doctors use radioactive isotopes as tracers for medical diagnoses as well as to kill cancer cells.

▶ The half-life of an isotope is the length of time that it takes half of a sample of that isotope to decay.

13. **MAIN Idea Explain** how half-life would help determine which isotopes might be useful for a medical test.

14. **Describe** what happens when beta particles pass through a Geiger counter.

15. **Explain** why background radiation can never be completely eliminated.

16. **Think Critically** Why is an archeologist unable to use carbon-14 to accurately date the age of a skeleton that is millions of years old?

Apply Math

17. **Use Percentages** What is the percentage of radioactive nuclei left after 3 half-lives pass?

18. **Use Fractions** If the half-life of iodine-131 is 8 days, how much of a 5.0-g sample is left after 32 days?

LAB   Transmutations

MODEL AND INVENT

Objectives
■ **Model** the decay of a radioactive isotope to a stable isotope.

Background: Imagine what would happen if the oxygen atoms around you began changing into nitrogen atoms. Without oxygen, most living organisms, including people, could not live. Fortunately, only a negligible percentage of oxygen atoms are radioactive and decay. Usually, when an unstable nucleus decays, an alpha or beta particle is thrown out of its nucleus, and the atom becomes a new element. A uranium-238 atom, for example, will undergo eight alpha decays and six beta decays to become lead. This process of one element changing into another element is called transmutation.

Question: *How could you create a model of a uranium-238 atom and the decay process it undergoes during transmutation?*

Preparation

Possible Materials

brown rice
white rice
colored candies
dried beans
dried seeds
glue
poster board

Safety Precautions

WARNING: *Never eat foods used in the lab.*

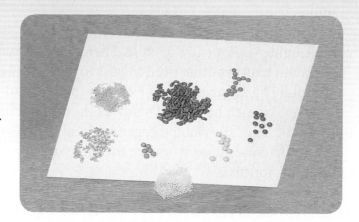

Make a Model

1. Read the procedure and safety information, and complete the lab form.

2. Choose two materials of different colors or shapes for the protons and neutrons of your nucleus model.

3. Choose a material for the negatively charged beta particle.

4. Decide how to model the transmutation process. Will you create a new nucleus model for each new element? How will you model an alpha or a beta particle leaving the nucleus?

5. Create a table like the one shown on the next page to show the results of each step of the transmutation of a uranium-238 atom into a lead-206 atom. A uranium-238 atom can undergo the following decay steps in order to transmute into a lead-206 atom: alpha decay, beta decay, beta decay, alpha decay, alpha decay, alpha decay, alpha decay, alpha decay, beta decay, beta decay, alpha decay, beta decay, beta decay, alpha decay.

6. Describe your model plan and transmutation table to your teacher, and ask how they can be improved.

7. Present your plan to the class. Ask classmates to suggest improvements.

? Inquiry   Lab

Data Table

Step	Identity of Element	Atomic Number	Mass Number	Type of Radiation Emitted
0	uranium	92	238	alpha particle
1				
2				
3				
4				
5				
6				
7				
8				
9				
10				
11				
12				
13				
14				

Test Your Model

1. Construct your model of a uranium-238 nucleus, showing the correct number of protons and neutrons.

2. Using your nucleus model, demonstrate the transmutation of a uranium-238 nucleus into a lead-206 nucleus by following the decay sequence outlined in step 5 of the previous section. Fill in your transmutation table as you proceed.

3. Show the emission of an alpha particle or a beta particle between each transmutation step.

Analyze Your Data

1. **Compare** how alpha decay and beta decay change the atomic number of an atom.

2. **Compare** how alpha decay and beta decay change the mass number of an atom.

Conclude and Apply

1. **Compare and contrast** your model with a real decay of uranium-238 into lead-206. How accurately does your model represent the decay of uranium-238? How is your model different from the decay of uranium-238?

2. **Calculate** the ratio of neutrons to protons in lead-206 and uranium-238. In which nucleus is the ratio closer to 3:2?

3. **Identify** Alchemists living during the Middle Ages spent much time trying to turn lead into gold. Identify the decay processes needed to accomplish this task.

Demonstrate how your model represents the transmutation of U-238 into Pb-206 to the class.

SCIENCE & HISTORY

THE ATOM BOMB

Nuclear fission was a well-researched phenomenon by January 1939. Newspapers around the world reported the "splitting of the atom." In late 1939, as Hitler's Germany threatened all of Europe, leading scientists in the U.S. urged the government to begin a top-secret atomic weapons research program. This program became known as the Manhattan Project.

War effort The Manhattan Project was a massive effort that involved more than 130,000 workers at its peak. From military officers and Nobel-prize winning scientists to thousands of engineers and factory workers, the quest involved dozens of facilities across the country. On July 16, 1945, the Manhattan project's work resulted in *Trinity,* the first test explosion of an atomic bomb.

Beyond testing By May of 1945, Germany had surrendered and the U.S. military was concentrating on defeating Japan. The Manhattan Project's atomic bombs, initially intended for use against Germany, would instead be used against Japan. **Figure 1** shows the devastation caused by the Hiroshima bomb. The explosion, resulting fires, and effects of radiation killed 200,000 people.

Figure 1 Hiroshima, Japan was the first city destroyed by an atomic bomb.

Figure 2 The US and the Soviet Union spent trillions of dollars building stockpiles of nuclear weapons.

The end of WW II Three days later, another atom bomb was dropped on Nagasaki, Japan. One day later, Japan surrendered. This action ended a conflict which killed more than 50 million people.

Controlling nuclear weapons The U.S. and U.S.S.R. built up growing stockpiles of nuclear weapons after WW II, as shown in **Figure 2.** Beginning in the 1970s, arms control agreements limited numbers of nuclear weapons and banned nuclear testing, anti-ballistic missile systems, and weapons in space. Such agreements had limited success. Over the years, negotiations continued, fewer new nuclear systems were developed, and nuclear stockpiles were reduced. Today, the United Nations and other organizations continue to seek to reduce the total number of nuclear weapons and work to prevent more countries from building these weapons of mass destruction.

WebQuest **Research** Robert Oppenheimer has been called the "American Prometheus." What was Robert Oppenheimer's role in the Manhattan Project, and why might Prometheus be an appropriate nickname? Write a short paragraph answering these questions.

THEME FOCUS Changes of Matter

One element can be changed into another element through radioactivity and nuclear reactions. The processes of fission, fusion, alpha decay, and beta decay all result in transmutations.

BIG Idea Forces inside the nucleus can cause changes that release particles and energy.

Section 1 The Nucleus

radioactivity (p. 620)
strong force (p. 618)

MAIN Idea The strong force holds an atomic nucleus together.

- Isotopes of an element have the same number of protons but different numbers of neutrons.
- The atomic number is the number of protons in a nucleus. The mass number is the number of protons and neutrons in a nucleus.
- A nucleus may decay due to its neutron to proton ratio.
- Radioactivity is the process of nuclear decay.

Section 2 Nuclear Decays and Reactions

alpha particle (p. 621)
beta particle (p. 622)
chain reaction (p. 624)
transmutation (p. 623)

MAIN Idea Incredibly high-energy particles are released during nuclear decays and reactions.

- Radioactivity can result in alpha particles, beta particles, or gamma rays.
- Nuclear fission occurs when a neutron strikes a nucleus, causing it to split into smaller nuclei.
- Nuclear fusion occurs when two nuclei combine to form another nucleus.
- A small amount of mass is the same as a tremendous amount of energy.

Section 3 Radiation Technologies and Applications

half-life (p. 632)
tracer (p. 630)

MAIN Idea Nuclear radiation can be dangerous, but it also has beneficial applications.

- Alpha and beta particles can be detected by Geiger counters and in wire chambers.
- Background radiation is low-level radiation emitted mainly by radioactive isotopes in Earth's rocks, soils, and atmosphere.
- Doctors use radioactive isotopes as tracers for medical diagnoses as well as to kill cancer cells.
- The half-life of an isotope is the length of time that it takes half of a sample of that isotope to decay.

Use Vocabulary

Compare and contrast the following pairs of terms.

19. radioactivity—half-life

20. tracer—beta particle

21. radioactivity—half-life

22. alpha particle—beta particle

23. beta particle—transmutation

24. strong force—radioactivity

Check Concepts

25. **BIG Idea** What keeps particles of a nucleus together?
    A) strong force
    B) weak force
    C) electromagnetic force
    D) gravity

26. What pushes the particles that make up a nucleus apart?
    A) strong force
    B) electromagnetic force
    C) gravity
    D) atomic repulsion

27. Which describes an isotope's half-life?
    A) a constant time interval
    B) a random time interval
    C) an increasing time interval
    D) a decreasing time interval

28. For which could carbon-14 dating be used?
    A) a fragment of wood from a Colonial house
    B) a marble column from ancient China
    C) dinosaur fossils that are millions of years old
    D) rocks that are billions of years old

Use the figure below to answer questions 29 and 30.

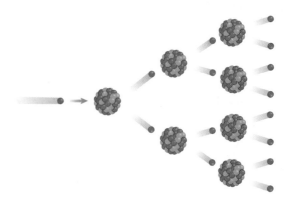

29. What is shown in the diagram above?
    A) decay reaction
    B) chain reaction
    C) beta emission
    D) alpha emission

30. What type of reaction is shown?
    A) alpha decay   C) nuclear fission
    B) beta decay   D) nuclear fusion

31. Which process is responsible for the tremendous energy released by the Sun?
    A) nuclear decay
    B) nuclear fission
    C) nuclear fusion
    D) alpha decay

32. Which describes all nuclei with more than 82 protons?
    A) radioactive   C) synthetic
    B) repulsive   D) stable

33. Which describes atoms with the same number of protons but a different number of neutrons?
    A) unstable   C) radioactive
    B) synthetic   D) isotopes

✓ **Assessment**   **Online Test Practice**

Concepts in Motion Interactive Concept Map

Interpret Graphics

34. Copy and complete the following concept map on radioactivity.

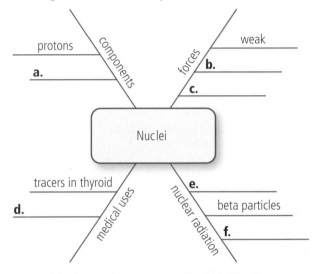

protons

components

a.

forces

weak

b.

c.

Nuclei

tracers in thyroid

medical uses

d.

nuclear radiation

e.

beta particles

f.

35. Copy and complete the following table.

Nuclear Reactions			
	Radioactive Decay	Nuclear Fission	Nuclear Fusion
Process	a.	d.	g.
Energy changes	b.	e.	h.
Safety issues	c.	f.	i.

Use the table below to answer question 36.

Isotope Half-Lives		
Isotope	Mass Number	Half-Life
Radon-222	222	4 days
Thorium-234	234	24 days
Iodine-131	131	8 days
Bismuth-210	210	5 days
Polonium-210	210	138 days

36. Graph the data in the table above with mass number on the *x*-axis and half-life on the *y*-axis. Infer from your graph whether there is a relation between the half-life and the mass number. If so, how does half-life depend on mass number?

Think Critically

37. **Explain** how properly sealing the basement of a house could help to reduce the background radiation in that house.

38. **Infer** how the mass of a nucleus changes when the nucleus emits only gamma radiation.

39. **Compare** Mass-energy equivalence becomes apparent during nuclear reactions but not during chemical reactions. What does this show about how the energies involved in nuclear reactions and chemical reactions compare?

40. **THEME FOCUS** **Infer** the type of nuclear radiation that is emitted in each of the following transmutations.
    a. uranium-238 to thorium-234
    b. boron-12 to carbon-12
    c. cesium-130 to cesium-130
    d. radium-226 to radon-222

41. **Predict** how the motion of an alpha particle would be affected if it passed between a positively-charged electrode and a negatively charged electrode. How would the motion of a gamma ray be affected?

42. **Infer** how the radiation that a person receives from cosmic rays changes when she or he goes skydiving.

Apply Math

43. **Calculate Number of Half-Lives** How many half-lives have elapsed when the amount of a sample's radioactive isotope is reduced to 3.125 percent of the original amount in the sample?

44. **Convert Units** Convert 75 kg to joules.

Standardized Test Practice

Multiple Choice

Record your answers on the answer sheet provided by your teacher or on a sheet of paper.

1. Which statement is true about all of the isotopes of an element?
   A. They have the same mass number.
   B. They have different numbers of protons.
   C. They have the same number of protons.
   D. They have the same number of neutrons.

2. Which happens in a nucleus during beta decay?
   A. The number of protons increases.
   B. The number of neutrons increases.
   C. The number of protons decreases.
   D. The number of protons plus the number of neutrons decreases.

Use the figure below to answer questions 3 and 4.

Helium-3   Helium-4

3. What does the illustration show?
   A. beta particles
   B. nuclear decay
   C. isotopes
   D. half-lives

4. Which is a true statement about both nuclei?
   A. They have the same atomic number.
   B. They have the same mass number.
   C. They have different numbers of electrons.
   D. They have different numbers of protons.

5. The atomic number for carbon is six. How many neutrons are in carbon-13?
   A. 6   C. 12
   B. 7   D. 13

6. If a radioactive material has a 10-y half-life, what fraction of the material will be left after 30 y?
   A. one-half
   B. one-third
   C. one-fourth
   D. one-eighth

7. Radioactive isotopes of which element are used to study the thyroid?
   A. uranium   C. iodine
   B. carbon   D. cadmium

8. To what is the atomic number of a nucleus equal?
   A. the number of neutrons
   B. the number of protons
   C. the number of neutrons and protons
   D. the number of neutrons minus the number of protons

Use the figure below to answer question 9.

$^{210}_{84}$Po $^{206}_{82}$Pb $^{4}_{2}$He

  +

9. What process is shown by this illustration?
   A. nuclear fusion
   B. chain reaction
   C. alpha decay
   D. beta decay

10. During which process does the mass number of an isotope not change?
   A. alpha decay
   B. beta decay
   C. nuclear fission
   D. nuclear fusion

✓ Assessment   Standardized Test Practice

Record your answers on the answer sheet provided by your teacher or on a sheet of paper.

11. What process contributes the most to the background radiation received by a person in the United States?

12. How much mass is equal to 9 billion J of energy?

Use the table below to answer questions 13–15.

Half-Lives of Isotopes	
Isotope	**Half-Life**
Carbon-14	5,730 years
Potassium-40	1.28 billion years
Iodine-131	8.04 days
Radon-222	4 days

13. Calculate how much of an 80 g sample of carbon-14 will be left after 17,190 years.

14. Potassium-40 decays to argon-40. What is the age of a rock in which 87.5 percent of the potassium-40 atoms have decayed to argon-40?

15. A sample containing which radioactive isotope will have about one-fourth of the isotope left after 16 days?

16. A radioactive tracer has a half-life of 2 h. Predict whether the tracer will be detectable after 24 h.

17. What is the resulting isotope when carbon-14 emits a beta particle?

Record your answers on a sheet of paper.

18. Compare the strength of the strong force on a proton and the strength of the electromagnetic force on a proton in a small nucleus and a large nucleus.

Use the figure below to answer questions 19 and 20.

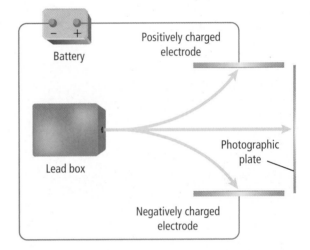

19. In the figure above, nuclear radiation is escaping from a small hole in the lead box. Which type of nuclear radiation is deflected toward the positively charged electrode? Why is this radiation deflected toward this electrode?

20. Explain why the radiation that struck the photographic plate was not deflected by the electrodes.

21. Describe the sequence of events that occur during a chain reaction.

NEED EXTRA HELP?																					
If You Missed Question . . .	1	2	3	4	5	6	7	8	9	10	11	12	13	14	15	16	17	18	19	20	21
Review Section . . .	1	2	1	1	1	3	3	1	2	2	3	2	3	3	3	3	2	1	2	2	2

UNIT 6

Applications of Chemistry

THEMES

Changes of Matter By understanding how matter undergoes change, we can engineer new materials.

Natural Resources Understanding the properties of our natural resources helps us find ways to manage and conserve.

Scientific Inquiry Unexpected results from scientific investigations can lead to new applications.

Structure and Properties of Matter New materials are being developed based on the properties of matter to meet specific needs.

Technology Access to raw materials can lead to new technologies, and new technologies can lead to the discovery of new materials.

WebQuest **STEM UNIT PROJECT**

Environment Research local recycling programs.
Develop a program that can be implemented in your school.

CHAPTER 21

Solutions

Your one-stop online resource
connectED.mcgraw-hill.com

- Video
- Audio
- Review
- Inquiry
- WebQuest
- Assessment
- Concepts in Motion
- Multilingual eGlossary

Launch Lab
Crystal Garden

Solutions surround you in your daily life. The air that you breathe, the steel bridge that you drive across, and the apple juice that you drink for breakfast are all solutions. In this lab, you will use a solution to grow a garden of crystals.

For a lab worksheet, use your StudentWorks™ Plus Online.

? Inquiry Launch Lab

FOLDABLES®

Make a three-tab Venn diagram. Label it as shown. Use it to organize your notes on solutions.

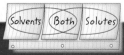

Solvents Both Solutes

THEME FOCUS Structures and Properties of Matter

The structures of atoms, ions, and molecules determine whether or not two substances will form a solution.

BIG (Idea A solution is a homogeneous mixture of a solute or solutes and a solvent or solvents.

Reading Preview

Essential Questions

▶ How do substances dissolve in a liquid?

▶ How do solid solutions and gas solutions form?

▶ What factors affect the rates at which solids dissolve in liquids?

Review Vocabulary

polar molecule: a molecule with a slightly positive end and a slightly negative end as a result of electrons being shared unequally

New Vocabulary

solute
solvent
alloy

g Multilingual eGlossary

How Solutions Form

MAIN ⟨Idea A solution forms when a solute or solutes and a solvent or solvents become evenly mixed.

Real-World Reading Link Solutions are all around you. For example, air is a solution of gases, and the ocean is a solution of salt water. Many structures, such as buildings or bridges, are also solutions made from different metals. Solutions are an important part of life.

What is a solution?

Hummingbirds are fascinating creatures. They can hover for long periods while they sip nectar from flowers through their long beaks. To attract hummingbirds, many people use feeder bottles containing a red liquid, as shown in **Figure 1.** The liquid is a solution of sugar and red food coloring in water.

Suppose you are making some hummingbird food. When you add sugar to water and stir, the sugar crystals disappear. When you add a few drops of red food coloring and stir, the color spreads evenly throughout the sugar water. Why does this happen?

Hummingbird food is one of many solutions. A solution is a homogeneous mixture, which means it has the same composition throughout the mixture. The reason why you no longer see the sugar crystals and the reason why the red dye spreads out evenly is that they have formed a solution. The sugar molecules and the red dye mixed evenly among the water molecules.

■ **Figure 1** Liquid solutions may contain gases, other liquids, and solids. For example, this hummingbird food contains sugar (solid), red food coloring (liquid), and oxygen and nitrogen (gases), all dissolved in water.

Gas phase

This diver breathes a solution of oxygen, nitrogen, and helium gases.

Solid phase

The bronze used to make this statue is a solid solution of copper and tin.

Solutes and Solvents

To describe a solution, you can say that one substance is dissolved in another. The substance being dissolved in a solution is the **solute.** The substance in which a solute is dissolved is the **solvent.** When a solid or gas dissolves in a liquid, the solid or gas is the solute and the liquid is the solvent. Thus, in salt water, salt is the solute and water is the solvent. In carbonated soft drinks, carbon dioxide gas is one of the solutes and water is the solvent. When a liquid dissolves in another liquid, the substance that is present in the larger amount is typically called the solvent.

✔ **Reading Check** **Explain** How do you know which substance is the solute in a solution?

Nonliquid solutions Solutions can also be gaseous or even solid. Examples of a gaseous solution and a solid solution are shown in **Figure 2.** Did you know that the air that you breathe is a solution? In fact, all mixtures of gases are solutions. Air is a solution of 78 percent nitrogen, 21 percent oxygen, and small amounts of other gases, such as argon, carbon dioxide, and water vapor.

The sterling silver and brass used in musical instruments are examples of solid solutions. Sterling silver contains 92.5 percent silver and 7.5 percent copper. Brass is a solution of copper and zinc metals. Solid solutions are known as alloys. An **alloy** is a mixture of elements that has metallic properties. Alloys are made by melting the solid solute and solvent together. Most coins, as shown in **Figure 3,** are alloys.

■ **Figure 2** Solutions can also be mixtures of solids or gases. Gases naturally diffuse within a common container. But solids, such as bronze, must be in a molten state to combine into a solution.

 **Inquiry** Virtual Lab

 **Inquiry** Video Lab

FOLDABLES
Incorporate information from this section into your Foldable.

VOCABULARY ·
SCIENCE USAGE V. COMMON USAGE
Solution
Science usage
a homogeneous mixture
Sugar water is a solution of solid sugar particles mixed with water.

Common usage
an action or process of solving a problem
The solution for getting the car from the mud was a tow truck. ·

FIGURE 3

Visualizing Metal Alloys

How do vending machines recognize whether the money being placed into them is the correct currency? They recognize coins by size, mass, and, sometimes, electrical conductivity.

The golden Sacagawea dollar was first issued in 2000 to replace the silver Susan B. Anthony dollar. It was the same size and mass as the Susan B. Anthony dollar, but the electrical conductivities were different.

An alloy's electrical conductivity is based on the composition of the alloy. Each alloy has its own specific electrical conductivity. If the Sacagawea dollar failed to match the conductivity of the Susan B. Anthony dollar, vending machines would have to be modified or replaced to take the Sacagawea dollar. By adding the right combination of zinc, manganese, and nickel to copper, the correct conductivity was achieved and the expensive task of replacing vending machines was avoided.

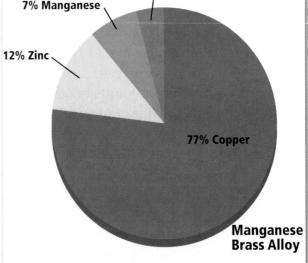

4% Nickel
7% Manganese
12% Zinc
77% Copper
Manganese Brass Alloy

After testing thousands of coins for conductivity, a manganese-brass alloy was chosen for the Sacagawea dollar.

Manganese brass alloy
Copper core
Manganese brass alloy

The Sacagawea dollar's copper core is half the coin's thickness. The outer layer is composed of the manganese brass alloy.

How Substances Dissolve

Fruity drinks made from powdered mixes, such as lemonade, are examples of solutions made by dissolving solids in liquids. Like the hummingbird food shown in **Figure 1,** they contain sugar as well as other substances, such as food coloring and additional flavors.

The dissolving of a solid in a liquid occurs at the surface of the solid. To understand how water solutions form, keep in mind two things that you have learned about water. Like the particles of any substance, water molecules are constantly moving, and water is a polar molecule.

✔ **Reading Check** **Identify** where a solid actually dissolves when placed in a liquid.

How it happens **Figure 4** shows the process of sugar dissolving in water. In step 1, the negative ends of water molecules are attracted to the positive ends of sugar crystals as water moves past the surface of the solid sugar. In step 2, the water molecules pull the sugar molecules into solution. Finally, in step 3, the water molecules and the sugar molecules mix evenly.

The process described in the three steps shown in **Figure 4** repeats as layer after layer of sugar molecules moves away from the crystal. The same three steps occur for any solid solute dissolving in a liquid solvent.

Dissolving liquids and gases A similar but more complex process takes place when liquids and gases dissolve. Liquid and gas particles move much more freely than particles of solids move. When gases dissolve in gases or when liquids and gases dissolve in liquids, particle movement eventually spreads solutes evenly throughout the solvent, resulting in a homogeneous mixture.

Dissolving solids in solids Solid particles do move a little, but this motion is not enough to spread particles evenly throughout a mixture. Solid metals are first melted and then mixed together while still molten. In the liquid state, the atoms can spread out evenly and will remain mixed after they have cooled.

■ **Figure 4** Dissolving sugar in water is a three-step process.

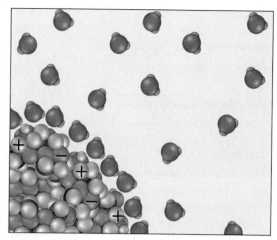

Step 1 At the surface of the sugar crystal, oppositely charged parts of the sugar and water molecules attract each other.

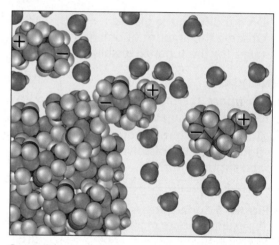

Step 2 Because the water molecules are moving in the liquid, they pull sugar molecules away from the crystal.

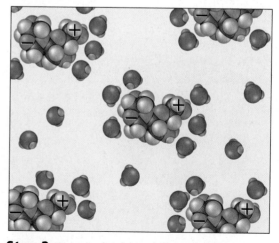

Step 3 Water molecules and sugar molecules continue to spread out until a homogeneous mixture forms.

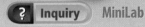
Observe the Effect of Surface Area

Procedure 🥽 🧤 🧹

1. Read the procedure and safety information, and complete the lab form.
2. Grind up two **sugar cubes.**
3. Place the ground sugar cubes into a **medium-sized glass,** and place two unground sugar cubes into a similar glass.
4. Add an equal amount of **water** at room temperature to each glass.
5. Do not disturb the glasses. Observe the amount of time it takes for the sugar to dissolve.

Analysis

1. **Compare** the times required to dissolve each sugar mixture.
2. **Conclude** What do you conclude about the dissolving rate and surface area?

■ **Figure 5** Surface area is the area of the exterior surface of an object, measured in square units. Increasing surface area increases the speed at which a solute is dissolved by a solvent.

Rate of Dissolving

If two substances form a solution, they will do so at a measureable rate. Sometimes the rate at which a solute dissolves into a solvent is fast. Other times it is slow. There are several things that you can do to speed up the rate of dissolving. Stirring, increasing the surface area of the solute, and increasing the temperature of the solvent are three of the most effective techniques.

Stirring Think about how you make a drink from a powdered mix. After you add the mix to water, you stir it. How can stirring speed up the dissolving process? Stirring a solution speeds up dissolving because it moves the solvent around, bringing more solvent into contact with the solute. The solvent attracts the particles of solute, causing the solid solute to dissolve faster.

Surface area Another way to speed up the dissolving of a solid in a liquid is to increase the surface area of the solute. Suppose you want to sweeten your water with a 5-g crystal of rock candy. If you put the whole crystal into a glass of water, it might take several minutes to dissolve, even with stirring. However, if you first grind the rock candy into a powder, it will dissolve in the same amount of water in only a few seconds.

Why does breaking up a solid cause it to dissolve faster? Breaking the solid into smaller pieces greatly increases its surface area, as you can see in **Figure 5.** Because dissolving takes place at the surface of the solid, increasing the surface area allows more solvent to come into contact with more solid solute. Therefore, the speed of the dissolving process increases.

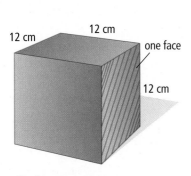

Surface area = 864 cm²

A face of a cube is the outer surface that has four edges. A cube has six faces of equal area.

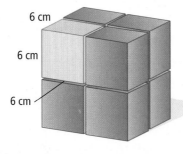

Surface area = 1,728 cm²

Pull apart the cube into eight smaller cubes of equal size. You now have a total of forty-eight faces.

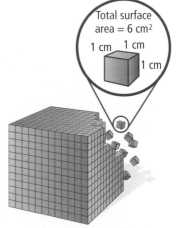

Total surface area = 6 cm²
1 cm   1 cm
1 cm

Surface area = 10,368 cm²

If you divide the cube into smaller cubes that are 1 cm on a side, you will have 1,728 cubes and 10,368 faces.

Calculate Surface Area Suppose the length, height, and width of a cube are each 1 cm. If the cube is cut in half to form two rectangular pieces, what is the total surface area of the new pieces?

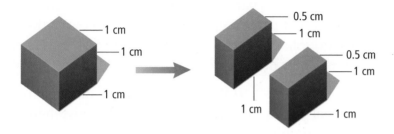

Identify the Unknown: Total surface area of the two new pieces

List Knowns: length = *l* = **1 cm**
height = *h* = **1 cm**
width = *w* = **0.5 cm**

Set Up the Problem: The rectangular solids each have six faces.
Surface area front and back = 2(***h*** × *w*)
Surface area left and right = 2(***h*** × *l*)
Surface area top and bottom = 2(*w* × *l*)
Surface area of one piece = 2(***h*** × *w*) + 2(***h*** × *l*) + 2(*w* × *l*)
Total surface area = Number of pieces × Surface area of one piece

Solve the Problem: Surface area of one piece =
2(**1 cm** × **0.5 cm**) + 2(**1 cm** × **1 cm**) + 2(**0.5 cm** × **1 cm**) = 4 cm²
The total surface area of the two new pieces = 2(4 cm²) = 8 cm²

Check the Answer: Total surface area of the original cube = 6(*w* × ***h***) = 6(**1 cm** × **1 cm**) = 6 cm²
Dividing the cube in two increased the surface area, which is reasonable.

Find **Additional Practice Problems** in the back of your book.

1. The length, height, and width of a cube are each 3 cm. If the cube is cut in half to form two rectangular pieces, what is the total surface area of the new pieces?

Review
Additional Practice Problems

2. If a cube that has a length, height, and width of 4 cm is broken down into 8 cubes of equal size, what is the surface area of the 8 new cubes?

3. **Challenge** A cube of salt with a length, height, and width of 5 cm each is attached along a face to another cube of salt with the same dimensions. How much surface area is lost by combining the cubes to form a rectangular solid?

■ **Figure 6** Hot water allows powdered chocolate mix to dissolve faster than cold water does. This means fewer solid clumps.

Temperature In addition to stirring and increasing the surface area of the solute, a third way to increase the rate at which most solids dissolve is to increase the temperature of the solvent. Think about making hot chocolate from a mix, as shown in **Figure 6.** The chocolate mix dissolves faster by mixing it with hot water instead of cold water. This is true of many solutions. Increasing the temperature of a solvent speeds up the movement of its particles. This temperature increase causes more solvent particles to come into contact with the solute. As a result, solute particles break loose and dissolve faster when the solvent is heated.

Controlling the process Think about how the three factors that you just learned affect the rate of dissolving. Can these factors combine to further increase the rate, or perhaps control the rate, of dissolving? Each technique of stirring, increasing the surface area, and heating, is known to increase the rate of dissolving by itself. When two or more techniques are combined, the rate of dissolving increases even more.

Consider a sugar cube placed in cold water. You know that the sugar cube will eventually dissolve. You can predict that heating the water will increase the rate by some amount. You can also predict that a combination of heating and stirring will further increase the rate. Finally, you can predict that increasing the surface area by crushing the cube combined with heating and stirring will result in the fastest rate of dissolving.

Section 1 Review

Section Summary

▶ A solution is a homogeneous mixture.

▶ Solutions are composed of solutes and solvents.

▶ Stirring, surface area, and temperature all affect the rate of dissolving.

4. **MAIN ‹Idea› Summarize** possible ways in which phases of matter could combine to form a solution.

5. **Draw** a diagram that shows how a solid dissolves in a liquid.

6. **Describe** how stirring, surface area, and temperature affect the rate of dissolving.

7. **Think Critically** Amalgams are sometimes used in tooth fillings and are made of mercury. Explain why an amalgam is a solution.

Apply Math

8. **Find Surface Area** Calculate the surface area of a rectangular solid with dimensions l = 2 cm, w = 1 cm, and h = 0.5 cm.

9. **Calculate Percent Increase** If the length of the rectangle in question 8 is increased by 10%, by what percentage will the surface area increase?

Reading Preview

Essential Questions

▶ How are the concentrations of solutions expressed?

▶ What is solubility?

▶ What are saturated, unsaturated, and supersaturated solutions?

▶ How do pressure and temperature affect the solubility of gases?

Review Vocabulary

solution: a homogeneous mixture that remains constantly and uniformly mixed and has particles that are so small that they cannot be seen with a microscope

New Vocabulary

concentration
solubility
saturated solution
unsaturated solution
supersaturated solution

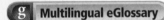 **Multilingual eGlossary**

Concentration and Solubility

MAIN Idea In a given amount of solvent, concentration is the amount of solute actually dissolved and solubility is the maximum amount of solute that can dissolve.

Real-World Reading Link Lemonade tastes a particular way because of the specific amounts of the lemon juice and sugar in the drink. For example, if it is not sweet enough, you can add sugar.

Concentration

Suppose you add one teaspoon of lemon juice to a glass of water to make lemonade. Your friend adds four teaspoons of lemon juice to the same amount of water in another glass. You could say that your glass of lemonade is diluted and your friend's lemonade is concentrated because your friend's drink now has more lemon flavor than your drink has. A concentrated solution is one in which a large amount of solute is dissolved in the solvent. A dilute solution is one that has a small amount of solute in the solvent. These are relative concentrations.

Precise concentrations How much real fruit juice is in an average fruit drink? *Concentrated* and *diluted* are not precise terms. However, concentrations of solutions can be described precisely. The **concentration** of a solution is the amount of solute actually dissolved in a given amount of solvent.

One way to state concentration precisely is to give the percentage by volume of the solute. The percentage by volume of the juice in the orange drink shown in **Figure 7** is 10 percent. Adding 10 mL of solute to 90 mL of the solvent makes 100 mL of a ten percent solution. This means that for a volume of 100 mL of juice drink, there is a volume of 10 mL of juice (the solute) and 90 mL of water (the solvent).

$$\frac{10 \text{ mL juice}}{10 \text{ mL juice} + 90 \text{ mL water}} \times 100 = 10\% \text{ by volume of juice}$$

■ **Figure 7** The concentrations of juice drinks often are given in percent by volume. Concentrations of juice drinks commonly range from 10 percent to 100 percent juice.

Identify *the product that has the highest concentration of orange juice.*

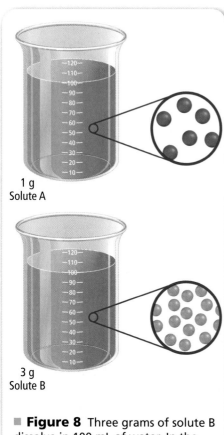

1 g
Solute A

3 g
Solute B

■ **Figure 8** Three grams of solute B dissolve in 100 mL of water. In the same temperature and volume of water, only one gram of solute A dissolves. This means that solute B is more soluble in water.

? Inquiry   Virtual Labs

How much can dissolve?

You can stir several teaspoons of sugar into lemonade, and the sugar will dissolve. However, if you continue adding sugar, a point is eventually reached when no more sugar dissolves and the excess sugar sinks to the bottom of the glass. This indicates how soluble sugar is in water. **Solubility** (sol yuh BIH luh tee) is the maximum amount of a solute that can be dissolved in a given amount of solvent at a given temperature. Solubility of substances dissolved in water is often expressed as grams of solute per 100 g of water (g/100 g water).

✓ **Reading Check Explain** What is solubility?

Comparing solubilities **Figure 8** shows two beakers with the same volume of water and two different solutes. In one beaker, one gram of Solute A dissolves completely, but additional solute does not dissolve and falls to the bottom of the beaker. In the other beaker, one gram of Solute B dissolves completely, and two more grams of Solute B also dissolve before additional solute begins to fall to the bottom of the beaker. If you assume that the temperature of the water is the same in both beakers, you can conclude that substance B is more soluble in water than substance A.

The solubilities of solutes in water vary. **Table 1** shows the solubilities of several substances in water at a temperature of 20°C. For solutes that are gases, such as hydrogen, oxygen, and carbon dioxide, the pressure must be given with the solubility, since the solubility can vary at different pressures.

Table 1	Solubility in Water at 20°C and Normal Atmospheric Pressure	
State of Substance	Substance	Solubility in g/100 g of Water
Solid	salt (sodium chloride)	35.9
	baking soda (sodium bicarbonate)	9.6
	washing soda (sodium carbonate)	21.4
	lye (sodium hydroxide)	109.0
	table sugar (sucrose)	203.9
Gaseous	hydrogen	0.00017
	oxygen	0.005
	carbon dioxide	0.16

Types of Solutions

How much solute can dissolve in a given amount of solvent? That depends on a number of factors, including the solubility of the solute. Here, you will examine three types of solutions that are defined by the amount of a solute dissolved in a solvent.

Saturated solutions If you add 35 g of copper(II) sulfate ($CuSO_4$) to 100 g of water at 20°C, only 32 g will dissolve. You have a saturated solution because no more copper(II) sulfate can dissolve. A **saturated solution** is a solution that contains all of the solute that it can hold at a given temperature. However, if you heat the mixture to a higher temperature, more copper(II) sulfate dissolves.

Generally, as the temperature of a liquid solvent increases, the amount of solid solute that can dissolve in it also increases. **Table 2** shows the amounts of a few solutes that can dissolve in 100 g of water at different temperatures. Each would form a saturated solution. Some of these are also compounds shown on the graph in **Figure 9.**

Solubility curves Each line on the graph from **Figure 9** is called a solubility curve for a particular substance. You can use a solubility curve to determine how much solute will dissolve at any temperature given on the graph. For example, about 79 g of potassium bromide (KBr) will form a saturated solution in 100 g of water at 50°C. How much sodium chloride (NaCl) will form a saturated solution with 100 g of water at the same temperature? You can also see that as temperature increases, so does the solubility of a substance.

Unsaturated solutions An **unsaturated solution** is any solution that can dissolve more solute at a particular temperature. Often times, when a saturated solution is heated to a higher temperature, it becomes unsaturated and is therefore able to dissolve more solute. The term *unsaturated* is not a precise term. If you look at **Table 2,** you will see that 35.9 g of sodium chloride (NaCl) forms a saturated solution in 100 g of water with a temperature of 20°C. However, an unsaturated solution of sodium chloride could be any amount less than 35.9 g in 100 g of water with a temperature of 20°C.

Table 2	Solubility of Compounds in g/100 g of Water		
Compound	0°C	20°C	100°C
Copper(II) sulfate	23.1	32.0	114
Potassium bromide	53.6	65.3	104
Potassium chloride	28.0	34.0	56.3
Potassium nitrate	13.9	31.6	245
Sodium chlorate	79.6	95.9	204
Sodium chloride	35.7	35.9	39.2
Sucrose (sugar)	179.2	203.9	487.2

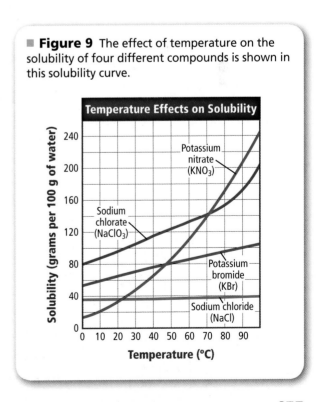

■ **Figure 9** The effect of temperature on the solubility of four different compounds is shown in this solubility curve.

A seed crystal of sodium acetate is added to a supersaturated solution of sodium acetate.

Excess solute immediately forms a solid.

A solid continues to form until the solution is saturated.

■ **Figure 10** A supersaturated solution is unstable.

Explain *why supersaturated solutions are unstable.*

Supersaturated solutions If you make a saturated solution of potassium nitrate (KNO_3) at 100°C and then let it cool to 20°C, part of the solute comes out of solution. At the lower temperature, the solvent cannot hold as much solute. Most other saturated solutions behave in a similar way when cooled.

However, if you cool a saturated solution of sodium acetate ($NaC_2H_3O_2$) from 100°C to 20°C without disturbing it, no solute comes out of solution. At this point, the solution is supersaturated. A **supersaturated solution** is one that contains more solute than a saturated solution at the same temperature. Supersaturated solutions are unstable. **Figure 10** shows that when a seed crystal of sodium acetate is dropped into the supersaturated solution, excess sodium acetate comes out of solution.

Solution Energy

The formations of some solutions are exothermic—they give off energy to the surrounding environment. One example is reusable heat packs. The heat packs contain a supersaturated solution of sodium acetate ($NaC_2H_3O_2$). The solution warms when Na^+ and $C_2H_3O_2^-$ ions interact with water molecules.

On the other hand, some substances must draw energy from the surroundings to dissolve. During this endothermic process, the solution becomes colder. Cold packs made of ammonium nitrate (NH_4NO_3) and water operate in this way. When the inner bag is broken, water mixes with ammonium nitrate. The interaction between ammonium nitrate and water draws energy from the surroundings, which causes the pack to cool.

Solubility of Gases

If you shake an opened bottle of a carbonated soft drink, it bubbles up and might squirt out. Shaking, stirring, or pouring a solution of a gas exposes more gas particles to the surface, where they escape from the liquid and come out of solution.

Figure 11 These carbonated soft drinks are bottled under pressure to keep carbon dioxide in solution. Opening the bottles reduces the pressure on the surface of the gas-liquid solution and carbon dioxide bubbles out of solution.

Pressure effects Carbonated soft drinks are bottled so that the pressure inside of the bottle is greater than the pressure outside of the bottle. This increases the amount of carbon dioxide dissolved in the liquid. When you open the bottle, the pressure inside of the bottle decreases and the carbon dioxide gas mixed in the soft drink escapes from the soft drink and bubbles out, as shown in **Figure 11.**

Temperature effects Another way to increase the amount of gas that dissolves in a liquid is to cool the liquid. This is just the opposite of what you do to increase the amounts of most solids dissolved in a liquid. For example, even more carbon dioxide will bubble out of a warm soft drink than a cold soft drink.

Section 2 Review

Section Summary

▶ Solubility curves help predict how much solute can dissolve at a particular temperature.

▶ Saturated, unsaturated, and supersaturated solutions are defined by how much solute can dissolve in a solvent.

▶ Solutions absorb or give off energy as they form.

▶ Temperature and pressure affect how much gas dissolves in a liquid.

10. **MAIN Idea Contrast** What is the difference between solubility and concentration?

11. **Compare and contrast** the difference between relative and precise concentrations. Give examples.

12. **Explain** Do all solutes dissolve to the same extent in the same solvent? How do you know?

13. **Identify** the type of solution that you have if solute continues to dissolve as you add more.

14. **Think Critically** Explain how keeping a carbonated beverage capped and refrigerated helps keep it from going flat.

Apply Math

15. **Calculate** By volume, orange drink is ten percent each of orange juice and corn syrup. A 1.5-L can of the drink costs $0.95. A 1.5-L can of orange juice is $1.49, and 1.5 L of corn syrup is $1.69. Per serving, does it cost less to make your own orange drink or to buy it?

Reading Preview

Essential Questions

▶ Why do some solutions conduct electricity?

▶ What are two ways that some solutes form ions in solution?

▶ How do solutes affect the freezing and boiling points of solvents?

Review Vocabulary

ion: charged particle that has greater or fewer electrons than protons

New Vocabulary

electrolyte
nonelectrolyte
ionization
dissociation

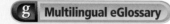 **Multilingual eGlossary**

Concepts in Motion

Animation

■ **Figure 12** Both hydrogen chloride (HCl) and water (H_2O) are polar molecules. Water surrounds the hydrogen chloride molecules and pulls them apart, forming positive hydrogen ions (H^+) and negative chloride ions (Cl^-). The water molecules attract the hydrogen ions and form hydronium ions (H_3O^+).

Particles in Solution

MAIN ◀Idea Dissolved particles can both lower the freezing point and raise the boiling point of a solvent.

Real-World Reading Link If you have ever made homemade ice cream, you know that one of the ingredients is salt. It might have struck you as strange that most of it was added to the ice, not the actual ice cream mixture. Salt helps to freeze the ice cream.

Ion Formation in Solution

Did you know that there are charged particles in your body that conduct electricity? In fact, you could not live without them. Some help nerve cells transmit messages. Each time that you blink your eyes or wave your hand, nerves control how muscles respond. Recall that these charged particles are called ions. Compounds that produce solutions of ions in water are known as **electrolytes.**

Solutions containing electrolytes conduct electricity. Some substances, such as sodium chloride, are strong electrolytes because they are entirely in the form of ions in solution. Strong electrolytes conduct a strong current. Other substances, such as acetic acid in vinegar, remain mainly in the form of molecules when they dissolve. They produce few ions, conduct a weak current, and they are called weak electrolytes. Substances that form no ions in water and do not conduct electricity are called **nonelectrolytes.** Fats and sugars are examples of nonelectrolytes.

Ionization Solutions of electrolytes form in two ways. One way applies to molecular solutions, such as hydrogen chloride in water. The molecules of hydrogen chloride are composed of neutral atoms. In order to form ions, the molecules must be broken apart so that the atoms take on a charge. The process in which molecular compounds dissolve in water and form charged particles is called **ionization.** The process is shown in **Figure 12,** using hydrogen chloride (HCl) as a model.

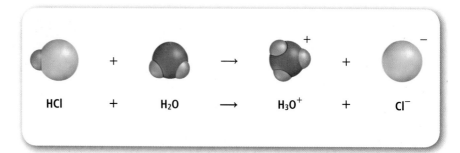

Dissociation The second way that solutions of electrolytes form is the separation of ions in ionic compounds. The ions already exist in the ionic compound. Polar water molecules surround the ionic compound and pull apart the compound into its individual ions. **Dissociation** is the process in which positive and negative ions of an ionic solid mix with the solvent to form a solution.

✓ **Reading Check Name** the two ways that solutions of electrolytes form.

A model of a sodium chloride (NaCl) crystal is shown in **Figure 13.** In the crystal, each positive sodium ion (Na^+) is attracted to six negative chloride ions (Cl^-). Each of the negative chloride ions is attracted to six positive sodium ions, and these attractions create a continuous pattern that exists throughout the crystalline structure.

When placed in water, the crystalline structure breaks apart. Remember that water molecules are polar, which means that the positive ends of the water molecules—the hydrogen atoms in a water molecule—are attracted to the negative chloride ions. Likewise, the negative end of the water molecule—the oxygen atom—is attracted to the positive sodium ions.

In **Figure 14,** water molecules approach the sodium ions (Na^+) and chloride ions (Cl^-) in the crystal. The water molecules break apart the crystalline structure by pulling the ions away and surrounding them in solution. The sodium and chloride ions have dissociated. The solution now consists of sodium and chloride ions mixed with water. The ions move freely through the solution and are capable of conducting an electric current.

✓ **Reading Check Compare and Contrast** What are the differences and similarities between dissociation and ionization?

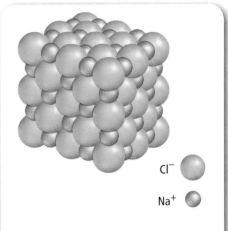

Cl⁻ ○
Na⁺ ◉

■ **Figure 13** This is a model of a sodium chloride crystal. Each chloride ion is surrounded by six sodium ions and vice versa.

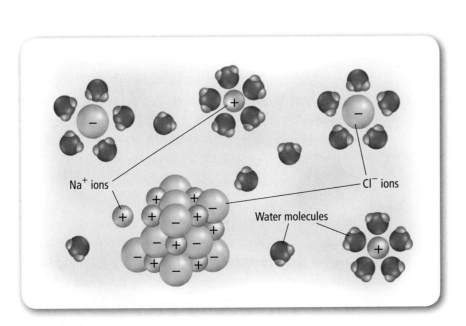

Na⁺ ions

Cl⁻ ions

Water molecules

Concepts in Motion

Animation

■ **Figure 14** Sodium chloride (NaCl) is an ionic compound, and water is a polar molecule. Sodium chloride dissociates as water molecules attract and pull the sodium and chloride ions from the crystal. Water molecules then surround and separate the Na⁺ and Cl⁻ ions.

Explain *Why will sodium chloride in solution conduct electricity?*

Effects of Solute Particles

All solute particles—polar and nonpolar, electrolyte and nonelectrolyte—affect the physical properties of the solvent. These effects can be useful. For example, adding antifreeze to water in a car radiator lowers the freezing point of the radiator fluid. Salt added to the ice and water mixture in an ice cream maker lowers the freezing point of the solution, and the ice cream freezes faster. The effect that a solute has on a solvent depends on the number of solute particles in solution and not on the chemical nature of the particles.

Lowering freezing point Adding a solute, such as antifreeze, to a solvent lowers the freezing point of the solvent. The amount that the freezing point lowers depends upon the concentration of the solute particles that you add.

As a substance freezes and changes state from a liquid to a solid, the particles arrange themselves in an orderly pattern. A solute interferes with the formation of this pattern, making it harder for the solvent to freeze, as shown in **Figure 15.** To overcome this, the temperature of the solvent must decrease to freeze the solution.

Animal antifreeze Animals that live in extremely cold climates have their own internal antifreeze. Caribou, for example, contain substances in the lower sections of their legs that prevent freezing in subzero temperatures. The caribou can stand for long periods of time in snow with no harm to their legs.

Fish living in polar waters also have a natural chemical antifreeze called glycoprotein (gli koh PROH teen) in their bodies. Glycoprotein prevents ice crystals from forming in moist tissues in fish.

Raising boiling point Surprisingly, antifreeze also raises the boiling point of the radiator fluid. How can it do this? Solute particles interfere with the solvent particles as they transition from liquid to gas at the surface of the solution. This lowers the vapor pressure. As a result, more energy is needed for the solvent to escape from the liquid surface. The boiling point of the solution will be higher than the boiling point of pure solvent. The amount the boiling point is raised depends upon the concentration of solute present.

■ **Figure 15** Freezing is the change from the liquid to the solid state. If a solute is added, it can prevent the liquid from forming an orderly solid at its usual freezing point.

Explain *how adding a solute affects the freezing point of a solution.*

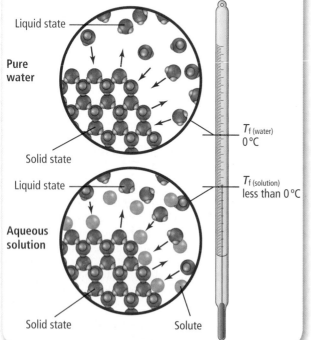

Car radiators The beaker in **Figure 16** on the left represents a car radiator when it contains water only—no antifreeze. Some of the water molecules on the surface will vaporize and change to the gaseous state. The number of water molecules that vaporize depends upon the temperature of the liquid water. As temperature increases, water molecules move more quickly and more molecules vaporize. Finally, when the pressure of the water vapor equals atmospheric pressure, the water begins to boil.

Figure 16 on the right shows the result of adding antifreeze to water. Particles of solute are evenly distributed throughout the solution, including at the surface. Now fewer water molecules can reach the surface and vaporize, making the vapor pressure of the solution lower than that of the pure solvent. This means that it will require a higher temperature to make the water boil. Antifreeze increases the boiling point of the water in a vehicle's radiator and helps prevent overheating.

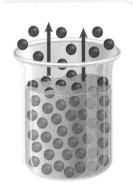 

In a beaker of pure water, water molecules vaporize freely from the surface.

Solute particles block part of the surface, making it more difficult for solvent to vaporize.

■ **Figure 16** Solute particles raise the boiling point of a solution.

Describe *how antifreeze in a car can help prevent both freezing and overheating.*

✔ **Reading Check Describe** How does antifreeze affect the vapor pressure of a pure solvent?

Section 3 Review

Section Summary

▶ Ionization and dissociation are two ways to form ions in solution.

▶ Solute particles lower the freezing point of a solution.

▶ Solute particles raise the boiling point of a solution.

16. **MAIN Idea Explain** how the concentration of a solute in a solution influences its boiling point and freezing point.

17. **Identify** what kinds of solute particles are present in water solutions of electrolytes and nonelectrolytes.

18. **Determine** whether ionization or dissociation has taken place if calcium phosphate ($Ca_3(PO_4)_2$) breaks into Ca^{2+} and PO_4^{3-}.

19. **Think Critically** People often put salt on ice that forms on sidewalks and driveways during the wintertime. The salt helps melt the ice, forming a saltwater solution. Explain why this solution resists refreezing.

20. **Graph** Use the data points (0, 12), (10, 8), (20, 4), and (30, 0) to graph the effect of a solute on the freezing point of a solvent. Label the *x*-axis *Solute (g)* and the *y*-axis *Freezing point (degrees Celsius).* Find the slope of the line that you graph.

Objective

■ **Determine** how adding salt to water affects the boiling point of water.

Background: Adding salt to water and adding antifreeze to a car radiator have a common result—increasing the boiling point.

Question: *How much can the boiling point of a solution be changed?*

Preparation

Materials

distilled water (400 mL)
Celsius thermometer
table salt ($NaCl$) (72 g)
ring stand
hot plate
250-mL beaker

Safety Precautions

Procedure

1. Read the procedure and safety information, and complete the lab form.

2. Copy the data table shown below. Bring 100 mL of distilled water to a gentle boil in a 250-mL beaker. Record the temperature in degrees Celsius. Do not touch the hot plate surface.

Effects of Solute on Boiling Point

Grams of NaCl Solute	Boiling Point (°C)
0	
12	
24	
36	

3. Dissolve 12 g of $NaCl$ in 100 mL of distilled water. Bring this solution to a gentle boil, and record its boiling point in the table.

4. Add 12 g more of $NaCl$ to bring the total $NaCl$ in solution to 24 g. Record the solution's boiling point in the table.

5. Repeat step 4, for a total of 36 g $NaCl$ in solution, and record the boiling point in the table.

6. Make a graph of your results. Place grams of $NaCl$ on the *x*-axis, and place boiling point on the *y*-axis.

Conclude and Apply

1. **Summarize** the results of steps 2–5.

2. **Explain** the differences between the boiling points of pure water and a saltwater solution.

3. **Predict** the effect of doubling the amount of water instead of the amount of $NaCl$ in step 4.

4. **Determine** the value for the boiling point temperatures of 6 g, 18 g, and 30 g of $NaCl$ in 100 mL of water as read from your graph.

5. **Predict** what would happen if you continued to add more salt. Would your graph continue in the same pattern or eventually level off? Design an experiment to test your prediction.

COMMUNICATE YOUR DATA

Determine Average your class's data and redraw the graph in procedure step 6 using the average data. As a class, discuss whether the average data is more accurate than the individual data.

? Inquiry   Lab

Reading Preview

Essential Questions

▶ What solutes do not dissolve well in water?

▶ How does polarity affect solubility?

▶ How does soap work?

Review Vocabulary

solvent: the substance in which the solute dissolves

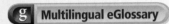 **Multilingual eGlossary**

Dissolving Without Water

MAIN Idea Polarity determines which solvent will dissolve a substance.

Real-World Reading Link Do you take vitamins? Some vitamins stay in your body and accumulate, and others quickly leave your body. What is the difference? The length of time that they will remain is determined by whether the vitamins dissolve in fat or in water.

When Water Will Not Work

Water often is referred to as the universal solvent because it can dissolve so many substances. However, there are some substances, such as oil, that cannot dissolve in water. Why?

As you learned in the first section, water molecules have positive and negative areas that allow them to attract polar or ionic solutes. Recall that polar molecules have positive areas and negative areas. However, nonpolar molecules have no separated positive and negative areas. Because of this, nonpolar molecules are not attracted to ionic or polar substances, such as water.

Nonpolar solutes Nonpolar molecules do not dissolve in water or only dissolve a very small amount. For example, the oil-contaminated water shown in **Figure 17** shows how the oil, which consists of nonpolar molecules, does not mix with seawater, a polar solution of water, salt, and nutrients.

Oils contain large molecules made of carbon and hydrogen atoms, which are called hydrocarbons. In hydrocarbons, carbon and hydrogen atoms share electrons in a nearly equal manner. The nonpolar oil molecules are not attracted to the polar water molecules in the ocean and will not dissolve.

■ **Figure 17** Oil is composed of nonpolar molecules, and water molecules are polar. Oil does not mix with water and remains on the surface of the water.

Nonpolar solvents Some substances around your house may be useful as nonpolar solvents. For example, mineral oil can dissolve candle wax from glass or metal candleholders. Mineral oil and wax are nonpolar substances. Oily peanut butter, a nonpolar solvent, is sometimes used to remove bubble gum, also a nonpolar substance, from hair.

Many nonpolar solvents are connected with specific jobs. Oil-based paints contain pigments that are dissolved in oil. In order to thin or remove such paints, a nonpolar solvent must be used. People who paint pictures using oil-based paints probably use the solvent turpentine. It comes from the sap of a pine tree. **Figure 18** shows how well turpentine dissolves nonpolar paint.

Dry cleaners also use nonpolar solvents. The word *dry* refers to the fact that no water is used in the process. Because molecules of a nonpolar solute can easily slip in among molecules of a nonpolar solvent, dry cleaning removes oil and grease stains.

Drawbacks of nonpolar solvents Although nonpolar solvents have many uses, they have some drawbacks, too. First, many nonpolar solvents are flammable. Also, some solvents are extremely toxic. These solvents are hazardous if they come into contact with skin or if their vapors are inhaled. For these reasons, you must always be careful when handling these substances and never use them in an enclosed area. Good ventilation is critical because nonpolar solvents tend to evaporate more readily than water. Even small amounts can produce high concentrations of harmful vapor in the air.

Versatile Molecules

Some substances are versatile because they have a nonpolar end and a polar end. For example, **Figure 19** shows that the alcohol ethanol has a polar and a nonpolar end. In addition, sodium stearate has a polar and a nonpolar end. As a result, ethanol and sodium stearate can dissolve polar and nonpolar substances. Sodium stearate is an important ingredient in soap.

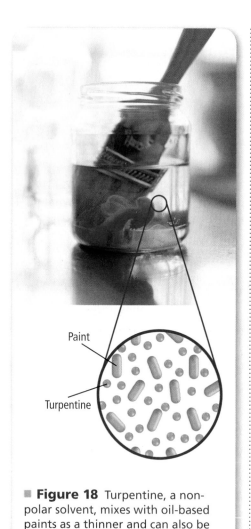

Figure 18 Turpentine, a nonpolar solvent, mixes with oil-based paints as a thinner and can also be used as a brush cleaner.

Paint
Turpentine

Figure 19 Ethanol and sodium stearate are examples of molecules that have both polar and nonpolar ends.

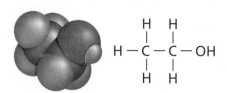

Ethanol (C₂H₅OH) has a polar −OH group and a nonpolar −C₂H₅ group.

The long end the hydrocarbon chain for sodium stearate is nonpolar. One end of the molecule is ionic.

How soap works The oils on human skin and hair keep them from drying out, but the oils can also attract and hold dirt. The oily dirt is a nonpolar mixture, so washing with water alone will not clean away the dirt. This is why soap is needed.

Soaps are substances that have polar and nonpolar properties. Soaps are salts of fatty acids. Fatty acids are long, hydrocarbon molecules with a nonpolar end and a carboxylic acid group –COOH at the other end. When a soap is made, the hydrogen atom of the acid group is removed, leaving a negative charge behind to form an ionic bond with a positive ion of sodium or potassium. For example, when Na^+ bonds with $-COO^-$, the salt, sodium stearate shown in **Figure 19** is made.

☑ **Reading Check Summarize** Why is soap required to clean oily dirt?

The ionic end of soap dissolves in water, and the long hydrocarbon portion dissolves in oily dirt. In this way, the dirt is removed from your skin, hair, or a fabric, suspended in the wash water, and washed away, as shown in **Figure 20.**

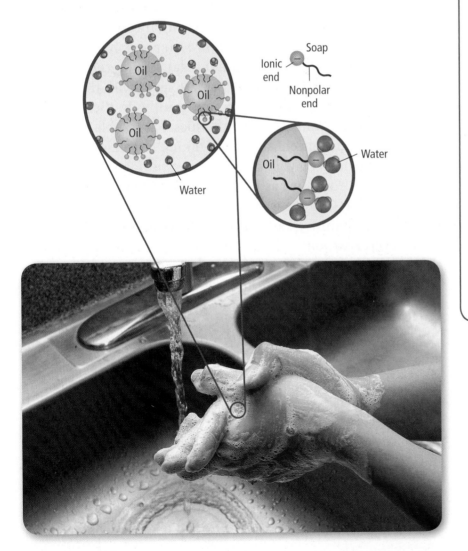

■ **Figure 20** Soap cleans because its nonpolar hydrocarbon part dissolves in oily dirt and its ionic part interacts strongly with water. The soap carries the oily dirt along as you use the water to rinse.

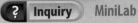

MiniLab

? Inquiry   MiniLab

Observe Clinging Molecules

Procedure 🖐️ 🥽 👕 🚫 💧

WARNING: *Rubbing alcohol is flammable.*

1. Read the procedure and safety information, and complete the lab form.
2. Lay **two clean pennies** side by side and heads up on a **paper towel.**
3. Slowly place drops of **water** from a **dropper** onto the head of one penny. Count each drop, and continue until the accumulated water spills off the edge of the penny.
4. With adult supervision, repeat step 3 using **rubbing alcohol** (a solution of approximately 70 percent isopropyl alcohol) and the other penny.

Analysis

1. **Identify** Which penny held the most drops before liquid spilled over the edge?
2. **Determine** How polar do you think isopropyl alcohol (C_3H_7OH) is compared to water?
3. **Conclude** How do the results of the investigation support the concept of polarity and the attraction between molecules?

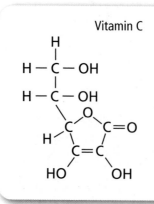

Figure 21 The structural formula of vitamin A shows a long hydrocarbon chain that makes it nonpolar. Foods such as liver, lettuce, cheese, eggs, carrots, sweet potatoes, and milk are good sources of this fat-soluble vitamin. Vitamins D, E, and K are also fat-soluble vitamins.

Vitamin A

Polarity and Vitamins

Taking the right types of vitamins in the correct doses is important to your health. Some of the vitamins that you need, such as vitamin A, as shown in **Figure 21,** are nonpolar and dissolve in fat, which is another nonpolar substance. Because fat and fat-soluble vitamins do not wash away with the water that is present in the cells throughout your body, the excess vitamins can accumulate in your tissues. Fat-soluble vitamins are toxic in high concentrations, so taking large doses that are not recommended by your physician can be dangerous.

Other vitamins, such as vitamins B and C, are polar molecules, meaning they are water soluble. When you look at the structure of vitamin C, as shown in **Figure 22,** you will see that it has several carbon-to-carbon bonds. This might make you think that it is nonpolar. But, if you look again, you will see that it also has several oxygen-to-hydrogen bonds that resemble those found in water. This makes vitamin C polar.

Figure 22 Although vitamin C has carbon-to-carbon bonds, it is water soluble because it also has polar groups. Foods that are good sources of vitamin C help heal wounds and help the body absorb iron.

Explain *Compare the number of oxygen atoms in vitamin C with the number in vitamin A in Figure 21. What effect does oxygen have in these two molecules?*

Vitamin C

Table 3	Sources of Vitamin C	
Food	Serving Size	Amount of Vitamin C (mg)
Orange juice, fresh	1 cup	124
Green peppers, raw	$\frac{1}{2}$ cup	96
Broccoli, raw	$\frac{1}{2}$ cup	70
Cantaloupe	$\frac{1}{4}$ melon	70
Strawberries	$\frac{1}{2}$ cup	42

Polar vitamins dissolve readily in the water that is present in your body. These vitamins do not accumulate in tissue because any excess vitamin is washed away. For this reason, you must replace water-soluble vitamins more quickly than fat-soluble vitamins by eating enough of the foods that contain them or by taking vitamin supplements. **Table 3** shows a variety of sources of vitamin C. In general, the best way to stay healthy is to eat a variety of healthy foods. Such a diet will supply the vitamins that you need with no risk of overdoses.

 Reading Check Restate Why is it necessary to replace water-soluble vitamins more quickly than fat-soluble vitamins?

VOCABULARY
WORD ORIGIN
Polar
comes from the Latin *polus* meaning *pole*
Polar solvents can dissolve polar solutes.

Section 4 Review

Section Summary

▶ Polar solvents dissolve polar solutes, and nonpolar solvents dissolve nonpolar solutes.

▶ Nonpolar solvents have many household and industrial uses.

▶ Some molecules have both polar and nonpolar parts and are versatile solvents.

21. **MAIN Idea Explain** how a polar solvent dissolves a polar solute and how a nonpolar solvent dissolves a nonpolar solute.

22. **Describe** polar and nonpolar molecules.

23. **Explain** how one solute can dissolve in both polar and nonpolar solvents.

24. **Draw** a diagram to explain how soap cleans your hands.

25. **Think Critically** What might happen to your skin if you washed too often?

Apply Math

26. **Calculate** If 60 mg of vitamin C in a multivitamin provides only 75 percent of the recommended daily dosage for children, how much is recommended?

27. **Interpret** To get the recommended dose of vitamin C, approximately how much fresh orange juice must you drink? (Refer to Table 3.)

✓ **Assessment** Online Quiz

Objective

- **Observe** the effects of temperature on the amount of solute that dissolves.

Background: Two major factors to consider when you are dissolving a solute in water are temperature and the ratio of solute to solvent. What happens to a solution as the temperature changes? To be able to draw conclusions about the effect of temperature, you must keep other variables constant. For example, you must be sure to stir each solution in a similar manner.

Question: *How does solubility change as temperature is increased?*

Preparation

Materials

water at room temperature
large test tubes
Celsius thermometer
table sugar
copper wire stirrer, bent into a spiral, as shown in the photo above, right
test-tube holder
graduated cylinder (25 mL)
beaker (250 mL) with 150 mL of water
electric hot plate
test-tube rack
ring stand and clamp
balance

Safety Precautions

WARNING: *Do not touch the test tubes or the hot plate surface when the hot plate is turned on or is cooling down. When heating a solution in a test tube, keep it pointed away from yourself and from others. Do not remove goggles until clean up, including washing hands, is completed.*

Procedure

1. Read the procedure and safety information, and complete the lab form.
2. Copy the data table below.
3. Place 20 mL of water in a test tube.
4. Add 30 g of sugar to the same test tube in step 3.
5. Stir the solution. Does the sugar dissolve?

Data Table	
Temperature (°C)	Total Grams of Sugar Dissolved

6. If it dissolves completely, add another 5 g of sugar to the test tube. Does it dissolve?

7. Continue adding 5 g amounts of sugar at a time until no more sugar dissolves.

8. Now place a beaker with 150-mL of water in it on the hot plate. Carefully hang the thermometer from the ring stand so that the bulb is immersed about halfway into the beaker, making sure that the bulb does not touch the sides or bottom. Record the starting temperature.

9. Using a test-tube holder, place the test tube into the water.

10. Turn on the hot plate. Gradually increase the temperature of the hot plate, while stirring the solution in the tube until all of the sugar dissolves.

11. Note the temperature at which all of the sugar dissolves.

12. Add another 5 g of sugar, and continue stirring. Note the temperature at which this additional sugar dissolves.

13. Continue in this manner until you have six data points. Turn off the hot plate. Note the total amount of sugar that has dissolved.

Analyze Your Data

1. **Graph** Analyze your data by constructing a line graph. Place grams of sugar per 100 g of water on the *y*-axis, and place temperature in degrees Celsius on the *x*-axis. Because you used only 20 mL of water, multiply the number of grams of sugar by five.

2. **Interpret** Using your graph, estimate the solubility of sugar at 100°C and at 0°C, the boiling and freezing points of water, respectively.

Conclude and Apply

1. **Determine** How did the saturation change as the temperature was increased?

2. **Evaluate** Compare your results with those given in **Table 2.** Calculate the percent error of your results for the solubility of sugar at 0°C, 20°C, and 100°C.

3. **Analysis** What might have caused the percent errors calculated in question 2?

Evaluate Many, but not all, solutes can form supersaturated solutions with water. Work with your class to develop a test to evaluate whether sugar can form a supersaturated solution in water.

SCIENCE & HISTORY

HIDDEN NOBEL PRIZE MEDALS

On April 9, 1940, chemist George de Hevesy faced a dilemma. Outside his laboratory in Copenhagen, Denmark, Nazi forces were occupying the city. Inside, he held in his hand two Nobel Prize medals for physics, belonging to Max von Laue and James Franck, German scientists who had protested Nazi policies. They had left their medals, which resembled the medal shown in **Figure 1,** in Copenhagen for safekeeping.

Dissolving gold De Hevesy decided that the only way to make sure the Nazis would not seize the medals was to hide them in plain sight. Gold is not a very reactive element. One of the only chemicals that can accomplish the task is aqua regia, a mixture of nitric acid and hydrochloric acid. De Hevesy reacted the Nobel Prize medals with aqua regia, put the resulting solution in a jar, and stored it on a shelf with other liquid solutions.

Figure 2 The precipitate lead iodide is formed when lead nitrate reacts with sodium iodide.

Although the Nazis searched the laboratory, they left the solution undisturbed. After the end of World War II, the gold was extracted from the solution and sent to the Royal Swedish Academy of Sciences, the group that awards Nobel Prizes. Franck and von Laue later received newly cast medals to replace the ones that de Hevesy had dissolved.

The process of precipitation In the Nobel Prize medal example, the gold and aqua regia reacted to form substances that dissolve in water. Together, they formed a solution. The gold was extracted from the solution by a process called precipitation, as shown in **Figure 2.** Precipitation occurs when a reaction in a solution forms a solid product, called a precipitate, that does not dissolve in the solvent. De Hevesy's gold formed a precipitate that could be filtered from the liquid, cleaned, dried, and molded into the replacement medals.

Figure 1 Nobel Prizes are awarded in several categories including chemistry, medicine, physics, literature, economics, and peace.

WebQuest **Create** Suppose you are a journalist present the day that Franck and von Laue received their recast Nobel Prize medals. Write a news article describing the scene and the interview that would take place with the scientists.

THEME FOCUS Structures and Properties of Matter
Solutions form as interactions between solute and solvent particles cause them to become mixed. The solutions can be described in terms of solubility and concentration.

BIG Idea A solution is a homogeneous mixture of a solute or solutes and a solvent or solvents.

Section 1 How Solutions Form

alloy (p. 647)
solute (p. 647)
solvent (p. 647)

MAIN Idea A solution forms when a solute or solutes and a solvent or solvents become evenly mixed.

- A solution is a homogeneous mixture.
- Solutions are composed of solutes and solvents.
- Stirring, surface area, and temperature all affect the rate of dissolving.

Section 2 Concentration and Solubility

concentration (p. 653)
saturated solution (p. 655)
solubility (p. 654)
supersaturated solution (p. 656)
unsaturated solution (p. 655)

MAIN Idea In a given amount of solvent, concentration is the amount of solute actually dissolved and solubility is the maximum amount of solute that can dissolve.

- Solubility curves help predict how much solute can dissolve at a particular temperature.
- Saturated, unsaturated, and supersaturated solutions are defined by how much solute can dissolve in a solvent.
- Solutions absorb or give off energy as they form.
- Temperature and pressure affect how much gas dissolves in a liquid.

Section 3 Particles in Solution

dissociation (p. 659)
electrolyte (p. 658)
ionization (p. 658)
nonelectrolyte (p. 658)

MAIN Idea Dissolved particles can both lower the freezing point and raise the boiling point of a solvent.

- Ionization and dissociation are two ways to form ions in solution.
- Solute particles lower the freezing point of a solution.
- Solute particles raise the boiling point of a solution.

Section 4 Dissolving Without Water

MAIN Idea Polarity determines which solvent will dissolve a substance.

- Polar solvents dissolve polar solutes, and nonpolar solvents dissolve nonpolar solutes.
- Nonpolar solvents have many household and industrial uses.
- Some molecules have both polar and nonpolar parts and are versatile solvents.

Standardized Test Practice

Multiple Choice

Record your answers on the answer sheet provided by your teacher or on a sheet of paper.

Use the graph below to answer questions 1–3.

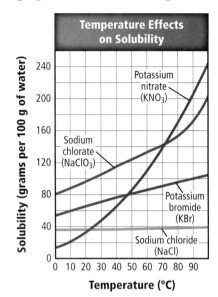

Temperature Effects on Solubility

1. How much potassium nitrate (KNO_3) will you have to add to 100 g of water at 40°C to prepare a saturated solution?
   A. 60 g
   B. 70 g
   C. 100 g
   D. 240 g

2. If 25 g of sodium chlorate ($NaClO_3$) are dissolved in 100 g of water at 70°C, how would you describe the solution?
   A. concentrated
   B. supersaturated
   C. saturated
   D. unsaturated

3. Which of the following will make a saturated solution if added to 100 g of water?
   A. 20 g of NaCl if the water is 50°C
   B. 100 g of KBr if the water is 90°C
   C. 80 g of $NaClO_3$ if the water is 30°C
   D. 60 g of KNO_3 if the water is 100°C

4. Which of the following statements about solubility is true as the temperature increases?
   A. The solubility of both gases and solids increases.
   B. The solubility of both gases and solids decreases.
   C. The solubility of gases increases, and the solubility of solids decreases.
   D. The solubility of gases decreases, and the solubility of solids increases.

Use the table below to answer question 5.

Solubility of Substances in Water at 20°C	
Substance	Solubility in g/100 g of Water
Sodium bicarbonate	9.6
Sodium carbonate	21.4
Sodium hydroxide	109.0
Sucrose	203.9

5. Which of the following is the most soluble in water at 20°C?
   A. sodium hydroxide
   B. sucrose
   C. sodium bicarbonate
   D. sodium carbonate

6. Which of the following statements about solute surface area is true?
   A. Grinding increases the surface area of the solute and slows down dissolving.
   B. Grinding increases the surface area of the solute and speeds up dissolving.
   C. Grinding decreases the surface area of the solute and slows down dissolving.
   D. Grinding decreases the surface area of the solute and speeds up dissolving.

✓ Assessment Standardized Test Practice

Short Response

Record your answers on the answer sheet provided by your teacher or on a sheet of paper.

7. When you take clothing to the dry cleaner, it is important to identify any stains that are on the clothing. Why does the dry cleaner need this information?

Use the figure below to answer question 8.

8. The drawing above shows carbon dioxide gas dissolved in water. Assuming temperature and volume are equal, which bottle could contain more dissolved CO_2? Explain.

9. Why does shaking or pouring a carbonated drink cause gas to come out of solution?

10. The solubility of potassium chloride in water is 34 g per 100 g of water at 20°C. A warm solution containing 100 g of potassium chloride in 200 g of water is cooled to 20°C. How many grams of potassium chloride will come out of solution? (Assume the solution is not supersaturated.)

11. Why is salt mixed with ice in an ice cream maker?

Extended Response

Record your answers on a sheet of paper.

12. Why can carp, catfish, and other fish with low-oxygen needs live in warmer waters than trout, which need more oxygen to survive?

Use the figure below to answer question 13.

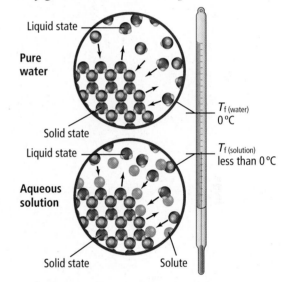

13. Explain what the figure shows. Describe this process.

14. You are given a clear-water solution containing potassium nitrate. How could you determine whether the solution is unsaturated, saturated, or supersaturated?

15. A solution conducts electricity. What do you know about the solution?

16. Why does antifreeze also raise the boiling point of water?

NEED EXTRA HELP?																
If You Missed Question . . .	1	2	3	4	5	6	7	8	9	10	11	12	13	14	15	16
Review Section . . .	2	2	2	1, 2	2	1	4	2	2	2	3	2	3	2	3	3

CHAPTER 22

Acids, Bases, and Salts

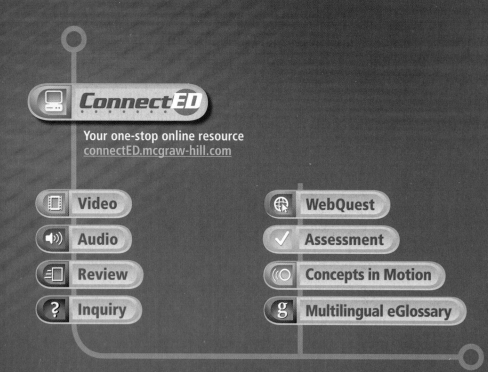

ConnectED

Your one-stop online resource
connectED.mcgraw-hill.com

- Video
- Audio
- Review
- Inquiry
- WebQuest
- Assessment
- Concepts in Motion
- Multilingual eGlossary

Launch Lab
The Effects of Acid Rain

Water mixes with rain and carbon dioxide in the atmosphere to create acid which can damage buildings and statues. Observe this reaction using soda water to represent acid rain and chalk to represent calcium carbonate, such as limestone and marble.

For a lab worksheet, use your StudentWorks™ Plus Online.

? Inquiry Launch Lab

FOLDABLES

Make a three-tab book. Label it as shown. Use it to compare and contrast your notes on acids, bases, and salts.

Acids   Salts   Bases

THEME FOCUS Structure and Properties of Matter

You can observe the properties of acids and bases in many everyday substances—from the sour taste of grapefruit juice to the slippery feel of soap.

BIG Idea Acids and bases can be defined in terms of hydrogen ions and hydroxide ions.

Reading Preview

Essential Questions

▶ What defines an acid or a base?
▶ How are common acids and bases used?
▶ How do acids and bases form ions in solution?

Review Vocabulary

electrolyte: compound that breaks apart in water, forming ions that can conduct electricity

New Vocabulary

acid
hydronium ion
indicator
hydroxide ion
base

 **Multilingual eGlossary**

 **Video** **BrainPOP**

Acids and Bases

MAIN Idea Acids produce hydronium ions (H_3O^+) in water, and bases produce hydroxide ions (OH^-) in water.

Real-World Reading Link You might have had orange juice this morning at breakfast. Then, you might have used dish soap to clean your glass. Orange juice and dish soap are just a few of the many everyday mixtures that can be classified as acids or bases.

Acids

What comes to mind when you hear the word *acid*? Do you think of a substance that can burn your skin or burn a hole through a piece of metal? Do you think about sour foods like those shown in **Figure 1?** Although some acids can burn and are dangerous to handle, the acids in foods are safe to eat.

Properties of acids An **acid** is a substance that produces hydrogen ions (H^+) in a water solution. The ability to produce these ions gives acids their characteristic properties. As acids dissolve in water, H^+ ions interact with water molecules to produce **hydronium ions** (H_3O^+) (hi DROH nee um • I ahnz).

Acids have several common properties. The sour taste of many foods is due to the presence of acids. However, taste should never be used to test for the presence of acids. Some acids can damage tissue by producing painful burns. Acids are corrosive. Some acids react strongly with certain metals, which seem to eat away the metals as metallic compounds and hydrogen gas form. An example of this type of reaction is zinc reacting with hydrochloric acid to produce zinc chloride and hydrogen gas. Acids also react with indicators to produce predictable changes in color. An **indicator** is an organic compound that changes color in the presence of acids and bases. For example, blue litmus paper is an indicator that turns red in acid.

■ **Figure 1** The acids in these common foods give them their distinctive sour tastes.

Name, Formula	Use	Other Information
Acetic acid (CH_3COOH)	food preservation, commercial organic syntheses	vinegar (about 5% acetic acid)
Acetylsalicylic acid ($HOOC–C_6H_4–OOCCH_3$)	pain relief, fever relief; to reduce inflammation	main component of aspirin
Ascorbic acid ($H_2C_6H_6O_6$)	antioxidant, vitamin	Vitamin C occurs naturally in some foods and is added to others.
Carbonic acid (H_2CO_3)	carbonated drinks	involved in cave formation and acid rain
Hydrochloric acid (HCl)	cleans steel in a process called pickling	Gastric juice in the stomach is a solution of HCl and water.
Nitric acid (HNO_3)	to make fertilizers	colorless, yet yellows when it is exposed to light
Phosphoric acid (H_3PO_4)	to make soft drinks, fertilizers, and detergents	Slightly sour but pleasant taste; detergents containing phosphates cause water pollution.
Sulfuric acid (H_2SO_4)	car batteries; to manufacture fertilizers and other chemicals	dehydrating agent that extracts water from air

Common acids Many foods contain acids. In addition to citric acid in citrus fruits, another example is lactic acid found in yogurt and buttermilk. It is also produced in your muscles during high levels of exercise when your muscles need oxygen. Any pickled food contains vinegar, also known as acetic acid. Your stomach uses hydrochloric acid to help digest your food.

At least four acids (sulfuric, phosphoric, nitric, and hydrochloric) play vital roles in industrial applications. All four of these acids are used in wastewater and water treatment. Sulfuric, phosphoric, and nitric acids are used in the production of fertilizers. Most of the nitric acid and sulfuric acid and approximately 90 percent of phosphoric acid produced are used for this purpose. Sulfuric acid is used to process pulp and make paper. Phosphoric acid is used in odor control. Nitric acid is involved in the production of nylon. Hydrochloric acid is part of the production of asphalt.

Reading Check   Name four acids that are important for industry.

Table 1 shows the names and formulas of a few acids, their uses, and some of their properties. Many acids can cause a chemical burn. For example, sulfuric acid reacts with skin cells and removes water as easily as it takes water from sugar, as shown in **Figure 2.**

■ **Figure 2** When sulfuric acid in the graduated cylinder is added to sugar in the beaker, a reaction occurs that produces black solid carbon and water.

MiniLab

? Inquiry MiniLab

Observe Acid Relief

WARNING: *Do not eat antacid tablets.*

Procedure 🥽 🧤 🚫 🧤 📋 ⚗️

1. Read the procedure and safety information, and complete the lab form.
2. Add 150 mL of **water** to a **250-mL beaker.**
3. Add three drops of **1*M* HCl** and 12 drops of **universal indicator.**
4. Observe the color of the solution.
5. Add an **antacid tablet** and observe for 15 minutes.

Analysis

1. **Describe** any changes that took place in the solution.
2. **Based** on the color of the mixture, how would you classify the HCl solution in step 3? How would you classify the solution at the end of step 5?
3. **Conclude** why the tablet is used to treat acid indigestion.

FOLDABLES
Incorporate information from this section into your Foldable.

■ **Figure 3** Bases are found in many cleaning products used in the home. For example, many glass cleaners contain ammonia (NH_3).

Identify *the property of bases that is evident in soaps.*

Bases

Most bases contain an OH^-, called a **hydroxide ion,** in their chemical formula. A **base** is a substance that produces hydroxide ions when it is dissolved in water. In addition, a base is any substance that accepts H^+ from acids. The definitions are related because the OH^- ions produced by some bases accept H^+ ions produced by some acids. Acids can be defined using this framework as well. An acid is any substance that donates H^+ to a base. In this definition, acids and bases are complements of one another—the base accepts the ion donated by the acid.

You might not be as familiar with bases as you are with acids. Although you can eat some foods that contain acids, you do not consume many bases. Some foods, such as egg whites, are slightly basic. Other examples of basic materials are baking powder and amines, organic compounds with a $-NH_2$ group, found in some foods. Some medicines, such as milk of magnesia and antacids, are also basic. Still, you come in contact with many bases every day. For example, each time you wash your hands using soap, you are using a base. Bases remove dirt and grime.

Bases also are important in many types of cleaning products, such as those shown in **Figure 3.** Bases are important in industry. For example, sodium hydroxide is used in the paper industry to separate fibers of cellulose from wood pulp. These fibers are made into paper.

Properties of bases Although acids and bases share some properties in common, bases have their own characteristic properties. In the pure, undissolved state, many bases are crystalline solids. In solution, bases feel slippery and have a bitter taste. Like strong acids, strong bases are corrosive and contact with skin can result in severe chemical burns. Therefore, taste and touch should never be used to test for the presence of a base. Finally, like acids, bases react with indicators to produce changes in color. Red litmus paper turns blue in bases.

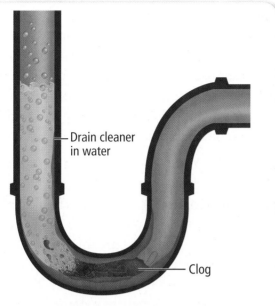

Some drain cleaners contain sodium hydroxide (NaOH), which dissolves grease, and small pieces of aluminum. The aluminum reacts with NaOH, producing hydrogen and dislodging solids, such as hair.

Aluminum hydroxide ($Al(OH)_3$) is a base that is used in water-treatment plants. Its sticky surface collects impurities, making them easier to filter from the water.

■ **Figure 4** Bases are used in a variety of applications, ranging from use in the home to use in industrial settings.

Common bases You are probably familiar with the bases in common cleaning products, but the uses of other bases, such as NaOH and $Al(OH)_3$, as shown in **Figure 4,** might be unfamiliar to you. **Table 2** includes some bases that have household and industrial uses and some additional information about them.

Table 2	Common Bases and Their Uses (Concepts in Motion) Interactive Table	
Name, Formula	**Use**	**Other Information**
Aluminum hydroxide ($Al(OH)_3$)	color-fast fabrics, antacid, water purification	sticky gel that collects suspended clay and dirt particles on its surface
Calcium hydroxide ($Ca(OH)_2$)	leather-making, mortar and plaster; lessen acidity of soil	slaked lime
Magnesium hydroxide ($Mg(OH)_2$)	laxative, antacid	called milk of magnesia when it is mixed with water
Sodium hydroxide (NaOH)	to make soap, oven cleaner, drain cleaner, textiles, and paper	called lye and caustic soda; generates heat when it is combined with water; reacts with metals to form hydrogen (exothermic reaction)
Ammonia (NH_3)	cleaners, fertilizer; to make rayon and nylon	irritating odor that is damaging to nasal passages and lungs

Solutions of Acids and Bases

Most of the products that rely on the chemistry of acids and bases are solutions, such as the cleaning products and food products mentioned previously. Because of its polarity, water is the main solvent in these products. Solutions of acids and bases produce ions that are capable of conducting an electric current. Thus, they are said to be electrolytes.

VOCABULARY
WORD ORIGIN
Acid
comes from the Latin *acidus* meaning sour
A grapefruit tastes sour because it contains ascorbic acid.

Ionization of acids You have learned that substances such as HCl, HNO_3, and H_2SO_4 are acids because of their abilities to produce hydrogen ions (H^+) in water. When an acid dissolves in water, the water molecules surround the neutral molecules of the acid, pulling them apart into ions. The positive hydrogen ions are attracted to the negative ends of the water molecules to form hydronium ions (H_3O^+). Therefore, an acid can be described more accurately as a compound that produces hydronium ions when dissolved in water. This process is shown in **Figure 5.**

Dissociation of bases Many bases are ionic compounds, formed from a positive metal ion and a negative hydroxide ion (OH^-). If you look at **Table 2,** you will find that most of the substances listed contain the letters OH in their chemical formulas. When bases that contain the letters OH dissolve in water, the negative areas of nearby water molecules attract the positive ion in the base. The positive areas of nearby water molecules attract the OH^- of the base. The base dissociates into a positive metal ion and a negative hydroxide ion (OH^-). This process is also shown in **Figure 5.** Unlike acid ionization, water molecules do not combine with the ions formed from the base.

$$NaOH(s) \xrightarrow{H_2O} Na^+(aq) + OH^-(aq)$$

Concepts in Motion
Animation

Figure 5 Acids and bases are classified by the ions that they produce when they are in an aqueous solution. Acids produce hydronium ions in water. Bases produce hydroxide ions in water.

When hydrogen chloride is added to water, a hydronium ion and a chloride ion are produced.

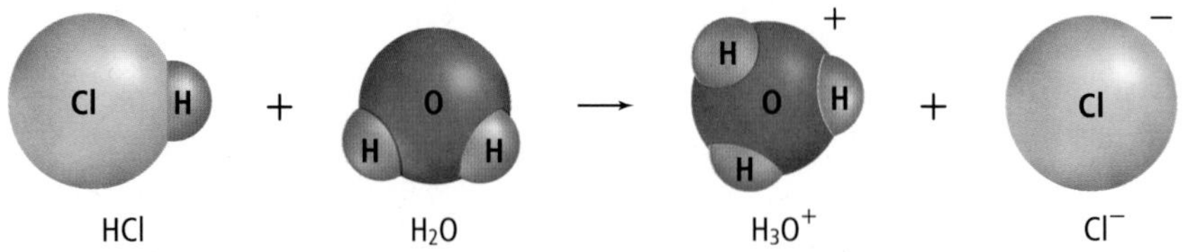
Cl
H
+
O
H
H
→
H
O
H
H
+
+
Cl
−
HCl
H₂O
H₃O⁺
Cl⁻

When sodium hydroxide dissolves in water, a sodium ion and a hydroxide ion are released.

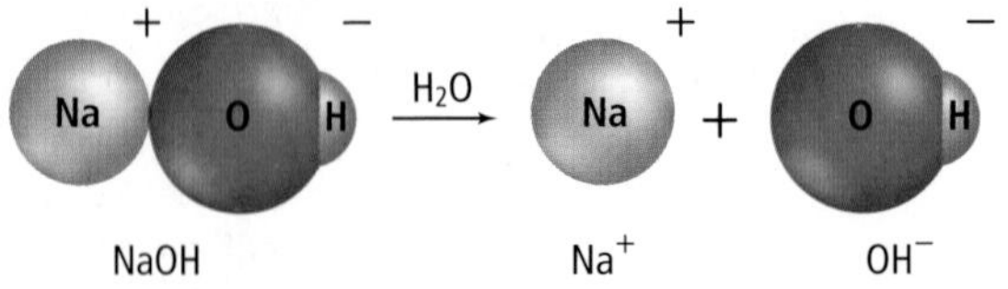
+
−
Na
O
H
H₂O
+
Na
+
−
O
H
NaOH
Na⁺
OH⁻

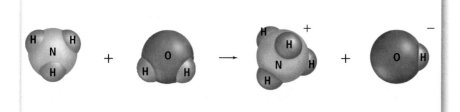

■ **Figure 6** Ammonia reacts with water to produce hydroxide ions in solution; therefore, it is a base.

Ionization of bases Recall that a base is also any substance that accepts an H^+ from acids. Ammonia (NH_3) is a base, even though it does not contain the letters OH in its chemical formula. Ammonia is a base because it accepts a hydrogen ion (H^+) from water. In a water solution, ionization takes place when the ammonia molecule attracts a hydrogen ion from a water molecule, forming an ammonium ion (NH_4^+). This produces a hydroxide ion (OH^-), as shown in **Figure 6.**

✓ **Reading Check** **Explain** how ammonia reacts in a water solution.

Ammonia is a common household cleaner. However, products containing ammonia should never be mixed with other chlorine-based cleaners (sodium hypochlorite), such as some bathroom bowl cleaners and bleach. A reaction between sodium hypochlorite and ammonia produces the toxic gases hydrazine and chloramine. Breathing these gases can severely damage lung tissues and cause death.

Section 1 Review

Section Summary

▶ Acids are sour tasting and corrosive, and they make blue litmus turn red.

▶ Bases exist as crystals in the solid state, are slippery, have a bitter taste, are corrosive, and make red litmus turn blue.

▶ The polar nature of water allows acids and bases to dissolve in water and form ions.

1. **MAIN Idea** **Describe** how an acidic solution forms when HCl is mixed in water and how a basic solution forms when NaOH is mixed in water.

2. **Explain** what an indicator is.

3. **Write** the formulas of three important acids and three important bases and describe their uses.

4. **Compare and contrast** how NH_3 and $Ca(OH)_2$ form OH^- ions in water.

5. **Think Critically** A friend asks you to get his favorite item from the kitchen, but he uses chemical formulas to ask for it. He asks for a drink which does not contain H_2CO_3, but does have $H_2C_6H_6O_6$. What might he be asking for?

Apply Math

6. **Calculate** the molecular mass of acetylsalicylic acid ($HOOC\text{-}C_6H_4\text{-}OOCCH_3$).

Section 1 • Acids and Bases **683**

Essential Questions

▶ What determines the strength of an acid or a base?

▶ How effectively do different acids and bases conduct electricity?

▶ What is the difference between strength and concentration?

Review Vocabulary

ionization: the process in which some molecules dissolve in water and separate into charged particles

New Vocabulary

strong acid
weak acid
strong base
weak base
pH
buffer

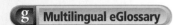

g **Multilingual eGlossary**

? **Inquiry**   **Video Lab**

■ **Figure 7** Only a few molecules of a weak acid, such as acetic acid, ionize in water. Fewer ions means lower conductivity, and therefore, the bulb is dimmer. Nearly all molecules of HCl, a strong acid, ionize in water. This provides greater electrical conductivity, so the bulb shines brighter.

Strength of Acids and Bases

MAIN ‹Idea› Strength describes the degree to which an acid or base ionizes or dissociates in water; concentration describes the amount of acid or base dissolved in a certain volume of water.

Real-World Reading Link During a basketball game, there are times when all ten players are concentrated in half the court. At other times, the players are spread out over the whole court. Similarly, an acid or a base would be considered dilute if a small amount were added to a large amount of water. Also, an acid or a base would be considered concentrated if that same amount were added to a small amount of water.

Strong and Weak Acids and Bases

Some acids must be handled with great care. For example, the sulfuric acid found in car batteries can burn your skin. Yet you drink acids, such as citric acid in orange juice and carbonic acid in soft drinks. Obviously, some acids are stronger than others.

The strength of an acid or a base depends on the degree to which acid or base particles form into ions in water. In a **strong acid,** all the acid ionizes upon dissolving in water. Hydrochloric acid, nitric acid, and sulfuric acid are examples of strong acids. In a **weak acid,** only a small fraction of the molecules ionize upon dissolving in water. Acetic acid and carbonic acid are examples of weak acids.

Ions in solution can conduct an electric current. The more ions that a solution contains, the more current it can conduct. The ability of a solution to conduct a current can be demonstrated by using a lightbulb connected to a battery with leads placed in the solution, as shown in **Figure 7.** The strong acid solution conducts more current, and the lightbulb burns brightly. The weak acid solution does not conduct as much current as a strong acid solution, and the bulb burns less brightly.

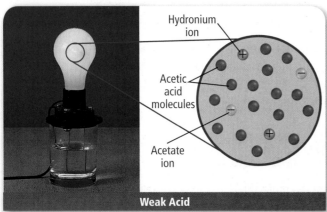

Weak Acid

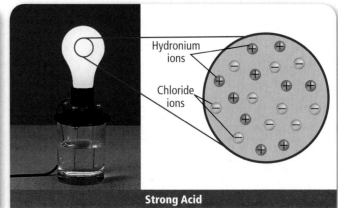

Strong Acid

Acid strength In strong acids, such as hydrochloric acid (HCl), nearly all the acid ionizes to form hydronium ions (H_3O^+) and chloride ions (Cl^-) and leaves almost no HCl molecules present. This is shown by writing the equation using a single arrow pointing toward the ions that are formed.

$$HCl(g) + H_2O(l) \rightarrow H_3O^+(aq) + Cl^-(aq)$$

Equations describing the ionization of weak acids, such as acetic acid (CH_3COOH), are written using double arrows pointing in opposite directions. This means that the reaction does not go to completion.

$$CH_3COOH(l) + H_2O(l) \rightleftharpoons H_3O^+(aq) + CH_3COO^-(aq)$$

An acetic acid solution is mostly made of CH_3COOH molecules. Only a relatively few hydronium ions (H_3O^+) and acetate ions (CH_3COO^-) are in solution.

Base strength Many bases are ionic compounds that dissociate to produce ions when they dissolve. A **strong base** dissociates completely upon dissolving in water. The dissociation of sodium hydroxide (NaOH) to form sodium ions (Na^+) and hydroxide ions (OH^-) is shown below.

$$NaOH(s) \rightarrow Na^+(aq) + OH^-(aq)$$

When ammonia (NH_3) is dissolved in water, an ammonia ion (NH_4^+) and a hydroxide ion (OH^-) form. This is shown using double arrows to indicate that not all of the molecules ionize. A **weak base** is one that does not ionize completely. Ammonia produces only a few ions in water, and most of the ammonia remains in the form of NH_3.

$$NH_3(aq) + H_2O(l) \rightleftharpoons NH_4^+(aq) + OH^-(aq)$$

Strength and concentration The terms *strength* and *concentration* can be confused when describing acids and bases. The terms *strong* and *weak* refer to the degree to which an acid or base ionizes or dissociates in solution. *Strong* acids ionize completely, and *strong* bases dissociate completely. *Weak* acids and bases ionize only partially. In contrast, the terms *dilute* and *concentrated* indicate the concentration of a solution, which is the amount of acid or base dissolved in the solution. It is possible to have dilute solutions of strong acids and bases and concentrated solutions of weak acids and bases, as shown in **Figure 8.**

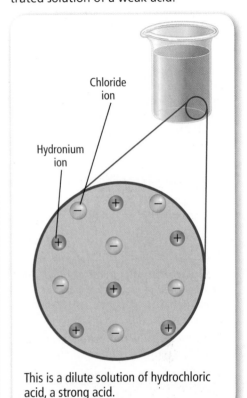

■ **Figure 8** You can have a dilute solution of a strong acid and a concentrated solution of a weak acid.

Chloride ion

Hydronium ion

This is a dilute solution of hydrochloric acid, a strong acid.

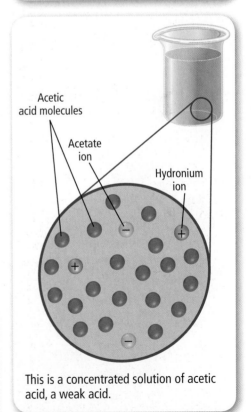

Acetic acid molecules

Acetate ion

Hydronium ion

This is a concentrated solution of acetic acid, a weak acid.

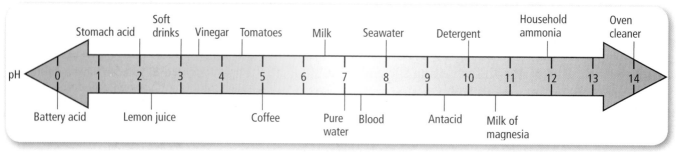

Soft drinks, Vinegar, Tomatoes, Milk, Seawater, Detergent, Household ammonia, Oven cleaner

Stomach acid

pH 0 1 2 3 4 5 6 7 8 9 10 11 12 13 14

Battery acid   Lemon juice   Coffee   Pure water / Blood   Antacid   Milk of magnesia

Figure 9 The pH scale helps classify solutions as acidic or basic. Each number on the pH scale represents a change in the H⁺ concentration of ten times. For example, a solution of pH = 3 has an H⁺ concentration that is ten times greater than a solution of pH = 4.

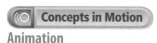

Concepts in Motion
Animation

Figure 10 The pH of a sample can be measured in several ways, two of which (indicator paper or a pH meter) are shown below.

Determine *which method provides a more accurate reading.*

Indicator Paper

pH Meter

pH of a Solution

If you have a swimming pool or a tropical fish aquarium, you know that the pH of the water must be controlled. Also, many products, such as shampoos, claim to control pH so it suits your type of hair. **pH** is a measure of the concentration of H⁺ (H_3O^+) ions in solution. The pH measures how acidic or basic a solution is. The greater the (H_3O^+) concentration is, the lower the pH and the more acidic the solution. A scale to indicate pH has been devised, as shown in **Figure 9.**

As the scale shows, solutions with pHs lower than 7 are described as acidic. The lower the value, the more acidic the solution. Solutions with pHs greater than 7 are basic. The higher the pH is, the more basic the solution. A solution that has a pH of exactly 7 has equal concentrations of (H_3O^+) ions and OH⁻ ions. These solutions are considered to be neutral. Pure water at 25°C has a pH of 7.

One way to determine pH is by using universal indicator paper. This paper undergoes color changes in the presence of H_3O^+ ions and OH⁻ ions in solution. The final color of the pH paper is matched with colors in a chart to find the pH, as shown in **Figure 10.**

An instrument called a pH meter is another tool to determine the pH of a solution. This meter is operated by immersing the electrodes in the test solution and reading the display. Small, battery-operated pH meters that have digital readouts are precise and convenient for use outside the laboratory when testing the pH of soils and streams, as shown in **Figure 10.**

Blood pH Your blood circulates throughout your body carrying oxygen, removing carbon dioxide, and absorbing nutrients from food that you have eaten. For blood to carry out its many functions properly, the pH of blood must remain between 7.0 and 7.8. The main reason for this is that enzymes, which are the protein molecules that act as catalysts for many reactions in the body, cannot work outside this pH range. Yet, you can eat foods that are acidic without changing the pH of your blood. How can this be? The answer is that your blood contains solutions called buffers that resist pH changes when small amounts of acids or bases are added.

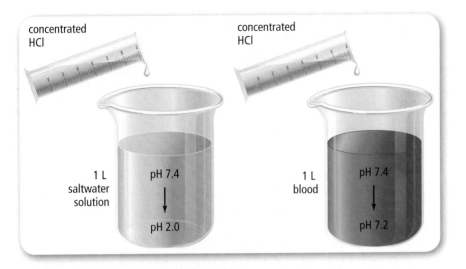

■ **Figure 11** This experiment shows how well buffers in blood work. Adding 1 mL of concentrated HCl to 1 L of salt water changes the pH from 7.4 to 2.0. Adding the same amount of concentrated HCl to 1 L of blood changes the pH from 7.4 to 7.2.

Buffers are solutions containing ions that react with acids or bases to minimize their effects on pH. One buffer system in blood involves a solution of carbonic acid (H_2CO_3) and bicarbonate ions (HCO_3^-). Because of these buffer systems, even small amounts of concentrated acids will not change the pH of blood much, as shown in **Figure 11.** Buffers help keep your blood close to a nearly constant pH of 7.4. If the pH of 7.4 is not maintained, a medical condition called acidosis can occur. This can occur during severe dehydration when excessive levels of acids build up in body fluids.

✔ **Reading Check Explain** what buffers are and how are they important for health.

Section 2   Review

Section Summary

▶ Strength refers to the degree to which an acid or base forms ions in water. Concentration refers to how much acid or base is present in solution.

▶ Acids and bases can conduct electricity in solution.

▶ Acids and bases are classified based on pH.

▶ Buffers are solutions that minimize the effects of the addition of an acid or a base on pH.

7. MAIN ⟨Idea⟩ **Compare and contrast** a dilute solution of a strong acid and a concentrated solution of a weak acid.

8. **Describe** two techniques used to measure the pH of a solution.

9. **Explain** how electricity can be conducted by acids and bases.

10. **Classify** pH values of 9.1, 1.2, and 5.7 as basic, acidic, or very acidic.

11. **Think Critically** The proper pH range for a swimming pool is between 7.2 and 7.8. Most pools use two substances, Na_2CO_3 and HCl, to maintain this range. How would you adjust the pH if you found it was 8.2? 6.9?

Apply Math

12. **Use Equations** To determine the difference in pH strength, calculate 10^n, where n is the difference between pHs. How much more acidic is a solution of pH 2.4 than a solution of pH 4.4?

✔ Assessment   Online Quiz

LAB Acid Concentrations

Objectives

■ **Determine** the relative concentrations of common acidic substances.

Background: The science of acids and bases is not investigated only in a high-tech laboratory by degreed scientists. You can investigate the acidic concentrations of things in your home using a simple homemade indicator solution.

Question: *How can you determine the relative concentration of the H⁺ ions in several acidic solutions?*

Preparation

Materials

homemade cabbage indicator (indicates acids and bases)
coffee filter
waxed paper
grease pencil or masking tape
teaspoons (3)
alum
cream of tartar
fruit preservative

Safety Precautions

Procedure

1. Read the procedure and safety information, and complete the lab form.

2. Use the grease pencil or masking tape and a pencil to label three areas on the waxed paper: *alum, cream of tartar,* and *fruit preservative.* These areas should be about 8 cm apart.

3. Place approximately $\frac{1}{2}$ teaspoon of each of the three powders on the waxed paper where labeled. Use a separate teaspoon for each substance.

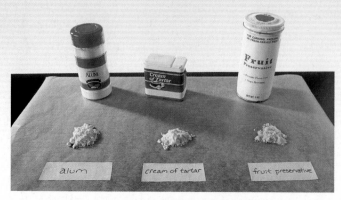

4. Cut three strips from the coffee filter, about 1 cm wide by 8 cm long.

5. Dip the end of one of the strips into the cabbage indicator solution, and then lay the wet end on top of the alum.

6. Wet a second strip and lay it in on top of the cream of tartar.

7. Wet the third strip and lay it on top of the fruit preservative.

8. Wait 5 minutes and then check the indicator strips. Record your observations.

Conclude and Apply

1. **Determine** if all three substances were acids. Did the indicator strips turn a similar color? Explain why each substance produced a different color.

2. **Propose** a possible rank of the H⁺ ion concentrations. Explain your reasoning.

3. **Predict** what you would have observed if you used sodium hydroxide instead of alum.

Compare your answer to question 3 above with other groups in the class. Come to an agreement on your prediction. As a group, design a procedure to test your prediction.

? Inquiry Lab

Section 3

Reading Preview

Essential Questions

▶ What is a neutralization reaction?

▶ What is a salt, and how does it form?

▶ What is the purpose of the indicator in a titration?

▶ How do soaps and detergents differ?

Review Vocabulary

nonpolar molecule: molecule that shares electrons equally and does not have oppositely charged ends

New Vocabulary

neutralization
salt
titration
soap

g Multilingual eGlossary

((O)) **Concepts in Motion**

Animation

■ **Figure 12** An antacid tablet containing the base sodium bicarbonate (NaHCO₃) reacts in your stomach much as it does in this dilute solution of HCl. Usually, people chew antacid tablets before swallowing them.

Explain *how chewing the tablet affects the rate of the reaction.*

Salts

MAIN ⟨Idea An acid and a base react to form a salt and water.

Real-World Reading Link The human body uses ions from salts to carry out many important functions, including prevention of muscle cramps, stabilization of irregular heartbeats, and sleep regulation.

Neutralization

Normally, stomach acid contains a dilute solution of hydrochloric acid and water. Too much acid can cause indigestion. Antacids contain bases or other compounds of sodium, potassium, calcium, magnesium, or aluminum that react with stomach acid to lower acid concentration. **Figure 12** shows antacid tablets reacting in a solution that is similar to stomach acid. The equation for this reaction is:

$$HCl(aq) + NaHCO_3(s) \longrightarrow NaCl(aq) + CO_2(g) + H_2O(l)$$

Neutralization is a chemical reaction between an acid and a base that forms a salt and water. In the above equation, HCl is neutralized by $NaHCO_3$.

Salt formation Hydrochloric acid can be neutralized by sodium hydroxide in another common neutralization reaction.

$$HCl(aq) + NaOH(aq) \longrightarrow NaCl(aq) + H_2O(l)$$

In the equation above, the hydrogen and hydroxide ion form water. Remaining ions form the salt sodium chloride (NaCl). A **salt** is a compound formed when negative ions from an acid combine with positive ions from a base.

Antacids contain bases that neutralize excess stomach acid.

Common Salts and Their Uses

Table 3   **Concepts in Motion** Interactive Table

Name, Formula	Common Name	Uses
Sodium chloride (NaCl)	salt	food, manufacture of chemicals
Sodium hydrogen carbonate ($NaHCO_3$)	sodium bicarbonate, baking soda	food, antacids
Calcium carbonate ($CaCO_3$)	calcite, chalk	manufacture of paint and rubber tires
Potassium nitrate (KNO_3)	saltpeter	fertilizers
Potassium carbonate (K_2CO_3)	potash	manufacture of soap and glass
Sodium phosphate (Na_3PO_4)	TSP	detergents
Ammonium chloride (NH_4Cl)	sal ammoniac	dry-cell batteries

Concepts in Motion
Animation

VOCABULARY ....
SCIENCE USAGE V. COMMON USAGE
Salt
Science usage
  a compound formed when negative ions from an acid combine with positive ions from a base
  A salt is a product of the reaction between an acid and a base.

Common usage
  an ingredient commonly used in cooking to provide flavor
  Carol added salt to the eggs. ....

Acid-base general equation The following general equation represents acid-base reactions in water.

$$acid + base \longrightarrow salt + water$$

Another example of a neutralization reaction occurs between HCl, an acid, and $Ca(OH)_2$, a base. This reaction produces the salt $CaCl_2$ and water.

$$2HCl(aq) + Ca(OH)_2(aq) \longrightarrow CaCl_2(aq) + 2H_2O(l)$$

Salts

There are many salts, a few of which are shown in **Table 3**. Most salts are composed of a positive metal ion and an ion with a negative charge, such as Cl^- or CO_3^{2-}. Ammonium salts contain the ammonium ion (NH_4^+), rather than a metal.

Salts are essential for many animals, large and small. Some domestic animals are supplied salts, as shown in **Figure 13**. Some animals find it at natural deposits. At these natural deposits, animals obtain ions, such as sodium ions (Na^+) and calcium ions (Ca^{2+}), that are important for health. Even insects, such as butterflies, need salts and are often found clustered on moist ground. You also need salts, especially because you lose them through perspiration. How humans obtain one salt, sodium chloride, is shown in **Figure 14**.

■ **Figure 13** Like many animals, horses need salts to help maintain body processes.

FIGURE 14

Visualizing Salt Production

Table salt (NaCl) is used for a variety of purposes: seasoning and preservation of food, deicing roadways, feeding animals and plants, and softening water. Salt is extracted from land and water. Three different technologies are commonly used.

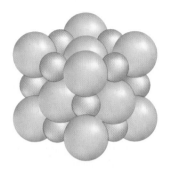

NaCl (Table salt)

Evaporation Salt Shallow ponds are filled with brine (salt water). The brine is moved from pond to pond to remove impurities and to concentrate it through evaporation. When the desired salt concentration has been reached, it is then pumped from evaporation ponds into crystallizing ponds. After the salt crystallizes, the remaining water is drained off. The crystals are washed, crushed, and dried.

Rock Salt Salt deposits exist due to the gradual evaporation of bodies of localized salt water. Salt can be found in deposits that have layers of sedimentary rocks, such as anhydrite, which is a calcium sulfate mineral. Salt deposits are found at relatively shallow depths, averaging 150 m to 600 m below Earth's surface. Salt is mined from these deposits by drilling and blasting.

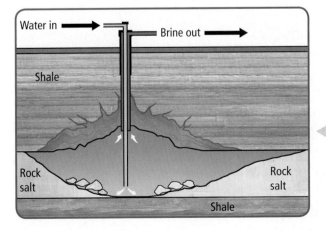

Solution Mining Solution mining is another technology used to obtain the table salt that we sprinkle on our food. A well pumps water into an underground salt deposit to dissolve the salt and to form a saturated solution called a brine. The brine is heated, and the salt crystallizes after the water boils away.

Titration

Sometimes you need to know the amount of acid or base in a solution, for example, to determine the purity of a commercial product. This can be done using titration (ti TRAY shun). **Titration** is the process in which a solution of known concentration is used to determine the concentration of another solution. **Figure 15** shows a titration procedure.

Titration involves a solution of known concentration called the standard solution. The standard solution is slowly added to a solution of unknown concentration, which contains an acid/base indicator. If the solution of unknown concentration is a base, a standard acid solution is used. If the solution of unknown concentration is an acid, a standard base solution is used.

The endpoint has a color change The titration shown in **Figure 15** shows how to determine the concentration of an acid solution. First, add a few drops of an indicator, such as phenolphthalein (fee nul THAY leen), to a carefully measured volume of the acidic solution of unknown concentration. Phenolphthalein is colorless in an acid, but it turns bright pink in the presence of a base.

Then, slowly and carefully add a basic solution of known concentration to this acid/indicator mixture. Toward the end of the titration, add the base drop by drop until one last drop of the base turns the solution pink and the color persists. The point at which the color persists is known as the endpoint, the point at which the acid is completely neutralized by the base. When the volume of base used is known, use that value, the known concentration of the base, and the volume of the acid to calculate the concentration of the acid solution.

VOCABULARY ....

ACADEMIC VOCABULARY

Process
  a series of actions or operations leading to an end
  To obtain accurate results, the process of titration must be slow. ....

■ **Figure 15** In this titration, a base of known concentration is being added to an acid of unknown concentration. The swirl of pink color shows that the end point is near but disappears until the endpoint has been reached. A permanent light-pink color marks the endpoint of this titration.

Explain *why you must add the base to the acid drop by drop near the end of the titration.*

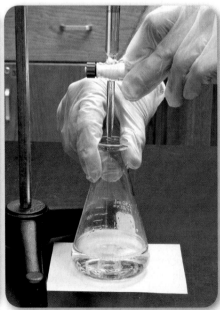

Figure 16 Natural indicators include red cabbage, radishes, and roses.

Many natural substances are acid-base indicators. In fact, the indicator litmus comes from a lichen, a combination of a fungus and an algae or a cyanobacterium. Flowers that are indicators include hydrangeas, which produce blue blossoms when the pH of the soil is acidic and pink blossoms when the soil is basic. The hydrangeas' color change is just the opposite of litmus.

Other natural indicators possess a range of colors. For example, the color of red cabbage varies from deep red at pH 1 to lavender at pH 7 and yellowish green at pH 10. **Figure 16** shows common substances that can be used as natural indicators. Grape juice is also an indicator, as you can learn by doing the MiniLab.

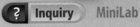

MiniLab

? Inquiry   MiniLab

Test a Grape Juice Indicator

Procedure

1. Read the procedure and safety information, and complete the lab form.
2. Add one-half cup of **water** to each of two **small glasses.**
3. Add 1 tablespoon of **purple grape juice** to each glass.
4. To one glass, add 1 teaspoon of **baking soda.** Stir.
5. To the other glass, add 1 teaspoon of **white vinegar.**
6. Note the color after each addition in steps 3, 4, and 5.

Analysis

1. **Describe** What happened when baking soda was added?
2. **Explain** Did the color change when you added vinegar? Why or why not?

Apply Science

How can you handle an upsetting situation?

At some time, most of us have experienced an upset stomach. Often, the cause is the excess acid within our stomachs. For digestive purposes, our stomachs contain dilute hydrochloric acid with a pH between 1.6 and 3.0. A doctor might recommend an antacid treatment for an upset stomach. What type of compound is "anti acid"?

Identify the Problem

You have learned that neutralization reactions of acids and bases form salts. Antacids typically contain small amounts of $Ca(OH)_2$, $Al(OH)_3$, or $NaHCO_3$, which are bases.

Whereas having an excess of acid lowers the pH of your stomach contents, antacid compounds raise the pH of your stomach contents. How does an antacid change pH to make you feel better?

Solve the Problem

1. Write the chemical equation for the reaction of HCl and $Mg(OH)_2$. Use it to explain why an antacid helps excess stomach acid.
2. Why is it important to have some acid in your stomach?
3. Design an experiment that compares how well antacid products neutralize acid.

Nonpolar hydrocarbon end — COO⁻ Na⁺ Ionic end

Soaps and Detergents

The next time that you are in a supermarket, go to the aisle with soaps and detergents. You will see all kinds of products—solid soaps, liquid soaps, and detergents for washing clothes and dishes. What are all these products? Do they differ from one another? Yes, they differ slightly in how they are made and in the ingredients included for color and aroma. Still, all these products are classified into two types—soaps and detergents.

Soaps The reason why soaps clean so well is explained by polar and nonpolar molecules. **Soaps** are salts. As shown in **Figure 17,** they have a nonpolar chain of carbon atoms on one end and a sodium or potassium salt of a carboxylic acid (kar bahk SIH lihk) group (–COOH) at the other end. The nonpolar, hydrocarbon end interacts with nonpolar oils and dirt so they can be removed readily, and the polar, ionic end ($-COO^-Na^+$ or $-COO^-K^+$) helps them dissolve in polar water.

To make an effective soap, the acid must contain 12 to 18 carbon atoms. If it contains fewer than 12 atoms, it will not be able to mix well to clean oily dirt. If it has too many carbon atoms, the salt will not be soluble in water. **Figure 18** shows how soap interacts with dirt particles to clean your hands.

Reading Check **Explain** why soaps must have polar and nonpolar ends.

Figure 18 Soaps are salts that have a nonpolar end and an ionic end.

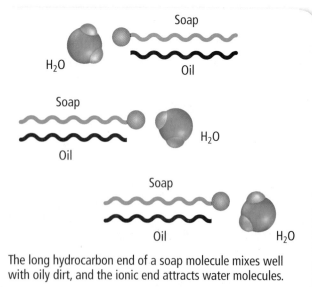

The long hydrocarbon end of a soap molecule mixes well with oily dirt, and the ionic end attracts water molecules.

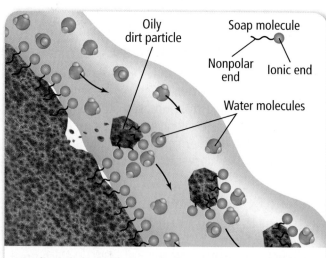

Dirt that is linked with the soap rinses away as water flows over it.

Commercial soaps One problem with all soaps is that the sodium ions and potassium ions can be replaced by ions of calcium, magnesium, and iron found in some water known as hard water. When this happens, the salts formed are insoluble and precipitate out of solution in the form of soap scum. Detergents were developed to avoid this problem.

✓ **Reading Check** **Explain** What is soap scum?

Detergents Similar to soaps, detergents have long hydro-carbon chains. But instead of a salt of a carboxylic acid group at the end, they may contain a salt of a sulfonic acid group. These acids form more soluble salts with the ions in hard water, lessening the problem of soap scum. Detergents can also be used in cold water. Most contain additional ingredients called builders and surfactants to enhance the ability to clean and to produce suds.

Despite solving the problem of cleaning in hard water, detergents are not the complete solution to our needs. Some detergents contained phosphates but are no longer produced because they cause water pollution. Certain sulfonic acid detergents also present problems in the form of excess foaming in water treatment plants and streams, as shown in **Figure 19**. These detergents do not break down easily and remain in the environment for long periods of time.

■ **Figure 19** Foam from detergents can build up in waterways. Phosphorus, which caused algal blooms, has been removed from detergents, but such pollution can still be harmful. For example, fish eggs can be destroyed by detergent pollution.

Section 3 Review

Section Summary

▶ Neutralization is a chemical reaction between an acid and a base.

▶ Salt is a dietary essential.

▶ Titration is a method used to determine the concentration of an acidic or a basic solution.

▶ Molecules of soaps and detergents have polar and nonpolar ends.

13. **MAIN Idea** **Write** the balanced chemical equation for one neutralization reaction. In your equation, which reactant contributed the salt's positive ion? Which one contributes the salt's negative ion?

14. **Identify** the purpose of an indicator in a titration experiment.

15. **Compare and contrast** the composition of detergents and soaps.

16. **Think Critically** Give the names and formulas of the salts formed in the following neutralizations: sulfuric acid and calcium hydroxide, nitric acid and potassium hydroxide, and carbonic acid and aluminum hydroxide.

Apply Math

17. **Calculate Ratios** In the following reaction, how many molecules of HCl are needed to produce four molecules of H_2O?

$$2HCl(aq) + Ca(OH)_2(aq) \rightarrow CaCl_2(aq) + 2H_2O(l)$$

Standardized Test Practice

Multiple Choice

Record your answers on the answer sheet provided by your teacher or on a sheet of paper.

Use the figures below to answer questions 1 and 2.

Acid Solution Data		
Solution	pH	Ionization
W	7	none
X	2	complete
Y	6	partial
Z	4	partial

1. Which solution has the lowest concentration of hydrogen ions?
   - A. solution W
   - B. solution X
   - C. solution Y
   - D. solution Z

2. Which solution contains a strong acid?
   - A. solution W
   - B. solution X
   - C. solution Y
   - D. solution Z

3. Hard water often contains various amounts of metallic substances. Which of the following ions does NOT contribute to hard water?
   - A. calcium
   - B. iron
   - C. magnesium
   - D. sodium

4. An unknown substance in solution is slippery to the touch, dissolves easily in water, and makes litmus paper turn blue. The substance is most likely
   - A. an acid.
   - B. a base.
   - C. a salt.
   - D. an indicator.

5. Which chemical formula below describes a hydronium ion?
   - A. H_3O^+
   - B. OH^-
   - C. $-COOH$
   - D. H_2O

Use the graph below to answer questions 6–8.

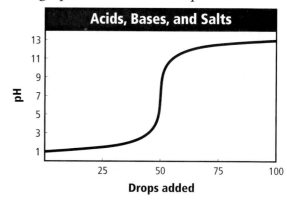

6. The titration curve above indicates the changes that happened to a solution of a strong acid as drops of a strong base were added. Before any drops were added, what was the pH of the solution?
   - A. 1
   - B. 3
   - C. 9
   - D. 13

7. At which drop count are the acid and base exactly neutralized?
   - A. 0
   - B. 25
   - C. 50
   - D. 75

8. At the instant of neutralization, what is in the beaker besides water?
   - A. acid only
   - B. base only
   - C. salt only
   - D. equal amounts of acid and base

9. The chemical equation for a reaction is $2HCl + Ca(OH)_2 \rightarrow CaCl_2 + 2H_2O$. How many water molecules are formed as x molecules of $CaCl_2$ form?
   - A. 2
   - B. twice as many, $2x$
   - C. half as many, $\frac{x}{2}$
   - D. an equal number, x

✓ Assessment   Standardized Test Practice

Short Response

Record your answers on the answer sheet provided by your teacher or on a sheet of paper.

Use the model below to answer questions 10 and 11.

10. What is the chemical formula for this substance?

11. How many hydronium ions can it form in water?

12. Write the chemical equation for the ionization reaction of hydrogen sulfide in water.

13. A conductivity apparatus is inserted into a beaker of water, but the lightbulb does not glow. Describe what will happen if NaCl crystals are added to the beaker and stirred.

14. At pH = 3.0, the hydronium ion concentration of a solution is $0.0010M$. When the pH of the solution drops to 1.0, the hydronium ion concentration increases by a factor of 100. What is the H_3O^+ concentration at pH = 1.0?

15. Compare and contrast the terms *strength* and *concentration* as they apply to acids and bases in solution.

16. Name the salt that is produced by each of the following acid-base pairs: HCl + NaOH, HCl + $Mg(OH)_2$, and HBr + KOH.

Extended Response

Record your answers on a sheet of paper.

Use the figure below to answer question 17.

17. One of the solutions has a pH of 10, and the other has a pH of 12. Explain why the lightbulbs glow with different intensities. Use the words *strong* and *weak* in your answer.

18. Describe how dishwashing liquid cleans dirty plates.

19. Na_2SO_4 is a soluble salt. Write the chemical equation describing its dissociation in water.

20. Identify the acid and base that form the salt Na_2SO_4 in a titration reaction.

21. Suppose you have an HCl solution of an unknown concentration. If 25.0 mL of this solution requires 50.0 mL of a known concentration of a NaOH solution to neutralize it, then how much more concentrated is the HCl solution than the NaOH solution?

NEED EXTRA HELP?																					
If You Missed Question . . .	1	2	3	4	5	6	7	8	9	10	11	12	13	14	15	16	17	18	19	20	21
Review Section . . .	2	2	3	1	1	2	2	3	3	1	1	2	2	2	2	2	3	2	3	3	3	2

CHAPTER 23
Organic Compounds

ConnectED

Your one-stop online resource
connectED.mcgraw-hill.com

- Video
- Audio
- Review
- Inquiry
- WebQuest
- Assessment
- Concepts in Motion
- Multilingual eGlossary

Launch Lab
Carbon, the Organic Element

The element carbon exists in several forms including dull, black charcoal; slippery, gray graphite; and bright, sparkling diamond. However, this is nothing compared with the millions of different compounds that carbon can form with other elements. In this lab, you will seek out the carbon hidden in two common substances.

For a lab worksheet, use your StudentWorks™ Plus Online.

? Inquiry) Launch Lab

FOLDABLES®

Make a vocabulary book. Label it with vocabulary from the chapter. Use it to organize your notes on vocabulary for organic compounds.

Organic compounds
Hydrocarbon
Isomer

THEME FOCUS Natural Resources
All organisms and many of our natural resources, including crude oil, natural gas, and timber, are made of organic compounds.

BIG (Idea Organic compounds contain the element carbon.

Chapter 23 • Organic Compounds   **705**

Essential Questions

▶ What is the difference between organic and inorganic compounds?

▶ Why can carbon form so many different compounds?

▶ What is the difference between a saturated and an unsaturated hydrocarbon?

▶ What are isomers and how do their properties vary?

Review Vocabulary

covalent bond: attraction formed between atoms when they share electrons

New Vocabulary

organic compound
hydrocarbon
saturated hydrocarbon
unsaturated hydrocarbon
isomer
benzene

g Multilingual eGlossary

■ **Figure 1** Organic compounds contain carbon. Carbon can form straight chains, branched chains, and rings.

Simple Organic Compounds

MAIN Idea Hydrocarbons are compounds made only of carbon and hydrogen atoms.

Real-World Reading Link How many structures could you make with only two types of blocks? Amazingly, a vast number of compounds can be made using only two atomic building blocks—hydrogen and carbon.

Organic Compounds

What do you have in common with gasoline, vanilla flavoring, and natural rubber? These items contain carbon compounds, as shown in **Figure 1.** You also contain carbon compounds. Most compounds containing the element carbon are **organic compounds.**

At one time, scientists thought that only living organisms made organic compounds, which is how they got their name. By 1830, scientists had made organic compounds in laboratories, but they still call them organic compounds. Of the millions of carbon compounds known today, more than 90 percent of them are organic compounds. The others, including carbon dioxide and the carbonates, are inorganic compounds.

Bonding You may wonder why carbon can form so many different organic compounds. A carbon atom has four electrons in its outer energy level. Therefore, a carbon atom can form four covalent bonds with atoms of carbon or with other elements. Recall that a covalent bond is formed when two atoms share a pair of electrons. The four available bonding sites allow carbon to form single, double, and triple bonds. As a result, carbon can make many types of compounds, including small compounds used as fuel, complex compounds found in medicines, and the long chains used in plastics.

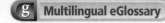

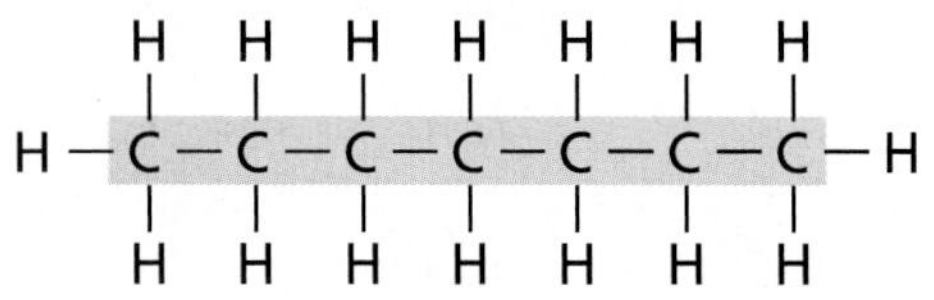

Heptane is a component of gasoline.

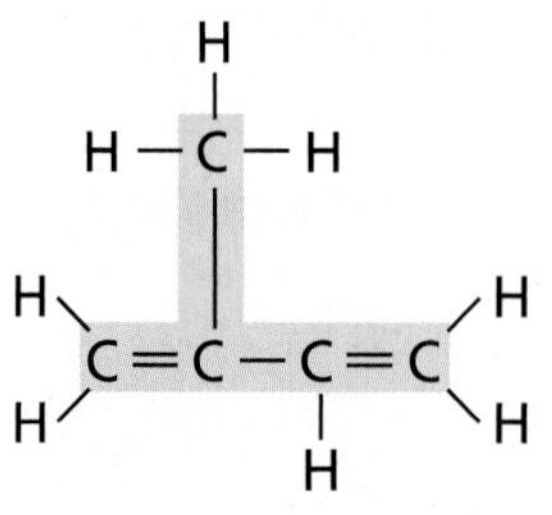

Isoprene exists in natural rubber.

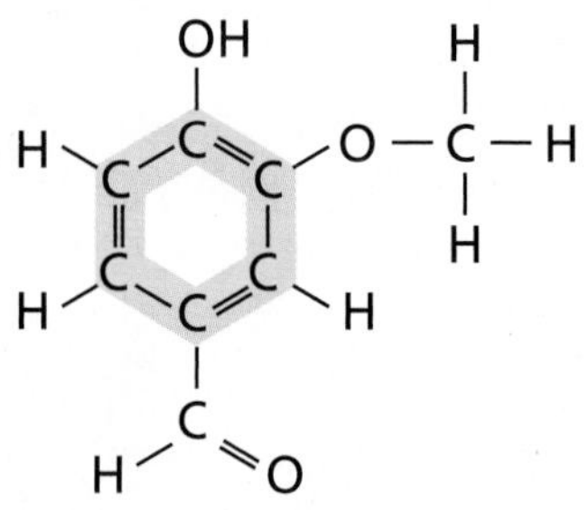

Vanillin is found in vanilla flavoring.

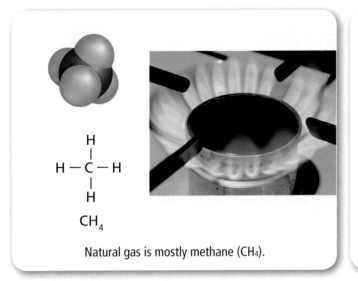

Natural gas is mostly methane (CH₄).

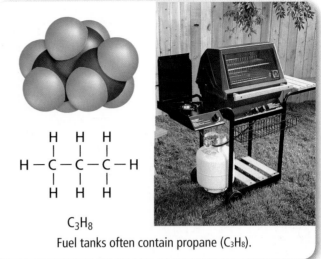

C₃H₈
Fuel tanks often contain propane (C₃H₈).

Arrangement Look back at **Figure 1.** Notice that carbon atoms can form straight chains, branched chains, and even rings. This variety of arrangements is another reason carbon can form so many compounds.

Representations Organic compounds are commonly represented in three ways, as shown in **Figure 2.** The simplest way is with the chemical formula. For example, a main component of natural gas used in homes is the organic compound methane. Methane's chemical formula is CH_4. The second way to represent the molecule is with the structural formula. The structural formula uses lines to show bonds between atoms. Each line between atoms represents a single covalent bond. The third way, the space-filling model, shows a more realistic picture of the relative size and arrangement of the atoms in the molecule. Most often, however, chemists use chemical and structural formulas to write about reactions.

✓ **Reading Check Identify** three ways to represent organic compounds.

Hydrocarbons

Carbon forms an enormous number of compounds with hydrogen alone. A compound composed of only carbon and hydrogen atoms is called a **hydrocarbon.** The natural gas methane is a hydrocarbon. Another hydrocarbon used as fuel is propane. Some stoves, most outdoor grills, and hot-air balloons burn propane. The structural formulas and space-filling models of propane and methane are shown in **Figure 2.**

Hydrocarbons produce more than 90 percent of the energy that humans use. Hydrocarbons are also important in medicines, foods, and clothing. To understand how hydrocarbons can play so many roles, you must examine how their structures differ.

■ **Figure 2** Organic molecules can be represented by their chemical formulas, by their structural formulas, or by space-filling models.

Compare and contrast *methane and propane.*

 Inquiry Virtual Lab

Table 1	Roots for Hydrocarbons
Root	**Number of Carbon Atoms**
Meth–	1
Eth–	2
Prop–	3
But–	4
Pent–	5
Hex–	6
Hept–	7
Oct–	8

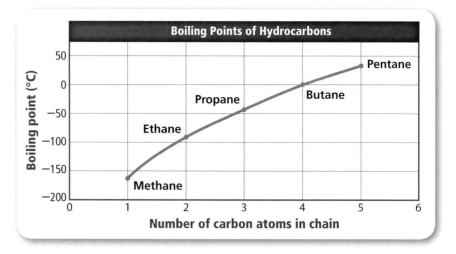

Figure 3 The root of a hydrocarbon's name indicates how many carbons are in the hydrocarbon. Boiling points of hydrocarbons increase as the number of carbon atoms in the chain increases.

Predict *the approximate boiling point of hexane.*

Concepts in Motion
Interactive Table

Hydrocarbon lengths The number of carbon atoms in a straight-chain hydrocarbon is indicated by the root of its name. For example, methane has one carbon atom, ethane has two, and propane has three. **Table 1** summarizes some of these roots. The length of the carbon chain affects the properties of the compound. For example, **Figure 3** shows a graph of the boiling points of some hydrocarbons. Notice the boiling point increases with the addition of carbon atoms.

Bonding in hydrocarbons Hydrocarbons can contain single, double, and triple bonds between carbon atoms. An easy way to know what type of bond that a hydrocarbon has is to look at the last three letters of the compound's name. The ending –*ane* indicates a single bonds only, the ending –*ene* indicates a double bond, and –*yne* indicates a triple bond.

Single bonds In some hydrocarbons, the carbon atoms are joined by single covalent bonds. Hydrocarbons containing only single-bonded carbon atoms are called **saturated hydrocarbons.** Saturated means that a compound holds as many hydrogen atoms as possible—it is saturated with hydrogen atoms. **Table 2** lists four saturated hydrocarbons. Each carbon atom can be considered a link in a chain connected by single covalent bonds.

Table 2	Some Saturated Hydrocarbons	Concepts in Motion	Interactive Table	
Name	**Methane**	**Ethane**	**Propane**	**Butane**
Chemical Formula	CH_4	C_2H_6	C_3H_8	C_4H_{10}
Structural Formula	H | H — C — H | H	H H | | H — C — C — H | | H H	H H H | | | H — C — C — C — H | | | H H H	H H H H | | | | H — C — C — C — C — H | | | | H H H H

Multiple bonds Hydrocarbons can also have double and triple bonds. Hydrocarbons that contain at least one double or triple bond are called **unsaturated hydrocarbons.** What would happen if the two carbon atoms in ethane (C_2H_6) shared two pairs of electrons and formed a double bond? The resulting compound would be ethene (C_2H_4). Ethene is sometimes called ethylene. Many fruits can form small quantities of ethylene gas, which aids in ripening. The hydrocarbon ethyne (C_2H_2) contains a triple bond in which three pairs of electrons are shared between the two carbon atoms. Ethyne has only two hydrogen atoms and is used in some welding torches. The differences between ethane, ethene, and ethyne are shown in **Figure 4.**

✓ **Reading Check** **State** the number of carbon atoms and hydrogen atoms in each molecule in **Figure 4.**

Isomers Pentane, isopentane, and neopentane have exactly the same chemical formula (C_5H_{12}). However, their carbon atoms are arranged very differently, as shown in **Figure 5.** In a molecule of pentane, the carbon atoms form a continuous chain. Isopentane has one branch, and neopentane has two branches. Pentane, isopentane, and neopentane are isomers. **Isomers** are compounds that have identical chemical formulas but different molecular structures and shapes.

Properties of isomers The arrangement of carbon atoms in each compound changes the shape of the molecule, which affects its physical properties. Generally, melting points and boiling points lower as the amount of branching in an isomer increases. You can see this pattern in **Figure 5.** You may notice that there is an exception to the pattern—the melting point of neopentane is higher than that of pentane or isopentane. In this case, the high melting point results from the symmetry of the molecule and its globular shape.

■ Figure 4 Saturated hydrocarbons, such as ethane, have only single bonds. Unsaturated hydrocarbons, such as ethene and ethyne, have double or triple bonds between carbon atoms.

 **Concepts in Motion** Animation

■ Figure 5 The amount of branching in the chain affects the compound's physical properties.

Compare *the number of carbon atoms in the isomers of pentane.*

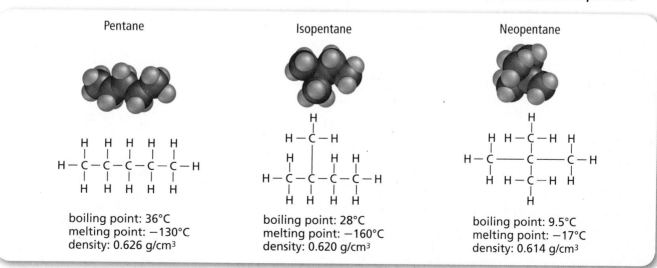

Pentane
boiling point: 36°C
melting point: −130°C
density: 0.626 g/cm³

Isopentane
boiling point: 28°C
melting point: −160°C
density: 0.620 g/cm³

Neopentane
boiling point: 9.5°C
melting point: −17°C
density: 0.614 g/cm³

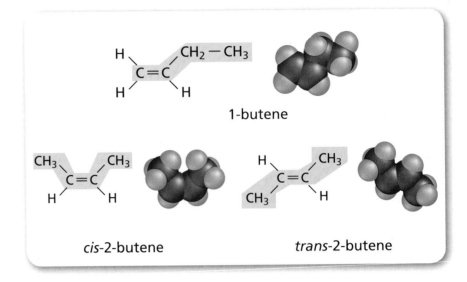

1-butene

cis-2-butene

trans-2-butene

Other isomers There are many other kinds of isomers in organic and inorganic chemistry. Examine the three non-branched isomers of butene shown in **Figure 6.** Notice that the double bond can be located in different places on the chain and that the chain can bend in different ways.

Another type of isomer differs only slightly in how the atoms are arranged in space. Such isomers form what are often called right-handed and left-handed molecules and look like mirror images. Two such isomers may have nearly identical physical and chemical properties.

Carbon Rings

Recall that carbon can also form rings. For example, cyclopropane has three carbon atoms joined into a ring by single bonds, as shown in **Figure 7.** *Cyclo–* means circular. Propane and cyclopropane are not isomers. Propane's chemical formula is C_3H_8, while cyclopropane's chemical formula is C_3H_6. Like straight and branched chains, cyclic carbon chains can have double and triple bonds. For example, cyclopentene (C_5H_8) has a double bond and cyclooctyne (C_8H_{12}) has a triple bond. These molecules are also shown in **Figure 7.**

MiniLab

? Inquiry   MiniLab

Model Hexane Isomers

Procedure 🥽 👕 ✂️ 🧤

1. Read the procedure and safety information, and complete the lab form.
2. To model hexane (C_6H_{14}), use soft **gumdrops** to represent carbon atoms, **raisins** to represent hydrogen atoms, and **toothpicks** for the chemical bonds. **WARNING:** *Never eat any food in the laboratory.*
3. Model the unbranched chain structure of hexane and draw its structural formula.
4. Make as many different branched formations of hexane as you can, and draw the structural formula of each.

Analysis

1. **Explain** How do you distinguish one hexane isomer from another?
2. **State** the total number of different hexane isomers found in your class.

■ **Figure 7** Cyclopropane, cyclopentene, and cyclooctyne are cyclic hydrocarbons.

Cyclopropane
C_3H_6

Cyclopentene
C_5H_8

Cyclooctyne
C_8H_{12}

Benzene **Figure 8** shows a special type of carbon ring called benzene. As you can see, the benzene molecule has six carbon atoms bonded into a ring. **Benzene** (C_6H_6) is a cyclic hydrocarbon with carbon atoms that are joined with alternating single and double bonds. The electrons in the double bonds are shared by all six carbon atoms in the ring. The equal sharing of electrons is represented by the benzene symbol: a circle in a hexagon.

Fused rings The sharing of these six electrons causes the benzene molecule to be very stable. The carbon atoms are bound in a rigid, flat structure. Benzene acts as a framework upon which new molecules can be built. Benzene rings can fuse together, like in the naphthalene (NAF thuh leen) molecule in **Figure 8**. Naphthalene is used in mothballs, which have a distinct odor. Many known compounds contain three or more rings fused together. Tetracycline (teh truh SI kleen) antibiotics are based on a fused ring system containing four fused rings.

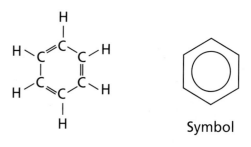

Symbol

Structural formula

Benzene can be represented in different ways.

Naphthalene

Naphthalene, used in mothballs, is a fused-ring system.

■ **Figure 8** Benzene (C_6H_6) is a very stable cyclic hydrocarbon. Benzene rings can fuse to make other compounds.

Section 1 Review

Section Summary

▶ Carbon can form many compounds because it has four electrons in its outer energy level.

▶ Hydrocarbons can be saturated or unsaturated.

▶ Isomers are compounds that have identical chemical formulas but different molecular structures.

▶ Benzene contains six carbon atoms bonded into a ring with alternating double and single bonds.

1. **MAIN Idea Define** the term *organic compounds* and explain how they got this name.

2. **Classify** each of the following compounds as organic or inorganic: C_4H_{10}, H_2O, FeO, CH_3COOH, and CaS.

3. **Compare and contrast** ethane, ethene, and ethyne.

4. **Explain** the term *saturated* in relation to hydrocarbons. What are these compounds saturated with?

5. **Describe** how boiling and melting points generally vary as branching in hydrocarbon isomers increases.

6. **Think Critically** Cyclobutane is a cyclic, saturated hydrocarbon containing four carbon atoms. Draw its structural formula. Are cyclobutane and butane isomers? Explain.

Apply Math

7. **Identify Trends** Adding one double bond to octane (C_8H_{18}) makes the hydrocarbon octene (C_8H_{16}). Write the formulas for adding one, two, and three more double bonds to octane. What is the decrease in the number of hydrogen atoms for each double bond added?

Reading Preview

Essential Questions

▶ What is a substituted hydrocarbon?
▶ What are the properties and uses of some common substituted hydrocarbons?
▶ What are aromatic compounds?

Review Vocabulary

acid: any substance that produces hydrogen ions (H⁺) in a water solution

New Vocabulary

substituted hydrocarbon
alcohol
ester
amine
aromatic compound

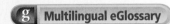 **Multilingual eGlossary**

Substituted Hydrocarbons

MAIN ‹Idea› Substituted hydrocarbons contain other elements besides carbon and hydrogen.

Real-World Reading Link Usually, a cheeseburger is a hamburger covered with melted American cheese and served on a bun. But you can make a cheeseburger with Swiss cheese and serve it on toast. Such substitutions would affect the taste of this cheeseburger. What would happen if you changed the atomic ingredients of a hydrocarbon?

Replacing Hydrogen

Chemists often change hydrocarbons into other compounds having different physical and chemical properties. They may include a double or triple bond or add different atoms or groups of atoms to a hydrocarbon. Some of these changed compounds are substituted hydrocarbons. A **substituted hydrocarbon** has one or more of its hydrogen atoms replaced by atoms, or groups of atoms, of other elements. The groups of atoms used in the substitution are called functional groups. Depending on what properties are needed, chemists decide what functional groups to add. Examples of substituted hydrocarbons are shown in **Figure 9**.

Substituting Oxygen Groups

Oxygen is found in the air, in water, and in many substituted hydrocarbons. Oxygen can form single and double bonds with carbon and single bonds with hydrogen. As a result, there are many compounds containing just carbon, hydrogen, and oxygen. These compounds include alcohols, organic acids, and esters.

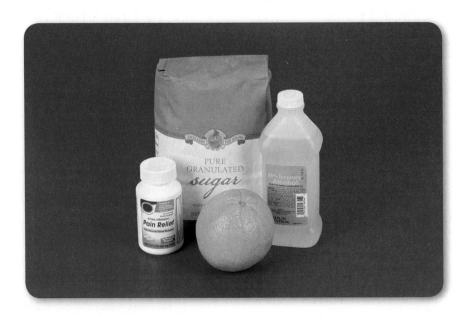

■ **Figure 9** By substituting elements such as oxygen, nitrogen, and chlorine for hydrogen atoms, a wide variety of organic compounds can be made. These diverse compounds have many uses.

Most ethanol (C_2H_5OH) is obtained from corn.

Ethanol
C_2H_5OH

Substituting –OH forms an alcohol

Ethane
C_2H_6

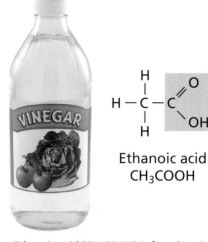

Ethanoic acid (CH_3COOH) is found in vinegar.

Ethanoic acid
CH_3COOH

Substituting –COOH forms an organic acid

Alcohols

Rubbing alcohol, used for rubbing aching muscles, is a substituted hydrocarbon. An **alcohol** is formed when –OH groups replace one or more hydrogen atoms in a hydrocarbon. Alcohols are an important group of organic compounds. They often serve as solvents and disinfectants and can be used as pieces to assemble larger molecules. **Figure 10** shows ethanol, an alcohol produced by the fermentation of sugar in grains and in fruit.

Organic acids

Organic acids form when a carboxyl group (–COOH) is substituted for one of the hydrogen atoms in a hydrocarbon. Ethanoic acid, also known as acetic acid, is an organic acid found in vinegar. As shown in **Figure 10,** the structures of ethanol and ethanoic acid are similar. You know some other organic acids, too—citric acid found in citrus fruits, such as oranges and lemons, and lactic acid found in sour milk.

✓ **Reading Check** **Explain** why organic acids are considered substituted hydrocarbons.

■ **Figure 10** Functional groups containing oxygen are found in alcohols and organic acids.

Describe *how ethane, ethanol, and ethanoic acid are related.*

 Inquiry   Video Lab

FOLDABLES

Incorporate information from this section into your Foldable.

The chemical reaction diagram at top:

H—C—C—C—C—OH + HO—C—C—H → H—C—C—C—C—O—C—C—H + H₂O

| Butyric acid | Ethyl alcohol | Ethyl butyrate | Water |

Figure 11 Butyric acid and ethyl alcohol combine to form the ester ethyl butyrate and water. Ethyl butyrate tastes like pineapple.

Esters Recall that mixing an acid and a base will yield water and a salt. In a similar way, alcohols and organic acids combine to form water and an ester. An **ester** is a substituted hydrocarbon with a –COOC– group. **Figure 11** shows the reaction of butyric (byew TIHR ihk) acid and ethyl alcohol to produce water and the ester ethyl butyrate, which is a component in pineapple flavor. Esters have many different applications. Esters of the alcohol glycerine are used to make commercial soaps. Other esters can be made into fibers for clothing, and still others are used in flavors and perfumes.

Esters for flavor and odors Many fruit-flavored soft drinks and desserts taste like real fruit. If you look at the label, you might be surprised to find that no fruit was used, only artificial flavor. Most likely, this artificial flavor contains esters.

The odor of some individual esters immediately makes you think of particular fruits, as shown in **Figure 12.** For example, octyl ethanoate smells like oranges, and both isopentyl ethanoate and butyl ethanoate smell like bananas. But many natural and artificial flavors contain a blend of many esters. Strawberry flavor, for example, may contain several esters. Making realistic synthetic flavors is an art in which chemists vary the blends of esters to achieve the desired taste.

Figure 12 Because of their strong, fruity aromas, esters are used as flavoring and scents.

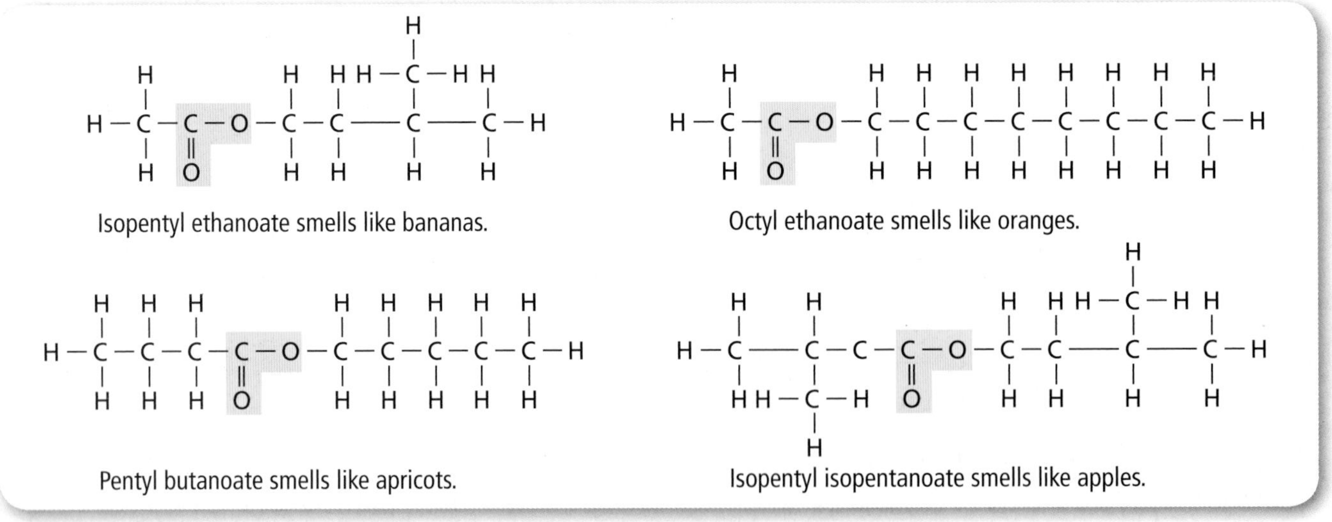

Isopentyl ethanoate smells like bananas.

Octyl ethanoate smells like oranges.

Pentyl butanoate smells like apricots.

Isopentyl isopentanoate smells like apples.

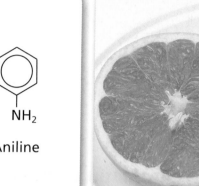

Aniline

Aniline is an amine used to make dyes.

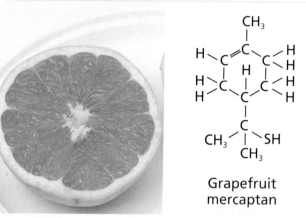

Grapefruit mercaptan

A mercaptan gives grapefruit its unique smell and taste.

Substituting Other Elements

Other functional groups can be added to hydrocarbons. Each group has unique properties. Three common groups are amines, mercaptans, and halocarbons.

Amines An **amine** forms when an amine group ($-NH_2$) replaces a hydrogen atom in a hydrocarbon. For example, aniline is formed when $-NH_2$ replaces a hydrogen atom on a benzene ring. Aniline, shown in **Figure 13,** is used to make dyes. Amines are also essential for life.

Mercaptans When the group $-SH$ replaces a hydrogen atom in a hydrocarbon, the resulting compound is a thiol. Thiols are commonly called mercaptans. Most mercaptans have unpleasant odors. This can be useful to animals such as skunks. Strangely, small concentrations of foul-smelling mercaptans are often found in pleasant-smelling substances. For example, the odor of grapefruits is due to the mercaptan shown in **Figure 13.**

Mercaptan odors are also powerful. You can smell skunk spray in concentrations as low as 0.5 parts per million. Such a powerful stink can be an asset to people, too. Natural gas has no odor of its own so it is impossible to smell a gas leak. For this reason, gas companies add small amounts of a mercaptan to the gas to make people aware of leaks.

Halocarbons A substituted hydrocarbon with a halogen, such as chlorine or bromine, in place of a hydrogen atom is called a halocarbon. For example, when four chlorine atoms replace four hydrogen atoms in ethene, the result is tetrachloroethene (teh truh klor uh eh THEEN), a solvent used in dry cleaning. Adding four fluorine atoms to ethene makes a compound that is a starting material for making nonstick coatings on cookware, as shown in **Figure 14.**

■ **Figure 13** Amines and mercaptans are two types of substituted hydrocarbons.

■ **Figure 14** Tetrafluoroethene is a halocarbon used to make non-stick coatings on pans.

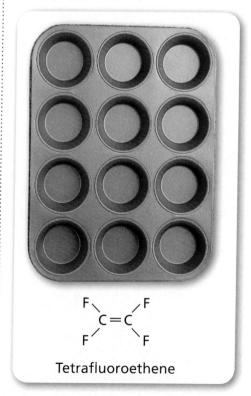

Tetrafluoroethene

Methyl salicylate gives these mints their wintergreen flavor.

Acetyl salicylic acid, commonly called aspirin, is a pain reliever.

■ **Figure 15** Aromatic compounds contain a benzene ring and often have strong scents and flavors.

Aromatic Compounds

Recall that benzene is very stable and makes a good building block for compounds. All compounds that contain a benzene structure are **aromatic compounds.** Benzene's hydrogen atoms can be replaced with various functional groups to make many different aromatic compounds.

Aromatic compounds and smell Aromatic compounds are so named because most of them have a distinctive smell. They contribute to the smell of cloves, cinnamon, and vanilla. Some aromatic compounds produce pleasant odors or tastes. For example, methyl salicylate, shown in **Figure 15,** produces a fresh wintergreen fragrance. Other aromatic compounds have less pleasant flavors and smells. Aspirin, also shown in **Figure 15,** is a sour-tasting aromatic compound. The different flavors are due to the different functional groups.

Section 2 Review

Section Summary

▶ In alcohols, the –OH group is substituted for a hydrogen atom.

▶ Organic acids contain the group –COOH.

▶ Esters are prepared by combining an alcohol and an organic acid.

▶ Amines contain the –NH₂ group and mercaptans contain the –SH group.

▶ Substituted hydrocarbons containing a halogen are called halocarbons.

▶ Many aromatic compounds have distinct odors.

8. **MAIN Idea** **Classify** each of the following as a hydrocarbon or a substituted hydrocarbon: ethyne, tetrachloroethene, ethanol, benzene, propane, and acetic acid.

9. **Identify** the structure that is present in all aromatic compounds.

10. **Explain** why chemists might want to prepare substituted hydrocarbons. Give two examples of possible substitutions.

11. **Identify** possible uses for each of the following types of substituted hydrocarbons: alcohols, esters, and halocarbons.

12. **Think Critically** Chloroethane (C_2H_5Cl) can be used as a spray-on anesthetic for localized injuries. How does chloroethane fit the definition of a substituted hydrocarbon? Diagram its structure.

Apply Math

13. **Use Percentages** The odor of mercaptans can be detected in concentrations as low as 0.5 parts per million. Express this concentration as a percent.

✓ **Assessment** Online Quiz

LAB

Alcohols and Organic Acids

Objectives

- **Initiate** a chemical reaction to produce a specific compound.
- **Gather** evidence to form conclusions about the identity of a new compound formed from a chemical reaction.

Background: Have you ever wondered how chemists change one substance into another? You have learned that changing the bonding between atoms holds the key to that process.

Question: *How can an alcohol change into an organic acid?*

Preparation

Materials

large test tube and stopper
$0.01M$ potassium permanganate solution (1 mL)
$6M$ sodium hydroxide solution (1 mL)
ethanol (3 drops)
10-mL graduated cylinder
universal indicator strips

Safety Precautions

WARNING: *Always handle chemicals carefully. Immediately flush any spill with water.*

Procedure

1. Read the procedure and the safety information and complete the lab form.
2. Pour 1 mL of $0.01M$ potassium permanganate solution and 1 mL of $6M$ sodium hydroxide solution into a test tube.
   WARNING: Potassium permanganate is an irritant. Sodium hydroxide is caustic.
3. Add 3 drops of ethanol to the test tube.

4. Stopper the test tube. Gently shake it for 1 minute. Observe and record any changes in the solution for 5 minutes.
5. Use a strip of universal indicator paper to determine the pH of the solution.

Conclude and Apply

1. **Draw** the structural formula of ethanol, and identify the part of the compound that makes it an alcohol.
2. **Identify** the part of a molecule that identifies a compound as an organic acid.
3. **Explain** how you know that a chemical change took place in the test tube.
4. **Predict** the formula of the acid produced when ethanol undergoes a chemical reaction in the presence of potassium permanganate.
5. **Identify** the acid produced by this reaction. (*Hint: This acid is found in vinegar.*)

COMMUNICATE YOUR DATA

Write an Article Suppose you are submitting a description of your investigation to a scientific journal. Write an article summarizing your procedure, your results, and your conclusions. Exchange articles with a classmate and peer review each other's articles.

Reading Preview

Essential Questions

▶ How are organic compounds obtained from petroleum?
▶ How do organic compounds combine to form polymers?
▶ What are some uses of polymers?

Review Vocabulary

condense: to change from gaseous to liquid state

New Vocabulary

monomer
polymer
depolymerization

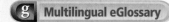 **Multilingual eGlossary**

Petroleum—A Source of Organic Compounds

MAIN ‹Idea› Petroleum is the source of organic compounds used to make fuels, polymers, and many other products.

Real-World Reading Link Suppose you fill your car's gas tank and buy a bottle of water. What do your purchases have in common? Both gasoline and plastic contain organic compounds that came from petroleum.

What is petroleum?

Do you carry a plastic comb in your pocket or purse? Do you know where that plastic originated? Chances are, it came from petroleum, a mixture of hydrocarbons and small amounts of other substances found deep within Earth. Because it is formed from the remains of fossilized material, it is a fossil fuel. The liquid part of petroleum is called crude oil. Crude oil is dark, flammable, and foul-smelling.

How can a thick, dark liquid like crude oil be transformed into a hard, brightly colored, useful object like a comb? The answer lies in the nature of crude oil. Crude oil is a mixture of thousands of organic compounds. To make items such as combs, specific compounds are refined from this mixture and then processed into consumer products.

Processing Crude Oil

The first step is to extract the crude oil from its underground source, as shown in **Figure 16.** Once the crude oil is obtained, chemists and engineers separate the crude oil into fractions containing compounds with similar boiling points. The separation process is known as fractional distillation and takes place in oil refineries.

■ **Figure 16** To obtain crude oil, large wells must be drilled deep into Earth.

The tower If you have ever driven past a refinery, you may have seen big, metal towers called fractionating towers. These towers, like the one shown in **Figure 17,** can rise as high as 35 m and be 18 m wide. Inside the tower is a series of metal plates arranged like the floors of a building. These plates have small holes through which vapors can pass. On the outside is a maze of scaffolding and pipes at various levels.

Separating organic compounds The tower separates crude oil into fractions containing compounds having a range of boiling points. Within a fraction, boiling points may range more than 100°C.

At the base of the tower, the crude oil is heated to more than 350°C. At this temperature, most hydrocarbons in the mixture become vapor and start to rise. A few compounds, such as those in asphalt, remain as liquids and are drained from the bottom of the tower. Compounds with higher boiling points condense on the lower plates, drain off through pipes on the sides of the tower, and are collected.

Fractions with lower boiling points may rise to the middle plates before condensing. Those with the lowest boiling points condense on the topmost plates. A few compounds never condense at all and are collected as gases at the top of the tower.

Figure 17 shows some typical fractions. Larger carbon chains have higher boiling points. They generally condense at the bottom of the tower, and smaller compounds condense at the top.

✔ **Reading Check** **Compare** the masses of compounds collected at the top of the tower to those collected at the bottom.

Why do the condensed liquids not fall back through the holes? The reason is that pressure from the rising vapors prevents this. In fact, the separation of the fractions is improved by the interaction of rising vapors with condensed liquid. The exact processes involved vary. For example, some towers add steam at the bottom to aid vaporization. The design and process used depend on the type of crude oil and on the fractions desired.

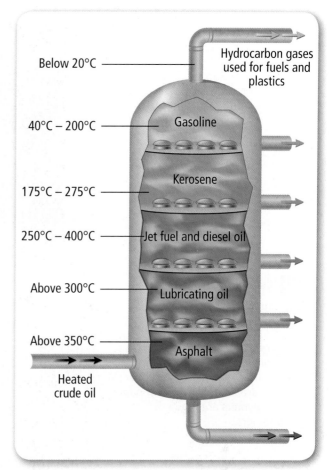

Below 20°C — Hydrocarbon gases used for fuels and plastics

40°C – 200°C — Gasoline

175°C – 275°C — Kerosene

250°C – 400°C — Jet fuel and diesel oil

Above 300°C — Lubricating oil

Above 350°C — Asphalt

Heated crude oil

■ **Figure 17** Fractions are separated in a fractionating tower by their boiling points. Larger hydrocarbons have high boiling points and condense on the bottom plates. The smallest hydrocarbons have much lower boiling points and are collected at the top of the tower as gases.

Infer *How might these fractions be further separated?*

 **Concepts in Motion** Animation

■ **Figure 18** Crude oil provides the organic compounds used in many common products. After crude oil has been refined, the organic compounds can be used to make various types of fuel, plastics, and synthetic fibers, as well as paint, dyes, and medicines.

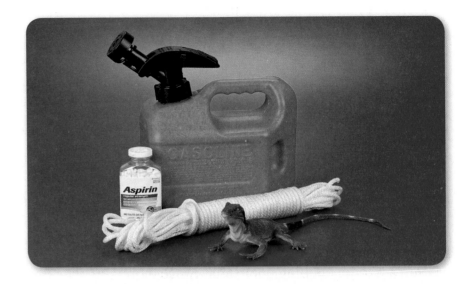

Concepts in Motion

Animation

■ **Figure 19** Both paper chains and polymers are long chains made up of smaller units. Paper chains are made of loops of paper and polymers are made of monomers.

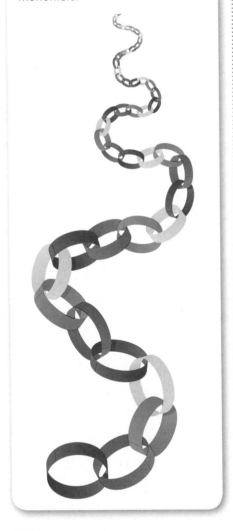

Uses for Petroleum Compounds

The lightest fractions from the top of the tower include butane and propane, which are used for fuel. The fractions that condense on the upper plates and contain from five to ten carbons are used for gasoline and solvents. Below these are fractions with 12 to 18 carbons that are used for kerosene and jet fuel. The bottom fractions go into lubricating oil, and the residue is used for paving asphalt.

Other petroleum products are obtained by further purifying crude oil fractions using different techniques to isolate individual compounds. After these are separated, they can be converted into substituted hydrocarbons. Chemists use these to make products ranging from medicines such as aspirin to insecticides, printers' ink, and flavorings, as shown in **Figure 18.** Aromatic dyes from crude oil have almost completely replaced natural dyes, such as indigo (a deep blue) and alizarin (a deep red). The first synthetic dye was a bright purple called mauve that was accidentally discovered in coal tar compounds.

Polymers

Did you ever loop together strips of paper to make paper chains for decorations, or have you ever strung paper clips together? A paper chain can represent the structure of a polymer, as shown in **Figure 19.** Some of the smaller molecules from petroleum are monomers that act like links in a chain. *Mono–* means one. A **monomer** is a small molecule that can combine with itself repeatedly to form a long chain. When these links are hooked together, they make new, extremely large molecules known as polymers. **Polymers** are long chains of monomers. Often, two or more different monomers, known as copolymers, combine to make one polymer molecule.

✅ **Reading Check** **Explain** how polymers are similar to paper chains.

Polymer properties The properties of polymers depend mostly on which monomers are used to make them. The amount of branching and the shape of the polymer also greatly affect its properties. Because of this diversity, polymers can be made light and flexible or strong enough to make plastic pipes, boats, and even some auto bodies. Other polymers can be spun into threads for use in clothing, suitcases, and backpacks.

Hydrocarbon polymers Some polymers are made entirely of carbon and hydrogen, as shown in **Table 3.** When ethene combines with itself repeatedly, it forms the common polymer polyethylene (pah lee EH thuh leen). Polyethylene is used widely in shopping bags and plastic bottles, as shown in **Table 3.** Another common polymer, polypropylene (pah lee PRO puh leen), is made of propene monomers and has many uses.

Versatile polystyrene Some polymers can take two completely different forms. Polystyrene (pah lee STI reen) forms brittle, transparent cases for CDs. But if a gas such as carbon dioxide is blown into the melted polystyrene as it is molded, bubbles will remain within the polymer when it cools. These bubbles make polystyrene the efficient insulator used in foam cups.

VOCABULARY ....
WORD ORIGIN
Polymer
from the Greek *poly–* meaning "many" and *–mer* meaning "part or segment"
Polymers are used to make plastic bags and foam cups. ....

Table 3	Hydrocarbon Polymers (⟨○ Concepts in Motion Interactive Table	
Polymer	**Monomer**	**Uses**
Polyethylene	$\left[CH_2 - CH_2 \right]$	• plastic bags • plastic bottles
Polypropylene	$\left[\begin{array}{c} CH_2 - CH \\ \quad\quad\ \ \| \\ \quad\quad CH_3 \end{array} \right]$	• glues • carpets • high-performance outdoor clothing
Polystyrene	$\left[\begin{array}{c} CH - CH_2 \\ \| \\ \bigcirc \end{array} \right]$	• foam packing • disposable food containers • CD cases

Table 4

Substituted Hydrocarbon Polymers Concepts in Motion Interactive Table

Polymer	Monomer	Uses
Polyurethane	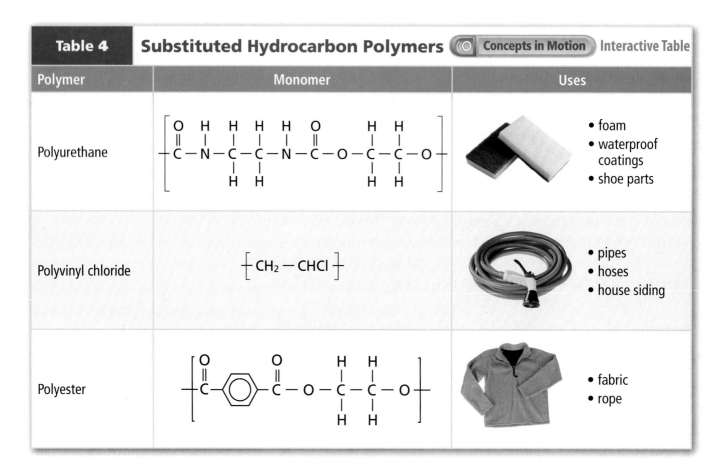	• foam • waterproof coatings • shoe parts
Polyvinyl chloride	$\left[\text{CH}_2 - \text{CHCl}\right]$	• pipes • hoses • house siding
Polyester		• fabric • rope

Substituted hydrocarbon polymers Polymers can contain elements other than carbon and hydrogen, as shown in **Table 4.** Polyurethane has functional groups that contain oxygen and nitrogen. The monomers of polyvinyl chloride, also known as PVC, are ethene monomers with chlorine substituted for one hydrogen atom. This substitution makes PVC harder and more heat resistant than polyethylene.

Polyesters Synthetic fibers called polyesters are made from an organic acid that has two –COOH groups and an alcohol that has two –OH groups, as shown in **Figure 20.** Polyester is strong because these chains are closely packed together. Many varieties of polyesters can be made, depending on what alcohols and acids are used. Polyesters can be woven or knitted into durable fabrics.

■ **Figure 20** Polyesters are formed when an organic acid with two –COOH groups combines with an alcohol that has two –OH groups.

Organic acid   Alcohol   Polyester (1 unit)   Water

■ **Figure 21** Plastics can be recycled into many structures, such as decks, gazebos, and this bench.

Depolymerization Disposing of polymers has become a problem because polymers have been used so widely and many polymers do not decompose. One way to combat this is by recycling, which recovers clean plastics for reuse in new products, such as the bench in **Figure 21.** Many communities recycle plastics.

Another approach is depolymerization. **Depolymerization** is a process that uses heat or chemicals to break the long polymer chain into its monomer fragments. These monomers can then be reused. However, each polymer requires a different depolymerization process and much research is needed to make this type of recycling economical.

Section 3 Review

Section Summary

▶ Crude oil is a dark, flammable liquid formed from fossilized materials.

▶ Organic compounds in crude oil can be separated using fractional distillation.

▶ Polymers are long chains of repeating chemical units called monomers.

▶ Polymers can be designed with specific properties.

▶ Depolymerization is the process of breaking a polymer into its components.

14. **MAIN Idea Identify** several items around your home that are made from organic compounds obtained from crude oil.

15. **Name** some of the fuels obtained from crude oil by fractional distillation.

16. **Describe** the process of fractional distillation.

17. **Explain** why polymers made from the same monomer can have physical properties that vary greatly.

18. **Describe** why depolymerization can be an expensive process.

19. **Think Critically** Based on the names of the polymers in this section, what do you think the polymer made from the monomer terpene is called?

Apply Math

20. **Calculate** If the mass of a monomer is 105 atomic mass units, find the mass of a polymer containing 122 monomers.

Essential Questions

▶ How are the structures of proteins, carbohydrates, lipids, and nucleic acids similar? How are they different?

▶ What types of polymers are found in the basic food groups?

▶ What is the function of DNA?

Review Vocabulary

base: any substance that forms hydroxide ions (OH⁻) in a water solution

New Vocabulary

protein
carbohydrate
lipid
nucleic acid
deoxyribonucleic acid (DNA)
nucleotide

g Multilingual eGlossary

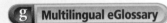

■ **Figure 22** Amino acids link together with peptide bonds. Here, glycine and cysteine combine to form the peptide glycyl cysteinate. Long chains of peptides are called proteins.

Biological Compounds

MAIN ‹Idea Proteins, carbohydrates, lipids, and nucleic acids are large, organic molecules made by plants and animals.

Real-World Reading Link You may have had cereal for breakfast this morning. Cereals, like all foods, contain biological compounds such as proteins and carbohydrates.

Biological Polymers

Like all polymers, biological polymers are huge molecules made of many monomers linked together. The monomers of biological polymers are usually larger and more complex in structure than those used to make plastics. Many of the important biological compounds in your body are polymers. Among them are the proteins and starches.

Proteins

Proteins are large organic polymers formed from organic monomers called amino acids. Even though only 20 amino acids are commonly found in nature, they can be arranged in so many ways that millions of different proteins exist. Proteins come in numerous forms and make up many of the tissues in your body, such as muscles, tendons, hair, and fingernails. In fact, proteins account for 15 percent of your total body weight.

Protein formation Amino acids, like glycine and cysteine, are the monomers that combine to form proteins. As shown in **Figure 22,** the amine group ($-NH_2$) of one amino acid combines with the carboxylic acid group ($-COOH$) of another amino acid to form a compound called a peptide. The bond joining them is known as a peptide bond. Peptides with about 50 or more amino acids are called proteins.

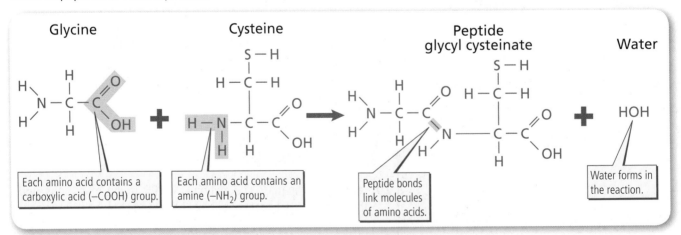

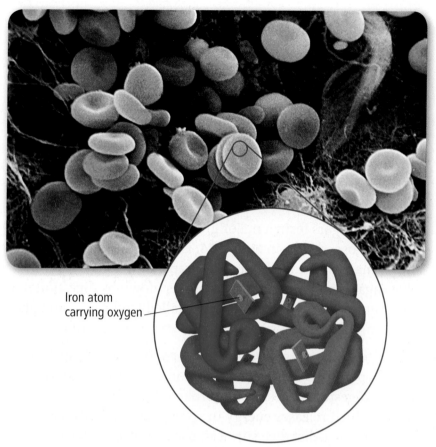

■ **Figure 23** Human blood contains disks called red blood cells. In red blood cells, hemoglobin has four protein chains coiled around each other. Each chain has an atom of iron, which carries oxygen.

Iron atom carrying oxygen

Protein structure Long protein molecules tend to twist and coil in a manner unique to each protein. For example, hemoglobin, which carries oxygen in your blood, has four chains that coil around each other, as shown in **Figure 23**. Each chain contains an iron atom that carries the oxygen. If you look closely, you can see all four iron atoms in hemoglobin.

When you eat foods that contain proteins, such as meat, dairy products, and some vegetables, your body breaks down the proteins into their amino acid monomers. Then your body uses these amino acids to make new proteins that form muscles, blood, and other body tissues.

☑ **Reading Check Explain** what happens when you eat foods containing protein.

Carbohydrates

If you hear the word *carbohydrate,* you may think of bread, cookies, or pasta. You may know someone who is on a low-carbohydrate diet to control weight. Or, you may have heard of carbohydrate loading by athletes. Runners, for example, often prepare for a long-distance race by eating, or loading up on, carbohydrates in foods such as vegetables and pasta. **Carbohydrates** are compounds containing carbon, hydrogen, and oxygen that have twice as many hydrogen atoms as oxygen atoms. Carbohydrates include the sugars and starches.

Test for Starch

Procedure 🥽 👕 🧤 🧼

WARNING: *Use caution when using iodine solution. Iodine solution can be a skin and eye irritant and is toxic if ingested or inhaled. Never eat any foods in the lab.*

1. Read the procedure and safety information, and complete the lab form.
2. Collect a variety of materials, including **cotton balls, paper, potato pieces, wood, apple pieces, laundry spray starch, bread,** and **metal.** Predict which items contain starch.
3. Use a **dropper** to place one or two drops of **iodine solution** on each material. For each item, record any color change.

Analysis

1. **Infer** What color indicates the presence of starch? (*Hint: Bread contains starch.*)
2. **Evaluate** How do your results compare with your predictions?

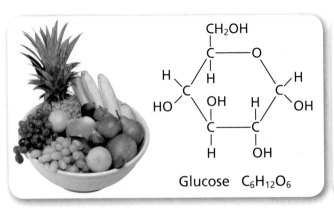

Glucose $C_6H_{12}O_6$

Sucrose $C_{12}H_{22}O_{11}$

■ **Figure 24** Fruits contain glucose and another simple sugar called fructose. Sucrose, commonly called table sugar, consists of a glucose molecule and a fructose molecule bonded together.

Explain *why sugars are carbohydrates.*

Sugars Sugars form a major group of carbohydrates. The sugar glucose, shown in **Figure 24,** is found in your blood and also in many sweet foods, such as grapes and honey. Common table sugar, known as sucrose, is broken down by digestion into two simpler sugars—fructose, often called fruit sugar, and glucose. Unlike starches, sugars provide energy soon after eating.

Starches Starch, shown in **Figure 25,** is a carbohydrate that is also a polymer. It is made of monomers of the sugar glucose. During digestion, the starch is broken down into similar sugars, which releases energy in your body cells. Athletes, especially long-distance runners, use starches to provide high-energy, long-lasting fuel for the body. The energy from starches can be stored in liver and muscle cells in the form of a compound called glycogen.

✓ **Reading Check Describe** the difference between sugars and starches.

Lipids

Fats, oils, and related compounds make up a group of organic compounds known as **lipids.** Lipids include animal fats, such as butter, and vegetable oils, such as corn oil. Lipids contain the same elements as carbohydrates but in different proportions. For example, lipids have fewer oxygen atoms and contain carboxylic acid groups. Lipids are long hydrocarbon chains, but they are not polymers.

■ **Figure 25** Starch, the major component of pasta, is a polymer made of glucose monomers.

Starch

Fats and oils Lipids are similar in structure to hydrocarbons. They can be classified as saturated or unsaturated according to the types of bonds in their carbon chains. Saturated fats, like saturated hydrocarbons, are saturated with hydrogen and contain only single bonds between carbon atoms. Unsaturated fats are like unsaturated hydrocarbons and have double or triple bonds. Unsaturated fats with one double bond are called monounsaturated, and those with two or more double bonds are called polyunsaturated.

Animal lipids, called fats, tend to be saturated and are solids at room temperature. Plant lipids, called oils, are unsaturated and are usually liquids, as shown in **Figure 26.** Sometimes, hydrogen is added to vegetable oils to form more saturated solid compounds called hydrogenated vegetable shortenings.

Have you heard that eating too much fat can be unhealthy? Evidence shows that too much saturated fat and cholesterol in the diet may contribute to some heart disease and that unsaturated fats may help to prevent heart disease. It appears that saturated fats are more likely to be converted to substances that can block the arteries leading to the heart. A balanced diet includes some fats, just as it includes proteins and carbohydrates.

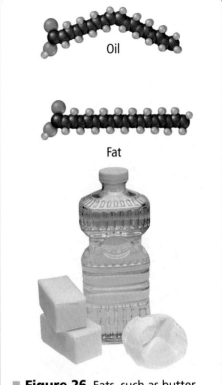

Oil

Fat

■ **Figure 26** Fats, such as butter and lard, are saturated lipids. Oils are unsaturated lipids.

Apply Science

Which foods should you choose?

What do you like to eat? You probably choose your foods by how good they taste. A better way might be to look at their nutritional values. Your body needs nutrients, such as proteins, carbohydrates, and fats, to give it energy and to help it build cells. Almost every food has some of these nutrients in it. The trick is to choose your foods so you do not get too much of one thing and not enough of another.

Identify the Problem

The table shows some basic nutrients for a variety of foods. The amount of the protein, carbohydrate, and fat is recorded as the number of grams in 100 g of the food. By examining these data, select the foods that best provide each nutrient.

Nutritional Values for Some Common Foods

Food (100 g)	Protein (g)	Carbohydrate (g)	Fat (g)
Cheddar cheese	25	1	33
Hamburger	17	23	17
Soybeans	13	11	7
Wheat	15	68	2
Potato chips	7	53	35

Solve the Problem

1. Using the table, list the foods that supply the most protein and carbohydrates. What might be the problem with eating too many potato chips?

2. In countries where meat and dairy products are not plentiful, people eat a lot of food made from soybeans. Give several reasons why people might wish to substitute meat and dairy products with soybean-based products.

■ **Figure 27** Each nucleotide looks like half of a ladder rung with an attached side piece. As you can see, each pair of nucleotides forms a rung on the ladder while the side pieces give the ladder a little twist that gives DNA its double helix shape.

Cholesterol Cholesterol is another lipid that is often in the news. It is found in meats, eggs, butter, cheese, and fish. Like fats and oils, cholesterol can be both helpful and harmful to your body. Some cholesterol is produced by the body to build cell membranes. It is also found in bile, a digestive fluid. However, too much cholesterol may cause serious damage to heart and blood vessels, similar to the damage caused by saturated fats.

Nucleic Acids

The nucleic acids are another important group of organic polymers that are essential for life. A **nucleic acid** is an organic polymer that controls the activities and reproduction of cells. You may have heard about DNA, which is one type of nucleic acid. **Deoxyribonucleic** (dee AHK sih ri boh noo klay ihk) **acid,** also called DNA, is an essential biological polymer found in the nuclei of cells that codes and stores genetic information. This information is known as the genetic code.

Nucleic acid monomers The monomers that make up DNA are called nucleotides. **Nucleotides** are complex molecules that make up DNA and that contain one of four organic bases, a sugar, and a phosphate unit. **Figure 27** shows the structure of DNA, along with a closer look at a nucleotide. DNA nucleotides form chains that are unique to an organism. Two nucleotide chains twist around each other, forming what resembles a twisted ladder called a double helix. The rungs of the ladder are paired organic bases. Your genetic code gives instructions for making other nucleotides and proteins needed by your body. Which proteins are made depends on the order of the bases in the DNA.

✓ **Reading Check** **Identify** the three components of a nucleotide.

Base pairs

Nucleotides

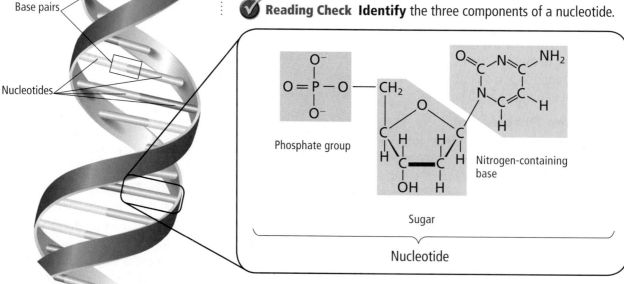

Phosphate group

Sugar

Nitrogen-containing base

Nucleotide

■ **Figure 28** A DNA profile analyzes the base pairs at several specific sites along the DNA strand.

DNA fingerprinting Human DNA contains more than five billion base pairs. The DNA of each person differs in some way from that of everyone else. The unique nature of DNA offers crime investigators a way to identify criminals from hair or fluids left at a crime scene. First, DNA from bloodstains or cells in saliva found on a soda bottle can be extracted in the laboratory. Then, chemists can break up the DNA into its nucleotide components and use radioactive and X-ray methods to obtain a picture of the nucleotide pattern. **Figure 28** shows a DNA profile similar to those used by investigators. Comparing this pattern to one made from the DNA of a suspect can link that suspect to the crime scene.

Section 4 Review

Section Summary

▶ Proteins are large, organic polymers that form muscles, blood, and other body tissues.

▶ Carbohydrates contain carbon, hydrogen, and oxygen.

▶ Sugars and starches are carbohydrates that provide energy to your body.

▶ Lipids include fats and oils.

▶ DNA is a nucleic acid that is found in the cell nucleus.

21. **MAIN ‹Idea Name** the monomers that make up the following biological polymers: proteins, nucleic acids, and starches.

22. **Identify** where your body gets the compounds that it needs to build proteins.

23. **Describe** the function of DNA.

24. **Explain** the difference between saturated and unsaturated fats and oils.

25. **Think Critically** Whole milk contains about 4 percent butterfat. Explain why you might choose to drink milk containing 2 percent fat.

Apply Math

26. **Use Percentages** You have read that your body is about 15 percent protein. Calculate the mass of protein in your body in kilograms.

LAB — Esters

Objectives

- **Prepare** an ester from an alcohol and an acid.
- **Detect** the results of the reaction by the odor of the product.

Background: Organic compounds known as acids and alcohols react to form another type of organic compound called an ester. Esters frequently produce a recognizable and often pleasant fragrance. Esters are responsible for many fruit flavors, such as apple, pineapple, pear, and banana. However, esters are not always aromatic in the chemical sense—they might not contain a benzene ring.

Question: *How do an acid and an alcohol react to produce a compound with different characteristics? Can the presence of the new compound formed be detected by its odor?*

Preparation

Materials

medium-size test tube
test-tube holder
250-mL beaker
10-mL graduated cylinder
water
hot plate
ring stand
thermometer
salicylic acid (1.0 g)
amyl alcohol (2 mL)
concentrated sulfuric acid (1 mL to be added by teacher)

Safety Precautions

Procedure

1. Read the procedure and safety information, and complete the lab form.

2. Add about 150 mL of water to the beaker, and heat it on the hot plate to 70°C.

3. Place approximately 1 g of salicylic acid in a test tube. Does this material have an odor? See the warning and illustration below for the proper way to detect odors in the laboratory.

   WARNING: *To detect an aroma safely, hold the container about 10 cm in front of your face and wave your hand over the opening to direct air currents to your nose.*

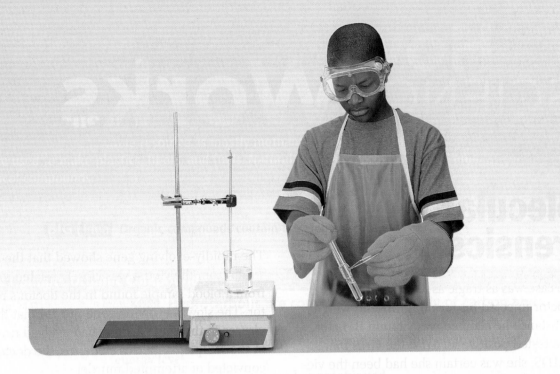

4. Check to see if amyl alcohol has an odor. If so, try to remember what it smells like.

5. Add 2 mL of amyl alcohol to the test tube.

6. Ask your teacher to carefully add 1 mL of concentrated sulfuric acid. **WARNING: *Sulfuric acid is caustic. Avoid all contact. Do not inhale fumes.***

7. Place the test tube in the hot water, and leave it untouched for about 12 to 15 minutes.

8. Remove the tube from the hot water using a test-tube holder, and allow it to cool. Check to see if you can detect a new aroma.

Analyze Your Data

1. **Identify** What did you smell after the reaction was complete?

2. **Infer** Look closely at the surface of the liquid in the test tube. Do you see any small droplets of an oily substance? What do you think it is?

Conclude and Apply

1. **Predict** This reaction formed the ester amyl salicylate. What esters would form if amyl alcohol were replaced by the following alcohols: methyl, ethyl, propyl, and isobutyl?

2. **Predict** Look at the equation for the reaction at the bottom of the page. One product is given. What do you think is the second product formed in this reaction? Explain.

3. **Design an Experiment** How might you modify the experiment to produce a different ester?

COMMUNICATE YOUR DATA

Poster Make a poster showing the reaction that took place. Use the poster to explain the formation of esters to students from another class.

Use Vocabulary

Complete each statement with the correct term from the Study Guide.

27. **BIG Idea** _____ are most compounds that contain the element carbon.

28. Amino acids combine to form large organic polymers known as _____.

29. _____ is the nucleic acid that contains your genetic information.

30. A(n) _____ is a compound containing the benzene-ring structure.

31. Sugars and starches are part of the group of organic compounds called _____.

32. Fats and oils are part of the group of organic compounds called _____.

33. _____ are compounds with identical chemical formulas but different structures.

Check Concepts

34. How would you describe a benzene ring?
    A) rare
    B) stable
    C) unstable
    D) saturated

35. What are the small units that make up polymers called?
    A) monomers   C) plastics
    B) isomers   D) carbohydrates

36. What type of compound is hemoglobin?
    A) carbohydrate   C) nucleic acid
    B) lipid   D) protein

37. What type of compounds form the DNA molecule?
    A) amino acids   C) polymers
    B) nucleotides   D) carbohydrates

38. Glucose and fructose both have the formula $C_6H_{12}O_6$. What are such compounds called?
    A) amino acids   C) isomers
    B) alcohols   D) polymers

39. If a carbohydrate has 16 oxygen atoms, how many hydrogen atoms does it have?
    A) 4   C) 16
    B) 8   D) 32

40. **THEME FOCUS** Which crude oil fractions are collected at the top of a fractionating tower?
    A) high boiling point, few carbon atoms
    B) high boiling point, many carbon atoms
    C) low boiling point, few carbon atoms
    D) low boiling point, many carbon atoms

Interpret Graphics

41. Copy and complete the following concept map about types of hydrocarbons.

Concepts in Motion   Interactive Concept Map

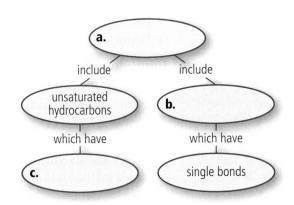

✓ **Assessment**   Online Test Practice

Use the figure below to answer question 42.

42. Write the chemical formula of each of these hydrocarbons, and identify which two are isomers.

43. Look at the fiber content of ten items of your clothing. Note the percentages of synthetic or natural fibers. Make a circle graph to compare the average percentages of natural and synthetic fibers. (*Hint: cotton, linen, wool, and silk are natural fibers.*)

Think Critically

44. Infer A healthy diet contains a variety of nutrients, including fats. However, saturated fats have some drawbacks. Based on this knowledge, how would you modify your diet to make it healthier? What general rule would you apply in making your choices?

45. Classify the following compounds as saturated, unsaturated, or substituted hydrocarbons: hexane (C_6H_{14}), isopropyl alcohol (C_3H_7OH), 2-chlorobutane (C_4H_9Cl), pentene (C_5H_{10}), and butyric acid (C_3H_7COOH).

46. Explain why the toughness and durability of many plastic polymers can be both an asset and a liability.

47. Describe how the structures of propyl alcohol and isopropyl alcohol might differ, although both have the formula C_3H_8O.

48. Draw a substituted hydrocarbon that has the following: six carbon atoms; single, double, and triple bonds; a hydroxyl group; and an amine group.

Apply Math

49. Solve One-Step Equations Although physicians disagree about what is a healthy level of blood cholesterol, many feel that levels above 200 mg/dL are harmful. A patient's blood cholesterol level measured 228 mg/dL. After two months on a low-fat diet, it dropped to 210 mg/mL. By what percent did the patient's cholesterol level decrease?

50. Use Percentages The label on a 500-mL bottle of vinegar states that it contains 6 percent acid by volume. How many milliliters of acid does this bottle contain?

Use the table below to answer question 51.

Nutritional Values for Some Common Foods			
Food (100 g)	Protein (g)	Carbohydrate (g)	Fat (g)
Cheddar cheese	25	1	33
Hamburger	17	23	17
Soybeans	13	11	7
Wheat	15	68	2
Potato chips	7	53	35

51. Calculate Percent The U.S. Food and Drug Administration recommends a maximum intake of 65 g of fat per day. What percent of the daily fat allowance is a 30 g serving of potato chips?

Standardized Test Practice

Multiple Choice

Record your answers on the answer sheet provided by your teacher or on a sheet of paper.

1. What atoms make up a hydrocarbon molecule?
   A. oxygen, carbon, and hydrogen
   B. nitrogen and carbon
   C. carbon and hydrogen
   D. oxygen and hydrogen

Use the figure below to answer questions 2 and 3.

2. What is the chemical formula of the compound shown above?
   A. C_3H_3
   B. CH_8
   C. C_6H_6
   D. C_3H_8

3. What is the name of this compound?
   A. propane   C. isoprene
   B. heptane   D. methane

4. Which of these contains carbon, hydrogen, and oxygen, and has twice as many hydrogen atoms as oxygen atoms?
   A. hydrocarbon
   B. carbohydrate
   C. alcohol
   D. isomer

5. Which of the following is not a polymer derived from petroleum?
   A. polypropylene
   B. acetylene
   C. polyethylene
   D. polystyrene

6. Which of the following is a type of recycling that breaks up the polymers into their original monomers?
   A. fractionation
   B. depolymerization
   C. isomerization
   D. saturation

Use the figure below to answer question 7.

Ethanoic acid   Ethanol   Tetrachloroethene
CH_3COOH C_2H_5OH C_2Cl_4

7. What is true of all three of these compounds?
   A. Their basic structural unit is a benzene ring.
   B. They are inorganic compounds.
   C. They are substituted hydrocarbons.
   D. They are polymers.

8. Which of these compounds is an alcohol that is often obtained from corn?
   A. ethanol
   B. acetic acid
   C. tetrachloroethene
   D. ethane

9. Which of these best shows the shape of the nucleic acid DNA?

A.   C.

B.    D.

✓ Assessment   Standardized Test Practice

Short Response

Record your answers on the answer sheet provided by your teacher or on a sheet of paper.

10. Describe the bonds between carbon atoms found in an organic compound.

Use the figure below to answer question 11.

11. These hydrocarbons are isomers. Write their chemical formulas.

12. Describe the general relationship between melting point, boiling point, and the amount of branching in an isomer.

13. Describe some properties and uses of alcohols.

14. Diagram the process used to separate petroleum compounds. On what physical property is this process based?

15. How are alcohols and organic acids similar? How are they different?

Extended Response

Record your answers on a sheet of paper.

16. Describe useful properties of polymers. Name several objects made of polymer material that would likely have been made of wood or metal in the past.

17. Identify the polymer material used to make CD cases and foam drinking cups. Explain the processes used to make these two different products.

Use the figure below to answer question 18.

18. How is this a good representation of a protein? Describe the importance of proteins in the human body.

19. When you read or hear about cholesterol in the news, it is usually associated with negative effects on the heart and blood vessels. Why does the body make a substance that can potentially damage the circulatory system?

20. Plastic polymers are inexpensive, but they do not decompose readily in landfills. Describe two ways to solve this problem.

NEED EXTRA HELP?																				
If You Missed Question . . .	1	2	3	4	5	6	7	8	9	10	11	12	13	14	15	16	17	18	19	20
Review Section . . .	1	1	1	4	3	3	2	2	4	1	1	1	2	3	2	3	3	4	4	3

CHAPTER 24

New Materials Through Chemistry

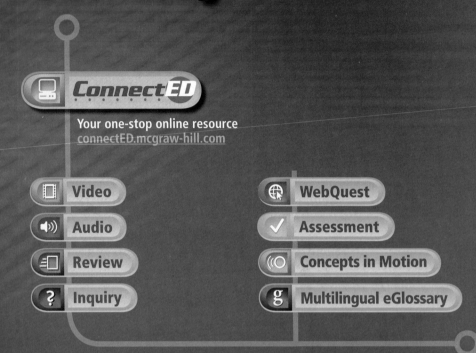

ConnectED

Your one-stop online resource
connectED.mcgraw-hill.com

- Video
- Audio
- Review
- Inquiry
- WebQuest
- Assessment
- Concepts in Motion
- Multilingual eGlossary

Launch Lab
Chemistry and Properties of Materials

When an engineer designs a vehicle, sports equipment, or even a structure like a bridge or building, the materials used for construction must be selected to match the function. Can the manufacturing process affect a material's performance?

For a lab worksheet, use your StudentWorks™ Plus Online.

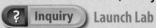 Launch Lab

FOLDABLES®

Make a four-door book using the labels shown. Use it to organize your notes on the classification of new materials.

Alloys	Ceramics
Composites	Polymers

THEME FOCUS Technology
Many everyday materials do not occur naturally, but are the products of chemistry.

BIG (Idea) Scientists can create synthetic materials with a wide range of properties and uses.

Section 1 • Alloys

Section 2 • Versatile Materials

Section 3 • Polymers and Composites

Reading Preview

Essential Questions

▶ What are alloys and how are they used?

▶ How do the properties of alloys determine their use?

Review Vocabulary

alloy: a mixture of elements that has metallic properties

New Vocabulary

luster
conductivity

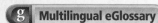
g Multilingual eGlossary

Alloys

MAIN ◀Idea An alloy is a mixture of elements that has metallic properties.

Real-World Reading Link Have you ever used a rusty fork? Probably not—most of the tableware we use everyday is made of a nonrusting mixture of iron and other elements called *stainless steel.* The addition of various elements changes the properties of iron.

Alloys Through Time

For ages, people have searched for better materials to use to make their lives more comfortable and their tasks easier. Ancient cultures used stone tools until methods for processing metals became known. Today, advances in processing and blending metals still occur as scientists continue to improve the art of making alloys. Recall that an alloy is a mixture of elements that has metallic properties. Alloys can produce materials with improved properties, such as greater hardness, strength, lightness, or durability.

The first alloy Historians think that bronze was accidently discovered more than 5,000 years ago by the ancient Sumerians. The Sumerians lived in the Tigris-Euphrates Valley (now Iraq). It is likely that Sumerians used rocks rich in copper and tin ore to keep their campfires from spreading. The hot campfire melted the copper and tin ores within the rocks, creating bronze. This first known mixture of metals became so popular and widely used that a 2,000-year span of history is known as the Bronze Age. The ancients did not have a chemical language for their discovery, but the bronze tools and objects, such as those shown in **Figure 1,** helped change the history of civilization.

■ **Figure 1** Artifacts from the Bronze Age prove that alloys of metal were used before 3,000 B.C. The development of bronze led to improved tools, weapons, and building materials and encouraged development of industry and trade.

■ **Figure 2** Copper is shiny, conducts thermal and electrical energy well, and can be easily drawn into a wire or hammered into sheets. Copper's malleability makes it ideal for use in pipes, wires, and tubing.

Explain *the difference between malleability and ductility.*

Materials change Bronze is still used today, but it is doubtful that ancient people would recognize it. The methods of processing this alloy have undergone many changes. Other alloys, such as brass, pewter, and steel, have been developed through the ages, giving people a large selection of materials to choose from today.

The Metallic Properties of Alloys

Alloys retain the metallic properties of the metals in them. Recall that metals have four distinct properties: luster, ductility, malleability, and conductivity. **Luster,** shown in **Figure 2,** is the property of a shiny appearance due to the ability to reflect light. The shiny appearance of aluminum foil and a new copper coin demonstrates the property of luster. The property that allows the material to be pulled into wires is known as ductility (duk TIH luh tee). The copper electrical wires in your home demonstrate the ductility of metals and alloys.

Recall that metals are malleable. The property that allows a material to be hammered or rolled into thin sheets is malleability (mal yuh BIH luh tee). Aluminum foil that is used in food preparation and food storage demonstrates the malleability of aluminum. Musical instruments like trombones or trumpets demonstrate the luster and malleability of brass. **Conductivity** (kahn duk TIH vuh tee) is the property of a material that allows thermal energy or electrical charges to move easily through it. Metals and alloys have high conductivities because some of their electrons are not tightly held by their atoms. Copper is used in electrical wiring because it is conductive as well as ductile.

✓ **Reading Check List** five other examples of items that you know have metallic properties.

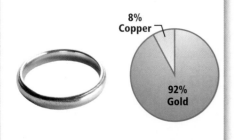
Choosing an alloy Why are gold alloys used in jewelry, but drill bits are made from steel? The alloy chosen to make an object depends upon how the object will be used. The characteristics desired in the final product must be considered before the product is constructed. For example, the manufacturer will have to consider how hard the alloy has to be to prevent the object from breaking when it is used. He or she must also consider if the object will be exposed to chemicals that will react with the alloy and cause the alloy to fail. These questions relate to the properties of the alloy and its intended use. This represents only two of the many possible questions that must be answered while a product is being designed.

Choosing a gold alloy Look at the characteristics of familiar objects, such as the gold jewelry, shown in **Figure 3.** The rings appear to be made of pure gold, but they are really made from gold-copper alloys. Gold is a bright, expensive metal that is soft and bends easily. Copper, on the other hand, is a relatively inexpensive metal that is harder than gold. When gold and copper are melted, mixed, and allowed to cool, an alloy forms with properties different from those of either pure gold or pure copper.

The properties of an alloy will vary depending upon the amount of each metal that is added. A ring made with a higher percentage of gold will bend easily due to gold's softness. This ring will be more valuable because it contains a higher percentage of gold, which is a more expensive metal. A ring with a higher percentage of copper will not bend as easily because copper is harder than gold. However, this ring will be less valuable because it contains more copper, a less-expensive metal.

✓ **Reading Check** **Compare and contrast** the properties of gold-copper alloys that are mostly gold with those that are mostly copper.

Uses of Alloys

If you see an object that looks metallic, it is most likely an alloy. Alloys are used in a variety of products, as shown in **Figure 4.** Alloys that are exceptionally strong are used to manufacture industrial machinery, construction beams, and railroad cars and rails. Automobile and aircraft bodies are constructed of alloys that are corrosion resistant and lightweight but are strong and able to carry heavy loads. Other types of alloys are used in products such as food cans, carving knives, and roller blades.

How are such a variety of alloys made? Scientists and engineers must find starting materials with some of the properties they need. Then, by adjusting the ratios of those materials, they can create a final product that suits their needs.

■ **Figure 3** Not all gold jewelry is the same. These rings vary in the amount of copper that has been added to the gold. A *karat* is a measure of gold's purity; pure gold is 24-karat.

Infer *How do the rings compare in hardness and malleability?*

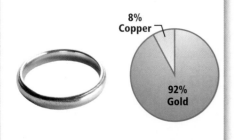

8%
Copper

92%
Gold

22-Karat

42%
Copper   58%
Gold

14-Karat

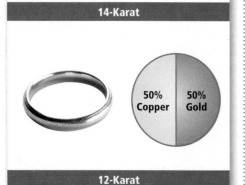

50%
Copper   50%
Gold

12-Karat

FIGURE 4

Visualizing Common Alloys

Most of the metal objects you use every day are made from alloys.

Stainless steel is an alloy of chromium, iron, and carbon. It is often used to make kitchenware. ▶

◀ Brass is an alloy of copper and zinc. It is used in musical instruments such as this trombone.

Pure silver is too soft to be very useful. Sterling silver, used in these pitchers, is an alloy of silver and copper. ▶

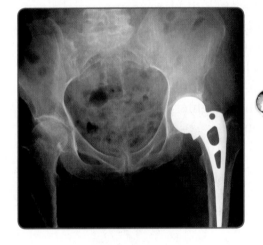

▼ Because they are lightweight and nonreactive, titanium alloys are often used in replacement joints.

Solder is an alloy of tin, silver, and copper that is used in electronics, such as this circuit board. ▼

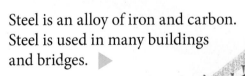

Steel is an alloy of iron and carbon. Steel is used in many buildings and bridges. ▶

Steel—an important alloy

Steel is an alloy of iron and carbon. Sometimes other elements are added to give steel specific properties. Steel is classified by the amount of carbon and other elements present, as well as by the manufacturing process that is used to refine the iron ore. The classes of steel, such as stainless, carbon, or high-strength steels, each have different properties and therefore different uses.

High-strength steel is a strong alloy and is used often if a great deal of strength is required. Office buildings have steel beams to support the weight of the structure. Bridges, overpasses, and streets also are reinforced with steel. Ship hulls, bedsprings, and automobile gears and axles are made from steel. Another class of steel, called stainless steel, is used in surgical instruments, cooking utensils, and food preparation. Stainless steel contains chromium as well as iron and carbon and gets it name from its resistance to rusting and corrosion.

✔ **Reading Check** **Explain** why steel is an important alloy.

Alloys in flight

Steel is not the only common type of alloy. Aluminum is familiar because it is used to make soda cans and cooking foil. Did you know that engineers also are using new aluminum and titanium alloys to build large commercial aircraft? The aircraft diagram in **Figure 5** shows how extensively alloys are used in new aircraft construction. The new alloys are strong and lightweight, and last longer than alloys used in the past. Some of the aluminum-zinc alloys are nearly as strong as steel, but a third of the weight. A lighter plane is less expensive to fly and produces less pollution.

Vocabulary . . . .
Academic Vocabulary
Structure
the arrangement of parts in a substance, body, or building
The house's structure was designed to withstand hurricane-force winds. . . . . . .

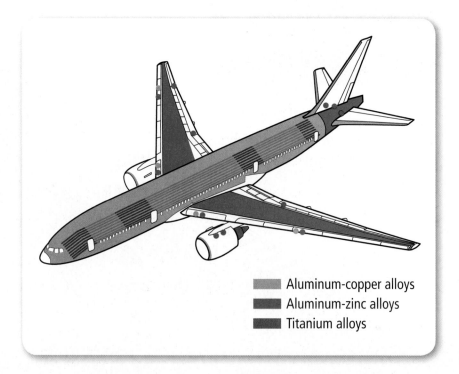

■ **Figure 5** Airplanes feature a number of alloys in their construction. Notice that the aircraft skin is made mostly of alloys.

Infer *what characteristics make alloys useful in airplane construction.*

- Aluminum-copper alloys
- Aluminum-zinc alloys
- Titanium alloys

Alloys in medicine Alloys that are resistant to tissue rejection and do not react chemically with the elements and compounds in the body can be used for a number of medical applications. One such alloy is surgical steel, a specific type of stainless steel made from iron, chromium, carbon, and nickel. Pins and screws made from surgical steel are used by doctors to connect broken bones and repair serious fractures, as shown in **Figure 6**. Surgical steel is also part of the metal plates used to repair damage to the skull. These plates protect the brain from injury and are safe to use inside the body.

If you have ever had a cavity, your dentist might have used a mercury-silver alloy to fill it, preventing further tooth decay. The alloy is malleable, durable, and helps to prevent bacterial growth and additional cavities in the same spot.

Shape memory alloys A special type of alloy commonly used in dentistry is called a shape memory alloy. A shape memory alloy can be bent and will return to its original shape when heat is applied. The wire in braces that helps to pull teeth into the correct position—called the archwire—is often made of shape memory alloys. Orthodontists bend the wire to fit a patient's mouth and then instruct the patient to drink warm liquids to help "activate" the archwire, which will cause it to return to its original shape and reposition teeth.

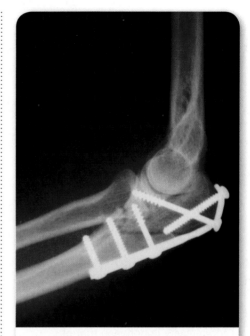

■ **Figure 6** Surgical steel screws and plates help to hold this elbow together as it heals. Because they are made of surgical steel, they are safe for use in the human body.

Identify *the properties that make surgical steel useful for implants.*

Section 1 Review

Section Summary

▶ Alloys have metallic properties.

▶ An alloy has characteristics that are different from, and often improved upon, the individual elements in it.

▶ Scientists and engineers can control the properties of an alloy by manipulating its composition.

▶ Alloy materials are used in industry, food service, medicine, and aeronautics.

1. **MAIN Idea** List the metallic properties of alloys.

2. **Identify** three different alloys in your home. Describe how each alloy is used and why the properties of the alloy make it suitable for that application.

3. **Describe** the importance of steel.

4. **Identify** two medical uses of alloys. Describe the desired characteristics in implements made from medical alloys

5. **Think Critically** If you were designing a skyscraper in an earthquake zone, what properties would the structural materials need?

Apply Math

6. **Calculate** 14-karat gold is 58 percent gold and 42 percent copper. Calculate the actual amount of gold in a 65-g, 14-karat gold chain.

7. **Find Mass** If a 7.6-g sample of copper can be hammered into a 2-cm × 2-cm sheet, calculate the number of grams necessary to hammer a 17-cm × 17-cm sheet under the same manufacturing conditions.

Essential Questions

▶ What properties of ceramics make them versatile?

▶ How are ceramic materials used?

▶ How do traditional and modern ceramic materials compare?

▶ What is a semiconductor and why it is useful?

Review Vocabulary

semiconductor: material having conductivity in between that of metals (good conductors) and nonmetals (insulators) and having electrical conductivities that can be controlled

New Vocabulary

ceramics
doping
integrated circuit

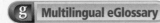

g Multilingual eGlossary

Versatile Materials

MAIN ‹ Idea Ceramics and semiconductors have conductivities that can range from highly insulating to superconductive.

Real-World Reading Link Think about the last time you had hot chocolate or tea. Did you use a mug? If you did, the mug was probably made from a ceramic material.

Ceramics

Do you think of floor tiles, pottery, or souvenir knickknacks when you see the word *ceramic*? By definition, **ceramics** are materials that are made from dried clay or claylike mixtures. Ceramics have been around for centuries—in fact, pieces of clay pottery from 16,000 B.C. have been found. The oldest known wall, built about 8,000 B.C., was around the Middle Eastern city of Jericho. The walls surrounding Jericho, as well as the homes inside the walls, were ceramic, consisting of bricks made from mud and straw that were baked in the Sun.

Around 1,500 B.C., the first glass vessels were made and kilns were used to fire and glaze pottery. By 50 B.C., the Romans had developed concrete and begun using it as a building material. Some of the structures built by the ancient Romans still stand today. About the same time that Romans were developing concrete, the Syrians were experimenting with glass-blowing techniques to make pitchers, bowls, cups, and jars. All of these items are ceramics. The china and glassware shown in **Figure 7** are also examples of ceramics.

■ **Figure 7** Glass, china, pottery, bricks, and tile are all examples of ceramics. Because of their wide range of properties, ceramics are used in a variety of different products.

Explain *how ceramics and alloys are similar.*

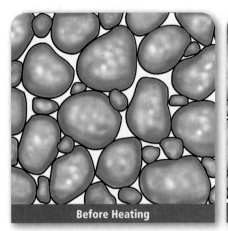
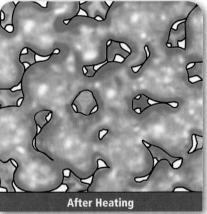

Before Heating

After Heating

Figure 8 When a ceramic object is fired, its particles merge together due to the loss of water, called dehydration. The object shrinks and the structure becomes more dense and stronger.

Making ceramics Traditional ceramics are made from easily obtainable raw materials—clay, silica (sand), and feldspar (crystalline rocks). These raw materials were used by ancient civilizations to make ceramic materials and are still used today. However, some of the more recent ceramics are made from compounds of metallic and nonmetallic elements, such as carbon, nitrogen, or sulfur.

After the raw materials are processed, ceramics usually are made by molding the ceramic into the desired shape, then heating it to temperatures between 1,000°C and 1,700°C. The heating process, called firing, causes the spaces between the particles to shrink as a result of the loss of water, called dehydration, as shown in **Figure 8.** The entire object shrinks as the spaces become smaller. This extremely dense internal structure gives ceramics their strength. Ceramics can be very durable. However, these same ceramics can also be brittle and will break if they are dropped or if the temperature changes too quickly.

✓ **Reading Check** **Summarize** the steps involved in making ceramics.

Properties of traditional ceramics Ceramics are known also for their chemical resistance to oxygen, water, acids, bases, salts, and strong solvents. These qualities make ceramics useful for applications where they may encounter these substances. For instance, ceramics are used for tableware because foods contain acids, water, and salts. Ceramic tableware is not damaged by contact with foods containing these substances.

Traditional ceramics also are used as insulators because they do not conduct heat or electricity. You may have seen electric wires attached to poles or posts with ceramic insulators. These insulators keep the current flowing through the wire instead of into the ground. Ceramic tiles were also used on the heat shield of the space shuttle to protect it from the extreme heat it experienced when it re-entered Earth's atmosphere.

MiniLab

? Inquiry MiniLab

Model a Ceramic

Procedure 🥽 ✋ 👕 🧤

1. Read the procedure and safety information, and complete the lab form.
2. Mix four tablespoons of **sand,** four tablespoons of **aquarium gravel or small pebbles,** and six tablespoons of **white glue** in a **paper cup.**
3. Add enough **water** to thoroughly mix the ingredients.
4. Stir the mixture until it is smooth.
5. Allow the mixture to sit for several days and observe.
6. Dispose of the cup as instructed by your teacher.

Analysis

1. **Draw** a picture of what your mixture looked like before you added the glue and water. Draw another picture after several days. Compare your drawings to those of your classmates.
2. **Explain** Is your mixture a ceramic? How is your mixture similar to a ceramic?
3. **Observe** What are some of the properties of your product?

Customizing ceramics The properties of ceramics can be customized, which makes them useful for a wide variety of applications. Changing the composition of the raw materials or altering the manufacturing process can change the properties of the ceramic. Manufacturing ceramics is similar to manufacturing alloys because scientists determine which properties are required and then attempt to create the ceramic material with those properties.

Ceramics in medicine Ceramics also have medical uses. Replacement hip sockets, for example, are made of ceramics because they are safe for use in the human body. Ceramics are strong and durable, but what properties make them safe for use in medicine? Like the alloys used in medicine, ceramics are resistant to body fluids, which can damage other materials. They are also relatively nonreactive and are resistant to rejection by the body. In the medical field, surgeons use ceramics in conjunction with alloys for the repair and replacement of joints such as hips, knees, shoulders, elbows, fingers, and wrists. Dentists use ceramics for braces as well as tooth replacement and repair.

 Reading Check **Explain** why ceramics are appropriate for medical applications.

Can you choose the right material?

Scientists continue to learn about atoms and how they interact. With this new knowledge, chemists today are able to create substances with a wide range of properties. This is especially evident in the production of specialized ceramics. Technical ceramics include ceramic products used in engineering applications, such as tiles on the outside of spacecraft, jet engine turbine blades, and gas burner nozzles.

Identify the Problem

As an engineer working on the design of a new car, you need to select the right ceramic materials to build parts of the car's engine and its onboard computer. The table at right shows the materials you have to choose from.

The ceramics listed in the table vary in their resistance to wear, electrical conductivity, chemical reactivity, and melting point. Using these properties, decide which materials should be used for the different components of the car.

Ceramic Properties

Material	Wear Resistant	Conducts Electricity	Reacts with Chemicals	Melting Point (°C)
A	highly	no	no	3,000
B	not at all	no	yes	100
C	moderately	yes	no	1,500
D	resistant	yes	no	500

Solve the Problem

1. Which of the above materials would you use when you build the engine? Explain the factors that you considered to make your decision.

2. Which of the above materials would you select when building the onboard computer? Explain your selection.

3. If you had to choose a material for building the car's bumper, what factors would you consider? Do you think that a ceramic material would be the best choice? Explain your answer.

Modern ceramics Ceramics can be customized to have nontraditional properties. Ceramics traditionally are used as insulators, but there are exceptions. For instance, chromium dioxide conducts electricity as well as most metals, and some copper-based ceramics have superconductive properties. One application of nontraditional ceramics uses a transparent, electrically conductive ceramic in aircraft windshields to keep them free of ice and snow.

Semiconductors

Another class of versatile materials is semiconductors. Recall that semiconductors are poorer conductors of electricity than metals but better conductors than nonmetals, and their electrical conductivities can be controlled. This property makes semiconductor devices useful and makes computers and other electronic devices possible. Two common semiconducting materials are the metalloids silicon (Si) and germanium (Ge).

☑ **Reading Check** **Define** the term *semiconductor.*

Doping Adding other elements to some metalloids can change their electrical conductivities. The process of adding impurities, such as other elements, to a semiconductor to modify conductivity is called **doping.** For example, the conductivity of silicon can be increased by replacing some of the silicon atoms with atoms of other elements, such as arsenic (As) or gallium (Ga), as shown in **Figure 9.**

Adding even a single atom of one of these elements to a million silicon atoms significantly changes the conductivity. By controlling the type and number of atoms added, the conductivity of silicon can vary over a wide range.

VOCABULARY ....
WORD ORIGIN
Ceramic
  comes from the Greek word
  keramos, meaning *potter's clay*
  In my ceramics class, I made a beauti-
  ful vase for my mother. ....

■ **Figure 9** Pure silicon is a poor conductor. Adding an element, such as gallium, as an impurity creates an area of fewer electrons called a hole. Therefore, the impurity changes the material's conductivity.

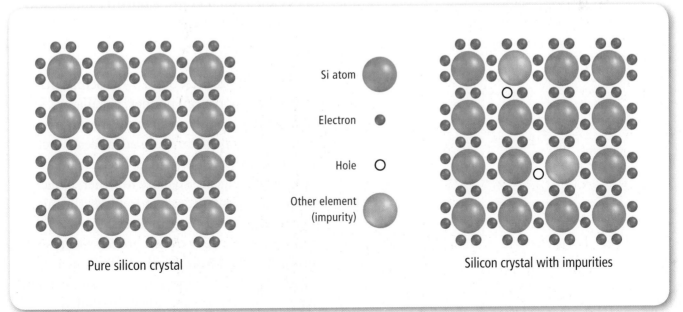

Si atom

Electron

Hole

Other element (impurity)

Pure silicon crystal

Silicon crystal with impurities

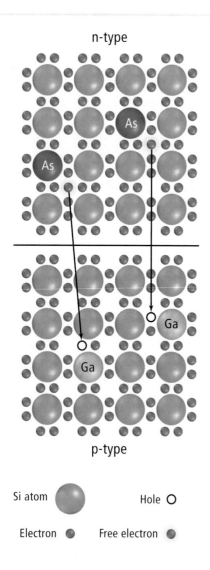

n-type

p-type

Si atom ⬤   Hole ○

Electron ⬤   Free electron ⬤

■ **Figure 10** Arsenic as an impurity adds free electrons. These electrons flow from the arsenic-doped silicon to the germanium-doped silicon, filling the available holes created by the addition of germanium. The electron flow is controlled by the sequence of n-type or p-type semiconductors.

Compare *How are n-type and p-type semiconductors different?*

Types of doping Depending on the element added, the overall number of electrons in the semiconductor can be either increased or decreased. If the impurity causes the overall number of electrons to increase, the semiconductor is called an n-type semiconductor. The extra electrons are called free electrons because they are very weakly attached to the impurity and will easily flow. If doping reduces the overall number of electrons, the semiconductor is called a p-type semiconductor. The silicon crystals will contain holes, or areas with fewer electrons. Electrons now can move from hole to hole across the crystal, increasing conductivity.

Controlling electron flow By placing n-type and p-type semiconductors together, semiconductor devices such as transistors and diodes can be made. These devices are used to control the flow of electrons in electrical circuits, as shown in **Figure 10**.

Integrated circuits During the 1960s, methods were developed for making these components extremely small. At the same time, the integrated circuit was developed. An **integrated circuit** is a small chip that contains many semiconducting devices. Integrated circuits as small as 1 cm on a side can contain millions of semiconducting devices. Because of their small size, integrated circuits are sometimes called microchips. **Figure 11** shows how small microchips can be.

Being able to pack so many circuit components onto a tiny integrated circuit was a technological breakthrough. Microchips make today's televisions, cell phones, calculators, and other devices smaller in size, cheaper to manufacture, and capable of more advanced functions than their predecessors. Also, because the circuit components are so close together, it takes less time for electric current to travel through the circuit. Therefore, computers, PDAs, and other electronic devices can process electronic signals rapidly. **Figure 11** illustrates how integrated circuits have given us faster, smaller, and more capable computers since the 1940s.

FIGURE 11

The History of Computers

The earliest computers were enormous and relied on vacuum tubes to process data. Today's computers use microchips, tiny flakes of silicon engraved with millions of circuit components.

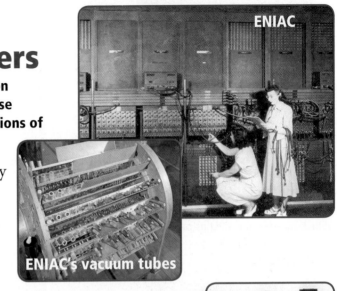

ENIAC

1946 One of the first computers was developed by the Army. It was called the Electronic Numerical Integrator and Computer, or ENIAC for short. ENIAC had more than 18,000 vacuum tubes and was the size of a room.

ENIAC's vacuum tubes

1953 The vacuum tubes of older computers were replaced by transistors, which used semiconductors to perform the same function in a much smaller space.

Transistor

A transistor-based computer

1961 The first commercially available integrated circuit (IC) combined 200,000 transistors on a single, miniature circuit called a "microchip."

A microchip

1977 The first home computers included keyboards and cables to connect them to a monitor or a television.

An early home computer

Today Modern computers run on processors that contain millions of transistors and perform operations measured in the billions each second.

A modern computer

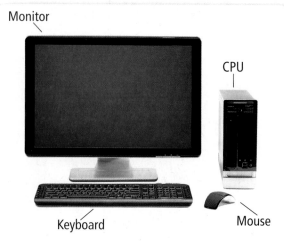

Monitor

CPU

Keyboard

Mouse

■ **Figure 12** Today's computers come in all shapes and sizes. This desktop computer has a monitor, keyboard, mouse, and CPU as part of its hardware.

Predict *what you think computers will look like in 20 years.*

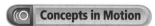
Concepts in Motion Animation

Computer components Computers receive and store information, follow instructions to perform tasks, and communicate information to the outside world. All three of these jobs require a combination of hardware and software components.

The term *computer hardware* refers to the major physical components of a computer, such as the keyboard, monitor, mouse, and central processing unit (CPU). These components are shown in **Figure 12**. All of these components contain semiconductors in integrated circuits. Inside the CPU, a special microchip called the processor acts as a control center. It is responsible for how all the components work together.

Software refers to the instructions that tell the computer what to do. The instructions are stored briefly in the computer's memory—on a special type of semiconductor chip—before being executed by the CPU. When a computer system is functioning properly, the hardware and software work together to perform tasks, sending and receiving millions of electronic signals each second.

Section 2 Review

Section Summary

▶ Ceramics are nonmetallic clay or claylike mixtures.

▶ The properties of ceramics can be customized by changing raw materials and manufacturing processes.

▶ Doping is the process of adding impurities to a semiconductor to change its conductivity.

▶ Integrated circuits are microchips that contain many semiconducting devices.

8. **MAIN Idea** **Describe** the electrical conductivities of traditional ceramics, modern ceramics, and semiconductors.

9. **List** five uses of ceramic materials. What properties make ceramics good choices for these applications?

10. **Describe** how modern ceramics are different from traditional ceramics.

11. **Explain** what semiconductors are and where they are used.

12. **Think Critically** Computers have changed the way businesses operate. If you operated a distribution center for a manufacturer, how would you use computers to assist you?

Apply Math

13. **Calculate** Ceramic A forms when heated to 1,400°C and has a density of 5.3 g/cm³. Ceramic B forms at a temperature 675°C cooler and is four times as dense. What temperature is required to form Ceramic B and what is its density?

14. **Solve a Problem** A developmental ceramic is designed to be 35 percent silica and 65 percent sulfur. If a researcher needs 75 g of this material, how many grams of each component will she need?

✓ **Assessment** Online Quiz

Reading Preview

Essential Questions

▶ How do natural and synthetic materials compare?

▶ What are polymers and how are they used?

▶ What is a composite and how is it used?

Review Vocabulary

polymer: a long chain of small, repeated molecules called monomers

New Vocabulary

synthetic
composite

 Multilingual eGlossary

Polymers and Composites

MAIN ‹Idea A huge variety of human-made products, from plastics to aircraft components, are made from polymers and composites.

Real-World Reading Link The tags in your clothes tell you the materials that are in them. Do you think those materials occurred naturally or were manufactured?

Polymers

The term *polymer* describes a class of substances that are composed of molecules arranged in large chains of simple, repeating units called monomers. Polymer chains can be very long. Polypropylene, for example, might have 50,000 to 200,000 monomers in its chain.

Some polymers, such as proteins, cellulose, and nucleic acids, occur naturally. Humans have used natural polymers for centuries. The ancient Egyptians soaked their burial wrappings in plant polymers called resins to help preserve their dead. Animal horns and turtle shells, which contain natural resins, were used to make combs and buttons for many years.

Making polymers In the 1800s, scientists began developing processes to improve natural polymers and to create new ones in the laboratory. Materials that do not occur naturally but are manufactured in a laboratory or chemical plant are called **synthetic.** In this section, the focus will be on synthetic polymers. Many synthetic polymers are designed to outperform their natural counterparts. Several examples of synthetic polymers are shown in **Figure 13.**

■ **Figure 13** Synthetic polymers are found in many materials we use every day, including computers, clothing, bedding, and wall paint.

History of synthetic polymers In 1839, Charles Goodyear, an American inventor, found that heating sulfur and natural rubber together made rubber no longer brittle when it became cold or soft when it became hot. In the 1850s Alexander Parkes created a plastic synthetic that would later become known as celluloid. It was used in applications such as umbrella handles and toys. These early synthetic polymers had many drawbacks, but they were the beginning of the development of a huge class of materials. **Figure 14** shows a time line of when some of these materials were created.

Today, synthetic polymers usually are made from the hydrocarbons found in fossil fuels, such as oil, coal, and natural gas. Recall that hydrocarbons are organic molecules made entirely of carbon and hydrogen.

Changing properties Polymers have a wide range of uses because their properties can be easily modified. If the composition or arrangement of monomers is changed, then the properties of the material will change. Changing the amount of branching, for example, is one way to change a polymer's properties. Low-density polyethylene (LDPE) has more branching than high-density polyethylene (HDPE).

Another way to change the properties of a polymer is to replace one or more of the hydrogen atoms in the monomer with another element or group. For example, if one of the hydrogen atoms in an ethylene monomer is replaced with a chlorine atom, the result is polyvinyl chloride (PVC). Because of the many ways to change polymers, the possibilities for creating new materials are almost limitless.

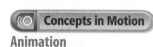
Concepts in Motion

Animation

■ **Figure 14** In the past 150 years, dozens of synthetic polymers have been developed for thousands of applications.

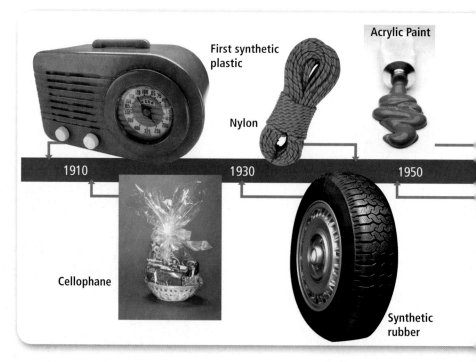

First synthetic plastic

Acrylic Paint

Nylon

1910   1930   1950

Cellophane

Synthetic rubber

Using Polymers

Because so many things today are made of polymers, some people call this "The Age of Plastics." However, plastics are only one group of polymers. Other groups include synthetic fibers, adhesives, surface coatings, and synthetic rubbers.

The plastics group Plastics are widely used for many products because they have desirable properties. They are usually lightweight, strong, impact resistant, waterproof, moldable, chemical resistant, and inexpensive. The properties of plastics vary widely. Some are clear, some melt at high temperatures, and some are flexible. These properties are determined by the composition of the polymer from which the plastic is made.

Synthetic fibers Nylon, polyester, acrylic, and polypropylene are examples of polymers that can be manufactured as fibers. Synthetic fibers can be mass-produced to almost any set of desired properties. For example, nylon is often used in wind and water-resistant clothing such as lightweight jackets. Polyester and polyester blended with natural fibers such as cotton often are used in clothing. Polyester fiber also is used to fill pillows and quilts. Polyurethane is the foam used in many mattresses and pillows.

Synthetic fibers called aramids are a family of nylons with special properties. Aramids are used to make fireproof clothing. Firefighters, military pilots, and race car drivers are examples of professionals that make use of this special fabric. Another aramid fiber is used to make bulletproof vests. Although they are lightweight, these aramids are five times stronger than steel.

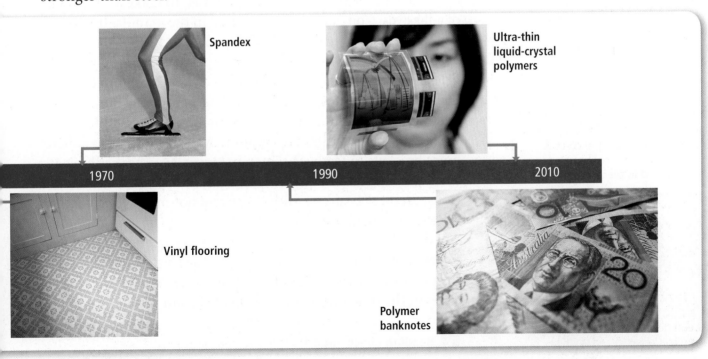

Spandex

Ultra-thin liquid-crystal polymers

1970   1990   2010

Vinyl flooring

Polymer banknotes

Silk Fibers **Nylon Fibers**

■ **Figure 15** Synthetic fibers are designed to mimic natural fibers, but they are not identical.

Describe *the similarities and differences between the nylon fiber and the silk fiber.*

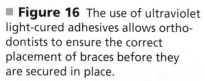

? **Inquiry** Video Lab

Taking a cue from nature Have you ever wondered where humans got the idea to create threads and fabrics? Spinning long fibers into threads is an idea not original to humans. Spiders spun fibers for their webs long before humans copied the idea and began spinning fibers themselves.

The idea behind nylon fiber is also borrowed from nature. It was produced in the laboratory as a possible substitute for silk. **Figure 15** shows the similarities and differences between silk and nylon fibers. The silkworm is a caterpillar that produces a strong and durable fiber when creating its cocoon. The silk is a highly desirable fiber that is woven into fabric for items such as blouses and stockings. Think of how many silkworm cocoons it would take to produce enough silk for a single blouse. Why do you think natural silk fabric is more expensive than nylon fabric?

Adhesives Synthetic polymers are used to make a wide range of adhesives. Contact cements are used in the manufacture of automobile parts, furniture, leather goods, and decorative laminates. They adhere instantly and the bond gets stronger after it dries. Structural adhesives are used in construction projects. One structural adhesive, called silicone, is used to seal windows and doors to prevent heat loss in homes and other buildings. Ultraviolet-cured adhesives are used by orthodontists to adhere brace brackets to teeth, as shown in **Figure 16.** These adhesives bond after exposure to ultraviolet light. Other types include transparent, pressure-sensitive tape and hot-melt adhesives that are used in hot glue guns.

Like other synthetic polymers, adhesives can be engineered to have specific properties. By altering the starting materials or the processes under which the adhesives are created, the properties of the final product can be made to suit any need.

✔ **Reading Check Identify** five uses of adhesives.

■ **Figure 16** The use of ultraviolet light-cured adhesives allows orthodontists to ensure the correct placement of braces before they are secured in place.

Surface coatings and elastic polymers Many surface coatings use synthetic polymers. Polyurethane is a popular polymer, used to protect and enhance wood surfaces. Many paints use synthetic polymers in their composition, too. For example, acrylic paints are useful because they are water-soluble when wet, but water-resistant when dry. Acrylic paints are therefore easy to clean off brushes, but the dried paint will not wash off surfaces.

Synthetic rubber is a synthetic elastic polymer. An elastic object is one that is capable of returning to its original size and shape after it is deformed, like a rubber band. Synthetic rubber is used to manufacture tires, gaskets, belts, and hoses. The soles of some shoes also are made from this rubber.

Composites

The properties of a synthetic polymer can be altered by using more than one material. A **composite** is a mixture of two or more materials—one embedded or layered in another. Embedding one material into another can produce a composite with the desired properties. For example, if a substance is light-weight but brittle, such as some plastics, embedding flexible fibers into it can alter the brittleness. After the substance has the flexible fibers embedded, the product is less brittle and can withstand greater forces before it breaks.

Glass fibers are used often to reinforce plastics because glass is inexpensive, but other materials can be used as well. Composite materials of plastic and glass fibers are used to construct boat and car bodies, as shown in **Figure 17**. These bodies are made of a mixture of small fibers of glass embedded in a plastic. The structure of the fiberglass reinforces the plastic, making a strong, lightweight composite.

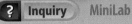

MiniLab

? Inquiry   MiniLab

Observe Drying

Procedure 🥽 🧤 🚰 🧪

1. Read the procedure and safety information, and complete the lab form.
2. Pour 400 mL of **water** into each of **four beakers.**
3. Submerge a **3 cm × 3 cm square of fabric** in each beaker and soak the squares for 3 minutes. Use **cotton, wool, polyester,** and **nylon cloth.**
4. Remove the fabric squares, lay them flat on several layers of **paper towels,** and place them in direct sunlight.
5. Check the dampness of each cloth square every 3 minutes for 15 minutes.

Analysis

1. **Describe** the results of your investigation.
2. **Infer** why athletes wear polyester or nylon clothing.

■ **Figure 17** Many cars, boats, and other vehicles have bodies that contain or are made entirely out of fiberglass, a composite of glass and plastic.

Infer *What parts of a car's body could be made of fiberglass?*

Figure 18 Satellites contain durable, light-weight composite materials.

Explain *why composites are used in artificial satellites.*

Composites in space Composite materials are used in the construction of satellites. Carbon fibers are used to strengthen the plastic body, creating a material that is more rigid and stronger than aluminum. In fact, carbon-fiber composites are 40 percent stronger than aluminum. Satellites made of carbon fiber composites, such as the one shown in **Figure 18,** are significantly lighter than those made of aluminum. Lighter-weight satellites are less expensive to launch into orbit, yet the structure is still able to withstand the stress of the launch.

The same can be said of composites used for air travel. Commercial aircraft made of composites also benefit from the strong yet lightweight properties of composite materials. Materials made of graphite composites are 13 percent lighter than those made with aluminum. The mass of a commercial aircraft can be reduced by more than 2,500 kg by using such advanced, lightweight composite materials. The lower mass results in cost savings by reducing the amount of fuel required to operate the aircraft. Some modern commercial aircraft contain 50 percent composite materials.

Reading Check Identify the advantages of using composite materials.

Section 3 Review

Section Summary

▶ The composition and chemistry of polymers allows almost limitless modifications.

▶ Polymers can be natural or synthetic.

▶ Polymers have a great variety of uses.

▶ Composite materials often are selected for products because they can be stronger and lighter than metals or alloys.

15. **MAIN Idea Discuss** reasons why chemists create new polymers and composites.

16. **Identify** four uses of synthetic polymers.

17. **Explain** the difference between natural and synthetic polymers and give an example of each.

18. **Explain** what a composite material is and give three examples of items that are made from composites.

19. **Think Critically** You are designing a new material for use in an airplane body. What properties should the material have?

20. Find Mass A telecommunications company launches 10,000-kg satellites. A new satellite made from composites promises to reduce that mass by 25%. What is the mass of the new satellite?

✓ Assessment   Online Quiz

LAB

Properties of New Materials

Objectives

- **Predict** the properties of a material.
- **Determine** possible uses for a material.

Background: This substance is fun to play with. But how do you describe its properties?

Question: *What are the properties of this new material and what can it be used for?*

Preparation

Materials

white glue
borax laundry soap
warm water
250-mL beaker or cup
100-mL beaker or cup
graduated cylinder
craft stick for mixing

Safety Precautions

Procedure

1. Read the procedure and safety information, and complete the lab form.

2. Prepare a data table to record your observations of the following: stretched slowly, stretched quickly, rolled into a ball and left alone, pressed onto newspaper ink, dropped on a hard surface.

3. Put about 100 mL of warm water in the larger beaker and add borax laundry soap until soap no longer dissolves.

4. Put 5 mL of water and 10 mL of white glue into the smaller beaker and mix completely.

5. Add 5 mL of the borax solution to the glue solution and continue mixing for a couple of minutes.

6. When the substance firms up, remove it from the container and continue to mix it by pressing with your fingers until it is like soft clay.

7. Examine the properties of this material and record them in your data table.

Conclude and Apply

1. **Identify** the properties of this material. Is your material a polymer? How do you know?

2. **Evaluate** Based on the properties of the material, what do you think it could be used for? How could the properties be modified?

COMMUNICATE YOUR DATA

Create Name your product and prepare an advertisement with text and graphics. Share your advertisement with other students in your class.

Standardized Test Practice

Multiple Choice

Record your answers on the answer sheet provided by your teacher or on a sheet of paper.

1. What property allows brass to be shaped into a musical instrument?
   A. conductivity
   B. ductility
   C. luster
   D. malleability

2. What element is combined with zinc to make brass?
   A. antimony
   B. copper
   C. silver
   D. tin

Use the figure below to answer question 3.

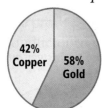

Composition of 14-Karat Gold

3. A 14-karat gold ring has a mass of 5.3 g. What is the mass of the gold used to make the ring?
   A. 2.2 g
   B. 3.1 g
   C. 5.0 g
   D. 5.3 g

4. An electrician uses copper electrical wire with a polymer coating. What properties are needed for the wire?
   A. ductility and malleability
   B. ductility and conductivity
   C. malleability and luster
   D. malleability and conductivity

Use the figure below to answer question 5.

5. How would the items in the photograph above be classified?
   A. ceramics   C. polymers
   B. composites   D. alloys

6. Which alloy contains iron and carbon?
   A. bronze   C. pewter
   B. brass   D. steel

7. Which of the following terms refers to substances and materials that are created in a laboratory or chemical plant?
   A. component
   B. composite
   C. integrated
   D. synthetic

8. An electrician uses copper electrical wire with a polymer coating. What property should the polymer coating on the wire have?
   A. low melting point
   B. high conductivity
   C. high malleability
   D. high resistivity

9. Which property of metals and alloys makes a French horn appear shiny?
   A. conductivity
   B. ductility
   C. luster
   D. malleability

✓ Assessment   Standardized Test Practice

Short Response

Record your answers on the answer sheet provided by your teacher or on a sheet of paper.

10. Define the term semiconductor and give an example of how it is used.

11. Explain how adding an impurity increases the conductivity of a semiconductor.

Use the figure below to answer question 12.

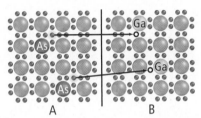

○ Si atom • Electron ○ Hole • Free electron

12. Which of the two semiconductors shown in the illustration above is an n-type? Which is a p-type? How can you tell?

13. The production of fibers from nylon is an idea that was borrowed from nature. Describe one advantage and one disadvantage of using nylon instead of natural fibers.

14. Name three properties of traditional ceramics that make them useful as food serving bowls.

15. Why do manufacturers frequently use alloys instead of metals when making different products?

Extended Response

Record your answers on a sheet of paper.

16. Describe the process of firing traditional ceramics. Explain how this process affects the properties of the ceramics.

17. List three groups of polymers and describe the uses of each group.

18. Explain what is meant by a hole in a semiconductor. Describe what happens to holes as current flows through the semiconductor.

19. Describe some properties of modern ceramics that traditional ceramics do not have. What are some uses of modern ceramics?

Use the figure below to answer question 20.

20. The objects in the figure above are both made from polyethylene. Identify what type of polyethylene was used for each object and explain why that material was chosen.

NEED EXTRA HELP?																				
If You Missed Question . . .	1	2	3	4	5	6	7	8	9	10	11	12	13	14	15	16	17	18	19	20
Review Section . . .	1	1	1	1	2	1	3	3	1	2	2	2	3	2	1	2	3	2	2	3

UNIT 7

Earth

THEMES

The Earth System Earth's system is an ever changing environment in which five spheres interact.

Energy All life depends on Earth's energy processes and resources.

Natural Resources Earth is a source of natural resources, such as fossil fuels, minerals, and water.

Scientific Inquiry Studying Earth's processes leads to a better understanding of Earth's dynamic past and future.

WebQuest STEM UNIT PROJECT

Career Research the field of climatology. Present
information on how ice cores can be used to determine
what climates existed in Earth's past.

CHAPTER 25

Earth's Internal Processes

ConnectED

Your one-stop online resource
connectED.mcgraw-hill.com

- Video
- Audio
- Review
- Inquiry
- WebQuest
- Assessment
- Concepts in Motion
- Multilingual eGlossary

Launch Lab
Global Jigsaw Puzzle

Alfred Wegener, a German scientist, noticed that the coastlines of continents appeared to fit together like pieces of a giant jigsaw puzzle. He suggested that the continents once formed a single large landmass. Use a map of the world to test this idea. Can you recreate this landmass by piecing together the continents?

For a lab worksheet, use your StudentWorks™ Plus Online.

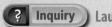 **Inquiry** Launch Lab

 **FOLDABLES**

Systems Create a tri-fold book to help you organize information about the different types of plate boundaries.

Plate Boundary Type	Description	Illustration
Divergent		
Convergent		
Transform		

THEME FOCUS Earth Materials and Processes
Differences in temperature and pressure within Earth's interior contribute to dynamic geologic change on Earth's surface.

BIG Idea Convection currents inside Earth cause plate motion.

Essential Questions

▶ What evidence led to the development of the continental drift hypothesis?

▶ What is the plate tectonics theory?

▶ How is mantle convection used to support the plate tectonics theory?

Review Vocabulary

hypothesis: a testable statement proposed to explain an observation or to answer a question

New Vocabulary

mid-ocean ridge
rift valley
convergent plate boundary
subduction
divergent plate boundary
transform plate boundary

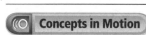
g Multilingual eGlossary

The Plate Tectonics Theory

MAIN ◀Idea Earth's surface is made of plates that move slowly and interact along plate boundaries.

Real-World Reading Link Have you ever been out in a canoe on a windy day? Even when you are not rowing, your canoe might move with water currents created by these strong winds. The water is not the only thing moving. Far below you, there are forces that cause continents to move. What sets the continents in motion?

Continental Drift

In the early twentieth century, there was no single explanation of how Earth processes related to each other. Much geologic research was done locally because transportation and communication were expensive. Based upon their observations, geologists developed hypotheses that emphasized vertical changes—for example, an erosion process that leveled high places and a mountain-building process that lifted these same areas up again.

Then in 1915, Alfred Wegener (VAY guh nur) proposed a hypothesis that suggested that Earth's continents were once part of a large supercontinent called Pangaea (pan GEE uh). Wegener proposed that about 200 million years ago the supercontinent separated into pieces that drifted over the surface of Earth like rafts on water, as shown in **Figure 1.** Wegener called his hypothesis continental drift. Other scientists, however, were skeptical of Wegener's continental drift hypothesis because he could not explain how the continents move.

 Concepts in Motion

Animation

■ **Figure 1** The supercontinent Pangaea began to break apart approximately 200 million years ago (mya). The continents have shifted to their present day positions over time.

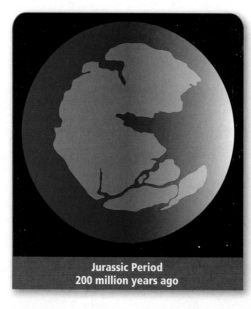

Jurassic Period
200 million years ago

Cretaceous Period
70 million years ago

Evidence for continental drift In addition to explaining how the continents move, Wegener needed evidence to prove that the continents were once connected. Wegener argued that the shapes of continents, as well as the types of fossils, rocks, and mountain ranges found on different continents, could be matched up. Matching these features on different continents provided evidence that the continents were once part of the supercontinent Pangaea.

Coastlines Wegener studied the shapes of continents to support his continental drift hypothesis. He observed that the eastern coastline of South America fit together with the western coastline of Africa. Also, the coastline of the eastern United States fit together with the northwestern coastline of Africa. When South America and Africa were placed together, their southern tips matched the coastline of the Weddell Sea in Antarctica.

Wegener's opponents argued that the coastlines are constantly wearing away due to wave action. How could someone compare coastlines in the past to coastlines today? Years later, during the revival of Wegener's hypothesis, oceanographers were able to show that the edges of the continental shelves, which extend beyond the coastline and far below sea level, are nearly a perfect match, as illustrated in **Figure 2.** Scientists used sonar to prove this point. This technology uses sound waves to determine ocean depth.

Fossils Wegener could not use just any ancient organism to support the continental drift hypothesis. For instance, animals that could fly or swim could appear in the fossil record in widely separated places due to their mobility, not because landmasses were necessarily connected. Massive land animals provided better evidence because they could not have crossed oceans. Animals such as Lystrosaurus or Cynognathus, which preceded the dinosaurs, supported a contiguous landmass. Glossopteris, a fern with large, heavy spores also supported continental drift. These living things were widely distributed, as shown in **Figure 3.**

✓ **Reading Check** **Identify** land-based animals used to support the continental drift hypothesis.

■ **Figure 2** Wegener used the analogy of a torn newspaper to explain Pangaea. To fit the pieces back together, you had to match up the shapes of the torn edges. While the coastlines of the continents matched reasonably well, the continental shelves fit together even better.

? **Inquiry**   Virtual Lab

■ **Figure 3** Wegener used fossil evidence to help support his continental drift hypothesis.

Explain *Why would an organism's ability to fly or swim eliminate fossil evidence from Wegener's support for continental drift?*

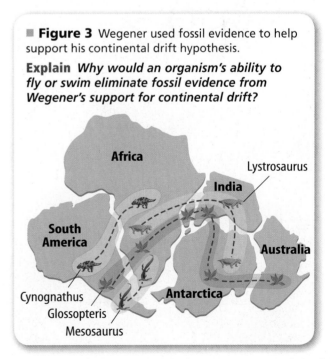

■ **Figure 4** Wegener's hypothesis showed that mountains separated by oceans were once part of the same mountain ranges. These mountains are similar in age, rock type, and structure.

Concepts in Motion Animation

■ **Figure 5** The discovery of the Mid-Atlantic Ridge helped Hess to explain continental drift.

Mid-Atlantic Ridge

Rocks and mountains Mountain ranges located on opposite sides of the ocean are, in some cases, similar in structure, rock type, and age, as shown in **Figure 4.** For decades, geologists attempted to explain the origin of these mountains as separate ranges. Wegener proposed that these mountain ranges were once part of one continuous range. He suggested that the mountains separated as the continents drifted apart over time.

The controversy Wegener, however, could not explain how the continents drifted. He suggested that Earth's rotation, the gravitational pull of the Sun and the Moon, and centrifugal force moved the continents. Physicists quickly concluded that these forces were unable to explain continental drift.

Wegener then proposed that the continents were pushing through the seafloor, but geologists rejected this idea, too. They argued that the seafloor was much too rigid and should break up, or fracture, as continents plowed through it. No evidence of any fractures existed. It was not until after Wegener's death in 1930 that new discoveries on the seafloor helped to support his hypothesis.

Seafloor Spreading Hypothesis

After World War II, Harry Hess revived Wegener's hypothesis. He used sonar to develop three-dimensional maps of the seafloor. Hess discovered a system of mountain ranges on the seafloor separated by valleys, which he called the **mid-ocean ridge** (MOR) system. The MOR system in the Atlantic Ocean, called the Mid-Atlantic Ridge, is shown in **Figure 5.**

Hess proposed that a MOR is produced by a process called seafloor spreading. He suggested that magma is buoyant and rises upward through cracks in the crust. This causes the crust to spread apart. A mid-ocean ridge forms as the seafloor spreads apart. A long, narrow depression called a **rift valley** forms in between peaks along the mid-ocean ridge. This process forms new oceanic crust as lava fills in the rift valley from below.

✓ **Reading Check Identify** How does a mid-ocean ridge form?

Rocks and sediments on the seafloor In addition to the discovery of the mid-ocean ridge system, more evidence was needed to support seafloor spreading. Scientists studied the age of sediments in drill cores from the seafloor. Layers of sediment in cores collected near the continents are old at the bottom of the core and younger near the top. However, sediment found in cores collected near a MOR are young. There are no old sediments near a MOR.

Additional evidence comes from comparing the ages of continental rocks with those of oceanic rocks. The oldest rocks on continents are approximately four billion years old. Rock samples from the seafloor are much younger—most oceanic rocks are less than 200 million years old. This difference in age can only be explained if rocks on the seafloor are continually being created at a mid-ocean ridge and are being destroyed at another location.

 Reading Check **Describe** How does the age of sediments change as you move away from a mid-ocean ridge?

EXAMPLE Problem 1

Calculate Spreading Distances The spreading rate along mid-ocean ridges varies. In the Atlantic Ocean, the spreading rate averages about 2.5 cm/y. Estimate how far, in kilometers, the Atlantic seafloor has spread over the past 100 million years.

Identify the Unknown: distance

List the Knowns:
time = 100,000,000 years
spreading rate = 2.5 cm/y

Set Up the Problem: distance = spreading rate × time

Solve the Problem: Substitute the known values into the equation:
distance = 2.5 cm/y × 100,000,000 y = 250,000,000 cm
250,000,000 cm × 1 km/100,000 cm = 2,500 km

Check the Answer: Rearrange the variables to check your answer. When you divide the distance you calculated by 100,000,000 years, can you reproduce the spreading rate?

PRACTICE Problems

Find **Additional Practice Problems** in the back of your book.

1. If the East Pacific Rise has a spreading rate of 12 cm/y, how far will it spread in 10 million years?

Additional Practice Problems

2. **Challenge** The Southwest Indian Ridge in the Indian Ocean has separated a distance of 272 km in 34 million years. Calculate the average spreading rate in cm/y.

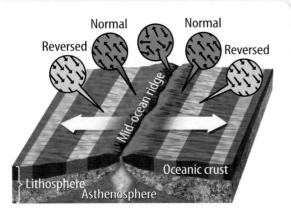

Figure 6 Iron-rich minerals in lava cool and crystallize on the seafloor. The minerals record the magnetic polarity of Earth at the time the rocks cooled and hardened.

FOLDABLES
Incorporate information from this section into your Foldable.

Figure 7 Earth's outermost layer is broken into tectonic plates, which are slabs of rigid rock that interact along plate boundaries.

Magnetic polarity of rocks

Additional evidence to support seafloor spreading comes from the magnetic properties of rocks on the seafloor. As the seafloor spreads along a MOR, lava erupts to form new oceanic crust. When lava cools, iron-rich minerals behave like compass needles and become oriented along Earth's magnetic field. When the rocks harden, this magnetic orientation, known as polarity, is locked in place.

Studies of rocks on the seafloor show that Earth's magnetic field has reversed direction many times. Scientists have discovered rock with alternating polarities on the seafloor that mirror each other on either side of a mid-ocean ridge. These bands, called magnetic stripes, result when Earth's magnetic field reverses direction, as shown in **Figure 6**.

The Plate Tectonics Theory

New discoveries on the seafloor in the 1960s led to the development of the plate tectonics theory. According to the plate tectonics theory, Earth's surface is made of separate slabs of rigid rock called plates that move slowly over Earth's upper mantle. The bottom part of a plate is composed of a solid layer of Earth's mantle. The top part of a plate is made of continental crust, oceanic crust, or both.

As plates move, they interact with each other along plate boundaries, as shown in **Figure 7**. There are three main types of plate boundaries: convergent, divergent, and transform.

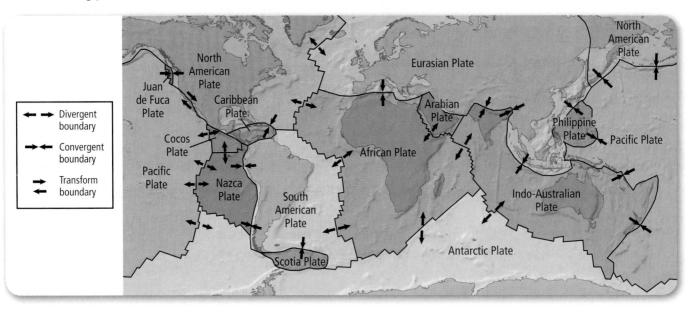

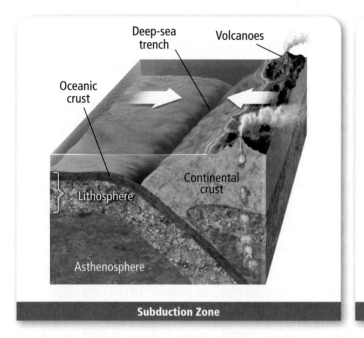

Deep-sea trench   Volcanoes

Oceanic crust

Lithosphere

Continental crust

Asthenosphere

Subduction Zone

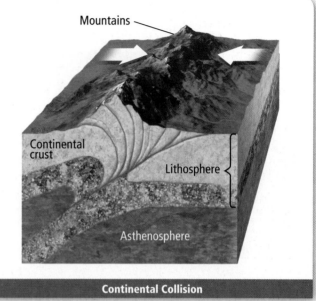

Mountains

Continental crust

Lithosphere

Asthenosphere

Continental Collision

Convergent plate boundaries The boundary where two plates collide is called a **convergent plate boundary.** Two different geologic processes result from collision along three types of convergent plate boundaries.

Subduction zones When a thick and buoyant continental plate meets a thin and dense oceanic plate, the denser plate dives beneath the continent in a process called **subduction.** Heat and fluids are transferred from the oceanic plate to the mantle, which then melts to form magma. This magma rises toward the surface and erupts as lava and ash to form a volcanic arc, which parallels the subduction zone, as shown to the left in **Figure 8.** This boundary also forms a deep-sea trench. The Andes Mountains in South America are an example of a continental-oceanic convergent plate boundary.

Convergent plate boundaries also occur where two oceanic plates collide. In this case, the oceanic plate that is older and colder, and therefore denser, subducts. Lava and ash that erupt here produce chains of volcanic islands called island arcs. Japan is an example of an oceanic-oceanic convergent plate boundary. The collision of two plates along a subduction zone can also cause large earthquakes capable of producing tsunamis.

Continental collision Along some convergent plate boundaries, two continental plates of equal density collide and do not subduct, as shown to the right in **Figure 8.** Because no subduction occurs, the plates collide and buckle upward to form a high range of folded mountains. Volcanic activity is noticeably absent here, and there is no deep-sea trench. The Himalaya Mountains in Asia are an example of a continental-continental convergent plate boundary.

■ **Figure 8** When two plates with different densities collide (left), the denser plate will subduct and form volcanoes along a convergent plate boundary. When two plates with the same density collide (right), neither plate subducts. Both plates will uplift and deform to create mountains.

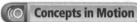 **Concepts in Motion**

Animation

VOCABULARY . . . .
ACADEMIC VOCABULARY
Parallel
    extending in the same direction, everywhere equidistant, and not meeting
    The hiking trail parallels the river for many kilometers. . . . .

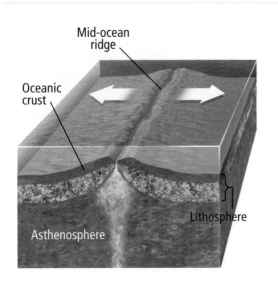

Figure 9 Two plates separate along a divergent plate boundary, such as a mid-ocean ridge. This process occurs where the seafloor spreads and a mid-ocean ridge forms.

Figure 10 Friction between two plates moving side by side along a transform plate boundary causes stress to build along faults. Faults are the site of brief, but rapid energy release called earthquakes.

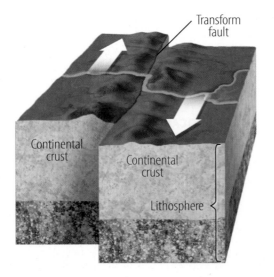

Divergent plate boundaries Two plates move apart along a mid-ocean ridge (MOR). The boundary between two plates that are moving apart is called a **divergent plate boundary.** At a divergent plate boundary, as shown in **Figure 9,** magma rises between the plates, erupts from a rift valley as lava, and then cools to form new crust. A mid-ocean ridge is one example of a divergent plate boundary. In some places, such as the East African Rift, divergent plate boundaries create intraplate rift valleys that form in the middle of a continent.

Transform plate boundaries Some plate boundaries have faults where horizontal movement takes place, as shown in **Figure 10.** In this case, no new rock is created, as in a divergent plate boundary. In addition, old rock is not being destroyed, as in a subduction zone. This type of boundary where plates slide horizontally past each other in opposite directions is called a **transform plate boundary.**

As rocks move along a transform plate boundary, stress accumulates in the crust and earthquakes result when energy is released as rocks move along a fault. The San Andreas Fault in California is an example of a transform plate boundary. While earthquakes are common along transform plate boundaries, they also occur along divergent and convergent plate boundaries.

Plates in Motion

Plate motion is caused by a combination of forces. Thermal energy from Earth's interior drives plate motion in a process called convection. As plates move and interact along plate boundaries, forces like slab pull, ridge push, and friction also contribute to the process.

Convection Thermal energy is both the residual heat from Earth's formation and the product of radioactive decay of elements in Earth's core. This thermal energy is transferred upward into the mantle. As mantle material gets hotter, it becomes more buoyant. Buoyant material rises to the base of the crust, where it cools and then sinks back down into the mantle. This continuous heating and cooling process produces convection currents in the mantle that move plates. Convection currents inside Earth are shown in **Figure 11.**

Slab pull, ridge push, and friction

Three other forces are thought to contribute to plate motion. When subduction occurs along a convergent plate boundary, a force called slab pull helps to move plates. You have probably experienced a similar effect at night. As you toss and turn in bed, some of the covers might fall off. Eventually, enough of the covers are on the floor that gravity pulls the rest of them off the bed. A similar effect occurs when an oceanic plate is pulled down along a subduction zone. This process can lead to more stress and separation along a divergent plate boundary.

In contrast to slab pull, ridge push moves plates along a mid-ocean ridge. Because the mid-ocean ridge is higher than the surrounding seafloor, gravity pulls material downward and away from the ridge. Friction between a plate and the mantle also has an affect on plate motion. For example, plates that drag continental material along with them are slower than those that drag oceanic material.

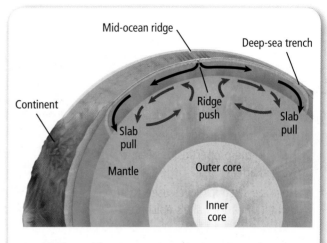

■ **Figure 11** Internal convection in the mantle is the driving force behind plate motion.

Section 1 Review

Section Summary

▶ Wegener used evidence from the shape of coastlines, types of fossils, rocks, and mountain ranges to support the continental drift hypothesis.

▶ The age and magnetic polarity of rocks and the thickness of sediments exposed on the seafloor are evidence for seafloor spreading.

▶ Earth's outermost layer is broken into plates of rigid rock that move and interact along plate boundaries.

▶ Convection in the mantle drives plate motion.

3. MAIN Idea **Compare and contrast** divergent, convergent, and transform plate boundaries.

4. **Explain** how coastlines, fossils, rocks, and mountains are evidence for continental drift.

5. **Describe** the process of mantle convection.

6. **Identify** patterns in the age and magnetic polarity of rocks on the seafloor.

7. **Explain** the plate tectonics theory.

8. **Think Critically** What would have to happen in Earth's core to stop plate motion?

Apply Math

9. **Calculate Distance** If two plates diverge at a rate of 1.3 cm/y, how much farther apart in km will the plates be after 200 million years?

10. **Estimate Time** The average distance across an ocean is 1,600 km. Two continents on either side of the ocean are converging at a rate of 10 cm/y. How long will it take for them to collide?

Essential Questions

▶ What causes earthquakes?

▶ How do the properties of seismic waves differ?

▶ How are seismic waves used to locate an earthquake's epicenter?

Review Vocabulary

friction: force that opposes the sliding motion between two materials

New Vocabulary

earthquake
fault
elastic rebound
focus
epicenter

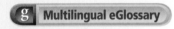 Multilingual eGlossary

 Video What's PHYSICAL and EARTH SCIENCE Got to Do With It?

■ **Figure 12** Most earthquakes occur along tectonic plate boundaries.

Earthquakes

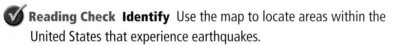 **MAIN ‹Idea** Earthquakes occur when blocks of rock move along faults, releasing energy.

Real-World Reading Link Imagine what it would be like to awaken to an earthquake. The room is shaking, furniture is moving, objects are falling off the shelves, and loud noises erupt in the chaos. People experience earthquakes around the world every day.

Earthquake Distribution

An **earthquake** is the sudden movement or vibration of the ground that occurs when rocks slip and slide along enormous cracks in Earth's crust called faults. For decades, scientists have known that earthquakes are not randomly distributed. Instead, they usually occur in well-defined zones. The zones where earthquakes occur correspond closely with tectonic plate boundaries. In fact, nearly 95 percent of all earthquakes occur along plate boundaries. Data from earthquakes have helped geologists to understand the structure and motion of Earth's plates. **Figure 12** shows the distribution of large earthquakes worldwide. Several million earthquakes occur each year, but the majority have very small magnitudes.

✓ **Reading Check Identify** Use the map to locate areas within the United States that experience earthquakes.

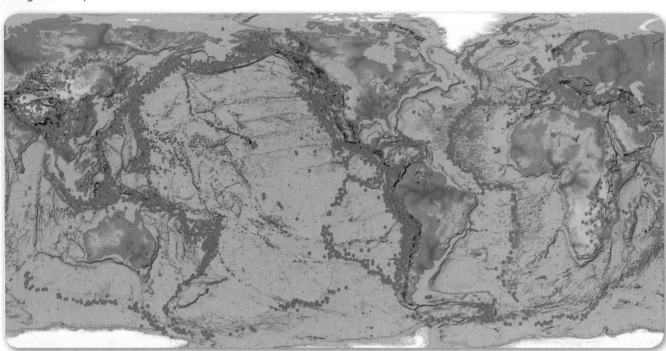

Figure 13 The depth at which an earthquake occurs, indicated by the yellow stars, depends on the type of plate boundary. Earthquakes tend to occur at shallow depths along divergent plate boundaries and at greater depths along convergent plate boundaries.

Earthquake depth The depths of earthquakes can provide information about plate boundaries. Earthquakes tend to occur at different depths along different types of plate boundaries, as shown in **Figure 13**. For example, earthquakes that occur along divergent and transform plate boundaries tend to be shallow, typically less than 70 km depth. However, earthquakes that occur along convergent plate boundaries commonly occur at depths greater than 70 km.

Causes of Earthquakes

When energy is released along a fault, an earthquake results. The shaking of the ground that occurs during an earthquake can cause buildings to collapse, as shown in **Figure 14**. The collapse of buildings usually causes most of the deaths and injuries that occur during an earthquake. What causes earthquakes?

Deformation Earthquakes are caused by forces applied to rocks along plate boundaries. A force applied to an object can cause the object to change its shape, or be deformed. The stress on an object is the force per unit area that acts on the object. There are four main types of stress: (1) compression stress, in which an object is squeezed or shortened; (2) tension stress, in which an object is stretched or lengthened; (3) shear stress, in which different parts of an object are moved in opposite directions along a plane; and (4) torsion stress, in which an object is twisted. These four types of stress cause objects, such as rocks along a fault, to change shape. A change in shape is classified as deformation.

Figure 14 This damage was caused when the buildings were shaken off their foundations during the May 2008 Sichuan Province earthquake in China.

Explain *How might this damage have been prevented?*

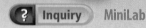

MiniLab

Model Stress

Procedure

WARNING: *Do not eat any food items in lab.*

1. Read the procedure and safety information, and complete the lab form.
2. Clasp a large **taffy bar** using both hands. The palms of your hands should face down at all times, with your thumbs touching and placed side-by-side.
3. Push one hand forward 2 cm while simultaneously pulling the other hand backward 2 cm toward your body. Return your hands to the original position.
4. Twist your hands in opposite directions. Return your hands to the original position.
5. Next, move your hands about 4 cm apart.
6. Finally, push your hands back together to the original position.

Analysis

1. **Identify** each type of stress that you modeled in steps 3–6 as one of the four main types identified.
2. **Describe** the shape of the taffy that results from each of the four types of deformation.

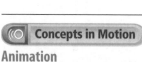

Concepts in Motion

Animation

■ **Figure 15** Rocks first move along a fault at a point of origin called the focus. The location on Earth's surface directly above the focus is called the epicenter.

Types of deformation Elastic deformation occurs when a material, such as rock, deforms as stress is applied but snaps back into its original shape when the stress is removed. Plastic deformation occurs when a material changes shape as a stress is applied and remains in the new shape when the stress is removed. Modeling clay can undergo plastic deformation. You might expect all rocks to be brittle, which means that they break when the applied stress is large enough. But deep inside Earth, where temperatures and pressures are high, rocks can undergo plastic deformation.

Energy release When an object is deformed, a form of energy called strain energy can be stored in the object. For example, when rock deforms along a crack in Earth's crust, strain energy builds up in the rock. As this energy is released, the rock moves. A **fault** is a crack in Earth's crust along which rock has moved. Earthquakes occur when a rock that has been stressed releases strain energy and moves along a fault. This sudden release of strain energy as rock moves along a fault is called **elastic rebound.** Elastic rebound produces energy called seismic waves that travel in Earth.

Seismic Waves

Seismic waves travel outward in all directions from the point where the strain energy is released. The point of origin for an earthquake is called a **focus.** The point on Earth's surface directly above the focus is called the **epicenter.** When you throw a stone into water, you see concentric rings of wave energy move out across the surface from the point of impact, similar to the way seismic waves move out from the focus in all directions, as shown in **Figure 15.** Seismic waves can be classified into three major types. Primary waves and secondary waves are body waves that travel through Earth. Surface waves travel across Earth's surface.

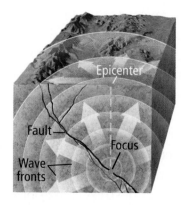

Seismic energy radiates outward in all directions from the earthquake focus.

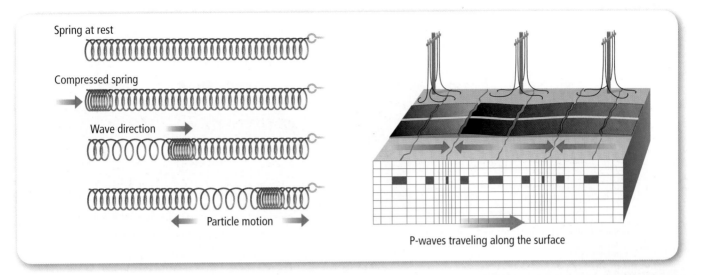

P-waves traveling along the surface

Primary waves One type of body wave is called a primary wave. Primary waves, which are also called P-waves, are similar to waves that travel along a coiled spring, as shown in **Figure 16**. Primary waves cause particles inside Earth to move back and forth in the same direction that the wave is traveling. The particles then return to their original positions after the wave passes. P-waves are the fastest seismic waves and can travel through Earth's interior with speeds between 5 km/s and 7 km/s. P-waves travel through both solids and liquids.

Secondary waves Secondary waves, or S-waves, are another type of body wave. Secondary waves are similar to energy that travels along the rope shown in **Figure 17**. These waves cause the particles inside Earth to move perpendicular to the direction that the wave is traveling. S-waves travel through Earth's interior more slowly than P-waves. As a result, S-waves lag behind P-waves as waves move away from an earthquake epicenter. The lag-time between the arrival of the first P-waves and the first S-waves can be used to locate an earthquake epicenter. Unlike P-waves, S-waves can travel only through solids.

■ **Figure 16** P-waves are compressional waves that travel through Earth's interior and are the first to arrive after an earthquake.

Concepts in Motion
Animation

■ **Figure 17** S-waves are transverse waves that move up and down perpendicular to the direction in which the wave travels.

Compare and contrast *P-waves and S-waves.*

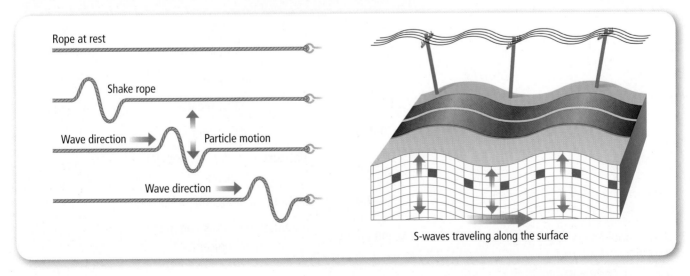

S-waves traveling along the surface

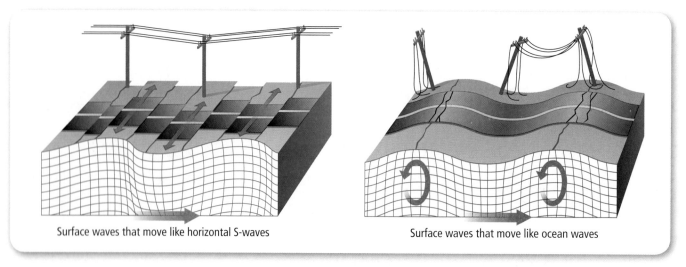

Surface waves that move like horizontal S-waves

Surface waves that move like ocean waves

Figure 18 Surface waves produce an up-and-down rolling motion similar to the motion caused by ocean waves. At the same time, the surface can shift from side-to-side.

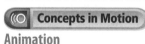

Animation

Surface waves

Surface waves only travel on Earth's surface. They move in a more complex manner, often in a rolling motion like ocean waves, as shown in **Figure 18**. As surface waves travel through material, they can exhibit an up-and-down rolling motion and also a side-to-side motion that parallels Earth's surface. Buildings, roads, and power lines are often damaged by the side-to-side rocking motion that results from surface waves.

Earthquake Measurement

Earthquake scientists called seismologists use two different scales to evaluate earthquake strength and intensity. The Richter magnitude scale, known as the Richter scale, measures the energy released during the earthquake, as shown in **Table 1.**

Table 1	Earthquake Magnitude and Frequency	Concepts in Motion Interactive Table
Richter Magnitude	**Description**	**Estimated Occurrence per Year**
< 2.0	recorded but not generally felt	3,000,000
2.0–2.9	potentially felt	1,300,000
3.0–3.9	felt by some	130,000
4.0–4.9	felt by most	13,000
5.0–5.9	damaging	1,319
6.0–6.9	destructive in densely populated areas	134
7.0–7.9	potential to inflict major damage	17
8.0 and above	potential to destroy communities near epicenter	1

Logarithmic scale The Richter scale is a logarithmic scale that measures the amplitude of the largest seismic wave recorded on a seismograph. Each increase of one unit on the Richter scale represents a 10-fold increase in wave amplitude. For example, seismic waves from a magnitude-8 earthquake are one hundred times larger in amplitude than those from a magnitude-6 earthquake

The Modified Mercalli scale, shown in **Table 2,** ranks earthquakes according to intensity or the amount of damage they cause. The scale ranges from I to XII. A rank of I on the scale represents an earthquake that is rarely felt by anyone. A rank of XII reflects earthquakes that cause the most severe damage.

VOCABULARY ....
WORD ORIGIN
Seismograph
from the Greek words *seismo* for earthquake and *graphein,* to write
The seismograph recorded ground vibrations during the earthquake. ......

Table 2	The Mercalli Scale of Earthquake Intensity Concepts in Motion Interactive Table
Level	**Description**
I	rarely felt by people
II	felt by resting people indoors; some hanging objects may swing
III	felt indoors by several; vibration like passing of a light truck
IV	felt indoors by many; vibration like passing of a heavy truck; standing autos rock; windows, dishes, and doors rattle; walls and frames may creak
V	felt by nearly everyone indoors and outdoors; small, unstable objects upset; dishes and glassware broken; swaying of tall objects noticed
VI	felt by all; walking is unsteady; many run outdoors; windows, dishes, and glassware broken; furniture overturned and plaster may crack
VII	difficult to stand; noticed by drivers of autos; furniture and chimneys broken; well-built buildings hardly damaged; poor structures sustain considerable damage
VIII	people frightened; ordinary buildings slightly damaged; driving of autos affected; tree limbs fractured; damage to tall objects; cracks in wet ground
IX	general panic; great damage in substantial buildings; some houses thrown off foundations; underground pipes broken; serious ground cracks
X	most masonry and frame structures destroyed; serious damage to dams, dikes, embankments; water splashed out of rivers, canals, lakes; rails bent
XI	few structures remain standing; bridges destroyed; broad fissures in the ground; slumps and landslides; rails bent greatly
XII	damage nearly total; waves seen on ground surfaces; lines of sight and level distorted; objects thrown into the air; large rock masses displaced

FIGURE 19

Locating an Earthquake Epicenter

Seismologists use P-wave and S-wave data from three different seismographs to locate an earthquake epicenter.

This is a typical seismograph record. Notice that the P-waves were first to arrive. The time that separates the arrival of the first P-wave from the arrival of the first S-wave is called lag time. The lag time can be used to determine the distance from a seismic station to the epicenter.

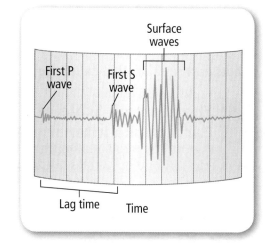

Seismologists use the lag time and a travel-time graph to determine the distance to an earthquake epicenter. Notice that the lag time is equal to the time that separates the S-wave curve from the P-wave curve on the graph.

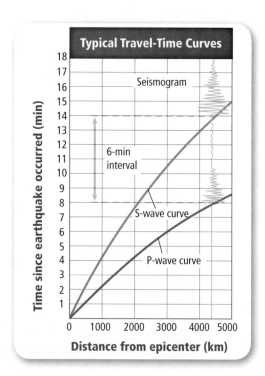

Once a distance is known, seismologists draw a circle with a radius equal to that distance around the seismic station where the data was collected. When data from three seismic stations are analyzed and circles are drawn, the circles will intersect at only one location—the earthquake's epicenter.

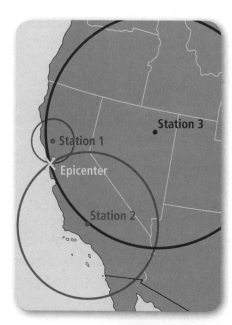

Earthquake Damage

The amount of damage caused by an earthquake is variable. Research has shown that buildings that do not meet construction codes in earthquake-active areas are the largest cause of damage and loss of life. Damage is greatest near the epicenter of an earthquake. In countries where there are poorly constructed buildings, it is common for tens of thousands of people to die in a large earthquake event. A large proportion of earthquake damage can be secondary, such as damage caused by landslides, fires, and earthquake-generated waves called tsunamis. While active earthquake zones are well known, predicting the precise times for earthquakes in those zones is not yet possible.

Earthquake-safe structures Scientists and engineers are finding ways to reduce the damage to structures during mild or moderate earthquakes. Damage can occur when old structures are shaken off their foundations, so securing a building to its foundation is important. One way to keep a building stable is to design a system that allows the whole structure to move as a unit. Base isolation systems use bearings that separate the building from the ground. These bearings are made of large rubber pads or giant metal springs. Large blocks of metal or concrete slide back and forth as the building sways in an earthquake using a design called active damping. Buildings can also be protected by using building materials that bend rather than break during an earthquake, as shown in **Figure 20.**

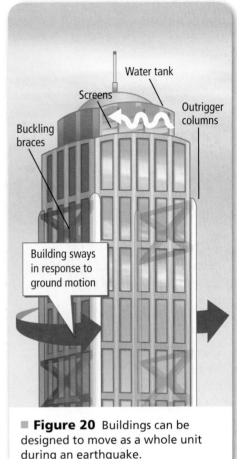

Water tank

Screens

Outrigger columns

Buckling braces

Building sways in response to ground motion

■ **Figure 20** Buildings can be designed to move as a whole unit during an earthquake.

Section 2 Review

Section Summary

▶ Most earthquakes occur along plate boundaries.

▶ Earthquakes result when energy is released as rocks move along a fault.

▶ Seismic waves travel at different speeds and in different directions and can be used to locate an earthquake epicenter.

11. **MAIN Idea Explain** what causes earthquakes.

12. **Compare and contrast** P-waves and S-waves.

13. **Summarize** the patterns of global earthquake distribution.

14. **Think Critically** Why are three sets of seismic data required to locate the epicenter of an earthquake?

Apply Math

15. **Calculate Time** If a P-wave travels at a rate of about 6 km/s through continental crust, how long will it take it to reach a seismic station located 1,200 km away?

16. **Evaluate Lag Time** If an S-wave travels at 4 km/s to the same seismic station in question 15, how much longer will secondary waves arrive after the primary waves?

Essential Questions

▶ How do geologists study Earth's interior?

▶ How does the composition and structure of each of Earth's layers differ?

Review Vocabulary

refraction: the bending of a wave as it changes speed in moving from one medium to another

New Vocabulary

discontinuity
shadow zone
lithosphere
asthenosphere

g Multilingual eGlossary

Earth's Interior

MAIN ‹Idea Earth's layers include the inner core, the outer core, the mantle, and the crust.

Real-World Reading Link When you cut into the inside of an apple, what do you see? There is a core, a layer of fleshy fruit, and an apple peel. Like an apple, Earth also has a layered structure. You can't cut a cross-section of Earth, but scientists can use other techniques to study Earth from the inside-out.

Earth From the Inside-Out

How do geologists study Earth's interior? Geologists use seismic waves to investigate Earth from the inside-out with a method similar to the method that doctors use to examine the inside of the human body. As ultrasound waves travel in the body, the direction and intensity of these waves change when waves pass through different types of body tissue. By recording these changes, an image of the inside of the human body can be produced. In a similar way, by studying changes in speed and direction of seismic waves as they travel in Earth, geologists can infer the structure of Earth's interior.

Refraction The direction in which seismic waves travel can change when the waves travel from one type of material into another. Waves usually change speed when they travel through different materials, as shown in **Figure 21.** However, as waves change speed, they can also change direction. Refraction occurs when a change in speed causes a wave to bend and change direction. The refraction and change of speed of seismic waves as they pass through Earth provides evidence of Earth's layered structure.

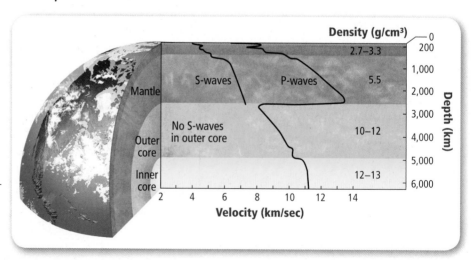

■ **Figure 21** The speeds of seismic waves change as waves travel through different materials within Earth's interior.

Speed and direction Why do the speed and the direction of seismic waves change as they travel through Earth? Earth's layers have different densities due to their different compositions. The boundary between two layers of material that have different densities is called a **discontinuity.** A discontinuity known as the Mohorovicic (moh huh ROH vee chich) discontinuity separates Earth's crust and mantle. Seismic waves can change both speed and direction when they encounter these discontinuities.

Shadow zones Discontinuities can be identified by studying seismic waves. Recall that after an earthquake, a seismograph records the arrival times of seismic waves at different locations. From these data, geologists can infer the paths that seismic waves traveled through Earth.

Examine the P-waves and S-waves produced by an earthquake in **Figure 22.** During an earthquake, some seismographs do not detect any seismic waves. **Shadow zones** are areas on Earth's surface where no seismic waves are recorded. Shadow zones occur for several reasons. One is that P-waves are refracted when they move into and out of the outer core and the inner core. This refraction causes the direction of P-waves to bend so that they do not reach the shadow zone.

The other reason the shadow zone occurs is that S-waves cannot travel through Earth's outer core. Recall that S-waves can travel through solids but cannot travel through liquids. Because S-waves do not travel in Earth's outer core, geologists have inferred that Earth's outer core is liquid.

The inner and outer cores The effects of the inner core on the movement of P-waves suggest that the inner core is solid. Why is the inner core a solid and the outer core a liquid? Deep inside Earth, the temperature and the pressure conditions are extreme. Despite high temperatures, high pressures at depth tend to keep Earth materials in a solid state. Temperatures are high enough in the outer core to overcome the effects of extreme pressure and melt material in the outer core. The pressure in the inner core is high enough to overcome the effects of extreme temperature to keep the inner core solid.

VOCABULARY...

SCIENCE USAGE V. COMMON USAGE

Focus

Science usage

the origin of an earthquake, where rocks first move along a fault

The earthquake focus was 10 km below Earth's surface along the San Andreas Fault.

Common usage

to concentrate

The student focused on completing his homework that was due the next day. ....

Concepts in Motion

Animation

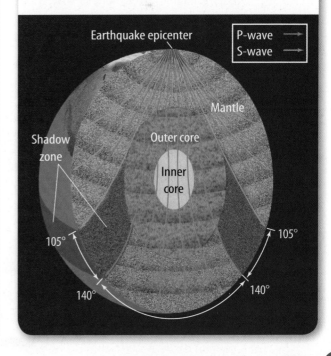

■ **Figure 22** S-waves cannot travel through Earth's outer core, but P-waves pass through both the liquid outer core and the solid inner core.

Identify *Which type of seismic waves will arrive at a seismic station on the opposite side of Earth from an earthquake epicenter?*

Earthquake epicenter

P-wave →
S-wave →

Mantle

Shadow zone

Outer core

Inner core

105°   105°

140°   140°

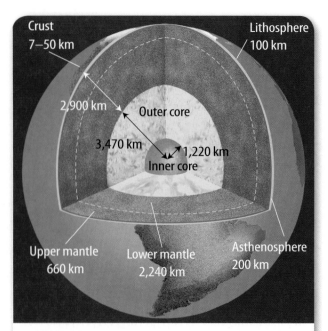

Crust 7–50 km
Lithosphere 100 km
2,900 km
Outer core
3,470 km
1,220 km
Inner core
Upper mantle 660 km
Lower mantle 2,240 km
Asthenosphere 200 km

■ **Figure 23** Layering of Earth is caused by heat and pressure. The densest materials are at the center and the less dense materials are near the crust.

Composition of Earth's Layers

Astronomers hypothesize that early Earth might have formed from meteorite-like material. Over hundreds of millions of years, this molten material separated into layers. As Earth cooled, the densest materials settled to the core and the lighter materials floated toward the mantle and crust. As a result, Earth's layers—the inner and outer core, the mantle, and the crust—shown in **Figure 23,** can be described based on their density.

The crust and uppermost mantle combine to form the **lithosphere** (LIH thuh spfihr). The lithosphere is made of solid, rocky material, mostly silicates that contain oxygen and silicon. It is broken into rigid slabs of rock called tectonic plates. Tectonic plates move on top of a plastic-like layer of the upper mantle called the **asthenosphere** (as THE nuh sfeer). The mantle below the asthenosphere is also composed of silicates that contain denser elements like calcium and aluminum.

Earth's outer and inner cores are composed of mostly metallic elements such as iron and nickel, with noticeable amounts of oxygen and sulfur also present. Earth's core has a composition similar to meteorites that have struck the surface since Earth's formation over 4.5 billion years ago.

 Reading Check **Describe** the difference between the lithosphere and the asthenosphere.

Section 3 Review

Section Summary

▶ Earthquake-generated seismic waves provide information about Earth's interior.

▶ Earth's interior has a layered structure.

▶ The crust and mantle are solid and composed mainly of silicates.

▶ The inner and outer cores have a high density and metallic composition. The inner core is solid and the outer core is liquid.

17. **MAIN ⟨Idea⟩ Describe** evidence for Earth's layered structure.

18. **Explain** the following points, using seismic evidence to support your argument.

   a. Earth has a non-uniform density.
   b. Earth has a layered structure.
   c. Earth has a liquid outer core.

19. **Compare and contrast** the inner and outer cores.

20. **Think Critically** Explain why it is difficult to determine what materials compose Earth's interior.

Apply Math

21. **Calculate** the percent of the mantle by depth that is classified as upper mantle, lower mantle, and asthenosphere.

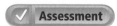 ✓ **Assessment** **Online Quiz**

Section 4

Reading Preview

Essential Questions

▶ What are the sources for molten material inside Earth?

▶ What determines the eruptive style of a volcano?

▶ What is the relationship between volcanoes and plate tectonics?

Review Vocabulary

melting point: temperature at which a solid turns into a liquid

New Vocabulary

viscosity
silica
cinder cone volcano
shield volcano
composite volcano

g Multilingual eGlossary

■ **Figure 24** There are 1,500 active volcanoes around the world today.

Volcanoes

MAIN ⟨Idea Volcanoes form when molten rock rises from deep within Earth and erupts from a crack in the crust.

Real-World Reading Link Have you ever dropped a can of soda and watched it erupt in a big sugary mess when you opened the can? Gases dissolved inside the soda can cause it to explode. Like this soda, volcanoes also store dissolved gas that contributes to their different eruptive styles.

Volcano Formation

Temperatures inside Earth range from 1,000°C at depths of 100 km to 7,000°C in the inner core. In spite of these temperatures, most of the rock within Earth's interior is solid. Recall that extreme pressures keep rock from melting. As pressure increases inside Earth, the melting point of rock also increases.

However, melting can occur within the asthenosphere. Molten, or liquid, rock stored inside Earth is called magma. Because magma is a liquid, it is less dense than the surrounding solid rock. As a result, magma is forced upward. As it rises, the pressure on magma decreases and melting can result. As the magma is forced upward into the lithosphere, it can create cracks in solid rock. These cracks provide channels for the magma to reach Earth's surface. A volcano is a feature that forms when magma reaches Earth's surface and erupts as either lava or ash. **Figure 24** shows the distribution of active volcanoes around the world.

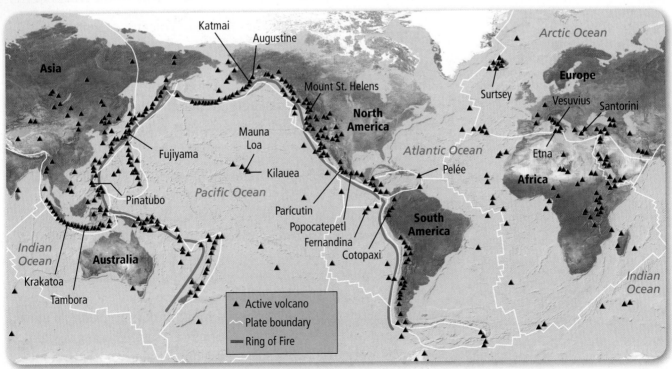

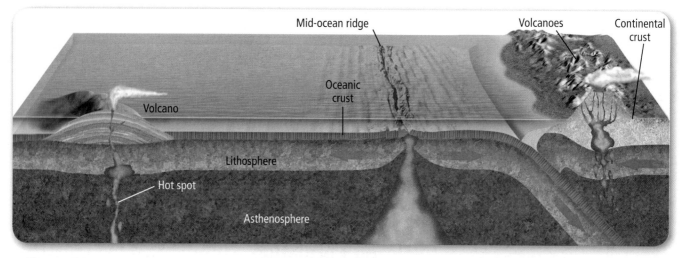

Mid-ocean ridge

Volcanoes

Continental crust

Volcano

Oceanic crust

Lithosphere

Hot spot

Asthenosphere

■ **Figure 25** Volcanic eruptions commonly occur at subduction zones along convergent plate boundaries and at rifts where plates separate along a divergent plate boundary.

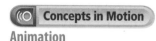

Concepts in Motion

Animation

Concepts in Motion

Animation

■ **Figure 26** Hot spots, such as the Hawaiian Islands, form where a plate moves slowly over a stationary plume of magma that originates deep within the mantle.

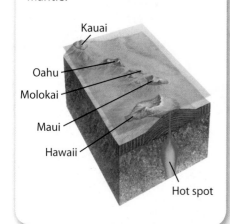

Kauai

Oahu

Molokai

Maui

Hawaii

Hot spot

Plate Boundaries and Hot Spots

Volcanic eruptions occur where plates separate along a mid-ocean ridge or continental rift or where plates collide along a subduction zone. Eruptions are also common at hot spots. Hot spots are areas deep within the mantle where the temperature is high enough to melt rock. Magma rises to the surface and erupts to form volcanoes. Hot spots often occur far away from plate boundaries.

Convergent plate boundaries Most of Earth's volcanoes are located along plate boundaries that surround the Pacific Ocean. This area is known as the Ring of Fire. Here, volcanoes form where tectonic plates collide along subduction zones, as seen in **Figure 25.** Earthquakes and volcanic eruptions often occur along these oceanic-continental and oceanic-oceanic convergent plate boundaries in the Ring of Fire.

Divergent plate boundaries Volcanic activity also occurs along divergent plate boundaries. Most of this activity goes unnoticed because it occurs under water at mid-ocean ridges, as shown in **Figure 25.** However, there are places where volcanic activity due to divergent plates occurs on land. The East African Rift Valley is an example of a divergent plate boundary exposed on land. It is also called a continental rift.

Hot spots Hot spots are areas of volcanic activity where magma moves toward Earth's surface in large, balloon-like plumes. Scientists hypothesize that these plumes originate near the core-mantle boundary. Hot spots are stationary, unlike the plates that move above them. When a hot spot exists under an oceanic plate, magma erupts to form a volcanic island chain. The Hawaiian Islands, depicted in **Figure 26,** are an example of an island chain. Yellowstone National Park is an example of a hot spot that exists under a continental plate. When a volcano moves away from a hot spot, it becomes inactive.

Eruptive Products

Have you ever seen photographs or video of an active volcano? What types of materials erupt? Eruptive products can include lava, chunks of solid materials, and superheated gas.

Lava Magma from a volcano or cracks called fissures may initially remain molten and erupt on Earth's surface as lava or ash. Magma can vary considerably in chemical composition, and this affects its physical properties, such as viscosity.

Viscosity is the measure of a fluid's resistance to flow. Fluids with low viscosities flow more easily than fluids with high viscosities. The viscosity of a fluid decreases as its temperature increases. For example, cold pancake syrup flows more slowly and with greater resistance than warm pancake syrup. This means cold pancake syrup has a higher viscosity than warm pancake syrup.

Another factor that affects the viscosity of magma is chemical composition. For example, mafic magmas have low viscosities and flow easily. Mafic magmas are rich in heavier elements like magnesium and iron. These magmas have low viscosities because they contain less silica than other varieties of magma. **Silica** is the chemical compound silicon dioxide, SiO_2, which is a common ingredient in most magma. Mafic magmas flow from cracks in Earth's crust, such as along mid-ocean ridges and oceanic hot spots like Hawaii. Magmas with high viscosities contain more silica. Intermediate and felsic magmas are rich in silicon, oxygen, aluminum, and potassium. These magma types typically erupt from explosive volcanoes along subduction zones and continental hot spots like Yellowstone.

The amount of dissolved gas in magma also affects its viscosity. For example, water vapor can break silicon-oxygen bonds in magma and decrease viscosity. Some magmas flow so easily that they explosively spew out when they erupt. These lavas with low viscosities can sometimes form a fire fountain where lava erupts into the air like water out of a faucet.

Pyroclastic material Any solid material that erupts from a volcano is classified as pyroclastic material. Often, lava erupts into the air as globules. These globules cool and solidify as they fall back to Earth. The smallest particles cool very quickly and form volcanic ash. Larger particles form volcanic cinders. Also, chunks of solid rock can be ripped away from the conduit of the volcano as it erupts. These chunks of rock form volcanic blocks. The larger the size of a pyroclastic material, the closer it will fall to the volcano. Blocks fall back to the ground near the volcano's main vent. Ash can be picked up and carried by the wind hundreds or even thousands of kilometers away.

? Inquiry   MiniLab

Model Lava Viscosity

Procedure 🥽🧤👕🍴🧼

WARNING: *Do not eat any food items in lab.*

1. Read the procedure and safety information, and complete the lab form.
2. Mix a small batch of batter from **pancake mix** according to the directions.
3. Mix a second small batch, but use 25 percent more **milk.**
4. Hold your finger under a **small funnel** and fill the funnel with the first batch of batter.
5. Have a partner hold a **watch** with a second hand. Remove your finger, and time how long it takes the funnel to empty. Record your data.
6. Clean the funnel, and repeat step 4 using the second batch of batter.
7. Gently heat a **small pan** on a **hot plate.**
8. Pour part of the first batch of batter on the pan and observe.
9. Repeat step 7 using the second batch of batter.

Analysis

1. **Identify** Which batter made the flattest pancakes? Explain why.
2. **Compare and contrast** the properties of each of the batters.
3. **Describe** Which batter modeled lava with a low viscosity?

? Inquiry   Video Lab

? Inquiry   Virtual Lab

Figure 27 Mount Redoubt in Alaska produced pyroclastic flows during a series of eruptions in May 2009. Pyroclastic flows are billowing clouds of superheated gas and ash that can travel at speeds in excess of 100 km/hr.

Explain *why volcanoes like Alaska's Mount Redoubt erupt explosively.*

Gases Volcanoes erupt a variety of gases, including water vapor, carbon dioxide, and sulfur dioxide. Water vapor is released more than any other gas. The sulfur dioxide can combine with oxygen and water in the atmosphere to form tiny droplets of sulfuric acid. Because these droplets are so small, they take a long time to fall back to Earth's surface. While in the atmosphere, the droplets can reduce the amount of solar radiation reaching Earth's surface. This can cause climatic cooling that might last for several years after a powerful eruption.

✓ **Reading Check Explain** How can sulfuric acid in the atmosphere affect climate?

Eruptive Styles

Volcanoes can erupt in a variety of ways, depending on magma composition and viscosity. Thick and sticky felsic magmas have high viscosities and resist eruption, causing the pressure inside a volcano to increase. When Earth's crust cracks under such high pressures, an explosive eruption occurs. This eruption style is often characterized by fast-moving, superheated pyroclastic flows, such as the one shown in **Figure 27.** In contrast, runny mafic lavas have low viscosities, and they erupt quietly. These eruptions are characterized by fluid lava flows, like the lava flow shown in **Figure 28.**

Eruptive style is strongly linked to both temperature and composition, factors that are hard to measure until after an eruption. Temperature and composition of magma can also be related to the type of plate boundary associated with an eruption. For example, high temperature mafic lavas tend to erupt from volcanoes along divergent plate boundaries and oceanic hot spots. Intermediate lava and ash is common along convergent plate boundaries. Low temperature felsic lava and ash erupt from continental hot spots like Yellowstone.

Figure 28 Volcanoes like Hawaii's Kilauea Volcano often produce basaltic lavas that erupt quietly.

Compare and contrast *the eruptions shown in Figures 27 and 28.*

Table 3	Comparison of Melt Properties		(((○ Concepts in Motion	Interactive Table
Composition	Silica Content	Gas Content	Viscosity	Volcano Type
Mafic	lowest	least (1–2%)	lowest	shield, fissure eruptions (such as MOR), cinder
Intermediate	intermediate	intermediate (3–4%)	intermediate	composite
Felsic	highest	highest (4–6%)	highest	volcanic dome

Types of Volcanoes

Volcanoes are classified according to their sizes, shapes, and the types of materials that they erupt. Recall that the eruptive products that form a volcano are related to the physical and chemical properties of magma. The temperature, composition, and gas content of magma help to determine the eruptive style and the type of volcano that results. **Table 3** summarizes the melt characteristics of the main types of volcanoes.

Cinder cone volcanoes When an eruption of high-temperature, gas-rich, mafic lava takes place, eruptive products often are ejected into the air explosively as a porous, dark colored rock called cinder. The cinders might accumulate near a volcanic vent, or a crack in Earth's crust. **Cinder cone volcanoes** tend to be small, have heights in the hundreds of meters range, and have short eruption cycles. When cinder cones form on the flanks of larger volcanoes, they are called parasitic cones. An example of a volcano with parasitic cones is Paricutin volcano in Mexico, as shown in **Figure 29.**

Shield volcanoes Because they form from high-temperature, gas-poor, mafic magmas, shield volcanoes erupt fluid lava flows that can travel far. **Shield volcanoes** are broad, flat structures composed of layer upon layer of lava flows. Think of pancake batter: if the batter is cold and sticky, it piles up and you get thick pancakes. You can add more milk to make a runny batter. In this case, you get thin pancakes as the batter easily flows across the skillet. Shield volcanoes, such as Kilauea and Mauna Loa volcanoes in Hawaii, are similar to pancakes made from runny batter. These volcanoes form from low viscosity magmas.

■ **Figure 29** Volcanoes, such as Paricutin in Mexico, have parasitic cones, which form from eruptions that take place away from the central vent.

Parasitic cone

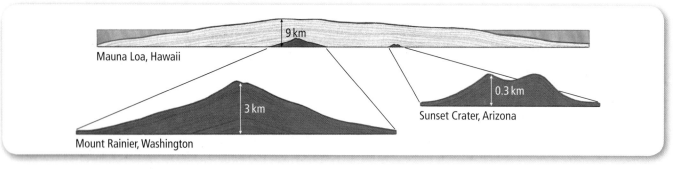

9 km
Mauna Loa, Hawaii

3 km
Mount Rainier, Washington

0.3 km
Sunset Crater, Arizona

■ **Figure 30** Shield volcanoes, such as Mauna Loa, have created some of the largest mountains on Earth. Cinder cones, such as Sunset Crater, are the smallest type of volcano. Some of the most famous volcanoes are composite volcanoes like Mount Rainier.

Composite volcanoes When volcanoes form along convergent plate boundaries, they tend to erupt intermediate and felsic magmas that have high viscosities and high silica contents. This is because as subduction of an oceanic plate takes place, water and sediment are exposed to higher temperatures. Partial melting of this material, in which the silica-rich portion of the rock and sediment melts first, produces magma with a high viscosity that erupts from volcanoes along a subduction zone. This results in explosive eruptions of pyroclastic material and very viscous lava flows.

Over time, lava and ash can accumulate to form a composite volcano. **Composite volcanoes** are large and steep-sided, often thousands of meters high and tens of kilometers across the base, and composed of layers of both lava and ash. Mount St. Helens and Mount Rainier in Washington are examples of composite volcanoes. **Figure 30** shows a size and shape comparison of cinder cone volcanoes, shield volcanoes, and composite volcanoes.

Section 4 Review

Section Summary

▶ Magma originates as molten rock material below the surface rises and erupts as pyroclastic material or lava on Earth's surface.

▶ Volcanoes are distributed worldwide along subduction zones, rift zones, and hot spots.

▶ The style of eruption, whether quiet or explosive, is related to magma composition.

22. **MAIN Idea Explain** how a volcano forms.

23. **Explain** why most volcanoes are located along plate boundaries.

24. **Compare and contrast** composite, cinder cone, and shield volcanoes.

25. **Describe** causes for variation in eruptive style of volcanoes.

26. **Think Critically** Identify some possible consequences if volcanic activity on Earth were to slow down or stop.

Apply Math

27. **Calculate** If a cinder cone is 540 m high and has a base diameter of 3 km, what is the volume of the volcanic cone in cubic meters? Use the formula: $V_{cone} = \frac{1}{3}\pi r^2 h$

28. **Estimate** A volcanic dome has a height of 12 meters and a diameter of 50 m. What is its volume? Use the formula: $V_{dome} = \frac{2}{3}\pi r^3$

✓ Assessment Online Quiz

LAB

A Case for Pacific Plate Motion

Objective

- **Describe** the overall motion of the Pacific Plate.

Background: To measure the rate of motion, you must have a starting and an ending point. You must also know the time it took to get from the start to the end. Volcanic activity associated with a hot spot beneath Hawaii gives geologists exactly what they need to measure Pacific Plate motion.

Question: *How can you determine the rate of plate motion?*

Preparation

Materials

ruler
scale map of the Hawaiian Islands
calculator

Procedure

1. Read the procedure and safety information, and complete the lab form.

2. Create a data table like the one shown.

3. Use the map scale to determine distances in km between each set of islands in the data table. Determine the distance to and from the central point for each island. Record distance (km) in the data table.

4. Refer to the average ages provided for each island on the map. Calculate and record the difference in age between each set of islands in the data table. The symbol *mya* represents age in millions of years ago.

5. Calculate rate of plate motion for each set of islands in km/y. Assume that the hot spot is stationary and that the Pacific Plate is moving over it.

Data Table

From/To	Distance (km)	Difference in Age (y)	Rate (km/y)
Hawaii to Maui	160		
Maui to Molokai			
Molokai to Oahu			
Oahu to Kauai			

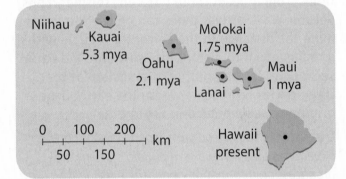

Conclude and Apply

1. **Describe** the overall motion of the Pacific Plate based on your data.

2. **Infer** why plate motion might change over geologic time.

3. **Identify** where the hot spot is located today. What type of geologic activity occurs in this area?

4. **Calculate** If the Pacific Plate continues to move at the same speed and in the same direction, how far away from the hot spot will the big island of Hawaii be in 1.5 million years?

Compare your plate motion calculations with classmates. What might explain any differences in the rates that you calculated for Pacific Plate motion?

LAB | Earthquake Epicenter

Objectives

- **Use** P-wave and S-wave arrival times to calculate the distance to an earthquake epicenter.

- **Determine** the location of an epicenter.

Background: Understanding earthquakes begins with location. To locate an earthquake epicenter, seismologists use P-wave and S-wave arrival times from at least three different seismographs. Seismic wave arrival times can be used to determine the distance to the epicenter. A circle with a radius equal to distance is drawn around each seismograph. The point of intersection of the three circles represents where the earthquake originated, as modeled in **Figure 1.**

Question: *You are on vacation in City A, and you experience an earthquake. The radio stations are broadcasting information about the earthquake. Your home is in City B. Is your home close to the earthquake epicenter?*

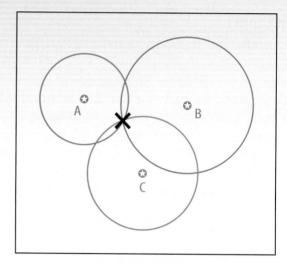

Figure 1

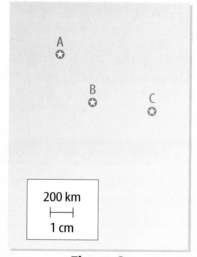

Figure 2

Preparation

Materials

plain white paper
graph paper
compass
metric ruler

Procedure

1. Read the procedure and safety information, and complete the lab form.

2. Draw a 20-cm × 15-cm rectangle on a piece of plain white paper. Arrange the rectangle so that the 20-cm edge is vertical. This rectangle will serve as your map.

3. Using a scale of 1 cm = 200 km, draw a distance scale just below the rectangle on the white paper. Place an arrow parallel to one of the 20-cm vertical edges, and label it *north*.

4. Within the rectangle, place City A 400 km from the north edge and 400 km from the west edge. Place City B 800 km from the north edge and 800 km from the east edge. Place City C 1,200 km from the south edge and 1,200 km from the west edge. Your map will look similar to **Figure 2.**

? **Inquiry** Lab

Data Table 1

City	P-wave Arrival	S-wave Arrival	Time Difference (min/sec)	Distance to Epicenter (km)	Distance on Map
A	08:40:00	08:43:00	3 min 0 sec		
B	08:41:15	08:45:00			
C	08:39:40	08:42:10			

5. The earthquake occurred at 08:37:00 PST. Copy and complete **Table 1** on your own paper. Subtract the P-wave and S-wave arrival times to find the time differences for Cities B and C.

6. Create a graph, and label it *Time Difference (min) versus Distance (km)*. Label the *y*-axis *Time Difference (min)* and the *x*-axis *Distance (km)*.

7. Plot the data provided in **Table 2** on your graph. Connect all points. Use this line to determine the distance from each city to the epicenter. Record this distance in **Table 1**.

8. Use the scale on your map to set the compass width equal to the distance from the epicenter to City A. **WARNING:** *Use caution when using a compass.*

9. Draw a circle on your map with a radius equal to this distance around City A. Your circle may go off the map, but do not worry. The epicenter is somewhere on the arc you can draw on the map. Repeat this process using the distance from Cities B and C to the earthquake epicenter.

Data Table 2

Time Difference Between P-waves and S-waves (min/sec)	Distance (km) Traveled by Waves
1 min 20 sec	500
2 min 12 sec	1,000
3 min 36 sec	2,000
4 min 30 sec	3,000
5 min 24 sec	4,000

Analyze Your Data

1. **Determine** the arrival time difference between P-waves and S-waves that traveled 2,500 km from the earthquake epicenter.

2. **Estimate** the distance that P-waves and S-waves travel in 3 minutes and 18 seconds.

3. **Calculate** how much closer to the epicenter is City A compared to City B.

Conclude and Apply

1. **Explain** the relationship between seismic wave travel time differences and the distance to an earthquake epicenter.

2. **Evaluate** the potential danger to your home in City B. How far is your city from the epicenter?

3. **Infer** how the arrival time difference between P-waves and S-waves changes for locations closest to an epicenter.

4. **Identify** the possible sources of error in the methods that you used to determine an epicenter. How can you minimize errors?

5. **Explain** why two sets of seismic data are not enough to locate an earthquake epicenter.

Compare your map with classmates to see whether your location matches their results. Hypothesize which cities will likely experience the greatest amount of damage during the earthquake.

In the Field
Yellowstone Supervolcano

It was a summer morning in 2003 when a park ranger was guiding tourists at Norris Geyser Basin in Yellowstone National Park. She noticed the smell of burning sap and inferred that the roots of the pine trees must be heating up. When she measured the temperature of the ground along the trail, she discovered that it was over 100°C. The park's geologist was alerted, and the area was marked for special monitoring.

Unsuspecting tourists Each year, millions of visitors enjoy Yellowstone's scenic waterfalls and mountainous landscape. The park also hosts 10,000 thermal features, including geysers such as Old Faithful, colorful hot springs, and gurgling mud pots. The geologists at Yellowstone Volcano Observatory know that beneath their feet lies the largest and most explosive volcano on Earth.

Trail blazed by the hot spot A hot spot has fueled Yellowstone for 17 million years. As the North American Plate moved over the hot spot, magma buoyed upward from beneath the crust. Parts of the crust melted adding gas to a large magma reservoir. As gases accumulated, so did the pressure. The crust expanded. Eventually the crust ruptured and the molten rock exploded in a series of super eruptions. The most recent eruptions occurred 2.1 million, 1.3 million, and 640,000 years ago.

Eruption of a supervolcano An eruption similar to the three super eruptions at Yellowstone has not been seen in human history. Consider the 1980 eruption of Mount St Helens in Washington, USA. Its eruption blasted out the north side of the mountain, ejecting a cubic kilometer of ash into the sky and leaving a crater 1.6-km-wide. The ash traveled as far east as the Rocky Mountains. This eruption pales in comparison to the extent of Yellowstone's ash deposits, as shown on the map in **Figure 1.**

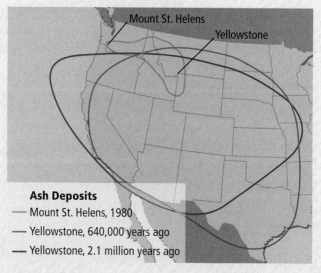

Ash Deposits
— Mount St. Helens, 1980
— Yellowstone, 640,000 years ago
— Yellowstone, 2.1 million years ago

Figure 1 An eruption of Yellowstone supervolcano would dwarf any eruption in human history.

In comparison to Mount St. Helens, the last three super eruptions at Yellowstone left a caldera measuring over 50 km wide. The most recent eruption of the Yellowstone supervolcano ejected 1,000 km³ of ash, enough to cover most of the Western United States under several feet of ash and darken the skies of the planet, resulting in a global winter.

Restless giant Geologists admit that it is not a matter of *if* the Yellowstone supervolcano will erupt again, but *when* it will erupt. So, they monitor the park using global positioning systems and satellite imagery to measure the ground movements. They have noticed that the land in the park bulges and subsides in cycles, sometimes fluctuating several meters at a time. The magma body that lies 10 km beneath Yellowstone continues to bulge, but geologists think an eruption at Yellowstone is highly unlikely for the next several thousand years.

🌐 **WebQuest**

Research Work with a partner to prepare a presentation about one of the largest volcanic eruptions that has occurred within the last decade.

THEME FOCUS Earth Materials and Processes

Thermal activity from within Earth's interior causes plates to collide, to separate, and to slide past each other along active plate boundaries.

BIG Idea Convection currents inside Earth cause plate motion.

Section 1 The Plate Tectonics Theory

convergent plate boundary (p. 777)
divergent plate boundary (p. 778)
mid-ocean ridge (p. 774)
rift valley (p. 774)
subduction (p. 777)
transform plate boundary (p. 778)

MAIN Idea Earth's surface is made of plates that move slowly and interact along plate boundaries.

- Wegener used evidence from the shape of coastlines, types of fossils, rocks, and mountain ranges to support the continental drift hypothesis.
- The age and magnetic polarity of rocks and the thickness of sediments exposed on the seafloor are evidence for seafloor spreading.
- Earth's outermost layer is broken into plates of rigid rock that move and interact along plate boundaries.
- Convection in the mantle drives plate motion.

Section 2 Earthquakes

earthquake (p. 780)
elastic rebound (p. 782)
epicenter (p. 782)
fault (p. 782)
focus (p. 782)

MAIN Idea Earthquakes occur when blocks of rock move along faults, releasing energy.

- Most earthquakes occur along plate boundaries.
- Earthquakes result when energy is released as rocks move along a fault.
- Seismic waves travel at different speeds and in different directions and can be used to locate an earthquake epicenter.

Section 3 Earth's Interior

asthenosphere (p. 790)
discontinuity (p. 789)
lithosphere (p. 790)
shadow zone (p. 789)

MAIN Idea Earth's layers include the inner core, the outer core, the mantle, and the crust.

- Earthquake-generated seismic waves provide information about Earth's interior.
- Earth's interior has a layered structure.
- The crust and mantle are solid and composed mainly of silicates.
- The inner and outer cores have a high density and metallic composition. The inner core is solid, and the outer core is liquid.

Section 4 Volcanoes

cinder cone volcano (p. 795)
composite volcano (p. 796)
shield volcano (p. 795)
silica (p. 793)
viscosity (p. 793)

MAIN Idea Volcanoes form when molten rock rises from deep within Earth and erupts from a crack in the crust.

- Magma originates as molten rock material below the surface rises and erupts as pyroclastic material or lava on Earth's surface.
- Volcanoes are distributed worldwide along subduction zones, rift zones, and hot spots.
- The style of eruption, whether quiet or explosive, is related to magma composition.

Review   Vocabulary eGames

Use Vocabulary

Match each phrase with the correct term from the Study Guide.

29. The _____ is the point of origin of an earthquake.

30. A crack or fracture in Earth's crust along which movement takes place is a _____.

31. A long, linear feature associated with a divergent plate boundary is a(n) _____.

32. A steep-sided explosive volcano that is built from alternating layers of lava and ash is a(n) _____.

33. A boundary marking an abrupt change in density within Earth's interior is a(n) _____.

Check Concepts

34. What process causes an object to permanently change its shape?
A) plastic deformation
B) elastic rebound
C) elastic deformation
D) brittle deformation

35. Which are characteristics of a shield volcano?
A) sticky, silica-rich magmas
B) great height compared to width
C) broad, gently-sloped base
D) steep-sided

36. What property was first used to identify Earth's layers?
A) temperature   **C)** density
B) composition   **D)** thickness

37. Which feature is common along divergent plate boundaries?
A) trenches   **C)** volcanic arcs
B) rift valleys   **D)** island arcs

38. Which evidence was NOT used by Wegener to support the continental drift hypothesis?
A) matching magnetic polarities
B) matching of continents
C) correlation of fossils among the continents
D) mountain-range matching among the continents

Use the figure below to answer question 39.

39. Which type of volcano shown in the image above is small with a height less than a few hundred meters?
A) Hawaiian volcano
B) cinder cone volcano
C) composite volcano
D) shield volcano

40. Which seismic waves travel the fastest within Earth's interior?
A) secondary waves
B) surface waves
C) primary waves
D) body waves

✓ Assessment   Online Test Practice

Interpret Graphics

Concepts in Motion   **Interactive Concept Map**

Use the figure below to answer question 41.

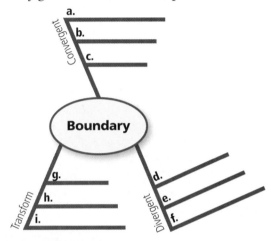

41. **THEME FOCUS** Copy and complete the concept map above describing divergent, convergent, and transform plate boundaries.

Use the table below to answer question 42.

Table 4	Magma Properties and Volcano Types		
General Composition	Relative Silica (SiO$_2$) Content	Relative Viscosity	Volcano Type
		lowest	
Intermediate	intermediate		
	highest		lava dome

42. Copy and complete the table of magma properties for the common volcano types.

Think Critically

43. **Explain** why the island of Kauai, often referred to as the "Garden Isle," has thicker soils and is better able to sustain agriculture than the island of Hawaii.

44. **BIG Idea** **Infer** why there are active volcanoes in the Andes Mountains and not in the Himalayas.

45. **Hypothesize** why continental rocks are significantly older than the oldest rocks exposed on the seafloor.

Apply Math

46. **Calculate** Suppose that hot spot volcanism produces an average of 76,000 m^3 of lava per day and does so continuously for 340 days. This lava flows across a region that has an area of 10 km^2. How thick is the resulting lava flow?

Use the figure below to answer question 47.

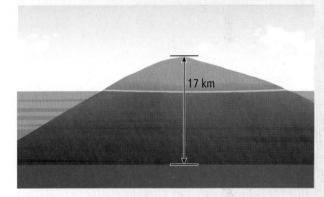

17 km

47. **Solve** Use the height of the shield volcano above from the seafloor to its summit, and the thickness that you calculated in question 46 to determine how many years it took the volcano to reach its current height.

48. **Evaluate** Mauna Loa rises 4 km above sea level. How many years did it take to grow from sea level to its present height?

49. **Calculate** An active volcano called Loihi lies just off the coast of the island of Hawaii. It is called a seamount because it still lies below the surface. If it grows at the same rate that you used in question 48 and lies 1,250 m below the surface, how long will it take for Loihi to break the ocean surface?

Standardized Test Practice

Record your answers on the answer sheet provided by your teacher or on a sheet of paper.

1. Which of the following is made of layers of ash and lava?
   A. shield volcano
   B. dome
   C. composite volcano
   D. cinder cone volcano

Use the figure below to answer question 2.

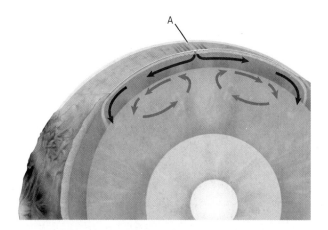

2. Suppose that the arrows in the figure above represent patterns of convection in Earth's mantle. Which type of plate boundary is most likely to occur along the region labeled "A"?
   A. transform plate boundary
   B. reverse plate boundary
   C. convergent plate boundary
   D. divergent plate boundary

3. Using sonar, how were oceanographers able to support the continental drift hypothesis?
   A. edges of the continental shelves matched
   B. edges of the continents matched
   C. fossils match similar fossils
   D. rocks match similar rocks

4. Which feature forms when two plates of different densities collide?
   A. mid-ocean ridge
   B. rift valley
   C. subduction zone
   D. transform boundary

Use the figure below to answer question 5.

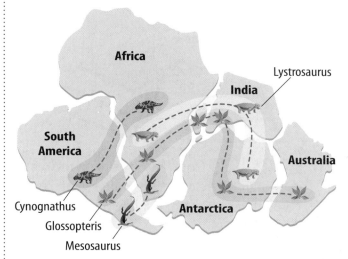

5. The figure above shows fossil evidence that supports the existence of Pangaea. Which of the following organisms was not used as evidence to support Wegener's continental drift hypothesis?
   A. cynognathus
   B. trilobite
   C. mesosaurus
   D. glossopteris

6. A shadow zone results when S-waves interact with which one of Earth's layers?
   A. crust
   B. mantle
   C. outer core
   D. inner core

Record your answers on the answer sheet provided by your teacher or on a sheet of paper.

Record your answers on a sheet of paper.

7. If two plates separate along a divergent plate boundary at a rate of 2.5 cm/y, how much farther apart will the two plates be in 230 years?

8. Explain why earthquakes can occur along any type of plate boundary.

Use the figure below to answer question 9.

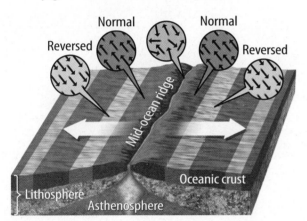

9. The illustration above shows magnetic polarity on the seafloor. How does magnetic polarity of rocks support seafloor spreading?

10. What is a hot spot? Why do volcanoes often form over hot spots?

11. Describe the three types of convergent plate boundaries.

12. How does a rift valley form along divergent boundaries?

13. Explain how geologists are able to infer that Earth's interior is layered.

14. Describe how the age and the thickness of sediments exposed on either side of a mid-ocean ridge vary away from the ridge.

15. What is the driving force behind plate motion? Explain this force.

16. Identify the three types of seismic waves produced during an earthquake. Describe their properties.

17. Why do earthquakes and volcanic eruptions occur along plate boundaries?

Use the figure below to answer question 18.

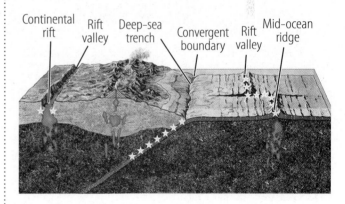

18. How does earthquake depth differ from one plate boundary to another? Why do deep earthquakes occur only at oceanic-oceanic convergent boundaries and oceanic-continental convergent boundaries?

NEED EXTRA HELP?																		
If You Missed Question . . .	1	2	3	4	5	6	7	8	9	10	11	12	13	14	15	16	17	18
Review Section . . .	4	1	1	1	1	3	1	2	1	4	1	1	3	1	1	2	2	2

CHAPTER 26
Earth Materials

ConnectED

Your one-stop online resource
connectED.mcgraw-hill.com

- Video
- Audio
- Review
- Inquiry
- WebQuest
- Assessment
- Concepts in Motion
- g Multilingual eGlossary

Launch Lab
Observe a Rock

Suppose you and a friend are hiking on a mountain trail. Upon reaching the top of a trail, you have an opportunity to look more closely at the rock that you have been hiking across all day. First, you notice that it sparkles in the Sun because of the silvery specks inside it. Looking closer, you also see clear glassy crystals and pink, irregular-shaped crystals. What is this rock made of? How did it get here?

For a lab worksheet, use your StudentWorks™ Plus Online.

? Inquiry Launch Lab

FOLDABLES

Textures Make a three-tab book to help you organize the types of textures that are common in igneous, sedimentary, and metamorphic rocks.

Texture

Igneous Rock | Sedimentary Rock | Metamorphic Rock

THEME FOCUS Earth Materials and Processes

Minerals and rocks are valuable natural resources that can be used to study geologic processes.

BIG Idea Internal and external processes on Earth continually change rocks from one type to another.

Essential Questions

▶ What are the common properties of minerals?

▶ How can physical properties be used to identify an unknown mineral?

▶ How do minerals form?

Review Vocabulary

ionic bond: attraction formed between oppositely charged ions in an ionic compound

New Vocabulary

mineral
streak
cleavage
fracture
hardness
magma

g Multilingual eGlossary

Minerals

MAIN ⟨Idea A mineral is a naturally occurring, inorganic solid with a crystalline form.

Real-World Reading Link It is 6 A.M. and the alarm sounds. Think about your morning routine. What types of rock and mineral resources do you use every morning? From cosmetics to deodorant to toothpaste, many of the products that you use every day are made from Earth materials.

Common Elements

Look at the periodic table of elements in the back of this textbook. Of the first 92 elements, 90 are found naturally in Earth. Only a small number of these 90 elements make up most of the common minerals in Earth's crust.

Elements in Earth's crust The crust, as shown in **Figure 1,** is Earth's rigid outermost layer. It includes all continental material and the material that forms the seafloor. The crust extends down tens of kilometers beneath the continents but is much thinner where it makes up the seafloor. Most minerals, whether on the continents or below the ocean surface, contain abundant oxygen and silicon.

In addition to oxygen and silicon, aluminum, iron, calcium, sodium, potassium, and magnesium are common elements in Earth's crust. **Figure 1** illustrates Earth's layers, including the lithosphere—the crust and a thin layer of the upper mantle. The composition of the materials that make up continental crust differ from materials that make up oceanic crust. Continental materials are less dense and contain a greater abundance of silicon, oxygen, aluminum, and potassium. Oceanic materials are dense and contain more iron and magnesium and less silicon and oxygen.

■ **Figure 1** The crust and a thin layer of the upper mantle make up the lithosphere. The lithosphere floats on a dense, plastic-like layer called the asthenosphere.

Identify *where the crust is thickest and where it is the thinnest.*

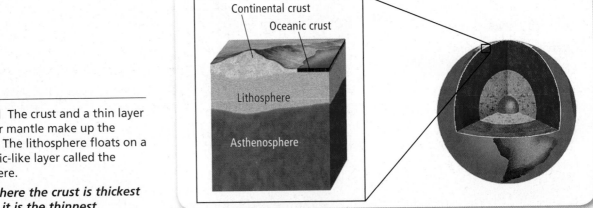

Continental crust
Oceanic crust
Lithosphere
Asthenosphere

The building blocks Picture Earth's most common elements as a set of interlocking blocks. In the set, you are given lots of common block types from which you can build thousands of different forms. But you are given only a few unique blocks. The bulk of the forms that you build consist of the most common blocks. Elements and the minerals that compose Earth's crust are similar to the block set. Common crustal building blocks include silicon and oxygen. **Figure 2** shows other common elements in Earth's crust that are the building blocks of most minerals.

What's a mineral?

Remember that atoms and ions can bond to form compounds. A **mineral** is a naturally occurring inorganic solid with a crystalline form. The term *inorganic* means that minerals are materials that are not living. The composition of any mineral is indicated by its chemical formula. For example, the element gold is a mineral with the chemical formula Au. Similarly, fluorite is a mineral that contains calcium ions (Ca^{2+}) and fluoride ions (F^-) that combine to form CaF_2, shown in **Figure 3.** The chemical composition of a mineral determines its physical properties.

Physical Properties of Minerals

Each mineral has a unique chemical composition. However, sometimes the composition of a mineral can vary slightly due to the presence of chemical impurities. These impurities cause some of the mineral's physical properties to change. For example, the gemstones ruby and sapphire are both the mineral corundum, even though they are different colors. A ruby is red because it contains more of the chemical impurity chromium than a sapphire, which is blue. Color is one of several physical properties that can be used to help identify minerals.

✓ **Reading Check Explain** why rubies are red and sapphires are blue.

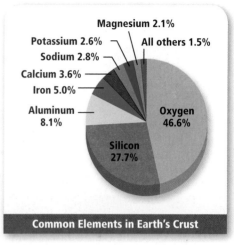

Figure 2 The two most abundant elements in Earth's crust are oxygen and silicon.

Common Elements in Earth's Crust
- Magnesium 2.1%
- Potassium 2.6%
- Sodium 2.8%
- All others 1.5%
- Calcium 3.6%
- Iron 5.0%
- Aluminum 8.1%
- Oxygen 46.6%
- Silicon 27.7%

Figure 3 Fluorite (CaF_2) is a mineral that contains calcium ions (Ca^{2+}) and fluoride ions (F^-) in a 1:2 ratio. This composition contributes to fluorite's unique set of physical properties.

Amethyst

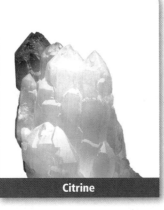
Citrine

Smoky Quartz

Rose Quartz

■ **Figure 4** Amethyst, citrine, smoky quartz, and rose quartz are the same mineral despite their different colors.

Inquiry Virtual Lab

■ **Figure 5** The mineral pyrite has a metallic luster. Talc is dull with a nonmetallic luster.

Metallic Luster

Nonmetallic Luster

Mineral Identification

You now know that the same mineral can contain chemical impurities that distort its color. Color can be one of the least reliable properties for mineral identification. The same mineral can occur naturally in a variety of colors. For example, quartz can be colorless, purple, yellow, gray, or rose-colored, as shown in **Figure 4**. In addition, two different minerals with different chemical compositions can be the same color. Olivine and pyroxene are both green despite their different chemical compositions. Because color alone is an unreliable property, other physical and chemical properties must be investigated to identify unknown minerals.

Luster The way a mineral reflects light is the physical property known as luster. There are two main types of luster—metallic and nonmetallic. As you might guess, minerals with metallic luster reflect light in a way that is similar to a metal surface, such as the shiny chrome on a car. Minerals that shine like glass or appear earthy or waxy exhibit nonmetallic luster, as shown in **Figure 5**.

Streak The color of a mineral in powdered form is called **streak.** A mineral's streak might not be the same color as the mineral itself. When a mineral occurs naturally in a variety of colors, the streak powder color generally stays the same, which helps to identify the mineral. A streak test is performed by rubbing a mineral on an unglazed, white porcelain tile. For example, when you rub the mineral graphite on a white porcelain tile, the streak color is gray.

Atomic arrangement Some physical properties are due to the orderly arrangement of atoms in a mineral's structure. This predictable pattern is what makes a mineral crystalline. The arrangement of atoms and ions and the bonds between them can reflect the way a mineral breaks, how hard it is, and what type of crystal shape it has.

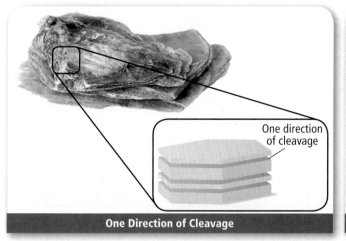

One Direction of Cleavage

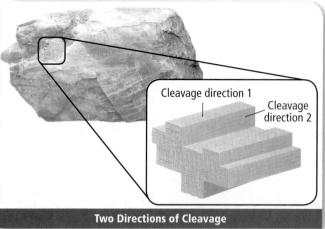

Two Directions of Cleavage

Cleavage When a mineral breaks along preferred planes of weakness creating sets of smooth parallel sides, it has **cleavage.** Any two parallel sets of planes define a single direction of cleavage. The mineral mica exhibits only one direction of cleavage and breaks in flat sheets, as shown in **Figure 6.** Other minerals, such as the feldspar shown in **Figure 6,** exhibit more than one direction of cleavage. Feldspar breaks into rectangles.

Fracture Some minerals do not separate along well-defined surfaces. In such cases, a mineral will break unevenly. Minerals that break with irregular surfaces exhibit **fracture.** Quartz is an example of a mineral that does not cleave; it fractures.

Hardness When you scratch a mineral with another material, the scratch forms as bonds in the mineral are broken. The physical property that measures a mineral's resistance to scratching is called **hardness.** Hardness depends on the strength of the chemical bonds between atoms or ions in a mineral. Today, you might be using a pencil that takes advantage of this physical property. The graphite in the core of some pencils has one plane of cleavage and very weak bonds between carbon atoms along this plane. The weak bonds make the graphite soft and easy to scratch. As you use the pencil, thin graphite plates are left behind on your paper. Other minerals with stronger bonds are not as easily broken and resist scratching.

You can compare the hardness of different materials by applying scratch tests and using Mohs hardness scale, shown in **Table 1.** When a hardness test is performed by rubbing two materials together, the softer of the two will be scratched. For example, quartz has a hardness of 7.0 and will scratch any object with a hardness less than 7.0. A copper penny has a hardness of 3.0 and will not scratch fluorite, but it will scratch both gypsum and talc. Your fingernail has a hardness of 2.5 and will scratch both gypsum and talc.

■ **Figure 6** Because mica has one direction of cleavage, it can be separated in layers. Feldspar is an example of a mineral with two directions of cleavage.

Identify *the two sets of cleavage planes in feldspar.*

Concepts in Motion
Interactive Table

Table 1	Mohs Scale of Hardness
Hardness	Mineral
1	talc
2	gypsum (fingernail 2.5)
3	calcite (copper penny 3.0)
4	fluorite
5	apatite (glass plate 5.5)
6	feldspar
7	quartz (streak plate 7.0)
8	topaz
9	corundum
10	diamond

■ **Figure 7** Minerals can be classified using six basic crystal systems that reflect the internal arrangement of atoms.

Compare and contrast *the cubic crystal system to the hexagonal crystal system.*

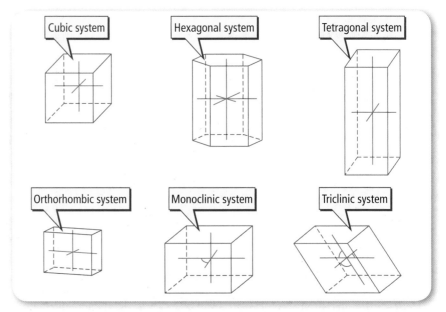

Cubic system | Hexagonal system | Tetragonal system

Orthorhombic system | Monoclinic system | Triclinic system

Crystal shape

Crystal shape The external crystal shape of a mineral is related to the orderly internal arrangement of its atoms or ions. Even though many unique shapes are exhibited by mineral crystals in nature, each can be sorted into one of the six crystal systems shown in **Figure 7.** Crystal shapes are used to determine the crystal system that a mineral belongs to. For example, lead and sulfur atoms in the mineral galena (PbS) bond to form cubes. Galena therefore belongs to the cubic crystal system.

Mineral Formation

A mineral crystal grows as atoms or ions bond to its surface. The types of atoms or ions that form bonds depend on the growing crystal's surroundings. Growth is also controlled by how fast atoms and ions can migrate to the crystal and by the temperature and pressure of the crystal's surroundings.

Mineral crystals can form in many different ways. Minerals can form when hot water, rich in dissolved atoms and ions, cools and crystallizes or when this hot water reacts with other rocks and minerals in its surroundings. Minerals can also form when molten rock cools and crystallizes or when water evaporates from solutions on or near Earth's surface.

Minerals from hot water Some minerals are produced from hot water that contains dissolved atoms and ions. You know that sugar dissolves fastest in hot water. As a hot water solution cools, its particles slow down and dissolved materials are able to crystallize out of solution. Minerals can also precipitate around the edges of hot springs where hot water reacts with surrounding rock, as shown in **Figure 8.** When hot water passes through cracks in cooler rock, minerals can crystallize within the cracks. Sometimes, veins or halos of concentrated minerals like gold, silver, or copper are produced in these cracks in rock.

MiniLab

? Inquiry MiniLab

Identify Salt's Crystal System

Procedure 🔬

1. Read the procedure and safety information, and complete the lab form.
2. Sprinkle a small quantity of **table salt** onto a **dark surface.**
3. Using a **magnifying lens,** observe the individual grains of salt.

Analysis

1. **Create** a sketch showing the shapes of several salt grains.
2. **Explain** Why are some of the grains more elongated than others?
3. **Infer** To which crystal system does salt belong? Support your answer.

Hydrothermal minerals form on the rims of hot springs.

Minerals can form as molten rock cools.

Minerals from molten rock Molten rock material inside Earth is called **magma**. When molten rock erupts on Earth's surface it is called lava, as shown in **Figure 8**. As magma cools, atoms and ions arrange themselves in an orderly pattern. When the temperature of the molten rock falls below the freezing point of a particular mineral, crystals of that mineral begin to form.

Minerals from evaporation Minerals can form from water solutions on or near Earth's surface. When water evaporates, dissolved mineral material can crystallize from a saturated solution. In a similar way, you might have seen films of mineral material left behind in a pan that was once filled with water. Dissolved minerals crystallize in the pan as the water evaporates.

Mineral Groups

About 3,800 minerals have been identified in nature. Some minerals are so common that they are called rock-forming minerals. Recall that only a few elements make up most of the Earth's crust. Similarly, only a few important mineral groups—the silicates and the nonsilicates—are common on Earth.

Silicates The chemical composition and arrangement of atoms or ions in minerals are used to classify minerals. The most abundant minerals in Earth's crust are silicate minerals.

The basic building block for the silicate minerals is a simple silicate ion (SiO_4^{4-}), composed of four oxygen atoms tightly bonded to a silicon atom. In silicate minerals, the elements silicon and oxygen bond together to form a geometric structure called a tetrahedron, as shown in **Figure 9**. Other metal ions are attracted to these silicate ions. Most silicate minerals contain silicon, oxygen, and a combination of one or more elements. One exception is quartz, which is pure silica (SiO_2). Silica is shorthand for silicon dioxide.

■ **Figure 8** Minerals form in a variety of different environments, such as hot springs, volcanoes, and saline lakes.

■ **Figure 9** The silicon tetrahedron, composed of one silicon ion attached to four oxygen ions, is the basic building block for all silicate minerals.

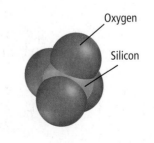

Oxygen

Silicon

Table 2	Common Minerals	
Silicates		
Olivine		O—Si Single tetrahedron
Pyroxene		Single chains
Amphibole		Double chains
Mica		Sheets
Feldspar		Three-dimensional networks
Quartz		
Nonsilicates		
Carbonates (calcite)		
Halides (halite)		
Oxides (hematite)		
Sulfides (pyrite)		
Sulfates (gypsum)		
Native element (sulfur)		

Silicate structures You might have seen houses being built in your neighborhood. The houses are constructed using similar materials. What makes them different is how these materials are put together. Similarly, different silicate minerals are formed when silicon-oxygen tetrahedrons are linked together in different ways. The simplest silicate structures have single silicon-oxygen tetrahedrons that are not linked. By linking silicon-oxygen tetrahedrons together in different combinations, chains, sheets, and three-dimensional networks can form. **Table 2** shows some common silicate minerals and their associated structures.

Minerals of the crust Earth's crust is composed of mainly silicate minerals, such as some of the common silicate minerals shown in **Table 2**. Minerals like quartz and potassium feldspar have low densities and form at low temperatures compared to other silicate minerals. Together, quartz and potassium feldspar group silicates make up most of Earth's continental crust. In contrast, Earth's oceanic crust is denser and contains a larger percentage of chain silicates, such as the pyroxene group. The oceanic crust is also composed of single tetrahedron silicates from the olivine group, along with the three-dimensional network silicates from the plagioclase feldspar group.

Nonsilicates The nonsilicates are classified as minerals that do not contain silicon. A few of the common nonsilicate minerals are also shown in **Table 2**. These include the carbonates, halides, oxides, sulfides, sulfates, and native elements. For example, the mineral calcite ($CaCO_3$) is a common carbonate mineral found in rocks like limestone and marble. Calcite does not contain silicon in its chemical composition. Hematite (Fe_2O_3) is a common oxide mineral. Pyrite (FeS_2), also known as fool's gold, is a common sulfide mineral, and sulfur (S) and gold (Au) are native elements.

The nonsilicate mineral groups are a source of many valuable ore minerals and building materials. To be an ore, a mineral must occur in large enough quantities to be economically recoverable. One example of an important iron ore is the mineral hematite, an iron oxide.

Mineral Uses

For centuries, humans have relied on minerals for their everyday needs. For example, humans have used salt to flavor and preserve food. Salt is processed from the mineral halite (NaCl). Civilizations have advanced with their mineral wealth. Consider all of the exploration and conflict that gold has caused. Humans have also spent vast amounts of money and have risked lives to search for and obtain gold.

People use minerals either directly as objects of wealth or as raw materials to make things. Civilizations in the regions fortunate enough to have large quantities of iron ore, called hematite, advanced rapidly. Consider all of the materials made from iron and steel. Without iron, there would be no machinery as we know it today for manufacturing goods.

✔ **Reading Check** **Identify** objects made from iron and steel.

Not all minerals need to provide metals to be valuable. Non-metallic minerals are valuable as well. For example, quartz is used to make glass and glass fibers. Glass fiber is used to make fiber optical cables. Most sands and gravels are largely quartz. When mixed with cement (calcite), quartz is used to make concrete, which is widely used in our society. Life as we know it would not be possible without mineral resources.

VOCABULARY ....
SCIENCE USAGE V. COMMON USAGE
Concrete
Science usage
  cemented particles that form a solid mass
  The concrete sidewalk was still wet when the little boy stepped on it.

Common usage
  a tangible object or idea
  A concrete plan was in place for the sale of the advertising agency. ....

Section 1 Review

Section Summary

▶ Few elements form the bulk of minerals in Earth's crust.

▶ Minerals can be classified based on their physical and chemical properties.

▶ Minerals can be sorted into two broad categories: silicates and nonsilicates.

▶ Minerals are valuable resources used every day.

1. **MAIN Idea** **Determine** whether diamonds produced in laboratories are considered to be minerals.

2. **Describe** two ways that minerals form.

3. **Compare and contrast** mineral cleavage and mineral fracture.

4. **Explain** why it is useful to test more than one physical property when attempting to identify a mineral.

5. **Think Critically** Why do you think some saw blades and drill bits are coated with diamond?

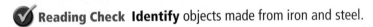

6. **Calculate Surface Area** Suppose you have a halite cube that measures 3 cm on each edge. What is its total surface area?

7. **Calculate Surface Area** Suppose you cleave the halite cube in question 6 exactly in half, perpendicular to one of its faces. What is the total surface area now?

✔ Assessment   Online Quiz

Be a Mineral Detective

Objectives

- **Observe** and record physical properties of minerals.

- **Determine** mineral names using your observations and identification keys.

Background: Detectives must gather facts and physical evidence to deduce the events that took place during a crime. Much like detectives, geologists gather physical and chemical evidence to better understand Earth processes. First, minerals are identified, and then the processes that formed them can be interpreted.

Question: *What physical properties can be used to distinguish similar-looking materials from each other?*

Preparation

Materials

mineral samples
unglazed porcelain tile
Mohs hardness scale
Reference Handbook, "Minerals"
penny
nail
glass plate

Safety Precautions

Procedure

1. Read the procedure and safety information, and complete the lab form.

2. Create a data table. Your table should have at least eight columns labeled as follows: Mineral Number, Color, Hardness, Streak, Cleavage, Fracture, Other Observations (such as smell, feel, or heft), and Mineral Name.

3. Obtain numbered mineral samples from your teacher. Observe each mineral and accurately record the data based on your tests for physical properties. Be descriptive in your observations.

4. Use the Reference Handbook "Minerals" at the back of your book to identify each unknown mineral. Your teacher may tell you when you are incorrect and ask you to try again.

5. To identify mineral hardness, recall that the hardness of your fingernail is roughly 2.5; penny, roughly 3.0; steel nail, 5–6; and glass, about 5–5.5.

6. To identify cleavage, look for sets of parallel sides. Fracture will appear as uneven surfaces.

Conclude and Apply

1. **Explain** why some minerals were difficult to identify. Why were some easier to identify?

2. **Determine** which physical properties were most reliable and which were least reliable for mineral identification.

3. **Describe** the physical tests that you used to distinguish between minerals.

Compare your results with your classmates. Discuss what kind of mineral evidence a detective might obtain from a crime scene and how it might be used to solve a crime.

? **Inquiry** Lab

Reading Preview

Essential Questions

▶ What are the common characteristics of rocks?

▶ Under what conditions do igneous rocks form?

▶ How are igneous rocks classified?

Review Vocabulary

mixture: combination of two or more substances that can be physically separated

New Vocabulary

rock
intrusive igneous rock
extrusive igneous rock

 **Multilingual eGlossary**

 **Video** **BrainPOP**

Igneous Rocks

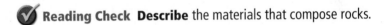 **MAIN Idea** Igneous rocks form when molten material cools and crystallizes.

Real-World Reading Link Did you ever pick up rocks and wonder where they came from? Some rocks might have formed on a beach. Others might have formed at the bottom of the ocean, while some might have erupted from a volcano. How can you tell one rock apart from another?

What's a rock?

A **rock** is a naturally formed mixture that can contain minerals, rock fragments, and volcanic glass bound together. Rocks are identified based on their texture and composition. The texture of a rock describes the size, shape, and arrangement of the rock's components. The composition describes the average chemistry of the rock. Rocks can undergo change through a cycle of different internal and external processes on Earth. Part of this cycle involves rock formed from molten material.

✓ **Reading Check** **Describe** the materials that compose rocks.

Intrusive Igneous Rocks

Igneous rocks are those formed from molten material that cools and crystallizes either underground or on Earth's surface. **Intrusive igneous rocks** are igneous rocks that form underground. Such rocks also are called *plutonic rocks,* after Pluto, the Roman god of the underworld. These rocks are intrusive because they form within Earth's crust. An example of an intrusive igneous rock is shown in **Figure 10.**

■ **Figure 10** The dark-colored veins of intrusive igneous rock formed when molten rock cut through surrounding rock layers and cooled slowly underground. Erosion of less resistant rock material around the veins eventually resulted in the spires of intrusive igneous rocks seen here.

Intrusive igneous rock

Magma composition Because magma is less dense than surrounding solid rock, it will rise up through cracks and fissures in Earth's crust. As magma rises and cools, different silicate minerals crystallize at different temperatures.

You are familiar with the freezing of liquid water to form ice at 0°C. Minerals freeze, or crystallize, at high temperatures, ranging from several hundred degrees Celsius to more than 1,000 degrees Celsius. Magmas must be cooled below the freezing temperature of a silicate mineral in order to crystallize.

Bowen's Reaction Series Silicate minerals crystallize in a predictable pattern. This pattern was first recognized by the Canadian geologist N.L. Bowen. Bowen observed that silicate minerals crystallize in two main sequences as magma cools. These sequences are now described by the process known as Bowen's reaction series. Bowen's reaction series is shown in **Figure 11.**

The two main sequences in Bowen's reaction series reflect simultaneous crystallization of silicate minerals with a decrease in temperature. On the left side of Bowen's reaction series, when the temperature of the magma is high, dense iron- and magnesium-rich minerals like olivine and pyroxine are first to crystallize. Due to their high densities, these minerals sink toward the bottom of the magma chamber. At the same time that olivine and pyroxene crystallize, calcium-rich feldspar also crystallizes, as shown on the right side of Bowen's reaction series. Late-forming, less dense minerals like potassium feldspar, muscovite mica, and quartz tend to crystallize at lower temperatures and rise to the top of the magma chamber.

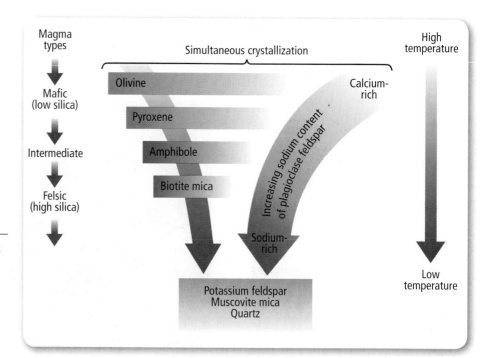

■ **Figure 11** Bowen's reaction series illustrates the sequence in which silicate minerals crystallize from magma at different temperatures.

Identify *two pairs of minerals that are likely to be found in the same rock.*

This piece of coarse-grained granite contains minerals that are visible without magnification, such as the minerals quartz (gray), potassium feldspar (pink), mica (clear), and hornblende (black).

This piece of fine-grained basalt contains mineral crystals that are too small to see, such as the minerals olivine, pyroxene, and plagioclase feldspar.

■ **Figure 12** Granite is a rock that cooled slowly inside Earth. Basalt is a rock that cooled quickly on Earth's surface.

FOLDABLES

Incorporate information from this section into your Foldable.

Three magma types Igneous rocks form from three types of magma—mafic magma, intermediate magma, and felsic magma. These magma types produce rocks that are classified into three groups—mafic rocks, intermediate rocks, and felsic rocks—based on their silica (SiO_2) composition.

Felsic rocks are silica-rich (65–75 percent SiO_2). These rocks have low densities and form much of the continental crust. Mafic rocks are silica-poor (45–55 percent SiO_2). These rocks are denser than those in the felsic group and form much of the oceanic crust. Finally, intermediate rocks have silica compositions between felsic and mafic rocks.

Intrusive rock textures Recall that texture describes the size, shape, and arrangement of all rock components. The size of the mineral crystals in a rock is called the grain size. Large grains are crystals that are big enough to see with the unaided eye. Rocks with large grains, like the granite shown in **Figure 12,** are called coarse-grained. Rocks will small grains, like the basalt in **Figure 12,** are called fine-grained. Based on the grain size of minerals in igneous rocks, you can determine where cooling and crystallization occurred.

Grain size depends on how quickly magma, lava, or volcanic ash cool and crystallize. An intrusive igneous rock forms when magma cools slowly inside Earth. As the magma slowly cools, crystals have time to grow. Intrusive igneous rocks therefore have a coarse-grained texture. A few examples of intrusive igneous rocks that cooled slowly underground and have coarse-grained textures include granite, diorite, and gabbro. In contrast, lava or ash that erupts onto Earth's surface, cools quickly and crystals have very little time to grow. Examples of igneous rocks that cooled quickly on Earth's surface include rhyolite, andesite, and basalt. These rock types have a fine-grained texture.

■ **Figure 13** Lava erupts from Kilauea volcano on the island of Hawaii. The cracks in Earth's crust through which magma flows to the surface at Kilauea volcano may be 60 km in depth.

Extrusive Igneous Rocks

Intrusive igneous rocks form when magma enters cracks within Earth's crust and then cools and crystallizes. Some cracks in the crust reach Earth's surface. Magma can flow through these cracks and erupt onto the surface as lava or volcanic ash. When lava or ash erupts onto the surface, it cools and crystallizes to form **extrusive igneous rocks.** For example, the lava flows in **Figure 13** will eventually cool to form new extrusive igneous rock.

Extrusive rock textures If a volcanic eruption occurs on land, lava comes in contact with air. If a volcanic eruption occurs on the seafloor, lava comes in contact with water. In either case, the temperature of the lava is much hotter than the temperature of the surrounding air or water. Because the temperature difference is large, the lava cools very quickly. When lava cools quickly, crystals do not have time to grow.

As a result of rapid cooling, the textures of extrusive igneous rocks differ from the textures of intrusive igneous rocks. Extrusive igneous rocks have small grain sizes and are called fine-grained. Fine-grained, extrusive igneous rocks often have grain sizes that are too small to be seen without a magnifying lens. Sometimes lava cools so quickly that almost no crystals are formed. The rocks form with a glassy texture. Obsidian is an example of an extrusive igneous rock with a characteristic glassy texture.

Other textures Extrusive igneous rocks can exhibit other textures. Pumice and scoria are examples of extrusive igneous rocks that have vesicular texture, as shown in **Figure 14.** These rocks formed as trapped gases escaped from a volcano during an eruption. As these gases escape, holes called vesicles are left behind in the rock. Another type of texture, called porphyritic texture, results when minerals cool at different rates and create a rock with two different-sized crystals, as shown in **Figure 14.**

■ **Figure 14** Pumice exhibits a vesicular texture as gases escape upon eruption. Andesite has a porphyritic texture with two distinct crystal sizes.

Vesicular Texture

Porphyritic Texture

Common Igneous Rocks

Despite differences in texture, extrusive igneous rocks are formed from the same three types of magma as intrusive igneous rocks. For example, silica-rich felsic magma forms coarse-grained granite when it cools slowly inside Earth's crust. The same magma forms a fine-grained rock called rhyolite when it cools quickly as lava or volcanic ash on Earth's surface. This means that the intrusive igneous rock granite and the extrusive igneous rock rhyolite have similar chemical compositions.

Reading Check **Identify** What is the main difference between granite and rhyolite?

As **Table 3** shows, it is the differences in crystal sizes that cause the granite and rhyolite to look dissimilar. Likewise, silica-poor mafic magma forms coarse-grained gabbro when it cools slowly underground, and it forms fine-grained basalt when it cools rapidly as lava on Earth's surface. Examine the common intrusive and extrusive igneous rocks identified in **Table 3.** What types of rock result from intermediate magmas that cool inside Earth's crust and on Earth's surface?

Table 3	Common Intrusive and Extrusive Igneous Rocks		
	Felsic Magma	**Intermediate Magma**	**Mafic Magma**
Intrusive	granite	diorite	gabbro
Extrusive	rhyolite	andesite	basalt

EXAMPLE Problem 1

Use Percentages The amount of silica (SiO_2) in an igneous rock and its texture can be used to classify the rock and determine its crystallization environment. Mafic rock contains 45–55 percent SiO_2, intermediate rock contains 55–65 percent SiO_2, and felsic rock contains 65–75 percent SiO_2. A 151-g rock is found to contain 89 g of SiO_2. Based on the percentage of SiO_2 in this rock, how would you classify it?

Identify the Unknown: percent SiO_2 = ? %

List the Knowns:
151 g of rock	$m_{rock} = 151$ g
89 g of SiO_2	$m_{silica} = 89$ g

Set Up the Problem: Divide the mass of SiO_2 by the total mass of the rock and multiply by 100.

$$\text{percent } SiO_2 = \frac{m_{silica}}{m_{rock}} \times 100$$

Solve the Problem: Substitute the known values.

$$\text{percent } SiO_2 = \frac{m_{silica}}{m_{rock}} \times 100 = \frac{89 \text{ g}}{151 \text{ g}} \times 100$$
$$= 59\%$$

The rock should be classified as intermediate.

Check the Answer: Is your answer correct? Check your answer by multiplying the percentage you calculated (expressed as a decimal) by the mass of the rock. The answer should be the mass of silica given in the problem.

PRACTICE Problems   Find **Additional Practice Problems** in the back of your book.

8. Solve How would you classify a 275-kg extrusive igneous rock from Mount Rainier that contains 189 kg of SiO_2?

Review

Additional Practice Problems

9. Calculate If the mass of an intrusive igneous rock discovered in Alaska is 89 kg and its percent SiO_2 is 63 percent, what is the mass of SiO_2 in the rock?

10. Challenge A 120-kg basaltic boulder found in Hawaii contains 47 percent SiO_2. How many kilograms of SiO_2 are in the boulder?

11. Challenge What is the greatest mass of SiO_2 that could be found in a 212 g rhyolite?

Gases Magma contains small amounts of dissolved gases, such as water vapor and carbon dioxide. As magma rises toward Earth's surface, the pressure exerted on the magma by the surrounding rock decreases. Under these lower pressure conditions, the gases form bubbles in the magma. When the magma rises through cracks in the crust, erupts on the surface, and cools, the bubbles can remain trapped in extrusive igneous rock. The rock that forms is filled with holes where the bubbles once were. Pumice is the product of a gas-rich eruption.

■ **Figure 15** This spindle-shaped volcanic bomb exploded from a volcano during a gas-rich eruption.

✔ **Reading Check** **Explain** how extrusive igneous rocks, such as pumice, form from gas-rich eruptions.

In some cases, dissolved gases can cause lava to erupt explosively. When an explosive eruption occurs, lava is thrown into the air. Large blobs of lava solidify as they travel through the air, forming irregularly-shaped rocks called volcanic bombs, like the one shown in **Figure 15**. Bombs are often streamlined in shape as lava cools mid-flight. The largest volcanic bombs can be several meters in diameter.

Section 2 Review

Section Summary

▶ Rocks are mixtures of various natural materials.

▶ Extrusive and intrusive igneous rocks form in different environments.

▶ There are three common types of magma compositions—felsic, intermediate, and mafic.

▶ Extrusive and intrusive igneous rocks formed from the same type of magma have similar compositions but different textures.

▶ Dissolved gases in magma can cause explosive volcanic eruptions.

12. **MAIN Idea** **Compare and contrast** extrusive and intrusive igneous rocks.

13. **Compare** the magma that forms the continental crust with the magma that forms the oceanic crust.

14. **Discuss** what happens to the dissolved gases in magma as the magma rises toward Earth's surface.

15. **Explain** why you would not expect to observe an igneous rock with quartz and olivine in it.

16. **Think Critically** Lava flows that are relatively low in SiO_2 tend to have low viscosities, which means they flow easily. Name an igneous rock that you would expect to form from a low-viscosity lava flow. Explain your answer.

Apply Math

17. **Calculate Granite Volume** Granite occurs in very large masses called batholiths. The Sierra Nevada in California is an example of a batholith made of many separate intrusions. Suppose the mountain range averages 8.33 km in thickness, extends 700 km from north to south, and averages 91.66 km wide. Assuming the intrusion is rectangular in shape, approximately how many cubic kilometers of granite reside in that mountain range?

Reading Preview

Essential Questions

▶ How do sedimentary rocks form?

▶ Where do sedimentary rocks form?

▶ Can sedimentary rocks be used to interpret Earth's history?

▶ How are sedimentary rocks classified?

Review Vocabulary

precipitate: a solid that results from a chemical reaction in solution

New Vocabulary

clast
pore space
compaction
cementation

 **Multilingual eGlossary**

Sedimentary Rocks

MAIN Idea Sedimentary rocks form from consolidated rock particles or crystallization from a solution.

Real-World Reading Link Think of all the tiny grains of sand on a beach. These grains of sand were once part of solid rock that weathered away over time. What types of weathering caused this change?

Clasts

Recall that rock is a mixture of minerals, rock fragments, or volcanic glass bound together. Some of these components could be bits and pieces of other rocks. Small rock and mineral fragments are called **clasts.** The word *clast* is from the Greek *klastos* which means "broken." Sedimentary rocks can form from the consolidation of clasts, the crystallization of minerals from saturated solutions, or the remains of living organisms.

Weathering What could reduce rocks into clasts? Inside Earth, rocks are exposed to high temperatures and pressures that protect them from breaking down. In contrast, rocks on Earth's surface are exposed to water, wind, and other forces. These forces break down rock into smaller pieces through a process called weathering. Over time, these pieces might be smashed by waves, polished in rivers and streams, or dissolved in the ocean. The Colorado River in **Figure 16** is weathering rock as it carves down through the Grand Canyon in Arizona.

■ **Figure 16** Rocks that tumble along river beds can be broken into smaller clasts. The Colorado River winds for 2,300 km carving canyons and breaking down rocks into clasts.

Transportation and deposition Mechanical weathering can occur when rocks collide and break into smaller clasts. Gravity, wind, water, and ice can carry these clasts to new locations. As the clasts are transported, they grind against each other often forming well-rounded grains.

Clasts can pile up when they are deposited. When clasts pile up, there are spaces between the individual clasts. The empty space between clasts is called **pore space.** All clasts, whether irregular or round, have pore space. Water, oil, and natural gas found beneath Earth's surface can be stored in these pore spaces.

Compaction and cementation Clasts can form sedimentary rocks, such as the sandstone shown in **Figure 17.** Following transportation and deposition, clasts can be buried in low-lying areas. As more material is deposited, clasts are buried even deeper. The weight of the overlying material forces the clasts together, reducing the pore space.

The process by which clasts stick together due to the weight of overlying material is called **compaction.** In contrast to the formation of clasts, which occurs on Earth's surface, compaction occurs beneath Earth's surface. Water moving between clasts carries dissolved minerals that can act as cement. Common minerals that cement clasts together include quartz, calcite, hematite, and clay minerals. Minerals precipitate slowly out of a water solution and crystallize in the spaces between clasts in a process called **cementation.** **Figure 18** shows sedimentary rock formed by compaction and cementation.

■ **Figure 17** Sandstone forms when sand grains are deposited, compacted, and cemented together.

■ **Figure 18** The weight of overlying material can force sediment grains together in a process called compaction. Minerals like quartz and calcite can then cement the grains together.

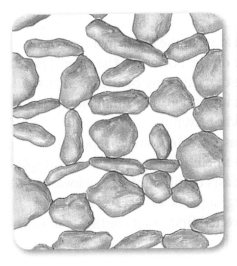

Essential Questions

▶ What are the physical conditions that cause metamorphism?

▶ Where does metamorphism occur?

▶ How are metamorphic rocks classified?

▶ What is the rock cycle?

Review Vocabulary

chemical reaction: process in which one or more substances are changed into new substances

New Vocabulary

foliated
rock cycle

g Multilingual eGlossary

Metamorphic Rocks and the Rock Cycle

MAIN ⟨**Idea**⟩ **Metamorphic rocks form when existing rocks are changed by heat, pressure, or chemical reactions.**

Real-World Reading Link It is time to leave for school. You toss a ham and cheese sandwich into your book bag and off you go. At lunchtime, you pull out your sandwich only to discover it has been crushed and the cheese has melted. What caused this change?

Metamorphic Rocks

Igneous and sedimentary rocks can be classified based on composition and texture. The same is true for metamorphic rocks. Metamorphic rocks, like the banded gneiss (NICE) shown in **Figure 22,** are rocks that have been changed by any combination of heat, pressure, and chemical reactions. As these agents begin to act on a rock, atoms within the rock rearrange and sometimes form new minerals. The word metamorphic is derived from *meta* (to change) and *morph* (form). Any igneous, sedimentary, or metamorphic rock can be changed through the process of metamorphism.

Regional and contact metamorphism Why do rocks experience changes in temperature, pressure, and chemical reactions? Regional movements of Earth's tectonic plates can cause rocks to be buried, resulting in an increase in temperature and pressure. Also, enormous forces are exerted on rocks when plates move, causing an increase in pressure. Fluids can also seep into rock and cause chemical reactions.

■ **Figure 22** Sharp folds in rock sometimes reflect intense changes in temperature and pressure that occur during large-scale regional metamorphism.

Metamorphic changes in rocks that occur over large areas are called regional metamorphism. Metamorphic changes can also occur over small areas. For example, when magma intrudes into rock, metamorphism can occur in areas that are in contact with the magma. These localized metamorphic changes are called contact metamorphism.

Composition Metamorphism is caused by changes in temperature, pressure, and chemical reactions. The presence of water within rock enables chemical reactions to occur. Some minerals in rock contain water molecules as part of their chemical composition. When changes in temperature and pressure occur during metamorphism, water can escape from these minerals. In other instances, water can flow into cracks or pores in rock and cause atoms in minerals to rearrange and change composition.

Reading Check **Identify** What are three agents of metamorphism?

Mineral composition Clay minerals, micas, and amphiboles are examples of minerals that contain water in their crystal structures. Clay minerals tend to lose water and form micas when metamorphism occurs. If the pressure and temperature within the rock continue to increase, the water can be removed from the rock completely. Minerals, such as garnet, that contain some of the same elements found in clays (aluminum, silicon) result. In this way, some new minerals form by dehydration, or loss of water, at high temperature and pressure.

Textures Metamorphic processes produce rocks with different textures. Recall that texture describes the size, shape, and arrangement of the crystals or grains in a rock. Some metamorphic rocks have foliated textures, as shown in **Figure 23**. **Foliated** rocks have crystals that are arranged in layers and bands. Foliated textures form under high pressure conditions. Metamorphic textures can also be nonfoliated, as shown in **Figure 23**, where crystals have a more random orientation.

(FOLDABLES)
Incorporate information from this section into your Foldable.

VOCABULARY ....
WORD ORIGIN
Foliation
  Comes from the Latin word *folio* which means "leaf"
  Foliated texture has the appearance of layered pages of a book. ....

Foliated Texture

Nonfoliated Texture

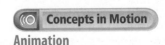
(((O)) **Concepts in Motion**)
Animation

■ **Figure 23** Foliated metamorphic rocks like gneiss have a layered texture. Nonfoliated rocks like quartzite are crystalline and uniform in color.

Mineral grains are randomly oriented when no direct force is involved.

Orientation of mineral grains is perpendicular to the direction of pressure caused by compression.

Mineral grains are parallel to the direction of shearing force. The rock has a foliated texture.

■ **Figure 24** The direction of applied force affects the orientation of mineral grains.

Describe *the texture that results after compression has occurred.*

Foliated rocks Shale and siltstone are common sedimentary rocks in Earth's crust. When metamorphism occurs, the clay minerals in these rocks change to mica minerals. As metamorphism continues, the grain size of the mica minerals increases.

When squeezed and heated, mica minerals align perpendicular to the direction of compression or parallel to the direction of shearing, as shown in **Figure 24.** For example, slate is a fine-grained, foliated metamorphic rock that splits easily along flat planes. Slate forms when shale undergoes metamorphism.

It is easiest to describe foliated metamorphic rocks based on the degree of metamorphism shown in **Table 6.** Slate is a fine-grained, foliated metamorphic rock that forms layers and exhibits rock cleavage. Rock cleavage results from the alignment of mineral grains in the rock. If the degree of metamorphism increases, some of the mineral grains in the slate grow and the rock phyllite forms. With continued grain growth, a rock with parallel mica sheets called schist (SHIHST) forms. Finally, at the highest degree of metamorphism, a banded black and white, coarse-grained, foliated rock called gneiss forms.

((○)) Concepts in Motion

Interactive Table

Table 6	Common Metamorphic Rocks								
Texture			**Composition**						**Rock Name**
Foliated	Layered		CHLORITE	MICA	QUARTZ				SLATE
		Fine-grained				FELDSPAR	AMPHIBOLE		PHYLLITE
		Coarse-grained							SCHIST
	Banded	Coarse-grained						PYROXENE	GNEISS
Nonfoliated		Fine- to coarse-grained	Quartz						QUARTZITE
			Calcite or dolomite						MARBLE

Nonfoliated rocks Similar in texture to intrusive igneous rocks, nonfoliated metamorphic rocks tend to have random crystal orientation. In general, crystal size tends to increase as the metamorphism increases. Unlike intrusive igneous rocks, nonfoliated metamorphic rocks are uniform in color. Picture a limestone rock. Under the influence of temperature, pressure, and in the presence of chemical fluids, this rock recrystallizes to form white, gray, or pink marble with coarse calcite crystals.

Classification Metamorphic rocks can be classified based on texture. Recall that metamorphic rocks can be foliated or nonfoliated. If they are foliated, they can be classified as slate-like, schistlike or gneisslike. Metamorphic rocks can also be classified according to their mineral composition. For example, a nonfoliated, quartz-rich metamorphic rock is called a quartz-ite. Marble is a nonfoliated metamorphic rock made of calcite. A foliated, schistlike rock made of garnet and mica is a garnet-mica schist.

The Rock Cycle

Rocks above and below Earth's surface are continually being changed into other types of rocks, as depicted in **Figure 25.** For example, high temperatures and pressures can change igneous and sedimentary rocks into metamorphic rocks. Sedimentary and metamorphic rocks can melt to form magma, which cools and crystallizes to form igneous rocks. Weathering, transportation, deposition, compaction, and cementation can change igneous and metamorphic rocks into sedimentary rocks. A rock can even be changed into a different rock of the same type. The continual changing of rocks into different types is called the **rock cycle.** An example of the rock cycle is shown in **Figure 26.**

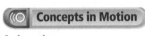
Concepts in Motion

Animation

■ **Figure 25** Rocks change from one type into another through a series of internal and external processes called the rock cycle.

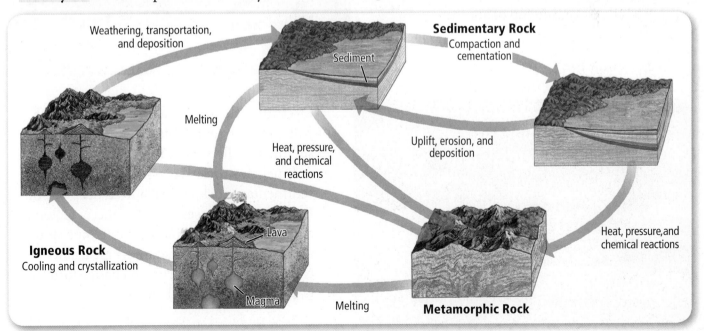

FIGURE 26

Visualizing the Rock Cycle

Rocks are continually changing form from one type to another through a series of processes that are part of the rock cycle.

When a volcano explosively erupts, the ash eventually cools and crystallizes to form igneous rock. Following an eruption, igneous rock exposed on the flanks of the steep-sided volcano will wear away. Water and other weathering agents such as snow and ice will transport these sediments downstream and into the valley below.

Some of these sediments are carried downslope by streams. As the stream flows toward the ocean, eventually these sediments are deposited. Time passes and sediments continue to pile up one on top of another.

Over time, the weight of overlying deposits compacts the sediments. Eventually these sediments will cement together to form a sedimentary rock. It is possible that the sedimentary rock will be buried deep within Earth as time passes or experience a tectonic event that causes an increase in temperature and pressure. These metamorphic agents cause the rock to once again change form to become a metamorphic rock.

Conservation of matter As rocks travel through the various phases of the rock cycle, the law of conservation of matter is upheld—matter is neither created nor destroyed. For example, minerals from granite are found in the clasts that form when those igneous rocks weather. These same minerals can become part of a clastic sedimentary rock such as a shale, a sandstone, or a conglomerate. When the minerals in rocks dissolve in water, the atoms and ions are not destroyed. These particles are instead distributed throughout the solution. The atoms and ions can then become part of the minerals that settle out of the solution when chemical sedimentary rocks form. The atoms and ions in any rock are never destroyed. They can be rearranged to form new Earth materials.

Figure 27 shows a few of the different ways rocks can be converted from one type to another. The igneous rocks in this image are weathering into clasts that will likely become sedimentary rock. Rocks can travel through the rock cycle and change forms many times throughout Earth's history, helping to tell a story about Earth's past.

■ **Figure 27** Water is eroding this igneous rock into sediments that will eventually be compacted and cemented together to form sedimentary rock.

Section 4 Review

Section Summary

▶ Metamorphic rocks can form from any pre-existing rocks.

▶ Metamorphic rocks are foliated or nonfoliated.

▶ Metamorphic rocks are classified based on composition and texture.

▶ Rocks are subject to many changes through a series of processes called the rock cycle.

23. **MAIN Idea Explain** the conditions necessary for metamorphism to occur.

24. **Compare and contrast** foliated and nonfoliated textures.

25. **Predict** what metamorphic rocks will form when the sedimentary rocks limestone and shale are exposed to heat and pressure.

26. **Identify** the metamorphic rock in question 25 that is foliated.

27. **Think Critically** A marble contains occasional layers of the mineral muscovite mica. What possible parent rock composition could have been metamorphosed to produce this rock?

Apply Math

28. **Calculate Mass** Suppose a sculptor wants to use marble with an average density of about 2.7 g/cm³. The block of marble is one meter on every side. What is the mass of the block? If the sculptor knows from experience that the finished product will only use 45 percent of the block and the rest will be waste, then what will the mass of the statue be?

LAB

Rock Identification

Objectives

- **Observe** physical properties of unknown rock samples.
- **Classify** each rock according to general type: igneous, sedimentary, or metamorphic.
- **Identify** each rock based on its composition and texture.

Preparation

Materials

rock samples
nail
penny (pre-1982)
magnifying lens or particle size chart
dilute HCl
paper towels
glass plate
unglazed porcelain tile
Resource Handbook, "Rocks"

Safety Precautions

Background: You've learned to observe and identify objects all around you. As a child, you learned by repetition and by constantly asking what things are called. By the time you were five years old, you could name many different objects. You probably haven't had the same experience with rock identification—at least not yet. In this lab, you will identify rocks using a more formal approach.

Question: *How can you determine the composition and texture of common rocks? What can a rock's composition and texture indicate to you about how it was formed?*

Procedure

1. Read the procedure and safety information, and complete the lab form.
2. Construct a horizontal data table with ten columns and four rows.

3. Label columns 1 through 10 across the top row as follows:

- Sample Number
- General Rock Type (igneous, sedimentary, metamorphic)
- Crystals (yes/no)
- Grain Size (large/small/none)
- Gas-Bubble Holes (yes/no)
- Fossils (yes/no)
- Particle Size (gravel/sand/silt/clay)
- Foliations (yes/no)
- Additional Observations
- Rock Name

4. Obtain eight rock samples from your teacher.

5. Observe physical and chemical characteristics of each rock. First, try to determine if the rock is igneous, sedimentary, or metamorphic. Then record all other characteristics you observe for each sample. Your data table columns will serve as a checklist to help organize these observations.

WARNING: *If you choose to use the acid test for calcite identification, use only one drop of HCl. A positive reaction to acid is the presence of tiny bubbles rising through the drop. Blot the liquid off the sample with a paper towel before leaving it. Be careful to keep acid away from skin and clothing.*

6. Identify the rock with the help of Resource Handbook, "Rocks," at the back of your textbook.

7. Confirm with your teacher that you have correctly identified each rock sample.

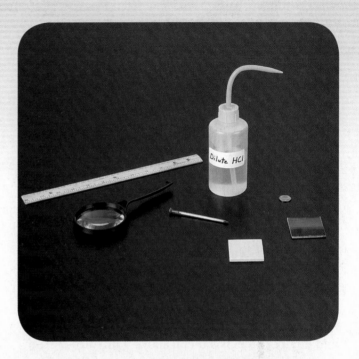

Analyze Your Data

1. **Identify** the physical characteristics that helped you distinguish between igneous and sedimentary rocks and between sedimentary and metamorphic rocks.

2. **Determine** Which rock samples required the fewest tests to identify? Why do you think that this is the case?

Conclude and Apply

1. **Describe** the steps that you followed to identify each rock type.

2. **Explain** which observations were useful for the largest number of rock samples.

3. **Evaluate** your data table in terms of including important observations. Explain whether there were any observations you left out of your data table that are critical for rock identification.

COMMUNICATE YOUR DATA

Present your findings to the class. Discuss any observations that were different from those of your classmates. Discuss the column headings you included in your data table and defend your choices.

Car Parts

Consider a compact car straight off the assembly line. Where did its many parts come from? If you trace each part back to its origins, you will eventually hit solid ground. A typical car is composed of no less than 30 minerals and many petroleum-based products, each of which was mined from Earth.

Iron and steel The mass of an average car, like any of those shown in **Figure 1,** is approximately 1,300 kg. Over 60 percent of the car's mass is iron and steel, which compose parts of the car's frame and engine. Steel is a metal alloy—a mixture of iron and trace amounts of other elements, such as tungsten and manganese. To produce the iron and steel for the car, 3,500 kg of raw ore is extracted from the mineral hematite. The average car also contains almost 130 kg of aluminum, most of which comes from recycled materials and the mineral bauxite.

Lead Galena is a lead sulfide mineral. Despite being composed of 86 percent lead, processing galena ore to remove lead is not efficient. It takes 210 kg of galena ore to make the 27 kg battery in a car. To process galena, the mineral must be separated and extracted from the surrounding rock, heated to isolate the lead from the sulfur, and then purified and poured into molds.

Precious metals Gold, copper, platinum, and nickel are also essential ingredients in many car parts. Gold was valued in prehistoric times because it does not rust, react with other elements, or break very easily. These characteristics make it invaluable for the electric systems in a car, such as the wiring of air bags, which protect you in an accident. In addition to gold, 20 kg of copper is used for wiring in the car, and 4 kg of nickel is used for plating the stainless steel on the car.

Figure 1 An average car is composed of over 30 different minerals and petroleum products.

Quartz Glass windshields are made from about 75 kg of industrial quartz, usually in the form of white sand. To make the glass, clean, white sand and additives, such as calcite, are heated to 1,675°C, which is hot enough to fuse quartz crystals together. The front windshield is a double paned layer of glass with a layer of vinyl in the middle, and it keeps you safe in an accident. The side windows are tempered glass, which is rapidly cooled to break into tiny pieces instead of deadly shards in an accident.

Petroleum products Three and a half barrels of petroleum-based products are used to make a car. Lightweight components of petroleum are extracted from crude oil and used to produce the plastics for the car body and the interior panels. Heavier components are used in rubber tires, paint ingredients, and synthetic fabric. Of course, some is also consumed as electricity for the assembly line that helped to build the car. A very small amount is also used to fill the gas tank and lubricate the engine.

🌐 **WebQuest**

Research Identify rock and mineral resources and petroleum-based products used in home construction. Prepare a poster or presentation with an illustration identifying these resources.

Chapter

THEME FOCUS
Rocks are made of minerals, which can be used to interpret Earth's geologic past and as important natural resources.

BIG Idea Internal and external processes on Earth continually change rocks from one type to another.

Section 1 Minerals

cleavage (p. 811)
fracture (p. 811)
hardness (p. 811)
magma (p. 813)
mineral (p. 809)
streak (p. 810)

MAIN Idea A mineral is a naturally occurring, inorganic solid with a crystalline form.

- Few elements form the bulk of minerals in Earth's crust.
- Minerals can be classified based on their physical and chemical properties.
- Minerals can be sorted into two broad categories: silicates and nonsilicates.
- Minerals are valuable resources used every day.

Section 2 Igneous Rocks

extrusive igneous rock (p. 820)
intrusive igneous rock (p. 817)
rock (p. 817)

MAIN Idea Igneous rocks form when molten material cools and crystallizes.

- Rocks are mixtures of various natural materials.
- Extrusive and intrusive igneous rocks form in different environments.
- There are three common types of magma compositions—felsic, intermediate, and mafic.
- Extrusive and intrusive igneous rocks formed from the same type of magma have similar compositions but different textures.
- Dissolved gases in magma can cause explosive volcanic eruptions.

Section 3 Sedimentary Rocks

cementation (p. 825)
clast (p. 824)
compaction (p. 825)
pore space (p. 825)

MAIN Idea Sedimentary rocks form from consolidated rock particles or crystallization from a solution.

- Sedimentary rocks are classified into three types: detrital, chemical, and biochemical.
- Detrital sedimentary rocks can be composed of a wide variety of different-sized clasts.
- Chemical sedimentary rock forms as minerals settle out from a solution rich in dissolved mineral matter.
- Biochemical sedimentary rocks form from the remains of organisms.

Section 4 Metamorphic Rocks and the Rock Cycle

foliated (p. 831)
rock cycle (p. 833)

MAIN Idea Metamorphic rocks form when existing rocks are changed by heat, pressure, or chemical reactions.

- Metamorphic rocks can form from any preexisting rocks.
- Metamorphic rocks are foliated or nonfoliated.
- Metamorphic rocks are classified based on composition and texture.
- Rocks are subject to many changes through a series of processes called the rock cycle.

Use Vocabulary

Match each phrase with the correct term from the Study Guide.

29. molten rock

30. fine-grained igneous rock

31. coarse-grained igneous rock

32. scratch resistance

33. uneven break

34. a mixture of consolidated minerals

35. rock or mineral fragment

36. layered texture in metamorphic rock

37. continual change of one rock type into another

38. empty space between clasts in a rock

39. a break along planes of weakness

40. natural, inorganic, crystalline solid

Check Concepts

41. What is the most common group of minerals?
  A) carbonates   **C)** sulfates
  B) silicates   **D)** oxides

42. What is a factor that controls grain size in igneous rocks?
  A) porosity   **C)** cooling rate
  B) cleavage   **D)** temperature

43. Which do all the silicate minerals have?
  A) the same hardness
  B) the same crystal structure
  C) silicon and oxygen
  D) silicon and carbon

44. Through which process does chemical sedimentary rock usually form?
  A) cementation of sand
  B) compaction of clasts
  C) evaporation of magma
  D) saturation of a solution

45. Which of these features are extrusive igneous rocks least likely to have?
  A) fossils
  B) small grains
  C) vesicular texture
  D) porphyritic texture

Use the figure below to answer question 46.

46. Which of these descriptions best describes the coarse-grained igneous rock shown above?
  A) extrusive and basaltic
  B) extrusive and granitic
  C) intrusive and basaltic
  D) intrusive and granitic

47. Which of the following is NOT an agent of metamorphism?
  A) lithification
  B) chemical reactions
  C) heat
  D) pressure

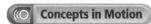 **Concepts in Motion** **Interactive Concept Map**

Interpret Graphics

48. **BIG Idea** Copy and complete the concept map on rocks.

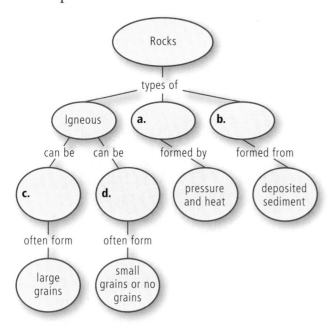

49. Copy the following table on a separate sheet of paper. Use Bowen's reaction series to fill in the blanks.

Crystallization Temperatures			
Temperature Range	Mineral 1	Mineral 2	Rock Name
Low	mica	a.	granite or rhyolite
Intermediate/high	pyroxene	calcium plagioclase	b.

Think Critically

50. **Explain** how cooling rate affects the texture of an igneous rock.

51. **List** the characteristics that help you identify sedimentary rocks.

52. **THEME FOCUS** **Determine** the identity of each rock from the following descriptions.
  a. quartz-rich, coarse interlocking grains, with noticeable amounts of potassium feldspar
  b. foliated mass of mica group minerals with some garnet
  c. nearly pure mass of fossil shells cemented by calcite
  d. dense black, fine-grained, with vesicular texture
  e. light colored, very low density, frothy (vesicular), glassy

Use the figure below to answer question 53.

53. **Describe** the types of forces applied and their relative directions which combined to form this foliated metamorphic rock.

Apply Math

54. **Calculate** If a miner were extracting rock that contained 0.015 percent gold by mass, how many kilograms of rock must be processed to obtain one kilogram of gold?

55. **Evaluate** The body of ore described in question 54 is estimated to contain 1.5×10^5 m^3 of ore and have an average density of 6.0 g/cm^3. How many kilograms of gold could potentially come from this body of ore?

Standardized Test Practice

Multiple Choice

*Record your answers on the answer sheet
provided by your teacher or on a sheet of paper.*

1. Which is the most common element by mass
   in Earth's crust?
   A. aluminum
   B. oxygen
   C. potassium
   D. silicon

Use the figure below to answer question 2.

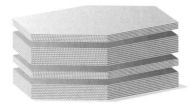

2. Which property of a mineral, used in mineral
   identification, is shown above?
   A. cleavage
   B. fracture
   C. hardness
   D. luster

3. Which property is a measure of a mineral's
   resistance to scratching?
   A. cleavage
   B. hardness
   C. luster
   D. streak

4. Why is graphite used in some pencils?
   A. Its hardness is high enough to leave a
      scratch on paper.
   B. Its hardness is low enough to leave a streak
      on paper.
   C. Its metallic luster gives it a dark color.
   D. Its metallic luster makes it shine on paper.

5. Which mineral is very important to the
   production of machinery for manufacturing
   goods?
   A. feldspar
   B. halite
   C. hematite
   D. quartz

Use the figure below to answer question 6.

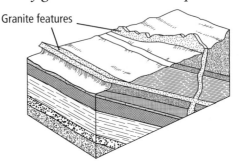

Granite features

6. The formation of which type of rock pro-
   duced the granite features shown above?
   A. foliated metamorphic rock
   B. extrusive igneous rock
   C. intrusive igneous rock
   D. nonfoliated metamorphic rock

7. Which is a dark-colored igneous rock with
   small grains?
   A. basalt   C. granite
   B. gabbro   D. rhyolite

8. What affect does cooling rate have on the
   grain size in intrusive igneous rocks?
   A. It forms fine-grained crystals.
   B. It forms coarse-grained crystals.
   C. It forms dark-colored crystals.
   D. It forms light-colored crystals.

9. Which is a coarse grained detrital sedimen-
   tary rock with well-rounded grains?
   A. sandstone   C. conglomerate
   B. shale   D. breccia

Short Response

Record your answers on the answer sheet provided by your teacher or on a sheet of paper.

10. If the percent by mass of silicon in Earth's crust is 27.7 percent and of iron is 5.0 percent, how much more silicon is present than iron?

Use the graph below to answer question 11.

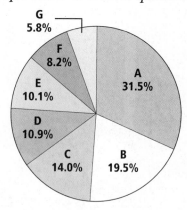

Abundance of Elements in Earth's Crust
(No Oxygen or Silicon)

11. The percent by mass of elements in Earth's crust are: 46.6% oxygen, 27.7% silicon, 8.1% aluminum, 5.0% iron, 3.6% calcium, 2.8% sodium, 2.6% potassium, 2.1% magnesium, and 1.5% all others. With oxygen and silicon not considered, which wedge represents the abundance of calcium?

12. How can studying sedimentary rocks help geologists interpret Earth's history?

13. What process does Bowen's reaction series illustrate? Explain.

Extended Response

Record your answers on a sheet of paper.

14. How does coal form? Describe the initial material, deposition environment, and geological conditions that lead to its formation.

15. Describe the rock cycle. Discuss what can happen to rock material within the cycle and the amount of time it might take for rocks to be transformed.

Use the illustration below to answer question 16.

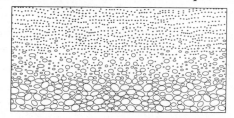

16. Describe how the sediment particles are arranged in the illustration above.

17. Describe the type of environment in which this arrangement of sediment particles might occur.

18. Generally, rocks consist of minerals. When molten rock cools down rapidly, it becomes a glass. Volcanic glass is an extrusive igneous rock. Assess whether this kind of rock contains minerals. Then explain your answer.

NEED EXTRA HELP?																		
If You Missed Question . . .	1	2	3	4	5	6	7	8	9	10	11	12	13	14	15	16	17	18
Review Section . . .	1	1	1	1	1	2	2	2	3	1	1	3	2	3	4	3	3	1,2

CHAPTER 27

Earth's Changing Surface

ConnectED

Your one-stop online resource
connectED.mcgraw-hill.com

- Video
- Audio
- Review
- Inquiry

- WebQuest
- Assessment
- Concepts in Motion
- Multilingual eGlossary

Launch Lab
Underground Water

Beneath your feet, there are vast amounts of water. Different types of rock absorb and store water in various amounts. For instance, sand soaks up water more easily than clay does. In this activity, you will model how water is stored by different rocks underground.

For a lab worksheet, use your StudentWorks™ Plus Online.

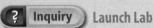
? Inquiry Launch Lab

FOLDABLES®

Surface Features
Make the following folded table to help you organize features that result from erosion and deposition on Earth's surface.

Surface-Changing Agent	Erosion	Deposition
Wind		
Water		
Glaciers		

THEME FOCUS Earth Materials and
Processes
**Surface processes, such as erosion and
deposition, can provide clues to interpret Earth's
geologic past.**

(BIG(Idea)) Earth's surface changes over time due to
weathering, erosion, and deposition.

Reading Preview

Essential Questions

▶ What is the role of weathering in recycling Earth materials?

▶ What are mechanical weathering and chemical weathering?

▶ What natural processes contribute to soil formation?

▶ What methods help to control soil loss?

Review Vocabulary

sediment: particles of rock deposited on Earth's surface that may become sedimentary rocks

New Vocabulary

weathering
soil

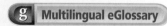

g **Multilingual eGlossary**

? **Inquiry**   Virtual Lab

□ **Video**   BrainPOP

Weathering and Soil

MAIN ◀Idea Weathering breaks down rocks on Earth's surface.

Real-World Reading Link Have you ever noticed the mud puddles that collect on sidewalks and streets after a powerful storm? The water might have drained down a hill or might have bubbled up from the saturated soil below. Water carries this mud and deposits it in low-lying areas.

Weathering

Rocks and other materials on Earth's surface are constantly changing. Chemical, physical, and biological processes act on these materials every day. These processes can cause physical and chemical changes. Physical changes include changes in the size or shape of a material. A chemical change occurs when the chemical composition of a material changes.

Rocks that form beneath Earth's surface break apart when exposed to changing surface conditions. **Weathering** is the process that involves the physical or chemical breakdown of materials on Earth's surface. Factors that influence weathering include the agent (such as water or air), the composition of the material being weathered, the climate, and time. Weathering rates vary from one area to another due to different surface conditions and climates.

Figure 1 shows granite that has been weathered. The minerals in this rock formed under high temperature and pressure conditions compared to rocks on the surface. As these minerals are exposed to water, air, or other agents, they break down.

Everyday weathering You can observe weathering processes in action every day. Paint on a house can crack or fade in color. Roadways develop cracks and potholes. Some metal items on cars can start to rust. Each of these events take place at different rates, depending on the climate where you live.

■ **Figure 1** The corners and edges of rocks, such as this granite, are more rapidly weathered than smooth rock faces. This often produces rounded forms like the granite shown here. This type of weathering is commonly called spheroidal weathering because of the spherelike shapes that result.

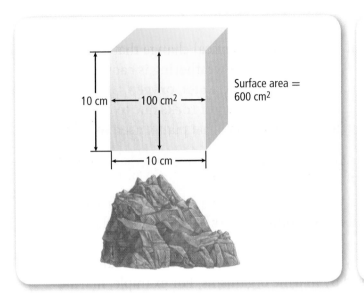
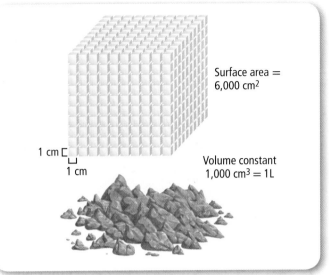

■ **Figure 2** Compared to the cube on the left, the cube on the right has a greater surface area due to its exposed inner surfaces.

Determine *which rock will weather at a faster rate.*

Mechanical Weathering

Mechanical weathering occurs when a physical change takes place. During mechanical weathering, rocks are broken down into smaller pieces without changing their chemical composition. Physical forces exerted on rocks can cause mechanical weathering. These forces result from collisions when rocks fall from a cliff or tumble through turbulent river rapids. Other processes, such as frost wedging and plant and animal activity, exert forces that break rocks down into smaller pieces.

As **Figure 2** shows, when a large particle breaks into smaller particles, the total surface area of all the particles increases. An increase in surface area causes the material to become even more susceptible to weathering.

Frost wedging Rocks on Earth's surface can be broken apart by frost wedging. Frost wedging occurs when water collects in the cracks of a rock and then freezes. When the water freezes, it expands and forces the crack apart. The water might melt and then refreeze, which forces the crack apart even further. After a number of freeze-thaw cycles, the crack can become large enough to permanently break the rock apart.

Biological activity Plant and animal activity can cause mechanical weathering. Water and nutrients seep into cracks in a rock. A plant can then root itself in these cracks. As the plant root grows, it expands and exerts a force that might eventually break the rock apart. Burrowing animals can also increase the size of a crack within a rock. As animals travel through or on top of rock and soil, they might loosen materials, resulting in additional weathering.

✓ **Reading Check** **Identify** What are three causes of mechanical weathering?

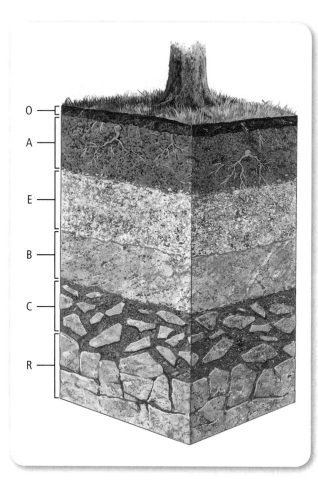

■ **Figure 5** This illustration provides a generalized view of a soil profile. There are six major soil horizons (O, A, E, B, C, and R) that vary in color and thickness. Some horizons do not appear in certain soils.

Explain *why the O and A horizons are darker in color.*

Concepts in Motion Animation

VOCABULARY ....

ACADEMIC VOCABULARY

Transport
  to move from one place to another
  Trains transport passengers from one station to the next. ....

Soil

One natural resource that is vital to life on Earth is soil. **Soil** is a mixture of weathered rock, organic matter, water, and air that is capable of supporting plant life. The organic matter in soil comes from plants, animals, and other organisms. Soil forms as the bedrock beneath it begins to break down due to mechanical or chemical weathering and is transported elsewhere.

✓ **Reading Check Describe** What are the primary components in soil?

Soil horizons If you dig a hole deep into the ground, you will see different soil layers called horizons. Each horizon has its own unique thickness, texture, color, and composition. **Figure 5** illustrates a collection of layered soil horizons called a soil profile.

Soils do not always contain every horizon. The horizons present depend on bedrock composition, climate, and the amount of organic matter available. Also, the shape of the surrounding landscape in an area affects the type of soil that forms there. On steeper surfaces, thin layers of soil with less organic matter are likely to develop. In flat, low-lying areas, thick soils with more organic matter can develop.

Figure 5 shows a complete set of soil horizons. From the surface to the bedrock, these horizons are O, A, E, B, C, and R. The O, for organic, and the A horizons are often referred to as topsoil. Plants absorb nutrients from the topsoil.

The E horizon, which lies directly below the topsoil, is a zone in which finer sediments and more soluble materials are transported deeper into the soil through a process called leaching. This is much like a coffeemaker that extracts the flavor from ground coffee. The B horizon collects materials from above and is usually darker than the E horizon. Together, the E and B horizon are called the subsoil. Roots from large trees sometimes anchor themselves and absorb water from the subsoil.

Collectively, horizons O, A, E, and B make up the true soil. The C horizon is partially weathered bedrock. The R horizon is unweathered bedrock.

| Arctic |
| Mountain |
| Desert |
| Prairie |
| Glacial |
| Wetlands |
| River |
| Temperate |
| Tropical |

45°N
40°N
35°N
30°N
60°N
20°N

Topsoil types

There are many different types of topsoil. Topsoils are classified based on climate, as shown in **Figure 6.** Topsoils also differ based on chemical and physical properties. Climate conditions, together with the rock type and the shape of the surrounding landscape, affect the type of topsoil that forms in an area.

Vegetation, if present, contributes important organic matter to topsoil. For example, forest topsoils of the eastern U.S. have a thin layer of organic matter, which affects soil acidity. Forest topsoils also have a sandy texture. In contrast, grassland topsoils of the western U.S. have thick topsoils because of long, well-rooted grasses. Grassland topsoils are enriched in calcite and white in color. In tropical areas, heavy rainfall washes away organic matter and leaves behind very thin, nutrient-poor topsoil. Tropical topsoil is rich in insoluble aluminum, iron oxides, and clay minerals. Finally, in the arctic or desert regions, where little or no vegetation grows, the topsoil is either thin or absent. Without organic matter, true topsoil cannot develop.

Parent material

The materials from which soil forms are called parent materials. Parent materials can be transported from distant sources. For example, rivers and glaciers help to transport rock and soil from one area to another. Some of the most fertile topsoils form when rivers flood and deposit new sediment on floodplains or when glaciers transport and deposit sediment. For thousands of years, the ancient Egyptians relied on the annual flooding of the Nile River to deposit nutrient-rich sediment. During the most recent ice advance, glaciers in North America transported materials that would later become some of the world's most fertile topsoils.

■ **Figure 6** Glacial topsoil in the northeastern U.S. is often composed of poorly sorted sediments deposited at the end of the most recent Ice Age. Heavy rainfall in the southeastern U.S. contributes to the formation of nutrient-rich wetland topsoils.

Describe *the topsoil type in the southwestern United States.*

Soil Conservation

Plants and animals remove the nutrients that they need from soil. Soil depletion is a serious agricultural problem in many areas of the world. For example, when the same crop is grown year after year in the same area, certain nutrients are used up. Soils in areas with heavy rainfall, such as rainforests, also tend to be low in certain nutrients. The recycling of organic matter in a rainforest might eventually return these compounds to the soil. Not for long, however—plants and animals and rain in the rainforest continue to strip away these nutrients from the soil.

Nutrients Farmers often add fertilizers containing nitrogen, phosphorous, and potassium to soil to supplement and replace nutrients. Farmers also use crop rotation to help preserve soil quality. This is because what one crop removes from the soil, another crop may return. Crop rotation also helps prevent erosion and reduce the risk of disease and invasion by unwanted pests. Another alternative is to allow the soil to go fallow, which means to rest, by not planting crops. This gives the soil time to replenish some of its nutrients.

Erosion prevention Most soil erosion occurs because the plant cover has been removed or because the land is steep. Farmers use several techniques to help prevent soil erosion. For example, contour plowing, as shown in **Figure 7,** reduces erosion by plowing furrows around a hill instead of up and down a slope. In the tie-ridging method, crops are planted on ridges and rainfall collects in basins.

■ **Figure 7** Farmers plow with the contour of the landscape. This makes the landscape less steep, and therefore, less prone to soil erosion.

Section 1 Review

Section Summary

▶ Weathering involves the breakdown of rock by physical or chemical processes.

▶ Mechanical weathering occurs when rocks break down physically due to collisions, frost wedging, and plant and animal activity.

▶ Chemical weathering occurs when rocks or minerals change composition.

▶ Soil is a mixture of organic matter, weathered rock material, water, and air.

1. **MAIN** ⟨**Idea**⟩ **Identify** examples of mechanical weathering and chemical weathering.

2. **Describe** the factors that affect soil formation.

3. **Explain** How does weathering recycle Earth materials?

4. **Think Critically** Outline the soil conservation methods that you might expect from any contractor building new roads in your community.

Apply Math

5. **Estimate** Suppose that soil erosion in your area averages 2.65 cm per year. The average soil profile is 3 m thick, and 45 percent of that is topsoil. Estimate how long it will take for the topsoil to erode.

Assessment   Online Quiz

FIGURE 22

Visualizing The Ogallala Aquifer

The Ogallala Aquifer, also known as the Great Plains Aquifer, is one of the world's largest aquifers. It supplies freshwater resources to eight states in the central United States, including South Dakota, Nebraska, Wyoming, Colorado, Kansas, Oklahoma, and New Mexico. The Ogallala Aquifer supplies water for residential, industrial, and agricultural use. The aquifer is used to irrigate about one-fifth of the annual agricultural output of the United States.

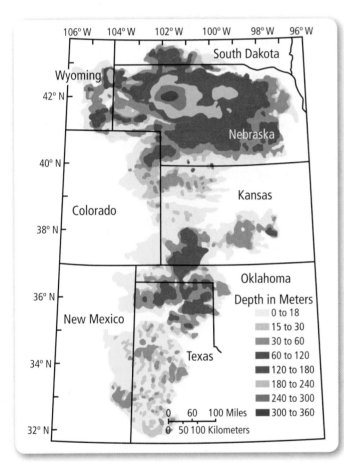

Based on studies of groundwater removal, it is possible that the Ogallala Aquifer will dry up in the next several decades. Freshwater resources have been withdrawn from the aquifer faster than they can be replenished. The depth of the Ogallala Aquifer and its subsurface area are shown here, with the largest freshwater reserve in Nebraska. The United States Department of Agriculture (USDA) in collaboration with Agricultural Resource Service and a collection of universities have funded a research initiative called The Ogallala Initiative.

The initiative was designed to help protect and preserve freshwater resources in the Great Plains. Through the initiative, scientists, engineers, farmers, and ranchers are committed to water conservation efforts. These efforts include the design and use of more efficient irrigation technologies, improved irrigation and precipitation management, studies of hydrology and climatology, and more effective crop and livestock maintenance.

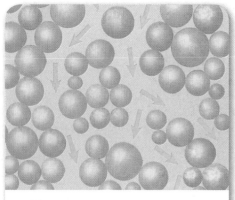

Figure 23 Fluids pass through interconnected pores but not through isolated pores.

Porosity Infiltration occurs as water passes through the pore spaces in sediments and rock, as shown in **Figure 23**. The **porosity** of a material is the percentage of the material's total volume that is pore space. Porosity depends on the variety and size of sediments in rock. Porosity can be calculated by dividing the total volume of all of the pore space by the total volume of the material and then multiplying by 100. In most materials, some pore spaces are isolated and some pore spaces are interconnected. Water passes through the interconnected pore spaces in a material. The more interconnected the pore spaces are, the more quickly water is able to pass through that material. A material that is highly permeable has a high porosity and contains many interconnected pore spaces. A material that has low porosity and lacks interconnected pore spaces prevents water from easily passing through the material.

 Reading Check **Explain** how porosity is calculated.

EXAMPLE Problem 1

Determine Aquifer Storage Suppose that an aquifer is 450 km long, 200 km wide, and 75 m thick. It has an average porosity of 7 percent. What volume of water can the aquifer potentially hold?

Identify the Unknown:	Volume the aquifer can hold (V)
List the Knowns:	**Length** of aquifer (L) = 450 km or **450,000 m**
	Width of aquifer (W) = 200 km, or **200,000 m**
	Thickness of aquifer (T) = 75 m
	Porosity = 7% = **0.07**
Set Up the Problem:	Total volume of aquifer = length × width × thickness
	$V = L \times W \times T$
	Water volume = **total volume × porosity**
Solve the Problem:	V = 450,000 m × 200,000 m × 75 m = **6.8 × 10¹² m³**
	Water volume = (6.8 × 10¹² m³) × 0.07 = 4.7 × 10¹¹ m³
Check the Answer:	Check the answer by dividing it by the porosity, thickness, and width values. The result should be the length of the aquifer in meters.

PRACTICE Problems

Find **Additional Practice Problems** in the back of your book.

11. The Dakota Sandstone aquifer is 925 km long, 560 km wide, averages 4,200 m in thickness, and has an average porosity of 9.3 percent. What is its water storage volume?

12. **Challenge** Suppose an aquifer is 100 km long and 50 km wide. What is the thickness of the aquifer if it has an average porosity of 5 percent and holds 25,000 cubic kilometers of water?

Review

Additional Practice Problems

Water Resources

You use water every day for drinking, bathing, washing dishes and clothes, and cooking your meals. About 50 percent of the population in the United States obtains their freshwater from the ground. Groundwater is also used for residential and industrial purposes as well as irrigation, as shown in **Figure 24.**

Springs and wells

Some groundwater flows naturally from springs, which form where the water table meets Earth's surface. Alternatively, wells can be drilled into rock beneath the surface. A well that comes in contact with the water table can be used to pump water to the surface. As water is pumped from the ground, the water table lowers, as shown in **Figure 25.** You might expect the water table's surface to remain horizontal as water is removed. This does not happen near the well because pumping changes the natural flow path of groundwater.

Much like surface water, groundwater flows downhill in response to gravity. The natural flow path of groundwater has both horizontal and vertical characteristics. As water is pumped out of the ground, the direction that the water flows is directed down toward the well, creating a cone of depression, as shown in **Figure 25.**

Depending on the pumping rate, the cone of depression can affect water levels in other area wells. As the water table lowers, wells have to be drilled deeper, and some wells even run dry.

Artesian wells

Wells drilled into pressurized aquifers are called artesian wells. Pressure causes water to flow up into the well. Artesian wells form when an aquifer is sandwiched between two aquitards. Water is able to flow from a higher elevation, such as a mountain, through the aquifer. One example of an artesian well might be porous sandstone positioned between two shale layers. The aquitards (shale layers) keep the water confined to the aquifer. When the aquifer is sloped, gravity acting on water adds pressure to the low side of the slope. Artesian wells are usually free flowing at the surface.

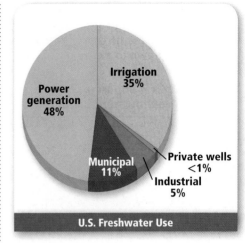

U.S. Freshwater Use

■ **Figure 24** Half of the U.S. population taps into the ground for their freshwater supplies. Municipal uses for freshwater include water used in homes and businesses.

Identify *the main sector for freshwater use.*

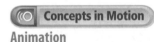 **Concepts in Motion**

Animation

■ **Figure 25** As areas become more populated, water wells are drilled more often. This lowers the water table, causing some existing water wells to go dry.

Infer *Why is a cone of depression created in the water table at the base of a water well?*

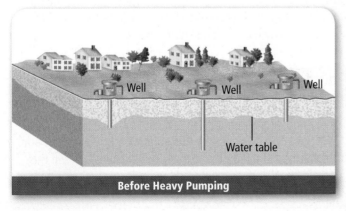

Before Heavy Pumping

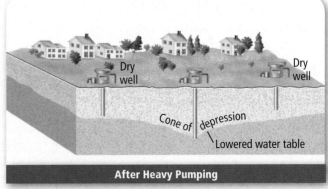

After Heavy Pumping

Section 3 • Groundwater **867**

Figure 26 Pollution can enter groundwater through infiltration. Pollutants can also be concentrated as they are drawn to a cone of depression.

Infer *Why is it important to prevent groundwater pollution?*

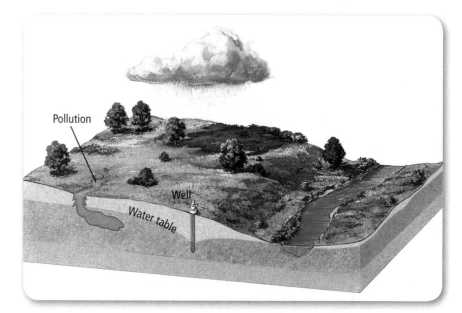

Pollution

Well

Water table

Pollution Both chemicals that are spilled or dumped on the ground and air pollutants washed from the air by precipitation enter groundwater through a process called infiltration, as shown in **Figure 26.** Soil can filter out some of the contaminants. But both natural pollutants, such as arsenic, and unnatural pollutants, like detergents and fertilizers, can infiltrate groundwater supplies. If groundwater becomes polluted, it is no longer useful until it is treated to remove all contaminants. Sources for water pollution include fossil fuel–burning power plants, factories, cars, oil and gas, and human waste. Groundwater pollution is regulated under the U.S. Environmental Protection Agency's Clean Water Act.

Section 3 Review

Section Summary

▶ Groundwater is a valuable natural resource.

▶ Only three percent of all water available on Earth is freshwater, most of which is stored underground.

▶ Freshwater travels from water to land to the atmosphere and back through a series of processes called the water cycle.

▶ Groundwater is replenished when surface water infiltrates into Earth.

▶ Rock and sediment, the surface features of an area, and the type and amount of vegetation control infiltration.

13. **MAIN Idea Explain** the importance of groundwater as a natural resource.

14. **Explain** why pore spaces must be interconnected for water to move through a material.

15. **Distinguish** between the unsaturated zone, the saturated zone, and the water table.

16. **Think Critically** In a rapidly growing neighborhood, streets, shopping centers, and homes are being built. Discuss the possible long-term effects of such growth on groundwater resources.

Apply Math

17. **Calculate** Suppose a community of 1,500 people uses water at a rate of 100 L per person per day. This community relies on an aquifer that contains 1.5 billion L of water. How many years will this water supply last if it is not replaced by the water cycle?

✓ Assessment   Online Quiz

Reading Preview

Essential Questions

▶ What are geologic epochs, periods, and eras?

▶ What are the characteristics of the main geologic eras?

▶ Which geologic laws are used to determine relative age?

▶ What methods are used to determine absolute age?

Review Vocabulary

radioactivity: the process that occurs when a nucleus decays and emits alpha, beta, and gamma radiation

New Vocabulary

absolute dating
relative dating
uniformitarianism
principle of superposition
unconformity
fossil

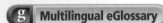
g Multilingual eGlossary

Geologic Time

MAIN ◀ Idea Relative dating and absolute dating are used to infer Earth's geologic past.

Real-World Reading Link Suppose you were asked to create a time line of your life for a history project. What events would you include on your time line? Geologists have created a time line to illustrate some significant geologic and biologic events in Earth's history.

Measuring Time

In your daily life, you have grown accustomed to the length of a year, a day, and an hour. These units of time were established based on astronomical observations. How do geologists measure time? Today, we know that the length of a day or a year is insignificant when compared to units of geologic time, as shown in **Figure 27.** Geologic time is subdivided into large units of time called eras, such as the Paleozoic, Mesozoic, and Cenozoic eras. These eras can be further subdivided into smaller units of time called periods and epochs.

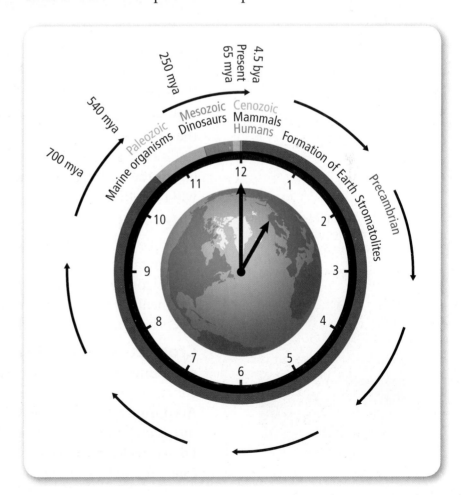

■ **Figure 27** Picture 4.5 billion years of geologic time converted to a 12-hour time line. Humans originated about 200,000 years ago, or at 11:59 and 58 seconds on the 12-hour clock. Units of geologic time are abbreviated as mya and bya for millions of years ago and billions of years ago, respectively.

MiniLab

Create a Time Line

Procedure

1. Read the procedure and safety information, and complete the lab form.
2. Obtain a 5-m length of **adding machine tape** and a **meterstick.**
3. Draw a horizontal line across one end of the tape, and label it *Present.*
4. Use a scale of 1 mm = 1 million years. Draw and label lines across the tape behind *Present* to mark the beginning of each of the following units of geologic time.

   Cenozoic: 65 mya

   Mesozoic: 248 mya

   Paleozoic: 543 mya

   Precambrian: 4,500 mya

Analysis

1. **Determine** where the first Homo sapiens (0.20 mya) would be labeled on the scale.
2. **Predict** What type of event might end one unit of geologic time and begin a new one?

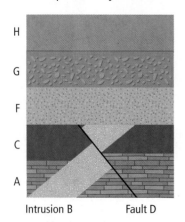

■ **Figure 28** This illustration shows an intrusion of igneous rock that was later separated by a fault.

H

G

F

C

A

Intrusion B   Fault D

Absolute and relative dating You can give your age in years, months, days, and even hours. To do this, you must have a starting point, your birth date. Birth records often give the exact date, hour, and minute of birth. This is an example of an absolute age. **Absolute dating** is the process of assigning an exact numerical age to an organism, an object, or an event.

Relative dating is the process of placing objects or events in their proper order in time. When you assign a relative age, you know which event happened first, second, and so on. The order is important. However, you will not know how much time has lapsed between events. For example, suppose your teacher has students in your class line up according to decreasing age. You would know that one student is older than the next student in line but you wouldn't know by how much.

Uniformitarianism For many centuries, Earth was thought to be thousands of years old. However, in the late-1700s, Scottish geologist James Hutton proposed that Earth was much older. Hutton's ideas were based on his principle of **uniformitarianism,** which states that the laws of nature operate today as they have in the past. Uniformitarianism is described by the phrase "the present is the key to the past."

Principles of Relative Dating

Geologists use several rules to help infer the relative age of rock layers. These rules were proposed in the mid-1600s by Nicolaus Steno, a Danish physician and geologist. Steno's rules help scientists determine the order that rock layers were deposited.

The principle of superposition One of Steno's rules is the principle of superposition. The **principle of superposition** states that in an undisturbed sequence of sedimentary rock layers, the youngest rocks will be at the top and the oldest rocks will be at the bottom.

Original horizontality Another of Steno's rules is the principle of original horizontality. The principle of original horizontality states that sedimentary rock layers are deposited as horizontal or nearly horizontal layers. In the Grand Canyon, sedimentary rock layers are horizontal. This means that the layers have not been disturbed since they were deposited.

Cross-cutting relationships Steno also observed that sometimes horizontal rock layers contain igneous rock that has cut across the layers. These igneous rocks are called intrusions. According to the principle of cross-cutting relationships, an intrusion or a fault is younger than the layers that it cuts across. For example, in **Figure 28,** the intrusion B is younger than the layers A and C. Fault D is younger than intrusion B.

Unconformities In some ways, rock layers are like a diary of Earth's past. If you have ever written in a diary, you know that there might be days when you skip an entry. Just as there are days when entries are missing from a diary, there are sometimes gaps of time missing from the geologic record. **Unconformities** are gaps in the rock record during which either erosion occurred or deposition was absent. A disconformity, shown in **Figure 29,** is an erosional surface between two horizontal layers of sedimentary rock. An angular unconformity is an erosional surface between rock layers that intersect at an angle. A nonconformity is an erosional surface between sedimentary rock and metamorphic or igneous rock.

Fossils

The remains or traces of organisms found in the geologic rock record are called **fossils.** Fossils can be the actual remains of an organism, such as a bone or a shell. Sometimes fossil remains, such as the inside of a shell, act as a mold. Over time, the mold fills with sediment that hardens into rock, preserving the structure of the original organism. In another process, called replacement, water containing dissolved minerals might replace the original shell or bone with new minerals. In other cases, only a trace impression, such as a footprint or burrow, is left for paleontologists to observe.

Fossil correlation While working as a canal engineer in England in the late 1800s, William Smith proposed the principle of faunal succession. Smith noticed that fossils appeared in the same order in rock layers in the canals, no matter where he excavated them. For example, if fossil Q were found in a sedimentary rock layer above fossil F, then Q also appeared above F in a different region. If you apply the law of superposition to this principle, you can conclude that fossil Q is younger than fossil F. The process of matching distinctive rock units from different regions is called correlation, as illustrated in **Figure 30.** Fossils that are most useful as global time markers are often from organisms that were widespread geographically but lived in only a narrow, well-defined period of time. Such fossils are called index fossils.

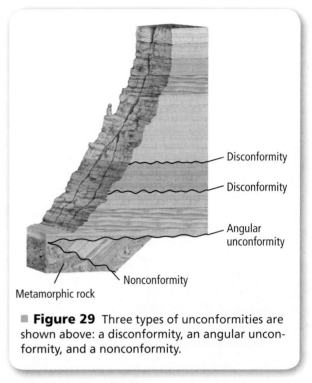

■ **Figure 29** Three types of unconformities are shown above: a disconformity, an angular unconformity, and a nonconformity.

Concepts in Motion Animation

Video BrainPOP

■ **Figure 30** Sometimes the relative age of rock units can be found by matching distinctive fossil-bearing rock units from different locations.

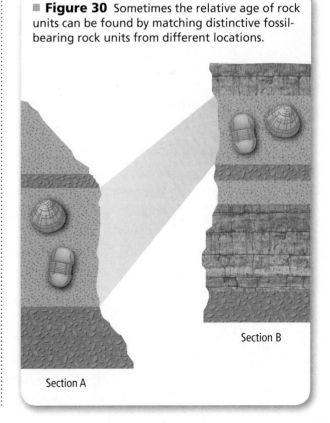

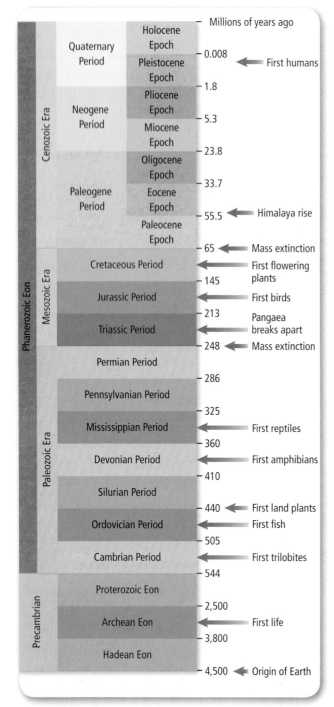

			Millions of years ago	
Phanerozoic Eon	Cenozoic Era	Quaternary Period	Holocene Epoch	
			Pleistocene Epoch	0.008 ← First humans
				1.8
		Neogene Period	Pliocene Epoch	
				5.3
			Miocene Epoch	
				23.8
		Paleogene Period	Oligocene Epoch	
				33.7
			Eocene Epoch	
				55.5 ← Himalaya rise
			Paleocene Epoch	
				65 ← Mass extinction
	Mesozoic Era	Cretaceous Period		First flowering plants
				145
		Jurassic Period		First birds
				213 Pangaea breaks apart
		Triassic Period		
				248 ← Mass extinction
	Paleozoic Era	Permian Period		
				286
		Pennsylvanian Period		
				325
		Mississippian Period		First reptiles
				360
		Devonian Period		First amphibians
				410
		Silurian Period		
				440 ← First land plants
		Ordovician Period		First fish
				505
		Cambrian Period		First trilobites
				544
	Precambrian	Proterozoic Eon		
				2,500
		Archean Eon		First life
				3,800
		Hadean Eon		
				4,500 ← Origin of Earth

■ **Figure 31** The geologic time scale is based on the appearance and disappearance of certain types of fossils. It is subdivided into four main eras.

 Concepts in Motion **Animation**

Geologic time scale Historians speak of the Industrial Revolution, and musicians refer to the Classical Period. These time periods began and ended with some event that marked a noticeable change in society or music. Similarly, scientists observe changes in the fossil record and relate them to time units within the geologic time scale, as shown in **Figure 31**.

Boundaries between units of time on the geologic time scale might also be established by evidence of a catastrophic geologic event. Such events, like a meteor impact or a large-scale volcanic eruption, might have caused environmental change that led to extinctions or changes in the life-forms present. Scientists revise the geologic time scale as new discoveries are made.

Absolute Dating

Absolute dating enables scientists to infer the numerical age of an object or event from the past. One type of absolute age dating process is based on the radioactive decay of unstable isotopes.

Recall that isotopes of an element have nuclei with the same number of protons but different numbers of neutrons. Some isotopes are radioactive and decay into other nuclei by emitting nuclear radiation. The isotope that decays is the parent isotope, and the isotope that forms is the daughter isotope. Some minerals in newly formed igneous rocks have radioactive nuclei that decay at the time the rock forms. Geologists can measure the ratio of the amount of parent isotope to the amount of daughter isotope to determine when the rock formed.

Half-life Every radioactive isotope has a half-life. The half-life is the time it takes for one half of a radioactive parent isotope to decay to its stable daughter isotope. After one half-life, half of the original parent isotope remains and the ratio of parent atoms to daughter atoms is 1:1. For example, if the half-life of a parent isotope is 500,000 years and a rock is analyzed with a daughter-to-parent ratio of 1:1, one half-life has elapsed and the age of that rock is 500,000 years.

✔ **Reading Check** **Determine** how much of the original parent isotope remains after one half-life.

Useful isotopes Generally, absolute ages are determined from unaltered minerals in igneous rocks or from organic materials. This is because sedimentary and metamorphic processes can involve water or other fluids. When these fluids interact with the rock, they can cause the amount of parent isotopes, daughter isotopes, or both to change. When this happens, the age of the rock is difficult to determine.

The type of isotope used for dating depends upon the composition and approximate age of the material being dated. For example, uranium-235 decays into lead-207 with a half-life of 713 million years. Geologists use U-235 to date igneous rocks that are tens of millions of years old. Carbon-14 decays to nitrogen-14 with a half-life of 5,730 years. C-14 occurs naturally in organic materials and can be used to date bone, wood, or peat younger than roughly 60,000 years old.

✔ **Reading Check** **Evaluate** Which isotope would be used to date a newly-discovered wooly mammoth fossil?

To understand how scientists combine relative and absolute age dating techniques to infer the geologic history of an area, look at **Figure 32.** You could conclude from the rock layers at Location A that Fossil C is between 2.1 and 2.4 million years old and Fossil A is older than 2.4 million years old. These conclusions use absolute age dating for the exact ages and superposition for the sequence of events in the area. What age relations can you determine for Fossils A, B, C, E, and F at Location B?

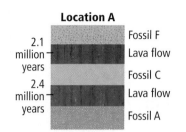

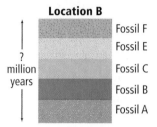

■ **Figure 32** Relative dating and absolute dating techniques can be combined to help determine the ages of some fossils.

Section 4 Review

Section Summary

▶ Uniformitarianism states that the present is the key to the past.

▶ Relative dating techniques are used to place objects or events in order from first to last.

▶ Fossils are the remains or traces of once-living organisms. Correlation is used to match rock formations from one area to another.

▶ Absolute dating techniques help to establish numerical ages for objects and events.

18. **MAIN Idea** **Discuss** how events in Earth's history determine boundaries between units of geologic time.

19. **Compare and contrast** absolute and relative age dating techniques.

20. **Describe** three different principles of relative dating.

21. **Think Critically** Suppose you are hiking down a canyon that has been deeply eroded, exposing horizontal rock layers. The rock layers are crosscut by an igneous intrusion. A fault cuts through both the intrusion and the layers of rock. Based on this description, place each geologic event in order.

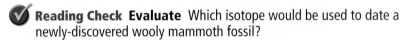

22. **Calculate an Age** Suppose the ratio of parent-to-daughter isotopes in an igneous rock is 1:7. The half-life of the parent isotope is 1.3 billion years. How old is the rock sample?

LAB

Stream Channel Formation

Objectives

- **Construct** a stream table to measure the effect of water speed on erosion.
- **Observe** how changes in water speed build new landforms.

Background: Large populations of people live in close proximity to rivers and streams.

Question: *How does the speed of a stream or river affect its ability to erode the surrounding landscape?*

Preparation

Materials

paint roller pan
sand
1-L beaker
rubber tubing (20 cm)
metric ruler
water
stopwatch
fine-mesh screen
wood block

Safety Precautions

WARNING: *Wipe up any water on the floor to prevent falls.*

Procedure

1. Read the procedure and safety information, and complete the lab form.
2. Place the mesh screen in the sink.
3. Pour 4–5 cm of moist sand into your pan, and smooth out the sand.

4. Set one end of the pan on the wood block, and hang the other end over the screen in the sink. Excess water will flow onto the screen.
5. Attach one end of the rubber tubing to the faucet, and place the other end in the beaker. Turn on the water so that it trickles into the beaker.

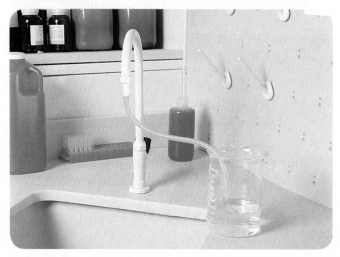

6. Measure the time it takes to fill the beaker with water to the 1-L mark. Record this time in your data table.
7. Without changing the water speed, hold the hose over the end of the pan that is resting on the wood block. Allow the water to flow into the sand for 2 minutes. At the end of two minutes, turn off the water.

? Inquiry Lab

Data Table

Time to Fill 1-L Beaker (s)	Depth of Channel (cm)	Length of Channel (cm)	Number of Channel Branches

8. Measure the depth and length of the eroded stream channel. Count the number of tributaries that formed. Record your measurements and any observations in the data table.

9. Empty the excess water from the tray, and smooth out the sand. Repeat steps 6–8 at the same water speed one more time. The thickness of the sand should be the same as in step 3.

10. Increase the water speed, and repeat steps 6–8.

Analyze Your Data

1. **Describe** how the shape of the channel and the number of tributaries changed when you increased the water speed.

2. **Predict** what might happen if you let the water run for longer periods of time.

3. **Predict** how your results would have differed if you raised the pan higher. Test your prediction.

Conclude and Apply

1. **Identify** the constants and variables in your experiment.

2. **Observe** Which water speed created the deepest and longest channel?

3. **Observe** Which water speed created the greatest number of tributaries?

4. **Explain** how the speed of a stream or river affects its ability to erode Earth's surface.

COMMUNICATE
YOUR DATA

Create a brochure for people buying homes in communities near rivers or streams. Outline some of the hazards of water erosion in the area.

Safe Water, A Global Concern

Safe water is a global concern. By 2030, it is estimated that nearly 50 percent of people worldwide will suffer from either insufficient water supplies or lack of access to safe water. Safe water is defined as water that is microbe-free, travels through pipes that keep the water clean, and is stored in sanitary containers or water treatment facilities.

Microbes and pollutants The most important aspect of safe water is the microbe content. Microbes carried by drinking water can cause serious epidemics, such as cholera and hepatitis, as well as many other diarrheal diseases. Water can also be carcinogenic if it has high concentrations of certain chemicals, such as nitrates and heavy metals. Nitrates in water can interfere with the body's ability to absorb oxygen. High arsenic levels in water can lead to the occurrence of cancer.

Water shortages Today, over 16 percent of the world's population lacks access to safe water, including many children. Experts in global water supplies believe that there is enough freshwater on Earth to meet the needs of every person. The struggle lies in making clean water accessible and affordable for all.

In some areas, seasonal, geographical and hydrological factors can lead to water shortages. Also, sometimes water demand in an area is not well-managed. The inadequate equipment and delivery subsystems fail, leaving communities with unreliable, and at times, unhealthy sources of drinking water. Some of the areas affected by water scarcity are shown in **Figure 1.**

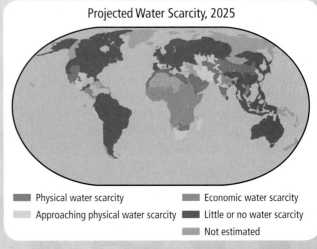

Projected Water Scarcity, 2025

- Physical water scarcity
- Approaching physical water scarcity
- Economic water scarcity
- Little or no water scarcity
- Not estimated

Figure 1 In areas that experience a physical water scarcity, more than 60 percent of surface water is allocated to agriculture, industries, or municipal use. In areas that experience an economic water scarcity, less than 25 percent of the surface water is withdrawn for human use.

The United Nations World Health Organization (WHO) has targeted safe water as part of a decade-long mission from 2005 to 2015. To make safe water available to more people, the United Nations WHO is focusing on global water issues on four levels: treating water and keeping it free of microbes and toxic chemicals, managing drinking water and irrigation water so that it is delivered safely, changing people's habits such as how they wash their hands, and managing natural resources and ecosystems.

WebQuest

Work In small groups, investigate one of the four global water initiatives outlined by the World Health Organization. Prepare a class discussion on the benefits of your group's initiative and how it helps to ensure safe drinking water worldwide.

BIG Idea Earth's surface changes over time due to weathering, erosion, and deposition.

Section 1 Weathering and Soil

soil (p. 850)
weathering (p. 846)

MAIN Idea Weathering breaks down rocks on Earth's surface.
- Weathering involves the breakdown of rock by physical or chemical processes.
- Mechanical weathering occurs when rocks break down physically due to collisions, frost wedging, and plant and animal activity.
- Chemical weathering occurs when rocks or minerals change composition.
- Soil is a mixture of organic matter, weathered rock material, water, and air.

Section 2 Shaping the Landscape

deposition (p. 854)
drainage basin (p. 855)
erosion (p. 854)
longshore current (p. 861)
sediment transport (p. 854)

MAIN Idea Earth's surface materials can be eroded, transported, and deposited, reshaping the landscape.
- Landforms that result from the erosion, transport, and deposition of sediment can provide clues about the natural processes that formed them.
- The age of a river can be determined from the shape of its channel.
- Glaciers carve solid rock as they move.
- Wind erosion creates landforms, such as desert pavement and dunes.
- Waves erode coastlines.
- Mass wasting occurs when material moves down a slope.

Section 3 Groundwater

aquifer (p. 864)
infiltration (p. 863)
porosity (p. 866)
water table (p. 864)

MAIN Idea Water can seep into the ground and be stored in porous rock as an important resource.
- Groundwater is a valuable natural resource.
- Only three percent of all water available on Earth is freshwater, most of which is stored underground.
- Freshwater travels from water to land to the atmosphere and back through a series of processes called the water cycle.
- Groundwater is replenished when surface water infiltrates into Earth.
- Rock and sediment, the surface features of an area, and the type and amount of vegetation control infiltration.

Section 4 Geologic Time

absolute dating (p. 870)
fossil (p. 871)
principle of superposition (p. 870)
relative dating (p. 870)
unconformity (p. 871)
uniformitarianism (p. 870)

MAIN Idea Relative dating and absolute dating are used to infer Earth's geologic past.
- Uniformitarianism states that the present is the key to the past.
- Relative dating techniques are used to place objects or events in order from first to last.
- Fossils are the remains or traces of once-living organisms. Correlation is used to match rock formations from one area to another.
- Absolute dating techniques help to establish numerical ages for objects and events.

Use Vocabulary

Match each phrase with the correct term from the Study Guide.

23. a rock layer that stores water underground

24. a technique of determining the order of geologic events

25. region that collects water for a stream system

26. idea that the laws of nature act today as they have in the past

27. a measure of the percentage of a material that is pore space

28. the chemical or physical breakdown of rocks and minerals

29. Earth's surface layer that consists of weathered rock, organic matter, water, and air

30. boundary separating the saturated zone and the unsaturated zone

31. a boundary representing a period of erosion or nondeposition

Check Concepts

32. **BIG Idea** What is the name of the process where surface materials are dropped by an erosional agent like water or wind?
   A) deposition
   B) differential weathering
   C) sediment transport
   D) infiltration

33. Which of the following is NOT a typical soil texture component?
   A) gravel   **C)** silt
   B) sand   **D)** clay

34. Identify the layer of material that slows or stops the movement of groundwater.
   A) aquifer   **C)** water table
   B) aquitard   **D)** saturated zone

35. Which time unit of the geologic time scale is the shortest?
   A) eon   **C)** period
   B) epoch   **D)** era

Use the figure below to answer questions 36 and 37.

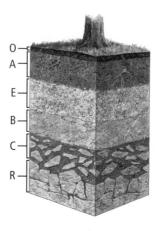

36. Which soil horizon contains the most organic matter?
   A) A horizon   **C)** C horizon
   B) B horizon   **D)** O horizon

37. Which soil horizon represents weathered bedrock?
   A) A horizon   **C)** C horizon
   B) B horizon   **D)** O horizon

Use the figure below to answer question 38.

38. **THEME FOCUS** Which glacial feature is shown in the figure above?
   A) V-shaped valley   **C)** flood plain
   B) moraines   **D)** deltas

✓ Assessment   Online Test Practice

Interpret Graphics

39. Copy and complete the table below.

Radioactive Decay		
Half-Lives	% Parent	% Daughter
a.	50	50
2	**b.**	**c.**
3	**d.**	**e.**
4	**f.**	**g.**

Use the figure below to answer question 40.

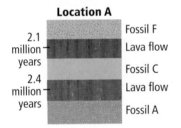

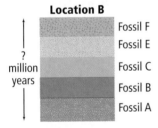

40. Describe how the absolute ages of the ancient lava flows could be used to determine the relative ages of fossils.

Think Critically

41. Explain why a soluble rock, such as limestone, can be preserved as cliffs in the western United States.

42. Describe how water can become contaminated.

43. Infer how lowering the water table can affect local wells.

44. Evaluate why building your home on coastal cliffs could cause problems.

45. Infer why rocks or fossils that are found lying on the ground are not useful for relative dating.

46. Apply The half-life of carbon-14 is 5,730 years. Give two reasons why carbon-14 could not be used to date a fossilized bone that was 30 million years old.

47. Explain why glacial deposits made of till are often poorly drained and swampy.

Apply Math

Use the figure below to answer question 48.

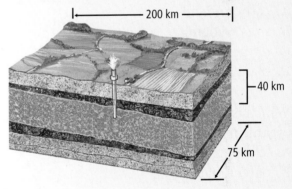

48. Evaluate Suppose an aquifer is known to be 200 km wide and 75 km long. If its average thickness is 40 km and it has an average porosity of 3 percent, estimate how many cubic kilometers of water could potentially be stored in the aquifer.

49. Calculate An igneous rock sample was dated using radioactive potassium. Measurement indicates that 1/32 of the original parent potassium is left in the sample. The half-life of potassium is 1.3 billion years. How old is this igneous rock?

Standardized Test Practice

Multiple Choice

Record your answers on the answer sheet provided by your teacher or on a sheet of paper.

Use the figure below to answer question 1.

1. Which type of chemical weathering agent created the landform shown in the image above?
   A. carbonic acid
   B. frost wedging
   C. oxygen
   D. plant and animal activity

2. Which is a mixture of weathered rock, organic matter, water, and air?
   A. feldspar
   B. parent material
   C. rock rust
   D. soil

3. Which process reflects the settling out of sediment from water, wind, or another erosional agent?
   A. deposition
   B. erosion
   C. transporting
   D. weathering

4. Which feature forms when the erosional force within a stream is exerted sideways?
   A. floodplain   C. source
   B. mouth   D. tributary

Use the figure below to answer question 5.

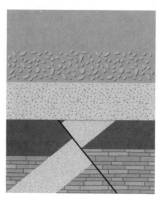

5. Which principle of relative dating is illustrated in the image above?
   A. crosscutting relationships
   B. fossil correlation
   C. unconformities
   D. uniformitarianism

6. Which is the name for the area where a river empties out into another body of water?
   A. floodplain   C. mouth
   B. source   D. tributary

7. Which feature forms at the front of a glacier?
   A. end moraine   C. lateral moraine
   B. ground moraine   D. medial moraine

8. Which feature forms when sand is moved by wind?
   A. currents   C. dunes
   B. delta   D. loess

9. Select the relative dating principle that states "In a sequence of undeformed sedimentary rocks, the oldest rock layer is on the bottom."
   A. original horizontality
   B. crosscutting relationships
   C. faunal succession
   D. superposition

✓ Assessment ▶ Standardized Test Practice

Record your answers on the answer sheet provided by your teacher or on a sheet of paper.

Record your answers on a sheet of paper.

Use the figure below to answer question 10.

Use the figure below to answer question 17.

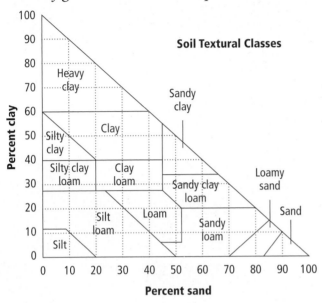

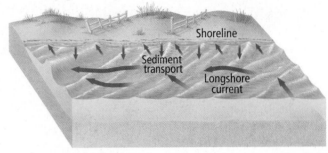

10. According to the graph, what is the greatest percentage of sand that can be in heavy clay soil?

11. What is the risk with over-pumping a well?

12. Why is vegetation important to the type of soil that forms?

13. Describe the drainage basin of a river.

14. Differentiate between permeability and porosity.

15. Describe the differences between relative dating and absolute dating.

16. How does a glacier erode land's surface?

17. **PART A** How does a longshore current form?
    PART B Use the image above to explain how a longshore current affects erosion and deposition along a shore.

18. If the half-life of a radioactive isotope is 7,000 years and the amount of the parent isotope present in an igneous rock is only one-fourth of the original amount, how old is the rock? Show your work.

19. Soil erosion in an area is 3.0 cm per year. The soil profile is 2.4 m thick and 25 percent of that is topsoil. How long will it take for the topsoil to erode away? Show your work.

20. Suppose you are a geologist and you discover rock layers exposed within a canyon similar to the Grand Canyon. Summarize the principles that you would use to determine the relative ages of rocks.

NEED EXTRA HELP?																				
If You Missed a Question . . .	1	2	3	4	5	6	7	8	9	10	11	12	13	14	15	16	17	18	19	20
Review Section . . .	1	1	2	2	4	2	2	2	4	1	3	1	2	3	4	2	2	4	1	4

CHAPTER 28

Weather and Climate

ConnectED

Your one-stop online resource
connectED.mcgraw-hill.com

- Video
- Audio
- Review
- Inquiry
- WebQuest
- Assessment
- Concepts in Motion
- g Multilingual eGlossary

Launch Lab
Atmospheric Pressure

Changes in atmospheric pressure are involved in producing winds and weather. You may not be aware of how much pressure the atmosphere exerts, but you will demonstrate its effects in this Launch Lab.

For a lab worksheet, use your StudentWorks™ Plus Online.

? Inquiry Launch Lab

THEME FOCUS Energy
Radiant energy from the Sun heats Earth's surface
and atmosphere unevenly, contributing to
changes in weather and climate.

BIG **Idea** Atmospheric properties, such as pressure,
temperature, and moisture content, determine the weather
and climate conditions in an area.

Reading Preview

Essential Questions
▶ What are the layers of the atmosphere?
▶ How is thermal energy transferred throughout the atmosphere?
▶ How do clouds form?
▶ What processes are involved in the water cycle?

Review Vocabulary
nucleus: a central point about which condensation takes place

New Vocabulary
troposphere
temperature inversion
greenhouse effect

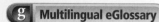 **g** Multilingual eGlossary

Earth's Atmosphere

MAIN ⟨Idea⟩ Earth's atmosphere can be divided into layers based on differences in composition and temperature.

Real-World Reading Link It is a late summer afternoon, and a storm approaches from the distance. The clouds move quickly across the sky. The sky darkens, the air cools, and raindrops begin to fall. From where are these storm clouds coming?

Composition of Earth's Atmosphere

When Earth formed 4.5 billion years ago, the early atmosphere was composed of mostly hydrogen and helium. As these gases escaped into space, volcanic eruptions formed an atmosphere made mostly of nitrogen, carbon dioxide, and water vapor. Increasing amounts of water vapor resulted in rainfall that slowly formed oceans. Collisions with asteroids and comets also added water to both the atmosphere and the oceans. Carbon dioxide dissolved in the oceans, greatly reducing the amount of carbon dioxide in the air. The evolution of photosynthetic organisms then increased atmospheric oxygen.

The atmosphere today Earth's atmosphere is a mixture of gases, including the water vapor and the oxygen that your body needs to survive. The amount of water vapor in the atmosphere varies from less than one percent to about four percent, depending on the location and the time of year. Dry air with less than one percent water vapor contains about 78 percent nitrogen and 21 percent oxygen. Most of the remaining one percent is argon. Other gases in the atmosphere are called trace gases because they are present in small amounts. Trace gases include carbon dioxide, ozone, neon, methane, and nitrogen oxides, as illustrated in **Figure 1.**

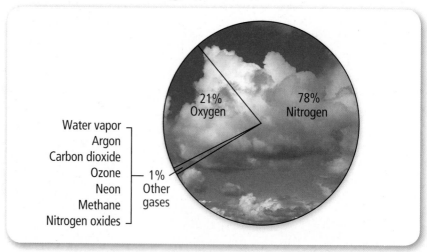

■ **Figure 1** Nearly 99 percent of Earth's atmosphere is composed of nitrogen (N_2) and oxygen (O_2).

Identify *other gases present in Earth's atmosphere.*

Life and the atmosphere Living organisms produce some of the gases present in the atmosphere today. The process of cellular respiration in these organisms produces water vapor and carbon dioxide. Photosynthetic organisms then convert this carbon dioxide and water into sugar and oxygen.

Carbon dioxide concentrations at different latitudes can change throughout the year. Due to an increase in plant productivity during the spring and summer, photosynthesis increases relative to respiration. The concentration of carbon dioxide in the atmosphere thus decreases until it reaches a minimum around October. In the fall and winter, the process is reversed and carbon dioxide increases until it reaches a maximum around May. Microorganisms in swamps, soil, and animals, such as termites and cows, produce other atmospheric gases including methane.

Structure of Earth's Atmosphere

Earth's atmosphere extends about 1,000 km above Earth's surface. Our weather takes place in the **troposphere,** a layer extending an average of 12 km above Earth's surface. In this layer, as shown in **Figure 2,** temperature decreases with altitude. Sometimes, however, temperature can increase with altitude. A **temperature inversion** occurs when air temperature increases with altitude and the air becomes stable. During a temperature inversion, air often resists the motion needed to form clouds and to disperse air pollution.

The stratosphere Above the troposphere is the stratosphere, which includes the ozone layer, as shown in **Figure 2.** Here, temperature always increases with altitude, creating a permanent temperature inversion. The place where this temperature inversion begins is called the tropopause. Temperatures within the stratosphere increase above the tropopause due to the ozone layer's ability to absorb ultraviolet radiation from the Sun. Temperature decreases in the mesosphere and increases in the uppermost layers of the atmosphere—the thermosphere and the exosphere. These layers are very low in density and do not affect weather conditions on Earth's surface.

VOCABULARY ·······························
WORD ORIGIN
Troposphere
comes from the Greek word *tropos,* meaning "a turn, change" and the Greek word *spharia,* meaning "sphere" *All weather occurs in the troposphere.* ···············

Concepts in Motion Animation

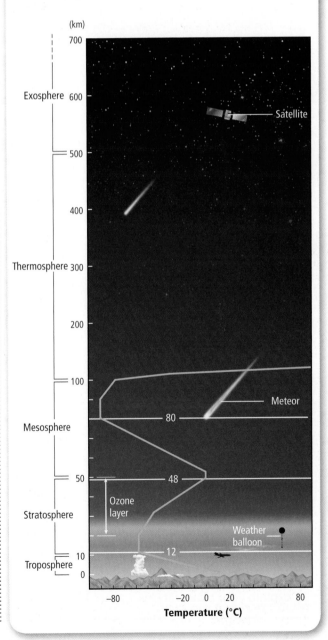

■ **Figure 2** Earth's atmosphere can be subdivided into five main layers. Each layer is classified based on composition, temperature, and altitude.

Identify *In which atmospheric layer is the ozone layer found?*

■ **Figure 3** Some gases in the atmosphere, such as water vapor and carbon dioxide, absorb infrared radiation and heat up the atmosphere. These gases then emit infrared waves in all directions. Earth's surface warms as it absorbs some of this infrared radiation.

Infer *How is Earth's atmosphere similar to being covered by a blanket at night?*

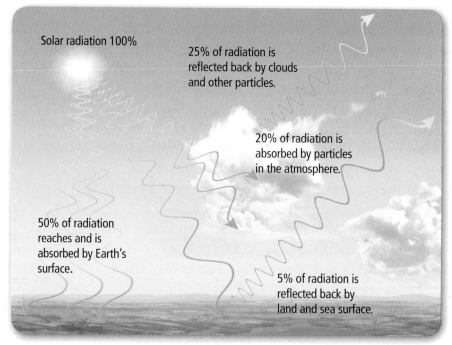

Solar radiation 100%

25% of radiation is reflected back by clouds and other particles.

20% of radiation is absorbed by particles in the atmosphere.

50% of radiation reaches and is absorbed by Earth's surface.

5% of radiation is reflected back by land and sea surface.

MiniLab

? Inquiry MiniLab

Model Convection

Procedure 🥽 👕 🧤

1. Read the procedure and safety information, and complete the lab form.
2. Place an **ice cube** in the center of a **large beaker.**
3. Add **room-temperature water** until the beaker is three-fourths full.
4. Place one drop of **food coloring** on the surface of the water.
5. Sketch what you observe.

Analysis

1. **Identify** where the water temperature was warmest and where it was the coldest in your sketch.
2. **Explain** Use your sketch to explain your observations.

Heating the Atmosphere

The energy that heats the atmosphere comes from the Sun. About 30 percent of this radiant energy is reflected back into space by the atmosphere, clouds, and Earth's surface. Another 20 percent is absorbed by the atmosphere. Some of this energy is absorbed by the ozone layer, which increases temperatures in the stratosphere.

Earth's surface directly absorbs roughly half of the radiant energy from the Sun, as shown in **Figure 3.** This energy heats Earth's surface. As it heats up, surface materials emit infrared radiation back into the atmosphere. Certain gases in the atmosphere, mainly water vapor, carbon dioxide, methane, and nitrous oxide, absorb and emit infrared waves, as shown in **Figure 3.** The **greenhouse effect** is a natural process in which certain gases in the atmosphere warm a planet as they absorb and emit infrared radiation. Without the greenhouse effect, Earth would not be warm enough to support life as we know it.

Other processes help to heat Earth's atmosphere. Convection transfers thermal energy as warm and cold air move from place to place. The release of latent heat also warms the atmosphere. Latent heat is the thermal energy released when water changes state from a gas to a liquid or from a liquid to a solid.

Heat Absorption by Earth Surfaces

Have you ever walked barefoot across the grass and onto an asphalt surface? Did you notice a change in temperature when your feet hit the asphalt? Surfaces covered by concrete, vegetation, snow, water, and ice absorb energy and warm the surface differently. This uneven heating of Earth's surface causes changes in the weather.

Water in the Atmosphere

The uneven heating of Earth's surface produces air currents that rise over warm areas. As warm air rises, it carries water vapor upward into the atmosphere. The air expands and cools as it rises. Clouds form when water vapor cools and condenses onto dust and other particles in the atmosphere called condensation nuclei. These cloud droplets are so small that air currents can keep them from falling to Earth as precipitation.

Types of clouds Meteorologists classify clouds based on their shapes and the altitudes where they form. The three basic cloud types are stratus, cumulus, and cirrus, as shown in **Figure 4**. Stratus clouds are layered, sheetlike clouds that form at altitudes below 2,000 m. Stratus clouds are often associated with rain. Cumulus clouds are puffy in shape and often occur with fair weather. Cumulus clouds typically form below 2,000 m. Under certain conditions, a cumulus cloud can grow upward and become a towering cumulonimbus cloud, sometimes called a thunderhead. This type of cloud can produce severe storms, lightning and thunder, and rain. Cirrus clouds are wispy clouds that form at altitudes above 6,000 m. Cirrus clouds are made of ice crystals.

Reading Check **Identify** the three cloud types.

Precipitation Cloud formation is often associated with precipitation. The main types of precipitation are rain, snow, sleet, and hail. For precipitation to occur, the tiny droplets that make up clouds must collide and stick together. When droplets become large enough, they fall to Earth as precipitation.

The type of precipitation that falls depends on air temperature. Rain falls when the air temperature is above freezing. When the air temperature is below freezing, ice crystals can clump together and fall as snow. When the air temperature is near freezing, rain can fall through layers of cold air to form sleet or freezing rain. Hail forms as layers of ice freeze onto droplets and air currents circulate these droplets inside a cloud.

■ **Figure 4** Different types of clouds form at different heights in the atmosphere.

Stratus clouds are layered and tend to form at low altitudes.

Cumulus clouds are puffy and can form at different altitudes.

Cirrus clouds have a wispy, icy appearance and form at high altitudes.

■ **Figure 5** In the water cycle, evaporation moves water from Earth's surface into the atmosphere. Precipitation returns water back to Earth's surface.

((◯ **Concepts in Motion**
Animation

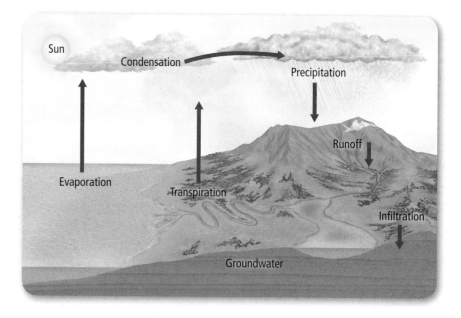

Global Water Cycle

Water is constantly cycling between Earth's surface and the atmosphere. This movement of water is called the water cycle, and it is shown in **Figure 5.** Energy from the Sun evaporates water from Earth's surface and forms water vapor. Most evaporation occurs from large bodies of water, such as oceans and lakes. Rising water vapor cools and condenses to form clouds. Condensation results in precipitation that falls over land and water. Some of the water that falls on land seeps into the ground, some becomes surface water that flows into rivers and streams, and some water enters the oceans. Water exposed to the Sun's energy evaporates again, and the cycle continues.

Section 1   Review

Section Summary

▶ Earth's atmosphere is divided into five main layers.

▶ The uneven heating of Earth's surface causes the movement of air.

▶ Clouds form from water vapor when rising air cools and condenses around small particles.

▶ Water is constantly moving between Earth's surface and the atmosphere in the water cycle.

1. MAIN ⟨Idea⟩ **Describe** the layers of the atmosphere and how temperature changes with altitude.

2. **Define** the greenhouse effect.

3. **Explain** why Earth's atmosphere today contains more oxygen than Earth's early atmosphere.

4. **Compare and contrast** cirrus, cumulus, and stratus clouds.

5. **Think Critically** How might an increase in average atmospheric temperatures affect the water cycle?

Apply Math

6. **Use Percentage** If the Southern Hemisphere contains 10 percent land and the Northern Hemisphere contains roughly 40 percent land, what percent of Earth is land?

✓ **Assessment** **Online Quiz**

LAB   Temperature Inversion

Objectives

■ **Make** a model that demonstrates a temperature inversion.

■ **Explain** what happens in the atmosphere during a temperature inversion.

Background: Temperature generally decreases with increasing altitude in the troposphere. Sometimes, a temperature inversion occurs near the ground and air becomes very stable and does not rise. This can result in fog or smog in cities. Does a liquid model the behavior of gases in the atmosphere?

Question: *How can a temperature inversion trap smog?*

Preparation

Materials

10-mL beaker containing 2–3 mL water
500-mL beaker containing 250 mL water
1,000-mL beaker containing 750 mL water
food coloring   ladle
long-stem dropper   thermal mitt
freezer   hot plate

Safety Precautions

Procedure

1. Read the procedure and safety information, and complete the lab form.

2. Chill 750 mL of water in a 1,000-mL beaker to near freezing.

3. Add one to two drops of food coloring to the 2–3 mL of water in the 10-mL beaker and let it stand at room temperature.

4. Heat 250 mL of water in a 500-mL beaker to near boiling.

5. Hold the ladle at the surface of the chilled water. Use the thermal mitt to pick up the 500-mL beaker. Slowly pour all the heated water into the ladle.

6. Allow the heated water to cascade slowly out of the ladle onto the surface of the chilled water. When you are finished, you should have a bottom layer of cold water and a top layer of hot water.

7. Use the long-stem dropper to inject a few drops of the colored water from step 3 into the cold water at the bottom of the 1,000-mL beaker.

Conclude and Apply

1. **Describe** what happened to the colored water.

2. **Explain** why this happened in terms of the temperatures of the water layers.

3. **Infer** how this is related to temperature inversions in the atmosphere.

4. **Suppose** the food coloring represents smog. Describe how this smog was affected by the temperature inversion.

COMMUNICATE YOUR DATA

Prepare a presentation of your observations. Include a labeled diagram showing the temperature layers and explain how your results relate to atmospheric processes.

Reading Preview

Essential Questions

▶ Why do differences in air pressure produce global winds?

▶ How do air masses interact along warm and cold fronts?

▶ What are the characteristics of severe storm systems?

Review Vocabulary

gradient: the rate of change of a quantity with distance

New Vocabulary

jet stream
Coriolis effect
air mass
weather front

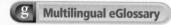
g Multilingual eGlossary

Video   BrainPOP

Weather

MAIN Idea When cold and warm air masses meet, the weather can change, sometimes resulting in severe storms.

Real-World Reading Link When you watch the weather forecast on the nightly news, you might see an L or H, cloudy skies or sunshine, and even signs of a severe storm on the map. Many people watch the weather because it impacts their daily lives.

Air Pressure

The atmosphere exerts a force on Earth's surface that is equal to the weight of the air above it. Air pressure is equal to this force divided by surface area. Differences in air pressure result from the uneven heating of Earth's surface. When air is heated, it expands and becomes less dense. This warm air weighs less and therefore exerts less pressure than an equal volume of cold air. Temperature differences, and consequently the differences in air pressure that result, create wind patterns. Wind is created when air moves from an area of high pressure to an area of low pressure.

Global winds and pressure systems Large-scale differences in air pressure create global wind belts. These belts are influenced by Earth's rotation and the unequal heating between the equator and the poles. Warm air rising near the equator and sinking over the poles creates general north-south wind circulation. Earth's rotation produces an east-west deflection of this circulation. **Figure 6** shows the major zones of high and low pressure air and the resulting global wind belts.

■ **Figure 6** In each hemisphere, there are four major pressure zones: the equatorial low, the subtropical high, the subpolar low, and the polar high. These pressure zones produce the trade winds, the westerlies, and the polar easterlies.

Explain *how the direction that winds blow differs for wind belts north and south of the equator.*

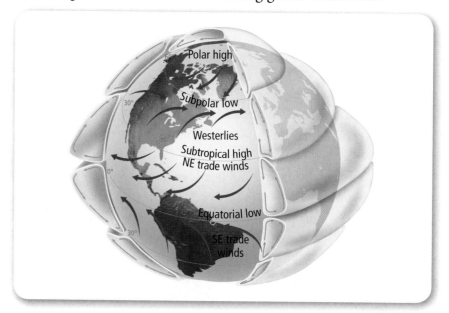

Jet streams Imbedded in the global wind belts are fast and powerful jet streams that can contribute to changes in the weather, such as storm development. A **jet stream** is a narrow band of fast-moving, high-altitude air. In the United States, the jet stream blows with the westerlies located about 12 km above the surface. Its speeds can exceed 500 km/h. Major storm tracks follow the jet stream as it moves north and south with changes in the seasons.

Pressure Systems

A low-pressure system, also known as a low (L), is a region in the troposphere where the air pressure is lower than the surrounding air. Because air flows from areas of high pressure to areas of low pressure, air flows toward the center of a low-pressure system. This inward air flow causes the air in the center of the low to rise, as shown in **Figure 7.** As this air rises, it expands and cools, and this results in cloud formation and precipitation.

A high-pressure system, also known as a high (H), is a region in the troposphere where the air pressure is higher than the surrounding air. As a result, air in a high pressure system flows away from the center. This outward flow causes the air in the center of a high to sink, as shown in **Figure 7.** Because sinking air does not lead to cloud formation, clear skies are often associated with highs.

Coriolis effect The direction of airflow is affected by Earth's rotation. The apparent deflection of an object due to Earth's rotation is called the **Coriolis effect.** In the Northern Hemisphere, the Coriolis effect causes winds to be deflected to the right. **Figure 7** shows how the Coriolis effect causes air to flow around highs and lows. Because of the Coriolis effect, air in the Northern Hemisphere flows counterclockwise around a low and clockwise around a high.

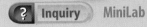

Create a Low-Pressure Center

Procedure 🥽 👕 🔥 🧪

1. Read the procedure and safety information, and complete the lab form.
2. Fasten a **birthday candle** firmly to the bottom of a **pie pan** or **plate** with **clay**.
3. Half-fill a **tall, narrow jar** with **water** and pour the water into the pan or plate.
4. Light the candle. Invert the jar over the candle. Set the jar mouth down into the water, and rest it on a **penny** to provide space for water to move into and out of the jar.
5. Write a brief description of what happens to the water level inside the jar when the candle burns out.

Analysis

1. **Infer** what happens to the air inside the jar when the candle is lit.
2. **Explain** what happens to air inside the jar when the candle goes out and why water rises in the jar when this happens.

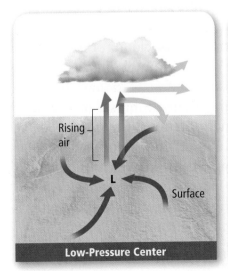

Low-Pressure Center

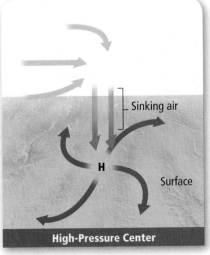

High-Pressure Center

■ **Figure 7** Air flowing into a low-pressure system rises, but air flowing out of a high-pressure system sinks. In the Northern Hemisphere, the Coriolis effect causes air to flow counterclockwise around a low and clockwise around a high.

Concepts in Motion Animation

■ **Figure 8** Weather fronts form when air masses of different temperatures and moisture content meet. The symbols shown next to each of these weather fronts are used by meteorologists to represent fronts on weather maps.

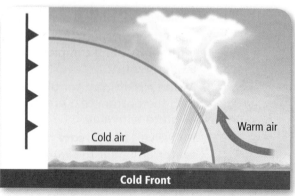

Cold air   Warm air

Cold Front

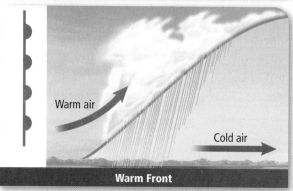

Warm air   Cold air

Warm Front

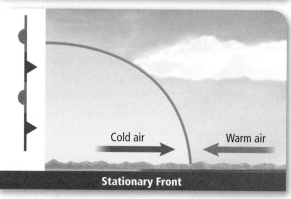

Cold air   Warm air

Stationary Front

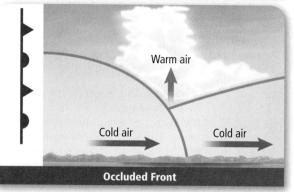

Warm air

Cold air   Cold air

Occluded Front

Air Masses and Weather Fronts

Weather around high-pressure and low-pressure systems is produced by the interaction of air masses. An **air mass** is a large volume of air with uniform moisture and temperature throughout. An air mass forms when air remains stationary for a time, such as in areas of high pressure. The air then takes on the characteristics of the area over which it forms.

Where air masses originate Air masses can be polar or tropical and continental or maritime. Continental air originates over land. It is relatively dry and can be extremely cold or extremely warm, depending on where it originates. For example, an air mass that forms in northern Canada will be cold and dry, whereas an air mass that forms over central Mexico will be warm and dry.

Maritime air masses are moist because they originate over the oceans. The maritime air masses affecting the United States originate over the Atlantic Ocean, the Pacific Ocean, and the Gulf of Mexico. The temperature of maritime air also depends upon latitude of the oceans over which it originates.

Where air masses meet Air masses interact in zones called **weather fronts.** Warm and cold fronts create different types of weather conditions, as shown in **Figure 8.** In a cold front, cold air forces warm air upward in a fast and chaotic manner, forming cumulus clouds. A cold front can also result in the formation of cumulonimbus clouds and severe storms. In a warm front, warm air rises gently above cold air, usually forming layered, stratus-type clouds or fog—a cloud with its base on the ground. Most layered clouds produce steady rainfall. Stationary fronts form where cold and warm air masses meet and neither front advances. Stationary fronts can last for days in an area and produce clouds and prolonged precipitation. Occluded fronts occur when a fast-moving cold front overtakes a slow warm front. Thunderstorms and strong winds are often associated with occluded fronts.

✔ **Reading Check** **Define** What is a weather front?

Severe Weather

Severe weather conditions, such as thunderstorms, downbursts, tornadoes, and hurricanes, occur almost every day in the United States. This is due, in part, to temperature differences between warm and cold air masses. Also, oceans provide a significant source of moisture to fuel such storms.

Thunderstorms A thunderstorm forms when warm and wet air rises rapidly. As the rising air cools, water vapor condenses to form water droplets or ice crystals. The latent heat of condensation warms the surrounding air, causing it to rise even higher. Rising air can form cumulonimbus clouds.

As cloud formation continues, tiny water droplets collide to form larger ones that fall toward Earth's surface. Raindrops cool the air around them, causing the air to sink. As air rises and sinks within the cloud, a charge separation occurs. Lightning is produced when oppositely charged particles in a single cloud, between two clouds, or between the cloud and the ground interact. The surrounding air is ionized, creating an electric field that generates a lightning bolt. Hotter than the surface of the Sun, the lightning bolt causes air to expand and collapse to produce thunder.

A typical cumulonimbus cloud, shown in **Figure 9,** contains ice crystals near its top. Water vapor can freeze onto these ice crystals. Repeated cycles of rising and sinking air add layers to the ice crystals and forms hailstones. Rain or hail produced during a thunderstorm can cool the surrounding air and cause it to sink. This sinking current of cold air spreads across Earth's surface as a downdraft.

Downbursts Downdrafts can produce severe winds called downbursts. Downbursts occur when cold air descends from a cloud and hits the ground. Sometimes the downdraft spreads out when it hits the ground, forming a series of gusty winds called squalls. The winds that result can be as fast as 260 km/h. The rapid change in wind speed and direction that results can be dangerous for aircraft during both takeoff and landing. Fortunately, automated warning systems inform pilots when these conditions are likely.

FOLDABLES
Incorporate information from this section into your Foldable.

Concepts in Motion Animation

■ **Figure 9** A tall, cumulonimbus cloud produces a thunderstorm. Rising air causes an updraft that helps pull moist air into the cloud. Condensing water vapor releases latent heat, which warms the surrounding air and helps maintain the updraft.

Infer *Why does this cumulonimbus cloud have an anvil-shaped top?*

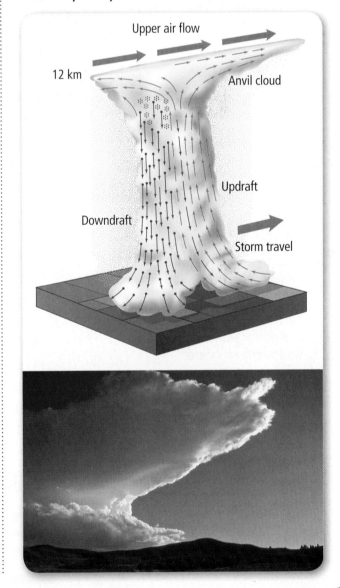

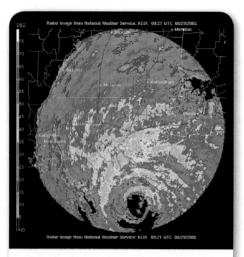

Figure 10 On August 29, 2005, Hurricane Katrina, the cyclonic storm shown here, made landfall in Louisiana with sustained wind speeds of 205 km/h, a 3 m storm surge, and torrential rain. The inundation of water caused 53 levees in New Orleans to fail, submerging the city in floodwater and debris.

Tornadoes and hurricanes Two types of severe storm systems that differ in where they form and how they develop are tornadoes and hurricanes. A tornado forms in cumulonimbus clouds where strong updrafts create a vortex effect. These twisting, funnel-shaped clouds can move across land at speeds of more than 65 km/h, creating a path of destruction between 150 m wide and 10 km long. Intense, circular winds in the funnel can reach speeds of 450–500 km/h.

Tornadoes are often short-lived and affect only small areas. However, they can cause tremendous amounts of damage, not only due to strong winds but also the extremely low pressure conditions in the center of the storm system. Tornadoes typically form when a cold and dry air mass meets a warm and wet air mass. This happens frequently in the Great Plains (or Tornado Alley), the lower Midwest, and in parts of the South. In the South, tornadoes often accompany hurricanes.

Hurricanes are tropical storms that cover large areas and last for days. Hurricanes that affect the eastern United States and Gulf of Mexico, as shown in **Figure 10,** often form as tropical depressions over the warm Atlantic waters near the coast of West Africa. As the tropical depression advances westward across the Atlantic, it is fueled by warm air and water. When wind speeds exceed 118 km/h, these tropical storms develop into hurricanes. Hurricanes are associated with strong winds, severe rainfall, and a storm surge that causes coastal flooding. Western Pacific hurricanes are called typhoons.

Section 2 Review

Section Summary

▶ Differences in air pressure and Earth's rotation produce wind belts in each hemisphere.

▶ Air flows counterclockwise around lows and clockwise around highs in the Northern Hemisphere.

▶ Air masses interact along weather fronts.

▶ Severe weather conditions include thunderstorms, tornadoes, and hurricanes.

7. **MAIN Idea Describe** the differences in the weather produced by warm fronts and cold fronts.

8. **Define** the Coriolis effect.

9. **Draw and describe** wind patterns around high pressure and low pressure systems.

10. **Compare and contrast** tornadoes and hurricanes.

11. **Describe** differences between continental air masses and maritime air masses.

12. **Think Critically** Use a map of the United States to explain why tornadoes are common in the Great Plains.

Apply Math

13. **Percentages** A tornado watch was issued on 25 days during one year in a Midwestern city. What percent of the year does this represent?

Assessment   Online Quiz

Reading Preview

Essential Questions

▶ What factors determine climate?

▶ What is a climate system?

▶ What are the dominant climate zones across the United States?

Review Vocabulary

lithosphere: rigid, outermost layer of Earth that includes the crust and upper mantle

New Vocabulary

maritime climate
continental climate
sea breeze

Climate

MAIN Idea Climate is determined by a location's latitude, altitude, and average temperature and precipitation.

Real-World Reading Link Suppose a foreign exchange student is going to stay with you for the school year. How would you describe the weather conditions to her, so that she is prepared for her study abroad?

Climate and Weather

Climate is often defined as the long-term average weather conditions for an area—temperature, precipitation, wind patterns and other factors that affect the weather. Climate also describes annual changes in the weather and its extremes.

Weather data collected every month for decades are analyzed to evaluate climate norms for an area. These norms do not describe daily weather conditions. They represent average conditions over a large area.

Factors that affect climate Climate is best considered as part of the Earth system. The Earth system can be visualized as five spheres that interact to create the environments in which we live, as shown in **Figure 11.** The atmosphere includes the air around us. The biosphere is all living organisms and the environments in which they live. The hydrosphere is liquid water in oceans, lakes, rivers, and underground. The cryosphere is frozen water in snow, ice, and glaciers. Finally, the lithosphere is Earth's outermost layer, including the crust and upper mantle.

■ **Figure 11** The five spheres of the Earth system create an interconnected ecosystem in which organisms live.

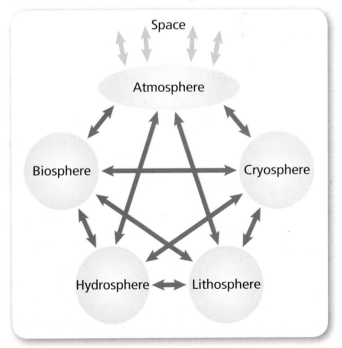

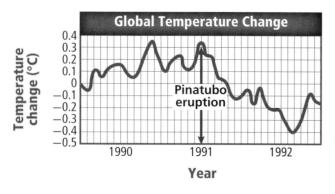

■ **Figure 12** Mount Pinatubo in the Philippines erupted in June 1991. During the eruption, more than 20 million tons of sulfur dioxide gas were ejected into the atmosphere, causing a worldwide decrease in average annual temperatures of 0.5°C for two years.

Global Temperature Change

Temperature change (°C)

Pinatubo eruption

1990   1991   1992

Year

Interactions within the Earth system Examine the volcanic eruption shown in **Figure 12**. A volcanic eruption is an excellent example of how Earth's spheres interact. Volcanoes erupt gas and ash from the lithosphere into the atmosphere. Gases can affect temperature, as shown in the graph in **Figure 12**. Ash in the atmosphere can cause clouds to form, which can produce precipitation that enters the hydrosphere. The area around the volcano can be covered with ash, which affects life in the biosphere. If enough gas and ash are erupted, it might also shield sunlight for days or months, affecting the climate of an area.

What factors determine climate?

The primary factor that influences the climate in any given area is latitude. Recall that the intensity of solar radiation received on Earth's surface is greatest at the equator (0°) and decreases toward the poles (90°N and 90°S). Also, the direction and strength of the prevailing winds depend on latitude. Other factors that affect climate include the distance to large bodies of water, such as an ocean, or an area's location relative to the mountains. Whether an area is on the east or west side of a continent also affects its climate.

Temperature The long-term average temperature in an area depends mainly on latitude. Because the intensity of solar radiation received at Earth's surface decreases with increasing latitude, average temperatures also decrease with increasing latitude. The intensity of solar radiation received is dependent on the angle with which the Sun's rays strike the surface. The Sun's most direct rays strike at an angle of 90° with the greatest intensity.

However, average summer temperatures do not decrease as much with latitude as average winter temperatures do. In the summer, the number of daylight hours increases from the equator to the poles. The longer period of daylight means that nearly as much solar energy is received during the summer at high latitudes as at lower latitudes.

VOCABULARY ····················
SCIENCE USAGE V. COMMON USAGE
Latitude
Science usage
   the angular distance in degrees measured north and south of the equator
   The Tropic of Cancer is located at 23.5°N latitude.

Common usage
   freedom to choose without any restriction
   The students were granted the latitude to select any project theme. ··········

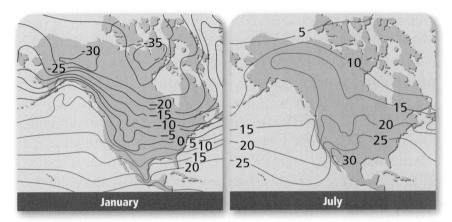

January   July

■ **Figure 13** The difference between average temperatures (°C) in high latitudes and low latitudes is not as large in summer as in winter.

Analyze *Why are warmer temperatures shifted northward during July?*

Figure 13 shows the difference in average temperatures in January and July over the Northern Hemisphere. Temperature differences between the southern United States and northern Canada are about 55°C in the winter, but only about 25°C in the summer. This difference is due in part to the amount of solar radiation different latitudes receive throughout the year.

The tropics receive the greatest amount of solar radiation because the Sun's rays are perpendicular (90°) to Earth's surface here most of the year. The tropics are between 23.5° north and 23.5° south of the equator. Temperatures in the tropics are warmer than those in higher latitudes, such as the temperate zone between 23.5° and 66.5° north and south of the equator. The polar zones are between 66.5° and 90° north and south of the equator. The poles receive the least amount of solar radiation and are colder than lower latitudes, as shown in **Figure 14**.

✓ **Reading Check** **Identify** the area that receives the greatest amount of solar radiation during the year.

■ **Figure 14** The amount of solar radiation that an area receives decreases from the equator to the poles.

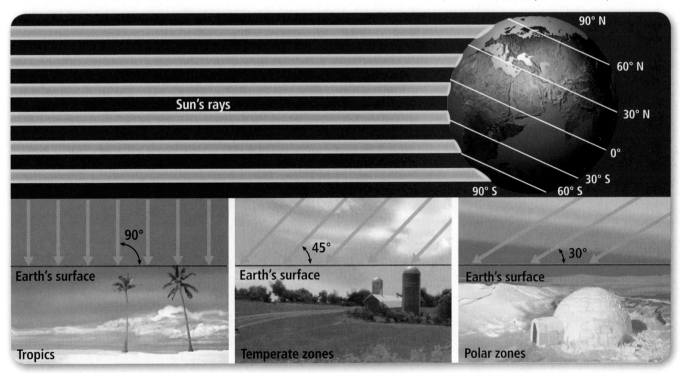

Sun's rays

90° N
60° N
30° N
0°
30° S
60° S
90° S

90°   Earth's surface   Tropics

45°   Earth's surface   Temperate zones

30°   Earth's surface   Polar zones

Ocean and land Oceans also affect climate. A climate that is strongly affected by an ocean is called a **maritime climate.** A climate that is not directly affected by an ocean is called a **continental climate.** Maritime climates are milder—summers are cooler, winters are warmer, and daily temperatures vary less throughout the year. Peoria, Illinois, and San Francisco, California, are at the same latitude—40°N. However, Peoria has a continental climate and San Francisco has a maritime climate. The temperature difference between the warmest and coldest months is 8°C in San Francisco and 30°C in Peoria.

Precipitation Wind and air pressure patterns affect the average precipitation of a region. Moisture is associated with low-pressure systems in the tropics and in mid-latitudes. Rainfall increases near the equator because the trade winds from both hemispheres converge there. Warm and wet air rises rapidly and rainstorms commonly develop due to this convergence. Dry climates are associated with high-pressure systems. Climate tends to be even drier in the subtropics east of the subtropical highs.

Precipitation is also affected by a region's location on a continent, as shown for North America in **Figure 15.** The west coast lies east of the subtropical high, which transports cold water currents and stable air. This creates the dry climates of California and the Southwest. The east coast lies west of the subtropical high, where southerly winds transport warm and unstable air from the Gulf of Mexico, increasing precipitation.

Another factor affecting precipitation is the prevailing wind. Because winds blow from the west in the mid-latitudes, the maritime climate is stronger on the west coast. This explains why San Francisco has developed a maritime climate, but Boston has more of a continental climate.

VOCABULARY

ACADEMIC

Region
a geographic area or subdivision of a larger area of interest
She studied crocodile habitats in a region near Everglades National Park.

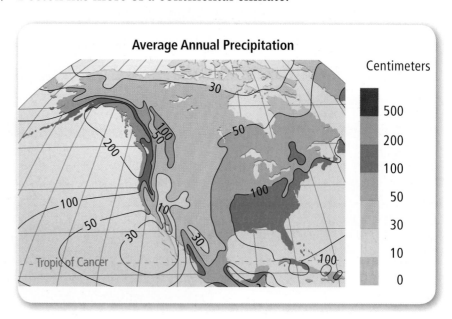

Average Annual Precipitation

■ **Figure 15** Average precipitation over North America tends to increase in coastal areas. Lines connecting points of equal precipitation are called isohyets.

Define *What areas in the United States receive the greatest amount of precipitation annually?*

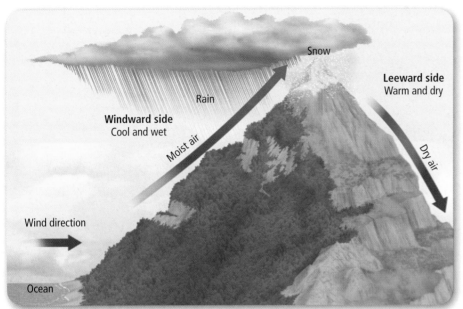

■ **Figure 16** Air on the windward side of a mountain rises and cools, resulting in precipitation. The climate on the windward side of a mountain tends to be cool and wet. Air on the leeward side of the mountain is drier and warms as it descends. The climate on the leeward side is often warm and dry.

Mountains Mountains can affect precipitation in an area. As an air mass approaches the mountains, it will rise up over them. Rising air cools, and condensation occurs. As a result, precipitation falls on the windward side of the mountains, as shown in **Figure 16.** The windward side of the mountain creates a rain shadow effect. On the opposite side of the mountains, called the leeward side, the air is drier and warms as it descends. Weather on the leeward side of the mountain is often warm and dry.

Water Coastlines and lakeshores can affect climate. For example, lake-effect snow often occurs around the Great Lakes in the winter, as shown in **Figure 17.** As cold Arctic air travels over warm lake water, the air mass absorbs both heat and moisture. When the air mass once again travels over cold land to the south and east of a lake, heavy snow falls.

Sea breezes are also caused by the influence of water. A **sea breeze** blows from over the water toward land in the afternoon, when the land is warmer than the water. Warm air rises over the land creating areas of low-pressure that allow cool, dense air to sink and take its place. The opposite occurs at night when the land is cooler than water. A land breeze occurs when cool, dense air over land creates an area of high-pressure that causes the air to blow from land toward the sea.

Small-scale climates Many small-scale changes in climate can occur within an area. Some are regional while others, termed microclimates, are local variations. For example, cities tend to be warmer than suburbs—a condition called the heat island effect. Building and pavement materials heat up more rapidly than land. On some clear, calm nights, downtown San Francisco can be as much as 8°C warmer than the surrounding area.

■ **Figure 17** Lake-effect snow occurs in areas east and south of the Great Lakes. Cold continental air masses travel over the warmer lake water, picking up moisture that condenses, freezes, and falls as snow.

Climate Zone	Vegetation
Cold Arid	tundra
Cold Dry winter	boreal evergreen forest
Warm Arid	desert
Warm Semi-arid	grassland
Wet winter Dry summer	Mediterranean forest
Wet summer Dry winter	temperate woodland
Warm Wet	subtropical deciduous forest
Warm Wet	tropical deciduous forest

■ **Figure 18** Climate zone influences the types of vegetation that will grow there.

Infer *What are two major factors that characterize climate types?*

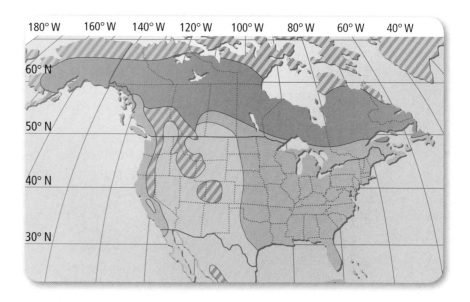

Types of Climates

Geographer Glenn Trewartha and meteorologist Lyle Horn designed a climate classification system, shown in **Figure 18.** It has three major divisions: cold and dry, arid and semi-arid, and warm and wet. The last designation includes temperate, subtropical, and tropical climates. The eastern United States is dominated by temperate and subtropical climates, whereas the western United States is arid and semi-arid with some cold, temperate, and subtropical climate zones. The types of vegetation grown in an area can be useful in climate classification.

Section 3 Review

Section Summary

▶ The distribution of solar radiation and the location of pressure and wind systems are factors that affect climate.

▶ Oceans, land masses, mountains, and large lakes can cause climate variations.

▶ Climates are classified based on average temperature and precipitation.

14. **MAIN Idea Explain** the relationship between climate and latitude.

15. **Describe** how weather patterns change from the windward to the leeward side of a mountain.

16. **Identify** the five spheres in the Earth system.

17. **Compare and contrast** continental climate and maritime climate.

18. **Think Critically** How does the daily temperature range vary between a maritime climate and a continental climate?

Apply Math

19. **Calculate** Mount Pinatubo ejected large volumes of gas and ash into the atmosphere during its eruption in June 1991. Scientists measured an average global temperature decline of about 2 percent during the next year. If the average temperature of an area was 25°C prior to the eruption, what was the average temperature the next year?

✓ **Assessment** Online Quiz

Reading Preview

Essential Questions

▶ How does climate change seasonally?

▶ What causes climate change?

▶ How do humans affect climate?

▶ What is El Niño's effect on weather?

Review Vocabulary

trace: a tiny, barely detectable amount present

New Vocabulary

global warming
El Niño
La Niña

g Multilingual eGlossary

Earth's Changing Climate

MAIN Idea Earth's climate undergoes seasonal and long-term changes.

Real-World Reading Link It is summertime, and your family has planned a camping trip to the mountains. Think about the alpine landscape, changes in vegetation and wildlife, and the temperature differences. Is it safe to assume that the climate will be the same?

Seasonal Changes

Seasonal changes occur as Earth revolves around the Sun. The hemisphere tilted toward the Sun experiences summer, and the hemisphere tilted away from the Sun experiences winter. During the summer, the intensity of solar radiation increases and temperatures rise. During the winter, the intensity of solar radiation decreases and temperatures drop. The equator experiences less seasonal change compared to the poles because the intensity of sunlight changes less at the equator when compared to higher latitudes.

Long-Term Changes

Over the past few million years, large ice sheets have both advanced and retreated. Large ice sheets covered much of Earth's surface during an ice age. The last ice age began 100,000 years ago and ended 10,000 years ago. **Figure 19** shows the peak of the last ice age, when ice sheets covered much of North America. By about 5,000 years ago, however, these ice sheets had melted almost completely. Climate patterns began to change into what they are today. The graph in **Figure 19** shows how temperatures have varied over the past 18,000 years.

■ **Figure 19** At the peak of the last ice age, about 18,000 years ago, thick ice sheets covered much of North America. The extent of the ice sheets is shown with a dashed, red line below.

Explain *the trend in temperature from 18,000 years ago to today.*

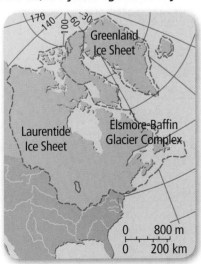

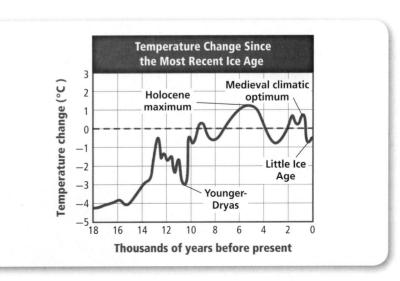

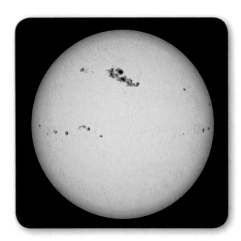

Figure 20 NASA and the European Space Agency (ESO) launched the Solar and Heliospheric Observatory in 1995. Since that time, the spacecraft has been monitoring the occurrence of sunspots and changes in space weather.

Causes of Climate Change

There are a number of factors that cause climate change. These factors affect climate on different time scales. For example, plate motion can affect climate over millions of years. Changes in ocean currents and snow or ice cover affect climate over a smaller scale, such as decades.

Periodic shifts in Earth's orbit around the Sun can also affect climate by changing the amount of solar radiation that Earth receives. The shape of Earth's orbit and the tilt and wobble of Earth's rotation axis have varied in predictable patterns over time. These patterns coincide with changes in climate.

The amount of solar radiation that Earth receives also changes. An example of this change can be observed through studies of sunspots, which are dark, cool spots on the Sun's surface. Sunspots reflect an increase in solar activity. Scientists have discovered a link between the occurrence of sunspots, as shown in **Figure 20,** and climate variations over time. From the mid-1600s to the early 1700s, almost no sunspots were visible. This meant that solar activity was at a minimum, and the amount of solar radiation emitted by the Sun was reduced. During this time, long winters and extreme cold occurred throughout Europe. In contrast, sunspot maximums have produced warmer climate conditions.

Catastrophic events, such as meteorite impacts and volcanic eruptions, can also affect climate. Recall that droplets of sulfuric acid form in the stratosphere as a result of volcanic eruptions. These droplets reflect solar radiation and cool Earth's surface. For example, the year following the 1815 eruption of Mt. Tambora in Indonesia was known as the year without a summer.

The human factor Human activities, such as energy consumption, deforestation, and industrial and agricultural practices, have affected Earth's atmosphere. These activities change the composition of the atmosphere and affect the water and carbon cycles. The carbon cycle is shown in **Figure 21.**

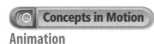
Animation

Figure 21 Humans influence the carbon cycle through practices that release carbon dioxide into the atmosphere. Carbon dioxide is removed from the atmosphere by plants and stored in rock, soil, and ocean reservoirs.

Describe *how trees affect levels of carbon dioxide and oxygen in the atmosphere.*

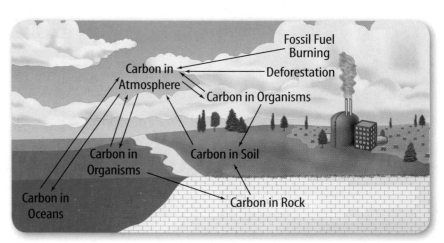

The carbon cycle The movement of carbon between the biosphere, the atmosphere, the hydrosphere, and the geosphere is called the carbon cycle. The carbon cycle makes life as we know it possible. For example, during photosynthesis, plants convert carbon dioxide and water into sugar and oxygen. Animals breathe in oxygen and release carbon dioxide back into the atmosphere during respiration. When plants and animals die, their remains are buried within layers of rock and sediment. Extreme heat and pressure within the geosphere causes these remains to be converted into fossil fuels, such as coal, petroleum, and natural gas. The process of fossil fuel formation takes millions of years. Humans extract these fossil fuels from the ground and use them as energy resources. When fossil fuels are burned, they release carbon dioxide, among other gases, back into the atmosphere and the carbon cycle continues.

 Reading Check Explain how carbon cycles through Earth's systems.

EXAMPLE Problem 1

Calculate Carbon Dioxide (CO_2) The concentration of CO_2 in the atmosphere (385 ppm in 2007) increased by 0.6 percent between 2007 and 2008. Assuming that this percent increase remains constant, you can predict the CO_2 concentration in the future by using the following formula:

CO_2 (year) = CO_2 (year) + CO_2 (year) × rate of CO_2 increase × years

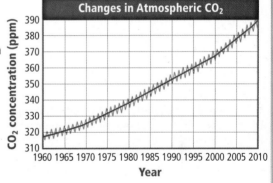

Changes in Atmospheric CO_2

Identify the Unknown: concentration of CO_2 in 2015 = ? ppm

List the Knowns: CO_2 (2007) = 385 ppm
rate of CO_2 increase = 0.6 percent

Set Up the Problem: CO_2 (2015) = CO_2 (2007) + CO_2 (2007) × rate of CO_2 increase × years

Solve the Problem: Substitute the known values into the equation.
concentration in 2015 = 385 ppm + 385 ppm × 0.006 × 8
concentration in 2015 = 403 ppm

Check the Answer: Multiply 385 ppm × 0.006 × 8, and subtract from 403 ppm. The result is the 2007 CO_2 concentration, 385 ppm, so your answer makes sense.

PRACTICE Problems

Find **Additional Practice Problems** in the back of your book.

20. Use the same method to determine the carbon dioxide concentration in 2025.

21. Challenge Assuming that conservation efforts reduce the rate of carbon dioxide increase to 0.3 percent per year, predict the concentration in 2015.

🔲 Review

Additional Practice Problems

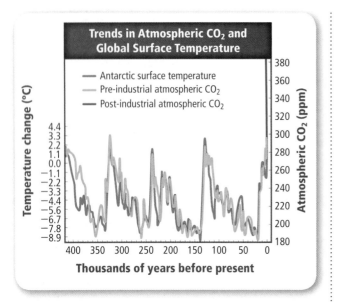

Figure 22 Greenhouse gas concentrations and average annual temperatures follow a similar trend that has fluctuated over the past 400,000 years.

Interpret *How does the trend in temperature illustrated for the past 400,000 years compare to changes in atmospheric carbon dioxide?*

Figure 23 A departure in average annual temperatures reflects an increase in temperature over the past 50 years. Eleven of the warmest years on record occurred between 1990 and 2009.

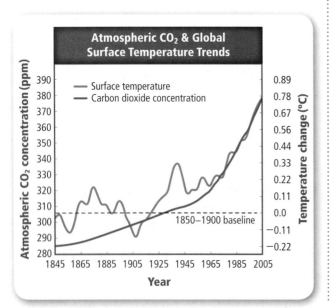

Greenhouse gases Water vapor, carbon dioxide, methane, and nitrogen oxide are all examples of heat-trapping greenhouse gases. These gases contribute to the greenhouse effect by absorbing radiant energy and emitting it back to Earth. Greenhouse gases help to make Earth habitable. Climate records indicate that greenhouse gas concentrations increase and decrease over time, as shown in **Figure 22.** Other observations suggest that atmospheric carbon dioxide has increased 30 percent since the late 1800s.

Human activities, such as the burning of fossil fuels and land development, contribute greenhouse gases to Earth's atmosphere. When fossil fuels burn, carbon in the fuels combines with oxygen in the air to form carbon dioxide.

In addition, carbon dioxide enters the atmosphere due to land-use changes. One type of land-use change is deforestation. Deforestation involves the clearing of large areas of forest land. As part of the carbon cycle, trees remove carbon dioxide from the atmosphere as photosynthesis occurs. With fewer trees, less carbon dioxide is removed from the atmosphere.

Global warming Scientists have discovered that this increase in greenhouse gas concentrations parallels an increase in the average annual temperatures on Earth. An increase in the average temperatures of Earth's near-surface air and oceans is called **global warming.** Climate data suggest that Earth's average temperature has increased by about 0.8°C in the last century, as shown in **Figure 23.** Some of this change can very likely be linked to an increase in atmospheric greenhouse gases, mainly carbon dioxide that results from human activity.

According to some estimates, carbon dioxide could cause global temperatures to increase between 1.8°C and 4°C by 2100. An increase in temperature is not the only cause for concern. In addition to warmer temperatures, climatologists forecast an increase in the occurrence of severe storms, a reduction in glacial ice, a rise in sea level, and a change in ocean circulation.

✓ **Reading Check Describe** the causes and effects of global warming.

El Niño and La Niña

El Niño is the warming of the Pacific Ocean off the coast of western South America that occurs every 3 to 10 years. During a normal year, trade winds push warm surface water westward in the Pacific Ocean, as shown in **Figure 24.** Off the western coast of South America, warm surface water is replaced by cold, nutrient-rich water that rises from below the surface. During an El Niño year, when the trade winds weaken, warm surface water is no longer pushed westward, as shown in **Figure 24.** The upwelling of cold, nutrient-rich water stops, and the surface water remains warm.

Fewer fish and other marine life can be supported by the nutrient-poor warm water. Rainfall in the western Pacific decreases, whereas heavy rain and flooding can occur on the normally dry coast of Peru. El Niño can dramatically alter weather around the world. For example, a strong El Niño can lead to flooding and mudslides in California and droughts in India, Australia, and parts of Africa.

The opposite of El Niño is **La Niña,** which occurs when trade winds in the Pacific are unusually strong and surface water is colder than normal. La Niña can cause drought in the southern United States and excessive rainfall in the northwestern United States.

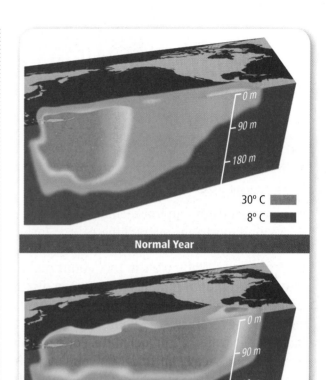

■ **Figure 24** El Niño causes a change in sea surface temperatures in the tropical Pacific that affects weather around the world.

Section 4 Review

Section Summary

▶ Climate can change over time scales ranging from millions of years to several years.

▶ Orbital changes, solar activity, plate movement, and catastrophic events can cause climate changes.

▶ Human activities can influence climate.

▶ El Niño is a warming in the Pacific Ocean that has global effects. La Niña is a cooling of the equatorial Pacific Ocean.

22. **MAIN** 〈**Idea**〉 **Identify** factors that influence climate on a long-term and a short-term scale.

23. **Identify** ways that humans might affect climate.

24. **Explain** how volcanic eruptions can cause climate to change.

25. **Think Critically** What could cause climate changes that occur on a time scale of 50,000 years?

Apply Math

26. **Calculate** Areas of the southeastern United States may receive 15 percent of their rain during hurricanes. If average rainfall for this area ranges from 100–200 cm, how much of this might be from hurricanes?

LAB Microclimates

DESIGN YOUR OWN

Objectives
- **Investigate** how environmental variables respond to various microclimates.

Background: While we can talk about global climate or regional climate, we also can discuss climate on a scale ranging from a few square meters to several square kilometers. Climate at this scale is called a microclimate. For example, because cold, heavy air flows downhill, valley fog sometimes forms in moist, low-lying areas. Along the California coast, coastal fog collects on the needles of redwood trees, drips to the ground, and is absorbed by the tree's shallow root system. This microclimate helps the coastal redwoods survive, nestled in the fog belt along the California coast.

Question: *How do variables such as proximity to water, differences in altitude, population growth, and urban development affect the climate of an area?*

Preparation

Possible Materials
directional compass
thermometer
rain gauge
meterstick
small plastic cups

Safety Precautions
🥽 👕 🧤

Form a Hypothesis

Choose a climate factor, such as sunlight, precipitation, or temperature. Form a hypothesis to explain how this factor influences local climate. Predict how this factor might change in neighboring microclimates. You can choose to compare temperature in a city to temperature in a nearby suburb, or rainfall amounts in the mountains to rainfall that occurs in the valley below, or even sunlight on opposite sides of a tall building.

Make a Plan

1. Decide which microclimate you will investigate and what climate factor you will monitor.
2. How will you measure the selected climate factor? What equipment will you need? How often will you make measurements?

❓ Inquiry Lab

3. Choose sites for making measurements. Consider the microclimate variable you wish to test.

4. Prepare a data table to record your measurements. Will you record the time of the measurements? Should you record the weather at the time of measurements?

5. Before you begin, list the steps of your procedure. Include all materials needed for each step. Does your procedure give you the data necessary to test your hypothesis?

Follow Your Plan

1. Read the procedure and safety information, and complete the lab form.

2. Be sure that your teacher approves your plan before you start.

3. Carry out your experiment as planned. Be sure to record all data in the appropriate places. Follow all appropriate safety precautions.

4. Record all observations and data.

Data Table	Microclimate Observations	
Time	Setting #1	Setting #2
7 am		
10 am		
1 pm		
4 pm		

Analyze Your Data

1. **Create** a graph of your data. Put the microclimate variable on the *y*-axis, and the climate variable on the *x*-axis. For example, if you decide to investigate temperature, place temperature of the microclimate on the *y*-axis and temperature of the climate on the *x*-axis.

2. **Discuss** any trends you see in the data based on your graph.

3. **Infer** any effects that the weather had on your microclimate variable.

Conclude and Apply

1. **Discuss** how your climate factor changed in your microclimate. Did your results support your hypothesis?

2. **Predict** how changes in global climate would affect your microclimate.

COMMUNICATE YOUR DATA

Compare your results with the data of other students who measured the same microclimate variable. Present the combined data to the class.

In the Field
Capturing a Tornado

Meteorologists forecasted severe thunderstorms in eastern Wyoming on June 5, 2009. As storm conditions intensified, most people went into their basements and storm cellars. While these people sought shelter, a group of tornado scientists were anxiously setting up their equipment with the hopes of capturing a tornado.

VORTEX2 Today, tornado scientists are participating in the largest single research experiment called VORTEX2, the successor to *Verification of Rotation in Tornadoes Experiment* (VORTEX) from the 1990s. VORTEX2 has recruited 200 scientists representing an international effort to study tornadoes in greater detail than ever before.

DOWs VORTEX2 equipment is carried on a fleet of 35 specialized vehicles. These vehicles include ten flat-bed trucks mounted with radar dishes, called *Doppler on Wheels* (DOWs), as shown in **Figure 1.** VORTEX2 also has 13 trucks outfitted as weather stations—called mobile mesonets—with four mobile balloon systems, and 24 sticknets. A sticknet is a network of tripods mounted on a vehicle, complete with sensors for measuring ground conditions around a tornado.

Chasing severe weather For five weeks of the 2009 tornado season, the VORTEX2 scientists traveled 16,000 km across nine states chasing after severe storm systems. Every morning, the scientists met to interpret regional weather patterns and to search for areas where tornadoes were most likely to occur. On days when the weather was calm, scientists practiced their tornado intercept strategy. Tornadoes develop suddenly, and the VORTEX2 scientists had to be prepared to navigate roads in the region and quickly position each piece of equipment in a specific area around the storm.

Figure 1 Doppler on Wheels is positioned close to a tornado, but at a distance where it is protected from rain, hail, and strong winds.

On June 5, 2009, VORTEX2 scientists made history when they detected rotation in a cumulonimbus cloud in Wyoming. Like they had done in their tornado intercept drills, they placed the DOWs both ahead of and behind the advancing storm system. They placed sticknets at equal intervals in the path of the storm, and they aligned the mobile mesonet trucks near the storm center.

VORTEX2 scientists watched as clouds began to rotate within the storm. These clouds eventually developed into a powerful tornado. The tornado traveled 10 km toward a team of VORTEX2 photographers, thinned, and eventually dissipated. The tornado lasted less than an hour. The data would take months for the scientists to analyze, but the June 5, 2009 Wyoming tornado has become one of the world's most actively studied tornadoes to date.

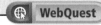
WebQuest

Research Ask students to research how tornadoes are classified and present their findings to the class.

THEME FOCUS Energy
Radiant energy from the Sun heats Earth's surface and atmosphere unevenly, contributing to changes in weather, such as the development of storm systems, and changes in climate, such as an ice age.

BIG Idea Atmospheric properties, such as pressure, temperature, and moisture content, determine the weather and climate conditions in an area.

Section 1 Earth's Atmosphere

greenhouse effect (p. 886)
temperature inversion (p. 885)
troposphere (p. 885)

MAIN Idea Earth's atmosphere can be divided into layers based on differences in composition and temperature.

- Earth's atmosphere is divided into five main layers.
- The uneven heating of Earth's surface causes the movement of air.
- Clouds form from water vapor when rising air cools and condenses around small particles.
- Water is constantly moving between Earth's surface and the atmosphere in the water cycle.

Section 2 Weather

air mass (p. 892)
Coriolis effect (p. 891)
jet stream (p. 891)
weather front (p. 892)

MAIN Idea When cold and warm air masses meet, the weather can change, sometimes resulting in severe storms.

- Differences in air pressure and Earth's rotation produce wind belts in each hemisphere.
- Air flows counterclockwise around lows and clockwise around highs in the Northern Hemisphere.
- Air masses interact along weather fronts.
- Severe weather conditions include thunderstorms, tornadoes, and hurricanes.

Section 3 Climate

continental climate (p. 898)
maritime climate (p. 898)
sea breeze (p. 899)

MAIN Idea Climate is determined by a location's latitude, altitude, and average temperature and precipitation.

- The distribution of solar radiation and the location of pressure and wind systems are factors that affect climate.
- Oceans, land masses, mountains, and large lakes can cause climate variations.
- Climates are classified based on average temperature and precipitation.

Section 4 Earth's Changing Climate

El Niño (p. 905)
global warming (p. 904)
La Niña (p. 905)

MAIN Idea Earth's climate undergoes seasonal and long-term changes.

- Climate can change over time scales ranging from millions of years to several years.
- Orbital changes, solar activity, plate movement, and catastrophic events can cause climate changes.
- Human activities can influence climate.
- El Niño is a warming in the Pacific Ocean that has global effects. La Niña is a cooling of the equatorial Pacific Ocean.

Review Vocabulary eGames

Use Vocabulary

Match each phrase with the correct term from the Study Guide.

27. area of interaction between air masses

28. climate with a strong ocean influence

29. air with uniform temperature and moisture content throughout

30. layer of the atmosphere where most weather occurs

31. global weather event(s) that involve changing sea surface temperatures

32. cool ocean air takes the place of warm air

33. fast-moving, high-altitude air currents

34. increase in average temperature of Earth

35. a region of very stable air that resists rising needed to form clouds and dispel pollution

36. warming of the atmosphere due to energy absorption by certain gases

Check Concepts

37. Which has a large range of temperatures and little ocean influence?
    A) continental climate
    B) El Niño
    C) La Niña
    D) maritime climate

38. Which describes the deflection of wind currents due to Earth's rotation?
    A) trade winds   C) greenhouse effect
    B) Coriolis effect   D) urban heat island

39. Which is the most important factor in determining climate at a given location?
    A) altitude   C) latitude
    B) continents   D) mountains

Use the image below to answer question 40.

40. Which influence on regional climate is shown above?
    A) continental location
    B) lake effect
    C) lee rain shadow
    D) maritime location

41. Which triggers droplet formation in clouds?
    A) evaporation   C) ozone
    B) particles   D) thermals

42. **BIG Idea** What type of weather is most closely associated with a warm front?
    A) drizzle or steady rain
    B) downbursts or windshear
    C) hurricanes or tornadoes
    D) thunderstorms

43. Which is a greenhouse gas?
    A) argon   C) oxygen
    B) nitrogen   D) water vapor

44. A heat island is an example of which type of climate?
    A) tropical climate   C) microclimate
    B) boreal climate   D) dry climate

✓ Assessment   Online Test Practice

Interpret Graphics

45. Illustrate Earth's major pressure belts and wind belts, with labels showing the westerlies and the trade winds.

46. **THEME FOCUS** **Create** a table comparing continental and maritime climates.

Use the table below to answer question 47.

City Temperature

Cities at °N Latitude	Average July Temperature (°C)	Average January Temperature (°C)
72	20	−45
41	21	9
45	29	−2
40	31	−10

47. Identify Based on the data in the table above, identify by latitude the cities that have maritime and continental climates.

Think Critically

48. Explain why the United States is prone to severe weather, such as tornadoes and hurricanes.

49. Explain why the release of latent heat enables an air mass to rise higher in a cumulonimbus cloud.

50. Propose an investigation to evaluate whether human activities have played a role in global warming.

51. Describe how the composition of Earth's atmosphere would be different than it is today if organisms in which photosynthesis occurs had never evolved.

52. Describe how the uneven heating of Earth's surface causes differences in air pressure.

Use the figure below to answer question 53.

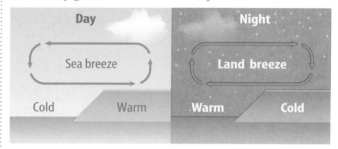

53. Explain why a sea breeze and a land breeze blow in opposite directions.

54. Explain how water droplets form in clouds and the conditions needed to make them fall as some form of precipitation.

55. Predict how a large forest fire might affect the formation of clouds over regions around the forest fire.

56. Explain why the increase of temperature with altitude in the stratosphere prevents air in the troposphere from rising into the stratosphere.

57. Describe what happens to ocean circulation off the western coast of South America during an El Niño event.

Apply Math

58. Convert The high temperature of a summer day in the United States might be 81°F. What would this temperature be in France where they use Celsius temperatures?
Hint: Use the formula $°C = (°F − 32) \times \frac{5}{9}$.

59. Compare City A had 175 cloudy days in 2004. In 2004, 56 percent of city B's days were cloudy. Which city had more cloudy days? Show the calculations you did to find your answer.

Standardized Test Practice

Multiple Choice

Record your answers on the answer sheet provided by your teacher or on a sheet of paper.

1. Which gas makes up most of Earth's atmosphere?
   - A. carbon dioxide
   - B. carbon monoxide
   - C. nitrogen
   - D. oxygen

Use the figure below to answer question 2.

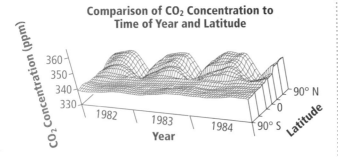

Comparison of CO₂ Concentration to Time of Year and Latitude

2. What causes the seasonal changes in atmospheric carbon dioxide in the Northern Hemisphere?
   - A. El Niño events
   - B. photosynthesis
   - C. weather fronts
   - D. greenhouse effect

3. Which chemical substance accumulated in the stratosphere over millions of years and now helps to shield Earth from ultraviolet radiation from the Sun?
   - A. argon
   - B. nitrogen
   - C. oxygen
   - D. ozone

4. Which is an example of climate?
   - A. today's high temperature
   - B. yesterday's rainfall
   - C. tomorrow's wind speed
   - D. average rainfall in an area over 30 years

5. Which type of air mass would most likely be moist and warm?
   - A. continental polar
   - B. continental tropical
   - C. maritime polar
   - D. maritime tropical

6. Which is a series of windy gusts formed when a downdraft hits Earth's surface with particularly strong force?
   - A. hail
   - B. downburst
   - C. thunderstorm
   - D. tornado

Use the figure below to answer question 7.

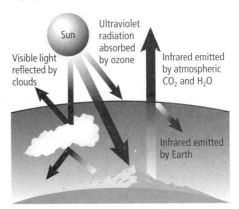

7. Which process illustrated in the image above results from the absorption of infrared radiation by atmospheric gases?
   - A. condensation
   - B. greenhouse effect
   - C. latent heat
   - D. precipitation

8. Which are fast and powerful wind currents embedded in global wind belts?
   - A. jet streams
   - B. sea breezes
   - C. subtropical highs
   - D. westerlies

✓ Assessment   Standardized Test Practice

Short Response

Record your answers on the answer sheet provided by your teacher or on a sheet of paper.

Use the table below to answer question 9.

Land Surface	Percent Reflected Solar Radiation
Bare soil	25
Snow	90
Desert	50
Dense forest	10

9. Which land surface reflects the greatest amount of solar radiation?

10. Atmospheric CO_2 at the beginning of the nineteenth century was 280 ppm. It is presently 385 ppm. What percentage of today's concentration existed at the beginning of the nineteenth century?

11. In the Southern Hemisphere, the Coriolis effect appears to deflect air currents to the left, instead of to the right. Infer the direction of air rotation around a high-pressure system in the Southern Hemisphere.

12. Describe what happens to a mass of air as it rises in the atmosphere.

13. What causes uneven patterns of heating on Earth's surface?

14. Which factors influence climate?

Extended Response

Record your answers on a sheet of paper.

Use the figure below to answer questions 15 and 16.

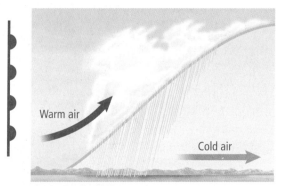

15. What are weather fronts?

16. Explain the interaction between the air masses shown in the image above. What type of front is this and what type of weather will it produce?

Use the graph below to answer question 17.

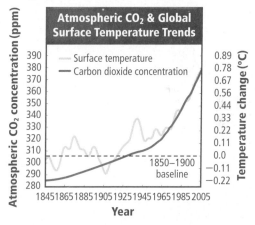

17. Use the graph above to explain the relationship of atmospheric CO_2 concentration and average temperature in °C. If the amount of atmospheric CO_2 decreases, what effect might this have on future temperatures?

NEED EXTRA HELP?																	
If You Missed Question . . .	1	2	3	4	5	6	7	8	9	10	11	12	13	14	15	16	17
Review Section . . .	1	1	1	3, 4	2	2	2	2	2	4	2	2	2	3	2	2	4

UNIT 8

Space

THEMES

Motion and Forces Life cycles of the galaxies depend on gravity and energy.

Scientific Inquiry Models allow scientists to study distant phenomena in the universe.

The Solar System and the Universe Our solar system is only a tiny part of a vast and ever-expanding universe.

Structure and Properties of Matter All the elements in the universe are made from matter formed in stars.

WebQuest **STEM UNIT PROJECT**

Technology Research resources required for astronauts to survive on the Moon. Plan and design a lunar base where astronauts could stay for extended periods of time.

Standardized Test Practice

Multiple Choice

Record your answers on the answer sheet provided by your teacher or on a sheet of paper.

Use the figure below to answer question 1.

1. Which theory explaining the Moon's origin is illustrated above?
   A. binary accretion theory
   B. capture theory
   C. fission theory
   D. giant impact theory

2. How far is Earth's magnetic axis tilted from its geographic axis?
   A. 5°
   B. 11.7°
   C. 15°
   D. 23.5°

3. What kind of eclipse occurs when Earth blocks light from reaching the Moon?
   A. solar
   B. new
   C. full
   D. lunar

4. In what way are Venus and Earth similar?
   A. atmospheric density
   B. liquid water oceans
   C. presence of greenhouse gases in atmosphere
   D. surface temperature

Use the figure below to answer question 5.

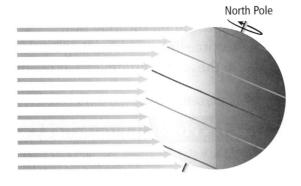

North Pole

5. What day is illustrated in the figure above?
   A. spring equinox
   B. fall equinox
   C. the summer solstice in the Northern Hemisphere
   D. the winter solstice in the Northern Hemisphere

6. Which is a group of constellations through which the Sun appears to move?
   A. ecliptic
   B. equinox
   C. solstice
   D. zodiac

7. Which Earth layer is responsible for generating the strong magnetic field around Earth?
   A. inner core
   B. outer core
   C. mantle
   D. crust

8. During which month of the year is Earth farthest from the Sun?
   A. January
   B. April
   C. July
   D. September

Short Response

Record your answers on the answer sheet provided by your teacher or on a sheet of paper.

9. How much longer is a sidereal month than a lunar phase cycle?

Use the figure below to answer question 10.

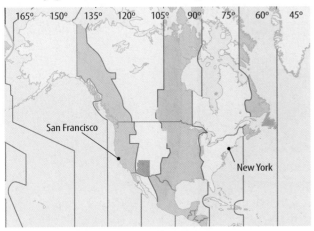

10. If it is 9:00 AM in New York City, what time is it in San Francisco?

11. If the collision of two planetary-sized objects (Earth and another object) formed the Moon, why is the Moon's solid core so small compared to Earth's iron-rich core?

12. What may have caused the CO_2 percentage difference between Earth's atmosphere compared to the atmospheres of Venus and Mars?

13. What is the International Date Line?

Extended Response

Record your answers on a sheet of paper.

Use the figure below to answer question 14.

14. **Part A** Which one of the features in the image above is older—the crater or the lava flow?

   Part B How does interpretation of the Moon's surface features help astronomers to understand the Moon's history?

15. How does the tilt of Earth's rotational axis affect the length of a day throughout the year?

16. How have recent Moon missions helped scientists to understand the Moon's formation?

NEED EXTRA HELP?																
If You Missed Question . . .	1	2	3	4	5	6	7	8	9	10	11	12	13	14	15	16
Review Section . . .	3	1	3	1	2	2	1	2	3	2	3	1	1	3	2	3

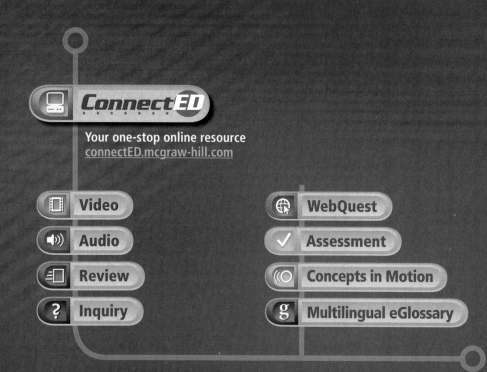

CHAPTER 30
The Solar System

ConnectED

Your one-stop online resource
connectED.mcgraw-hill.com

- Video
- Audio
- Review
- Inquiry
- WebQuest
- Assessment
- Concepts in Motion
- Multilingual eGlossary

Launch Lab
A Scale Model of the Solar System

Even if you consider only the planets, the solar system is enormous compared to objects that you see in your daily life. You might be able to better understand the size of the solar system by using a scale model. Distances between the orbits of the planets are measured in astronomical units (AUs). One AU equals 150 million km, the average distance between Earth and the Sun.

For a lab worksheet, use your StudentWorks™ Plus Online.

 **Inquiry** Launch Lab

FOLDABLES®

The Solar System
Make a layered-look book to help you organize information about the planets.

Neptune
Uranus
Saturn
Jupiter
Mars
Earth
Venus
Mercury

Planets

THEME FOCUS The Solar System and the Universe

The solar system originated as an interstellar cloud of dust and gas collapsed to form the Sun, planets, dwarf planets, and other objects.

BIG Idea Gravitational forces formed the solar system and cause the planets to orbit the Sun.

Section 1

Reading Preview

Essential Questions

▶ How do geocentric and heliocentric models of the solar system differ?

▶ What is each planet's position in the solar system?

▶ How are planets classified?

Review Vocabulary

ellipse: an oblong, closed curve drawn around two foci

New Vocabulary

geocentric model
heliocentric model
extrasolar planet

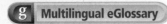
g Multilingual eGlossary

■ **Figure 1** Ptolemy proposed his geocentric model of the solar system in 140 A.D. In the geocentric model of the solar system, the Sun, the Moon, and the other planets revolve around Earth. Ptolemy's model accounted for the backward motion of Mars, which is described today as retrograde motion.

Planet Motion

MAIN ◀Idea **All objects that orbit the Sun move in elliptical orbits.**

Real-World Reading Link You are going for a run on your high school track. There are eight lanes. Which lane will you choose? If you choose the outermost lane—lane eight—you will run farther than if you choose lane one. The planets also travel different distances depending on their positions relative to the center of our solar system.

Models of the Solar System

Think of how difficult it would be to construct a model of the solar system if you did not know that Earth rotates. At night, planets appear to move across the sky in a path around Earth. It is easy to see why many early scientists thought that Earth was the center around which everything that they saw in the sky revolved.

Geocentric model In the **geocentric model** of the solar system, Earth is considered the center and everything else revolves around it. Scientists explained this model by saying that the planets, the Sun, and the Moon were embedded in separate spheres that rotated around Earth. They thought that the stars were imbedded in another sphere that also rotated around Earth. The sphere bearing the stars moved in a regular, predictable way, but those bearing the planets seemed to move erratically against this background. For this reason, they were called *planasthai*, a Greek word that means "to wander." Our word *planet* comes from this meaning. These ideas led to the model of the solar system proposed by the Greek philosopher Ptolemy, as shown in **Figure 1**.

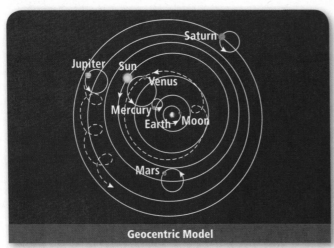

Geocentric Model

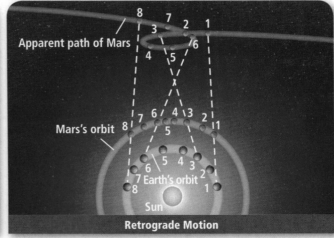

Retrograde Motion

Geocentric modifications Although Ptolemy's model of the solar system was accepted and used for over a thousand years, it had many problems. One problem was the fact that planets periodically appeared to move in a retrograde, or backward, motion when viewed against a background of stars, as shown in **Figure 1.** To account for this retrograde motion, Ptolemy's model described a complicated system of circular paths that the planets followed.

Heliocentric model In spite of its complexity, the Earth-centered model of the solar system endured for centuries. However, in 1543, Polish astronomer Nicholas Copernicus proposed a different model based on the ideas of an early Greek philosopher named Aristarchus. The **heliocentric model** stated that the Moon revolves around Earth and that Earth and the other planets revolve around the Sun. In this model, the apparent motion of the planets, the stars, and the Sun is due to Earth's rotation. The Sun is considered the center of the solar system in the heliocentric model, as shown in **Figure 2.** Copernicus placed the known planets in the correct order in circular orbits around the Sun.

✓ **Reading Check Compare and contrast** heliocentric and geocentric models of the solar system.

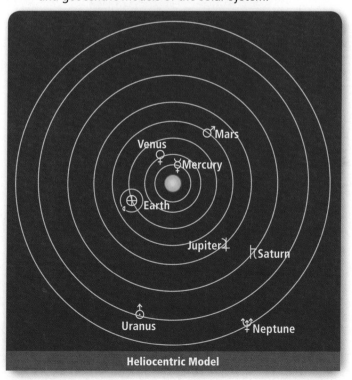
Heliocentric Model

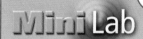
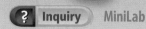

? Inquiry MiniLab

Model Retrograde Motion

Procedure
1. Read the procedure and safety information, and complete the lab form.
2. On a piece of **graph paper,** label the x-axis *Right ascension.* Separate it into 15-minute increments of time, from 10:00 P.M. to 12:00 A.M.
3. Label the y-axis *Declination.* Separate it into 2° increments, from 0° to 20°.
4. Plot and label these positions for Jupiter: Oct. 1, 2003 (10h 38m, 10°); Jan. 1, 2004 (11h 21m, 6°); April 1, 2004 (10h 51m, 9°); May 1, 2004 (10h 44m, 9°); Aug. 1, 2004 (11h 20m, 6°).
5. Connect the plotted positions in chronological order with a solid line.

Analysis
1. **Interpret** What do you notice about Jupiter's direction of motion?
2. **Explain** Using a heliocentric model of the solar system, explain why Jupiter appears to move as it does.

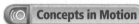
Concepts in Motion

Animation

■ **Figure 2** In the heliocentric model of the solar system, Earth and the other planets revolve around the Sun.

Describe *the model of the solar system that is accepted today.*

Galileo Using his telescope, Italian astronomer Galileo Galilei discovered evidence to support Copernicus' heliocentric model of the solar system. He observed that Venus has phases like the Moon has phases. These phases could be explained only if Venus orbited the Sun and passed between Earth and the Sun on each orbit. He also discovered moons in orbit around Jupiter. These observations convinced him that everything did not revolve around Earth. He concluded that Earth and Venus orbit the Sun and that the Sun is the center of the solar system.

Understanding the Solar System

The heliocentric model of the solar system is now accepted as fact. Earth and the other planets revolve around the Sun. The apparent motion of the planets and stars in the night sky is due to Earth's rotation as it travels around the Sun.

In the early 1600s, the German astronomer Johannes Kepler discovered that the planets move around the Sun in ellipses rather than perfect spheres. Kepler also found that the planets are located at different orbital distances relative to the Sun. The closer a planet is to the Sun, the faster it moves. As a result, planets farther from the Sun take longer to orbit than the innermost planets. For example, Mercury takes just 88 Earth days to complete an orbit while Neptune takes 164 Earth years.

Apply Science

How can you use heliocentric longitude (H.L.) to chart planet position?

A planet's heliocentric longitude is the position of the planet on any day of any year. This position is plotted by measuring the angle between a line of the spring equinox in the Northern Hemisphere and the position of the planet on its orbit. This measurement is done counterclockwise because that is how planets appear to move when viewed from above the North Pole of Earth.

Identify the Problem
What were the approximate positions of Mercury, Venus, Earth, Mars, and Jupiter on March 4, 2004?

Solve the Problem
Using this diagram as a model, make a larger drawing on a piece of paper. The H.L.s for Mercury, Venus, Earth, Mars, and Jupiter on March 4, 2004, are 344°, 103°, 164°, 84°, and 164°, respectively.

Use a protractor to measure the H.L. angles, and mark the position of each planet on its orbit in the diagram. Label it with its name or planet symbol: Mercury, Venus, Earth, Mars, and Jupiter.

1. **Compare and contrast** your model of planetary positions for Mercury, Venus, Earth, Mars, and Jupiter on March 4, 2004, to their positions on January 1, 2009. The heliocentric longitudes on January 1, 2009, for Mercury, Venus, Earth, Mars, and Jupiter are 6°, 49°, 101°, 269°, and 302°, respectively.

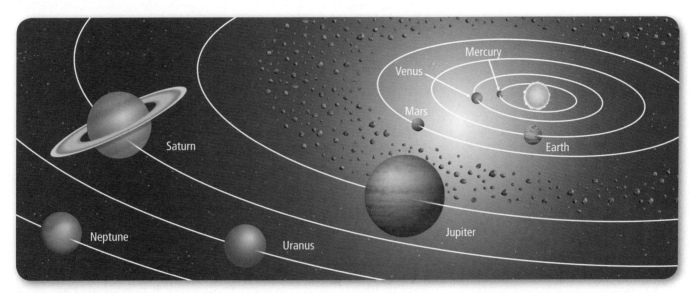

Saturn

Venus

Mercury

Mars

Earth

Jupiter

Neptune

Uranus

Classifying the planets

The planets are classified in several ways. Planets similar to Earth in size, structure, and composition are called terrestrial planets. Terrestrial is from the Latin word *terra,* meaning Earth. The giant planets are called *Jovian* planets from the English name for Jupiter, *Jove*.

Planets are also classified by location. The system used most often classifies planets whose orbits are between the Sun and the asteroid belt as the inner planets and those beyond the asteroid belt as the outer planets, as shown in **Figure 3.** A third system classifies planets whose orbits are between Earth's orbit and the Sun as inferior planets, and those whose orbits are beyond Earth's orbit as superior planets.

Origin of the solar system

The composition and motion of the planets suggest that the solar system formed when a large cloud of gas and dust called a nebula contracted under the force of gravity. Many scientists believe that this contraction was triggered by the explosion of a nearby star about 4.6 billion years ago. The nebula would have been spinning slowly initially and sped up as contraction continued.

Density within the solar nebula increased. Gravity pulled dust and gas toward the center, and the cloud contracted even more. It also began to rotate faster, much like a figure skater's rotation speed increases as she pulls her arms in. The nebula flattened into a disk with a dense center called a solar nebula.

Temperature within the solar nebula increased. When the temperature soared above ten million degrees Celsius, nuclear reactions began to fuse hydrogen atoms into helium atoms and the Sun formed at the center of the rotating disk. Viewed from above the north pole of the Sun, the disk rotated in a counterclockwise direction. The remaining matter condensed to form the planets and other objects that travel in counterclockwise orbits around the Sun.

■ **Figure 3** Planets in the solar system are classified as terrestrial or Jovian planets. They are also classified as inner or outer planets. The Jovian planets are sometimes called the gas giants.

Identify *the Jovian planets and the outer planets.*

VOCABULARY ....
ACADEMIC VOCABULARY
Contract
to be drawn together or to shrink in size
Your diaphragm expands and contracts as you breathe. ....

Other Solar Systems

Until recently, our solar system was the only one of its kind known. Now we know that many other stars in the universe have planets that revolve around them. **Figure 4** shows an artist's rendition of a gas giant orbiting a star. **Extrasolar planets** are planets in orbit around stars other than the Sun. These planets are helping astronomers infer how planetary systems like our solar system formed.

As of summer 2010, astronomers have discovered more than 350 extrasolar planets that orbit stars like the Sun. Many more likely exist. Some of these planetary systems have been discovered around stars that are similar in size, age, and structure to the Sun. Most of the planets that have been found are large and massive like Jupiter. If large planets exist, it is possible that smaller and less massive planets like Earth also exist in these planetary systems.

New satellites and space probes developed by the United States' National Aeronautics and Space Administration (NASA) and other international space agencies continue to search for Earth-like planets around stars. It is difficult because less massive planets located light-years away from Earth are more difficult to find than giant planets. Even so, some of the planets that have been discovered are comparatively low in mass and a few of these have masses less than Neptune's mass. Scientists hypothesize that this indicates that many more planets with lower masses exist.

■ **Figure 4** NASA has discovered over 350 extrasolar planets that orbit stars similar in size, age and structure to the Sun.

Section 1 Review

Section Summary

▶ A geocentric model places Earth at the center of the solar system, whereas a heliocentric model places the Sun at the center.

▶ Planets travel in elliptical orbits around the Sun.

▶ Planets are classified by their locations and characteristics.

▶ Extrasolar planets have been discovered orbiting Sun-like stars.

1. **MAIN Idea** **Summarize** Johannes Kepler's observations of planetary motion.

2. **Compare and contrast** geocentric and heliocentric models.

3. **Classify** planets in two ways using their positions in the solar system.

4. **Define** the term *terrestrial planet*.

5. **Describe** how the solar system formed.

6. **Think Critically** Would a year on Venus be longer or shorter than an Earth year? Explain.

Apply Math

7. **Calculate** Use the equation (planet's period in years)^2 = (planet's distance in AUs)^3 to find the orbital period of a planet that is 19.2 AU from the Sun.

Reading Preview

Essential Questions

▶ What are the important surface features of the inner planets?

▶ How does Earth's atmosphere compare to the atmospheres of Mercury, Venus, and Mars?

▶ What are the successes and failures of various missions to Mars?

Review Vocabulary

robot lander: part of a spacecraft that lands on another planet or other object

New Vocabulary

Mercury
Venus
Earth
Mars

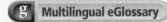
g Multilingual eGlossary

? Inquiry   Virtual Lab

■ **Figure 5** Two probes have flown by Mercury: *Mariner 10* from 1974–1975 and *MESSENGER* from 2008–2009. These pictures from the *MESSENGER* reveal a cratered surface including a crater resembling a giant paw print.

Identify *another object in space that has similar surface features.*

The Inner Planets

MAIN ◄Idea The inner planets are mainly made of rock surrounding iron cores.

Real-World Reading Link Picture yourself in the produce aisle at the grocery store. What types of fruit do you see? Apples and pears have similar structures. Both have a core and seeds, but their colors, textures, and tastes are different. Like these fruits, the inner planets have similar structures but different compositions and surface features.

Planets Near the Sun

The solar system formed when a nebula collapsed about 4.6 billion years ago. Gradually, planets, dwarf planets, moons, asteroids, meteoroids, and comets began to form from the flattened disk. Close to the young Sun, temperatures were high enough to prevent light gases in the disk from condensing. As a result, planets close to the Sun are made of heavier elements, such as iron, silicon, and oxygen. These inner planets have iron cores surrounded by solid rock.

Mercury The smallest and closest planet to the Sun is **Mercury.** Mercury revolves around the Sun in 88 days, faster than any other planet in the solar system. Mercury is covered by craters, as shown in **Figure 5.**

Mariner 10 was the first U.S. space probe sent to explore Mercury. *Mariner 10* discovered a magnetic field around Mercury, suggesting that Mercury has a much larger iron core than expected and is missing some lighter materials in its mantle. One theory to explain Mercury's interior is that two similarly sized bodies collided. They merged to form one large iron core, and lighter material vaporized and was lost to space.

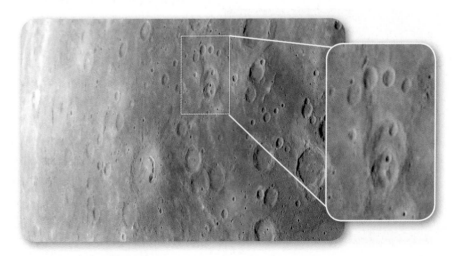

■ **Figure 6** Venus has an extreme greenhouse effect due to large concentrations of carbon dioxide in the atmosphere.

FOLDABLES

Incorporate information from this section into your Foldable.

■ **Figure 7** Radar images of volcanoes with visible lava flows on Venus were captured by the *Magellan* probe.

Mercury's surface Mercury's relatively large core and thin outer layers resulted in some extreme differences in its surface topography. For example, Mercury's surface is covered with deep craters and steep cliffs up to 3 km high. NASA's *MESSENGER* spacecraft completed three flybys of Mercury and created a mosaic of nearly 90 percent of Mercury's surface from January 2008 through September 2009. The spacecraft has completed a gravity assist and will enter Mercury's orbit for a more detailed study of the closest planet to the Sun in 2011.

Is there an atmosphere around Mercury? Mercury is small with a weak gravitational pull. Its surface experiences temperature extremes, suggesting that the planet has no atmosphere. However, data collected by *Mariner 10* hinted at an atmosphere. It was soon discovered that these gases were deposited by the solar wind and vapors that diffused upward through the crust. Mercury has no true atmosphere. The lack of an atmosphere, along with Mercury's close proximity to the Sun, produces temperature extremes ranging from 430°C to –170°C. Mercury, however, is not the hottest planet in the solar system. Its closest planetary neighbor, Venus, is the solar system's hottest planet.

Venus The second closest planet to the Sun is **Venus.** Venus is often referred to as Earth's sister planet. Their sizes and masses are almost identical. However, one major difference is that the entire surface of Venus is blanketed by a dense atmosphere. The atmosphere has 92 times the surface pressure of Earth's atmosphere at sea level. It is composed mostly of carbon dioxide. The clouds in the atmosphere also contain droplets of sulfuric acid and water vapor which help create distinct layers, as shown in the false color ultraviolet image in **Figure 6.**

The greenhouse effect Clouds on Venus are so dense that only about two percent of sunlight reaches the planet's surface. This has made studies of Venus's surface features virtually impossible. Heat radiated from Venus's surface is absorbed by the carbon dioxide in the atmosphere, producing an extreme greenhouse effect. Due to the greenhouse effect, temperatures on the surface of Venus are extreme, between 450°C and 475°C. Venus is the hottest planet in the solar system.

In the 1970s, the former Soviet Union sent the *Venera* landers to Venus. One of the landers photographed and mapped the surface before being destroyed by intense heat and pressure. The *Venera* landers discovered that Venus is hot and dry. The landers photographed many surface features on Venus including mountains, valleys, and inactive volcanoes. In 1995, the *Magellan* probe was the first U.S. spacecraft to send back detailed radar images of these landforms, as shown in **Figure 7.**

Earth The third planet from the Sun is **Earth.** Unlike other planets in the solar system, temperatures on Earth allow water to exist as a solid, a liquid, and a gas on its surface. Water makes life possible here. Earth's atmosphere is composed of mostly nitrogen and oxygen with trace amounts of carbon dioxide, water vapor, argon, and other gases. Earth has a greenhouse effect, although less extreme than the greenhouse effect on Venus, with an average temperature of 15°C. Without a greenhouse effect, average temperatures on Earth would be below freezing. The presence of ozone (O_3) in the atmosphere protects life from intense solar radiation.

Life on Earth Life has been discovered in extreme environments on Earth where temperatures, pressures, and even acidity and alkalinity can vary. For example, an entire ecosystem of exotic marine organisms has been discovered living under great pressure near volcanic vents on the seafloor. Organisms have also been found living under the ice in deep polar waters and in acidic hot springs near volcanoes, as shown in **Figure 8.** These discoveries have inspired scientists to search for life elsewhere in the solar system.

■ **Figure 8** Life exists in extreme environments on Earth. Microorganisms thrive in acidic hot springs with average temperatures of 75°C.

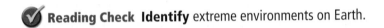 **Reading Check** **Identify** extreme environments on Earth.

Mars The fourth planet from the Sun is **Mars.** Mars is also called the red planet because of the presence of iron oxide minerals in some of the weathered rocks on its surface gives it a red color, as shown in **Figure 9.** Mars's surface color also varies seasonally due to the common occurrence of severe dust storms. Mars rotates on a tilted axis at an angle of 25°, which is very close to Earth's tilt. Because of this, Mars also experiences changes in seasons as it orbits the Sun, much like Earth does. The average temperature on Mars is about –60°C. The polar ice caps on Mars, composed of frozen carbon dioxide and frozen water, grow during winter and shrink during summer.

■ **Figure 9** Iron oxide minerals present in weathered rocks exposed on Mars's surface produce the planet's distinct red color. For this reason, Mars is often referred to as the red planet.

Reading Check **Describe** the composition of polar ice caps on Mars.

Changes in coloration and the size of ice caps were observed from Earth long before rovers were sent to explore the planet. Some people saw these changes and thought that life-forms like lichen might survive on Mars. They hypothesized that these life-forms grew more during the summer, making the surface look darker. We know now that what actually causes this seasonal change in coloration is wind blowing lighter colored dust off the darker layers below. Despite this fact, the search for evidence of water and past life on Mars still continues today.

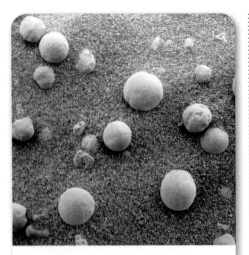

■ **Figure 10** These spherical concretions found on Mars provide evidence of Mars's wet past.

? **Inquiry** **Video Lab**

Mars's atmosphere The Martian atmosphere is much thinner than Earth's atmosphere and is composed of mostly carbon dioxide with some nitrogen, argon, and traces of oxygen and water vapor. The thin atmosphere does not filter out intense solar radiation like Earth's atmosphere, and surface temperatures on Mars range from –140°C to 20°C. This temperature difference produces strong winds that can cause global dust storms. As the seasons change during the Martian year, winds blow dust around on the planet's surface. When the wind blows dust off one area, it may appear darker.

Martian moons Mars has two small, heavily cratered moons called Phobos and Deimos. These names are from the Greek words for "fear" and "panic." Phobos is slowly spiraling inward toward Mars and is expected to hit the Martian surface in about 50 million years. Deimos, which is one of the smallest known moons in the solar system, orbits far above the surface.

Was Mars once wet? Currently, Mars appears to be a very dry world. But has it always been dry? In late 2003, NASA sent the Martian landers *Spirit* and *Opportunity* in search of liquid water. **Figure 10,** an image taken by *Opportunity,* shows round deposits on Mars that have been labeled as "blueberries." These spherical concretions are composed of hematite, an iron oxide mineral that forms in the presence of liquid water on Earth.

Figure 11, an image taken by the *Mars Reconnaissance Orbiter* in August 2009, shows sediment layering and gullies on Mars. These surface features look like they were made by running water. Wind might have formed the layering of sediment and gullies, but more evidence, including sediment deposits discovered in gullies, suggests that liquid water was once present on Mars. Scientists today know that water ice remains frozen all year in the polar ice caps.

■ **Figure 11** This photograph taken by the *Mars Reconnaissance Orbiter* reveals evidence that suggests water flowed on the Martian surface at one time in the past.

Identify *features that indicate water may have carved this landscape.*

NASA on Mars

Mars has been a priority of NASA planetary exploration for several decades. Although not all missions to Mars have been successful, several missions have returned a tremendous amount of data. One of the earliest missions to Mars was the *Mariner 9* space probe that orbited Mars from 1971–1972. This probe revealed long channels carved by flowing water. *Valles Marineris* (the valley of the mariners), as shown in a computer-generated image in **Figure 12,** is a large canyon that was discovered by *Mariner 9.* It is so large that it would stretch from San Francisco to New York City if on Earth. It probably formed when Mars was still geologically active. *Mariner 9* also discovered large, extinct volcanoes on Mars. One of these volcanoes, Olympus Mons, is the largest known volcano in the solar system. It is over three times higher than Mount Everest.

The Viking probes In 1976, the *Viking 1* and *Viking 2* orbiters photographed the entire surface of Mars. Each spacecraft consisted of an orbiter and a lander. While their landers conducted meteorological, chemical, and biological experiments on the planet's surface, the orbiters traveled around the planet. Although the initial results of biological tests provided data that seemed to indicate the presence of life, scientists soon realized that this data could also indicate chemical reactions not associated with life. So far, experiments have found no conclusive evidence of life in the soil.

VOCABULARY ....
WORD ORIGIN
Volcano
comes from the Italian word *vulcano* after the Roman god of fire
The volcano ejected a cloud of ash that darkened the sky for days. ....

■ **Figure 12** Mars's *Valles Marineris* is six to seven times deeper, ten times longer, and twenty times wider than the Grand Canyon in Arizona. It is a series of rift valleys, formed sometime in Mars's past by geologic activity rather than by water erosion.

■ **Figure 13** The Mars Exploration Rovers *Spirit* and *Opportunity* are equipped with panoramic cameras that have photographed features on the Martian surface for over six years.

Opportunity, shown here in an artistic rendering of the rover, discovered evidence of a shallow, salty sea bed at its landing site on the *Meridiani Planum.*

These light-colored volcanic rocks and fine-grained soils were photographed by *Spirit* in Gusev Crater. Cracks within these rocks indicate evidence that water once flowed there.

This photograph was taken by *Opportunity* in an area called Victoria Crater.

Global Surveyor, Pathfinder,* and *Odyssey Onboard cameras of NASA's 1996 *Global Surveyor* spacecraft showed that the walls of *Valles Marineris* have distinct layers similar to those of the Grand Canyon in Arizona. They also showed that a large plateau, resembling a dried-up seabed or mudflat covers a large area of Mars's northern hemisphere.

Evidence indicates that water is frozen within Mars's crust in the form of permafrost. In 1996, NASA's *Mars Pathfinder* and its rover, *Sojourner,* collected data that indicates the iron oxide minerals in Mars's crust may have been leached out by groundwater. Launched in 2001, the *Mars Odyssey* orbiter discovered frost stored within a thin layer of soil near Mars's poles.

Mars Exploration Rover mission

Until 2003, most of the information about Mars, its atmospheric structure and composition, its geology, and its surface features came from *Mariner 9,* the *Viking* probes, *Mars Global Surveyor, Mars Pathfinder,* and *Odyssey.* The most recent data comes from NASA's *Mars Reconnaissance Orbiter* and the Mars Exploration Rover mission and its two landers, *Spirit* and *Opportunity,* which actively collected data on Mars for more than six years.

Recall that the round concretions, called blueberries, indicate water might have once existed on the surface of Mars. These concretions were discovered by the *Opportunity* rover in *Meridiani Planum. Opportunity* confirmed that the concretions were hematite deposits, or iron oxide minerals. The fact that the hematite was deposited on top of other surface features suggests that it was deposited in a standing body of water. Other deposits of hematite have been found by the *Spirit* rover in Gusev Crater.

✔ **Reading Check** **Define** What are blueberries?

The Mars Exploration rovers continue to provide a great deal of evidence that suggests liquid water existed on Mars in the past. Photographs of the Martian surface taken by *Spirit* and *Opportunity* are shown in **Figure 13.**

960 **Chapter 30** • The Solar System

Martian Meteorites

Space probes sent to Mars are not the only way we have learned about our nearest planetary neighbor. Meteorites have been discovered on Earth that originated on Mars. In 1984, a group of scientists discovered a meteorite from Mars in Antarctica which they named ALH84001. The meteorite was ejected by an impact event on Mars about 16 million years ago. However, an interesting discovery was announced in 1996 that initiated debate over the possibility of life on Mars. Scientists discovered long, microscopic structures in the meteorite, shown in **Figure 14,** that they believed were fossilized evidence of ancient life.

■ **Figure 14** Scientists found strange microscopic structures in a Martian meteorite fragment found in Antarctica.

✓ **Reading Check Identify** What structures were discovered in the meteorite ALH84001?

This discovery began a debate in which many scientists suggested other possible explanations for the microscopic structures. The majority of scientists do not believe these structures are evidence of past life. Evidence indicates that processes unrelated to living organisms likely formed these structures. Research continues today to determine whether life exists, or may have once existed, on Mars.

Section 2 Review

Section Summary

▶ Mercury has an extreme range of temperatures, many craters, and long, steep cliffs.

▶ Venus has an extreme greenhouse effect.

▶ Surface temperatures on Earth allow water to exist as a solid, a liquid, and a gas.

▶ Earth is the only planet known to support life as we know it.

▶ Mars has seasons like Earth, and there is evidence to suggest that water flowed on its surface sometime in the past.

8. **MAIN Idea Compare and contrast** the size, shape, surface features, and atmospheric conditions of the four inner planets.

9. **Describe** the atmosphere of Venus.

10. **Explain** why scientists believe that Mars was once a wetter planet.

11. **Infer** why the Mars Exploration Rover mission is considered a success.

12. **Think Critically** Explain why Mars is a more likely planet for humans to visit than Venus.

Apply Math

13. **Use Percentages** If the diameter of Mars is 53.3 percent of Earth's diameter (12,756 km), determine its diameter in kilometers.

14. **Convert Units** If Earth takes 365.25 days and Mars takes 686.98 Earth days to orbit the Sun, how many Earth years does Mars take to orbit once around the Sun?

✓ **Assessment** Online Quiz

Section 2 • The Inner Planets **961**

LAB   Surface Features on Mars

Objectives

- **Identify** surface features shown in a photograph of Mars.
- **Interpret** what processes were involved in the formation of these surface features.

Background: Scientists study Mars using photographs and data collected by spacecraft that have explored the Martian surface. Comparing surface features on Earth to those discovered on Mars enables scientists to hypothesize about what processes might have formed these features. You can interpret some of these Martian surface features using NASA photographs.

Question: *How can you infer surface processes that have shaped the Martian surface over time?*

Preparation

Materials

photograph of the Martian surface from NASA
clear, plastic sheet
scissors
erasable marker
transparent removable tape

Safety Precautions

Procedure

1. Read the procedure and safety information, and complete the lab form.
2. Cut the clear, plastic sheet to fit as an overlay on the photograph. Tape it over the photograph.
3. Use the erasable marker to mark, trace, or circle the features that you can recognize, such as craters, lava flows, flow structures (channels), terraces, and sediment layers.

4. Indicate the direction that a flow of lava or water may have traveled.

Conclude and Apply

1. **Create** a table to show the data you will collect in this lab.
2. **Identify** and describe all of the features that you were able to recognize in the photograph in your data table.
3. **Identify** the process that might have formed each feature in your data table.
4. **Infer** from which direction sunlight was shining on the Martian surface when the photograph was taken.
5. **Evaluate** Which questions cannot be answered with a simple photo analysis of Martian surface features?

Data Table	
Surface Feature	**Formation Process**

Explain to friends, classmates, or your family why a comparison between surface features on Earth and Mars can help NASA scientists to interpret surface processes on Mars.

? Inquiry   Lab

Reading Preview

Essential Questions

▶ How do the outer planets compare and contrast?

▶ What are some important characteristics of each outer planet?

▶ What are the successes of NASA missions Galileo and Cassini to explore the outer solar system?

Review Vocabulary

space probe: an instrument that is sent to space to gather information

New Vocabulary

Jupiter
Saturn
Uranus
Neptune
dwarf planet
comet
asteroid
meteoroid
Sedna

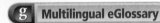 **Multilingual eGlossary**

The Outer Planets

MAIN ⟨Idea The outer planets are mainly made of light elements and simple compounds.

Real-World Reading Link Picture different types of melons, such as cantaloupe and honeydew, in the produce aisle of the grocery store. These fruits have a mushy core, and the thickness of their fruit and rinds differs from apples and pears. The melons are like the outer planets, larger in size with a very different composition from the inner planets.

Planets Far from the Sun

The inner and the outer planets are very different. The outer planets formed far from the Sun, where temperatures were colder. As a result, compounds like water (H_2O) exist only as ice. Closer to the Sun, H_2O exists as a gas, with Earth and Mars as the only exceptions. Because of the lack of heavy elements and the presence of water-ice and gases, such as hydrogen, helium, methane, and ammonia, the outer planets grew fast.

Jupiter The fifth planet from the Sun is **Jupiter.** Jupiter is the largest planet and contains more than twice the mass of all of the other planets combined. It is composed of about 90 percent hydrogen and 10 percent helium with trace amounts of ammonia, methane, and water vapor. Below Jupiter's gaseous atmosphere, shown in **Figure 15,** there is a layer of liquid metallic hydrogen subject to extreme pressure. Scientists believe that a metallic core, 5–15 times the mass of Earth, resides beneath this layer. Temperatures in the core could be as high as 40,000°C.

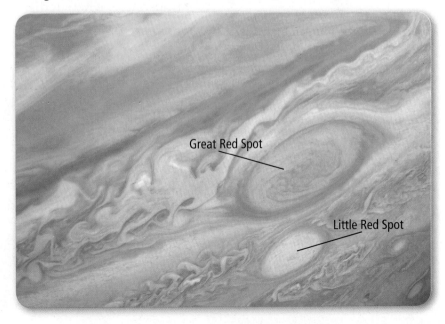

Great Red Spot

Little Red Spot

■ **Figure 15** The Great Red Spot on Jupiter is an atmospheric storm that has raged for more than 300 years. The Little Red Spot has wind speeds that rival its larger and older counterpart.

Explain *What is the Great Red Spot?*

MiniLab

Compare Diameters of the Outer and Inner Planets

Procedure

1. Read the procedure and safety information, and complete the lab form.
2. Research the outer and inner planets to find their diameters in kilometers.
3. Select a scale to draw the planets. The scale must be such that the small planets can be seen and the larger planets can fit onto your piece of **paper.**
4. Using your scale, draw a circle with the diameter of the smallest planet.
5. Using the same scale, draw circles representing each of the other planets.

Analysis

1. **Compare** How do the sizes of the inner planets compare to those of the outer planets?

■ **Figure 16** More than 60 moons orbit Jupiter. The four largest moons are Ganymede, Callisto, Io, and Europa.

You have probably seen pictures of Jupiter's colorful bands of clouds. Its atmosphere has bands of white, red, and brown clouds. The Great Red Spot, shown in **Figure 15,** is the most spectacular of these cloud formations, although Jupiter's Little Red Spot, discovered in 2005, rivals its older relative.

The *Voyager* probes and *Galileo* *Voyager 1* and *Voyager 2* flew past Jupiter in 1979. The *Galileo* space probe reached Jupiter in 1995. These probes studied the composition, structure, and motion of Jupiter's atmosphere and many features of its moons, including the active volcanoes on Io. The probes also searched for new moons and discovered faint rings around Jupiter.

Jupiter's moons Jupiter has 63 known moons. Many are small, rocky bodies that could be captured asteroids. However, four moons are large enough to be considered small planets. Italian astronomer Galileo was the first to discover these moons with his telescope. The four Galilean moons shown in **Figure 16** are Ganymede, Callisto, Io, and Europa. Ganymede is the largest moon in the solar system. Ganymede is larger than Mercury.

Io experiences a constant tug-of-war between the gravities of Jupiter and Europa. This heats up Io's interior and causes it to be the most volcanically active object in the solar system. Io's volcanoes were first seen in photographs from the *Voyager* probes.

Data sent by the *Galileo* space probe suggest that the three other Galilean moons most likely have oceans of water trapped beneath their ice-rock crusts. Using gravity and magnetic field measurements, NASA scientists have generated a model that shows the existence of water beneath the surface on Europa, Ganymede, and Callisto. Some scientists speculate that Europa's ocean is very large and could possibly support life. When *Galileo's* mission ended, NASA crashed the probe into Jupiter's atmosphere, where it disintegrated during its descent.

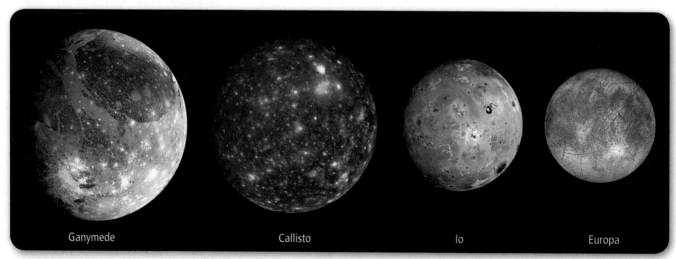

| Ganymede | Callisto | Io | Europa |

Saturn The sixth planet from the Sun is **Saturn.** It is the second-largest and has the lowest density of any planet in the solar system—even less dense than water. Like Jupiter, Saturn is composed of hydrogen and helium with traces of methane, ammonia, and water vapor. Its interior has a similar structure to Jupiter. Below the atmosphere is a thick layer of liquid hydrogen and likely a rocky core. Saturn has the largest and most complex ring system of any planet, shown in **Figure 17.**

Saturn's rings Saturn's rings are made of particles of water-ice. These particles range in size from millimeters to tens of meters in diameter. Each ring particle orbits Saturn. The rings appear as seven broad bands, each composed of thousands of thin ringlets. Even though the ring system is over 280,000 km wide, the rings are less than 3 km thick. The rings are so thin that they seem to disappear when seen on edge. Most scientists believe the rings resulted from a collision that caused a moon to break apart 4 billion years ago.

The structure of the rings is caused by the gravitational interaction between ring particles and Saturn's moons. These interactions cause the divisions within the rings. The gravitational forces due to some moons, called shepherd moons, keep ring particles from spreading outward. As a result, the rings seem to have sharp edges.

Cassini-Huygens The *Cassini-Huygens* spacecraft approached Saturn in July 2004. Since then, the *Cassini* orbiter has collected data about Saturn, its ring system, and its many moons. Saturn is now known to have at least 60 moons. A number of these moons were discovered by *Cassini's* cameras.

Cassini-Huygens is actually two spacecrafts in one. *Cassini* completed its four-year mission orbiting Saturn. The second spacecraft, the *Huygens* probe, was released from *Cassini* as it passed by Saturn's largest moon, Titan. In January 2005, *Huygens* descended to Titan's surface. It discovered methane lakes and hydrocarbon sand dunes on Titan, as shown in **Figure 18,** and evidence for a liquid ammonia lake beneath the surface. Hydrocarbons are compounds that contain hydrogen and carbon, the building blocks of life.

■ **Figure 17** *Cassini-Huygens* provided detailed views of Saturn and its rings. Saturn's largest ring was discovered by NASA's *Spitzer Space Telescope* in October 2009, orbiting six million kilometers away from the planet.

■ **Figure 18** The *Huygens* probe descended through Titan's atmosphere and photographed diverse surface features that scientists believe could include both continents and oceans.

Uranus

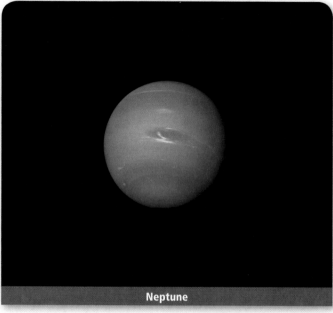
Neptune

■ **Figure 19** Uranus and Neptune are about the same size. Both Uranus and Neptune have faint cloud bands and a ring system.

Compare and contrast characteristics of Uranus and Neptune.

VOCABULARY · · · · · · · · · · · · · · ·
SCIENCE USAGE V. COMMON USAGE
Ring
Science usage
(planetary ring) ice and rock particles that orbit a planet
The Cassini *spacecraft photographed Saturn's ring system.*

Common usage
a band of jewelry worn around your finger.
Her wedding ring sparkled in the summer sunlight. · · · · · · · · · · · · · · ·

Uranus The seventh planet from the Sun is **Uranus.** The only space probe to study Uranus was *Voyager 2*, which completed its flyby in January 1986. Uranus is four times larger in diameter than Earth and has about 27 moons and a thin ring system made of eleven rings. Unlike Earth and the other planets, Uranus's axis of rotation is tilted so that it is nearly parallel to the plane of the planet's orbit, as shown in **Figure 19.**

Uranus's atmosphere contains about 83 percent hydrogen, 15 percent helium, 2 percent methane, and other trace gases. Methane in the upper atmosphere absorbs red light from the Sun, and the planet's clouds reflect blue-green light. This produces Uranus's blue-green color. Unlike Jupiter and Saturn, Uranus probably has a high-pressure ocean of liquid water, methane, and ammonia on its surface and a rocky core.

Neptune The eighth planet from the Sun is **Neptune.** Neptune, as shown in **Figure 19,** is about the same size as Uranus. It was discovered in 1846 after studies of Uranus indicated that its orbit was being affected by the gravity of another unseen planet. *Voyager 2* completed its flyby in 1989 and is also the only space probe to have ever visited Neptune.

Neptune's atmosphere contains about 3 percent methane, making it appear bluer than Uranus. Neptune's atmosphere is fueled by internal heat, creating the fastest winds (2,000 km/h) in the solar system. Neptune has a rocky core surrounded by an icy mantle and a layer of liquid metallic hydrogen at the surface. Similar to Uranus, Neptune also has a faint ring system. Neptune has at least 13 moons. Triton is the largest moon with a diameter of 2,700 km and a thin atmosphere composed of mostly nitrogen, although some methane is also present. *Voyager 2* detected nitrogen geysers erupting on Triton.

Dwarf Planets

Prior to 2005, Pluto was recognized as the ninth planet in the solar system. However, in 2005, astronomers discovered Eris (ER ihs), an object in our solar system larger than Pluto. This discovery led to a change in the definition of a planet by the International Astronomical Union (IAU) in August 2006. Pluto, Eris, and Ceres are now classified as dwarf planets. **Dwarf planets** are nearly round objects in orbit around the Sun that are not satellites and that have not cleared the debris in their orbits. In 2008, the IAU recognized two additional dwarf planets, with about a dozen added to the list of potential candidates. The newest dwarf planets classified include Makemake and Haumea.

Pluto, Eris, and Ceres Discovered in 1930, Pluto has a diameter of 2,300 km, a thin atmosphere, and a surface of solid, icy rock. Pluto takes about 248 Earth years to orbit the Sun. Pluto's three moons are shown in **Figure 20.** Eris, discovered in 2005, is about 2,400 km in diameter and has one moon. Eris orbits the Sun about once every 557 Earth years. With a diameter of 940 km, Ceres was discovered in the asteroid belt in 1801. A year on Ceres is equal to about 4.6 Earth years.

Comets and Other Objects

A **comet** is composed of dust and rock particles, frozen water, methane, and ammonia. As a comet approaches the Sun, it begins to vaporize due to heat from the Sun. The released dust and gases form a bright cloud called a coma around the nucleus, which is the solid part of the comet. The solar wind pushes on the vaporized coma, forming a tail that always points away from the Sun, as shown in **Figure 21.** Most comets originate from a vast disk of icy objects beyond Neptune's orbit—called the Kuiper Belt—or from the Oort Cloud, a cloud of comets that surrounds the solar system.

Asteroids Rocky objects formed from material similar to the material in planets' composition are called **asteroids.** Most asteroids are found in between the orbits of Mars and Jupiter. Asteroids vary in size from tiny particles to roughly 500 km in diameter.

Meteoroids Sand- to boulder-sized objects in the solar system are called **meteoroids.** Meteoroids are smaller than asteroids—so small they are not observable until they enter Earth's atmosphere. When a meteoroid enters the atmosphere, it heats up and emits light. Most burn up completely, and we see them as meteors, or "shooting stars." Those that impact Earth's surface are called meteorites.

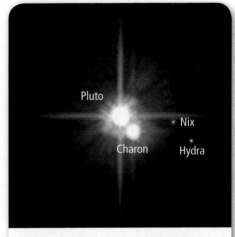

■ **Figure 20** Pluto's three moons are Charon, Nix, and Hydra. Some astronomers argue that with three moons in orbit, Pluto should be promoted back to planetary status. The dwarf planet debate rages on today.

■ **Figure 21** Some comets have predictable orbits, such as Halley's comet, which passes near Earth every 76 years.

Figure 22 In this artistic rendition of Sedna, the distant planetoid orbits the Sun in the Kuiper Belt beyond Pluto.

Sedna Another object in the solar system that has puzzled astronomers is Sedna, shown in **Figure 22. Sedna** has been labeled a distant planetoid. With a diameter of 1,200–1,700 km, it is smaller than Pluto but larger than comets in the Kuiper Belt. Also, it has an elliptical orbit; Sedna travels within the range of 76–950 AU from the Sun. These greater distances are far beyond the Kuiper Belt but much closer than the Oort Cloud. Recall that one astronomical unit (AU) is equal to 150 million kilometers.

Some scientists suggest that the Sedna's orbit was severely affected by a passing star less than 100 million years after the Sun formed. This supports the idea that our Sun formed as part of a star cluster. A close encounter with another star so soon after the Sun's formation could only be explained if the Sun formed within a cluster.

Another puzzle is Sedna's apparent rate of rotation of 40 days. This is best explained if the object has another object in orbit around it. Even if Sedna has no companion, there may be other similar objects out there, some even larger than Pluto. At present, objects this far away with orbits this elliptical can be detected only when they are closest to the Sun.

Section 3 Review

Section Summary

▶ Jupiter is the largest planet in the solar system and has more than sixty moons.

▶ Saturn has the most complex system of rings.

▶ Uranus's axis of rotation is tilted on its side, and Neptune's atmosphere generates the strongest winds ever measured.

▶ Pluto, Eris, and Ceres are dwarf planets.

▶ Along with planets and dwarf planets, asteroids, comets, and meteoroids orbit around the Sun.

15. **MAIN Idea Compare and contrast** the compositions and structures of Jupiter, Saturn, Uranus, and Neptune.

16. **Describe** the composition of Saturn's rings.

17. **Identify** three discoveries made by the *Galileo* space probe.

18. **Think Critically** Explain what would happen to the density of Saturn if its atmospheric layers were compressed toward the center of the planet.

Apply Math

19. **Use Numbers** If the diameters of Earth and Jupiter are 12,756 km and 142,984 km respectively, how many Earths could fit side-by-side across Jupiter's diameter?

20. **Use Decimals** On average, if Earth is 1 AU from the Sun and Venus is 0.71 AU from the Sun, what is the distance from Earth to Venus in kilometers? (1 AU = 150,000,000 km)

✓ Assessment   Online Quiz

Reading Preview

Essential Questions

▶ What does life in extreme environments on Earth suggest about the possibility of life elsewhere in the solar system?

▶ What conditions do scientists look for when evaluating the possibility of life on a planet or moon?

▶ Where is NASA searching for life beyond planet Earth?

Review Vocabulary

fossil: remains, imprints, or traces of past life

New Vocabulary

extraterrestrial life

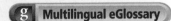 **g** Multilingual eGlossary

Life in the Solar System

MAIN ◀ Idea Conditions on Mars and the moons Europa and Titan might have supported life.

Real-World Reading Link For decades, sci-fi movies and books have hinted at the possibility of life beyond Earth. It is possible to find strange life-forms in extreme environments here on Earth. Is it possible that life exists elsewhere within our solar system?

Life As We Know It

Life exists all around us—in the air, on rocks, and in the sea. It even thrives in extreme environments on Earth, as shown in **Figure 23.** Does it exist anywhere else in the solar system, and if so, where? Life on Earth is carbon-based and requires water to survive. Does water exist elsewhere in the solar system in sufficient amounts to make life possible?

Think about what you consider to be life. You might think of yourself, your friends, or your pets. There is so much diversity to life on Earth. There is life that we can see, such as grass, trees, and animals. There are also life-forms that we cannot see. For example, there are microorganisms that are so small that you need a microscope to view them.

If you were asked to search for evidence of tiny life-forms in a soil sample, what would you do? You might look for moving things, but motion is not always indicative of life. You might look for residues or traces of life that once lived in the soil, such as fossils. You might check to see if gases are being emitted from the soil. Think about what type of tests you could perform on Martian soil samples to search for life.

■ **Figure 23** Tube worms and other life-forms living near hydrothermal vents on the seafloor get their energy from bacteria that consume methane and hydrogen sulfide. Predators, such as crabs, feed on the community.

Infer *How are conditions on the seafloor different from other areas of the ocean?*

■ **Figure 24** Heat- and acid-tolerant microorganisms survive in geothermal environments, like the hot springs and mud pots in Yellowstone National Park.

■ **Figure 25** Mono Lake contains large amounts of dissolved salts and lye. Despite its alkalinity, Mono Lake is home to a wide variety of plants and animals.

Exotic life on Earth Before we rule out the possibility of life on other planets or moons, we should take a closer look at life on Earth, where there are exotic organisms living in the most extreme environments. These organisms provide clues as to how to search for life elsewhere in our solar system.

In the 1970s, scientists piloting a research submarine named *Alvin* discovered some interesting life-forms while exploring volcanic vents on the seafloor. These life-forms included unusual species of crabs, clams, and tube worms. The species were living under extreme conditions, thriving in high-pressure, deep, dark, and cold environments.

Alvin discovered bacteria colonies thriving off of superheated, chemically-enriched fluids erupting from volcanic vents on the seafloor. These organisms use thermal and chemical energy to survive and are the foundation for a deep-sea food chain. They produce energy by a process called chemosynthesis. Chemosynthetic organisms are able to produce energy in the presence of certain chemical compounds, such as hydrogen sulfide and methane, and in the absence of any sunlight.

A surprising amount of life has been found surviving in other extreme environments on Earth—boiling hot springs, arctic sea ice, salty dried-up sea beds, and dark, cold, and inhospitable caves. For example, there are microscopic life-forms that thrive around geothermal features, like hot springs and mud pots at Yellowstone National Park, Wyoming, shown in **Figure 24.** Three types of bacteria thrive in the alkaline mud of Mono Lake, California, as shown in **Figure 25,** where the environment is dark, lacks oxygen, and contains a large amount of salt and lye. What does this tell us about the ability of life to exist in extreme environments on other planets or moons?

Can life exist on other worlds?

Think about the conditions on Earth that make life possible. None of the planets or moons you have studied is identical to Earth, but some share characteristics in common with exotic environments on Earth. Can you say for certain that life does not exist elsewhere in our solar system?

The high temperatures found on Mercury and Venus and the lack of a solid surface on all of the gas giants make finding life as we know it unlikely. This does not rule out the possibility, however. Life beyond Earth is called **extraterrestrial life.** There are places where scientists called exobiologists are searching for extraterrestrial life.

Mars During most of Earth's history, single-celled organisms were the only life-forms on Earth. Could such life-forms be present on Mars today, or could they have existed on Mars in the past when the planet was wet? Evidence from recent missions to Mars suggests that the planet had large amounts of liquid water on its surface in its past. These missions have also revealed evidence that water-ice exists on the Martian surface today, as shown in **Figure 26.**

The *Viking* landers and the *Mars Exploration Rovers* performed tests designed to detect life on Mars. The first tests showed positive results, but further evidence suggested that these results could be explained just as well by chemical reactions that did not require living organisms. Ultimately, the *Viking* experiments were unsuccessful. However, ground- and space-based telescopes have discovered a small amount of atmospheric methane on Mars. The presence of methane indicates that something must be producing it, most likely a geologic process, but possibly a biologic process. Whether life existed on Mars in the past or if it is thriving today, the search for extraterrestrial life is an ongoing quest.

Europa Data from the *Galileo* probe sent to explore Jupiter and its moons indicate that three of the Galilean moons have ice-covered oceans. Ganymede and Callisto's subsurface oceans might be partially frozen. However, surface cracking, as shown in **Figure 27,** tidal bulges, and apparent movement of rafts of ice on the surface, support the idea of a liquid ocean on Europa.

Europa's ocean could hold more than twice the amount of water that Earth's oceans hold. Some scientists speculate that Europa's ocean could harbor life. One reason is that it might be deep and long-lasting, possibly even liquid water that is warmed by tides from Jupiter's gravitational pull on Europa.

On Earth, we find life everywhere that there is water. If life can exist in superheated, chemically-enriched fluids erupting from volcanic vents on the seafloor, could it survive in a similar environment on Europa? Ocean currents have been detected on Europa. Today, some scientists consider Europa to be the most promising location to look for extraterrestrial life, as shown in **Figure 28.**

✓ **Reading Check** **Identify** two places where NASA is searching for life.

■ **Figure 26** The *Mars Reconnaissance Orbiter* discovered debris aprons on the Martian surface in 2005. Astronomers believe that these aprons are made rock and sediment deposited by water-ice.

■ **Figure 27** Surface cracking on Europa may indicate that a liquid ocean lies beneath.

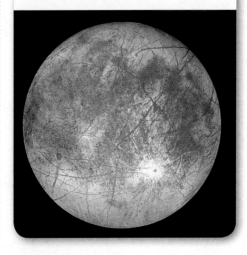

FIGURE 28

Visualizing Life on Europa

NASA scientists are searching for extraterrestrial life on Europa. Although Europa is far from the Sun, it is believed that Jupiter's strong gravitational pull warms the planet from the inside-out. Despite its cold, icy exterior, liquid water beneath the ice on Europa could sustain life. Scientists are searching for clues in Lake Vostok, Antarctica, to support their claims.

Lake Vostok is a freshwater lake that resides, under high pressure conditions, nearly two miles beneath the Antarctic ice sheet. Scientists have discovered communities of microbes called extremophiles living in deep, cold, and dark environments miles beneath the ice.

Scientists extract ice cores from above Lake Vostok in search for extremophiles. Life here could lead to the discovery of extraterrestrial life-forms in extreme environments elsewhere in our solar system. Lake Vostok in Antarctica is, therefore, an ideal place to study the possibility of extraterrestrial life on Jupiter's moon Europa.

Scientists believe that a liquid water ocean exists beneath Europa's cracked icy surface. Considering Lake Vostok in Antarctica, if life can exist in deep, dark, and cold environments here on Earth, is it possible that life exists on Europa?

Titan Saturn's moon Titan is larger than Mercury and has an atmosphere composed of mostly nitrogen with some argon and methane. Surface features on Titan indicate this moon has an interesting geologic history. Darker areas contain relatively pure water-ice, and brighter areas are composed of hydrocarbons. A bright cloud of the hydrocarbon methane is visible near Titan's south pole. Evidence suggests that a dynamic atmosphere exists on Titan, as shown in **Figure 29.**

The presence of hydrocarbons on the surface of Titan interests exobiologists. Also, the apparent absence of large impact craters indicates that Titan has experienced internal geologic activity. This activity could provide the energy needed for organic molecules to develop into the building blocks of life.

However, when the *Huygens* probe landed on Titan in 2005, it found that the surface at the landing area was dry. Although there was evidence for liquid having been present in the past, there seemed to be no seas of liquid hydrocarbons.

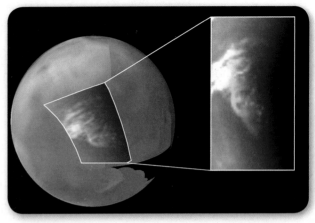

■ **Figure 29** This infrared image of Titan's atmosphere taken by *Cassini* in June 2009 shows visible cloud formations. Atmospheric gases include mostly nitrogen, argon, and methane, which is of interest to exobiologists.

Section 4 Review

Section Summary

▶ Life on Earth is carbon-based and requires liquid water to survive.

▶ Life exists in extreme environments on Earth.

▶ Mars and Europa have conditions that may be conducive to extraterrestrial life.

21. **MAIN ⟨Idea⟩ Define** extraterrestrial life.

22. **Describe** how discoveries of life in extreme environments on Earth support the idea that life may exist elsewhere in the solar system.

23. **Explain** how superheated, chemically-enriched fluids from volcanic vents on Earth's seafloor provide energy to support life.

24. **Identify** locations in the solar system where life might have once existed.

25. **Think Critically** Explain the similarities between extreme environments on Earth and on Jupiter's moon Europa.

Apply Math

26. **Use Ratios** Atmospheric pressure on Venus is 92 times Earth's atmospheric pressure at sea level. If atmospheric pressure on Earth at sea level is 101.3 kPa, find the atmospheric pressure on the surface of Venus in kPa.

27. **Use Percentages** The average composition of Mars's atmosphere is 95 percent carbon dioxide, 3 percent nitrogen, and 2 percent argon. Approximately what fraction of Mars's atmosphere is nitrogen?

LAB

Life in Soil

DESIGN YOUR OWN

Objectives

- **Identify** characteristics that you might see if life existed in an unknown sand or soil sample.
- **Demonstrate** procedures you could use to search for life in the unknown sample.
- **Describe** evidence of life that you could observe in each of your samples.
- **Compare and contrast** observations that may or may not indicate that life could exist in your samples.

Background: What is life? You know that living organisms need water and food to survive. The type of food depends on the life-form. You consume food for energy every day. Other life-forms require different sources of energy. You know that many plants make their own food through a process called photosynthesis. You have learned that life-forms on Earth can also obtain their energy from a process called chemosynthesis. Suppose you were given a soil sample from an extreme environment here on Earth. How might you determine if life is present in the sample? You might look for evidence that an organism was taking in one gas and expelling others. Or you might look for energy being produced.

Question: *How could you determine if an unknown sample of sand or soil contained life?*

Preparation

Possible Materials

clear plastic cups (3)
sand or soil
antacid
dry yeast
warm water
magnifying glass
thermometer

Safety Precautions

Form a Hypothesis

Form a hypothesis about what evidence might indicate that a sand or soil sample contains life. Have your teacher approve your hypothesis before you begin your experiment.

Make a Plan

1. Read the procedure and safety information, and complete the lab form.
2. As a group, agree upon the hypothesis and decide how to test it. Identify which results will confirm your hypothesis.
3. Define the steps in detail that you will need to test your hypothesis. Include a control run.
4. Identify the materials that you will need to perform the experiment. Determine who will prepare your unknown samples.
5. Prepare a data table where you will record your observations.

6. Read the entire experiment to make sure all the steps will be performed in a logical order.

7. Identify all constants, variables, and controls of the experiment. Keep in mind that the person who prepares your unknowns must follow your instructions carefully and not divulge which sample(s) contain life.

Follow Your Plan

1. Make sure that your teacher approves your plan and oversees the preparation of your unknown samples before you begin.

2. Conduct your experiment as planned.

3. While performing the experiment, record your observations in your data table.

Analyze Your Data

1. **Create** drawings of each unknown sand or soil sample.

2. **Label** your drawings with all of your observations, such as production of heat or change in odor.

3. **Interpret** your data. Evaluate any sources for error that might exist in the design of this investigation.

Conclude and Apply

1. **Examine** Do the illustrations show the observations that you expected when following your procedure? Describe any unexpected observations.

2. **Determine** Do your observations indicate that life exists in any of your unknown samples? Describe the conclusive evidence for life and any inconclusive observations that you make.

3. **Evaluate** Did your results support your initial hypothesis? Explain why or why not.

4. **Explain** Design a plan to test for the existence of life in extreme environments on Earth. Describe the types of environments that you might include in this investigation.

COMMUNICATE YOUR DATA

Compare your data with those of other students. Discuss how your experiments may have differed and how this could have changed the observations that you made.

SCIENCE & TECHNOLOGY

Killer Asteroids

On June 30, 1908, a meteor with a 37 m diameter entered Earth's atmosphere and exploded high above the Tunguska region of Siberia, Russia. Although the meteor did not impact Earth, the force of the explosion—equal to 185 Hiroshima-size nuclear bombs—flattened forests for 2,100 km^2.

Past meteor impacts Asteroids have done some serious damage on Earth in the past. Most scientists agree that the meteor that struck Earth off the coast of present-day Mexico caused the dinosaur extinction event about 65 mya. In October 2009, researchers discovered a second meteor crater over 480 km in diameter off the west coast of present-day India. This impact event occurred about 300,000 years after the meteor struck in Mexico, and likely contributed to the same mass extinction event.

The Spaceguard Survey Currently, there are no large asteroids on a collision course with Earth. Scientists say the probability of a large meteor striking Earth in the near future is highly unlikely. In 1998, NASA established the Spaceguard Survey, a program that uses telescopes to search for near-Earth asteroids (NEAs). NEAs, such as the asteroid shown in **Figure 1,** are asteroids that have orbits that direct their course toward Earth because of gravity exerted by planets in our solar system.

Deflecting asteroids The Spaceguard Survey has identified more than 90 percent of NEAs that are 1 km or larger in diameter, large enough to cause a catastrophic impact. New asteroids are discovered every day, so these estimates will continue to change. It is possible that scientists will discover an asteroid that is on a collision course with planet Earth. Is there any way to avoid an impact?

Figure 1 This near-Earth asteroid named Mathilde is approximately 50 km in diameter.

One option that scientists have considered—firing missiles at an asteroid in order to destroy it—is risky. Instead of being destroyed, the asteroid might break into pieces. One of the pieces could still enter Earth's orbit. Other pieces might enter an orbit that could still cause them to strike Earth at a later time.

Another asteroid-deflecting strategy is called a "gravity tractor." In this scenario, a large spacecraft hovers near an asteroid for several months or years. Gravity attracts the asteroid to the spacecraft and slowly alters the asteroid's orbit. The spacecraft uses thrusters to avoid being drawn in by the asteroid's gravity. This method would be effective if scientists had at least several years' warning of a future asteroid impact. Scientists have also discussed simply slamming a spacecraft into an asteroid in order to change its orbit.

WebQuest

Research near-Earth asteroids. Prepare a NASA special report that summarizes details of the most recent near-Earth asteroid flyby to present to the class.

THEME FOCUS The Solar System and the Universe
Gravitational forces caused a collapsing nebula to form the Sun, the planets, dwarf planets, and many other celestial objects in our solar system.

BIG Idea Gravitational forces formed the solar system and cause the planets to orbit the Sun.

Section 1 Planet Motion

extrasolar planet (p. 954)
geocentric model (p. 950)
heliocentric model (p. 951)

MAIN Idea All objects that orbit the Sun move in elliptical orbits.

- A geocentric model places Earth at the center of the solar system, whereas a heliocentric model places the Sun at the center.
- Planets travel in elliptical orbits around the Sun.
- Planets are classified by their locations and characteristics.
- Extrasolar planets have been discovered orbiting Sun-like stars.

Section 2 The Inner Planets

Earth (p. 957)
Mars (p. 957)
Mercury (p. 955)
Venus (p. 956)

MAIN Idea The inner planets are mainly made of rock surrounding iron cores.

- Mercury has an extreme range of temperatures, many craters, and long, steep cliffs.
- Venus has an extreme greenhouse effect.
- Surface temperatures on Earth allow water to exist as a solid, a liquid, and a gas.
- Earth is the only planet known to support life as we know it.
- Mars has seasons like Earth, and there is evidence to suggest that water flowed on its surface sometime in the past.

Section 3 The Outer Planets

asteroid (p. 967)
comet (p. 967)
dwarf planet (p. 967)
Jupiter (p. 963)
meteoroid (p. 967)
Neptune (p. 966)
Saturn (p. 965)
Sedna (p. 968)
Uranus (p. 966)

MAIN Idea The outer planets are mainly made of light elements and simple compounds.

- Jupiter is the largest planet in the solar system and has more than sixty moons.
- Saturn has the most complex system of rings.
- Uranus's axis of rotation is tilted on its side, and Neptune's atmosphere generates the strongest winds ever measured.
- Pluto, Eris, and Ceres are dwarf planets.
- Along with planets and dwarf planets, asteroids, comets, and meteroids orbit around the Sun.

Section 4 Life in the Solar System

extraterrestrial life (p. 971)

MAIN Idea Conditions on Mars and the moons Europa and Titan might have supported life.

- Life on Earth is carbon-based and requires liquid water to survive.
- Life exists in extreme environments on Earth.
- Mars and Europa have conditions that may be conducive to extraterrestrial life.

Use Vocabulary

Compare and contrast the following pairs of terms.

28. Uranus, Neptune

29. extrasolar planet, extraterrestrial life

30. dwarf planet, Sedna

31. Venus, Earth

32. Mercury, Jupiter

33. asteroid, Sedna

34. Jupiter, Saturn

35. Earth, Mars

Check Concepts

36. Who proposed the modern heliocentric model of the solar system?
    A) Copernicus   C) Kepler
    B) Galileo   D) Ptolemy

37. Which planet has the most complex ring system?
    A) Jupiter   C) Saturn
    B) Neptune   D) Uranus

38. Which planet is the red planet?
    A) Jupiter   C) Neptune
    B) Mars   D) Venus

39. Which chemical or gas produces a green-blue to blue color on Uranus and Neptune?
    A) carbon dioxide
    B) methane
    C) oxygen
    D) sulfur dioxide

40. Which planet has the highest surface temperature?
    A) Earth   C) Mercury
    B) Mars   D) Venus

41. Which phrase describes all Jovian planets?
    A) inner planets
    B) outer planets
    C) lie beyond Saturn's orbit
    D) lie beyond the Kuiper belt

42. Which Galilean moon has active volcanoes?
    A) Callisto   C) Ganymede
    B) Europa   D) Io

Use the figure below to answer question 43.

43. Which planet is shown?
    A) Jupiter
    B) Mars
    C) Saturn
    D) Venus

44. What is a reason that Europa might be a likely environment for extraterrestrial life?
    A) liquid methane
    B) presence of organic molecules
    C) rust-colored rocks
    D) liquid ocean

✓ Assessment   Online Test Practice

Interpret Graphics

45. Make a table summarizing the gases contained in the atmospheres of the planets.

Use the figure below to answer question 46.

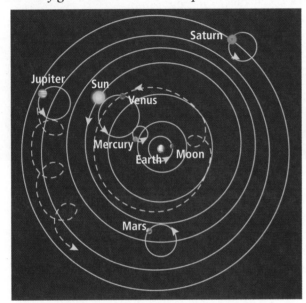

46. **BIG Idea** What is the difference between the geocentric and heliocentric models of the solar system? Which is shown above?

Use the table below to answer questions 47–48.

Planet Orbits			
Planet	Distance to Sun (AU)	Planet	Distance to Sun (AU)
Mercury	0.39	Jupiter	5.20
Venus	0.72	Saturn	9.54
Earth	1.00	Uranus	19.19
Mars	1.52	Neptune	30.07

47. Which planet is a little over nine times farther from the Sun than Earth is from the Sun?

48. How far is the orbit of Mars from that of Jupiter?

Think Critically

49. **THEME FOCUS** **Describe** how the solar system formed.

50. **Explain** why Venus has a higher surface temperature than Mercury, even though Mercury is closer to the Sun.

51. **Develop** a hypothesis about where the oxygen that combined with the iron in rocks on Mars originated.

52. **Explain** why Mars and Europa interest scientists searching for extraterrestrial life. Do Mars and Europa resemble any extreme environments on Earth today?

Apply Math

Use the diagram below to answer question 53.

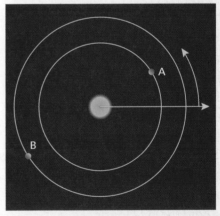

53. **Use Angles** If the Heliocentric Longitude (H.L.) of planet A is 33 degrees, what is the approximate H.L. of planet B?

54. **Calculate** Use the equation $V = \frac{4}{3}\pi r^3$ to determine the volume of Venus, which has a radius of 6,052 km.

Standardized Test Practice

Multiple Choice

Record your answers on the answer sheet provided by your teacher or on a sheet of paper.

Use the figure below to answer question 1.

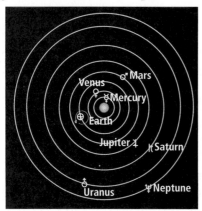

1. Which modern astronomer proposed this model of the solar system?
   A. Ptolemy
   B. Kepler
   C. Galileo
   D. Copernicus

2. Which object's gravity holds the planets in their orbits?
   A. Earth
   B. Jupiter
   C. the Sun
   D. Mercury

3. Which did Galileo observe about Venus?
   A. It goes through phases.
   B. It has two moons.
   C. It is covered by clouds.
   D. Its surface temperature is over 450°C.

4. Who determined that planets orbit the Sun in elliptical orbits?
   A. Brahe
   B. Copernicus
   C. Kepler
   D. Ptolemy

Use the figure below to answer question 5.

5. Identify the comet's tail in the image above.
   A. 1
   B. 2
   C. 3
   D. 4

6. Which is composed of dust and rock particles mixed with frozen water, methane, and ammonia?
   A. asteroid
   B. comet
   C. meteoroid
   D. meteorite

7. Which gas giant is the second largest planet?
   A. Jupiter
   B. Saturn
   C. Uranus
   D. Neptune

8. Which planet revolves around the Sun the fastest?
   A. Mercury
   B. Venus
   C. Saturn
   D. Earth

✓ Assessment   Standardized Test Practice

Short Response

Record your answers on the answer sheet provided by your teacher or on a sheet of paper.

9. Saturn takes 29.5 Earth years to complete one orbit. How many years will it take Saturn to complete 195° of its orbit?

10. The orbital radius of Jupiter is 5.2 AU. What fraction of that distance is Earth's orbital radius?

Use the figure below to answer question 11.

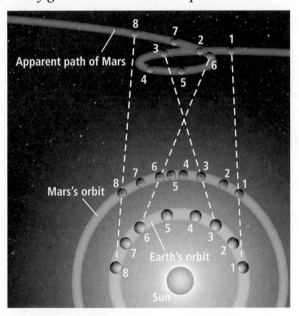

11. How does this image support the heliocentric model of the solar system?

12. What causes Mars to have a reddish color?

13. What is generally true of all Jovian planets?

14. Why does it appear unlikely that Venus is capable of supporting life as we know it?

Extended Response

Record your answers on a sheet of paper.

Use the figure below to answer question 15.

15. **PART A** What surface features shown above did the NASA rovers *Spirit* and *Opportunity* discover on Mars?
    PART B Why is this discovery so important?

16. Explain why scientists believe that life might exist beneath Europa's ice-covered surface?

17. Would a year on Venus be longer or shorter than a year on Earth? Explain.

NEED EXTRA HELP?																	
If You Missed Question . . .	1	2	3	4	5	6	7	8	9	10	11	12	13	14	15	16	17
Review Section . . .	1	1	2	1	3	3	3	2	1	1	1	2	3	2	4	4	1, 2

CHAPTER 31
Stars and Galaxies

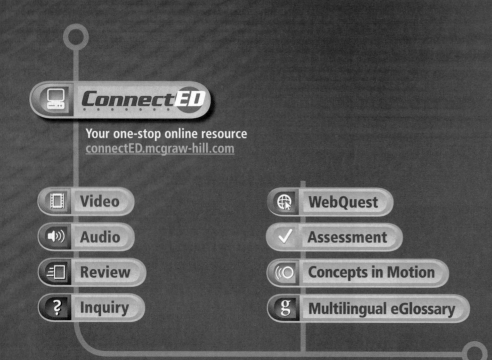

ConnectED

Your one-stop online resource
connectED.mcgraw-hill.com

- Video
- Audio
- Review
- ? Inquiry
- WebQuest
- ✓ Assessment
- Concepts in Motion
- g Multilingual eGlossary

Launch Lab
Stars in the Sky

Have you ever looked up at the night sky and been amazed by the number of stars that you see? The number of stars that you see depends on your location. In this lab, you will explore a quick way to estimate how many stars you can see in different parts of the night sky.

For a lab worksheet, use your StudentWorks™ Plus Online.

? Inquiry   Launch Lab

FOLDABLES

Cosmology Construct a three-tab book to help identify what you know, what you want to know, and what you learn about stars, galaxies, and cosmology.

Know? | Like to Know? | Learned?

THEME FOCUS The Solar System and the Universe

Our solar system is a subsystem of the universe, which includes all celestial objects.

BIG Idea Stars and galaxies form as a result of gravity.

Reading Preview

Essential Questions

▶ What is a constellation?

▶ How do optical telescopes differ?

▶ What are the similarities and differences between radio telescopes and optical telescopes?

Review Vocabulary

electromagnetic spectrum: the entire range of wavelengths of electromagnetic energy

New Vocabulary

constellation
radio telescope
light-year
spectroscope

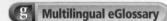

g Multilingual eGlossary

Observing the Universe

MAIN ⟨**Idea**⟩ Telescopes form images by collecting and focusing the electromagnetic radiation emitted by distant objects.

Real-World Reading Link Sometimes when you watch clouds drift by, you can identify shapes and patterns in them. One cloud might look like a ship; another might resemble a rabbit or a bear. In a similar way, people long ago looked for patterns in the stars.

Constellations

People of long ago named the patterns in stars in the night sky after characters in stories, animals, and even tools. Many of the names given to these groups of stars by ancient cultures remain today and are called **constellations.**

Astronomers use constellations to locate and name stars. Ancient cultures used constellations to navigate from place to place, to create calendars, and to predict changes in the seasons. For example, Orion, shown in **Figure 1,** is one of the most recognizable constellations in the Northern Hemisphere during the winter months. Orion was a great hunter who had two hunting dogs, Canis Major (big dog) and Canis Minor (little dog). The constellations visible in the sky change throughout the year.

From Earth, the stars in a constellation appear relatively close together. How do astronomers study star distance and other characteristics like the composition, the temperature, and the age of a star? They use tools, such as telescopes and spectroscopes, to study the stars.

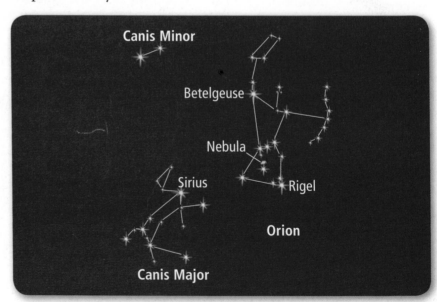

■ **Figure 1** Canis Major and Orion contain some of the brightest stars visible from the Northern Hemisphere—Sirius, Betelgeuse, and Rigel.

Telescopes

Many stars are visible with the unaided eye. However, to study stars and other objects in space, scientists and amateur astronomers use different types of telescopes. Optical telescopes detect the visible light that these objects produce. Radio telescopes detect the radio waves that these objects produce.

Optical telescopes Optical telescopes collect more light from distant objects than the unaided eye, producing brighter and clearer images. There are two types of optical telescopes—refracting telescopes and reflecting telescopes.

Refracting telescopes A refracting telescope uses a convex lens, which is curved outward like a ball, to collect and focus visible light. The objective lens collects the light and the eyepiece lens magnifies the image, as shown in **Figure 2**. However, there is a limit to the size of a refracting telescope. Because the objective lens can only be supported along its edges, it can sag in the middle if it is too large and blur the image. As a result, the largest refracting telescope has an objective lens that is about one meter in diameter.

Reflecting telescopes A reflecting telescope uses mirrors to collect and focus visible light. As light passes through the open end of a reflecting telescope, it strikes a concave mirror, as shown in **Figure 2**. Often, a smaller mirror is used to reflect light into a camera or an eyepiece. In very large reflecting telescopes, the astronomer can sit inside of the telescope to view distant objects through an eyepiece lens. Astronomers can also operate reflecting telescopes remotely, using cameras to collect images. Because mirrors can be supported from behind, reflecting telescopes with mirrors that are several meters in diameter can be built.

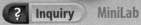
■ **Figure 2** A refracting telescope uses a convex lens to collect and focus visible light and to form an image at the focal point. A reflecting telescope uses a concave mirror to collect and focus visible light.

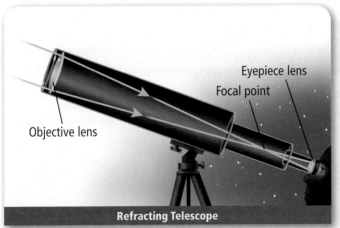

Eyepiece lens
Focal point
Objective lens

Refracting Telescope

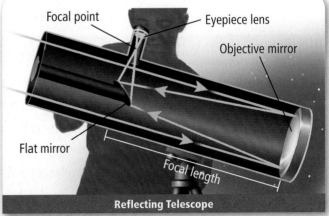

Focal point — Eyepiece lens
Objective mirror
Flat mirror
Focal length

Reflecting Telescope

Focal point and focal length The focal point of a lens or a mirror is the point at which parallel light rays converge to produce a clear image. This image can then be magnified by an eyepiece lens. The distance from a lens or a mirror to its focal point is called the focal length. The magnifying power (M_p) of a telescope is equal to the focal length of the objective lens or mirror (f_o) divided by the focal length of the eyepiece lens (f_e).

Magnification

$$\text{Magnifying power} = \frac{\text{focal length of the objective lens (mm)}}{\text{focal length of the eyepiece lens (mm)}}$$

$$M_p = f_o / f_e$$

EXAMPLE Problem 1

Calculate Magnification Find the magnifying power of a telescope with a focal length of 1,200 mm using eyepieces of 20 mm and 6 mm.

Identify the Unknown: the magnifying power of the telescope: M_p

List the Knowns: The focal length of the objective is 1,200 mm.

$$f_o = \textbf{1,200 mm}$$

The focal lengths of the eyepieces are 20 mm and 6 mm.

$$f_e = \textbf{20 mm}; f_e = \textbf{6 mm}$$

Set Up the Problem: $M_p = f_o / f_e$

Solve the Problem: Substitute the known values into the equation for magnifying power:

$M_p = \textbf{1,200 mm} / \textbf{20 mm} = 60$ $M_p = \textbf{1,200 mm} / \textbf{6 mm} = 60$

Notice that the units cancel, and that magnifying power has no unit.

Check the Answer: Does the answer seem reasonable? Multiply the focal length of each eyepiece by the magnifying power of each lens. Does this reproduce the focal length of the objective?

PRACTICE Problems

Find **Additional Practice Problems** in the back of your book.

1. Find the magnifying power of a telescope with a focal length of 2,500 mm when using eyepieces with focal lengths of 50 mm and 10 mm.

Review

Additional Practice Problems

2. Find the magnifying power for a telescope with a focal length of 900 mm when using an eyepiece with a focal length of 12 mm.

3. **Challenge** The magnifying power of a telescope is 400. Determine the focal length of the objective lens when the focal length of the eyepiece lens is 10 mm.

Adaptive Optics If you look at the night sky, you will notice that the stars seem to twinkle. This twinkle effect occurs as a result of temperature variations in Earth's atmosphere. Twinkling blurs images produced by telescopes on Earth. Some telescopes use a system called adaptive optics to make images sharper. In an adaptive optics system, the light from the objective mirror strikes a small, deformable mirror before it is focused. A sensor samples the light from the objective mirror to determine how distorted the images are. The surface of the deformable mirror is then adjusted many times a second to reduce the distortion.

Radio telescopes Like visible light, radio waves are a form of electromagnetic radiation emitted by stars and other objects in space. Radio waves can be detected even during the day, when the Sun makes it impossible to see the faint visible light from other stars, and on cloudy days when the light is obstructed. A **radio telescope** collects and amplifies radio waves. **Figure 3** shows an image produced by radio telescopes in the Very Large Array (VLA) near Socorro, New Mexico. Like optical telescope images, Earth's atmosphere can blur the images that radio telescopes produce.

✓ **Reading Check** **Explain** why radio telescopes are useful.

Space telescopes Astronomers can avoid the blurring effects of Earth's atmosphere altogether by placing telescopes in space. In 1990, the *Hubble Space Telescope* (HST), shown in **Figure 4,** was placed in orbit around Earth by the Space Shuttle. Because of its location outside of Earth's atmosphere, the HST has obtained clearer images of space objects than those obtained by most ground-based telescopes.

Earth's atmosphere absorbs other types of electromagnetic radiation, such as ultraviolet waves, X-rays, and infrared radiation. Some space telescopes, such as the *Chandra X-Ray Observatory* and the *Spitzer Space Telescope,* produce images using wavelengths that are absorbed by the atmosphere. NASA plans to launch a new space telescope—the James Webb Space Telescope—in 2014 to investigate the origins of our universe.

■ **Figure 3** The VLA has 27 radio telescopes. The telescopes detect radio waves emitted from stars and produce images like this one of distant galaxies.

■ **Figure 4** The *Hubble Space Telescope,* launched in 1990, orbits Earth at an altitude of 610 km. NASA astronauts, shown below, made upgrades and repairs to the telescope in May 2009.

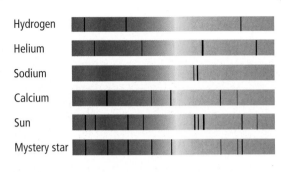

Hydrogen
Helium
Sodium
Calcium
Sun
Mystery star

■ **Figure 5** The dark lines in stellar spectra show astronomers what elements are present in the stars being studied.

Determine *which elements are contained in the mystery star.*

Concepts in Motion
Animation

The light-year

The light-year Light travels at a speed of 300,000 km/s in space. At this speed, light from the Sun takes 8 minutes and 20 seconds to travel to Earth, a distance of approximately 150 million kilometers. Some objects are so far away it takes millions or billions of years for light from these objects to reach Earth. To measure such astronomical distances, a unit called the light-year was quantified. A **light-year** is the distance light travels in one year. A light-year is equivalent to about 9.5 trillion kilometers.

Spectroscopes

The visible light from stars can provide information about the star's composition, its surface temperature, and even how fast it is moving toward or away from Earth. The light produced by stars and other objects in space is made of different wavelengths of visible light. The wavelengths produced by a star, for example, depend on the star's temperature and composition.

Astronomers use a spectroscope to analyze the different wavelengths of light collected by optical telescopes. A **spectroscope** uses a prism or diffraction grating to separate light into its component wavelengths. The separated wavelengths are called the spectrum of the star. The spectrum can be used to determine the temperature and composition of the star. **Figure 5** shows a spectrum from a star produced by a spectroscope.

Section 1 Review

Section Summary

▶ Constellations are patterns of stars that are used to name and locate stars.

▶ Reflecting and refracting telescopes collect and focus visible light to produce a clear image that can be magnified.

▶ A radio telescope collects and amplifies radio waves.

▶ Spectroscopes are used to determine the composition of a star.

4. **MAIN Idea Compare and contrast** optical telescopes and radio telescopes.

5. **Describe** how astronomers use spectroscopes to study the stars.

6. **Explain** why telescopes are launched into Earth's orbit.

7. **Define** the term *light-year.*

8. **Think Critically** What would be an advantage of building a telescope on a high mountain?

Apply Math

9. **Calculate** If the magnifying powers (M_p) of refracting and reflecting telescopes are 20 and 100, respectively, how much greater is the M_p of the reflecting telescope?

✓ Assessment Online Quiz

Essential Questions

▶ How do stars form?

▶ How do astronomers classify stars?

▶ How did the Sun form?

▶ Does the Sun continue to evolve over time?

Review Vocabulary

nuclear fusion: a nuclear reaction in which atomic nuclei join together

New Vocabulary

nebula
main sequence
giant
white dwarf
supernova
neutron star
black hole
photosphere
sunspots

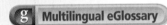

Evolution of Stars

MAIN Idea The life cycle of a star depends on its mass.

Real-World Reading Link At about 150 million kilometers away, the Sun is Earth's closest star. In comparison, the distance between Earth and the Moon is 384,000 kilometers. That means that the Sun is 440 times farther away from Earth than it is from the Moon.

How do stars form?

Stars form from a large cloud of gas, ice, and dust called a **nebula.** Particles within the cloud exert gravitational forces on each other, causing the cloud to contract, and the temperature to increase. A protostar forms at the center of the cloud when the temperature there exceeds 1 million Kelvin (K). When the temperature exceeds 10 million K, nuclear fusion of hydrogen into helium begins. A star is born.

H-R diagram In the early 1900s, Einjar Hertzsprung and Henry Russell discovered that the hottest stars produced the most light and energy. The relationship between the brightness and temperature of stars can be inferred from the Hertzsprung-Russell (H-R) diagram, as shown in **Figure 6.** About 90 percent of all stars are classified as main sequence stars. **Main sequence** stars include the broad band of stars from the hot, bright stars in the upper left corner of the diagram to the cool, dim stars in the lower right corner. All other stars generally fall into three areas on the H-R diagram.

■ **Figure 6** Scientists use temperature and brightness to classify stars. Most stars plot within the main sequence. Supergiants, giants, and white dwarfs plot in different areas on the H-R diagram.

Identify *the dimmest stars with the highest temperatures on the H-R diagram.*

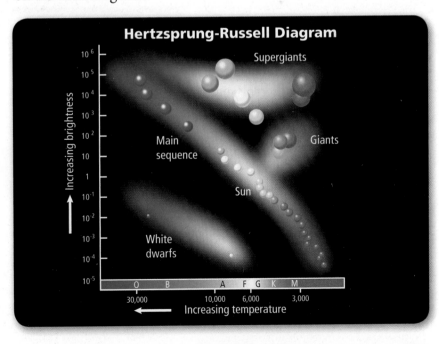

Section 2 • Evolution of Stars   989

■ **Figure 7** Only a small portion of the giant star Antares is shown here. Supergiant stars, such as Betelgeuse, are too large to be shown at all on this scale.

How do stars evolve?

Recall that the protostar, which forms in the center of a cloud fragment, continues to contract until nuclear fusion begins. Then the energy from the fusion of hydrogen into helium exerts an outward pressure on the material in the star. Meanwhile, the gravitational forces inside the star continue to pull material inward toward the star's center. Equilibrium is reached when the outward pressure exerted by the fusion reactions equals the inward pull of gravity.

Main sequence Once equilibrium is reached, the star becomes a main sequence star. As long as the star is fusing hydrogen into helium, it remains on the main sequence. Stars tend to maintain equilibrium for most of their existences. The Sun has been a main sequence star for about five billion years and will continue to be a main sequence star for another five billion years. **Figure 7** shows the relative size of the Sun compared to the stars that form as equilibrium comes to an end.

Equilibrium ends When a star uses up all of the hydrogen in its core, it is no longer in a state of equilibrium. The star enters a new evolutionary stage. What happens to the star depends on its mass. A star with an average mass, like the Sun, will become a red giant, then a white dwarf, and finally, a black dwarf.

Stars more massive than the Sun can become supergiants and eventually neutron stars or black holes. Stars much lower in mass than the Sun could remain on the main sequence as red dwarfs for up to 16 trillion years.

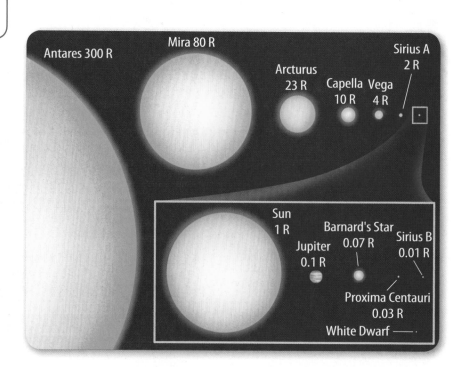

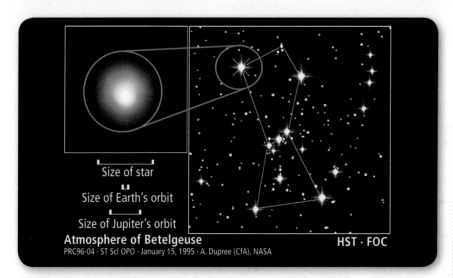

■ **Figure 8** Betelgeuse is located in the constellation of Orion. Its diameter is larger than the diameter of Jupiter's orbit around the Sun.

Classify *What type of star is Betelgeuse?*

Size of star
Size of Earth's orbit
Size of Jupiter's orbit
Atmosphere of Betelgeuse
PRC96-04 · ST Scl OPO · January 15, 1995 · A. Dupree (CfA), NASA

HST · FOC

Giants and white dwarfs As equilibrium ends, the star's core contracts, the temperature increases, and the star evolves into a **giant star.** As the core heats up, the star's outer layers expand and cool. In about five billion years, our Sun will expand to become a giant star like Antares.

A giant star's core continues to contract and increase in temperature. When the core temperature reaches 100 million K, helium fuses to form carbon. Eventually, the core uses up its helium. Now the star is enormous and its surface is cooler. A **white dwarf** forms as the core of a giant star no longer supports fusion and the star's outer layers escape into space, leaving a hot and dense core.

Supergiants, neutron stars, and black holes Stars that are over eight times more massive than our Sun evolve differently. Their core temperatures are hot enough to fuse heavier elements. The star expands into a supergiant, such as Betelgeuse shown in **Figure 8.** Fusion reactions end when iron accumulates in the star's core.

Supernovas As iron accumulates in the core of a supergiant star, energy is absorbed and the star can no longer support itself. The core collapses, forcing electrons into the nuclei of iron atoms. This collapse releases a huge amount of energy and very fast-moving neutrons in the form of a supernova. A supernova is a gigantic explosion in which the temperature within the collapsing star is 10 billion K and the atomic nuclei in the core split into neutrons and protons. When the star explodes, it leaves behind a ball of neutrons called a neutron star.

Black holes Very massive stars, 25 times more massive than the Sun, face a different end. In this case, the final collapse of the core continues past the neutron star stage and forms a black hole. A **black hole** is an area in space that is so dense that nothing can escape its pull of gravity.

VOCABULARY
ACADEMIC VOCABULARY
Collapse
to fall down, give way, or cave in
The runner collapsed at the finish line after winning the race.

FIGURE 9

Visualizing H-R Diagrams

The Pleiades star cluster blazed into existence 100 million years ago, when dinosaurs roamed Earth and mammals were just gaining ground. Today, astronomers study star clusters because they offer valuable insights into how the universe evolved.

The Pleiades

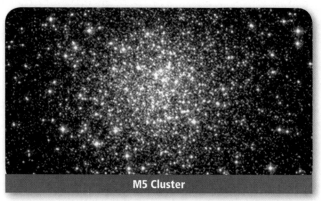

M5 Cluster

The Pleiades, located in the constellation Taurus, is among one of the earliest celestial objects to be discovered by our ancient ancestors. It is 440 light-years from Earth.

M5 is a globular star cluster located 24,500 light-years from Earth. Stars within this cluster are more abundant and more compact than stars within the Pleiades cluster. The presence of cool, red giant stars indicates that the M5 cluster is an older star cluster than the Pleiades.

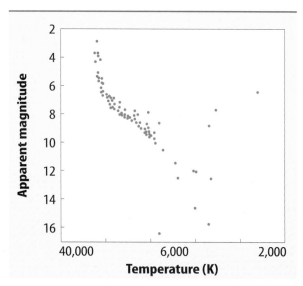

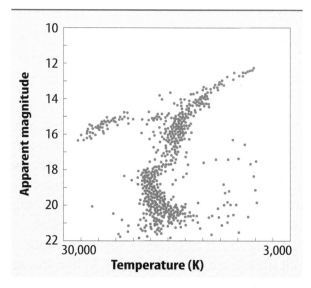

The H-R diagram for stars within the Pleiades star cluster shows a well-defined main sequence trend. Stars in the upper left portion of the diagram are hot, blue, and massive. As stars age, they burn hydrogen and eventually become red giants. The presence of hot, blue stars and lack of red giants indicate that the Pleiades is a young star cluster.

The H-R diagram for M5 cluster stars indicates that these stars were once hot and blue, but they are now cool, red giants. Sun-sized stars evolve from main sequence stars, like those identified in the young Pleiades cluster, into red giants, like those identified in the older M5 cluster.

The Sun—A Main Sequence Star

Although the Sun is an average-sized star, it is by far the largest object in our solar system. The Sun is about 1.4 million kilometers in diameter—109 times larger than Earth and even 10 times larger than Jupiter. Like all stars, the Sun is made almost entirely of hydrogen and helium. About 70 percent of the Sun's mass is hydrogen and about 28 percent is helium. Much of the remaining 2 percent of the Sun's mass is oxygen and carbon. The Sun contains 99 percent of all the mass in the solar system.

Reading Check **Identify** What elements make up most of the Sun's mass?

The Sun's interior The Sun's interior can be divided into several distinct layers or zones: the core, the radiation zone, and the convection zone. These zones are shown in **Figure 10.**

The core The core is the Sun's innermost layer, with a radius of 140,000 km. Nuclear fusion occurs in the core. Fusion reactions change hydrogen nuclei into helium nuclei. The temperature in the Sun's core is about 15 million K—so hot that atoms have been stripped of their electrons. The core is a gas made of charged particles known as a plasma.

The radiation zone The radiation zone lies above the core and extends from the core to 500,000 km from the star's center. In the radiation zone, the thermal energy produced by nuclear fusion is transferred from the core by electromagnetic radiation. Temperatures are extreme in this zone, but they decrease as distance from the core increases from approximately 7 million K to approximately 2 million K toward the top of the radiation zone.

The convection zone The Sun's outermost layer is the convection zone. Here, thermal energy is transferred from the top of the radiation zone to the photosphere by the process of convection. Columns of hot material form convection cells as they rise to the surface, cool, and then sink back to the bottom of the convection zone.

Reading Check **Describe** the three layers of the Sun's interior.

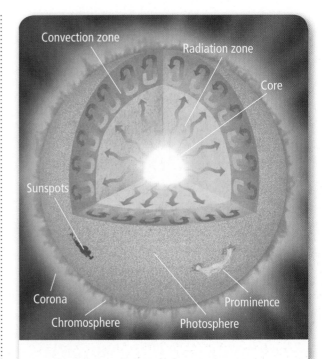

■ **Figure 10** The Sun's interior is divided into three layers: the core, the radiation zone, and the convection zone.

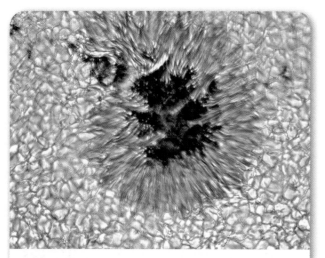

■ **Figure 11** The large, dark area is a sunspot. The smaller, surrounding cells are called granules. Sunspots are often 2,000 K cooler than granules.

■ **Figure 12** This prominence of dense plasma appears suspended in the Sun's chromosphere as it loops back to the surface.

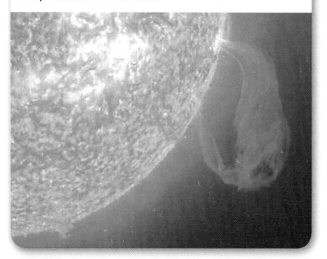

Surface features on the Sun

The **photosphere** is the layer of the Sun that emits light into space. Surrounding the photosphere is the Sun's atmosphere, which can be subdivided into two layers. The inner layer of the atmosphere is the chromosphere, and the outer layer is the corona.

Granules The photosphere has a grainy appearance, as shown in **Figure 11**. The individual grains are called granules and are about 1,000 km across. Granules are the tops of the convection cells. The bright areas are hot material that is rising, and the dark areas are cooler material that is sinking.

Sunspots Sunspots are dark, cool areas in the photosphere where the Sun's magnetic field has weakened. Temperatures in the center of a sunspot can be 2,000 K cooler than its surroundings. Sunspots move as the Sun rotates. By monitoring sunspots, astronomers have discovered that the Sun rotates faster at its equator than at its poles.

Sunspots are not permanent features on the Sun's surface. They appear and disappear over periods of days, weeks, or months. The number of sunspots changes in a regular pattern, called a sunspot cycle, which occurs every 11.2 years.

Prominences Intense magnetic fields associated with sunspots can produce huge arching columns of ionized gas (plasma) called prominences, as shown in **Figure 12**. Prominences result as ionized gases flow from the convection zone toward the photosphere. When the magnetic field strength is significant, dense plasma loops from one area on the Sun's surface to another in the form of a prominence. Some prominences blast material from the Sun's surface into space at speeds of more than 1,000 km/s.

Flares Gases near a sunspot sometimes brighten suddenly, shooting gas outward at high speeds in what are classified as solar flares. Temperatures within these solar flares can reach 100 million K. Particles produced in a flare possess so much energy that the Sun's magnetic field cannot hold onto them as it can with prominences, and particles escape into space instead.

CMEs Sometimes large bubbles of plasma are emitted from the Sun, as shown in **Figure 13.** These are known as coronal mass ejections (CMEs). During sunspot cycle minimums there is usually one CME per week, but during a sunspot cycle maximum, there are two to three per day. When a CME is released in the direction of Earth, it appears as a halo around the Sun.

☑ **Reading Check** **Describe** a coronal mass ejection.

As a coronal mass ejection passes Earth, the planet is exposed to a sudden shock wave of solar wind. Earth's atmosphere protects us, but occurrences of auroras increase. When astronomers observe a CME, they post an alert to watch for auroras at latitudes lower than normal.

Auroras are colorful ribbons of light that result when high-energy particles in coronal mass ejections and the solar wind encounter Earth's magnetic field. This generates large electric currents that flow toward Earth's poles. These electric currents ionize gases in Earth's atmosphere.

When atmospheric gases are ionized, the ions can combine with electrons and produce light. This light is called the aurora borealis, or northern lights, when it occurs in the Northern Hemisphere. In the Southern Hemisphere, it is called the aurora australis. While CMEs present little danger to life on Earth, the most powerful solar winds can disrupt Earth's magnetosphere. These disruptions can interfere with orbiting satellites, radio signals, and can even cause power outages.

☑ **Reading Check** **Identify** What is the aurora borealis?

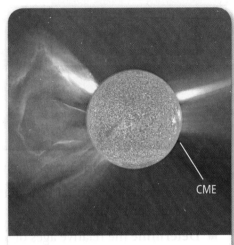
■ **Figure 13** Coronal mass ejections can hurl a million tons of matter into space each second. They damage satellites and can endanger astronauts.

Section 2 Review

Section Summary

▶ A large cloud of gas, ice, and dust is called a nebula.

▶ An H-R diagram relates the temperature and brightness of stars.

▶ Once nuclear fusion begins, a star spends most of its existence as a main sequence star.

▶ A main sequence star can evolve into a giant or supergiant.

▶ The Sun's energy is produced by nuclear fusion.

10. **MAIN Idea** **Describe** the life cycle of stars as illustrated on the H-R diagram.

11. **Explain** how stars form.

12. **Diagram** and describe the Sun's layers.

13. **Predict** what will happen to the Sun as it burns hydrogen fuel.

14. **Think Critically** Explain how equilibrium changes as a main sequence star evolves into a giant star.

Apply Math

15. **Calculate** Eighty percent of all stars are red dwarfs. Out of a random sample of 2,000 stars in the galaxy, how many will plot on the H-R diagram as red dwarfs?

■ **Figure 15** This image was taken with the *Hubble Space Telescope* and shows the giant elliptical galaxy ESO 325-G004. This galaxy is more than 450 million light-years from Earth and contains several hundred billion stars. The bright crosses are images of nearby stars and are not part of the galaxy.

Calculate *How many kilometers are there in 450 million light-years?*

Concepts in Motion   **Animation**

■ **Figure 16** This image shows the irregular galaxy NGC 4449. This galaxy is about 20,000 light-years across and is about 12 million light-years away.

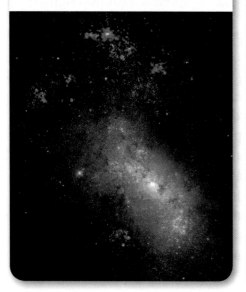

Elliptical galaxies Unlike spiral galaxies, elliptical galaxies are not disk-shaped and they do not have radiating arms. Instead, elliptical galaxies have shapes that range from nearly spherical to football-shaped. **Figure 15** shows the elliptical galaxy identified by astronomers as ESO 325-G004. Elliptical galaxies contain less ice, dust, and gas than spiral galaxies. As a result, there is less star formation than occurs in a spiral galaxy. For this reason, elliptical galaxies contain mostly older stars.

Elliptical galaxies differ in size more than spiral galaxies. The largest galaxies in the universe are elliptical galaxies. The largest elliptical galaxies might contain over 10 trillion stars and can be over 1 million light-years in diameter. The smallest elliptical galaxies are called dwarf elliptical galaxies and can be only a few thousand light-years in diameter. Dwarf elliptical galaxies usually contain less than 10 million stars.

Irregular galaxies Galaxies that do not have an elliptical or a spiral shape are classified as irregular galaxies. Irregular galaxies, such as NGC 4449 shown in **Figure 16,** are usually larger than dwarf elliptical galaxies but smaller than spiral galaxies. Irregular galaxies range in size from about 3,000 to 30,000 light-years in diameter, and they can contain between 100 million and 10 billion stars.

The smallest irregular galaxies are called dwarf irregular galaxies. Dwarf irregular galaxies might be some of the most common types of galaxies. They contain high percentages of young stars and large amounts of ice, dust, and gas.

Two of the closest galaxies to the Milky Way are the Large Magellanic Cloud (LMC) and the Small Magellanic Cloud (SMC). Both are small, irregular galaxies that orbit the Milky Way galaxy at a distance of about 200,000 light-years.

✓ **Reading Check   Compare and contrast** spiral galaxies, elliptical galaxies, and irregular galaxies.

The Local Group Just as groups of stars form galaxies, galaxies also form groups. The Milky Way is part of a larger group of galaxies called the **Local Group,** which includes about fifty galaxies spread out over a distance of 10 million light-years across.

Most of the galaxies in the Milky Way are dwarf elliptical galaxies and dwarf irregular galaxies. There are only three spiral galaxies in the Local Group, including the Milky Way galaxy, the Andromeda Galaxy, and a galaxy named M33. The Milky Way galaxy and the Andromeda Galaxy are the largest galaxies in the Local Group. Many of the dwarf and irregular galaxies in the Local Group orbit the Milky Way or the Andromeda Galaxy.

How do galaxies form?

Astronomers hypothesize that the first galaxies began to form about 13.7 billion years ago as enormous clouds of gas began to contract due to gravity. These clouds of gas were much larger than the gas clouds from which stars form. As these galactic clouds contracted, smaller regions formed where the density of gas was even higher. Stars then formed in these high-density regions.

Astronomers believe that the first galaxies that formed tended to be irregular in shape and were smaller and closer together than galaxies are today. Astronomers hypothesize that many of the galaxies seen today formed when the earliest galaxies merged with each other.

Galaxies interact When galaxies are close to each other, the gravitational forces between them can change their shapes. For example, some irregular galaxies could have passed close enough to each other for a large group of stars to be ripped away from one of the galaxies. If galaxies pass close to each other, gravitational forces between the galaxies can also cause them to merge. For example, large elliptical galaxies can form when two spiral galaxies merge. However, spiral galaxies do not change shape when they merge with much smaller galaxies.

Figure 17 shows an interaction between two large spiral galaxies. These galaxies have been moving toward each other for about 40 million years. The centers of the two galaxies are about 100,000 light-years apart. The two galaxies will continue to move around each other for billions of years, eventually merging to form a single large galaxy without any spiral structure. You might think that some of the stars in these galaxies are colliding with each other. However, stars in galaxies are so far apart that when galaxies collide, the stars just move past each other.

FOLDABLES

Incorporate information from this section into your Foldable.

■ **Figure 17** The gravitational forces that these two galaxies exert on each other have slowly begun to distort their shapes. These galaxies are about 114 million light-years away from Earth.

■ **Figure 18** This figure shows a drawing of what a side view of the Milky Way galaxy might look like. The Milky Way has a nuclear bulge in its center, a thin, flat galactic disk, and a spherical halo of globular clusters.

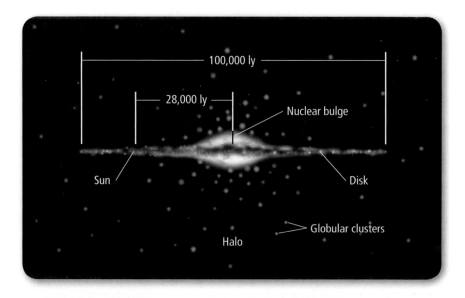

The Milky Way

Recall that the Sun is one of possibly 200–400 billion stars in the Milky Way galaxy. A side view of the Milky Way galaxy is shown in **Figure 18.** Like most spiral galaxies, the Milky Way has three parts. As shown in **Figure 18,** these parts include the galactic disk, the nuclear bulge, and the halo.

The galactic disk The Milky Way has a galactic disk about 100,000 light-years in diameter. The galactic disk includes spiral arms that radiate outward. Stars form within these spiral arms where the concentration of dust and gas is high.

The Sun is located in a spiral arm about 28,000 light-years from the center of the galaxy. Because the Milky Way is rotating, the Sun orbits the center of the galaxy at a speed of about 220 km/s. During its lifetime, the Sun has completed about 22 orbits around the center of the Milky Way. Although the Sun is about 5 billion years old, the Milky Way galaxy is much older. The oldest stars in the disk are thought to be 9–10 billion years old.

The nuclear bulge Stars are much closer together in the central region of a spiral galaxy compared to stars within the galactic disk. For example, in the vicinity of the Sun, the stars are typically 200 times farther apart than they are in the center of the Milky Way. The center of the Milky Way galaxy where stars are grouped more closely together is called the nuclear bulge.

Most of the stars in the nuclear bulge are older stars. In some spiral galaxies, the nuclear bulge is stretched so that it forms the shape of a bar across the center of the galaxy. These galaxies are called barred spiral galaxies. Observations by radio telescopes indicate that the Milky Way galaxy might be more accurately classified as a barred spiral galaxy.

The halo The halo is a roughly spherical region that surrounds the nuclear bulge and the galactic disk. Its diameter is approximately 200,000 light-years across. The halo is made of globular clusters, which are groups of very old stars. Globular clusters typically contain about one million stars. Most of the oldest stars in the Milky Way galaxy are found in the halo and are about 13 billion years old.

The galactic center The center of the Milky Way cannot be seen from Earth because clouds of dust and gas prevent visible light from passing through. However, radio waves can pass through these clouds. Data obtained by radio telescopes, such as the image shown in **Figure 19,** show that there is a dense object at the center of the galaxy that is emitting tremendous amounts of energy. This object, called Sgr A* (saj ay star), is smaller than the solar system and has a mass that is about 2.6 million times the mass of the Sun. Sgr A* is a black hole created by merging stars and clouds of gas and dust in the galaxy's center. Energy is emitted as hot gas spirals into the black hole.

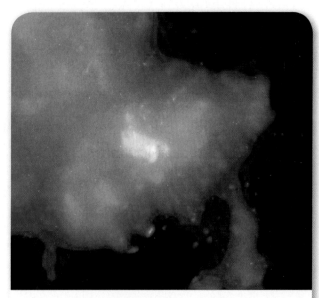
■ **Figure 19** This image shows the intensity of radio waves emitted from the center of the Milky Way galaxy. A black hole might be at the center of the bright yellow area.

Section 3 Review

Section Summary

▶ A galaxy is a large group of stars, gas, and dust held together by gravity.

▶ The three main types of galaxies are elliptical, irregular, and spiral.

▶ The attractive gravitational forces between galaxies can cause them to merge.

▶ The Milky Way galaxy contains 200–400 billion stars and has a spiral shape.

▶ Spiral galaxies usually have a galactic disk, a nuclear bulge, and a halo of star clusters.

16. **MAIN Idea** **Compare and contrast** the Milky Way galaxy to other galaxies in the universe.

17. **Describe** the shapes of the three main types of galaxies.

18. **Draw** the overall structure of the Milky Way galaxy, and indicate where the Sun is located.

19. **Describe** the Local Group of galaxies.

20. **Think Critically** How might the Sun be affected if the Andromeda Galaxy and the Milky Way galaxy collide?

Apply Math

21. **Calculate** Assume there are 400 billion stars in the Milky Way galaxy, plus or minus 200 billion. Based on this estimate, what is the range of the number of stars that might exist in the Milky Way?

22. **Use Percentages** A dwarf elliptical galaxy has fewer than 1 million stars and a small, irregular galaxy contains 100 million stars. What percent of the number of stars found in a small, irregular galaxy are found in a dwarf elliptical galaxy?

Reading Preview

Essential Questions

▶ What is the most accepted theory used to explain the origins of the universe?

▶ What evidence is used to support an expanding universe?

▶ What is dark matter?

▶ How does dark energy affect the expansion of the universe?

Review Vocabulary

universe: the space that contains all known matter and energy

New Vocabulary

cosmology
big bang theory
cosmic background radiation
dark matter
dark energy

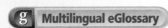 Multilingual eGlossary

 Inquiry Video Lab

Cosmology

MAIN ⟨Idea⟩ The universe has been expanding for billions of years.

Real-World Reading Link Ancient cultures used the sky to forecast changes in the seasons and to construct calendars. A calendar is a type of time line. Is it possible to construct a time line for the universe?

The Expanding Universe

The study of how the universe began, how it continues to change, and what it is made of is called **cosmology.** An important discovery about the universe was made in 1929 by the astronomer Edwin Hubble. Hubble discovered that almost all galaxies are moving away from Earth at speeds that are dependent on the their distance from Earth. Galaxies that are farther from Earth move faster. Hubble's results can only be explained if the universe is expanding.

Evidence of the Big Bang

Hubble's results and other data support an explanation for how the universe formed called the big bang theory. According to the **big bang theory,** all matter and energy in the universe were compressed into a single point, which began to expand outward about 13.7 billion years ago. This was not an explosion, but instead, an expansion of space itself. Telescopes collect evidence of this expansion, such as galaxies that formed roughly 3.7 billion years after the big bang, as shown in **Figure 20.**

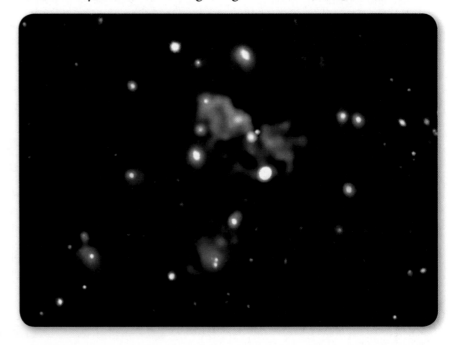

■ **Figure 20** X-ray radiation detected 10 billion light-years from Earth indicate that these galaxies formed approximately 3.7 billion years after the big bang.

Cosmic microwave background radiation In 1965, scientists discovered cosmic microwave radiation dispersed in space. **Cosmic background radiation** is the residual microwave radiation emitted when the universe was very young. Its discovery supports the big bang theory for the origins of the universe.

Recently, detailed measurements of the cosmic background radiation have been made by an orbiting observatory called the *Wilkinson Microwave Anisotropy Probe* (WMAP). A map of these measurements is shown in **Figure 21.** The bright areas on the map show regions where the density of matter was high and galaxies first formed. The WMAP data and other data indicate that the big bang occurred about 13.7 billion years ago.

The Doppler effect The speed and direction that galaxies move can be determined by the behavior of light and the Doppler effect. As a light source approaches you, wavelengths become shorter and frequencies are higher. As the light source moves away, wavelengths become longer and frequencies are lower. Hubble determined the universe was expanding by measuring the Doppler effect in the light emitted from galaxies, as shown in **Figure 22.** He found that the light from galaxies outside of the Local Group exhibit a redshift. This suggests that these galaxies are moving away from the Milky Way galaxy.

VOCABULARY · · · · · · · · · · · · · · · ·
WORD ORIGIN
Cosmic
   derived from the Greek word *kosmos*, which means "universe"
   The existence of cosmic background radiation dispersed throughout the universe supports the big bang theory. · · ·

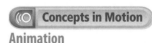
Animation

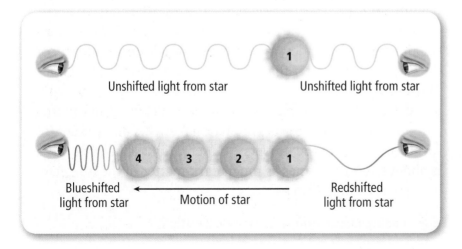

Unshifted light from star   Unshifted light from star

Blueshifted   ←   Redshifted
light from star   Motion of star   light from star

■ **Figure 22** The observer on the left experiences a blue shift as wavelengths of light emitted by an approaching object are shortened. The observer on the right experiences a redshift as wavelengths emitted by an object moving away are stretched.

■ **Figure 23** This image was obtained by the *Hubble Space Telescope.* Almost all the objects in this image are galaxies. The smallest and reddest galaxies are far from Earth and thought to be some of the oldest known.

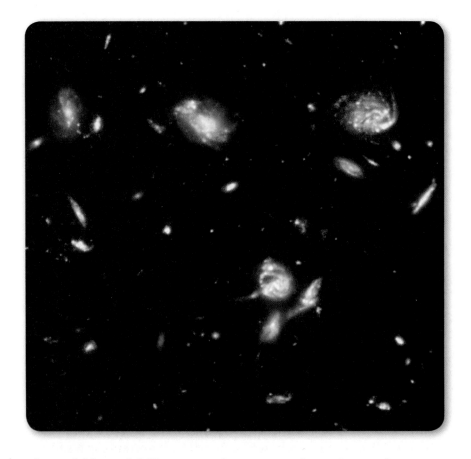

FOLDABLES
Incorporate information from this section into your Foldable.

The Hubble redshift Many observations have shown that the light from all distant galaxies exhibits a redshift. However, the redshift of very distant galaxies is not due to the Doppler effect. Instead, it is caused by the stretching of space as the universe expands. The stretching of space causes the wavelengths of light waves to stretch and the visible light to have longer wavelengths. This shift is known as the cosmological (or Hubble) redshift. All the galaxies you can see in **Figure 23** are moving away from the Milky Way due to the expansion of the universe.

What is the universe made of?

Stars and galaxies are made almost entirely of hydrogen and helium gas, with small amounts of heavier elements. Hydrogen, helium, and these heavier elements form stars and galaxies that can be detected with various types of telescopes. However, over the past several decades, astronomers have found evidence for another type of matter that cannot be found with a telescope. **Dark matter** is an invisible form of matter that does not emit any detectable electromagnetic energy of its own. Dark matter can only be detected by its gravitational effects on ordinary forms of matter in the universe. Observations show that there is about five to six times as much dark matter in the universe as there is ordinary matter.

☑ **Reading Check Define** What is dark matter?

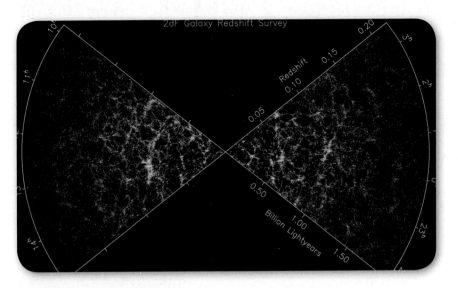

■ **Figure 24** These maps of sections of the universe show areas densely populated with galaxies and other areas almost empty.

Dark energy All matter in the universe exerts a gravitational force on all other matter, including dark matter. This attractive force might be strong enough to slow down the expansion of the universe. However, in the late 1990s, astronomers discovered that the expansion rate was actually speeding up. **Dark energy** is an invisible form of energy that creates a repulsive force causing the universe to expand at a faster rate.

The structure of the universe Astronomers have mapped the positions of thousands of galaxies in the universe. **Figure 24** shows that galaxies tend to form dense clusters. There are also enormous regions of space called voids where almost no galaxies exist.

Section 4 Summary

Section Summary

▶ The big bang theory is the accepted theory for the formation of the universe.

▶ The universe is estimated to be 13.7 billion years old.

▶ A redshift in the light emitted by stars indicates that the universe is expanding.

▶ The universe contains mostly dark matter and dark energy.

23. **MAIN Idea Explain** how the expansion of space could cause a redshift in the light from distant galaxies.

24. **Define** dark matter and dark energy.

25. **Summarize** the big bang theory.

26. **Describe** How does the Doppler effect support the big bang theory?

27. **Think Critically** How could the presence of a repulsive force cause the universe to expand forever?

Apply Math

28. **Calculate** The currently accepted age of the universe is 13.7 billion years, plus or minus one percent. How much time does this one percent represent?

29. **Use Percentages** What percent of the universe's age is the first 400,000 years?

LAB Expansion of the Universe

MODEL AND INVENT

Objectives

- **Model** the shift in wavelengths of light caused by the expansion of the universe.
- **Measure** the shift in wavelengths as the model universe expands.
- **Determine** increases in distances between galaxy clusters illustrated in your model.

Background: You have read that the universe is expanding. In fact, recent calculations indicate that it is expanding more quickly now than it has in the past. Astronomers are able to measure this expansion because wavelengths of light coming from distant objects lengthen as a result.

Question: *Can you create a model that demonstrates the expansion of the universe and its effect on light?*

Preparation

Materials

round balloon
permanent marker (black or dark blue)
medium-sized binder clip (3 cm)
flexible plastic ruler

Safety Precautions

Make the Model

1. Read the procedure and safety information, and complete the lab form.
2. Work in teams of two or more. Collect all necessary materials.
3. Slightly inflate the balloon.
4. Use the binder clip to hold air in the balloon long enough for you to draw on the balloon and make measurements. If the binder clip does not hold, have one student in your team hold it closed while another student makes measurements.
5. Model the positions of four galaxy clusters by placing four dots in different locations and at different distances from each other on the surface of the balloon.
6. Mark the locations *R* (for reference), *A*, *B*, and *C*.
7. Sketch an image on the slightly inflated balloon with grid lines to show the relative positions *R*, *A*, *B*, and *C*.
8. Create a data table to record the distance between each galaxy cluster (*A*, *B*, and *C*) and the reference galaxy (*R*) on your balloon.
9. Obtain your teacher's approval of your sketch and data table before proceeding.

Data Table

Location	1st Distance Measurement	2nd Distance Measurement	Change in Distance	1st Wavelength Measurement	2nd Wavelength Measurement	Change in Wavelength
A						
B						
C						

? Inquiry Lab

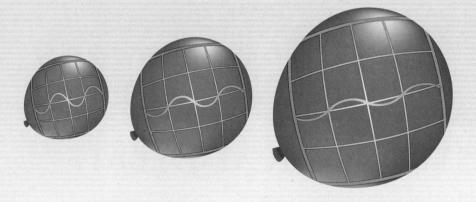

Test the Model

1. Inflate your balloon slightly, so that it is not very big.

2. Use the binder clip to temporarily seal the end of the balloon (or have another student in your team hold it closed). Do not tie it.

3. Measure the distance from your reference galaxy (R) to each of the three galaxy clusters, and record these distances on your data table.

4. Draw a wavy line to represent wavelengths of light traveling from your reference galaxy to each of the three galaxy clusters.

5. Measure the wavelength for each wavy line, and record these distances on your data table.

6. Continue to inflate your balloon (be careful not to inflate it too much). Replace the binder clip, or tie the end of the balloon.

7. Measure and record the distances from your reference galaxy to each of the three galaxy clusters.

8. Measure and record the wavelength of each wavy line on your inflated model.

Analyze Your Data

1. **Calculate** the change in distance and the change in wavelength from the reference galaxy to each galaxy cluster as you inflated the balloon. Record these values in your data table.

2. **Analyze** whether objects moved on your model or your entire model expanded.

Conclude and Apply

1. **Explain** any changes in distance or wavelength noted in your model.

2. **Conclude** whether the objects moved apart because of their individual motions or due to expansion in your model.

3. **Infer** how measurements on your model can be related to the expansion of the universe.

COMMUNICATE
YOUR DATA

Develop a spreadsheet that summarizes the data from all of the teams of students. Compare your data to the class data. Discuss reasons for differences in the data that different teams of students collected.

SCIENCE & HISTORY

AN ASTRONOMICAL DEBATE

In 1930, Subrahmanyan Chandrasekhar, shown in **Figure 1,** traveled for three weeks from India to graduate school in Cambridge, England. While in route, Chandrasekhar, also known as Chandra, made the most of his time. He observed that the life cycle of stars could be better understood by applying the ideas of Albert Einstein and his contemporaries to models of how stars evolve.

Famous colleagues Arthur Stanley Eddington was a well-respected scientist who had been at Cambridge for 20 years before Chandra arrived. Eddington hypothesized that every star within the main sequence eventually evolves into a white dwarf. His hypothesis was not universally accepted.

Trouble looms Chandra finished his PhD and stayed at Cambridge as a faculty member. While he continued researching the life cycle of stars, Eddington visited him daily to ask about his work. Chandra shared his discovery that some stars do develop into white dwarves, while others do not. He thought Eddington was pleased with these results.

Conflict between colleagues Chandra looked forward to presenting his discovery at the Royal Astronomical Society in 1935. Two days prior to the meeting, Chandra saw the meeting agenda. Eddington was scheduled to speak right after him, on the same exact subject. Why hadn't Eddington mentioned this to him in person? The night before the meeting, Eddington approached Chandra and told him that he had requested Chandra to be given extra time to make his presentation.

Figure 1 Chandra discovered that a star's mass determines how it will evolve.

Discovery is discredited At the meeting, Chandra explained that stars having a mass close to that of the Sun become white dwarves, but that there is a limit to this mass. He explained that more massive stars will collapse in on themselves. Scientists today recognize that Chandra had discovered the existence of black holes. Sadly, Eddington followed Chandra's presentation with a harsh critique of the younger scientist's discoveries. No one in the room questioned Eddington's arguments.

Science was stalled by disagreement Chandra was crushed. His work from the past five years was discredited in one day. Most of his colleagues respected Eddington too much to disagree with him. Although some of the leading scientists at the time privately agreed with Chandra's discovery, they were not willing to confront or question Eddington's authority. Chandra's work was not fully credited until he was awarded the Nobel Prize for physics nearly 50 years later. Today, the Chandra X-Ray Observatory is named in Subrahmanyan Chandrasekhar's honor.

 WebQuest **Research** other great debates in astronomy.

THEME FOCUS The Solar System and the Universe
Our solar system resides in the Milky Way galaxy, one of many galaxies in an expanding universe.

BIG Idea Stars and galaxies form as a result of gravity.

Section 1 Observing the Universe

constellation (p. 984)
light-year (p. 988)
radio telescope (p. 987)
spectroscope (p. 988)

MAIN Idea Telescopes form images by collecting and focusing the electromagnetic radiation emitted by distant objects.

- Constellations are patterns of stars that are used to name and locate stars.
- Reflecting and refracting telescopes collect and focus visible light to produce a clear image that can be magnified.
- A radio telescope collects and amplifies radio waves.
- Spectroscopes are used to determine the composition of a star.

Section 2 Evolution of Stars

black hole (p. 991)
giant star (p. 991)
main sequence (p. 989)
nebula (p. 989)
neutron star (p. 991)
photosphere (p. 994)
sunspots (p. 994)
supernova (p. 991)
white dwarf (p. 991)

MAIN Idea The life cycle of a star depends on its mass.

- A large cloud of gas, ice, and dust is called a nebula.
- An H-R diagram relates the temperature and brightness of stars.
- Once nuclear fusion begins, a star spends most of its existence as a main sequence star.
- A main sequence star can evolve into a giant or supergiant.
- The Sun's energy is produced by nuclear fusion.

Section 3 Galaxies and the Milky Way

galaxy (p. 997)
Local Group (p. 998)
Milky Way (p. 997)

MAIN Idea Galaxies are classified into three main groups based on shape.

- A galaxy is a large group of stars, gas, and dust held together by gravity.
- The three main types of galaxies are elliptical, irregular, and spiral.
- The attractive gravitational forces between galaxies can cause them to merge.
- The Milky Way galaxy contains about 200–400 billion stars and has a spiral shape.
- Spiral galaxies usually have a galactic disk, a nuclear bulge, and a halo of star clusters.

Section 4 Cosmology

big bang theory (p. 1002)
cosmic background radiation (p. 1003)
cosmology (p. 1002)
dark energy (p. 1005)
dark matter (p. 1004)

MAIN Idea The universe has been expanding for billions of years.

- The big bang theory is the accepted theory for the formation the universe.
- The universe is estimated to be 13.7 billion years old.
- A redshift in the light emitted by stars indicates that the universe is expanding.
- The universe contains mostly dark matter and dark energy.

Use Vocabulary

Match each phrase with the correct term from the Study Guide.

30. patterns of stars

31. telescope that detects radio waves

32. the distance light travels in one year

33. plotted from the upper left to the lower right on the H-R diagram

34. star in which the core contracts and outer layers expand and cool

35. layer of the Sun from which light is emitted

36. dark, cooler areas on the Sun's photosphere

37. **BIG** **Idea** large group of stars, gas, and dust held together by gravity

38. spiral galaxy that contains our solar system

39. study of the evolution of the universe

Check Concepts

40. Which telescope uses lenses to collect and focus light and form an image?
A) adaptive
B) radio
C) reflecting
D) refracting

41. Which form of electromagnetic waves are optical telescopes used to study?
A) infrared radiation
B) radio waves
C) visible light
D) X-rays

42. Which magnifies the image in a telescope?
A) eyepiece   C) focus
B) focal length   D) objective

43. Which forms when the core of a giant star can no longer support fusion?
A) giant
B) neutron
C) red dwarf
D) white dwarf

44. Which forms from a star that is over 25 times the mass of the Sun?
A) black hole
B) giant star
C) neutron star
D) white dwarf

45. Which is a feature of the Sun that can reach 100 million K?
A) CME   C) prominence
B) flare   D) sunspot

Use the figure below to answer question 46.

46. Which type of galaxy is illustrated above?
A) dwarf elliptical
B) spiral
C) elliptical
D) irregular

47. Which part of a spiral galaxy contains the oldest stars?
A) halo
B) disk
C) nuclear bulge
D) photosphere

✓ **Assessment**   **Online Test Practice**

Interpret Graphics

Use the figure below to answer question 48.

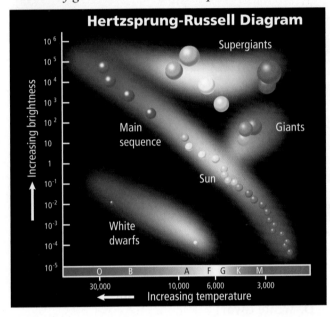

Hertzsprung-Russell Diagram

48. **Make** a table summarizing both the brightness and temperatures of stars on this H-R diagram.

49. **Draw** and label scale models of each of the following stars: the Sun; Antares, 500 times larger than the Sun; and Sirius B, 100 times smaller (in terms of diameter) than the Sun.

50. **Create** a concept map to explain the life cycle of a star, such as the Sun.

51. **Draw,** label, and describe the parts in a reflecting telescope.

Think Critically

52. **Explain** how energy released in the core of the Sun is eventually emitted from the photosphere.

53. **Compare and contrast** elliptical, irregular, and spiral galaxies.

54. **Discuss** the benefits of using a radio telescope.

55. **Explain** why high sunspot activity on the Sun can affect Earth's magnetic field.

56. **THEME FOCUS** **Describe** how the Sun's position in the Milky Way affects how we perceive our galaxy.

Apply Math

57. **Calculate** Use the equation $M_p = f_o / f_e$ to determine the magnifying power of a telescope in which the focal lengths of the objective lens and the eyepiece lens are 1500 mm and 9 mm, respectively.

Use the figure below to answer question 58.

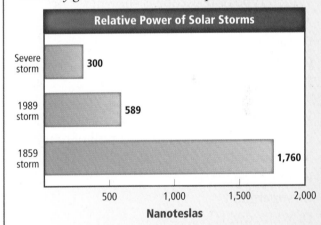

Relative Power of Solar Storms

Nanoteslas

58. **Solve** A solar storm that took place in 1859 is called "the perfect storm" because of its great power. The diagram above shows the relative power of some historic solar storms. Using this diagram, calculate approximately how much more powerful was the solar storm of 1859 than the one of 1989.

Standardized Test Practice

Multiple Choice

Record your answers on the answer sheet provided by your teacher or on a sheet of paper.

1. What is used as an objective in a reflecting telescope?
   A. antenna
   B. camera
   C. lens
   D. mirror

Use the figure below to answer question 2.

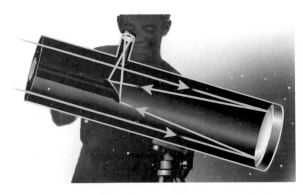

2. Which type of telescope is shown?
   A. optical
   B. radio
   C. ultraviolet
   D. X-ray

3. Which group of stars has a balanced outward pressure of fusion and inward pull of gravity?
   A. giant
   B. main sequence
   C. supergiant
   D. white dwarf

4. What does the existence of cosmic microwave background radiation help support?
   A. critical density
   B. Hubble redshift
   C. the inflationary model
   D. the big bang theory

Use the figure below to answer question 5.

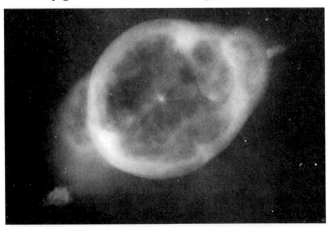

5. Which stage of stellar evolution shown in the image above?
   A. black hole
   B. giant
   C. main sequence
   D. white dwarf

6. What occurs inside a main sequence star?
   A. Energy from fusion exceeds gravity.
   B. Fusion shuts down.
   C. Gravity exceeds energy from fusion.
   D. It attains stellar equilibrium.

7. Which may be responsible for the accelerating expansion of the universe?
   A. dark energy
   B. dark matter
   C. regular energy
   D. regular matter

8. What is the correct order for stellar evolution?
   A. nebula, main sequence, red giant, white dwarf
   B. nebula, red giant, white dwarf, main sequence
   C. main sequence, red giant white dwarf, nebula
   D. nebula, main sequence, white dwarf, red giant

Short Response

Record your answers on the answer sheet provided by your teacher or on a sheet of paper.

9. If the focal length of an objective lens is 2,400 mm and the focal length of the eyepiece lens is 20 mm, what is the magnifying power of the telescope?

10. What is the area in square meters of one of the four 8.2-meter reflectors in the Very Large Telescope? Use the equation $A = \pi r^2$.

Use the figure below to answer question 11.

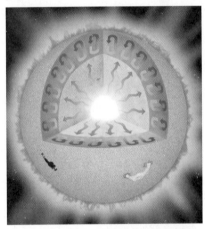

11. Describe the structure of the Sun's interior.

12. What are coronal mass ejections?

13. How do astronomers think that galaxies like the Milky Way formed?

14. What does Hubble's redshift describe?

Extended Response

Record your answers on a sheet of paper.

15. How can a spectroscope be used to determine the composition of a star?

16. Explain how galaxies such as the Milky Way galaxy formed.

Use the figure below to answer questions 17 and 18.

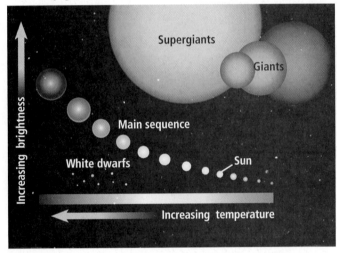

17. Use the illustration above to explain the evolutionary stages of a star like the Sun. Include a description of the temperature and composition of the star as it evolves from one stage to the next.

18. What are the evolutionary stages of a star more than eight times the mass of the Sun? How does this differ from the evolutionary stages of a star that is less massive than the Sun?

NEED EXTRA HELP?																		
If You Missed Question . . .	1	2	3	4	5	6	7	8	9	10	11	12	13	14	15	16	17	18
Review Section . . .	1	1	2	4	2	2	4	2	1	1	2	2	3	4	1	3	2	2

Student Resources

Table of Contents

Make Comparisons

Why learn this skill?

Suppose you want to buy a portable MP3 music player, and you must choose between three models. You would probably compare the characteristics of the three models, such as price, amount of memory, sound quality, and size in order to determine which model is best for you. In the study of chemistry, you often make comparisons between the structures of elements and compounds. You will also compare scientific discoveries or events from one time period with those from another period.

Learn the Skill

When making comparisons, you examine two or more items, groups, situations, events, or theories. You must first decide what will be compared and which characteristics you will use to compare them. Then identify any similarities and differences.

For example, comparisons can be made between the two models in the illustration on this page. The structure of the hydrogen atom can be compared to the structure of the oxygen atom. By reading the labels, you can see that both atoms have protons, neutrons, and electrons.

Practice the Skill

Create a table with the heading *Hydrogen and Oxygen Atoms*. Make three columns. Label the first column *Protons*. Label the second column *Neutrons*. Label the third column *Electrons*. Make two rows. Label the first row *Hydrogen*. Label the second row *Oxygen*. List the number of protons for each atom in the first column. Fill in the number of neutrons and electrons for each atom in the remaining columns. When you have finished the table, answer the following questions.

1. What is being compared? How are they being compared?
2. What do hydrogen and oxygen atoms have in common?
3. Describe how the differences between these two atoms affect the number of energy levels that each atom has.

Apply the Skill

Make Comparisons On page 180 you will find illustrations of a dry cell and an wet cell. Compare these two illustrations carefully. Then, identify the similarities and the differences between the two cells.

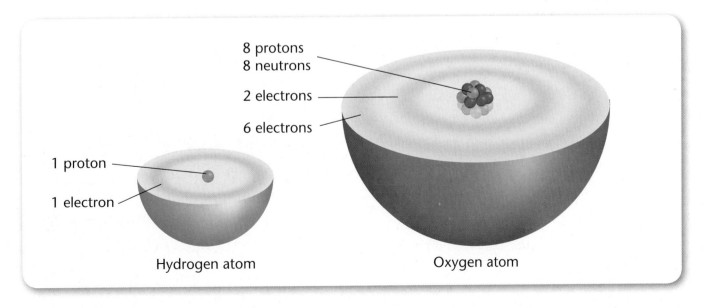

8 protons
8 neutrons
2 electrons
6 electrons

1 proton
1 electron

Hydrogen atom   Oxygen atom

Take Notes and Outline

Why learn this skill?

One of the best ways to remember something is to write it down. Taking notes—writing down information in a brief and orderly format—not only helps you remember, but also makes studying easier.

Learn the Skill

There are several styles of note taking, but all put information in a logical order. As you read, identify and summarize the main ideas and details that support them and write them in your notes. Paraphrase—that is, state in your own words—the information rather then copying it directly from the text. Using note cards or developing a personal "shorthand"—using symbols to represent words—can help.

You might also find it helpful to create an outline when taking notes. When outlining material, first read the material to identify the main ideas. In textbooks, section headings provide clues to main topics. Then, identify and fill in the subheadings and subdetails.

Main Topic
- I. First Idea or Item
  - A. First Detail
    - 1. Subdetail
    - 2. Subdetail
  - B. Second Detail
- II. Second Idea or Item
  - A. First Detail
  - B. Second Detail
    - 1. Subdetail
    - 2. Subdetail
- III. Third Idea or Item

Practice the Skill

Read the following excerpt from Chapter 12. Use the steps you just read about to take notes or create an outline. Then answer the questions that follow.

Look around your darkened room at night. After your eyes adjust to the darkness, you can see some familiar objects. Brightly colored objects look gray or black in the dim light. If you turn on the light, however, you might see all of the objects in the room, including their colors. What you see depends on the amount of light in the room and the color of the objects. To see an object, it must reflect some light back to your eyes.

Objects can absorb, reflect, and transmit light. Objects that transmit light allow light to pass through them. An object's material determines the amount of light it absorbs, reflects, and transmits. The material in the first candleholder in Figure 1 is opaque. Opaque materials only absorb and reflect light; no light passes through them. As a result, you cannot see the candle.

Materials such as the second candleholder in Figure 1 are described as translucent. Translucent materials transmit light but also scatter it. You cannot see clearly through translucent materials, and objects appear blurry. The third candleholder in Figure 1 is transparent. Transparent materials transmit light without scattering it, so you can see objects clearly through them.

1. What is the main topic of the excerpt?
2. What are the first, second, and third ideas?
3. Name one detail for each of the ideas.

Apply the Skill

Take Notes and Outline Go to Section 1 of Chapter 15 and take notes by paraphrasing and using shorthand or by creating an outline. Use the section title and headings to help you create your outline. Summarize the section using only your notes.

Analyze Media Sources

Why learn this skill?

To stay informed, people use a variety of media sources, including print media, broadcast media, and electronic media. The Internet has become an especially valuable research tool. It is convenient to use, and the information contained on the Internet is plentiful. Whichever media source you use to gather information, it is important to analyze the source to determine its accuracy and reliability.

Learn the Skill

There are a number of issues to consider when analyzing a media source. Most important is to check the accuracy of the source and content. The author and publisher or sponsors should be credible and clearly indicated. To analyze print media or broadcast media, ask yourself the following questions:

- Is the information current?

- Are the resources revealed?

- Is more than one resource used?

- Is the information biased?

- Does the information represent both sides of an issue?

- Is the information reported firsthand or secondhand?

Practice the Skill

To analyze print media, choose two articles—one from a newspaper and the other from a news magazine—about an issue on which public opinion is divided. Then answer these questions:

1. What points are the articles trying to make? Were the articles successful? Can the facts be verified?

2. Did either article reflect a bias toward one viewpoint or another? List any unsupported statements.

3. Was the information reported firsthand or secondhand? Do the articles seem to represent both sides fairly?

4. How many resources can you identify in the articles? List them.

To analyze electronic media, visit a Web site as instructed by your teacher. Read the information on that Web site, and then answer these questions.

1. Who is the author or sponsor of the Web site?

2. What links does the Web site contain? How are they appropriate to the topic?

3. What resources were used for the information on the Web site?

Apply the Skill

Analyze Sources of Information Think of an issue on which public opinion is divided. Use a variety of media resources to read about this issue. Which news source more fairly represents the issue? Which news source has the most reliable information? Can you identify any biases?

Debate Skills

New research leads to new scientific information. There are often opposing points of view on how this research is conducted, how it is interpreted, and how it is communicated. Some of the features in your book offer a chance to debate a current controversial topic. Here is an overview on how to conduct a debate.

Choose a Position and Research

First, choose a scientific issue that has at least two opposing viewpoints. The issue can come from current events, your textbook, or your teacher. These topics could include human cloning or environmental issues. Topics are stated as affirmative declarations, such as "Cloning human beings is beneficial to society."

One speaker will argue the viewpoint that agrees with the statement, called the positive position, and another speaker will argue the viewpoint that disagrees with the statement, called the negative position. Either individually or with a group, choose the position for which you will argue. The viewpoint that you choose does not have to reflect your personal belief. The purpose of debate is to create a strong argument supported by scientific evidence.

After choosing your position, conduct research to support your viewpoint. Use resources in your media center or library to find articles, or use your textbook to gather evidence to support your argument.

Hold the Debate

You will have a specific amount of time, determined by your teacher, in which to present your argument. Organize your speech to fit within the time limit: explain the viewpoint that you will be arguing, present an analysis of your evidence, and conclude by summing up your most important points. Try to vary the elements of your argument. Your speech should not be a list of facts, a reading of a newspaper article, or a statement of your personal opinion, but an analysis of your evidence in an organized manner. It is also important to remember that you must never make personal attacks against your opponent. Argue the issue. You will be evaluated on your overall presentation, organization and development of ideas, and strength of support for your argument.

Additional Roles There are other roles that you or your classmates can play in a debate. You can act as the timekeeper. The timekeeper times the length of the debaters' speeches and gives quiet signals to the speaker when time is almost up (usually a hand signal).

You can also act as a judge. There are important elements to look for when judging a speech: an introduction that tells the audience what position the speaker will be arguing, strong evidence that supports the speaker's position, and organization. The speaker also must speak clearly and loudly enough for everyone to hear. It is helpful to take notes during the debate to summarize the main points of each side's argument.

Scientific Methods

Scientists use an orderly approach called the scientific method to investigate problems. This includes organizing and recording data so others can understand them. Scientists use many variations in this method when they perform an investigation. Although there is variation, there are six common steps to the scientific methods, as shown in **Figure 1.**

Identify a Question

The first step in a scientific investigation or experiment is to identify a question to be answered or a problem to be solved. For example, you might ask which gasoline is the most efficient.

Gather and Organize Information

After you have identified your question, begin gathering and organizing information. There are many ways to gather information, such as researching in a library, interviewing those knowledgeable about the subject, testing and working in the laboratory and field. Fieldwork is investigations and observations done outside of a laboratory.

Researching Information Before moving in a new direction, it is important to gather the information that already is known about the subject. Start by asking yourself questions to determine exactly what you need to know. Then you will look for the information in various reference sources. Some sources may include textbooks, encyclopedias, government documents, professional journals, science magazines, and the Internet. Always list the sources of your information.

Evaluate Sources of Information Not all sources of information are reliable. You should evaluate all of your sources of information, and use only those you know to be dependable.

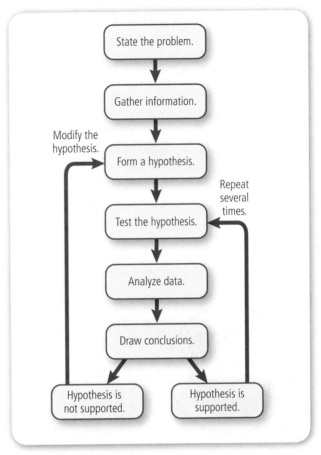

Figure 1 This flowchart can help you visualize scientific methods.

For example, if you are researching ways to make homes more energy efficient, a site written by the U.S. Department of Energy would be more reliable than a site written by a company that is trying to sell a new type of weatherproofing material. Also, remember that research always is changing. Consult the most current resources available to you. For example, a 1985 resource about saving energy would not reflect the most recent findings.

Sometimes scientists use data that they did not collect themselves, or conclusions drawn by other researchers. This data must be evaluated carefully. Ask questions about how the data were obtained, if the investigation was carried out properly, and if it has been duplicated exactly with the same results. Only when you have confidence in the data can you believe it is true and feel comfortable using it.

Interpret Scientific Illustrations As you research a topic in science, you will see drawings, diagrams, and photographs to help you understand what you read. Some illustrations are included to help you understand an idea that you can't see easily by yourself, like the tiny particles in an atom in **Figure 2.** A drawing helps many people to remember details more easily and provides examples that clarify difficult concepts or give additional information about the topic you are studying. Most illustrations have labels or a caption to identify or to provide more information.

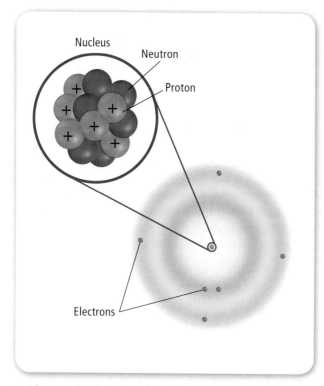

Figure 2 This drawing shows an atom of carbon with its six protons, six neutrons, and six electrons.

Concept Maps One way to organize data is to draw a diagram that shows relationships among ideas (or concepts). A concept map can help make the meanings of ideas and terms more clear, and it can help you both understand and remember what you are studying. Concept maps are useful for breaking large concepts down into smaller parts, which makes learning easier.

Network Tree A type of concept map that shows how related ideas branch out from a central concept is a network tree, shown in **Figure 3.** In a network tree, the words are written in the ovals, while the description of the type of relationship is written across the connecting lines.

When constructing a network tree, write down the topic and all major topics on separate pieces of paper or note cards. Then arrange them in order from general to specific. Branch the related concepts from the major concept and describe the relationship on the connecting line. Continue to more specific concepts until finished.

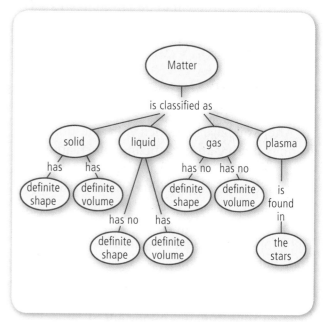

Figure 3 A network tree shows how concepts or objects are related.

Events Chain Another type of concept map is an events chain. Sometimes called a flowchart, it models the order or sequence of items. An events chain can be used to describe a sequence of events, the steps in a procedure, or the stages of a process.

When making an events chain, first find the one event that starts the chain. This event is called the initiating event. Then, find the next event and continue until the outcome is reached, as shown in **Figure 4.**

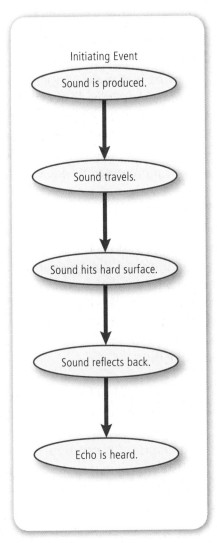

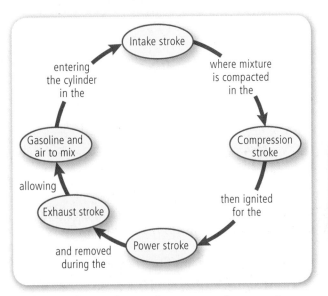

■ **Figure 5** A cycle map shows events that occur in a cycle.

Spider Map A spider map is a type of concept map that you can use for brainstorming. When you have a central idea, you might find that you have a jumble of ideas that relate to it but are not necessarily clearly related to each other. The spider map on sound in **Figure 6** shows that if you write these ideas outside the main concept, then you can begin to separate and group unrelated terms so they become more useful.

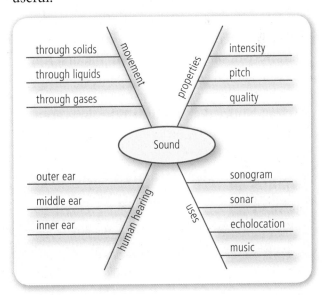

■ **Figure 6** A spider map allows you to list ideas that relate to a central topic but not necessarily to one another.

■ **Figure 4** Events-chain concept maps show the order of steps in a process or event. This concept map shows how a sound makes an echo.

Cycle Map A cycle map is a specific type of events chain. It is used when the series of events do not produce a final outcome, but instead relate back to the beginning event, such as in **Figure 5.** Therefore, the cycle repeats itself.

To make a cycle map, first decide what event is the beginning event. This is also called the initiating event. Then list the next events in the order that they occur, with the last event relating back to the initiating event. Words can be written between the events that describe what happens from one event to the next. The number of events in a cycle map can vary, but usually contain three or more events.

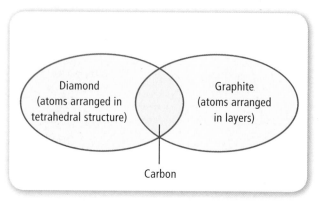

■ **Figure 7** This Venn diagram compares and contrasts two substances made from carbon.

Venn Diagram To illustrate how two subjects compare and contrast you can use a Venn diagram. You can see the characteristics that the subjects have in common and those that they do not, as shown in **Figure 7.**

To create a Venn diagram, draw two overlapping ovals that are big enough to write in. List the characteristics unique to one subject in one oval, and the characteristics of the other subject in the other oval. The characteristics in common are listed in the overlapping section.

Make and Use Tables One way to organize information so it is easier to understand is to use a table. Tables can contain numbers, words, or both. To make a table, list the items to be compared in the first column and the characteristics to be compared in the first row. The title should clearly indicate the content of the table, and the column or row heads should be clear. Notice that in **Table 1** the units are included.

Table 1	Recyclables Collected During Week		
Day of Week	Paper (kg)	Aluminum (kg)	Glass (kg)
Monday	5.0	4.0	12.0
Wednesday	4.0	1.0	10.0
Friday	2.5	2.0	10.0

Make a Model One way to help you better understand the parts of a structure, the way a process works, or to show things too large or small for viewing is to make a model. For example, an atomic model made of a plastic-ball nucleus and pipe-cleaner electron shells can help you visualize how the parts of an atom relate to each other. Other types of models can by devised on a computer or represented by equations.

Form a Hypothesis

A possible explanation based on previous knowledge and observations is called a hypothesis. After researching gasoline types and recalling previous experiences in your family's car you form a hypothesis—our car runs more efficiently because we use premium gasoline. To be valid, a hypothesis has to be something you can test by using an investigation.

Predict When you apply a hypothesis to a specific situation, you predict something about that situation. A prediction makes a statement in advance, based on prior observation, experience, or scientific reasoning. Scientists test predictions by performing investigations. Based on previous observations and experiences, you might form a prediction that cars are more efficient with premium gasoline. The prediction can be tested in an investigation.

Design an Experiment A scientist needs to make many decisions before beginning an investigation. Some of these include: how to define the variables, how to carry out the investigation, what steps to follow, and how to record the data. It also is important to address any safety concerns.

Test the Hypothesis

Now that you have formed your hypothesis, you need to test it. Using an investigation, you will make observations and collect data, or information. This data might either support or not support your hypothesis. Scientists collect and organize data as numbers and descriptions.

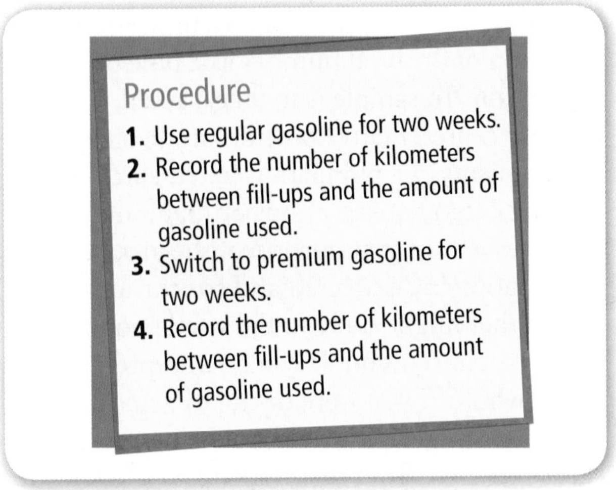

Procedure

1. Use regular gasoline for two weeks.
2. Record the number of kilometers between fill-ups and the amount of gasoline used.
3. Switch to premium gasoline for two weeks.
4. Record the number of kilometers between fill-ups and the amount of gasoline used.

■ **Figure 8** A procedure tells you what to do step by step.

Follow a Procedure In order to know what materials to use, as well as how and in what order to use them, you must follow a procedure. **Figure 8** shows a procedure you might follow to test your hypothesis.

Identify and Manipulate Variables and Controls In any experiment, it is important to keep everything the same except for the item you are testing. The one factor you change is called the independent variable. The change that results is the dependent variable. Make sure you have only one independent variable, to assure yourself of the cause of the changes you observe in the dependent variable. For example, in your gasoline experiment the type of fuel is the independent variable. The dependent variable is the fuel economy.

Many experiments also have a control—an individual instance or experimental subject for which the independent variable is not changed. You can then compare the test results to the control results. To design a control you can have two cars of the same type. The control car uses regular gasoline for four weeks. After you are done with the test, you can compare the experimental results to the control results. All other factors in an experiment must remain constant. **Table 2** summarizes the types of variables that are used in an experiment.

Table 2	Types of Variables
Variable	**Description**
Dependent	Changes according to the changes of the independent variable
Independent	The variable that is changed to test the effect on the dependent variable
Constant	A factor that does not change when other variables change
Control	The standard by which the test results can be compared

Collect Data

Whether you are carrying out an investigation or a short observational experiment, you will collect data, as shown in **Figure 9.** Scientists collect data as numbers and descriptions and organize it in specific ways.

Observe Scientists observe items and events, then record what they see. When they use only words to describe an observation, it is called qualitative data. Scientists' observations also can describe how much there is of something. These observations use numbers, as well as words, in the description and are called quantitative data. For example, if a sample of the element gold is described as being "shiny and very dense" the data are qualitative. Quantitative data on this sample of gold might include "a mass of 30 g and a density of 19.3 g/cm^3."

■ **Figure 9** Collecting data is one way to gather information directly.

■ **Figure 10** Record data neatly and clearly so it is easy to understand.

When you make observations you should examine the entire object or situation first, and then look carefully for details. It is important to record observations accurately and completely. Always record your notes immediately as you make them, so you do not miss details or make a mistake when recording results from memory. Never put unidentified observations on scraps of paper. Instead they should be recorded in a notebook, like the one in **Figure 10.** Write your data neatly so you can easily read it later.

At each point in the experiment, record your observations and label them. That way, you will not have to determine what the figures mean when you look at your notes later. Set up any tables that you will need to use ahead of time, so you can record any observations right away. Remember to avoid bias when collecting data by not including personal thoughts when you record observations. Record only what you observe.

Estimate Scientific work also involves estimating. To estimate is to make a judgment about the size or the number of something without measuring or counting. This is important when the number or size of an object or population is too large or too difficult to accurately count or measure.

Sample Scientists may use a sample or a portion of the total number as a type of estimation. To sample is to take a small, random representative portion of the objects or organisms of a population for research. By making careful observations or manipulating variables within that portion of the group, information is discovered and conclusions are drawn that might apply to the whole population. A poorly chosen sample can be unrepresentative of the whole. If you were trying to determine the rainfall in an area, it would not be best to take a rainfall sample from under a tree.

Measure You use measurements every day. Scientists also take measurements when collecting data. When taking measurements, it is important to know how to use measuring tools properly. Accuracy also is important.

Length To measure length, the distance between two points, scientists use meters. Smaller measurements might be measured in centimeters or millimeters.

Length is measured using a metric ruler or meter stick. When using a metric ruler, line up the 0-cm mark with the end of the object being measured and read the number of the unit where the object ends. Look at the metric ruler shown in **Figure 11.** The centimeter lines are the long, numbered lines, and the shorter lines are millimeter lines. In this instance, the length would be 4.50 cm.

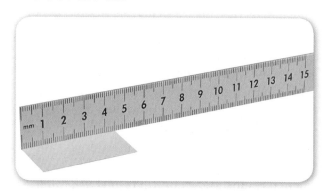

■ **Figure 11** This metric ruler has centimeter and millimeter divisions.

Mass The SI unit for mass is the kilogram (kg). Scientists can measure mass using units formed by adding metric prefixes to the unit gram (g), such as milligram (mg). To measure mass, you might use a triple-beam balance similar to the one shown in **Figure 12.** The balance has a pan on one side and a set of beams on the other side. Each beam has a rider that slides on the beam.

When using a triple-beam balance, place an object on the pan. Slide the largest rider along its beam until the pointer drops below zero. Then move it back one notch. Repeat the process for each rider proceeding from the larger to smaller until the pointer swings an equal distance above and below the zero point. Sum the masses on each beam to find the mass of the object. Move all riders back to zero when finished.

Instead of putting materials directly on the balance, scientists often take a tare of a container. A tare is the mass of a container into which objects or substances are placed for measuring their masses. To mass objects or substances, find the mass of a clean container. Remove the container from the pan, and place the object or substances in the container. Find the mass of the container with the materials in it. Subtract the mass of the empty container from the mass of the filled container to find the mass of the materials you are using.

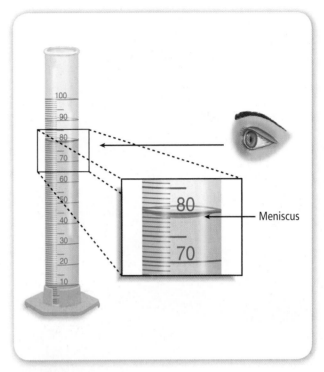

■ **Figure 13** Graduated cylinders measure liquid volume.

Liquid Volume To measure liquids, the unit used is the liter. When a smaller unit is needed, scientists might use a milliliter. Because a milliliter takes up the volume of a cube measuring 1 cm on each side it also can be called a cubic centimeter ($cm^3 = cm \times cm \times cm$).

You can use graduated cylinders to measure liquid volume. A graduated cylinder, shown in **Figure 13,** is marked from bottom to top in milliliters. In lab, you might use a 10-mL graduated cylinder or a 100-mL graduated cylinder. When measuring liquids, notice that the liquid has a curved surface. Look at the surface at eye level, and measure the bottom of the curve. This is called the meniscus. The graduated cylinder in **Figure 13** contains 79.0 mL, or 79.0 cm^3, of a liquid.

Temperature Scientists often measure temperature using the Celsius scale. Pure water has a freezing point of 0°C and boiling point of 100°C. The unit of measurement is degrees Celsius. Two other scales often used are the Fahrenheit and Kelvin scales.

■ **Figure 12** A triple-beam balance is used to determine the mass of an object.

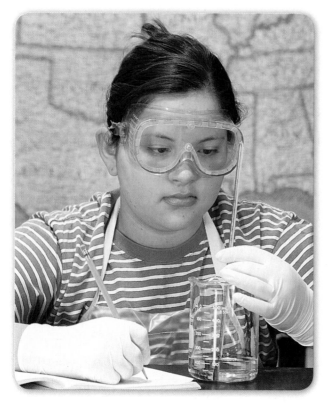

■ **Figure 14** A thermometer measures the temperature of an object.

Scientists use a thermometer to measure temperature. Most thermometers in a laboratory are glass tubes with a bulb at the bottom end containing a liquid such as colored alcohol. The liquid rises or falls with a change in temperature. To read a glass thermometer like the thermometer in **Figure 14,** rotate it slowly until a red line appears. Read the temperature where the red line ends.

Form Operational Definitions An operational definition defines an object by how it functions, works, or behaves. For example, when you are playing hide and seek and a tree is home base, you have created an operational definition for a home base.

Objects can have more than one operational definition. For example, a ruler can be defined as a tool that measures the length of an object (how it is used). It can also be a tool with a series of marks used as a standard when measuring (how it works).

Analyze the Data

To determine the meaning of your observations and investigation results, you will need to look for patterns in the data. Then you must think critically to determine what the data mean. Scientists use several approaches when they analyze the data they have collected and recorded. Each approach is useful for identifying specific patterns.

Interpret Data The word interpret means "to explain the meaning of something." When analyzing data from an experiment, try to find out what the data show. Identify the control group and the test group to see whether or not changes in the independent variable have had an effect. Look for differences in the dependent variable between the control and test groups.

Classify Sorting objects or events into groups based on common features is called classifying. When classifying, first observe the objects or events to be classified. Then select one feature that is shared by some members in the group, but not by all. Place those members that share that feature in a subgroup.

You can classify members into smaller and smaller subgroups based on characteristics. Remember that when you classify, you are grouping objects or events for a purpose. Keep your purpose in mind as you select the features to form groups and subgroups.

Compare and Contrast Observations can be analyzed by noting the similarities and differences between two more objects or events that you observe. When you look at objects or events to see how they are similar, you are comparing them. Contrasting is looking for differences in objects or events.

Recognize Cause and Effect A cause is a reason for an action or condition. The effect is that action or condition. When two events happen together, it is not necessarily true that one event caused the other. Scientists must design a controlled experiment to recognize the exact cause and effect.

Draw Conclusions

When scientists have analyzed the data they collected, they proceed to draw conclusions about the data. These conclusions are sometimes stated in words similar to the hypothesis that you formed earlier. They may confirm a hypothesis, or lead you to a new hypothesis.

Infer Scientists often make inferences based on their observations. An inference is an attempt to explain observations or to indicate a cause. An inference is not a fact, but a logical conclusion that needs further investigation. For example, you may infer that a fire has caused smoke. Until you investigate, however, you do not know for sure.

Apply When you draw a conclusion, you must apply those conclusions to determine whether the data supports the hypothesis. If your data do not support your hypothesis, it does not mean that the hypothesis is wrong. It means only that the result of the investigation did not support the hypothesis. Maybe the experiment needs to be redesigned, or some of the initial observations on which the hypothesis was based were incomplete or biased. Perhaps more observation or research is needed to refine your hypothesis. A successful investigation does not always come out the way you originally predicted.

Avoid Bias Sometimes a scientific investigation involves making judgments. When you make a judgment, you form an opinion. It is important to be honest and not to allow any expectations of results to bias your judgments. This is important throughout the entire investigation, from researching to collecting data to drawing conclusions.

Communicate

The communication of ideas is an important part of the work of scientists. A discovery that is not reported will not advance the scientific community's understanding or knowledge. Communication among scientists also is important as a way of improving their investigations.

Scientists communicate in many ways, from writing articles in journals and magazines that explain their investigations and experiments, to announcing important discoveries on television and radio. Scientists also share ideas with colleagues on the Internet or present them as lectures, like the student is doing in **Figure 15.**

■ **Figure 15** A student communicates to her peers about her investigation.

Safety Guidelines in the Laboratory

The laboratory is a safe place to work if you are aware of important safety rules and if you are careful. You must be responsible for your own safety and for the safety of others. The safety rules given here and the first aid instruction in **Table 3** will protect you and others from harm in the lab. While carrying out procedures in any of the **Launch Labs, MiniLabs,** or **Labs,** notice the safety symbols and warning statements. The safety symbols are explained in the chart on the next page.

1. Always obtain your teacher's permission to begin a lab.
2. Study the procedure. If you have questions, ask your teacher. Be sure you understand all safety symbols shown.
3. Use the safety equipment provided for you. Goggles and a safety apron should be worn when any lab calls for using chemicals.
4. When you are heating a test tube, always slant it so the mouth points away from you and others.
5. Never eat or drink in the lab. Never inhale chemicals. Do not taste any substance or draw any material into your mouth.
6. If you spill any chemical, wash it off immediately with water. Report the spill immediately to your teacher.
7. Know the location and proper use of the fire extinguisher, safety shower, fire blanket, first aid kit, and fire alarm.

8. Keep all materials away from open flames. Tie back long hair.
9. If a fire should break out in the classroom, or if your clothing should catch fire, smother it with the fire blanket or a coat, or get under a safety shower. **NEVER RUN.**
10. Report any accident or injury, no matter how small, to your teacher.

Follow these procedures as you clean up your work area.
1. Turn off the water and gas. Disconnect electrical devices.
2. Return materials to their places.
3. Dispose of chemicals and other materials as directed by your teacher. Place broken glass and solid substances in the proper containers. Never discard materials in the sink.
4. Clean your work area.
5. Wash your hands thoroughly after working in the laboratory.

Table 3	First Aid in the Laboratory
Injury	**Safe Response**
Burns	Apply cold water. Notify your teacher immediately.
Cuts and bruises	Stop any bleeding by applying direct pressure. Cover cuts with a clean dressing. Apply cold compresses to bruises. Notify your teacher immediately.
Fainting	Leave the person lying down. Loosen any tight clothing and keep crowds away. Notify your teacher immediately.
Foreign matter in eye	Flush with plenty of water. Use eyewash bottle or fountain. Notify your teacher immediately.
Poisoning	Note the suspected poisoning agent and call your teacher immediately.
Any spills on skin	Flush with large amounts of water or use safety shower. Notify your teacher immediately.

These safety symbols are used in laboratory and investigations in this book to indicate possible hazards. Learn the meaning of each symbol and refer to this page often. *Remember to wash your hands thoroughly after completing lab procedures.*

SAFETY SYMBOLS	HAZARD	EXAMPLES	PRECAUTION	REMEDY
DISPOSAL	Special disposal procedures need to be followed.	certain chemicals, living organisms	Do not dispose of these materials in the sink or trash can.	Dispose of wastes as directed by your teacher.
BIOLOGICAL	Organisms or other biological materials that might be harmful to humans	bacteria, fungi, blood, unpreserved tissues, plant materials	Avoid skin contact with these materials. Wear mask or gloves.	Notify your teacher if you suspect contact with material. Wash hands thoroughly.
EXTREME TEMPERATURE	Objects that can burn skin by being too cold or too hot	boiling liquids, hot plates, dry ice, liquid nitrogen	Use proper protection when handling.	Go to your teacher for first aid.
SHARP OBJECT	Use of tools or glassware that can easily puncture or slice skin	razor blades, pins, scalpels, pointed tools, dissecting probes, broken glass	Practice common-sense behavior and follow guidelines for use of the tool.	Go to your teacher for first aid.
FUME	Possible danger to respiratory tract from fumes	ammonia, acetone, nail polish remover, heated sulfur, moth balls	Make sure there is good ventilation. Never smell fumes directly. Wear a mask.	Leave foul area and notify your teacher immediately.
ELECTRICAL	Possible danger from electrical shock or burn	improper grounding, liquid spills, short circuits, exposed wires	Double-check setup with teacher. Check condition of wires and apparatus.	Do not attempt to fix electrical problems. Notify your teacher immediately.
IRRITANT	Substances that can irritate the skin or mucous membranes of the respiratory tract	pollen, moth balls, steel wool, fiberglass, potassium permanganate	Wear dust mask and gloves. Practice extra care when handling these materials.	Go to your teacher for first aid.
CHEMICAL	Chemicals that can react with and destroy tissue and other materials	bleaches such as hydrogen peroxide; acids such as sulfuric acid, hydrochloric acid; bases such as ammonia, sodium hydroxide	Wear goggles, gloves, and an apron.	Immediately flush the affected area with water and notify your teacher.
TOXIC	Substance may be poisonous if touched, inhaled, or swallowed.	mercury, many metal compounds, iodine, poinsettia plant parts	Follow your teacher's instructions.	Always wash hands thoroughly after use. Go to your teacher for first aid.
FLAMMABLE	Open flame may ignite flammable chemicals, loose clothing, or hair.	alcohol, kerosene, potassium permanganate, hair, clothing	Avoid open flames and heat when using flammable chemicals.	Notify your teacher immediately. Use fire safety equipment if applicable.
OPEN FLAME	Open flame in use, may cause fire.	hair, clothing, paper, synthetic materials	Tie back hair and loose clothing. Follow teacher's instructions on lighting and extinguishing flames.	Always wash hands thoroughly after use. Go to your teacher for first aid.

Eye Safety Proper eye protection should be worn at all times by anyone performing or observing science activities.

Clothing Protection This symbol appears when substances could stain or burn clothing.

Animal Safety This symbol appears when safety of animals and students must be ensured.

Radioactivity This symbol appears when radioactive materials are used.

Handwashing After the lab, wash hands with soap and water before removing goggles

Math Review

Use Fractions

A fraction compares a part to a whole. In the fraction $\frac{2}{3}$, the 2 represents the part and is the numerator. The 3 represents the whole and is the denominator.

Reduce Fractions To reduce a fraction, you must find the largest factor that is common to both the numerator and the denominator, the greatest common factor (GCF). Divide both numbers by the GCF.

Example Twelve of the 20 chemicals in the science lab are in powder form. What fraction of the chemicals are in powder form?

Step 1 Write the fraction.
$$\frac{\text{part}}{\text{whole}} = \frac{12}{20}$$

Step 2 To find the GCF of the numerator and denominator, list all of the factors of each number.

Factors of 12: 1, 2, 3, 4, 6, 12 (the numbers that divide evenly into 12)
Factors of 20: 1, 2, 4, 5, 10, 20 (the numbers that divide evenly into 20)

Step 3 List the common factors.
1, 2, 4

Step 4 Choose the greatest factor in the list. The GCF of 12 and 20 is 4.

Step 5 Divide the numerator and denominator by the GCF.
$$\frac{12 \div 4}{20 \div 4} = \frac{3}{5}$$

$\frac{3}{5}$ of the chemicals are in powder form.

Practice Problem At an amusement park, 66 of 90 rides have a height restriction. What fraction of the rides, in its simplest form, has a height restriction?

Add and Subtract Fractions To add or subtract fractions with the same denominator, add or subtract the numerators and write the sum or difference over the denominator. After finding the sum or difference, find the simplest form for your fraction.

Example 1 In the forest outside your house, 18 of the animals are rabbits, $\frac{3}{8}$ are squirrels, and the remainder are birds and insects. How many are mammals?

Step 1 Add the numerators.
$$\frac{1}{8} + \frac{3}{8} = \frac{(1 + 3)}{8} = \frac{4}{8}$$

Step 2 Find the GCF.
$$\frac{4}{8} \text{ (GCF, 4)}$$

Step 3 Divide the numerator and denominator by the GCF.
$$\frac{4}{4} = 1, \frac{8}{4} = 2$$

$\frac{1}{2}$ of the animals are mammals.

Example 2 If $\frac{7}{16}$ of a region is covered by freshwater, and $\frac{1}{16}$ of that is in glaciers, how much freshwater is not frozen?

Step 1 Subtract the numerators.
$$\frac{7}{16} - \frac{1}{16} = \frac{(7 - 1)}{16} = \frac{6}{16}$$

Step 2 Find the GCF.
$$\frac{6}{16} \text{ (GCF, 2)}$$

Step 3 Divide the numerator and denominator by the GCF.
$$\frac{6}{2} = 3, \frac{16}{2} = 8$$

$\frac{3}{8}$ of the freshwater is not frozen.

Practice Problem A bicycle rider is riding at a rate of 15 km/h for $\frac{4}{9}$ of his ride, 10 km/h for $\frac{2}{9}$ of his ride, and 8 km/h for the remainder of the ride. How much of his ride is he riding at a rate greater than 8 km/h?

Unlike Denominators To add or subtract fractions with unlike denominators, first find the least common denominator (LCD). This is the smallest number that is a common multiple of both denominators. Rename each fraction with the LCD, and then add or subtract. Find the simplest form if necessary.

Example 1 A chemist makes a paste that is $\frac{1}{2}$ table salt (NaCl), $\frac{1}{3}$ sugar ($C_6H_{12}O_6$), and the remainder is water (H_2O). How much of the paste is a solid?

Step 1 Find the LCD of the fractions.
$\frac{1}{2} + \frac{1}{3}$ (LCD, 6)

Step 2 Rename each numerator and each denominator with the LCD.
$1 \times 3 = 3, 2 \times 3 = 6$
$1 \times 2 = 2, 3 \times 2 = 6$

Step 3 Add the numerators.
$\frac{3}{6} + \frac{2}{6} = \frac{(3+2)}{6} = \frac{5}{6}$

$\frac{5}{6}$ of the paste is a solid.

Example 2 The average precipitation in Grand Junction, CO, is $\frac{7}{10}$ inch in November and $\frac{3}{5}$ inch in December. What is the sum of these averages?

Step 1 Find the LCD of the fractions.
$\frac{7}{10} + \frac{3}{5}$ (LCD, 10)

Step 2 Rename each numerator and each denominator with the LCD.
$7 \times 1 = 7, 10 \times 1 = 10$
$3 \times 2 = 6, 5 \times 2 = 10$

Step 3 Add the numerators.
$\frac{7}{10} + \frac{6}{10} = \frac{(7+6)}{10} = \frac{13}{10}$

$\frac{13}{10}$ inches total precipitation, or $1\frac{3}{10}$ inches.

Practice Problem On an electric bill, about $\frac{1}{8}$ of the energy is from solar energy and about $\frac{1}{10}$ is from wind power. How much of the total bill is from solar energy and wind power combined?

Example 3 In your body, $\frac{7}{10}$ of your muscle contractions are involuntary (cardiac and smooth muscle tissue). Smooth muscle makes $\frac{3}{15}$ of your muscle contractions. How many of your muscle contractions are made by cardiac muscle?

Step 1 Find the LCD of the fractions.
$\frac{7}{10} - \frac{3}{15}$ (LCD, 30)

Step 2 Rename each numerator and each denominator with the LCD.
$7 \times 3 = 21, 10 \times 3 = 30$
$3 \times 2 = 6, 15 \times 2 = 30$

Step 3 Subtract the numerators.
$\frac{21}{30} - \frac{6}{30} = \frac{(21-6)}{30} = \frac{15}{30}$

Step 4 Find the GCF.
$\frac{15}{30}$ (GCF, 15)

$\frac{1}{2}$ of all muscle contractions are cardiac muscle.

Example 4 Tony wants to make cookies that call for $\frac{3}{4}$ of a cup of flour, but he only has $\frac{1}{3}$ of a cup. How much more flour does he need?

Step 1 Find the LCD of the fractions.
$\frac{3}{4} - \frac{1}{3}$ (LCD, 12)

Step 2 Rename each numerator and each denominator with the LCD.
$\frac{3 \times 3}{4 \times 3} = \frac{9}{12}$
$\frac{1 \times 4}{3 \times 4} = \frac{4}{12}$

Step 3 Subtract the numerators.
$\frac{9}{12} - \frac{4}{12} = \frac{(9-4)}{12} = \frac{5}{12}$

$\frac{5}{12}$ of a cup of flour

Practice Problem Using the information provided to you in Example 3 above, determine how many muscle contractions are voluntary (skeletal muscle).

Multiply Fractions To multiply with fractions, multiply the numerators and multiply the denominators. Find the simplest form if necessary.

Example Multiply $\frac{3}{5}$ by $\frac{1}{3}$.

Step 1 Multiply the numerators and denominators.

$$\frac{3}{5} \times \frac{1}{3} = \frac{(3 \times 1)}{(5 \times 3)} = \frac{3}{15}$$

Step 2 Find the GCF.

$$\frac{3}{15} \text{ (GCF, 3)}$$

Step 3 Divide the numerator and denominator by the GCF.

$$\frac{3 \div 3}{15 \div 3} = \frac{1}{5}$$

$\frac{3}{5}$ multiplied by $\frac{1}{3}$ is $\frac{1}{5}$.

Practice Problem Multiply $\frac{3}{14}$ by $\frac{5}{16}$.

Find a Reciprocal Two numbers whose product is 1 are called multiplicative inverses, or reciprocals.

Example Find the reciprocal of $\frac{3}{8}$.

Step 1 Inverse the fraction by putting the denominator on top and the numerator on the bottom.

$$\frac{8}{3}$$

The reciprocal of $\frac{3}{8}$ is $\frac{8}{3}$.

Practice Problem Find the reciprocal of $\frac{4}{9}$.

Divide Fractions To divide one fraction by another fraction, multiply the dividend by the reciprocal of the divisor. Find the simplest form if necessary.

Example 1 Divide $\frac{1}{9}$ by $\frac{1}{3}$.

Step 1 Find the reciprocal of the divisor. The reciprocal of $\frac{1}{3}$ is $\frac{3}{1}$.

Step 2 Multiply the dividend by the reciprocal of the divisor.

$$\frac{\frac{1}{9}}{\frac{1}{3}} = \frac{1}{9} \times \frac{3}{1} = \frac{(1 \times 3)}{(9 \times 1)} = \frac{3}{9}$$

Step 3 Find the GCF.

$$\frac{3}{9} \text{ (GCF, 3)}$$

Step 4 Divide the numerator and denominator by the GCF.

$$\frac{3 \div 3}{9 \div 3} = \frac{1}{3}$$

$\frac{1}{9}$ divided by $\frac{1}{3}$ is $\frac{1}{3}$.

Example 2 Divide $\frac{3}{5}$ by $\frac{1}{4}$.

Step 1 Find the reciprocal of the divisor. The reciprocal of $\frac{1}{4}$ is $\frac{4}{1}$.

Step 2 Multiply the dividend by the reciprocal of the divisor.

$$\frac{\frac{3}{5}}{\frac{1}{4}} = \frac{3}{5} \times \frac{4}{1} = \frac{(3 \times 4)}{(5 \times 1)} = \frac{12}{5}$$

$\frac{3}{5}$ divided by $\frac{1}{4}$ is $\frac{12}{5}$ or $2\frac{2}{5}$.

Practice Problem Divide $\frac{3}{11}$ by $\frac{7}{10}$.

Use Ratios

A ratio compares two different numbers. For example, if you have 3 dogs and 5 cats, the ratio of dogs to cats can be written "3 to 5," 3:5. Notice that you have eight total animals in this example—3 dogs and 5 cats.

Ratios can represent one type of probability, called odds. This is a ratio that compares the number of ways a certain outcome occurs to the number of possible outcomes. For example, if you flip a coin 100 times, what are the odds that it will come up heads? There are two possible outcomes, heads or tails, so the most likely outcome of 100 flips is 50 heads and 50 tails, or 50:50. Like fractions, ratios can be written in simplest form—50:50 becomes 1:1. To write this as a fraction, you would say $\frac{1}{2}$ the flips are heads and $\frac{1}{2}$ are tails.

Example 1 A chemical solution contains 40 g of salt and 64 g of baking soda. What is the ratio of salt to baking soda as a fraction in simplest form?

Step 1 Write the ratio.

salt:baking soda = 40:64

Step 2 Express the fraction in simplest form.
The GCF of 40 and 64 is 8.
$40 \div 8 = 5$, and $64 \div 8 = 8$

The ratio of salt to baking soda in the sample is 5:8.

Example 2 Sean rolls a 6-sided die 6 times. Predict how many times the side with a 3 will show in six rolls.

Step 1 Write the ratio as a fraction.
$$\frac{\text{number of sides with a 3}}{\text{number of total sides}} = \frac{1}{6}$$

Step 2 Multiply by the number of attempts.
$\frac{1}{6} \times 6$ attempts $= \frac{6}{6} = 1$

In six rolls, Sean will likely roll one 3.

Practice Problem Two metal rods measure 100 cm and 144 cm in length. What is the ratio of their lengths in simplest form?

Use Decimals

A fraction with a denominator that is a power of ten can be easily written as a decimal. The decimal point separates the ones place from the tenths place. For example, $\frac{27}{100}$ means 0.27.

Any fraction can be written as a decimal using division. For example, the fraction $\frac{5}{8}$ can be written as a decimal by dividing 5 by 8. Written as a decimal, it is 0.625.

Add or Subtract Decimals When adding and subtracting decimals, line up the decimal points before carrying out the operation.

Example 1 Find the sum of 47.68 and 7.80.

Step 1 Line up the decimal places when you write the numbers.

47.68
+7.80

Step 2 Add the decimals.

47.68
+7.80
55.48

The sum of 47.68 and 7.80 is 55.48.

Example 2 Find the difference of 42.17 and 15.85.

Step 1 Line up the decimal places when you write the numbers.

42.17
−15.85

Step 2 Subtract the decimals.

42.17
−15.85
26.32

The difference of 42.17 and 15.85 is 26.32.

Practice Problem Find the sum of 1.245 and 3.842.

Multiply Decimals To multiply decimals, multiply the numbers like you multiply numbers without decimals. Count the decimal places in each factor. The product will have the same number of decimal places as the sum of the decimal places in the factors.

Example Multiply 2.4 by 5.9.

Step 1 Multiply the factors like two whole numbers.
$$24 \times 59 = 1416$$

Step 2 Find the sum of the number of decimal places in the factors. Each factor has one decimal place, so the sum is two decimal places.

Step 3 The product will have two decimal places.
14.16

The product of 2.4 and 5.9 is 14.16.

Practice Problem Multiply 4.6 by 2.2.

Divide Decimals When dividing decimals, change the divisor to a whole number. To do this, multiply both the divisor and the dividend by the same power of ten. Then place the decimal point in the quotient directly above the decimal point in the dividend. Then divide as you do with whole numbers.

Example Divide 8.84 by 3.4.

Step 1 Multiply both factors by 10.
$$3.4 \times 10 = 34, 8.84 \times 10 = 88.4$$

Step 2 Divide 88.4 by 34.

```
     2.6
34)88.4
   -68
    204
   -204
      0
```

8.84 divided by 3.4 is 2.6.

Practice Problem Divide 75.6 by 3.6.

Use Proportions

An equation that shows that two ratios are equivalent is a proportion. The ratios $\frac{2}{4}$ and $\frac{5}{10}$ are equivalent, so they can be written as $\frac{2}{4} = \frac{5}{10}$. This equation is a proportion. When two ratios form a proportion, the cross products are equal. To find the cross products in the proportion $\frac{2}{4} = \frac{5}{10}$ multiply the 2 and the 10, and the 4 and the 5. Therefore $2 \times 10 = 4 \times 5$, or $20 = 20$.

Because you know that both ratios are equal, you can use cross products to find a missing term in a proportion. This is known as solving the proportion.

Example The heights of a tree and a pole are proportional to the lengths of their shadows. The tree casts a shadow of 24 m when a 6-m pole casts a shadow of 4 m. What is the height of the tree?

Step 1 Write a proportion.
$$\frac{\text{height of tree}}{\text{height of pole}} = \frac{\text{length of tree's shadow}}{\text{length of pole's shadow}}$$

Step 2 Substitute the known values into the proportion. Let h represent the unknown value, the height of the tree.
$$\frac{h}{6 \text{ m}} = \frac{24 \text{ m}}{4 \text{ m}}$$

Step 3 Find the cross products.
$$h \times 4 \text{ m} = 6 \text{ m} \times 24 \text{ m}$$

Step 4 Simplify the equation.
$$4h = 144 \text{ m}$$

Step 5 Divide each side by 4.
$$\frac{(4 \text{ m}) h}{4} = \frac{144 \text{ m}}{4}$$

The height of the tree is 36 m.

Practice Problem The ratios of the weights of two objects on the Moon and on Earth are proportional. A rock weighing 3 N on the Moon weighs 18 N on Earth. How much would a rock that weighs 5 N on the Moon weigh on Earth?

Use Percentages

The word *percent* means "out of one hundred." It is a ratio that compares a number to 100. Suppose you read that 77 percent of the Earth's surface is covered by water. That is the same as reading that the fraction of the Earth's surface covered by water is $\frac{77}{100}$. To express a fraction as a percent, first find the equivalent decimal for the fraction. Then, multiply the decimal by 100 and add the percent symbol.

Example Express $\frac{13}{20}$ as a percent.

Step 1 Find the equivalent decimal for the fraction.

$$\begin{array}{r} 0.65 \\ 20\overline{)13.00} \\ 12.0 \\ \hline 1.00 \\ 1.00 \\ \hline 0 \end{array}$$

Step 2 Rewrite the fraction $\frac{13}{20}$ as 0.65.

Step 3 Multiply 0.65 by 100 and add the % symbol.
$$0.65 \times 100 = 65 = 65\%$$

So, $\frac{13}{20} = 65\%$

This also can be solved as a proportion.

Example Express $\frac{13}{20}$ as a percent.

Step 1 Write a proportion.
$$\frac{13}{20} = \frac{x}{100}$$

Step 2 Find the cross products.
$$1300 = 20x$$

Step 3 Divide each side by 20.
$$\frac{1300}{20} = \frac{20x}{20}$$

$$65\% = x$$

Practice Problem In one year, 73 of 365 days were rainy in one city. What percent of the days in that city were rainy?

Solve One-Step Equations

A statement that two expressions are equal is an equation. For example, A = B is an equation that states that A is equal to B.

An equation is solved when a variable is replaced with a value that makes both sides of the equation equal. To make both sides equal, the inverse operation is used. Addition and subtraction are inverses, and multiplication and division are inverses.

Example 1 Solve the equation x - 10 = 35.

Step 1 Find the solution by adding 10 to each side of the equation.
$$x - 10 = 35$$
$$x - 10 + 10 = 35 + 10$$
$$x = 45$$

Step 2 Check the solution.
$$x - 10 = 35$$
$$45 - 10 = 35$$
$$35 = 35$$

Both sides of the equation are equal, so $x = 45$.

Example 2 In the formula $a = bc$, find the value of c if $a = 20$ and $b = 2$.

Step 1 Rearrange the formula so the unknown value is by itself on one side of the equation by dividing both sides by b.
$$a = bc$$
$$\frac{a}{b} = \frac{bc}{b}$$
$$\frac{a}{b} = c$$

Step 2 Replace the variables a and b with the values that are given.
$$\frac{a}{b} = c$$
$$\frac{20}{2} = c$$
$$10 = c$$

Step 3 Check the solution.
$$a = bc$$
$$20 = 2 \times 10$$
$$20 = 20$$

Both sides of the equation are equal, so $c = 10$ is the solution when $a = 20$ and $b = 2$.

Practice Problem In the formula $h = gd$, find the value of d if $g = 12.3$ and $h = 17.4$.

Math Skill Handbook

Use Statistics

The branch of mathematics that deals with collecting, analyzing, and presenting data is statistics. In statistics, there are three common ways to summarize data with a single number—the mean, the median, and the mode.

The mean of a set of data is the arithmetic average. It is found by adding the numbers in the data set and dividing by the number of items in the set.

The median is the middle number in a set of data when the data are arranged in numerical order. If there were an even number of data points, the median would be the mean of the two middle numbers.

The mode of a set of data is the number or item that appears most often.

Another number that often is used to describe a set of data is the range. The range is the difference between the largest number and the smallest number in a set of data.

A frequency table shows how many times each piece of data occurs, usually in a survey. **Table 1** below shows the results of a student survey on favorite color.

Table 1	Student Color Choice	
Color	Tally	Frequency
Red	IIII	4
Blue	HHH	5
Black	II	2
Green	III	3
Purple	HHH II	7
Yellow	HHH I	6

Based on the frequency table data, which color is the favorite?

Example The speeds (in m/s) for a race car during five different time trials are 39, 37, 44, 36, and 44.

Find the mean:

Step 1 Find the sum of the numbers.

$$39 + 37 + 44 + 36 + 44 = 200$$

Step 2 Divide the sum by the number of items, which is 5.

$$200 \div 5 = 40$$

The mean is 40 m/s.

Find the median:

Step 1 Arrange the measures from least to greatest.

36, 37, 39, 44, 44

Step 2 Determine the middle measure.

36, 37, <u>39</u>, 44, 44

The median is 39 m/s.

Find the mode:

Step 1 Group the numbers that are the same together.

44, 44, 36, 37, 39

Step 2 Determine the number that occurs most in the set.

<u>44</u>, <u>44</u>, 36, 37, 39

The mode is 44 m/s.

Find the range:

Step 1 Arrange the measures from greatest to least.

44, 44, 39, 37, 36

Step 2 Determine the greatest and least measures in the set.

<u>44</u>, 44, 39, 37, <u>36</u>

Step 3 Find the difference between the greatest and least measures.

$$44 - 36 = 8$$

The range is 8 m/s.

Practice Problem Find the mean, median, mode, and range for the data set 8, 4, 12, 8, 11, 14, 16.

Use Geometry

The branch of mathematics that deals with the measurement, properties, and relationships of points, lines, angles, surfaces, and solids is called geometry.

Perimeter The perimeter (P) is the distance around a geometric figure. To find the perimeter of a rectangle, add the length and width and multiply that sum by two, or $2(l + w)$. To find perimeters of irregular figures, add the lengths of the sides.

Example 1 Find the perimeter of a rectangle that is 3 m long and 5 m wide.

Step 1 You know that the perimeter is 2 times the sum of the width and length.

$$P = 2(3\ m + 5\ m)$$

Step 2 Find the sum of the width and length.

$$P = 2(8\ m)$$

Step 3 Multiply by 2.

$$P = 16\ m$$

The perimeter is 16 m.

Example 2 Find the perimeter of a shape with sides measuring 2 cm, 5 cm, 6 cm, 3 cm.

Step 1 You know that the perimeter is the sum of all the sides.

$$P = 2\ cm + 5\ cm + 6\ cm + 3\ cm$$

Step 2 Find the sum of the sides.

$$P = 2\ cm + 5\ cm + 6\ cm + 3\ cm$$
$$P = 16\ cm$$

The perimeter is 16 cm.

Practice Problem Find the perimeter of a rectangle with a length of 18 m and a width of 7 m.

Practice Problem Find the perimeter of a triangle measuring 1.6 cm by 2.4 cm by 2.4 cm.

Area of a Rectangle The area (A) is the number of square units needed to cover a surface. To find the area of a rectangle, multiply the length by the width, or $l \times w$. When finding area, the units also are multiplied. Area is given in square units.

Example Find the area of a rectangle with a length of 1 cm and a width of 10 cm.

Step 1 You know that the area is the length multiplied by the width.

$$A = (1\ cm \times 10\ cm)$$

Step 2 Multiply the length by the width. Also multiply the units.

$$A = 10\ cm^2$$

The area is 10 cm².

Practice Problem Find the area of a square whose sides measure 4 m.

Area of a Triangle To find the area of a triangle, use the formula:

$$A = \frac{1}{2}(\text{base} \times \text{height})$$

The base of a triangle can be any of its sides. The height is the perpendicular distance from a base to the opposite endpoint, or vertex.

Example Find the area of a triangle with a base of 18 m and a height of 7 m.

Step 1 You know that the area is $\frac{1}{2}$ the base times the height.

$$A = \frac{1}{2}(18\ m \times 7\ m)$$

Step 2 Multiply $\frac{1}{2}$ by the product of 18×7. Multiply the units.

$$A = \frac{1}{2}(126\ m^2)$$
$$A = 63\ m^2$$

The area is 63 m².

Practice Problem Find the area of a triangle with a base of 27 cm and a height of 17 cm.

Circumference of a Circle The diameter (*d*) of a circle is the distance across the circle through its center, and the radius (*r*) is the distance from the center to any point on the circle. The radius is half of the diameter. The distance around the circle is called the circumference (*C*). The formula for finding the circumference is:

$$C = 2\pi r \ \text{ or } \ C = \pi d$$

The circumference divided by the diameter is always equal to 3.1415926... This nonterminating and nonrepeating number is represented by the Greek letter π (pi). An approximation often used for π is 3.14.

Example 1 Find the circumference of a circle with a radius of 3 m.

Step 1 You know the formula for the circumference is 2 times the radius times π.
$$C = 2\pi \,(3 \text{ m})$$

Step 2 Multiply 2 times the radius.
$$C = \pi \,(6 \text{ m})$$

Step 3 Multiply by π.
$$C = 19 \text{ m}$$

The circumference is 19 m.

Example 2 Find the circumference of a circle with a diameter of 24.0 cm.

Step 1 You know the formula for the circumference is the diameter times π.
$$C = \pi \,(24.0 \text{ cm})$$

Step 2 Multiply the diameter by π.
$$C = 75.4 \text{ cm}$$

The circumference is 75.4 cm.

Practice Problem Find the circumference of a circle with a radius of 19 cm.

Area of a Circle The formula for the area of a circle is: $A = \pi r^2$

Example 1 Find the area of a circle with a radius of 4.0 cm.

Step 1 $\quad A = \pi \,(4.0 \text{ cm})^2$

Step 2 Find the square of the radius.
$$A = 16\pi \text{ cm}^2$$

Step 3 Multiply the square of the radius by π.
$$A = 50 \text{ cm}^2$$

The area of the circle is 50 cm².

Example 2 Find the area of a circle with a radius of 225 m.

Step 1 $\quad A = \pi \,(225 \text{ m})^2$

Step 2 Find the square of the radius.
$$A = 50{,}625\pi \text{ m}^2$$

Step 3 Multiply the square of the radius by π.
$$A = 158{,}962.5 \text{ m}^2$$

The area of the circle is 158,962 m².

Example 3 Find the area of a circle whose diameter is 20.0 mm.

Step 1 You know the formula for the area of a circle is the square of the radius times π and that the radius is half of the diameter.
$$A = \pi \left(\frac{20.0 \text{ mm}}{2}\right)^2$$

Step 2 Find the radius.
$$A = \pi \,(10.0 \text{ mm})^2$$

Step 3 Find the square of the radius.
$$A = 100\pi \text{ mm}^2$$

Step 4 Multiply the square of the radius by π.
$$A = 314 \text{ mm}^2$$

The area is 314 mm².

Practice Problem Find the area of a circle with a radius of 16 m.

Volume The measure of space occupied by a solid is the volume (V). To find the volume of a rectangular solid, multiply the length by the width by the height, or $V = l \times w \times h$. It is measured in cubic units, such as cubic centimeters (cm^3).

Example Find the volume of a rectangular solid with a length of 2.0 m, a width of 4.0 m, and a height of 3.0 m.

Step 1 You know the formula for volume is the length times the width times the height.
$$V = 2.0 \text{ m} \times 4.0 \text{ m} \times 3.0 \text{ m}$$

Step 2 Multiply the length by the width by the height.
$$V = 24 \text{ m}^3$$

The volume is 24 m³.

Practice Problem Find the volume of a rectangular solid that is 8 m long, 4 m wide, and 4 m high.

To find the volume of other solids, multiply the area of the base by the height.

Example 1 Find the volume of a prism that has two triangular bases with lengths of 8.0 m and heights of 7.0 m. The height of the entire solid is 15.0 m.

Step 1 You know that the base is a triangle, and the area of a triangle is $\frac{1}{2}$ the base times the height, and the volume is the area of the base times the height.
$$V = \left[\frac{1}{2}(b \times h)\right] \times 15 \text{ m}$$

Step 2 Find the area of the base.
$$V = \left[\frac{1}{2}(8 \text{ m} \times 7 \text{ m})\right] \times 15 \text{ m}$$
$$V = \left(\frac{1}{2} \times 56 \text{ m}^2\right) \times 15 \text{ m}$$

Step 3 Multiply the area of the base by the height of the solid.
$$V = 28 \text{ m}^2 \times 15 \text{ m}$$
$$V = 420 \text{ m}^3$$

The volume is 420 m³.

Example 2 Find the volume of a cylinder that has a base with a radius of 12.0 cm and a height of 21.0 cm.

Step 1 You know that the base is a circle, and the area of a circle is the square of the radius times π, and the volume is the area of the base times the height.
$$V = (\pi r^2) \times 21 \text{ cm}$$
$$V = \pi (12 \text{ cm})^2 \times 21 \text{ cm}$$

Step 2 Find the area of the base.
$$V = 144\pi \text{ cm}^2 \times 21 \text{ cm}$$
$$V = 452 \text{ cm}^2 \times 21 \text{ cm}$$

Step 3 Multiply the area of the base by the height of the solid.
$$V = 9,490 \text{ cm}^3$$

The volume is 9,500 cm³.

Example 3 Find the volume of a cylinder that has a diameter of 15 mm and a height of 4.8 mm.

Step 1 You know that the base is a circle with an area equal to the square of the radius times π. The radius is one-half the diameter. The volume is the area of the base times the height.
$$V = (\pi r^2) \times 4.8 \text{ mm}$$
$$V = \left[\pi \left(\frac{1}{2} \times 15 \text{ mm}\right)^2\right] \times 4.8 \text{ mm}$$
$$V = \pi (7.5 \text{ mm})^2 \times 4.8 \text{ mm}$$

Step 2 Find the area of the base.
$$V = 56.25\pi \text{ mm}^2 \times 4.8 \text{ mm}$$
$$V = 176.63 \text{ mm}^2 \times 4.8 \text{ mm}$$

Step 3 Multiply the area of the base by the height of the solid.
$$V = 847.8 \text{ mm}^3$$

The volume is 847.8 mm³.

Practice Problem Find the volume of a cylinder with a diameter of 7 cm in the base and a height of 16 cm.

Science Applications

Measure in SI

The metric system of measurement was developed in 1795. A modern form of the metric system, called the International System (SI), was adopted in 1960 and provides the standard measurements that all scientists around the world can understand.

The SI system is convenient because unit sizes vary by powers of 10. Prefixes are used to name units. Look at **Table 2** for some common SI prefixes and their meanings.

Table 2	Common SI Prefixes		
Prefix	Symbol	Meaning	
kilo–	k	1,000	thousand
hecto–	h	100	hundred
deka–	da	10	ten
deci–	d	0.1	tenth
centi–	c	0.01	hundredth
milli–	m	0.001	thousandth

Example How many grams equal one kilogram?

Step 1 Find the prefix *kilo–* in **Table 2.**

Step 2 Using **Table 2,** determine the meaning of *kilo–*. According to the table, it means 1,000. When the prefix kilo– is added to a unit, it means that there are 1,000 of the units in a "kilounit."

Step 3 Apply the prefix to the units in the question. The units in the question are grams. There are 1,000 grams in a kilogram.

Practice Problem Is a milligram larger or smaller than a gram? How many of the smaller units equal one larger unit? What fraction of the larger unit does one smaller unit represent?

Dimensional Analysis

Convert SI Units In science, quantities such as length, mass, and time sometimes are measured using different units. A process called dimensional analysis can be used to change one unit of measure to another. This process involves multiplying your starting quantity and units by one or more conversion factors. A conversion factor is a ratio equal to one and can be made from any two equal quantities with different units. If 1,000 mL equal 1 L, then two ratios can be made.

$$\frac{1,000 \text{ mL}}{1 \text{ L}} = \frac{1 \text{ L}}{1,000 \text{ mL}} = 1$$

One can convert between units in the SI system by using the equivalents in **Table 2** to make conversion factors.

Example 1 How many cm are in 4 m?

Step 1 Write conversion factors for the units given. From **Table 2,** you know that 100 cm = 1 m. The conversion factors are

$$\frac{100 \text{ cm}}{1 \text{ m}} \text{ and } \frac{1 \text{ m}}{100 \text{ cm}}$$

Step 2 Decide which conversion factor to use. Select the factor that has the units you are converting from (m) in the denominator and the units you are converting to (cm) in the numerator.

$$\frac{100 \text{ cm}}{1 \text{ m}}$$

Step 3 Multiply the starting quantity and units by the conversion factor. Cancel the starting units with the units in the denominator. There are 400 cm in 4 m.

$$4 \text{ m} \times \frac{100 \text{ cm}}{1 \text{ m}} = 400 \text{ cm}$$

Practice Problem How many milligrams are in one kilogram? (Hint: You will need to use two conversion factors from **Table 2.**)

Table 3	Unit System Equivalents	
Type of Measurement	Equivalent	
Length	1 in. = 2.54 cm 1 yd = 0.91 m 1 mi = 1.61 km	
Mass and weight (measured in standard Earth gravity)	1 oz = 28.35 g 1 lb = 0.45 kg 1 ton (short) = 0.91 tonnes (metric tons) 1 lb = 4.45 N	
Volume	1 in.^{3} = 16.39 cm^3 1 qt = 0.95 L 1 gal = 3.78 L	
Area	1 in.^{2} = 6.45 cm^2 1 yd^2 = 0.83 m^2 1 mi^2 = 2.59 km^2 1 acre = 0.40 hectares	
Temperature	$°C = \frac{(°F - 32)}{1.8}$ $K = °C + 273$	

Convert Between Unit Systems **Table 3** gives a list of equivalents that can be used to convert between English and SI units.

Example If a meterstick has a length of 100 cm, how long is the meterstick in inches?

Step 1 Write the conversion factors for the units given. From **Table 3**, 1 in. = 2.54 cm.
$\frac{1 \text{ in.}}{2.54 \text{ cm}}$ and $\frac{2.54 \text{ cm}}{1 \text{ in.}}$

Step 2 Determine which conversion factor to use. You are converting from cm to in. Use the conversion factor with cm on the bottom.
$\frac{1 \text{ in.}}{2.54 \text{ cm}}$

Step 3 Multiply the starting quantity and units by the conversion factor. Cancel the starting units with the units in the denominator. Round your answer to the nearest tenth.
$100 \text{ cm} \times \frac{1 \text{ in.}}{2.54 \text{ cm}} = 39.37 \text{ in.}$

The meterstick is about 39.4 in. long.

Practice Problem A book has a mass of 5 lbs. What is the mass of the book in kg?

Practice Problem Use the relationship 1 in. = 2.54 cm to show how 1 in.^{3} = 16.39 cm^3.

Precision and Significant Figures

When you make a measurement, the value you record depends on the precision of the measuring instrument. This precision is represented by the number of significant figures recorded in the measurement. When counting the number of significant figures, all figures are counted except zeros at the end of a number with no decimal point such as 2,050, and zeros at the beginning of a decimal such as 0.03020. When adding or subtracting numbers with different precision, round the answer to the smallest number of decimal places of any number in the sum or difference. When multiplying or dividing, the answer is rounded to the smallest number of significant figures of any number being multiplied or divided.

Example The lengths 5.28 and 5.2 are measured in meters. Find the sum of these lengths and record your answer using the correct number of significant figures.

Step 1 Find the sum. [align the decimal points]
5.28 m   2 digits after the decimal
+ 5.2 m   1 digit after the decimal
10.48 m

Step 2 Round to one digit after the decimal because the least number of digits after the decimal of the numbers being added is 1.

The sum is 10.5 m.

Practice Problem How many significant figures are in the measurement 7,071,301 m? How many significant figures are in the measurement 0.003010 g?

Practice Problem Multiply 5.28 and 5.2 using the rule for multiplying and dividing. Record the answer using the correct number of significant figures.

Scientific Notation

Oftentimes the numbers used in science are very small or very large. Because these numbers are difficult to work with, scientists use scientific notation. To write numbers in scientific notation, move the decimal point until only one non-zero digit remains on the left. Then count the number of places you moved the decimal point and use that number as a power of ten. For example, the average distance from the Sun to Mars is 227,800,000,000 m. In scientific notation, this distance is 2.278×10^{11} m. Because you moved the decimal point to the left, the number is a positive power of ten.

The mass of an electron is about 0.000 000 000 000 000 000 000 000 000 000 911 kg. Expressed in scientific notation, this mass is 9.11×10^{-31} kg. Because the decimal point was moved to the right, the number is a negative power of ten.

Example Earth is 149,600,000 km from the Sun. Express this in scientific notation.

Step 1 Move the decimal point until one non-zero digit remains on the left. 1.496 000 00

Step 2 Count the number of decimal places you have moved. In this case, eight.

Step 3 Show that number as a power of ten, 10^8. Earth is 1.496×10^8 km from the Sun.

Practice Problem Express each of the following in scientific notation: 0.005835 g, 300,000 m/s, 15,000,000 K, 0.00020 cm.

Make and Use Graphs

Data in tables can be displayed in a graph—a visual representation of data. Common graph types include line graphs, bar graphs, and circle graphs.

Line Graph A line graph shows a relationship between two variables that change continuously. The independent variable is changed and is plotted on the x-axis. The dependent variable is observed and is plotted on the y-axis.

Example Draw a line graph of the data in **Table 4** from a cyclist in a long-distance race.

Table 4	Bicycle Race Data
Time (h)	Distance (km)
0	0
1	8
2	16
3	24
4	32
5	40

Step 1 Determine the x-axis and y-axis variables. Time varies independently of distance and is plotted on the x-axis. Distance is dependent on time and is plotted on the y-axis.

Step 2 Determine the scale of each axis. The x-axis data ranges from 0 to 5. The y-axis data ranges from 0 to 50.

Step 3 Using graph paper, draw and label the axes. Include units in the labels.

Step 4 Draw a point at the intersection of the time value on the x-axis and corresponding distance value on the y-axis. Connect the points and label the graph with a title, as shown in **Figure 1.**

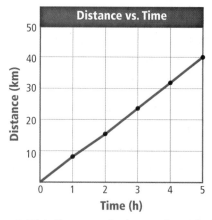

■**Figure 1** This line graph shows the relationship between distance and time during a bicycle ride.

Practice Problem A puppy's shoulder height is measured during the first year of her life. The following measurements were collected: (3 mo, 52 cm), (6 mo, 72 cm), (9 mo, 83 cm), (12 mo, 86 cm). Graph this data.

Find a Slope The slope of a straight line is the ratio of the vertical change, rise, to the horizontal change, run.

$$\text{slope} = \frac{\text{vertical change (rise)}}{\text{horizontal change (run)}} = \frac{\text{change in } y}{\text{change in } x}$$

Example Find the slope of the graph in Figure 20.

Step 1 You know that the slope is the change in *y* divided by the change in *x*.

$$\text{slope} = \frac{\text{change in } y}{\text{change in } x}$$

Step 2 Determine the data points you will be using. For a straight line, choose the two sets of points that are the farthest apart.

$$\text{slope} = \frac{(40 - 0) \text{ km}}{(5 - 0) \text{ h}}$$

Step 3 Find the change in *y* and *x*.

$$\text{slope} = \frac{40 \text{ km}}{5 \text{ h}}$$

Step 4 Divide the change in *y* by the change in *x*.

$$\text{slope} = 8 \text{ km/h}$$

The slope of the graph is 8 km/h.

Bar Graph To compare data that does not change continuously, you might use a bar graph. A bar graph uses bars to show the relationships between variables. The *x*-axis variable is divided into parts. The parts can be numbers, such as years, or a category, such as a type of animal. The *y*-axis is a number and increases continuously along the axis.

Example A recycling center collects 4.0 kg of aluminum on Monday, 1.0 kg on Wednesday, and 2.0 kg on Friday. Create a bar graph of this data.

Step 1 Select the *x*-axis and *y*-axis variables. The measured numbers (the masses of aluminum) should be placed on the *y*-axis. The variable divided into parts (collection days) is placed on the *x*-axis.

Step 2 Create a graph grid like you would for a line graph. Include labels and units.

Step 3 For each measured number, draw a vertical bar above the *x*-axis value up to the *y*-axis value. For the first data point, draw a vertical bar above Monday up to 4.0 kg.

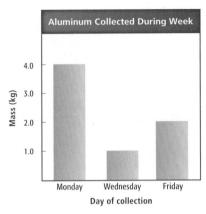

Practice Problem Draw a bar graph of the gases in air: 78% nitrogen, 21% oxygen, 1% other gases.

Circle Graph A circle graph is a circle divided into sections that represent the relative size of each piece of data. The entire circle represents 100%, half represents 50%, and so on.

Example Air is made up of 78% nitrogen, 21% oxygen, and 1% other gases. Display the composition of air in a circle graph.

Step 1 Multiply each percent by 360° and divide by 100 to find the angle of each section in the circle.

$$78 \times \frac{360°}{100} = 280.8°$$

$$21 \times \frac{360°}{100} = 75.6°$$

$$1 \times \frac{360°}{100} = 3.6°$$

Step 2 Use a compass to draw a circle and to mark the center of the circle. Draw a straight line from the center to the edge of the circle.

Step 3 Use a protractor and the angles you calculated to divide the circle into parts.

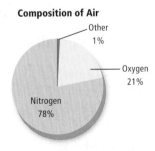

Practice Problem Draw a circle graph to represent the amount of aluminum collected during the week shown in the bar graph above.

Formulas

Chapter 1 The Nature of Science

$$\text{Density} = \frac{\text{mass}}{\text{volume}}$$

$$\text{Kelvin} = {}^\circ\text{Celsius} + 273$$

$$\% \text{ Error} = \left| \frac{\text{(Accepted value} - \text{Experimental value)}}{\text{Accepted value}} \right| \times 100$$

Chapter 2 Motion

$$\text{Speed} = \frac{\text{distance}}{\text{time}}$$

$$\text{Acceleration} = \frac{\text{change in velocity}}{\text{time}}$$

$$\text{Change in velocity} = \text{final velocity} - \text{initial velocity}$$

Chapter 3 Forces and Newton's Laws

$$\text{Acceleration} = \frac{\text{net force}}{\text{mass}}$$

$$\text{Force} = \text{mass} \times \text{acceleration}$$

$$\text{Gravitational force} = \text{mass} \times \text{(acceleration due to gravity)}$$

$$\text{Weight} = \text{mass} \times \text{gravity}$$

$$\text{Momentum } (p) = \text{mass} \times \text{velocity}$$

$$\text{Force} = \frac{(mv_f - mv_i)}{\text{time}}$$

$$\text{Change in position} = \text{initial velocity (change in time)} + \frac{1}{2} \text{acceleration (change in time)}^2$$

$$\text{Average velocity} = \frac{\text{change in position}}{\text{change in time}}$$

$$\text{Average acceleration} = \frac{\text{change in velocity}}{\text{change in time}}$$

Chapter 4 Work and Energy

$$\text{Kinetic energy} = \frac{1}{2} (\text{mass}) \times (\text{velocity})^2$$

$$\text{Gravitational potential energy (GPE)} = \text{mass} \times \text{gravity} \times \text{height}$$

$$\text{Mechanical energy} = \text{gravitational potential energy} + \text{kinetic energy}$$

$$\text{Work} = \text{force} \times \text{distance}$$

$$\text{Power} = \frac{\text{work}}{\text{time}}$$

$$\text{Efficiency} = \left(\frac{\text{work}_{out}}{\text{work}_{in}} \right) \times 100\%$$

$$\text{Ideal mechanical advantage (IMA)} = \frac{\text{length of effort arm}}{\text{length of resistance arm}} = \frac{L_e}{L_r}$$

Ideal mechanical advantage (IMA) $= \dfrac{\text{radius of wheel}}{\text{radius of axle}} = \dfrac{r_w}{r_a}$

(IMA) $= \dfrac{\text{effort distance}}{\text{resistance distance}} = \dfrac{\text{length of slope}}{\text{height of slope}} = \dfrac{l}{h}$

Chapter 5 Thermal Energy

Change in thermal energy = mass × change in temperature × specific heat   or

$Q = m \times (T_{final} - T_{initial}) \times C_p$

Chapter 6 Electricity

Electric current $= \dfrac{\text{voltage difference}}{\text{resistance}}$ or $I = \dfrac{V}{R}$

Electric power = current × voltage difference   or $P = I \times V$

Electric energy = power × time   or $E = P \times t$

Series Circuits

$I_t = I_1 = I_2 = I_3 = \dots$

$V_t = V_1 + V_2 + V_3 + \dots$

$R_t = R_1 + R_2 + R_3 + \dots$

Parallel Circuits

$I_t = I_1 + I_2 + I_3 + \dots$

$V_t = V_1 = V_2 = V_3 = \dots$

$\dfrac{1}{R_t} = \dfrac{1}{R_1} + \dfrac{1}{R_2} + \dfrac{1}{R_3} + \dots$

Chapter 9 Introduction to Waves

Wave velocity = wavelength × frequency

or $v_w = \lambda \times f$

Chapter 12 Light

Index of refraction $= \dfrac{\text{speed of light in a vacuum}}{\text{speed of light in a substance}}$ or $n = \dfrac{c}{v}$

Chapter 14 Solids, Liquids, and Gases

Pressure $= \dfrac{\text{force}}{\text{area}}$ or $P = \dfrac{F}{A}$

Boyle's law $P_1 \times V_1 = P_2 \times V_2$

Charles's law $\dfrac{V_1}{T_1} = \dfrac{V_2}{T_2}$

Chapter 21 Solutions

Surface area of a rectangular solid $= 2(h \times w) + 2(h \times l) + 2(w \times l)$

Additional Practice Problems

Chapter 1
The Nature of Science

1. How many centimeters are in four meters?

2. How many deciliters are in 500 mL?

3. How many liters are in 2,540 cm^3?

4. A young child has a mass of 40 kg. What is the mass of the child in grams?

5. Iron has a density of 7.9 g/cm^3. What is the mass in kg of an iron statue that has a volume of 5.4 L?

6. A 2-L bottle of soda has a volume of 2,000 cm^3. What is the volume of the bottle in cubic meters?

7. A big summer movie has a running time of 96 minutes. What is the movie's running time in seconds?

8. The temperature in space is approximately 3 K. What is this temperature in degrees Celsius?

9. The x-axis of a certain graph is distance traveled in meters and the y-axis is time in seconds. Two points are plotted on this graph with coordinates (2, 43) and (5, 68). What is the elapsed time between the two points?

10. A circle graph has labeled segments of 57%, 21%, 13%, and 6%. What percentage does the unlabeled segment have?

11. A car can travel 14 km on 1 L of gasoline. What percent of its fuel efficiency does another car have if it travels 10 km on 1 L of gasoline?

12. What is the fuel efficiency of a car if it gets 45 percent of the fuel efficiency of another car that travel 15 km on 1 L of gasoline?

13. Data from a new Web site takes 17 min to download. How many seconds does it take?

14. A circle graph shows the comparative effects of five new technologies. The circle graph has segments labeled 45 percent, 32 percent, 12 percent, and 3 percent. What is the correct label of the fifth segment?

15. If a new Internet connections reduces a download time of 17 min by 85 percent, how much time (in minutes and seconds) will it take now?

16. Because of a new health center in a town, the cases of influenza are 25 percent fewer. How many cases should you expect during the next month if the town usually has 300 cases per month?

17. If a person can sell 10 computers per day and a new technology allows her to triple her sales, how many computers can she expect to sell the next day?

18. A technologically advanced engine runs at twice the efficiency of an older engine. If the older engine allows a car to travel at a fuel efficiency of 8 km/L, what is the fuel efficiency of the advanced engine?

Chapter 2
Motion

1. John rides his bike 2.3 km to school. After school, he rides an additional 1.4 km to the mall in the opposite direction. What is his total distance traveled?

2. A squirrel runs 4.8 m across a lawn, stops, then runs 2.3 m back in the opposite direction. What is the squirrel's displacement from its starting point?

3. An ant travels 75 cm in 5 s. What was the ant's speed?

4. It takes a car one minute to go from rest to 30 m/s east. What is the acceleration of this car?

5. It took you 6.5 h to drive 550 km. What was your average speed?

6. A bus leaves at 9 A.M. with a group of tourists. They travel 350 km before they stop for lunch. Then they travel an additional 250 km until the end of their trip at 3 p.m. What was the average speed of the bus?

7. Halfway through a cross-country meet, a runner's speed is 4 m/s. In the last stretch, she increases her speed to 7 m/s. What is her change in speed?

8. You are in a car traveling an average speed of 60 km/h. The total trip is 240 km. How long does the trip take?

9. You are riding in a train that is traveling at a speed of 120 km/h. How long will it take to travel 950 km?

10. A car goes from rest to a velocity of 108 km/h north in 10 s. What is the car's acceleration in m/s²?

11. A cart rolling south at a speed of 10 m/s comes to a stop in 2 s. What is the cart's acceleration?

12. A car with a mass of 1,200 kg has a velocity of 30 m/s west. What is the car's momentum?

13. If a 5,000-kg mass is moving at a velocity of 40 m/s south, what is its momentum?

14. How fast must a 50-kg mass travel to have a momentum of 1,500 kg·m/s east?

Chapter 3
Forces and Newton's Laws

1. If you are pushing on a box with a force of 20 N and there is a force of 7 N on the box due to sliding friction, what is the net force on the box?

2. A weight lifter is trying to lift a 1,500-N weight but can apply a force of only 1,200 N on the weight. One of his friends helps him lift it at a constant velocity. What force was applied to the weight by the weight lifter's friend?

3. During a tug-of-war, Team A pulls with a force of 5,000 N while Team B pulls with a force of 8,000 N. What is the net force applied to the rope?

4. A 80-kg mass has an acceleration of 5.5 m/s² north. What is the net force applied?

5. A force of 3,200 N west is applied to a 160-kg mass. What is the acceleration of the mass?

6. A 2.5-kg object is dropped from a height of 1,000 m. What is the force of air resistance on the object when it reaches terminal velocity?

7. How much force is needed to lift a 25-kg mass at a constant velocity?

8. A person is on an elevator that moves downward with an acceleration of 1.8 m/s². If the person weighs 686 N, what is the net force on the person?

9. What is the net force on a 4,000-kg car that doubles its velocity from 15 m/s west to 30 m/s west over 10 seconds?

10. A book with a mass of 1 kg is sliding to the left on a table. If the frictional force on the book is 5 N, calculate the book's acceleration. Is it speeding up or slowing down?

Chapter 4
Work and Energy

1. When moving a couch, you exert a force of 400 N and push it 4.0 m. How much work have you done on the couch?

2. How much work is needed to lift a 50-kg weight 3.0 m?

3. By applying a force of 50 N, a pulley system can lift a box with a mass of 20 kg. What is the mechanical advantage of the pulley system?

4. How much energy do you save per hour if you replace a 60-watt lightbulb with a 55-watt lightbulb?

5. What is the efficiency of a machine if your work on the machine is 1200 J and the machine's output work is 300 J?

6. What power is used by a machine to perform 800 J of work in 25 s?

7. A person pushes a box up a ramp that is 3 m long and 1 m high. If the box has a mass of 20 kg and the person pushes with a force of 80 N, what is the efficiency of the ramp?

8. A lever has a mechanical advantage of 5. How large would a force need to be to lift a rock with a mass of 100 kg?

9. What is the kinetic energy from the motion of a 5.0-kg object moving at 7.0 m/s?

10. An object has 600 J of kinetic energy from its motion and a speed of 10 m/s. What is its mass?

11. If you throw a 0.4-kg ball at a speed of 20 m/s, what is the kinetic energy from the ball's motion?

12. If you have a mass of 80 kg and you are standing on a platform 3.0 m above the ground, what is the gravitational potential energy between you and Earth relative to the ground?

13. A 2.0-kg book is moved from a shelf that is 2.0 m off the ground to a shelf that is 1.5 m off the ground. What is the change in GPE?

14. A car moving at 30 m/s has 900 kJ of kinetic energy from its motion. What is the car's mass?

15. A car with a mass of 900 kg is traveling at a speed of 25 m/s. What is the kinetic energy from the car's motion?

16. If your weight is 500 N, and you are standing on a floor that is 20 m above the ground, what is the gravitational potential energy between you and Earth, relative to the ground?

Chapter 5
Thermal Energy

1. Water has a specific heat of 4,184 J/(kg·°C). How much energy is needed to increase the temperature of a kilogram of water 5.0°C?

2. The temperature of a block of iron, which has a specific heat of 450 J/(kg·°C), increases by 3°C when 2,700 J of energy are added to it. What is the mass of this block of iron?

3. How much energy is needed to heat 1.0 kg of sand, which has a specific heat of 664 J/(kg·°C), from 30°C to 50°C?

4. 1 kg of liquid water (specific heat = 4,184 J/(kg·°C)) is heated from freezing (0°C) to boiling (100°C). What is the water's change in thermal energy?

5. A concrete statue (specific heat = 600 J/(kg·°C)) sits in sunlight and warms up to 40°C. Overnight, it cools to 15°C and transfers 90,000 J of thermal energy to its surroundings. What is its mass?

6. A substance with a mass of 10.0 kg transfers 106.5 kJ of thermal energy to its surroundings when its temperature drops 15°C. What is this substance's specific heat?

7. How much heat is needed to raise the temperature of 100 g of water by 50°C, if the specific heat of water is 4,184 J/kg·°C?

8. A calorimeter contains 1.0 kg of water (specific heat = 4,184 J/(kg·°C)). An object with a mass of 4.23 kg is added to the water. If the water temperature increases by 3.0°C and the temperature of the object decreases by 1.0°C, what is the specific heat of the object?

9. A sample of an unknown metal has a mass of 0.5 kg. Adding 1,985 J of thermal energy to the metal raises its temperature by 10°C. What is the specific heat of the metal?

Chapter 6
Electricity

1. A circuit has a resistance of 4.0 Ω. What voltage difference will produce a current of 1.4 A in the circuit?

2. How many amperes of current will there be in a circuit if the voltage difference is 9.0 V and the resistance in the circuit is 3.0 Ω?

3. If a voltage difference of 3.0 V causes a 1.5 A current in a circuit, what is the resistance in the circuit?

4. The current in an appliance is 3.0 A and the voltage difference is 120 V. How much power is being supplied to the appliance?

5. What is the current into a microwave oven that requires 700 W of power if the voltage difference is 120 V?

6. What is the voltage difference in a circuit that uses 2,420 W of power when the current through the circuit is 11 A?

7. How much energy is converted when a 110 kW appliance is used for 3.0 hours?

8. How much does it cost to light six 100-W lightbulbs for six hours if the price of electrical energy is $0.09/kWh?

9. An electric clothes dryer uses 4 kW of electric power. How long did it take to dry a load of clothes if electric power costs $0.09/kWh, and the cost of using the dryer was $0.27?

10. What is the resistance of a lightbulb that draws 0.50 amp of current when plugged into a 120-V outlet?

11. How large is the current through a 100-W lightbulb that is plugged into a 120-V outlet?

12. The current through a hair dryer connected to a 120-V outlet is 8 A. How much electrical power does the hair dryer use?

Chapter 7
Magnetism and Its Uses

1. How many turns are in the secondary coil of a step-down transformer that reduces a voltage from 900 V to 300 V and has 15 turns in the primary coil?

2. A step-down transformer reduces voltage from 2,400 V to 120 V. What is the ratio of the number of turns in the primary coil to the number of turns in the secondary coil of the transformer?

3. The current produced by an AC generator switches direction twice for each revolution of the coil. How many times does a 110-Hz alternating current switch direction each second?

4. What is the output voltage from a step-down transformer with 200 turns in the primary coil and 100 turns in the secondary coil if the input voltage was 800 V?

5. The coil of a 60-Hz generator makes 60 revolutions each second. How many revolutions does the coil make in five minutes?

6. What is the output voltage from a step-up transformer with 25 turns in the primary coil and 75 turns in the secondary coil if the input voltage was 120 V?

7. How many turns are in the primary coil of a step-down transformer that reduces a voltage from 400 V to 100 V and has 80 turns in the secondary coil?

8. How many turns are in the secondary coil of a step-up transformer that increases voltage from 30 V to 150 V and has seven turns in the primary coil?

9. If a generator coil makes 6,000 revolutions in two minutes, how many revolutions does it make each second?

Chapter 8
Energy Sources and the Environment

1. A gallon of gasoline contains about 2,800 g of gasoline. If burning one gram of gasoline releases about 48 kJ of energy, how much energy is released when a gallon of gasoline is burned? (1 kJ = 1,000 J)

2. An automobile engine converts the energy released by burning gasoline into mechanical energy with an efficiency of about 25%. If burning 1 kg of gasoline releases about 48,000 kJ of energy, how much mechanical energy is produced by the engine when 1 kg of gasoline is burned?

3. Refer to the pie chart *Energy Sources* in **Figure 2.** According to the pie chart, petroleum represents 38 percent of the consumable energy sources in the United States. If 39.7 quadrillion BTUs of petroleum are consumed each year, what amount of energy (in quadrillion BTUs) does coal use account for?

4. Refer to **Figure 8,** a graph of carbon dioxide concentration (ppm) vs. time (years). If the average carbon dioxide concentration in 1960 was 315 parts per million and the average carbon dioxide concentration in 2010 is 385 parts per million, what is the percentage change in the concentration of carbon dioxide over the past 50 years?

5. A nuclear reactor contains 100,000 kg of enriched uranium. About 4% of the enriched uranium is the isotope uranium-235. What is the mass of uranium-235 in the reactor core?

6. Suppose the number of uranium-235 nuclei that are split doubles at each stage of a chain reaction. If the chain reaction starts with one nucleus split in the first stage, how many nuclei will have been split after six stages?

7. From 1970 to 2010, the carbon dioxide concentration in Earth's atmosphere increased from about 325 parts per million to about 385 parts per million. What is the percentage change in the concentration of carbon dioxide over the past 40 years?

8. About 85% of the energy used in the U.S. comes from fossil fuels. How many times greater is the amount of energy used from fossil fuel than the amount used from all other energy sources?

Chapter 9
Introduction to Waves

1. What is the wavelength of a wave with a frequency of 0.4 kHz traveling at 16 m/s?

2. What is the wavelength of a wave with a frequency of 5 Hz traveling at 15 m/s?

3. A wave has a wavelength of 250 cm and a frequency of 4 Hz. What is its speed?

4. Two waves are traveling in the same medium with a speed of 340 m/s. What is the difference in frequency of the waves if the one has a wavelength of 5.0 m and the other has a wavelength of 0.2 m?

5. What is the speed of a wave that has a wavelength of 6.0 m and a frequency of 3.0 Hz?

6. What is the frequency of a wave with a wavelength of 7 m traveling at 21 m/s?

7. A light ray strikes a plane mirror. The angle between the incident light ray and the normal to the mirror is 55°. What is the angle between the reflected ray and the normal?

Chapter 10
Sound

1. What is the wavelength of a 440-Hz sound wave traveling with a speed of 347 m/s?

2. A sound wave with a frequency of 440 Hz travels in steel with a speed of 5,200 m/s. What is the wavelength of the sound wave?

3. A wave traveling in water has a wavelength of 750 m and a frequency of 2 Hz. How fast is this wave moving?

4. At 0°C sound travels through air with a speed of about 331 m/s and through aluminum with a speed of 4,877 m/s. How many times longer is the wavelength of a sound wave in aluminum compared to the wavelength of a sound wave in air if both waves have the same frequency?

5. The speed of sound in air at 0°C is 331 m/s, and at 20°C is 344 m/s. What is the percentage change in the speed of sound at 20°C compared to 0°C?

6. What is the frequency of the first overtone of a 440-Hz wave?

7. The wreck of the *Titanic* is at a depth of about 3,800 m. A sonar unit on a ship above the *Titanic* emits a sound wave that travels at a speed of 1,500 m/s. How long does it take a sound wave reflected from the *Titanic* to return to the ocean surface?

8. A sonar unit on a ship emits a sound wave. The echo from the ocean floor is detected two seconds later. If the speed of sound in water is 1,500 m/s, how deep is the ocean beneath the ship?

9. One flute plays a note with a frequency of 443 Hz, and another flute plays a note with a frequency of 440 Hz. What is the frequency of the beats that the flute players hear?

10. A sound wave has a wavelength of 50 m and a frequency of 22 Hz. What is the speed of the sound wave?

Chapter 11
Electromagnetic Waves

1. Express the number 20,000 in scientific notation.

2. An electromagnetic wave has a wavelength of 0.054 m. What is the wavelength in scientific notation?

3. Earth is about 4,500,000,000 years old. Express this number in scientific notation.

4. The speed of electromagnetic waves in air is 300,000 km/s. What is the frequency of electromagnetic waves that have a wavelength of 5×10^{-3} km?

5. Radio waves with a frequency of 125,000 Hz have a wavelength of 1.84 km when traveling in ice. What is the speed of the radio waves in ice?

7. Two cylinders contain pistons that are connected by fluid in a hydraulic system. A force of 1,300 N is exerted on one piston with an area of 0.05 m². What is the force exerted on the other piston that has an area of 0.08 m²?

8. A gas-filled weather balloon floating in the atmosphere has an initial volume of 850 L. The weather balloon rises to a region where the pressure is 56 kPa, and its volume expands to 1,700 L. If the temperature remains the same, what was the initial pressure on the weather balloon?

9. In a hydraulic system, a force of 7,500 N is exerted on a piston with an area of 0.05 m². If the force exerted on a second piston in the hydraulic system is 1,500 N, what is the area of this second piston?

10. A gold bar weighs 17.0 N. If the density of gold is 19.3 g/cm³, what is the volume of the gold bar?

11. A book is sitting on a desk. If the surface area of the book's cover is 0.05 m² and atmospheric pressure is 100.0 kPa, what is the downward force of the atmosphere on the book?

Chapter 15
Classification of Matter

1. The size of particles in a solution is about 1 nm (1 nm = 0.000000001 m). Write 0.000000001 m in scientific notation.

2. A chemical reaction produces two new substances, one with a mass of 34 g and the other with a mass of 39 g. What is the total mass of the reactants?

3. The human body is about 65% oxygen by mass. If a person has a mass of 75.0 kg, what is the mass of oxygen in his body?

4. Two solutions, one with a mass of 450 g and the other with a mass of 350 g, are mixed. A chemical reaction occurs and 125 g of solid crystals are produced that settle on the bottom of the container. What is the mass of the remaining solution?

5. Carbon reacts with oxygen to form carbon dioxide according to the following equation: $C + O_2 \rightarrow CO_2$. When 120 g of carbon reacts with oxygen, 440 g of carbon dioxide are formed. How much oxygen reacted with the carbon?

6. Salt water is distilled by boiling it and condensing the vapor. After distillation, 1,164 g of water have been collected and 12 g of salt are left behind in the original container. What was the original mass of the salt water?

7. Calcium carbonate, $CaCO_3$, decomposes according to the reaction: $CaCO_3 \rightarrow CaO + CO_2$. When 250 g of $CaCO_3$ decompose completely, the mass of CaO is 56% of the mass of the products of this reaction. What is the mass of CO_2 produced?

8. Water breaks down into hydrogen gas and oxygen gas according to the reaction: $2H_2O \rightarrow 2H_2 + O_2$. In this reaction, the mass of oxygen produced is eight times greater than the mass of hydrogen produced. If 36 g of water form hydrogen and oxygen gas, what is the mass of hydrogen gas produced?

9. A 112-g serving of ice cream contains 19 g of fat. What percentage of the serving is fat?

10. The mass of the products produced by a chemical reaction is measured. The reaction is repeated five times, with the same mass of reactants used each time. The measured product masses are 50.17 g, 50.12 g, 50.17 g, 50.10 g, and 50.14 g. What is the average of these measurements?

Chapter 16
Properties of Atoms and the Periodic Table

1. The boron atom has a mass number of 11 and an atomic number of 5. How many neutrons are in the boron atom?

2. A magnesium atom has 12 protons and 12 neutrons. What is its mass number?

3. Iodine-127 has a mass number of 127 and 74 neutrons. What percentage of the particles in an iodine-127 nucleus are protons?

4. How many neutrons are in an atom of phosphorus-31?

5. What is the ratio of neutrons to protons in the isotope radium-234?

6. About 80% of all magnesium atoms are magnesium-24, about 10% are magnesium-25, and about 10% are magnesium-26. What is the average atomic mass of magnesium?

7. The half-life of the radioactive isotope rubidium-87 is 48,800,000,000 years. Express this half-life in scientific notation.

8. The radioactive isotope nickel-63 has a half-life of 100 years. How much of a 10.0-g sample of nickel-63 is left after 300 years?

9. A sample of the radioactive isotope cobalt-62 is prepared. The sample has a mass of 1.00 g. After three minutes, the mass of cobalt-62 remaining is 0.25 g. What is the half-life of cobalt-62?

10. A neutral phosphorus atom has 15 electrons. How many electrons are in the third energy level?

Chapter 17
Elements and Their Properties

1. In seawater the concentration of fluoride ions, F^-, is 1.3×10^{-3} g/L. How many liters of seawater would contain 1.0 g of F^-?

2. There are three isotopes of hydrogen. The isotope deuterium, with one proton and one neutron in the nucleus, makes up 0.015% of all hydrogen atoms. Of every million hydrogen atoms, how many are deuterium?

3. A vitamin and mineral supplement pill contains 1.0×10^{-5} g of selenium. According to the label on the bottle, this amount is 18% of the recommended daily value. What is the recommended daily value of selenium in g?

4. The density of silver is 10.5 g/cm^3 and the density of copper is 8.9 g/cm^3. What is the difference in mass between a piece of silver with a volume of 5 cm^3 and a piece of copper with a volume of 5 cm^3?

5. A person has a mass of 68.3 kg. If 18% of the mass of a human body is carbon, what is the mass of carbon in this person's body?

6. A gold ore produces about 5 g of gold for every 1,000 kg of ore that is mined. If one ounce = 28.3 g, how many kg of ore must be mined to produce an ounce of gold?

7. A metal bolt with a mass of 26.6 g is placed in a 50-mL graduated cylinder containing water. The water level in the cylinder rises from 27.0 mL to 30.5 mL. What is the density of the bolt in g/cm^3?

8. On a circle graph showing the percentage of elements in the human body, the wedge representing nitrogen takes up 10.8°. What is the percentage of nitrogen in the human body?

9. The melting point of aluminum is 660.0°C. What is the melting point of aluminum on the Fahrenheit temperature scale?

10. The synthetic element hassium-269 has a half-life of 9.3 s. The synthetic element fermium-255 has a half-life of 20.1 h. How many times longer is the half-life of fermium-255 than the half-life of hassium-261?

Chapter 18
Chemical Bonds

1. What is the formula of the compound formed when ammonium ions (NH_4^+) and phosphate ions (PO_4^{3-}) combine?

2. Show that the sum of positive and negative charges in a unit of calcium chloride ($CaCl_2$) equals zero.

3. What is the formula for iron(III) oxide?

4. How many hydrogen atoms are in three molecules of ammonium phosphate, $(NH_4)_3PO_4$?

5. The overall charge on the polyatomic phosphate ion (PO_4^{3-}) is 3−. What is the oxidation number of phosphorus in the phosphate ion?

6. The overall charge on the polyatomic dichromate ion ($Cr_2O_7^{2-}$) is 2−. What is the oxidation number of chromium in this polyatomic ion?

7. What is the formula for lead(IV) oxide?

8. What is the formula for potassium chlorate?

9. What is the formula for carbon tetrachloride?

10. What is the name of NaF?

11. What is the name of Al_2O_3?

12. What percentage of the mass of a sulfuric acid molecule (H_2SO_4) is sulfur?

Chapter 19
Chemical Reactions

1. Lithium reacts with oxygen to form lithium oxide according to the equation: $4Li + O_2 \rightarrow 2Li_2O$. If 27.8 g of Li react completely with 32.0 g of O_2, how many grams of Li_2O are formed?

2. What coefficients balance the following equation: $\_Zn(OH)_2 + \_H_3PO_4 \rightarrow \_Zn_3(PO_4)_2 + \_H_2O$?

3. Aluminum hydroxide, $Al(OH)_3$, decomposes to form aluminum oxide, Al_2O_3, and water according to the reaction: $2Al(OH)_3 \rightarrow Al_2O_3 + 3H_2O$. If 156.0 g of $Al(OH)_3$ decompose to from 102.0 g of Al_2O_3, how many grams of H_2O are formed?

4. In the following balanced chemical reaction one of the products is represented by the symbol X: $BaCO_3 + C + H_2O \rightarrow Ba(OH)_2 + 2X$. What is the formula for the compound represented by X?

5. When propane (C_3H_8) is burned, carbon dioxide and water vapor are produced according to the following reaction: $C_3H_8 + 5O_2 \rightarrow 3CO_2 + 4H_2O$. How much propane is burned if 160.0 g of O_2 are used and 132.0 g of CO_2 and 72.0 g of H_2O are produced?

6. Increasing the temperature usually causes the rate of a chemical reaction to increase. If the rate of a chemical reaction doubles when the temperature increases by 10°C, by what factor does the rate of reaction increase if the temperature increases by 30°C?

7. When acetylene gas (C_2H_2) is burned, carbon dioxide and water are produced. Find the coefficients that balance the chemical equation for the combustion of acetylene:
_C_2H + _O_2 → _CO_2 + _H_2O.

8. What coefficients balance the following equation:
_CS_2 + _O_2 → _CO_2 + _SO_2?

9. When methane, CH_4, is burned, 50.1 kJ of energy per gram are released. When propane, C_3H_8, is burned, 45.8 kJ of energy are released. If a mixture of 1.0 g of methane and 1.0 g of propane is burned, how much energy is released per gram of mixture?

10. A chemical reaction produces 0.050 g of a product in 0.18 s. In the presence of a catalyst, the reaction produces 0.050 g of the same product in 0.0070 s. How much faster is the rate of reaction in the presence of the enzyme?

Chapter 20
Radioactivity and Nuclear Reactions

1. How many protons are in the nucleus $^{81}_{36}$ Kr?

2. How many neutrons are in the nucleus $^{56}_{26}$ Fe?

3. What is the ratio of neutrons to protons in the nucleus $^{241}_{95}$ Am ?

4. How many alpha particles are emitted when the nucleus $^{222}_{86}$ Rn decays to $^{218}_{84}$ Po?

5. An alpha particle is the same as the helium nucleus $^{4}_{2}$ He. What nucleus is produced when the nucleus $^{226}_{88}$ Ra decays by emitting an alpha particle?

6. A sample of $^{38}_{71}$ Cl is observed to decay to 25% of the original amount in 74.4 minutes. What is the half-life of $^{38}_{71}$ Cl?

7. How many beta particles are emitted when the nucleus $^{40}_{19}$ K decays to the nucleus $^{40}_{20}$ Ca?

8. How long will it take a sample of $^{194}_{84}$ Po to decay to 1/8 of its original amount if $^{194}_{84}$ Po has a half-life of 0.70 s?

9. The half-life of $^{131}_{53}$ I is 8.04 days. How much time would be needed to reduce 1.00 g of $^{131}_{53}$ I to 0.25 g?

10. A sample of radioactive carbon-14 sample has decayed to 12.50% of its original amount. If the half-life of carbon-14 is 5730 years, how old is this sample?

11. Recall that objects in motion have kinetic energy. How fast would a 1,500-kg car need to travel in order to increase its mass by 1.0 kg?

12. The Kashiwazaki-Kariwa nuclear power plant is capable of converting approximately 8 billion J of nuclear energy into electrical energy every second. How much energy does the Kashiwazaki-Kariwa nuclear power plant convert to electrical energy in 1 year? How much mass is this equivalent to?

Chapter 21
Solutions

1. A cup of orange juice contains 126 mg of vitamin C and $\frac{1}{2}$ cup of strawberries contain 42 mg of vitamin C. How many cups of strawberries contain as much vitamin C as one cup of orange juice?

2. A Sacagawea dollar coin is made of manganese brass alloy that is $\frac{1}{25}$ nickel. Express this number as a percentage.

3. What is the total surface area of a 2-cm cube?

4. A cube has 2-cm sides. If it is split in half, what is the total surface area of the two pieces?

5. What is the increase in surface area when a cube with 2-cm sides is divided into eight equal parts?

6. How much surface area is lost if two 4-cm cubes are attached at one face?

7. At 20°C, the solubility in water of potassium bromide (KBr) is 65.3 g/100 mL. What is the maximum amount of potassium bromide that will dissolve in 237 mL of water?

8. At 20°C, the solubility of sodium chloride (NaCl) in water is 35.9 g/100 mL. If the maximum amount of sodium chloride is dissolved in 500 mL of water at 20°C, the mass of the dissolved sodium chloride is what percentage of the mass of the solution?

9. At 60°C, the solubility of sucrose (sugar) in water is 287.3 g/100 mL. At this temperature, what is the minimum amount of water needed to dissolve 50.0 g of sucrose?

10. A fruit drink contains 90% water and 10% fruit juice. How much fruit juice does 500 mL of fruit drink contain?

Chapter 22
Acids, Bases, and Salts

1. The difference between the pH of an acidic solution and the pH of pure water is 3. What is the pH of the solution?

2. The pH of rain that fell over a region had measured values of 4.6, 5.1, 4.8, 4.5, 4.5, 4.9, 4.7, and 4.8. What was the mean value of the measured pH?

3. A molecule of acetylsalicylic acid, or aspirin, has the chemical formula $COOHC_6H_4COOCH_3$. What is the mass of a molecule of acetylsalicylic acid in amu?

4. If 5.5% of 473.0 mL of vinegar is acetic acid, how many milliliters of acetic acid are there?

5. The difference between the pH of a basic solution and the pH of pure water is 2. What is the pH of the solution?

6. On the pH scale, a decrease of one unit means that the concentration of H^+ ions increases 10 times. If the pH of a solution changes from 6.5 to 4.5, how has the concentration of H^+ ions changed?

7. Write the balanced chemical equation for the neutralization of H_2SO_4, sulfuric acid, by KOH, potassium hydroxide.

8. Write the balanced chemical equation for the neutralization of hydrobromic acid (HBr) by aluminum hydroxide ($Al(OH)_3$).

9. Write the equation for the reaction when nitric acid (HNO_3) ionizes in water.

10. When sodium oxide (Na_2O) reacts with water, the base sodium hydroxide (NaOH) is formed. Write the balanced equation for this reaction.

Chapter 23
Organic Compounds

1. Fats supply 9 Calories per gram; carbohydrates and proteins each supply 4 Calories per gram. If 100 g of potato chips contain 7 g of protein, 53 g of carbohydrates, and 35 g of fats, how many Calories are in 100 g of potato chips?

2. The basal metabolism rate (BMR) is the amount of energy required to maintain basic body functions. The BMR is approximately 1.0 Calories/hr per kilogram of body mass. For a person with a mass of 65 kg, how many Calories are needed each day to maintain basic body functions?

Additional Practice Problems

3. The hydrocarbon octane, C_8H_{18}, has a boiling point of 259°F. What is its boiling point on the Celsius temperature scale?

4. Four molecules of a saturated hydrocarbon contain carbon atoms and 56 hydrogen atoms. What is the formula for a molecule of this hydrocarbon?

5. For saturated hydrocarbons, the number of hydrogen atoms in a molecule can be calculated by the formula $N_H = 2N_C + 2$, where N_H is the number of hydrogen atoms and N_C is the number of carbon atoms in the molecule. If a molecule of the saturated hydrocarbon decane has 22 hydrogen atoms, how many carbon atoms does a decane molecule contain?

6. A food Calorie is an energy unit equal to 4,184 joules. If a person uses 2,070 Calories in one day, what is the power being used? Express your answer in watts.

7. In each 100 g of cheddar cheese, there are 33 g of fat. Calculate how many grams of fat are in 250 g of cheddar cheese.

8. A car gets 25 miles per gallon of gas. If the car is driven 12,000 miles in one year and gasoline costs $2.55 per gallon, what was the cost of the gasoline used in one year?

Chapter 24
New Materials Through Chemistry

1. A 14-karat gold earring has a mass of 10 g. What is the mass of gold in the earring?

2. In 1997, about 6,400,000,000 kg of polyvinyl chloride were used in the United States. About 6% of the PVC used was for packaging. Express in scientific notation how many kilograms of PVC were used for packaging in 1997.

3. A stainless steel spoon contains 30.0 g of iron, 6.8 g of chromium, and 3.2 g of nickel. What percentage of the stainless steel is chromium?

4. The molecules in a sample of polypropylene have an average length of 60,000 monomers. The monomer of polypropylene has the formula CH_2CHCH_3. Express in scientific notation the mass, in amu, of a polypropylene molecule made of 60,000 monomers.

5. A certain process for manufacturing integrated circuits packs 47,600,000 transistors into an area of 340 mm². If this process is used to produce an integrated circuit with an area of 1 cm², express in scientific notation the number of transistors in this integrated circuit.

6. The melting points of five different samples of a new aluminum alloy have measured values of 631.5°C, 632.3°C, 636.1°C, 637.4°C, and 630.2°C. What is the mean of these measurements?

7. The measured values of the copper content of seven bronze buttons found at an archaeological site are 83%, 90%, 91%, 72%, 79%, 87%, and 89%. What is the median of these measurements?

8. The number of transistors and other components per mm² on an integrated circuit has doubled, on average, every two years. If integrated circuits contained 100,000 transistors in 1992, estimate how many transistors an integrated circuit of the same size contained in 2008.

9. A car contains 200 kg of plastic parts instead of steel parts. The density of steel is twice the density of plastic. If the volume of the plastic parts equals the volume of the same parts made of steel, how much less is the mass (kg) of the car by using plastic parts instead of steel?

Chapter 25
Earth's Internal Processes

1. How long after an earthquake will a seismograph 3,000 km away from the epicenter record S-waves that travel at 3.6 km/s?

2. If it takes 48 min 20 s for P-waves traveling at 6.0 km/s to reach and be recorded by a seismograph, how far away is the epicenter?

3. If S-waves lag behind P-waves by 1 min 52 s for every 1,000 km of distance from the earthquake epicenter, how far away from the earthquake epicenter is a seismograph that measures a time lag of 5 min 30 s between the arrivals of P-waves and S-waves?

4. With a P-wave–S-wave lag time of 1 min 52 s for every 1,000 km of distance form the earthquake epicenter, how much difference between the arrival times of P-waves and S-waves would be measured at 3,000 km? At 3,500 km?

5. If surface waves travel at 3.2 km/s, when would you expect them to arrive at a seismograph that is 2,500 km away from the earthquake epicenter?

6. If S-waves travel at 3.6 km/s and P-waves travel at 6.0 km/s through Earth's crust, what percentage of the speed of P-waves would you expect S-waves to travel at other depths inside Earth?

7. The volume of a cone is $1/3\pi r^2 h$. If Paricutín is 424 m high and has a base 2.8 km across, what is the volume of the cinder cone?

8. If two places diverge at a rate of 1.3 cm/year, how much farther apart, in kilometers, will the tow plates be after 200 million years?

9. How many times faster are plates moving at 7.3 cm/year than are plates moving at 1.3 cm/year?

10. The volume of a sphere is $4/3(\pi r^3)$. The radius of Earth is 6,378 km, and the radius of its core (including both outer and inner cores) is 3,486 km. What percent of the total volume of Earth is its core?

Chapter 26
Earth Materials

1. If oxygen makes up 46.6% of the mass of Earth's crust and silicon makes up 27.7%, what is the total percent of the crust's mass for oxygen and silicon?

2. If oxygen and silicon make up 74.3% of the mass of Earth's crust, what percent of this number is silicon's percentage?

3. How many total atoms of aluminum and oxygen combined make up one molecule of corundum (Al_2O_3)?

4. What is the ratio of silicon to oxygen in a molecule of olivine (($Mg,Fe)_2SiO_4$)?

5. Other than the common eight elements that make up Earth's crust, all other elements make up only 1.5%. If oxygen makes up 46.6% of Earth's crust, how many times greater is the amount of oxygen in Earth's crust than those other elements?

6. The Mohs scale of hardness consists of ten categories, each associated with a mineral standard. How many categories are between hardness 7-quartz and hardness 2-gypsum?

7. The volume of a sphere is $4/3(\pi r^3)$. If a particle of gravel has a radius of 2 mm and a sand grain has a radius of 1 mm, how many times bigger is the volume of the gravel?

8. If a sedimentary rock has a porosity of 15%, how much volume do the mineral grains and cement take up?

9. If you have a halite cube that measures 4 cm on each side, what is the total surface area of the cube?

10. What is the ratio of oxygen atoms to potassium atoms in one molecule of K-feldspar, ($KAlSiO_8$)?

Chapter 27
Earth's Changing Surface

1. If soil erosion averages 2.5 cm per year and the average soil profile is 3.2 m thick with 40% topsoil, how long will it take for the topsoil to erode away?

2. The average soil profile in your area is 2.1 m thick. Topsoil erodes at 2.0 cm per year. What percent of the soil profile is topsoil if it erodes in 14 years?

3. Soil erosion in your area averages 3.5 cm per year. The average soil profile is 3.7 m thick and 35% of that is topsoil. Soil replacement though weathering is 0.2 cm per year. How long will it take for the topsoil to erode?

4. Observations taken at the mouth of a stream include a flow rate of 120 m^3/s and a suspended sediment load of 1.8 kg/m^3. How many kilograms of sediment potentially could drop out of this stream each day?

5. Observations taken at the mouth of a stream include a flow rate of 85 m^3/s and a suspended sediment load of 1.6 kg/m^3 . How many kilograms of sediment potentially could drop out of this stream each day?

6. An aquifer is 400 km in length, 185 km in width, and 80 km thick. Sampling of the aquifer material shows it has an average porosity of 15%. What is the volume of porosity?

7. Suppose a community with a human population of 800 has a water consumption rate of 900 L per person per day. This community relies on an aquifer that is thought to contain 1.2 billion L of water. Assuming no change in average water consumption or significant recharge of the aquifer, how many years will this water supply last?

8. An aquifer has dimensions of 350 km in length, 175 km in width, and 65 km thick. Sampling of the aquifer material shows it has an average porosity of 10%. What is the volume of porosity?

9. If three half-lives for an isotope have passed, what fraction of the original isotope would be present in the igneous rock?

10. If the half-life of an isotope is 12,000 years and the amount of the isotope present is only $\frac{1}{16}$ of the original amount present, how old is the igneous rock containing the isotope?

Chapter 28
Weather and Climate

1. If a snowy surface reflects 90% of the solar radiation that strikes it and bare soil reflects 30%, how many times more solar radiation is reflected from a snowy surface than from bare soil?

2. The amount of rainfall over five days is 4 cm, 2 cm, 0.4 cm, 0.2 cm, and 1.3 cm. What is the average rainfall per day?

3. Air pressure at Earth's surface is 101.3 kPa. What is the air pressure at 16 km if it is $\frac{1}{10}$ the air pressure at Earth's surface?

4. Air pressure at Earth's surface is 101.3 kPa. If air pressure in an average car tire is equal to two atmospheres, what is the pressure in an average car tire?

5. Much of northern Florida receives 1,300 mm of rain on average per year, and the southern part of the Everglades receives 1,650 mm of rain on average per year. What percent of the rainfall in the southern part of the Everglades is the rainfall in northern Florida?

6. If the tilt of Earth's axis has varied from 21.5° to 24.5° over time, what is the value of the range of Earth's axial tilt?

7. Most of Earth's atmosphere is within 30 km of Earth's surface. If Earth's atmosphere extends about 10,000 km above Earth's surface, what percent of the atmosphere's depth contains most of Earth's atmosphere?

8. If a tornado travels at 50 km/h and cuts a path of destruction 10 km long, how many minutes did it take to do this?

9. If there have been eight glacial periods during the past 200,000 years, what is the rate at which these have occurred?

10. The three main gases in Earth's atmosphere are nitrogen: 78%, oxygen: 21%, and argon and trace gases: 1%. If a circle graph were drawn to represent these data, how many degrees would represent each section?

Chapter 29
The Earth-Moon-Sun System

1. Earth's diameter is 12,714 km from pole to pole and 12,756 at the equator. How much less is Earth's diameter from pole to pole than at the equator?

2. Earth's circumference is 40,075 km at the equator and 40,008 km through the two poles. How much greater is Earth's circumference around the equator?

3. If a day on Earth lasts 23 h and 56 min, how many minutes are in one day?

4. Earth's average density is 5.52 g/cm³. If the Moon's average density is 3.31 g/cm³, what percent of Earth's density is the Moon's density?

5. Earth is one AU (149,600,000 km) from the Sun. If Jupiter is 5.2 AU from the Sun, how many kilometers is Jupiter from the Sun?

6. If Earth rotates 15° each hour, how many degrees will it rotate in three hours?

7. Earth is tilted 23.5°. If the Sun is 50° above the southern horizon when it is directly over the equator, how high in the sky will the Sun be on the first day of summer?

8. The Sun has a diameter of 1,392,000 km and is about 400 times larger than the Moon. What is the approximate diameter of the Moon?

9. If there are 29.5 days in one synodic month, how many synodic months are there in one year of 365 days?

10. The South Pole-Aitken Basin on the Moon is 12 km deep and 2,500 km wide. How many times wider is the basin than it is deep?

Chapter 30
The Solar System

1. What fraction of a complete orbit (360°) would a planet move through if its H.L. changes from 32° to 302°?

2. What fraction of a complete orbit (360°) would a planet move through if its H.L. changes from 152° to 242°?

3. Earth's atmospheric pressure is 101.3 kPa. What is the atmospheric pressure on the surface of Titan if it is 1.5 times that of Earth?

4. What is the atmospheric pressure on the surface of Mars if it is 0.6 percent of Earth's?

5. If the atmospheric pressure of Triton's atmosphere is 0.002 percent of Earth's, what is the atmospheric pressure on the surface of Triton?

6. If a planet has a H.L. of 73°, what percentage of the total orbit (360°) would this represent?

7. If the diameters of Earth and Uranus are 12,756 km and 51,118 km respectively, approximately how many Earths could fit across Uranus's diameter?

8. The volume of a sphere is $4/3(\pi r^3)$. If the radius of Earth is 6,378 km and the radius of Jupiter is 71,492 km, how many Earths would fit inside Jupiter?

9. The volume of a sphere is $4/3(\pi r^3)$. If the radius of Mars is 3,394 km and the radius of Earth is 6,378 km, what percentage of Earth's volume is Mars's volume?

10. On average, Earth is 150 million km (1 AU) from the Sun. If Saturn is 9.53 AU from the Sun, what is the distance from the Sun to Saturn in kilometers?

11. If Venus takes 0.62 years to orbit the Sun and Saturn takes 29.42 years to orbit the Sun, how many times will Venus orbit the Sun during each orbit of Saturn?

12. If Earth is tilted 23.5° and Uranus is tilted 97.9°, how many times greater is the axial tilt of Uranus than that of Earth?

Chapter 31
Stars and Galaxies

1. If the focal length of a telescope's objective is 900 mm and the focal length of its eyepiece is 10 mm, what is the magnifying power of this telescope?

2. If the focal length of a telescope's objective is 700 mm and the magnifying power of the telescope is 40 times, what is the focal length of an eyepiece?

3. If it takes 25 days for a sunspot to travel once around the Sun, about how many days would it take to travel across the face of the Sun?

4. If a solar prominence blasts material from the Sun at a speed of 600 km/s, how long in hours and minutes would it take for material to arrive at Earth 150 million km away?

5. How many times larger in area is a 250-mm–diameter objective than a 100-mm–diameter objective?

6. How many times larger in area is an 8-m–diameter objective than a 5-m–diameter objective?

7. How many times wider is an elliptical galaxy that is nine million light-years across than one that is 3,000 light-years across?

8. A telescope's objective has a focal length of 1,200 mm. It is used with two eyepieces that have focal lengths of 12 mm and 18 mm. How many times greater is the magnifying power of the telescope when the 12-mm eyepiece is used?

9. A telescope's objective has a focal length of 1,500 mm. It is used with two eyepieces that have focal lengths of 20 mm and 12.5 mm. What percent of the magnifying power obtained using the 12.5-mm eyepiece is achieved using the 20-mm eyepiece?

10. If the age of the universe was thought to be 20 billion years but is now considered to be 13.7 billion years, how much younger is the universe now previously thought to be?

Reference Handbook

Table 1	Minerals with Metallic Luster

Mineral (Formula)	Color	Streak	Hardness	Specific Gravity	Crystal System	Breakage Pattern	Uses and Other Properties
Bornite (Cu_5FeS_4)	bronze, tarnishes to dark blue purple	gray-black	3	4.9–5.4	tetragonal	uneven fracture	source of copper called "peacock ore" because of the purple shine when it tarnishes
Chalcopyrite ($CuFeS_2$)	brassy to golden yellow	greenish black	3.5–4	4.2	tetragonal	uneven fracture	main ore of copper
Chromite ($FeCr_2O_4$)	black or brown	brown to black	5.5	4.6	cubic	irregular fracture	ore of chromium, stainless steel, metallurgical bricks
Copper (Cu)	copper red	copper red	3	8.5–9	cubic	hackly	coins, pipes, gutters, wire, cooking utensils, jewelry, decorative plaques; malleable and ductile
Galena (PbS)	gray	gray to black	2.5	7.5	cubic	cubic cleavage perfect	source of lead, used in pipes, shields for X-rays, fishing equipment sinkers
Gold (Au)	pale to golden yellow	yellow	2.5–3	19.3	cubic	hackly	jewelry, money, gold leaf, fillings for teeth, medicines; does not tarnish
Graphite (C)	black to gray	black to gray	1–2	2.3	hexagonal	basal cleavage (scales)	pencil lead, lubricants for locks, rods to control some small nuclear reactions, battery poles
Hematite (specular) (Fe_2O_3)	black or reddish brown	red or reddish brown	6	5.3	hexagonal	irregular fracture	source of iron; roasted in a blast furnace, converted to "pig" iron, made into steel
Magnetite (Fe_3O_4)	black	black	6	5.2	cubic	conchoidal fracture	source of iron, naturally magnetic, called lodestone
Pyrite (FeS_2)	light, brassy yellow	greenish black	6.5	5.0	cubic	uneven fracture	source of iron, "fool's gold," alters to limonite
Pyrrhotite ($Fe_{1-x}S$)* *contains one less atom of Fe than S	bronze	gray-black	4	4.6	hexagonal	uneven fracture	an ore of iron and sulfur; may be magnetic
Silver (Ag)	silvery white, tarnishes to black	light gray to silver	2.5	10–12	cubic	hackly	coins, fillings for teeth, jewelry, silver plate, wires; malleable and ductile

Table 2   Minerals with Nonmetallic Luster

Mineral (Formula)	Color	Streak	Hardness	Specific Gravity	Crystal System	Breakage Pattern	Uses and Other Properties
Augite ((Ca, Na)(Mg, Fe, Al)(Al, Si)$_2$O$_6$)	black	colorless	6	3.3	monoclinic	2-directional cleavage	square or 8-sided cross section
Corundum (Al$_2$O$_3$)	colorless, blue, brown, green, white, pink, red	colorless	9	4.0	hexagonal	fracture	gemstones: ruby is red, sapphire is blue; industrial abrasive
Feldspar (orthoclase) (KAlSi$_3$O$_8$)	colorless, white to gray, green, yellow	colorless	6	2.5	monoclinic	two cleavage planes meet at 90° angle	insoluble in acids; used in the manufacturing of porcelain
Feldspar (plagioclase) (NaAlSi$_3$O$_8$)(CaAl$_2$Si$_3$O$_8$)	gray, green, white	colorless	6	2.5	triclinic	two cleavage planes meet at 86° angle	used in ceramics; striations present on some faces
Fluorite (CaF$_2$)	colorless, white, blue, green, red, yellow, purple	colorless	4	3–3.2	cubic	cleavage	used in the manufacturing of optical equipment; glows under ultraviolet light
Garnet (Mg, Fe, Ca, Mn)$_3$(Al, Fe, Cr)$_2$(SiO$_4$)$_3$	deep yellow-red, green, black	colorless	7.5	3.5	cubic	conchoidal fracture	used in jewelry; also used as an abrasive
Hornblende Ca$_2$Na(Mg, Fe^2)$_4$(Al, Fe$_3$, Ti)$_3$Si$_8$O$_{22}$(O, OH)$_2$	green to black	gray to white	5–6	3.4	monoclinic	cleavage in two directions	will transmit light on thin edges; 6-sided cross section
Limonite (hydrous iron oxides)	yellow, brown, black	yellow, brown	5.5	2.7–4.3	N/A	conchoidal fracture	source of iron; weathers easily, coloring matter of soils
Olivine ((Mg, Fe)$_2$SiO$_4$)	olive green	colorless	6.5	3.5	orthorhombic	conchoidal fracture	gemstones, refractory sand
Quartz (SiO$_2$)	colorless, various colors	colorless	7	2.6	hexagonal	conchoidal fracture	used in glass manufacture, electronic equipment, radios, computers, watches, gemstones
Topaz (Al$_2$SiO$_4$(F, OH)$_2$)	colorless, white, pink, yellow, pale blue	colorless	8	3.5	orthorhombic	basal cleavage	valuable gemstone

Table 3	Rocks	
Rock Type	**Rock Name**	**Characteristics**
Igneous (intrusive)	granite	large mineral grains of quartz, feldspar, hornblende, and mica; usually light in color
	diorite	large mineral grains of feldspar, hornblende, and mica; less quartz than granite; intermediate in color
	gabbro	large mineral grains of feldspar, hornblende, augite, olivine, and mica; no quartz; dark in color
Igneous (extrusive)	rhyolite	small or no visible grains of quartz, feldspar, hornblende, and mica; light in color
	andesite	small or no visible grains of quartz, feldspar, hornblende, and mica; less quartz than rhyolite; intermediate in color
	basalt	small or no visible grains of feldspar, hornblende, augite, olivine, and mica; no quartz; dark in color; vesicles may be present
	obsidian	glassy texture; no visible grains; volcanic glass; fracture is conchoidal; color is usually black, but may be red-brown or black with white flecks
	pumice	frothy texture; floats; usually light in color
Sedimentary (clastic)	conglomerate	coarse-grained; gravel- or pebble-sized grains
	sandstone	sand-sized grains 1/16 to 2 mm in size; varies in color
	siltstone	grains smaller than sand but larger than clay
	shale	smallest grains; usually dark in color
Sedimentary (chemical or biochemical)	limestone	major mineral is calcite; usually forms in oceans, lakes, rivers, and caves; often contains fossils; effervesces in dilute HCl
	coal	occurs in swampy, low-lying areas; compacted layers of organic material, mainly plant remains
Sedimentary (chemical)	rock salt	commonly forms as seawater evaporates
Metamorphic	gneiss	well-developed banding because of alternating layers of different minerals, usually of different colors; common parent rock is granite
	schist	well-developed parallel arrangement of flat, sheetlike minerals, mainly micas; common parent rocks are shale and phyllite
	phyllite	shiny or silky appearance; may look wrinkled; common parent rocks are shale and slate
	slate	harder, denser, and shinier than shale; common parent rock is shale
Metamorphic (nonfoliated)	marble	interlocking calcite or dolomite crystals; common parent rock is limestone
	soapstone	composed mainly of the mineral talc; soft with a greasy feel
	quartzite	hard and well-cemented with interlocking quartz crystals; common parent rock is sandstone

Weather Map Symbols

Sample Plotted Report at Each Station

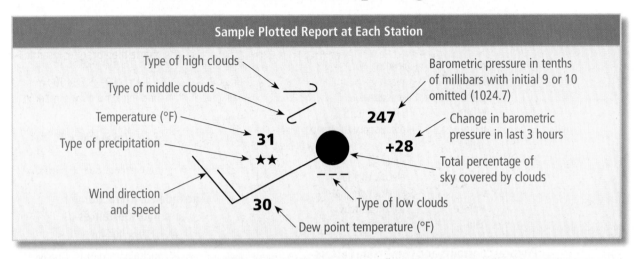

Type of high clouds

Type of middle clouds

Temperature (°F)

Type of precipitation

Wind direction and speed

Barometric pressure in tenths of millibars with initial 9 or 10 omitted (1024.7)

247

Change in barometric pressure in last 3 hours

+28

Total percentage of sky covered by clouds

31

★★

Type of low clouds

30

Dew point temperature (°F)

Symbols Used in Plotting Report

Precipitation	Wind Direction and Speed	Sky Coverage	Fronts and Pressure Systems
≡ Fog	○ 0 calm	○ No cover	(H) or High — Center of high- or
★ Snow	╱ 1–2 knots	◐ 1/10 or less	(L) or Low — low-pressure system
● Rain	⟍ 3–7 knots	◓ 2/10 to 3/10	▲▲▲▲ Cold front
⊺ Thunderstorm	⟍ 8–12 knots	◒ 4/10	●●●● Warm front
	⟍ 13–17 knots	◑ 1/2	▲▲●● Occluded front
❟ Drizzle	⟍ 18–22 knots	◓ 6/10	▲●▲● Stationary front
▽ Showers	⟍ 23–27 knots	◕ 7/10	
	⟍ 48–52 knots	◖ Overcast with openings	
	1 knot = 1.852 km/h	● Completely overcast	

Clouds

Some Types of High Clouds	Some Types of Middle Clouds	Some Types of Low Clouds
⌐⊃ Scattered cirrus	∠ Thin altostratus layer	⌂ Cumulus of fair weather
⌐⊃ Dense cirrus in patches	⫽ Thick altostratus layer	⏝ Stratocumulus
⌐⌐ Veil of cirrus covering entire sky	⫽ Thin altostratus in patches	- - - Fractocumulus of bad weather
⌐⌐ Cirrus not covering entire sky	⫽ Thin altostratus in bands	— Stratus of fair weather

Physical Science Reference Tables

Standard Units

Symbol	Name	Quantity
m	meter	length
kg	kilogram	mass
Pa	pascal	pressure
K	kelvin	temperature
mol	mole	amount of a substance
J	joule	energy, work, quantity of heat
s	second	time
C	coulomb	electric charge
V	volt	electric potential
A	ampere	electric current
Ω	ohm	resistance

Physical Constants and Conversion Factors

Acceleration due to gravity	g	9.8 m/s/s or m/s^2
Avogadro's Number	N_A	6.02×10^{23} particles per mole
Electron charge	e	1.6×10^{-19} C
Electron rest mass	m_e	9.11×10^{-31} kg
Gravitation constant	G	6.67×10^{-11} N $\times$ m^2/kg^2
Mass-energy relationship		1 u (amu) $= 9.3 \times 10^2$ MeV
Speed of light in a vacuum	c	3.00×10^8 m/s
Speed of sound at STP		331 m/s
Standard Pressure		1 atmosphere
		101.3 kPa
		760 Torr or mmHg
		14.7 lb/in.^{2}

Wavelengths of Light in a Vacuum

Violet	$4.0 - 4.2 \times 10^{-7}$ m
Blue	$4.2 - 4.9 \times 10^{-7}$ m
Green	$4.9 - 5.7 \times 10^{-7}$ m
Yellow	$5.7 - 5.9 \times 10^{-7}$ m
Orange	$5.9 - 6.5 \times 10^{-7}$ m
Red	$6.5 - 7.0 \times 10^{-7}$ m

The Index of Refraction for Common Substances
($\lambda = 5.9 \times 10^{-7}$ m)

Air	1.00
Alcohol	1.36
Canada Balsam	1.53
Corn Oil	1.47
Diamond	2.42
Glass, Crown	1.52
Glass, Flint	1.61
Glycerol	1.47
Lucite	1.50
Quartz, Fused	1.46
Water	1.33

Heat Constants

	Specific Heat (average) (kJ/kg × °C) (J/g × °C)	Melting Point (°C)	Boiling Point (°C)	Heat of Fusion (kJ/kg) (J/g)	Heat of Vaporization (kJ/kg) (J/g)
Alcohol (ethyl)	2.43 (liq.)	−117	79	109	855
Aluminum	0.90 (sol.)	660	2467	396	10500
Ammonia	4.71 (liq.)	−78	−33	332	1370
Copper	0.39 (sol.)	1083	2567	205	4790
Iron	0.45 (sol.)	1535	2750	267	6290
Lead	0.13 (sol.)	328	1740	25	866
Mercury	0.14 (liq.)	−39	357	11	295
Platinum	0.13 (sol.)	1772	3827	101	229
Silver	0.24 (sol.)	962	2212	105	2370
Tungsten	0.13 (sol.)	3410	5660	192	4350
Water (solid)	2.05 (sol.)	0	–	334	–
Water (liquid)	4.18 (liq.)	–	100	–	–
Water (vapor)	2.01 (gas)	–	–	–	2260
Zinc	0.39 (sol.)	420	907	113	1770

Standard Units

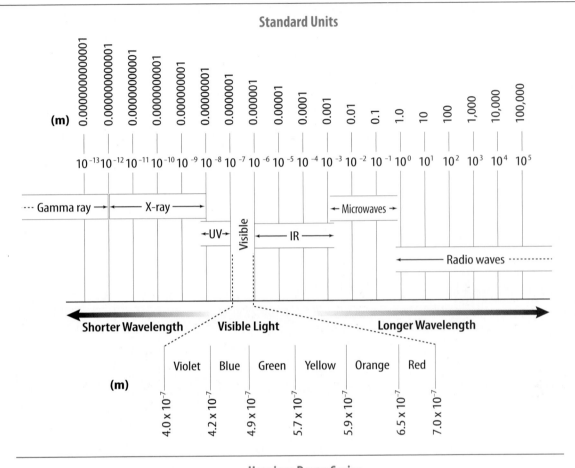

(m)

0.0000000000001 0.000000000001 0.00000000001 0.0000000001 0.000000001 0.00000001 0.0000001 0.000001 0.00001 0.0001 0.001 0.01 0.1 1.0 10 100 1,000 10,000 100,000

10^{-13} 10^{-12} 10^{-11} 10^{-10} 10^{-9} 10^{-8} 10^{-7} 10^{-6} 10^{-5} 10^{-4} 10^{-3} 10^{-2} 10^{-1} 10^{0} 10^{1} 10^{2} 10^{3} 10^{4} 10^{5}

--- Gamma ray → ←——— X-ray ———→

←UV→ Visible ←——— IR ———→

← Microwaves →

←——— Radio waves ·········

← Shorter Wavelength   **Visible Light**   Longer Wavelength →

(m)

Violet	Blue	Green	Yellow	Orange	Red
4.0×10^{-7}	4.2×10^{-7}	4.9×10^{-7}	5.7×10^{-7}	5.9×10^{-7}	6.5×10^{-7}	7.0×10^{-7}

Uranium Decay Series

Atomic number and chemical symbol

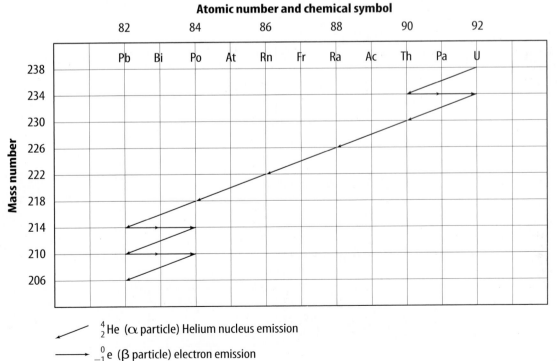

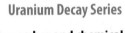
$^{4}_{2}$He (α particle) Helium nucleus emission

$^{0}_{-1}$e (β particle) electron emission

PERIODIC TABLE OF THE ELEMENTS

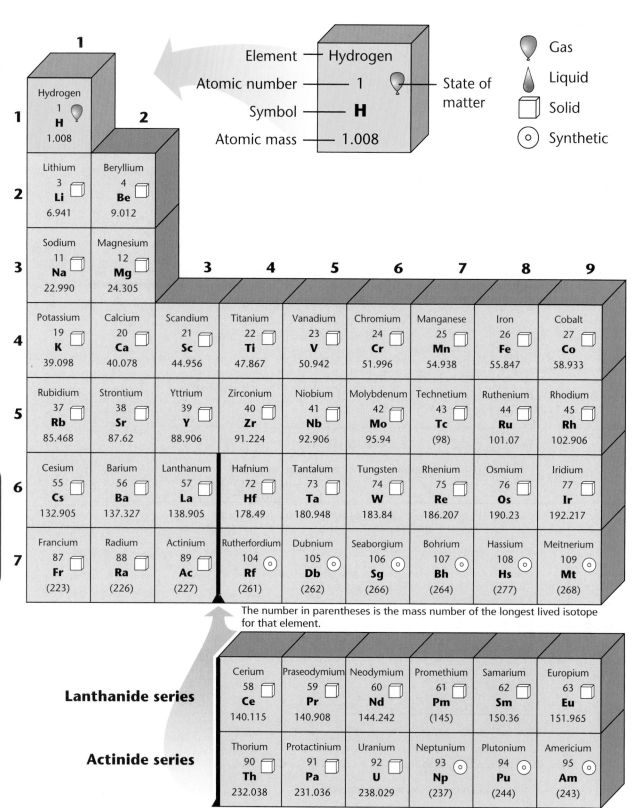

The number in parentheses is the mass number of the longest lived isotope for that element.

Legend

- Metal
- Metalloid
- Nonmetal
- Recently observed

			13	14	15	16	17	18
								Helium 2 He 4.003
			Boron 5 B 10.811	Carbon 6 C 12.011	Nitrogen 7 N 14.007	Oxygen 8 O 15.999	Fluorine 9 F 18.998	Neon 10 Ne 20.180
10	**11**	**12**	Aluminum 13 Al 26.982	Silicon 14 Si 28.086	Phosphorus 15 P 30.974	Sulfur 16 S 32.066	Chlorine 17 Cl 35.453	Argon 18 Ar 39.948
Nickel 28 Ni 58.693	Copper 29 Cu 63.546	Zinc 30 Zn 65.39	Gallium 31 Ga 69.723	Germanium 32 Ge 72.61	Arsenic 33 As 74.922	Selenium 34 Se 78.96	Bromine 35 Br 79.904	Krypton 36 Kr 83.80
Palladium 46 Pd 106.42	Silver 47 Ag 107.868	Cadmium 48 Cd 112.411	Indium 49 In 114.82	Tin 50 Sn 118.710	Antimony 51 Sb 121.757	Tellurium 52 Te 127.60	Iodine 53 I 126.904	Xenon 54 Xe 131.290
Platinum 78 Pt 195.08	Gold 79 Au 196.967	Mercury 80 Hg 200.59	Thallium 81 Tl 204.383	Lead 82 Pb 207.2	Bismuth 83 Bi 208.980	Polonium 84 Po 208.982	Astatine 85 At 209.987	Radon 86 Rn 222.018
Darmstadtium 110 Ds (281)	Roentgenium 111 Rg (272)	Copernicium 112 Cn (285)	Ununtrium ★ 113 Uut (284)	Ununquadium ★ 114 Uuq (289)	Ununpentium ★ 115 Uup (288)	Ununhexium ★ 116 Uuh (291)		Ununoctium ★ 118 Uuo (294)

★ The names and symbols for elements 113, 114, 115, 116, and 118 are temporary. Final names will be selected when the elements' discoveries are verified.

Gadolinium 64 Gd 157.25	Terbium 65 Tb 158.925	Dysprosium 66 Dy 162.50	Holmium 67 Ho 164.930	Erbium 68 Er 167.259	Thulium 69 Tm 168.934	Ytterbium 70 Yb 173.04	Lutetium 71 Lu 174.967
Curium 96 Cm (247)	Berkelium 97 Bk (247)	Californium 98 Cf (251)	Einsteinium 99 Es (252)	Fermium 100 Fm (257)	Mendelevium 101 Md (258)	Nobelium 102 No (259)	Lawrencium 103 Lr (262)

Glossary

g Multilingual eGlossary

A multilingual science glossary at connectED@mcgraw-hill.com includes Arabic, Bengali, Chinese, English, Haitian Creole, Hmong, Korean, Portuguese, Russian, Tagalog, Urdu, and Vietnamese.

Como usar el glosario en español:
1. Busca el termino en ingles que desees encontrar.
2. El termino en espanol, junto con la definicion, se encuentran en la columna de la derecha.

Pronunciation Key

Use the following key to help you sound out words in the glossary.

a	back (BAK)	ew	food (FEWD)
ay	day (DAY)	yoo	pure (PYOOR)
ah	father (FAH thur)	yew	few (FYEW)
ow	flower (FLOW ur)	uh	comma (CAHM uh)
ar	car (CAR)	u (+con)	rub (RUB)
e	less (LES)	sh	shelf (SHELF)
ee	leaf (LEEF)	ch	nature (NAY chur)
ih	trip (TRIHP)	g	gift (GIHFT)
i (i+con+e)	idea, life (i DEE uh, life)	j	gem (JEM)
oh	go (GOH)	ing	sing (SING)
aw	soft (SAWFT)	zh	vision (VIHZH un)
or	orbit (OR but)	k	cake (KAYK)
oy	coin (COYN)	s	seed, cent (SEED, SENT)
oo	foot (FOOT)	z	zone, raise (ZOHN, RAYZ)

English	A	Español

absolute dating: process of assigning a numerical age to organisms, objects or events. **(p. 870)**

datación absoluta: asignar edad a organismos, objetos o eventos **(p. 870)**

acceleration: rate of change of velocity; can be calculated by dividing the change in the velocity by the time it takes the change to occur. **(p. 56)**

aceleración: tasa de cambio de la velocidad; se calcula dividiendo el cambio en la velocidad por el tiempo que toma para que ocurra el cambio. **(p. 56)**

acid: substance that produces hydrogen ions (H^+) in a water solution. **(p. 678)**

ácido: sustancia que produce iones de hidrógeno, (H^+), en una solución de agua. **(p. 678)**

acid precipitation: water with a pH below 5.6 that falls to Earth as rain or snow and can harm plants and animals and corrode buildings. **(p. 260)**

precipitación ácida: el agua con un pH inferior a 5,6 que cae a la Tierra como lluvia o nieve y puede dañar a las plantas y los animales, y oxida a los edificios. **(p. 260)**

acoustics: the study of sound. **(p. 324)**

acústica: el estudio del sonido. **(p. 324)**

air mass: a large volume of air that has the characteristics of the area over which it forms. **(p. 892)**

masa de aire: gran volumen de aire con las características de la zona en que se forma. **(p. 892)**

air resistance: force that opposes the motion of objects that move through the air. **(p. 87)**

resistencia del aire: fuerza que se opone al movimiento de los objetos que se mueven por el aire. **(p. 87)**

alcohol: substituted hydrocarbon, such as ethanol, that is formed when –OH groups replace one or more hydrogen atoms in a hydrocarbon. **(p. 713)**

alcohol: hidrocarbonados sustituidos, como el etanol, que se forma cuando grupos –OH reemplazan a uno o más átomos de hidrógeno en un hidrocarburo. **(p. 713)**

allotropes: different molecular structures of the same element **(p. 533)**

alótropos: estructuras moleculares diferentes de un mismo elemento. **(p. 533)**

alloy: a mixture of elements that has metallic properties. **(p. 647)**

aleación: una mezcla de elementos que tiene propiedades metálicas. **(p. 647)**

alpha particle: particle consisting of two protons and two neutrons that is emitted from a decaying atomic nucleus. **(p. 621)**

alternating current (AC): electric current that reverses its direction of flow in a regular pattern. **(p. 220)**

amine: a substituted hydrocarbon with an amine group ($-NH_2$). **(p. 715)**

amplitude: a measure of the size of the disturbance of a wave, related to the energy that it carries. **(p. 283)**

analog signal: an electric signal whose values changes smoothly over time. **(p. 354)**

aquifer: rock unit that can store water and transfer it through its pore space. **(p. 864)**

aromatic compound: an organic compound that contains the benzene ring structure, most have a distinctive smell. **(p. 716)**

asteroid: rocky object formed from material similar to that of the planets, orbiting the Sun. **(p. 967)**

asthenosphere: the plastic-like layer of Earth made of partially-molten rock material directly beneath the tectonic plates. **(p. 790)**

astronomical unit: about 150 million km; equal to the average distance from Earth to the Sun, used to measure distances within the solar system. **(p. 968)**

atom: the smallest particle of an element that still retains the properties of the element. **(p. 489)**

atomic number: number of protons in an atom's nucleus. **(p. 495)**

average atomic mass: weighted-average mass of an element's isotopes according to their natural abundance. **(p. 497)**

average speed: total distance an object travels divided by the total time it takes to travel that distance. **(p. 48)**

partícula alfa: partícula compuesta por dos protones y dos neutrones y que es emitida por un núcleo atómico en descomposición. **(p. 621)**

corriente alterna (CA): corriente eléctrica que invierte su dirección de flujo en un patrón regular. **(p. 220)**

amina: un hidrocarburo sustituido con un grupo amino ($-NH_2$). **(p. 715)**

amplitud: medida del tamaño del desplazamiento por una onda, indica la cantidad de la energía que transporta. **(p. 283)**

señal analógica: una señal eléctrica cuya values cambia fluidamente a lo largo del tiempo. **(p. 354)**

roca acuífera: formación rocosa que puede contener agua y transferirla por medio de sus poros. **(p. 864)**

compuesto aromático: compuesto orgánico que contiene la estructura del anillo bencénico, la mayoría tiene una fragancia característica. **(p. 716)**

asteroide: objeto rocoso formado a partir de material similar a la de los planetas, en órbita solar. **(p. 967)**

astenofera: capa maleable o suave de la tierra hecha de material rocoso parcialmente fundido y que está directamente debajo de las placas tectónicas. **(p. 790)**

unidad astronómica: alrededor de 150 millones de kilómetros; igual a la distancia media entre la Tierra y el Sol, se usa para medir distancias en el sistema solar. **(p. 968)**

átomo: la partícula más pequeña de un elemento que mantiene las propiedades del elemento. **(p. 489)**

número atómico: número de protones en el núcleo de un átomo. **(p. 495)**

masa atómica promedio: masa media ponderada de los isótopos de un elemento en función de su abundancia natural. **(p. 497)**

velocidad promedio: distancia que recorre un objeto dividida por el tiempo que dura en recorrer dicha distancia. **(p. 48)**

English	B	Español

balanced chemical equation: chemical equation with the same number of atoms of each element on both sides of the equation. **(p. 586)**

base: a substance that produces hydroxide ions (OH^-) in a water solution. (p. 680)

benzene: (C_6H_6) cyclic hydrocarbon whose carbon atoms are joined with alternating single and double bonds. **(p. 711)**

beta particle: high-energy electron that is emitted when a neutron decays into a proton. **(p. 622)**

bias: occurs when a scientist's expectations change how the results of an experiment are viewed. **(p. 11)**

ecuación química equilibrada: ecuación química con el mismo número de átomos de cada elemento en ambos lados de la ecuación. **(p. 586)**

base: sustancia que forma iones de hidróxido, OH^-, en una solución de agua. **(p. 680)**

benceno: (C_6H_6) hidrocarburos cíclico cuyos átomos de carbono están unidos con una alternancia de enlaces simples y dobles. **(p. 711)**

partícula beta: electrón de alta energía que se emite cuando un neutrón se convierte en un protón. **(p. 622)**

predisposición: ocurre cuando las expectativas de un científico cambian la forma en que son vistos los resultados de un experimento. **(p. 11)**

Glossary

big bang theory: theory that about 13.7 billion years ago, the entire universe was contained in a single point that began expanding outward. **(p. 1002)**

binary compound: compound that is composed of two elements. **(p. 566)**

biomass: renewable organic matter from plants and animals, such as wood and animal manure, that can be burned to provide thermal energy **(p. 253)**

black hole: a region in space that is so dense that nothing can escape its inward pull of gravity. **(p. 991)**

boiling point: the temperature at which the pressure of the vapor of a liquid is equal to the external pressure acting on the surface of the liquid. **(p. 435)**

Boyle's law: states that the volume and pressure of a gas are related, such that if the temperature of a gas remains constant, an increase in volume causes a proportional decrease in the pressure. **(p. 448)**

buffer: solution that resists changes in pH when limited amounts of acid or base are added. **(p. 687)**

buoyancy: ability of a fluid, which include liquids and gases, to exert an upward force on an object immersed in it. **(p. 441)**

teoría del Big Bang: teoría de que cerca de 13.7 mil millones de años, el universo entero estaba contenido en un solo punto que comenzó a expandirse. **(p. 1002)**

compuesto binario: compuesto conformado por dos elementos. **(p. 566)**

biomasa: materia orgánica renovable que proviene de plantas y animales, tales como madera y estiércol animal, que puede ser incinerada para proveer energía térmica. **(p. 253)**

agujero negro: zona en el espacio que es tan denso que nada puede escapar a su atracción de la gravedad. **(p. 991)**

punto de ebullición: temperatura a la cual la presión del vapor de un líquido es igual a la presión externa que actúa sobre la superficie del líquido. **(p. 435)**

ley de Boyle: establece que el volumen y la presión de un gas son relacionados de manera que si la temperatura se mantiene constante, un aumento el volumen causaría una disminución proporcional en la presión. **(p. 448)**

buffer: solución que se resiste a los cambios en el pH al añadir cantidades limitadas de ácido o base. **(p. 687)**

fuerza de flotación: capacidad de un fluido, líquido o gas, para ejercer una fuerza ascendente sobre un objeto inmerso en el fluido. **(p. 441)**

English	C	Español

carbohydrate: group of biological compounds containing carbon, hydrogen, and oxygen with twice as many hydrogen atoms as oxygen atoms. **(p. 725)**

carrier wave: specific frequency that a radio station is assigned and uses to broadcast signals. **(p. 352)**

carrying capacity: maximum number of individuals of a given species that the environment can support. **(p. 255)**

catalyst: substance that speeds up a chemical reaction without being permanently changed itself. **(p. 601)**

cementation: process of minerals precipitating out of solution into the spaces between clasts. **(p. 825)**

centripetal acceleration: acceleration of an object toward the center of a curved or circular path. **(p. 59)**

centripetal force: a force that is directed toward the center of a curved or circular path. **(p. 90)**

ceramics: versatile materials made from dried clay or clay-like mixtures with customizable properties; produced by a process in which an object is molded and then heated to high temperatures, increasing its density. **(p. 746)**

chain reaction: series of fission reactions caused by the release of additional neutrons in every step. **(p. 624)**

charging by contact: the transferring of electrical charge between objects by touching or rubbing. **(p. 174)**

carbohidrato: grupo de compuestos biológicos que contienen carbono, hidrógeno y oxígeno que contienen el doble de átomos de hidrógeno que de oxígeno. **(p. 725)**

onda transportadora: frecuencia específica que se le asigna a una estación de radio y que la usa para emitir señales. **(p. 352)**

capacidad de carga: máximo número de individuos de una especie que el medio ambiente puede apoyar. **(p. 255)**

catalizador: sustancia que acelera una reacción química sin cambiar el mismo permanentemente. **(p. 601)**

cementación: es un proceso por el cual la materia mineral llevada en solución por las aguas, se deposita en las clásticas para mantenerlos unidos. **(p. 825)**

aceleración centrípeta: aceleración de un objeto dirigida hacia el centro de un trayecto curvo o circular. **(p. 59)**

fuerza centrípeta: fuerza dirigida hacia el centro de un trayecto curvo o circular. **(p. 90)**

cerámicas: materiales versátiles hechos con arcilla seca o mezclas parecidas a la arcilla con propiedades adaptables, producidos mediante un proceso en el cual un objeto es moldeado y luego sujeto a altas temperaturas, aumentando su densidad. **(p. 746)**

reacción en cadena: serie continua de reacciones de fisión causado por la liberación de neutrones en cada paso. **(p. 624)**

cargar por contacto: transferir carga eléctrica entre objetos por contacto o frotación. **(p. 174)**

Glossary

charging by induction: the rearranging of electrons on a neutral object caused by bringing a charged object close to it. **(p. 174)**

Charles's law: states that the temperature and volume of a gas are related such that, if the pressure is constant, an increase in temperature will produce a proportionate increase in the volume. **(p. 450)**

chemical bond: force that holds atoms together in a compound. **(p. 556)**

chemical change: change of one substance into a new substance. **(p. 473)**

chemical equation: shorthand method used to describe chemical reactions using chemical formulas and other symbols. **(p. 584)**

chemical formula: chemical shorthand that uses symbols to tell what elements are in a compound and their ratios. **(p. 553)**

chemical potential energy: energy that is due to chemical bonds. **(p. 117)**

chemical property: any characteristic of a substance, such as flammability, that can be observed that produces a new substance. **(p. 472)**

chemical reaction: process in which one or more substances are changed into new substances. **(p. 582)**

cinder cone volcano: small steep-sloped volcano with a short eruption cycle, composed of cinder, formed at vents in Earth's crusts, often around the central vent of a larger volcano. **(p. 795)**

circuit: closed conducting loop through which an electric current can flow. **(p. 201)**

clasts: small rock and mineral fragments that can become part of another rock. **(p. 824)**

cleavage: manner in which a mineral breaks along planes of weakness, creating sets of smooth parallel sides; determined by the arrangement of atoms in the mineral's structure. **(p. 811)**

cochlea: spiral-shaped, fluid-filled structure in the inner ear that contains tiny hair cells and converts sound waves to nerve impulses. **(p. 310)**

coefficient: number in a chemical equation that represents the number of units of each substance taking part in a chemical reaction. **(p. 584)**

coherent light: light of one wavelength that travels in one direction with a constant distance between the corresponding crests of the waves. **(p. 382)**

collision model: explains why certain factors affect reaction rates, states that particles must collide in order to react. **(p. 599)**

colloid: heterogeneous mixture whose particles never settle. **(p. 466)**

cargar por inducción: redistribuir los electrones de un objeto neutro debido al acercarle a un objeto con carga. **(p. 174)**

ley de Charles: establece que la temperatura y el volumen de un gas son relacionados de tal manera que sin un cambio de la presión, un aumento de temperatura producirá un aumento proporcional en el volumen. **(p. 450)**

enlace químico: fuerza que mantiene a los átomos juntos en un compuesto. **(p. 556)**

cambio químico: transformación de una sustancia en una nueva sustancia. **(p. 473)**

ecuación química: método simplificado utilizado para describir reacciones químicas por medio de fórmulas químicas y otros símbolos. **(p. 584)**

fórmula química: nomenclatura química que usa símbolos para expresar cuales elementos están en un compuesto y en cuales proporciones. **(p. 553)**

energía química potencial: energía debido a los enlaces químicos. **(p. 117)**

propiedad química: cualquier característica de una sustancia, como por ejemplo la combustibilidad, que puede ser observado y produce un nuevo sustancia. **(p. 472)**

reacción química: proceso en el cual una o más sustancias son cambiadas por nuevas sustancias. **(p. 582)**

volcán de cono de ceniza: volcán pequeño e inclinado con un ciclo de erupciones corto, compuesto de fragmentos volcánicos acumulados en una chimenea volcánica en la superficie de la tierra, se pueden encontrar alrededor de volcanes más grande. **(p. 795)**

circuito: circuito conductor cerrado a través del cual puede fluir una corriente eléctrica. **(p. 201)**

clástica: roca o fragmentos minerales que pueden hacerse parte de otra roca. **(p. 824)**

exfoliación: manera en la que un mineral se quiebra a lo largo de sus planos creando lados paralelos lisos, se produce por la forma los átomos se disponen en la estructura del mineral **(p. 811)**

cóclea: estructura en forma de espiral, llena de líquido en el oído interno que contiene células diminutas del pelo y convierte las ondas sonoras en impulsos nerviosos. **(p. 310)**

coeficiente: número en una ecuación química que representa el número de unidades de cada una de las sustancias que participan en una reacción química. **(p. 584)**

luz coherente: luz de una sola longitud de onda que viaja en una sola dirección con una distancia constante ente las crestas de las olas. **(p. 382)**

modelo de colisión: explica por qué ciertos factores afectan las velocidades de reacción, afirma que las partículas deben chocar para que reaccionen. **(p. 599)**

coloide: mezcla heterogénea cuyas partículas nunca se sedimentan. **(p. 466)**

combustion reaction: a type of chemical reaction that occurs when a substance reacts with oxygen to produce energy in the form of heat and light. (p. 590)

comet: object in elliptical orbit around the Sun composed of dust and particles, frozen water, methane and ammonia; vaporized due to heat from Sun, producing a coma and tail. (p. 967)

compaction: process by which clasts stick together due to the weight of overlying material. (p. 825)

composite: mixture of two materials, one of which is embedded or layered in the other. (p. 757)

composite volcano: large and steep-sided volcano composed of layers of lava and ash. (p. 796)

compound: substance in which the atoms of two or more elements are combined in a fixed proportion. (p. 464)

compound machine: machine that is a combination of two or more simple machines. (p. 109)

compression: denser region of a longitudinal wave. (p. 279)

concave lens: a lens that is thicker at the edges than in the middle; causes light rays to diverge and forms reduced, upright, virtual images; often used in combination with other lenses. (p. 410)

concave mirror: a reflective surface that curves inward and can magnify objects or create real images. (p. 402)

concentration: the amount of solute actually dissolved in a given amount of solvent. (p. 653)

conduction: transfer of thermal energy by collisions between the particles that make up matter (p. 144)

conductivity: property of metals and alloys that allows heat or electrical charges to pass through the material easily. (p. 741)

conductor: material, such as copper wire, through which electrons can move easily. (p. 173)

constant: in an experiment, a variable that does not change. (p. 9)

constellation: star pattern that appears to form images, is used by astronomers to locate and name stars, and often is named for a mythological figure. (p. 984)

continental climate: climate not affected by an ocean, characterized by extremes in temperatures. (p. 898)

control: standard used for comparison of test results in an experiment. (p. 10)

convection: transfer of thermal energy in a fluid by the movement of warmer and cooler fluid from one place to another. (p. 145)

reacción de combustión: un tipo de reacción química que ocurre cuando una sustancia reacciona con oxígeno y produce energía en forma de calor y luz. (p. 590)

cometa: objeto en órbita elíptica alrededor del Sol, compuesta de polvo y partículas, agua congelada, metano y amoníaco, vaporizado por el calor del sol, produce una coma y una cola. (p. 967)

compactación: proceso por el que las clásticas se unen, debido al peso de materiales sobrepuestos. (p. 825)

material compuesto: mezcla de dos materiales, uno de los cuales está integrado por capas o fundido en el otro. (p. 757)

volcán compuesto: Volcanes grandes e inclinados formados por capas de lava y ceniza. (p. 796)

compuesto químico: sustancia en la que los átomos de dos o más elementos se combinan en una proporción fija. (p. 464)

máquina compuesta: máquina compuesta por dos o más máquinas simples. (p. 109)

compresión: la región densa de una onda longitudinal. (p. 279)

lente cóncavo: lente que es más grueso en los bordes que en el centro; hace que los rayos de luz se desvíen y forma imágenes reducidas, verticales y virtuales, frecuentemente se utiliza en combinación con otros lentes. (p. 410)

espejo cóncavo: superficie reflexiva que se curva hacia el interior puede aumentar imágenes o crear imágenes reales. (p. 402)

concentración: la cantidad de soluto que se disuelve en una cantidad dada de disolvente. (p. 653)

conducción: transferencia de energía térmica por colisiones entre las partículas que componen una materia. (p. 144)

conductividad: propiedad de los metales y aleaciones que permite fácilmente el paso de calor o cargas eléctricas a través del material. (p. 741)

conductor: material, como el alambre de cobre, a través del cual los electrones se pueden pasar con facilidad. (p. 173)

constante: en un experimento, una variable que no cambia. (p. 9)

constelación: patrón de estrellas que aparece formar imágenes, utilizada por los astrónomos a localizar y nombrar estrellas, a menudo tienen nombre de una figura mitológica. (p. 984)

clima continental: clima no afectado por un océano, se caracteriza por temperaturas extremas. (p. 898)

control: estándar usado para la comparación de resultados de pruebas en un experimento. (p. 10)

convección: transferencia de energía térmica en un fluido por el movimiento de fluidos con mayores y menores temperaturas de un lugar a otro. (p. 145)

convergent plate boundary: boundary where two plated collide, produces either subduction zones or continental collisions. **(p. 777)**

convex lens: a lens that is thicker in the middle than at the edges and can form real or virtual images. **(p. 408)**

convex mirror: a reflective surface that curves outward, away from the viewer, and forms a reduced, upright, virtual image. **(p. 405)**

Coriolis effect: the effect of rotation on the motion of any object or fluid; on Earth, air traveling north or south from the equator appears to deflect right or left respectively; the combination of the Coriolis effect and Earth's heat imbalance creates the trade winds, polar easterlies, and prevailing westerlies. **(p. 891)**

cornea: transparent covering on the eyeball through which light enters the eye. **(p. 411)**

cosmic background radiation: residual radiation from the formation of the universe; predicted by the big bang theory and later detected in 1965. **(p. 1003)**

cosmology: study of how the universe began, what it is made of, and how it continues to evolve. **(p. 1002)**

covalent bond: attraction formed between atoms when they share electrons. **(p. 561)**

crest: the highest point on a transverse wave. **(p. 279)**

límite de placas convergentes: frontera en la que dos placas chocan, produce una zona de subducción o un colisión continental. **(p. 777)**

lente convexo: lente que es más grueso en el centro que en los bordes y que puede formar imágenes reales o virtuales. **(p. 408)**

espejo convexo: una superficie reflectante que se curva hacia afuera, lejos del espectador, y forma una imagen virtual, reducida, en posición vertical. **(p. 405)**

efecto Coriolis: efecto de la rotación en el movimiento de cualquier objeto o líquido; en la tierra, el aire que viaja hacia el norte o hasta el sur desde el ecuador parece desviar a la derecha o izquierda, respectivamente; la combinación del efecto de Coriolis y el desequilibrio de calor de la Tierra crea los vientos alisios, Brisas polares, y predominantes del oeste. **(p. 891)**

córnea: cubierta transparente del globo ocular a través de la cual entra la luz al ojo. **(p. 411)**

radiación de fondo cósmico: radiación residual de la formación del universo, predicho por la teoría del Big Bang y, posteriormente detectado en 1965. **(p. 1003)**

cosmología: el estudio de cómo empezó el universo, de lo que está hecho, y cómo sigue evolucionando. **(p. 1002)**

enlace covalente: atracción formada entre átomos que comparten electrones. **(p. 561)**

cresta: la punta más alta en una onda transversal. **(p. 279)**

English	D	Español

dark energy: energy that might be causing the accelerating expansion of the universe. **(p. 1005)**

dark matter: type of matter that cannot be seen and can only be detected by its gravitational effects. **(p. 1004)**

decibel: unit for sound intensity; abbreviated dB. **(p. 313)**

decomposition reaction: chemical reaction in which one substance breaks down into two or more substances. **(p. 591)**

density: mass per unit volume of a material. **(p. 18)**

deoxyribonucleic acid: also called DNA, a type of essential biological compound found in the nuclei of cells that codes and stores genetic information. **(p. 728)**

dependent variable: factor that changes as a result of changes in the other variables. **(p. 9)**

depolymerization: process using heat or chemicals to break a polymer chain into its monomer fragments. **(p. 723)**

deposition: dropping of sediment previously in transport as the erosional agent slows down, or in the case of ice, melts. **(p. 854)**

energía oscura: energía que posiblemente es la causa de la expansión acelerada del universo. **(p. 1005)**

materia oscura: tipo de materia que no puede ser visto y sólo se puede detectar por sus efectos gravitatorios. **(p. 1004)**

decibel: unidad que mide la intensidad del sonido; se abrevia dB. **(p. 313)**

reacción de descomposición: reacción química en la cual una sustancia se descompone en dos o más sustancias. **(p. 591)**

densidad: masa por unidad de volumen de un material. **(p. 18)**

ácido desoxirribonucleico: también conocido como ADN, compuesto biológico esencial encontrado en el núcleo de las células, codifica y almacena información genética. **(p. 728)**

variable dependiente: factor que varía como resultado de los cambios en las otras variables. **(p. 9)**

despolimerización: proceso en el que se utilizan calor o químicos para descomponer una cadena de polímeros en sus fragmentos de monómeros. **(p. 723)**

deposición: el depósito de sedimentoss, previamente transportados por la erosión cuando esta se retrasa, o en caso de que el hielo se derrita. **(p. 854)**

diatomic molecule: a molecule that consists of two atoms of the same element, joined by a covalent bond. (p. 527)

diffraction: the bending of waves around an obstacle; can also occur when waves pass through a narrow opening. (p. 290)

diffusion: spreading of particles throughout a given volume until they are uniformly distributed. (p. 479)

digital signal: an electric signal with only two possible values: ON and OFF. (p. 354)

direct current (DC): electric current that flows in only one direction. (p. 220)

discontinuity: boundary between two layers of material that have different densities. (p. 788)

displacement: distance and direction of an object's change in position from the starting point. (p. 45)

dissociation: process in which an ionic compound separates into its positive and negative ions in solution. (p. 659)

distance: the length an object travels, measured in SI units of meters. (p. 45)

distillation: process that can separate two substances in a mixture by evaporating a liquid and recondensing its vapor. (p. 472)

divergent plate boundary: the boundary between two plates that are moving apart. (p. 778)

doping: process of adding impurities to a semiconductor to modify its conductivity. (p. 749)

Doppler effect: change in frequency that occurs when a source is moving relative to an observer. (p. 315)

double-displacement reaction: reaction in which two ionic compounds in solution are combined, can produce a precipitate, water, or a gas. (p. 592)

drainage basin: land area that gathers water for a major river or several major river systems. (p. 855)

ductile: property of metals and alloys that allows them to be drawn into wires. (pp. 570, 741)

dwarf planet: nearly-spherical object in orbit around the Sun that is not a satellite, and has not cleared the debris in its orbit. (p. 967)

molécula diatómica: molécula que consta de dos átomos del mismo elemento, unidas por un enlace covalente. (p. 527)

difracción: un cambio de dirección de las ondas al rozar el borde de un obstáculo, la cual también puede ocurrir cuando éstas pasan a través de una abertura angosta. (p. 290)

difusión: propagación de partículas en la totalidad de un volumen determinado hasta que se distribuyen de manera uniforme. (p. 479)

señal digital: una señal eléctrica, con sólo dos valores posibles: ON y OFF. (p. 354)

corriente directa (CC): corriente eléctrica que fluye en una sola dirección. (p. 220)

discontinuidad: límite entre dos capas de material que tienen diferentes densidades. (p. 788)

desplazamiento: distancia y dirección del cambio de posición de un objeto desde el punto inicial. (p. 45)

disociación: proceso en el cual un compuesto iónico se separa en sus iones positivos y negativos en una solución. (p. 659)

distancia: la longitud de un objeto de viajes, medido en unidades SI de metros. (p. 45)

destilación: proceso que puede separar dos sustancias de una mezcla por medio de la evaporación de un líquido y la recondensación de su vapor. (p. 472)

límite de placas divergentes: frontera entre dos placas tectónicas que se están separando. (p. 778)

dopaje: proceso que consiste en añadir impurezas a un semiconductor para modificar a su conductividad. (p. 749)

efecto Doppler: cambio en la frecuencia que se produce cuando una fuente se mueve respecto a un observador. (p. 315)

reacción de doble desplazamiento: reacción química en cual se combinan dos compuestos iónicos en una solución, puede producir un precipitado, agua o gas. (p. 592)

cuenca de drenaje: área que recolecta el agua para un rio importante o para varios sistemas de ríos importantes (p. 855)

ductibilidad: propiedad de los metales y aleaciones que les permiten ser convertidos en alambres. (pp. 570, 741)

planeta enano: objeto casi-esférico en órbita alrededor del Sol que no es un satélite, y no ha limpiado la vecindad de su órbita. (p. 967)

English	E	Español

eardrum: tough membrane in the outer ear that is about 0.1 mm thick and transmits sound vibrations into the middle ear. (p. 309)

Earth: third planet from the Sun; the only planet in the solar system known to have life, and where water exists as a solid, a liquid, and a gas. (p. 957)

tímpano: membrana fuerte del oído externo que tiene aproximadamente 0.1 mm de grueso y transmite las vibraciones del sonido al oído medio. (p. 309)

Tierra: tercer planeta desde el Sol, el único planeta del sistema solar donde se ha encontrado vida, y donde el agua existe como sólido, líquido y gas. (p. 957)

earthquake: sudden movement or vibration of the ground that occurs when rocks slip and slide along enormous cracks in Earth's crust. **(p. 780)**

ecliptic: path of Earth's orbit around the Sun; from Earth, seen as the path that the Sun travels across the zodiac. **(p. 924)**

echolocation: process by which objects are located by emitting sounds and interpreting the sound waves that are reflected from those objects. **(p. 324)**

efficiency: ratio of the output work done by the machine to the input work done on the machine, expressed as a percentage. **(p. 110)**

El Niño: periodic warming of the Pacific Ocean off the coast of western South America that changes global weather patterns. **(p. 905)**

elastic potential energy: energy that is stored by compressing or stretching an object. **(p. 117)**

elastic rebound: the sudden release of strain energy from rock as it moves along a fault. **(p. 782)**

electrical power: rate at which electrical energy is converted to another form of energy; expressed in watts (W). **(p. 188)**

electric circuit: a closed path that electric current follows. **(p. 179)**

electric current: the net movement of electric charges in a single direction, measured in amperes (A). **(p. 178)**

electric field: a region surrounding every electric charge in which a force of attraction or repulsion is exerted on other electric charges. **(p. 172)**

electric motor: device that converts electrical energy to mechanical energy by using the magnetic forces between an electromagnet and a permanent magnet to make a shaft rotate. **(p. 213)**

electrolyte: compound that breaks apart in water, producing charged particles (ions) that can conduct electricity. **(p. 658)**

electromagnet: temporary magnet created when there is a current in a wire coil. **(p. 210)**

electromagnetic force: the attractive or repulsive force between electric charges and magnets. **(p. 210)**

electromagnetic induction: process by which electric current is produced in a wire loop by a changing magnetic field. **(p. 216)**

electromagnetic wave: waves created by vibrating electric charges; consists of vibrating electric and magnetic fields, and can travel through a vacuum or through matter. **(p. 338)**

electromagnetism: the interaction between electric charges and magnets. **(p. 210)**

electron: particle with an electric charge of 1−, surrounds the nucleus of an atom. **(p. 489)**

terremoto: movimiento brusco o vibración de la tierra que ocurre cuando roca se desliza a lo largo de una grieta enorme en la corteza de la tierra. **(p. 780)**

eclíptica: sendero de la órbita de la Tierra alrededor del Sol; desde la Tierra, vista como el camino que el Sol recorre por todo el zodiaco. **(p. 924)**

ecolocalización: proceso por el cual los objetos son localizados por emitir sonidos e interpretando las ondas de sonido que se reflejan. **(p. 324)**

eficiencia: relación del trabajo efectuado por una máquina y el trabajo hecho en ésta, expresada en porcentaje. **(p. 110)**

El Niño: calentamiento periódico del Océano Pacífico frente a la costa occidental de América del Sur, causa cambios globales en los patrones climáticos. **(p. 905)**

energía elástica potencial: energía almacenada por compresionar o estrechar un objeto. **(p. 117)**

recuperación elástica: la liberación repentina de energía de las rocas a lo largo de una falla. **(p. 782)**

potencia eléctrica: velocidad a la cual la energía eléctrica se convierte en otra forma de energía; se expresa en vatios (W). **(p. 188)**

circuito eléctrica: un camino cerrado que sigue la corriente electrica. **(p. 179)**

corriente eléctrica: movimiento neto de cargas eléctricas en una sola dirección, medido en amperios (A). **(p. 178)**

campo eléctrico: una región en torno a toda carga eléctrica en la que una fuerza de de atración o repulsión se ejerce hacia otras cargas eléctricas. **(p. 172)**

motor eléctrico: dispositivo que convierte la energía eléctrica en energía mecánica por las fuerzas magnéticas entre un electroimán y un imán permanente, y hace que un eje gire. **(p. 213)**

electrolito: compuesto que se descompone en agua y así produce partículas cargadas (iones) que pueden conducir electricidad. **(p. 658)**

electroimán: imán temporal crea cuando hay una corriente en una bobina de alambre.**(p. 210)**

fuerza electromagnética: fuerza de atracción o repulsión entre cargas eléctricas y imanes. **(p. 210)**

inducción electromagnética: proceso mediante el cual se produce la corriente eléctrica en un aro de alambre por un campo magnético variable. **(p. 216)**

ondas electromagnéticas: olas creadas por la vibración de cargas eléctricas; consta de vibrar los campos eléctricos y magnéticos, y puede viajar a través del vacío o través de la materia. **(p. 338)**

electromagnetismo: la interacción entre las cargas eléctricas y los imanes. **(p. 210)**

electrón: partícula que rodea con carga eléctrica de 1−, rodea el núcleo de un átomo. **(p. 489)**

electron cloud: area around the nucleus of an atom where the atom's electrons are most likely to be found. **(p. 493)**

electron dot diagram: uses the symbol for an element and dots representing the number of electrons in the element's outer energy level. **(p. 504)**

electroscope: a device, sometimes consisting of two leaves of metallic foil, used to detect electric charge. **(p. 176)**

element: substance with atoms that are all alike. **(p. 462)**

ellipse: elongated, closed curve with two foci; shape of Earth's orbit around the Sun. **(p. 920)**

endergonic reaction: chemical reaction that requires energy input in the form of light, thermal energy or electricity in order to proceed. **(p. 596)**

endothermic reaction: chemical reaction that requires thermal energy in order to proceed. **(p. 596)**

energy: the ability to cause change, measured in joules. **(p. 114)**

epicenter: the point on Earth's surface directly above the focus of an earthquake. **(p. 782)**

equilibrium: state in which forward and reverse reactions or processes occur at equal rates. **(p. 602)**

equinox: occurs twice a year in March and September, when Earth's rotational axis is perpendicular to a line connecting the center of Earth to the center of the Sun. **(p. 927)**

erosion: removal of surface material through the process of weathering. **(p. 854)**

ester: a substituted hydrocarbon with a –COOC– group. **(p. 714)**

exergonic reaction: chemical reaction that releases some form of energy, such as light or thermal energy. **(p. 595)**

exothermic reaction: chemical reaction in which energy is primarily given off in the form of thermal energy. **(p. 595)**

experiment: organized procedure for testing a hypothesis; tests the effect of one thing on another under controlled conditions. **(p. 9)**

extrasolar planets: planets in orbit around another star. **(p. 954)**

extraterrestrial life: life beyond Earth, object of the search by exobiologists. **(p. 970)**

extrusive igneous rock: rock that formed from lava or ash that solidified on Earth's surface. **(p. 820)**

nube de electrones: área alrededor del núcleo de un átomo en donde hay más probabilidad de encontrar los electrones de los átomos. **(p. 493)**

diagrama de punto de electrones: usa el símbolo de un elemento y puntos que representan el número de electrones en el nivel de energía externo del elemento. **(p. 504)**

electroscopio: un aparato, a veces con hojas metales, que detecta a carga eléctrica. **(p. 176)**

elemento: sustancia en la cual todos los átomos son iguales. **(p. 462)**

elipse: curva cerrada, alargada, con dos focos; la forma de la órbita terrestre alrededor del sol. **(p. 920)**

reacción endergónica: reacción química que requiere entrada de energía en la forma de luz, energía térmica o electricidad para proceder. **(p. 596)**

reacción endotérmica: reacción química que requiere energía de energía térmica para proceder. **(p. 596)**

energia: la habilidad para efectuar un cambio, medida en julio **(p. 114)**

epicentro: punto en la superficie de la tierra localizada directamente arriba del foco del terremoto. **(p. 782)**

equilibrio: estado en que reacciones o procesos avancen y retrocesan a tasas iguales. **(p. 602)**

equinoccio: producido dos veces al año, en marzo y en septiembre, cuando el eje de rotación de la Tierra es perpendicular a una línea que conecta el centro de la Tierra en el centro del sol. **(p. 927)**

erosión: eliminación del material de la superficie de la tierra mediante el proceso de intemperización. **(p. 854)**

ester: un hidrocarburo sustituido con un –COOC–. **(p. 714)**

reacción exergónica: reacción química que libera una forma de energía, tal como luz o energía térmica. **(p. 595)**

reacción exotérmica: reacción química en la cual la energía es inicialmente emitida en forma de energía térmica. **(p. 595)**

experimento: procedimiento organizado para probar una hipótesis; prueba el efecto de una cosa sobre otra bajo condiciones controladas. **(p. 9)**

extrasolar planets: planetas extrasolares: planetas en órbita alrededor de otra estrella. **(p. 954)**

vida extraterrestre: vida más allá de la Tierra, objeto de la búsqueda por exobiólogos. **(p. 970)**

roca ígnea estrusiva: roca que se ha formado de la lava o de ceniza solidificada en la superficie de la tierra. **(p. 820)**

English	F	Español

fault: crack in Earth's crust along which rock has moved. **(p. 782)**

field: a region of space in which every point has a physical quantity, such as a force. **(p. 77)**

falla: grieta en la corteza de la Tierra a lo largo de cual la roca se ha movido. **(p. 782)**

campo: una región en cual cada punto se puede expresar por una cantidad, por ejemplo, una fuerza. **(p. 77)**

filter: a transparent material that selectively transmits light. **(p. 375)**

first law of thermodynamics: states that if the mechanical energy of a system is constant, the increase in the thermal energy of the system equals the sum of the thermal energy transferred into the system and the work done on the system. **(p. 155)**

fission: process of splitting an atomic nucleus into two or more nuclei with smaller masses, releasing large amounts of energy. **(p. 241)**

fluorescent light: light generated by using phosphors to convert ultraviolet radiation to visible light. **(p. 378)**

focal length: distance from the center of a lens or mirror to the focal point. **(p. 402)**

focal point: the point on the optical axis of a curved mirror or lens where light rays that are initially parallel to the optical axis converge after striking the mirror or lens. **(p. 402)**

focus: point of origin for an earthquake, the point from which seismic waves originate. **(p. 782)**

foliated: texture of some metamorphic rocks in which crystals are arranged in layers and bands as a result of high pressure conditions. **(p. 831)**

force: a push or pull exerted on an object. **(p. 72)**

fossil: remains or traces of ancient organism, found in the geologic rock record. **(p. 871)**

fossil fuel: oil, natural gas, and coal; formed from the remains of ancient plants and animals that were buried and altered over millions of years. **(p. 235)**

fracture: manner in which a mineral without cleavage will break, producing uneven, irregular surfaces. **(p. 811)**

frequency: the number of wavelengths that pass a fixed point each second; is expressed in hertz (Hz). **(p. 280)**

free fall: describes the fall of an object on which only the force of gravity is acting. **(p. 89)**

friction: force that opposes the sliding motion between two touching surfaces. **(p. 74)**

fusion: reaction in which two or more atomic nuclei form a nucleus with a larger mass, releasing large amounts of energy **(p. 241)**

filtro: un material transparente que transmite la luz de forma selectiva. **(p. 375)**

primera ley de la termodinámica: establece que si la energía mecánica del sistema es constante, el aumento de la energía térmica del sistema es igual a la suma de la energía térmica transferida al sistema y el trabajo realizado en el sistema. **(p. 155)**

fisión: proceso de división en el cual un núcleo atómico de divide entre dos o más núcleos con masas más pequeñas, produciendo grandes cantidades de energía. **(p. 241)**

luz fluorescente: luz generada mediante el uso de fósforo para convertir la radiación ultravioleta a la luz visible. **(p. 378)**

longitud focal: distancia desde el centro de un lente o espejo al punto focal. **(p. 402)**

punto focal: el punto en el eje óptico de un espejo o lente curvo en el cual los rayos de luz, que inicialmente son paralelos al eje óptico, convergen después de chocar al espejo o lente. **(p. 402)**

foco: punto de origen de un terremoto, punto desde donde se originan las ondas sísmicas. **(p. 782)**

rocas foliadas: textura en ciertas rocas metamórficas en la que los cristales están alineadas en hojas y bandas como resultado de condiciones de gran presión. **(p. 831)**

fuerza: impulso o tracción sobre un objeto. **(p. 72)**

fósil: restos o rastros de organismos encontrados en las rocas. **(p. 871)**

combustibles fósiles: petróleo, gas natural y carbón; formado a partir de los restos de antiguas plantas y animales que fueron enterrados y alterados durante millones de años. **(p. 235)**

fractura: forma en que un mineral sin exfoliación se romperá, produciendo superficies desiguales y irregulares. **(p. 811)**

frecuencia: el número de longitudes de onda que pasan por un punto fijo en un segundo; se expresa en hercios (Hz). **(p. 280)**

caída libre: describe la caída de un objeto sobre cual la única fuerza que le actúa es la de la gravedad. **(p. 89)**

fricción: fuerza que se opone al movimiento deslizante entre dos superficies en contacto. **(p. 74)**

fusión: proceso en la cual dos o más núcleos atómicos forman un núcleo con mayor masa, produciendo grandes cantidades de energía. **(p. 241)**

English	G	Español

galaxy: large group of stars, dust, and gas; held together by gravity. **(p. 997)**

galvanometer: a device that uses an electromagnet to measure electric current. **(p. 213)**

gamma ray: electromagnetic wave with a wavelength less than about 100 trillionths of a meter; usually emitted from a decaying atomic nucleus. **(pp. 351, 622)**

galaxia: grupo grande de estrellas, polvo y gas; se mantienen unidos por la gravedad. **(p. 997)**

galvanómetro: un dispositivo que utiliza y un electroimán para medir la corriente eléctrica. **(p. 213)**

rayo gama: onda electromagnética con longitud de onda menor a cien trillonésimas de un metro, generalmente emitido por un núcleo atómico en descomposición. **(pp. 351, 622)**

Glossary

Geiger counter: radiation detector that produces a click or a flash of light when a charged particle is detected. **(p. 629)**

generator: device that uses electromagnetic induction to convert mechanical energy to electrical energy. **(p. 217)**

geocentric model: Earth-centric model of the solar system. **(p. 950)**

geothermal energy: thermal energy contained in and around magma; can be converted by a power plant into electrical energy. **(p. 252)**

giant star: late stage in a star's life cycle that occurs when its hydrogen fuel is depleted, its core contracts, and its outer layers expand and cool. **(p. 991)**

Global Positioning System (GPS): a system of satellites and ground monitoring stations that enable a receiver to determine its location at or above Earth's surface. **(p. 357)**

global warming: increase in average temperatures of Earth's near-surface air and oceans. **(p. 904)**

graph: visual display of information or data. **(p. 21)**

gravitational potential energy (GPE): energy that is due to the gravitational force between objects. **(p. 118)**

gravity: attractive force between two objects that depends on the masses of the objects and the distance between them. **(p. 76)**

greenhouse effect: natural process in which certain gases in the atmosphere absorb and emit infrared radiation, warming the planet. **(p. 886)**

group: vertical column in the periodic table. **(p. 502)**

contador Geiger: detector de radiación que produce un sonido seco o un destello de luz al detectar una partícula cargada. **(p. 629)**

generador: dispositivo que usa inducción electromagnética para convertir energía mecánica en energía eléctrica. **(p. 217)**

modelo geocéntrico: modelo del sistema solar con la Tierra en el centro. **(p. 950)**

energía geotérmica: energía térmica en y alrededor del magma, la cual se puede convertir mediante una planta industrial en energía eléctrica. **(p. 252)**

estrella gigante: fase avanzada del ciclo de vida de una estrella que ocurre cuando se agote su hidrógeno, su núcleo se contrae, y sus capas externas se expanden y enfrían. **(p. 991)**

Sistema de Posicionamiento Global (GPS): sistema de satélites y estaciones de monitoreo en tierra que permiten que un receptor determine su ubicación en o sobre la superficie terrestre. **(p. 357)**

calentamiento global: aumento de las temperaturas promedias del aire cerca de la superficie de la Tierra y los océanos. **(p. 904)**

gráfica: presentación visual de información que puede suministrar una forma rápida de comunicar gran cantidad de información. **(p. 21)**

energía gravitacional potencial: energía debida a la fuerza gravitoria entre objetos **(p. 118)**

gravedad: fuerza de atracción entre dos objetos que depende de las masas de los objetos y de la distancia entre ellos. **(p. 76)**

efecto invernadero: proceso natural en donde ciertos gases en la atmosfera absorben y emiten radiaciones infrarrojas, calentando el planeta. **(p. 886)**

grupo: columna vertical en la tabla periódica. **(p. 502)**

English	H	Español

half-life: amount of time it takes for half the nuclei in a sample of a radioactive isotope to decay. **(p. 632)**

hardness: measure of a mineral's resistance to scratching, described by Mohs scale of hardness. **(p. 811)**

hazardous waste: wastes that are poisonous, cause cancer, or can catch fire. **(p. 257)**

heat: energy that is transferred between objects due to a temperature difference between those objects. **(p. 140)**

heat engine: device that converts some thermal energy into mechanical energy. **(p. 156)**

heat of fusion: amount of energy required to change a substance from the solid phase to the liquid phase. **(p. 435)**

heat of vaporization: the amount of energy required for a liquid at its boiling point to become a gas. **(p. 435)**

vida media: tiempo requerido para que se descomponga la mitad de los núcleos de una muestra de isótopo radiactivo. **(p. 632)**

dureza: medida de la resistencia de un mineral a ser rasguñado; descrita por la escala de dureza de Mohs. **(p. 811)**

residuo peligroso: residuo tóxico, provoca cáncer, o puede incendiarse. **(p. 257)**

calor: la energía que se transfiere entre los objetos debido a una diferencia de temperatura entre esos objetos. **(p. 140)**

motor de calor: dispositivo que convierte térmica en energía mecánica. **(p. 156)**

calor de fusión: cantidad de energía necesaria para cambiar una sustancia del estado sólido al líquido. **(p. 435)**

calor de vaporización: cantidad de energía necesaria para que un líquido en su punto de ebullición se convierta en gas. **(p. 435)**

Glossary

heliocentric model: Sun-centered model of the solar system. (p. 951)

heterogeneous mixture: substance in which its different components are easily distinguished. (p. 465)

holography: technique that produces a complete three-dimensional photographic image of an object. (p. 385)

homogeneous mixture: substance containing two or more components that are blended uniformly so that individual components are indistinguishable with a microscope. (p. 467)

hydrate: compound that has water chemically attached to its atoms and written into its chemical formula. (p. 570)

hydrocarbon: compound containing only carbon and hydrogen atoms. (p. 707)

hydroelectricity: electricity produced from the energy of moving water. (p. 250)

hydronium ion: H_3O^+ ion, forms when an acid dissolves in water and H^+ ions interact with water. (p. 678)

hydroxide ion: OH^- ion, forms when a base dissolves in water. (p. 680)

hypothesis: possible explanation for a problem using what is known and what is observed. (p. 9)

modelo heliocéntrico: modelo del sistema solar con el sol en el centro. (p. 951)

mezcla heterogénea: sustancia en la que los diferentes componentes se distinguen fácilmente (p. 465)

holografía: técnica que produce una imagen fotográfica tridimensional completa de un objeto. (p. 385)

mezcla homogénea: sustancia que contiene dos o más componentes que se combina de manera uniforme a fin de que los componentes individuales no se pueden distinguir con un microscopio. (p. 467)

hidrato: compuesto que contiene agua químicamente apegado a sus átomos y escrita en su fórmula química. (p. 570)

hidrocarburo: compuesto que contiene únicamente átomos de carbono e hidrógeno. (p. 707)

hidroelectricidad: electricidad producida por la energía de movimiento de agua. (p. 250)

ion de hidronio: ion H_3O^+, se forma cuando un ácido se disuelve en agua y el ion H^+ interactúa con el agua. (p. 678)

hydroxide ion: ion OH^-, forma cuando una base disuelve en agua. (p. 680)

hipótesis: explicación posible para resolver un problema al utilizar lo que conoce y lo que observa. (p. 9)

English		Español

incandescent light: light produced by heating a piece of metal, usually tungsten, until it glows. (p. 378)

incoherent light: light that can contain more than one wavelength, travel in more than one direction, and has varying distances between the corresponding crests of the waves. (p. 382)

independent variable: factor that, as it changes, affects the measure of another variable. (p. 9)

index of refraction: property of a material indicating how much the speed of light is reduced in the material compared to the speed of light in a vacuum. (p. 370)

indicator: organic compound that changes color in acids and bases. (p. 678)

inertia: tendency of an object to resist any change in its motion. (p. 80)

infiltration: process by which water enters Earth and becomes groundwater, controlled by topography, surface materials, and type and amount of vegetation. (p. 863)

infrared wave: electromagnetic wave with a wavelength between about 1 mm and 700 billionths of a meter. (p. 348)

inhibitor: substance that slows down a chemical reaction or prevents it from occurring by combining with a reactant. (p. 601)

luz incandescente: luz que se produce al calentar una pieza de metal, generalmente tungsteno, hasta que brille. (p. 378)

luz incoherente: luz que puede contener más de una longitud de onda, viaja en más de una dirección, y tiene distancias variables entre las crestas correspondientes de las olas. (p. 382)

variable independiente: factor que, a medida que cambia, afecta la medida de otra variable. (p. 9)

índice de refracción: propiedad de un material que indica cuanto la velocidad de la luz se reduce en el material en comparación con la velocidad de la luz en el vacío. (p. 370)

indicador: compuesto orgánico que cambia de color en presencia de ácidos y bases. (p. 678)

inercia: tendencia de un objeto a resistir cualquier cambio en su movimiento.(p. 80)

infiltración: proceso por el cual el agua entra en la tierra y se convierte en agua subterránea, controlado por la topografía, materiales superficiales, y tipo y cantidad de vegetación. (p. 863)

onda infrarroja: onda electromagnética que tiene una longitud de onda entre aproximadamente 1 mm y 700 billonésimas de metro. (p. 348)

inhibidor: sustancia que reduce una reacción química o previene que ocurra por combinar con un reactivo. (p. 601)

Glossary

instantaneous speed: speed of an object at a given point in time; is constant for an object moving with constant speed, and changes with time for an object that is slowing down or speeding up. **(p. 49)**

insulator: material in which electrons and thermal energy are not able to move easily. **(p. 173)**

integrated circuit: tiny chip that can contain millions of transistors, diodes, and other components. **(p. 750)**

intensity: amount of energy that flows through a certain area in a specific amount of time. **(p. 312)**

interference: the process of two or more waves overlapping and combining to form a new wave. **(p. 292)**

internal combustion engine: heat engine that burns fuel inside the engine in chambers or cylinders. **(p. 157)**

intrusive igneous rock: rock that formed from magma that solidified within Earth's crust, also called plutonic rock. **(p. 817)**

ion: charged particle that has either more or fewer electrons than protons. **(pp. 558, 676)**

ionic bond: the force of attraction between the opposite charges of the ions in an ionic compound. **(p. 560)**

ionization: process in which electrolytes dissolve in water and separate into charged particles. **(p. 658)**

isomers: compounds with identical chemical formulas but different molecular structures and shapes. **(p. 709)**

isotope: atom of an element that has a specific number of neutrons. **(p. 496)**

velocidad instantánea: velocidad de un objeto en un punto dado en el tiempo; es constante para un objeto que se mueve a una velocidad constante y cambia con el tiempo en un objeto que está reduciendo o aumentando su velocidad. **(p. 49)**

aislador: material a través del cual los electrones y la energía térmica no se pueden transferir con facilidad. **(p. 173)**

circuito integrado: pedazo minúsculo que puede contener millones de transistores, diodos y otros componentes. **(p. 750)**

intensidad: cantidad de energía que fluye a través de cierta área en un tiempo específico. **(p. 312)**

interferencia: el proceso de dos o más ondas se superponen y se combinan para formar una nueva ola. **(p. 292)**

motor de combustión interna: motor de calor que quema combustible en su interior en cámaras o cilindros. **(p. 157)**

roca ígnea intrusiva: roca ígnea que se ha formado en el magma solidificándose dentro de la corteza terrestre; también llamada roca plutónica. **(p. 817)**

ion: partícula cargada que tiene ya sea más o menos electrones que protones. **(pp. 558, 676)**

enlace iónico: fuerza de atracción entre las cargas opuestas de los iones en un compuesto iónico. **(p. 560)**

ionización: proceso en el cual los electrolitos se disuelven en agua y se separan en partículas cargadas. **(p. 658)**

isómeros: compuestos con fórmulas químicas idénticas pero con estructuras moleculares diferentes. **(p. 709)**

isótopo: átomo de un elemento que tiene un cierto número de neutrones. **(p. 496)**

English	J	Español

jet stream: narrow band of powerful, fast-moving, high-altitude air embedded in the global wind belts. **(p. 891)**

joule: SI unit of work and energy. **(p. 107)**

Jupiter: fifth planet from the Sun; the largest planet with continuous, swirling, high-pressure gas storms. **(p. 963)**

corriente de chorro: flujo de aire de gran alcance, rápido movimiento, gran altura que se introduce en los cinturones de viento mundiales. **(p. 891)**

julio: unidad del SI de trabajo y energía. **(p. 107)**

Júpiter: quinto planeta desde el Sol, el planeta más grande con tormentas continuas de gas de alta presión. **(p. 963)**

English	K	Español

kinetic energy: energy a moving object has because of its motion; determined by the mass and speed of the object. **(p. 116)**

kinetic theory: explanation of the behavior of particles in gases; states that matter is made of constantly moving particles that collide without losing energy. **(p. 432)**

energía cinética: energía que tiene un cuerpo debido a su movimiento, se determina por la masa y velocidad del objeto. **(p. 116)**

teoría cinética: explicación del comportamiento de las partículas en un gas, la cual establece que las sustancias son compuestas de partículas en constante movimiento que se chocan sin perder energía. **(p. 432)**

English	L	Español

La Niña: periodic cooling of equatorial ocean surface temperature that modify global weather patterns. **(p. 905)**

law of conservation of charge: states that charge can be transferred from one object to another but it cannot be created or destroyed. **(p. 171)**

law of conservation of energy: states that energy cannot be created or destroyed. **(p. 120)**

law of conservation of mass: states that the mass of all substances present before a chemical change equals the mass of all the substances remaining after the change. **(p. 475)**

law of conservation of momentum: states that if no external forces act on a group of objects, their total momentum does not change. **(p. 91)**

Le Chatelier's principle: states that if a stress is applied to a reaction at equilibrium, the reaction shifts in the direction opposite of the stress. **(p. 603)**

lever: simple machine consisting of a bar free to pivot about a fixed point called the fulcrum. **(p. 109)**

light-year: distance light travels in one year—about 9.5 trillion km. **(p. 988)**

linearly polarized light: light whose magnetic field vibrates in only one direction. **(p. 384)**

lipid: group of biological compounds that contains the same elements as carbohydrates but in different proportions, includes saturated and unsaturated fats and oils. **(p. 726)**

lithosphere: the layer of Earth made of rocky material broken up into tectonic plates, consists of Earth's crust and uppermost mantle. **(p. 790)**

Local Group: group of about 50 galaxies including the Milky Way. **(p. 998)**

longitudinal wave: a wave in which the matter in the medium moves back and forth along the direction that the wave travels. **(p. 277)**

longshore current: movement of water, and often, sediment, parallel to the shoreline. **(p. 861)**

loudness: human perception of sound volume, depends primarily on intensity. **(p. 312)**

lunar eclipse: occurs during full moon, when the Moon enters Earth's umbra and Earth casts a curved shadow on the Moon's surface. **(p. 934)**

luster: property of metals and alloys that describes having a shiny appearance. **(p. 741)**

La Niña: enfriamiento periódico de temperatura superficial del océano ecuatorial; modifica a los patrones climáticos globales. **(p. 905)**

ley de la conservación de carga: Estados de que se puede ser transferida de un objeto a otro, pero no puede ser creada o destruida. **(p. 171)**

ley de la conservación de energía: establece que la energía no puede crearse ni destruirse. **(p. 120)**

ley de conservación de la masa: establece que la masa de todas las sustancias presente antes de un cambio químico es igual a la masa de todas las sustancias resultantes después del cambio. **(p. 475)**

ley de conservación del moméntum: establece que si no actúan fuerzas externas sobre un grupo de objetos, su cantidad de movimiento total no cambia. **(p. 91)**

principio de Le Chatelier: establece que si una tensión se aplica a una reacción en el equilibrio, la reacción se acelera en la dirección opuesta de la tensión. **(p. 603)**

palanca: máquina simple que consiste de una barra que puede girar sobre un punto fijo llamado pivote. **(p. 109)**

año luz: distancia que recorre la luz en el periodo de un año; alrededor de 9,5 mil de billón km. **(p. 988)**

luz polarizada lineal: luz cuya campo magnético vibra en una sola dirección. **(p. 384)**

lípido: grupo de compuestos biológicos que contiene los mismos elementos que los hidratos de carbono, pero en diferentes proporciones, incluye grasas saturadas e insaturadas y aceites. **(p. 726)**

litosfera: La capa de la tierra hecha de material rocoso dividida en placas tectónicas, consta de la Corteza terrestre y el manto superior. **(p. 790)**

Grupo Local: grupo de alrededor de 50 galaxias inclusivo a la Vía Láctea. **(p. 998)**

onda longitudinal: onda por la cual la materia en el medio se mueve para adelante y para atrás en la dirección en que viaja la onda. **(p. 277)**

corriente costera: movimiento de agua y a menudo sedimento, paralelo a la línea de playa. **(p. 861)**

volumen de sonido: percepción humana de la fuerza del sonido, depende primariamente en la intensidad. **(p. 312)**

eclipse lunar: producido durante una luna llena, cuando la Luna entra la penumbra de la Tierra y la Tierra proyecta una sombra curva sobre la superficie de la Luna. **(p. 934)**

lustre: propiedad de los metales y aleaciones que describe la manera come brilla. **(p. 741)**

English	M	Español

machine: device that makes doing work easier by increasing the force applied to an object, changing the direction of an applied force, or increasing the distance over which a force can be applied. (p. 109)

magma: molten rock material inside Earth. (p. 813)

magnetic domain: group of atoms in a magnetic material in which the magnetic poles of the atoms are aligned in the same direction. (p. 207)

magnetic field: region surrounding a magnet that exerts a force on other magnets and objects made of magnetic materials. (p. 203)

magnetic pole: region on a magnet where the magnetic force exerted by a magnet is strongest; like poles repel and opposite poles attract. (p. 204)

magnetism: the properties and interactions of magnets. (p. 202)

main sequence: section of the H-R diagram that is plotted from the upper left to the lower right and contains 90 percent of all known stars. (p. 989)

malleable: property of metals and alloys that allows them to be hammered or rolled into thin sheets. (pp. 518, 741)

maria: relatively flat, dark-colored regions on the Moon's surface. (p. 936)

maritime climate: climate strongly influenced by presence of an ocean, tends to be mild due to moisture. (p. 898)

Mars: fourth planet from the Sun; called the red planet because of high concentrations of iron oxide. (p. 957)

mass: amount of matter in an object. (p. 18)

mass number: sum of the number of protons and neutrons in an atom's nucleus. (p. 495)

matter: anything that takes up space and has mass. (p. 18)

mechanical advantage (MA): ratio of the output force exerted by a machine to the input force applied to the machine. (p. 111)

mechanical energy: sum of the potential energy and kinetic energy of the objects in a system. (p. 121)

mechanical wave: a wave that can only travel through matter. (p. 276)

medium: matter through which a wave travels. (p. 276)

melting point: temperature at which a solid begins to liquefy. (p. 434)

Mercury: smallest and closest planet to the Sun. (p. 955)

metal: element that is shiny, malleable, ductile, and a good conductor of heat and electricity. (p. 518)

máquina: artefacto que facilita la ejecución del trabajo por aumentar la fuerza que se aplica a un objeto, cambiar la dirección de una fuerza aplicada o aumentar la distancia sobre la cual se puede aplicar una fuerza. (p. 109)

magma: material de roca fundida dentro de la tierra. (p. 813)

dominio magnético: grupo de átomos en un material magnético en el cual los polos magnéticos de los átomos están alineados en la misma dirección. (p. 207)

campo magnético: región que rodea a un imán que ejerce una fuerza sobre otros imanes y objetos hechos de materiales magnéticos. (p. 203)

polo magnético: zona en un imán en donde la fuerza magnética ejercida por un imán es la más fuerte; los polos-iguales se repelen y los polos-opuestos se atraen. (p. 204)

magnetismo: propiedades e interacciones de los imanes. (p. 202)

secuencia principal: sección del diagrama HR que se traza desde la parte superior izquierda a la inferior derecha y contiene el 90 por ciento de todas las estrellas conocidas. (p. 989)

maleable: la propiedad de los metales y aleaciones que les permite ser martillados o enrollados en láminas delgadas. (pp. 518, 741)

mar lunar: uno de varios regiones en la superficie de la luna relativamente plano y de color oscuro. (p. 936)

maritime climate clima marítimo: clima fuertemente influenciada por la presencia de un océano, tiende a ser templado debido a la humedad. (p. 898)

Marte: cuarto planeta desde el Sol, se llama el planeta rojo debido a las altas concentraciones de óxido de hierro. (p. 957)

masa: cantidad de materia en un objeto. (p. 18)

número de masa: suma del número de protones y neutrones en el núcleo de un átomo. (p. 495)

materia: todo lo que ocupa espacio y tiene masa. (p. 18)

ventaja mecánica (MA): relación de la fuerza ejercida por una máquina y la fuerza aplicada a dicha máquina. (p. 111)

energía mecánica: suma de la energía potencial y energía cinética de los objetos en un sistema. (p. 121)

onda mecánica: una ola que sólo puede viajar a través de la materia. (p. 276)

medio: materia a través de la cual viaja una onda. (p. 276)

punto de fusión: temperatura a la cual un sólido comienza a licuarse. (p. 434)

Mercurio: el planeta más pequeño y más cercano al sol. (p. 955)

metal: elemento que por lo general es brillante, maleable, dúctil, y un buen conductor del calor y la electricidad. (p. 518)

metallic bonding: occurs because some electrons move freely among a metal's positively charged ions, explains properties such as ductility and the ability to conduct electricity. **(p. 519)**

metalloid: element that shares some properties with metals and some with nonmetals. **(p. 532)**

meteoroid: sand-to-boulder sized rocky object orbiting the Sun. **(p. 967)**

microscope: instrument that uses two convex lenses to magnify small, close objects. **(p. 418)**

microwave: electromagnetic wave with wavelength between about 0.1 mm and 30 cm. **(p. 347)**

mid-ocean ridge (MOR): a system of mountain ranges with a rift valley between them that extends around Earth on the seafloor; formed where oceanic plates spread apart due to magma rising from Earth's mantle. **(p. 774)**

Milky Way: spiral galaxy that is about 100,000 light-years in diameter and contains from 200 to 400 billion stars, including the Sun. **(p. 997)**

mineral: naturally occurring, inorganic solid with a crystalline form. **(p. 809)**

mirage: image of a distant object produced by the refraction of light through air layers of different densities. **(p. 372)**

model: can be used to represent an idea, object, or event that is too big, too small, too complex, or too dangerous to observe or test directly. **(p. 12)**

modulation: process of adding a signal to a carrier wave by altering the carrier wave's amplitude, frequency, or other properties. **(p. 352)**

molar mass: the mass in grams of one mole of a substance. **(p. 589)**

mole: SI unit for quantity equal to 6.022×10^{23} units of that substance. **(p. 588)**

molecule: a neutral particle that forms as a result of electron sharing among atoms. **(p. 561)**

momentum: property of a moving object that equals its mass times its velocity. **(p. 86)**

monomer: small molecule that can combine with itself repeatedly to form a long chain. **(pp. 720, 771)**

Moon phase: changing appearance of the Moon as viewed from Earth, depending on the relative positions of the Sun, the Moon, and Earth. **(p. 931)**

motion: a change in an object's position relative to a reference point. **(p. 44)**

music: collection of sounds deliberately used in a regular pattern. **(p. 317)**

enlace metálico: ocurre debido a que algunos electrones se mueven libremente entre los iones de cargados positiva de un metal y explica propiedades tales como la ductibilidad y la capacidad para conducir electricidad. **(p. 519)**

metaloide: elemento que tiene algunas propiedades de los metales y algunas de los no metales. **(p. 532)**

meteorito: objeto de piedra de tamaño entre arena hasta tamaño de piedra en órbita solar. **(p. 967)**

microscopio: instrumento que usa dos lentes convexos para amplificar objetos pequeños y cercanos. **(p. 418)**

microonda: onda electromagnética con longitud de onda entre aproximadamente 0.1 mm y 30 cm. **(p. 347)**

dorsal oceánica-(MOR): un sistema de cadenas de montañas marcado con un valle larga y extendido alrededor de la Tierra en el fondo marino; formado por las placas oceánicas que se separan debido a la emergencia de magma. **(p. 774)**

Vía Láctea: galaxia espiral que es de unos 100.000 años-luz de diámetro y contiene 200 a 400 mil millones de estrellas, incluyendo el sol. **(p. 997)**

mineral: sólido inorgánico de forma cristalina que se producen naturalmente. **(p. 809)**

espejismo: imagen de un objeto distante producida por la refracción de la luz a través de capas de aire de diferentes densidades. **(p. 372)**

modelo: puede ser usado para representar una idea, objeto o evento que es demasiado grande, demasiado pequeño, demasiado complejo o demasiado peligroso para ser observado o probado directamente. **(p. 12)**

modulación: proceso de agregar una señal a una onda portadora mediante la alteración de la amplitud de la onda portadora, la frecuencia, u otra propiedad. **(p. 352)**

masa molar: la masa en gramos de un mol de una sustancia. **(p. 589)**

mol: unidad SI de cantidad igual a $6,022 \times 10^{23}$ unidades de dicha sustancia. **(p. 588)**

molécula: partícula neutra que se forma al compartir electrones entre átomos. **(p. 561)**

moméntum: propiedad de un objeto en movimiento que es igual a su masa por su velocidad. **(p. 86)**

monómero: pequeña molécula que se puede formar una cadena por combinar consigo misma repetidamente. **(pp. 720, 771)**

fase de la Luna: el cambio de la apariencia de la luna como se ve desde la Tierra, depende de la posición relativa entre el Sol, la Luna y la Tierra. **(p. 931)**

movimiento: un cambio de puesto en relación con un punto de referencia. **(p. 44)**

música: colección de sonidos que se usan deliberadamente en un patrón regular. **(p. 317)**

English	N	Español

nebula: interstellar cloud of gas, ice and dust. **(p. 989)**

Neptune: eighth planet from the Sun; has storms similar to Jupiter's and is blue due to methane in the atmosphere. **(p. 966)**

net force: sum of all of the forces that are acting on an object. **(p. 73)**

neutralization: chemical reaction that occurs when the H_3O^+ ions from an acid react with the OH^- ions from a base to produce water molecules and a salt. **(p. 689)**

neutron: electrically neutral particle inside the nucleus of an atom. **(p. 489)**

neutron star: produced by a collapsing star when protons and electrons in the star's core collide to form neutrons. **(p. 991)**

Newton's first law of motion: states that an object moving at a constant velocity keeps moving at that velocity unless an unbalanced force acts on it. **(p. 80)**

Newton's second law of motion: states that the acceleration of an object is in the same direction as the net force on the object, and that the acceleration equals the net force exerted on it divided by its mass. **(p. 82)**

Newton's third law of motion: states that when one object exerts a force on a second object, the second object exerts a force on the first object that is equal in strength and in the opposite direction. **(p. 84)**

node: a point in a standing wave at which the interfering waves always cancel. **(p. 294)**

nonelectrolyte: substance that does not ionize in water and cannot conduct electricity. **(p. 658)**

nonmetal: element that usually is a gas or brittle solid at room temperature, is not malleable or ductile, is a poor conductor of heat and electricity, and typically is not shiny. **(p. 526)**

nonpolar bond: a covalent bond in which electrons are shared equally by both atoms. **(p. 562)**

nonpolar molecule: molecule that shares electrons equally and does not have oppositely charged ends. **(p. 563)**

nonrenewable resources: natural resources, such as fossil fuels, that cannot be replaced by natural processes as quickly as they are used. **(p. 240)**

nuclear reactor: an apparatus in which controlled nuclear chain reactions generate electricity. **(p. 242)**

nuclear waste: radioactive by-product that results when radioactive materials are used. **(p. 246)**

nucleic acid: essential organic polymer that controls the activities and reproduction of cells. **(p. 728)**

nebula: nube interestelar de gas, hielo y polvo. **(p. 989)**

Neptuno: octavo planeta desde el Sol, tiene tormentas similares a las de Júpiter y es de color azul debido al metano en la atmósfera. **(p. 966)**

fuerza neta: suma de las fuerzas que actúan sobre un objeto. **(p. 73)**

neutralización: reacción química que ocurre cuando los iones H_3O^+ de un ácido reaccionan con los iones OH^- de una base para producir moléculas de agua. **(p. 689)**

neutrón: partícula con carga eléctrica neutral del núcleo de un átomo. **(p. 489)**

estrella de neutrones: producida por una estrella que se colapsa cuando los protones y los electrones colisionaron y se formaron neutrones en el núcleo de la estrella. **(p. 991)**

primera ley del movimiento de Newton: establece que un objeto con una velocidad constante continuará a menos que una fuerza neta se interpone. **(p. 80)**

segunda ley de movimiento de Newton: establece que la aceleración de un objeto es en la misma dirección que la fuerza neta del objeto y que la aceleración es igual a la fuerza neta dividida por su masa. **(p. 82)**

tercera ley de movimiento de Newton: establece que cuando un objeto ejerce una fuerza sobre un segundo objeto, el segundo objeto ejerce una fuerza igual de fuerte sobre el primer objeto y en dirección opuesta. **(p. 84)**

nodo: un punto en una onda estacionaria en la que las olas siempre interfiriendo cancelar. **(p. 294)**

no electrolito: sustancia que no se ioniza en el agua y no puede conducir electricidad. **(p. 658)**

no metal: elemento que por lo general es un gas o un sólido frágil a temperatura ambiente, no es maleable o dúctil, es mal conductor del calor y la electricidad, y por lo general no es brillante. **(p. 526)**

enlace no polar: enlace covalente en el cual los electrones son compartidos por igual por ambos átomos. **(p. 562)**

molécula no polar: molécula que comparte equitativamente los electrones y que no tiene extremos con cargas opuestas. **(p. 563)**

recursos no renovables: recursos naturales, tales como combustibles fósiles, que no son reemplazados por procesos naturales tan pronto como son usados. **(p. 240)**

reactor nuclear: aparato en lo que las reacciones nucleares controladas generan la electricidad. **(p. 242)**

desperdicio nuclear: subproducto radioactivo que resulta del uso de materiales radiactivos. **(p. 246)**

ácido nucleico: polímero orgánico esencial que controla las actividades y la reproducción de las células. **(p. 728)**

Glossary

nucleotide: complex, organic molecule that makes up DNA; contains an organic base, a phosphoric acid unit, and a sugar. (p. 728)

nucleus: the small, positively charged center of an atom, contains protons and neutrons. (p. 489)

nucleótido: molécula orgánica compleja que compone el ADN, contiene una base orgánica, una unidad de ácido fosfórico y un azúcar. (p. 728)

núcleo: el pequeño centro de carga positiva de un átomo, contiene protones y neutrones. (p. 489)

English	O	Español

Ohm's law: states that the current in a circuit equals the voltage difference divided by the resistance. (p. 182)

opaque: material that absorbs or reflects all light and does not transmit any light. (p. 368)

optical axis: imaginary straight line that is perpendicular to the surface at the center of a mirror or lens. (p. 402)

optical scanner: device that reads intensities of reflected light and converts the information to digital signals. (p. 388)

organic compound: one of a large number of compounds that contain the element carbon. (p. 706)

overtone: vibration whose frequency is a multiple of the fundamental frequency. (p. 318)

oxidation: the loss of electrons from the atoms of a substance in a chemical reaction. (p. 593)

oxidation number: positive or negative number that indicates how many electrons an atom has gained, lost, or shared to become stable. (p. 565)

ley de Ohm: establece que la corriente en un circuito es igual a la diferencia de voltaje dividida entre la resistencia. (p. 182)

opaco: material que absorbe o refleja toda la luz pero no la transmite. (p. 368)

eje óptico: línea recta imaginaria que es perpendicular a la superficie en el centro de un espejo o lente. (p. 402)

escáner óptico: dispositivo que lee la intensidad de la luz reflejada y convierte la información en señales digitales. (p. 388)

compuesto orgánico: una de un gran número de compuestos que contiene el elemento carbono. (p. 706)

sobretono: vibración cuya frecuencia es un múltiplo de la frecuencia fundamental. (p. 318)

oxidación: la pérdida de electrones de los átomos de una sustancia en una reacción quimica. (p. 593)

número de oxidación: número positivo o negativo que indica cuántos electrones ha ganado, perdido o compartido un átomo para alcanzar la estabilidad. (p. 565)

English	P	Español

parallel circuit: circuit in which electric current has more than one path to follow. (p. 186)

pascal: SI unit of pressure. (p. 443)

period: horizontal row in the periodic table. (p. 502); the amount of time it takes one wavelength to pass a fixed point; is expressed in seconds. (p.280)

periodic table: organized list of all known elements that are arranged by increasing atomic number and by changes in chemical and physical properties. (p. 498)

petroleum: liquid fossil fuel formed from decayed remains of ancient organisms; can be refined into fuels and used to make plastics. (p. 236)

pH: a measure of the concentration of hydronium ions in a solution using a scale ranging from 0 to 14, with 0 being the most acidic and 14 being the most basic. (p. 686)

photochemical smog: the ozone-containing pollution that results from the reaction between sunlight and vehicular or industrial exhaust. (p. 260)

circuito paralelo: circuito en el cual la corriente eléctrica tiene más de una trayectoria para seguir. (p. 186)

pascal: unidad SI de presión. (p. 443)

período: fila horizontal en la tabla periódica. (p. 502); El tiempo que requiere para que una longitud de onda pase un punto fijo; se expresa en segundos. (p. 280)

tabla periódica: lista organizada de todos los elementos conocidos y ordenados de manera ascendente por número atómico y por cambios en sus propiedades químicas y físicas. (p. 498)

petróleo: combustible fósil líquido que se forma a partir de residuos en descomposición de organismos ancestrales y que puede ser refinado para producir combustibles y fabricar plásticos. (p. 236)

pH: medida de la concentración de iones de hidronio en una solución, usando una escala de 0 a 14, en la cual 0 es la más ácida y 14 la más básica. (p. 686)

smog fotoquímica: la contaminación que incluye ozono y que resulta de la reacción entre la luz solar y el escape vehicular o industriales. (p. 260)

photon: massless energy-containing particle that electromagnetic waves sometimes behave like; the frequency of the electromagnetic wave increases with the energy of the particle. **(p. 342)**

photosphere: layer of the Sun that emits light into space. **(p. 994)**

photovoltaic cell: device that converts solar energy into electricity; also called a solar cell. **(p. 248)**

physical change: any change in size, shape, or state of matter in which the identity of the substance remains the same. **(p. 471)**

physical property: any characteristic of a material, such as size or shape, that can be observed without changing the identity of the material. **(p. 469)**

pigment: colored material that is used to change the color of other substances. **(p. 376)**

pitch: perception of how high or low a sound is; related to the frequency of the sound waves. **(p. 314)**

plane mirror: flat, smooth mirror that reflects light to form upright, virtual images. **(p. 401)**

plasma: matter with enough energy to overcome the attractive forces within its atoms, composed of positively and negatively charged particles. **(p. 436)**

polar bond: a covalent bond in which the electrons are not shared equally, resulting in a slightly positive end and a slightly negative end. **(p. 563)**

polar molecule: a neutral molecule in which unequal electron sharing results a slightly positive end and a slightly negative end. **(p. 563)**

pollutant: any substance that contaminates the environment. **(p. 256)**

polyatomic ion: positively or negatively charged, covalently bonded group of atoms. **(p. 569)**

polyethylene: polymer formed from a chain containing many ethylene units; often used in plastic bags and plastic bottles. **(p. 721)**

polymer: class of natural or synthetic substances made up of many smaller, simpler molecules, called monomers, arranged in large chains. **(pp. 720, 771)**

population: the total number of individuals of one species occupying the same area. **(p. 255)**

pore space: empty space between clasts, can contain water, oil and natural gas. **(p. 825)**

porosity: percentage of a material's total volume of pore space, calculated by dividing total volume of pore space by the total volume of the material. **(p. 866)**

potential energy: energy that is stored due to the interactions between objects. **(p. 117)**

power: the rate at which energy is converted; measured in watts (W). **(p. 126)**

precipitate: insoluble compound that is formed in a solution during a double-displacement reaction. **(p. 592)**

fotón: partícula que contiene energía como la cual algunas veces se comportan las ondas electromagnéticas; la frecuencia de la onda electromagnética aumenta con la energía del fotón. **(p. 342)**

fotosfera: capa del Sol que emite luz hacia el espacio. **(p. 994)**

célula fotovoltaica: dispositivo que convierte la energía solar en electricidad; también llamada celda solar. **(p. 248)**

cambio físico: cualquier cambio en tamaño, forma o estado de una sustancia en la cual la identidad de la sustancia sigue siendo la misma. **(p. 471)**

propiedad física: cualquier característica de un material, tal como tamaño o forma, que se puede observar sin cambiar la identidad del material. **(p. 469)**

pigmento: material de color que se usa para cambiar el color de otras sustancias. **(p. 376)**

tono: percepción de qué tan alto o bajo es un sonido; relacionado a la frecuencia de las ondas sonoras. **(p. 314)**

espejo plano: espejo plano y liso que refleja la luz y forma imágenes verticales y virtuales. **(p. 401)**

plasma: materia con la energía suficiente para superar las fuerzas de atracción entre sus átomos, consiste de partículas con cargas positivas y negativas. **(p. 436)**

enlace polar: enlace en el que los electrones no se comparten por igual, resultando en un lado ligeramente positivo y un lado ligeramente negativo. **(p. 563)**

molécula polar: molécula con un extremo ligeramente positivo y otro ligeramente negativo como resultado de un compartir desigual de los electrones. **(p. 563)**

contaminante: cualquier sustancia que ensucia al medioambiente. **(p. 256)**

ion poliatómico: grupo de átomos enlazados covalentemente, con carga positiva o negativa. **(p. 569)**

polietileno: polímero formado por una cadena que contiene varias unidades de etileno; es comúnmente usado en la fabricación de bolsas y envases plásticos. **(p. 721)**

polímero: clase de sustancias naturales o sintéticas compuestas por muchas moléculas más simples y pequeñas, llamadas monómeros, ordenadas en largas cadenas. **(pp. 720, 771)**

población: el número total de individuos de una especie que ocupa la misma zona. **(p. 255)**

espacio poroso: espacio entre las clásticas, que pueden contener agua, aceite o gas natural. **(p. 825)**

porosidad: porcentaje de espacio poroso en el total del volumen de un material; calculado dividiendo el total del espacio poroso por el volumen total del material. **(p. 866)**

energía potencial: energía almacenada que un objeto tiene debido a las interacciones entre objetos. **(p. 117)**

potencia: tasa de cambio en que la energía se convierte; medida en vatios (W). **(p. 126)**

precipitado: compuesto insoluble que resulta en una solución por medio de una reacción de doble desplazamiento. **(p. 592)**

pressure: amount of force exerted per unit area; SI unit is the pascal (Pa). **(p. 443)**

principle of superposition: states that in an undisturbed sequence of sedimentary rock layers, the youngest rocks will be at the top, and the oldest will be at the bottom. **(p. 870)**

product: in a chemical reaction, the new substance or substances formed. **(p. 582)**

protein: large, complex, biological polymer formed from amino acid units; make up many body tissues such as muscles, tendons, hair, and fingernails. **(p. 724)**

proton: particle in the nucleus with an electric charge of 1+. **(p. 489)**

presión: cantidad de fuerza ejercida por unidad de área; la unidad SI es el pascal (Pa). **(p. 443)**

principio de superposición: establece que en las capas de sedimentos de rocas las capas más recientes estarán arriba y las capas más antiguas abajo. **(p. 870)**

producto: en una reacción química, la sustancia formada o las sustancias formados. **(p. 582)**

proteína: polímero biológico extenso y complejo formado por unidades de aminoácidos; conforma muchos tejidos del cuerpo como los músculos, los tendones, el pelo y las uñas. **(p. 724)**

protón: partícula en el núcleo con una carga eléctrica de 1+. **(p. 489)**

English	Q	Español

quark: particle of matter that makes up protons and neutrons. **(p. 489)**

quark: partícula de materia que constituye los protones y neutrones. **(p. 489)**

English	R	Español

radiant energy: energy carried by an electromagnetic wave. **(p. 357)**

radiation: transfer of energy by electromagnetic waves. **(p. 147)**

radioactive element: element, such as radium, whose nucleus breaks down and emits particles and energy. **(p. 520)**

radioactivity: process that occurs when a nucleus decays and emits matter and energy. **(p. 620)**

radio telescope: telescope that collects and magnifies radio waves. **(p. 987)**

radio wave: electromagnetic wave with wavelength longer than about 10 cm, used for communications. **(p. 345)**

rarefaction: the less-dense region of a longitudinal wave. **(p. 279)**

reactant: in a chemical reaction, the substance that reacts. **(p. 582)**

reaction rate: the rate at which reactants change into products in a chemical reaction. **(p. 598)**

real image: an image that appears at a certain location as a result of rays of light converging at that location. **(p. 403)**

reduction: the gain of electrons by the atoms of a substance in a chemical reaction. **(p. 593)**

reflecting telescope: uses mirrors and lenses to collect and focus light from distant objects. **(pp. 416, 985)**

energía radiante: energía transportada por una onda electromagnética. **(p. 357)**

radiación: transferencia de energía mediante ondas electromagnéticas. **(p. 147)**

elemento radiactivo: elemento, como el radio, cuyo núcleo se divide y emite partículas y energía. **(p. 520)**

radiactividad: proceso que ocurre cuando un núcleo se descompone y emite materia y energía. **(p. 620)**

radio telescopio: telescopio que recoge y amplifica las ondas de radio. **(p. 987)**

onda de radio: onda electromagnética con longitud de onda más larga de aproximadamente 10 cm y que se usa en las comunicaciones. **(p. 345)**

rarefacción: la región menos densa de una onda longitudinal. **(p. 279)**

reactante: la sustancia que reacciona en una reacción química. **(p. 582)**

velocidad de reacción: velocidad en que se transforman los reactivos en productos en una reacción química. **(p. 598)**

imagen real: imagen que aparece en un cierto lugar como consecuencia de rayos de luz que convergen en ese lugar. **(p. 403)**

reducción: la obtención de electrones por los átomos de una sustancia en una reacción química. **(p. 593)**

telescopio reflexivo: usa espejos y lentes para recolectar y enfocar la luz proveniente de objetos distantes. **(pp. 416, 985)**

Glossary

refracting telescope: uses lenses to gather and focus light from distant objects. **(pp. 416, 985)**

refraction: the bending of a wave caused by a change in its speed as it travels from one medium to another. **(p. 288)**

regolith: layer of debris on the Moon's surface formed by the accumulation of meteoric material. **(p. 936)**

relative dating: process of placing objects or events in order in time according to which occurred or formed first, second, and so on. **(p. 870)**

renewable resource: energy source that is replaced by natural processes faster than it is used. **(p. 248)**

resistance: tendency for a material to oppose electron flow and to convert electrical energy into other forms of energy, such as thermal energy and light; measured in ohms (Ω). **(p. 181)**

resonance: the process by which an object is made to vibrate by absorbing energy at its natural frequencies. **(p. 295)**

resonator: hollow, air-filled chamber that amplifies sound when the air inside it vibrates. **(p. 319)**

retina: inner lining of the eye that has cells which convert light images into electric signals for interpretation by the brain. **(p. 411)**

reversible reaction: a reaction that can proceed in both the forward and the reverse directions. **(p. 601)**

revolution: motion of Earth around the Sun, used to measure time in years. **(p. 924)**

rift valley: long narrow depression formed in between the peaks along the mid-oceanic ridge. **(p. 774)**

rock: naturally formed mixture of minerals, rock fragments or volcanic glass, bound together, identified based on composition and texture. **(p. 817)**

rock cycle: the continual changing of rocks into different types through processes such as high temperature and pressure, weathering, erosion and sedimentation. **(p. 833)**

rotation: spinning of Earth on its axis; used to measure time in days. **(p. 924)**

telescopio refractivo: usa lentes para reunir y enfocar la luz proveniente de objetos distantes. **(pp. 416, 985)**

refracción: el cambio en dirección de una ola debido a un cambio en su velocidad por viajar de un medio a otro. **(p. 288)**

regolito: capa de escombros en la superficie de la Luna, formada por la acumulación de material meteórico. **(p. 936)**

datación relativa: proceso de ordenar cronológicamente objetos o eventos según estos ocurrieron ya sea primero segundo etc. **(p. 870)**

recursos renovables: fuente de energía que es se reemplaza más rápido de que se consume. **(p. 248)**

resistencia: tendencia de un material de oponerse al fluido de los electrones y convertir la energía eléctrica en energía térmica y luz; se mide en ohmios (Ω). **(p. 181)**

resonancia: el proceso por el cual un objeto vibra al absorber energía en sus frecuencias naturales. **(p. 295)**

resonador: cámara hueca, llena de aire, que amplifica el sonido cuando vibra el aire en su interior. **(p. 319)**

retina: capa interna del ojo que posee células que convierten imágenes iluminadas en señales eléctricas para que el cerebro las interprete. **(p. 411)**

reacción reversible: una reacción que puede proceder tanto en el avance que en la dirección retroceso. **(p. 601)**

revolución: movimiento de la Tierra alrededor del Sol, sirve para medir el tiempo en años. **(p. 924)**

valle del rift: surco central formado entre los picos a lo largo de la cresta de la dorsal oceánica. **(p. 774)**

roca: mezcla naturalmente formada de minerales, fragmentos de roca o vidrio volcánico mezclado, identificable por medio su composición y textura. **(p. 817)**

ciclo de las rocas: el continuo proceso de cambio de las rocas a través de diferentes procesos como las altas temperaturas, presión, meteorización, erosión y sedimentación. **(p. 833)**

rotación: giro de la Tierra en su eje, se utiliza para medir el tiempo en días. **(p. 924)**

English	S	Español

salt: compound formed when negative ions from an acid combine with positive ions from a base. **(pp. 580, 689)**

saturated hydrocarbon: hydrocarbon, such as propane or methane, in which all the carbon atoms are connected by single covalent bonds. **(p. 708)**

saturated solution: any solution that contains all the solute it can hold at a given temperature. **(p. 655)**

sal: compuesto iónico que se forma cuando un halógeno adquiere un electrón de un metal. **(pp. 580, 689)**

hidrocarburo saturado: hidrocarburo, como el propano y el metano, en la que todos los átomos de carbono están unidos por enlaces covalentes simples. **(p. 708)**

solución saturada: cualquier solución que contiene todo el soluto que puede retener a una temperatura determinada. **(p. 655)**

Saturn: sixth planet from the Sun; the second-largest planet with the most complex ring system. **(p. 965)**

scientific law: statement about what happens in nature that seems to be true all the time; does not explain why or how something happens. **(p. 13)**

scientific method: pattern of investigation procedures that can include stating a problem, forming a hypothesis, researching and gathering information, testing a hypothesis, analyzing data, and drawing conclusions. **(p. 8)**

sea breeze: local wind created by temperature and pressure differences between ocean and land; wind blows toward land in the afternoon, and away from land at night. **(p. 899)**

second law of thermodynamics: states that energy spontaneously spreads from regions of higher concentration to regions of lower concentration. **(p. 155)**

sediment transport: process of moving eroded materials from one place to another through the movement of water, ice, wind and due to gravity. **(p. 854)**

Sedna: distant planetoid, smaller than Pluto, larger than comets, with a very elliptical orbit. **(p. 968)**

semiconductor: material that conducts an electric current under certain conditions. **(pp. 534)**

series circuit: circuit in which electric current has only one path to follow. **(p. 185)**

shadow zone: area on Earth's surface where no seismic waves from a given Earthquake are recorded. **(p. 788)**

shield volcano: large, broad, flat volcano composed of layer upon layer of basaltic lava flows. **(p. 795)**

SI: International System of Units—the improved, universally accepted version of the metric system that is based on multiples of ten and includes the meter (m), liter (L), and kilogram (kg). **(p. 15)**

silica: chemical compound, silica dioxide (SiO_2), a common ingredient in most magma and much of Earth's crust. **(p. 793)**

simple machine: machine that does work with only one movement; examples include lever, pulley, wheel and axle, inclined plane, screw, and wedge. **(p. 109)**

single-displacement reaction: chemical reaction in which one element replaces another element in a compound. **(p. 591)**

sliding friction: frictional force that opposes the motion of two surfaces sliding past each other. **(p. 75)**

soap: organic salt with a nonpolar, hydrocarbon end that interacts with oils and dirt and a polar end that causes it to dissolve in water. **(pp. 694)**

Saturno: sexto planeta desde el Sol; el segundo planeta más grande con el sistema de anillos más complejos. **(p. 965)**

ley científica: enunciado acerca de lo que ocurre en la naturaleza, lo cual parece ser cierto en todo momento, no explica cómo o por qué algo ocurre. **(p. 13)**

método científico: patrón organizado de procedimientos de investigación que puede incluir el planteamiento de un problema, formulación de una hipótesis, investigación y recopilación de información, comprobación de la hipótesis, análisis de datos y elaboración de conclusiones. **(p. 8)**

brisa del mar: viento local creado por las diferencias de temperatura y de presión entre el océano y la tierra; el viento sopla hacia la tierra en la tarde, y fuera de la tierra en la noche. **(p. 899)**

segunda ley de la termodinámica: afirma que la energía de forma espontánea se extiende desde las regiones de mayor concentración a regiones de menor concentración. **(p. 155)**

transporte de sedimento: movimiento del material erosionado de un lugar a otro por medio del los movimientos de agua, hielo, viento o gravedad. **(p. 854)**

Sedna: planetoide lejano, más pequeño que Plutón, más grande que los cometas, con una órbita muy elíptica. **(p. 968)**

semiconductor: los material que conduce la corriente eléctrica bajo ciertas condiciones. **(pp. 534)**

circuito en serie: circuito en el cual la corriente eléctrica tiene una sola trayectoria para seguir. **(p. 185)**

zona de sombra: área en la superficie de la tierra en donde no se registran ondas sísmicas de un terremoto dado. **(p. 788)**

volcán en escudo: volcán grande y amplio, formado por numerosas capas de lava. **(p. 795)**

SI: Sistema Internacional de Unidades: la versión del sistema métrico mejorada y aprobada universalmente por científicos que se basa en múltiples de diez e incluye el metro (m), el litro (L) y el kilogramo (Kg). **(p. 15)**

sílice: compuesto químico, dióxido de silicio (SiO_2), es abundante en el magma y en la mayoría de la corteza terrestre. **(p. 793)**

máquina simple: máquina que realiza el trabajo con un solo movimiento; ejemplos incluye palanca, polea, rueda y eje, plano inclinado, tornillo y cuña. **(p. 109)**

reacción de un solo desplazamiento: reacción química en la cual un elemento reemplaza a otro elemento en un compuesto. **(p. 591)**

fricción deslizante: fuerza de fricción que se opone al movimiento de dos superficies que se deslizan entre sí. **(p. 75)**

jabón: sal orgánica con un extremos de hidrocarburo no polar que interactúa con aceites y suciedad, y un extremo polar que ayuda a disolverlo en agua. **(pp. 694)**

Glossary

Glossary

society: group of people that share similar values and beliefs. **(p. 30)**

soil: mixture of weathered rock, organic matter, water, and air that is capable of supporting plant life. **(p. 850)**

solar collector: device used in an active solar heating system that transforms radiant energy from the Sun into thermal energy. **(p. 153)**

solar eclipse: occurs during the new moon, when Earth enters the Moon's umbra and casts a shadow on the Earth's surface. **(p. 934)**

solenoid: a cylindrical coil of wire, used to produce a magnetic field when an electrical current passes through the wire. **(p. 210)**

solstice: occurs twice a year, in June and December, when Earth's rotational axis is tilted the most toward, or the most away, from the Sun. **(p. 926)**

solubility: maximum amount of a solute that can be dissolved in a given amount of solvent at a given temperature. **(p. 654)**

solute: in a solution, the substance being dissolved. **(p. 647)**

solution: homogenous mixture, remains constantly and uniformly mixed and has particles that are so small they cannot be seen with a microscope. **(pp. 467)**

solvent: in a solution, the substance in which the solute is dissolved. **(p. 647)**

sonar: system that uses the reflection of sound waves to detect objects underwater. **(p. 326)**

sound quality: result of the differences between sounds having the same pitch and loudness. **(p. 318)**

specific heat: amount of heat needed to raise the temperature of 1 kg of a material 1°C. **(p. 141)**

spectroscope: device that disperses light into its component wavelengths, using a prism or diffraction grating. **(p. 988)**

speed: distance an object travels per unit of time. **(p. 46)**

sphere: three-dimensional, round object whose surface is the same distance from the center in every direction. **(p. 918)**

standard: exact, agreed-upon quantity used for comparison. **(p. 14)**

standing wave: a wave pattern that forms when waves of equal wavelength and amplitude, but traveling in opposite directions, continuously interfere with each other; does not appear to be travelling. **(p. 294)**

static electricity: the accumulation of excess electric charge on an object. **(p. 170)**

static friction: frictional force that prevents two surfaces from sliding past each other. **(p. 75)**

sociedad: grupo de personas que comparten valores y creencias similares. **(p. 30)**

suelo: mezcla de piedras, materia orgánica agua y aire; capaz de mantener vida vegetal. **(p. 850)**

recolector solar: dispositivo usado en un sistema de calefacción solar activa que transforma la energía radiante del Sol en energía térmica. **(p. 153)**

eclipse solar: se produce durante la luna nueva, cuando la Tierra entra en la sombra de la Luna y proyecta una sombra sobre la superficie de la Tierra. **(p. 934)**

solenoide: bonina cilíndrica de alambre, utilizado para producir un campo eléctrico cuando pasa una corriente eléctrica. **(p. 210)**

solsticio: producido dos veces al año, en junio y en diciembre, cuando el eje de la rotación de la Tierra se inclina más hacia, o la más lejos, al sol. **(p. 926)**

solubilidad: máxima cantidad de soluto que puede ser disuelto en una cantidad dada de solvente a una temperatura determinada. **(p. 654)**

soluto: en una solución, la sustancia que está disuelta. **(p. 647)**

solución: mezcla homogénea, permanece constante y uniformemente mezclada y tiene partículas tan pequeñas que no pueden ser vistas en un microscopio. **(pp. 467)**

solvente: en una solución, la sustancia en la cual se disuelve el soluto. **(p. 647)**

sonar: sistema que usa la reflexión de las ondas sonoras para detectar objetos bajo el agua. **(p. 326)**

timbre: resulto de las diferencias entre sonidos del mismo tono e intensidad sonora. **(p. 318)**

calor específico: cantidad de calor necesaria para aumentar la temperatura un grado centígrado en un kilogramo de material. **(p. 141)**

espectroscopio: dispositivo que dispersa la luz por sus longitudes de onda, utiliza un prisma o una rejilla de difracción. **(p. 988)**

velocidad: distancia que recorre un objeto por unidad de tiempo. **(p. 46)**

esfera: objeto redondo en tres dimensiones, cuya superficie es la misma distancia del centro en todas direcciones. **(p. 918)**

estándar: cantidad exacta y acordada, usada para hacer comparaciones. **(p. 14)**

onda estacionaria: patrón de una onda que se forma cuando ondas con la misma longitud de onda y amplitud, pero que viajan en direcciones opuestas, interfieren continuamente entre sí; parece que no se mueve. **(p. 294)**

electricidad estática: la acumulación del exceso de carga eléctrica en un objeto. **(p. 170)**

fricción estática: fuerza que evita que dos superficies en contacto se deslicen una sobre otra. **(p. 75)**

streak: color of a mineral in its powdered form. **(p. 810)**

strong acid: any acid that dissociates almost completely in solution. **(p. 684)**

strong base: any base that dissociates completely in solution. **(p. 685)**

strong force: attractive force that acts between protons and neutrons in an atomic nucleus. **(p. 618)**

subduction: the movement of a dense oceanic plate under a buoyant continental plate. **(777)**

sublimation: the process of a solid changing directly to a vapor without forming a liquid. **(p. 435)**

substance: element or compound that cannot be broken down into simpler components without losing the properties of the original substance. **(p. 462)**

substituted hydrocarbon: hydrocarbon with one or more of its hydrogen atoms replaced by atoms, or groups of atoms, of other elements. **(p. 712)**

sunspots: darker, cooler areas of the Sun's photosphere. **(p. 994)**

supernova: gigantic explosion of a star in which the temperature within the collapsing star reaches 10 billion K, can evolve into a neutron star. **(p. 991)**

supersaturated solution: any solution that contains more solute than a saturated solution at the same temperature. **(p. 656)**

suspension: heterogeneous mixture containing a liquid, and in which visible particles slowly settle due to gravity. **(p. 466)**

synthesis reaction: chemical reaction in which two or more substances combine to form a different substance. **(p. 591)**

synthetic: a material that is made in a laboratory or chemical plant and does not occur naturally. **(p. 753)**

system: a region or set of regions around which a boundary can be defined. **(p. 114)**

raya: el color del polvo de un mineral. **(p. 810)**

ácido fuerte: cualquier ácido que se disocie casi por completo en una solución. **(p. 684)**

base fuerte: cualquier base que se disocie completamente en una solución. **(p. 685)**

interacción nuclear fuerte: fuerza de atracción que mantiene juntos los protones y neutrones en un núcleo atómico. **(p. 618)**

subducción: el movimiento de una placa oceánica bajo una densa placa continental boyante. **(777)**

sublimación: proceso mediante el cual un sólido se convierte directamente en vapor sin pasar por el estado líquido. **(p.435)**

sustancia: elemento o compuesto que no se puede descomponer en componentes más simples sin perder las propiedades de la sustancia original. **(p. 462)**

hidrocarburo sustituido: un hidrocarburo en el cual uno o más de sus átomos de hidrógeno son reemplazados por un átomo, o grupo de átomos, de otros elementos. **(p. 712)**

manchas solares: áreas más oscuro, más frías de la fotosfera del sol. **(p. 994)**

supernova: explosión de una estrella gigantesca en la que la temperatura dentro de la estrella que se colapsa alcanza 10 mil millones K, puede convertirse en una estrella de neutrones. **(p. 991)**

solución sobresaturada: cualquier solución que contenga más soluto que una solución saturada a la misma temperatura. **(p. 656)**

suspensión: mezcla heterogénea que contiene un líquido, y en el cual las partículas visibles lentamente se sedimentan. **(p. 466)**

reacción síntesis: reacción química en la cual se combinan dos o más sustancias y forman una sustancia diferente. **(p. 591)**

sintético: un material que se realiza en un laboratorio of fábrica de productos químicos y no se produce de forma natural. **(p. 753)**

sistema: una región o conjunto de regiones alrededor de lo cual se puede distinguir con un límite. **(p. 114)**

English	T	Español

technology: application of science to benefit people. **(p. 22)**

temperature: measure of the average kinetic energy of all the particles that make up an object. **(p. 139)**

temperature inversion: describes the increasing temperature of air with increasing altitude. **(p. 885)**

terminal velocity: the maximum speed an object will reach when falling through a substance, such as air. **(p. 88)**

tecnología: aplicación de la ciencia para el beneficio de la población. **(p. 22)**

temperatura: medida de la energía cinética promedio de todas las partículas que componen un objeto. **(p. 139)**

inversión térmica: describe el incremento de la temperatura del aire con el incremento de la altitud. **(p. 885)**

velocidad límite: la velocidad máxima que un objeto puede alcanzar cuando pasa en caída libre por una sustancia como aire. **(p. 88)**

Glossary

Glossary

theory: explanation of things or events based on knowledge gained from many observations and investigations. **(p. 13)**

thermal energy: sum of the kinetic and potential energy of the particles that make up an object. **(p. 139)**

thermal expansion: increase in the volume of a substance when the temperature is increased. **(p. 437)**

thermal insulator: a material through which thermal energy moves slowly. **(p. 149)**

thermodynamics: study of the relationship between thermal energy, heat, and work. **(p. 154)**

tide: the periodic rise and fall of sea level caused by the gravitational attraction among Earth, the Moon, and the Sun. **(p. 930)**

time zone: any one of 24 longitudinal strips on Earth about 15° wide on Earth's surface in which local time is the same. **(p. 923)**

titration: process in which a solution of known concentration is used to determine the concentration of another solution. **(p. 692)**

total internal reflection: the complete reflection of light at a boundary that occurs when light strikes at an angle greater than the critical angle, and light travels faster in the second medium than in the first medium. **(p. 386)**

tracer: radioactive isotope, such as iodine-131, that can be detected by the radiation it emits after it is absorbed by a living organism. **(p. 630)**

transceiver: device that transmits radio signals at one frequency and receives radio signals at a different frequency, allowing a user to talk and listen at the same time. **(p. 355)**

transform plate boundary: tectonic plate boundary in which plates slide horizontally past each other in opposite directions. **(p. 778)**

transformer: device that uses electromagnetic induction to increase or decrease the voltage of an alternating current. **(p. 220)**

transition elements: elements in groups 3 through 12 of the periodic table; occur in nature as uncombined elements and include the iron triad and coinage metals. **(p. 522)**

translucent: material that transmits and scatters light so that objects viewed through it appear blurry. **(p. 368)**

transmutation: process of changing one element to another through radioactive decay. **(p. 623)**

transparent: material that transmits light without scattering so that objects are clearly visible through it. **(p. 368)**

transuranium elements: elements having more than 92 protons, all of which are synthetic and unstable. **(p. 539)**

teoría: explicación de las objetos o eventos que se basa en el conocimiento obtenido a partir de numerosas observaciones e investigaciones. **(p. 13)**

energía térmica: suma de la energía cinética y potencial de las partículas que componen un objeto. **(p. 139)**

expansión térmica: aumento del volumen de una sustancia al aumentar la temperatura. **(p. 437)**

aislante térmico: material a través del cual se mueve la energía térmica muy lentamente. **(p. 149)**

termodinámica: estudio de la relación entre la energía térmica, el calor y el trabajo. **(p. 154)**

marea: los ascensos y descensos periódicos del nivel del mar causado por la atracción gravitatoria entre la Tierra, la Luna y el Sol. **(p. 930)**

zona horaria: uno de 24 franjas de la Tierra alrededor de 15° de ancho en la superficie de la Tierra en el que la hora local es la misma. **(p. 923)**

titulación: proceso mediante el cual una solución con una concentración conocida es usada parea determinar la concentración de otra solución. **(p. 692)**

reflexión interna total: la reflexión completa de la luz en un límite que se produce cuando la luz incide en un ángulo mayor que el ángulo crítico, y la luz viaja más rápido en el segundo medio que en el primer medio. **(p. 386)**

indicador radiactivo: isótopo radioactivo, tal como el yodo-131, que se detecta por la radiación que emite después de ser absorbido por un organismo vivo. **(p. 630)**

radio transmisor-receptor: dispositivo que transmite señales de radio a una frecuencia y recibe señales de radio en una frecuencia diferente, lo que permite al usuario hablar y escuchar al mismo tiempo. **(p. 355)**

límite de transformación de placa: borde tectónica en la que las placas se deslizan horizontalmente entre sí en direcciones opuestas. **(p. 778)**

transformador: dispositivo que usa inducción electromagnética para aumentar o disminuir el voltaje de una corriente alterna. **(p. 220)**

elementos de transición: los elementos de los grupos 3 al 12 de la tabla periódica que se encuentran en la naturaleza como elementos sin combinar e incluyen la tríada de hierro y los metales con los que se fabrican las monedas. **(p. 522)**

translúcido: material que transmite y dispersa la luz para que los objetos vistos a través de ella se vean borrosos. **(p. 368)**

transmutación: proceso de cambio de un elemento a otro mediante la descomposición radioactiva. **(p. 623)**

transparente: material que transmite luz sin dispersarse de manera que los objetos son claramente visibles a través de ella. **(p. 368)**

elementos transuránicos: elementos con más de 92 protones, que son sintéticos e inestables. **(p. 539)**

transverse wave: wave in which the matter in the medium moves at right angles to the direction of the wave, has crests and troughs. **(p. 276)**

troposphere: lowest layer of the atmosphere, extending 12 km above Earth's surface, layer where weather takes place. **(p. 885)**

trough: the lowest point on a transverse wave. **(p. 279)**

turbine: large wheel that rotates when pushed by steam, wind, or water and provides mechanical energy to a generator. **(p. 218)**

Tyndall effect: tendency for a beam of light to scatter as it passes through a colloid. **(p. 467)**

onda transversal: onda por la cual la materia en el medio se mueve en ángulos rectos con respecto a la dirección en que viaja la onda; tiene crestas y depresiones. **(p. 276)**

troposfera: capa más baja de la atmosfera; se extiende por 12 km sobre la superficie de la tierra, es la capa donde ocurre el clima. **(p. 885)**

valle: el punto más bajo de una onda transversal. **(p. 279)**

turbina: rueda grande que gira al ser impulsada por vapor, viento o agua y que suministra energía mecánica a un generador. **(p. 218)**

efecto Tyndall: tendencia de un rayo de luz para dispersar al pasar a través de un coloide. **(p. 467)**

English	U	Español

ultrasound: sound waves with frequency above 20,000 Hz; cannot be heard by humans. **(p. 326)**

ultraviolet wave: electromagnetic wave with wavelength between about 400 billionths and 10 billionths of a meter. **(p. 349)**

unconformity: gap in the rock record representing a time during with exposed rock was eroded before new layers were deposited. **(p. 871)**

uniformitarianism: states that the laws of nature operate today as they have in the past. **(p. 870)**

unsaturated hydrocarbon: hydrocarbon, such as ethene or ethyne, that contains at least one double or triple bond between carbon atoms. **(p. 709)**

unsaturated solution: any solution that can dissolve more solute at a given temperature. **(p. 655)**

Uranus: seventh planet from the Sun; appears blue-green due to methane in the atmosphere; its axis of rotation is tilted on its side. **(p. 966)**

ultrasonido: onda de sonido con frecuencia superior a 20,000 Hz, no percibido por seres humanos. **(p. 326)**

onda ultravioleta: ondas electromagnética con longitud de onda entre aproximadamente 10 y 400 billonésimas de metro. **(p. 349)**

disconformidad: evidencia que la roca fue expuesta a erosión antes de que nuevas capas fuesen depositadas. **(p. 871)**

uniformitarismo: establece que las leyes de la naturaleza operan hoy de la misma forma que operaban en el pasado. **(p. 870)**

hidrocarburo no saturado: hidrocarburo, como el etileno, que contiene al menos un enlace doble o triple entre los átomos de carbono. **(p. 709)**

solución no saturada: cualquier solución que puede disolver más soluto a una temperatura determinada. **(p. 655)**

Urano: séptimo planeta desde el Sol, aparece de color azul-verde debido al metano en la atmósfera, su eje de rotación inclina a su lado. **(p. 966)**

English	V	Español

variable: quantity that can have more than a single value, can cause a change in the results of an experiment. **(p. 9)**

velocity: the speed and direction of a moving object. **(p. 51)**

Venus: second planet from the Sun; has a dense atmosphere of mostly carbon dioxide and very high surface temperatures. **(p. 956)**

virtual image: an image formed by diverging light rays that is perceived by the brain, even though the light rays do not actually originate from the place where the image appears to be located. **(p. 402)**

viscosity: a fluid's resistance to flowing. **(pp. 446, 793)**

variable: una cantidad que puede tener más de un valor, puede causar un cambio en los resultados de un experimento. **(p. 9)**

velocidad direccional: la rapidez y dirección de un objeto en movimiento. **(p. 51)**

Venus: segundo planeta desde el Sol, tiene una densa atmósfera compuesta principalmente por dióxido de carbono y tiene las temperaturas de superficie muy elevada. **(p. 956)**

imagen virtual: una imagen formada por las distintas rayos de luz que son percibidas por el cerebro, a pesar de que los rayos de luz en realidad no originan donde la imagen parece que se encuentra. **(p. 402)**

viscosidad: resistencia de un fluido al flujo. **(pp. 446, 793)**

Glossary

Glossary

visible light: electromagnetic waves with wavelengths of 700 to 400 billionths of a meter that can be detected by human eyes. **(p. 348)**

voltage difference: related to the force that causes electric charges to flow; measured in volts (V). **(p. 178)**

volume: amount of space occupied by an object. **(p. 17)**

luz visible: ondas electromagnéticas con longitudes de onda entre 400 y 700 billonésimas de metro y que pueden ser detectadas por el ojo humano. **(p. 348)**

diferencia de voltaje: relacionados con la fuerza que hace que las cargas eléctricas a fluir; se mide en voltios (V). **(p. 178)**

volumen: cantidad de espacio ocupado por un objeto. **(p. 17)**

English	W	Español

water table: the upper boundary of the saturated zone in an aquifer. **(p. 864)**

wave: a repeating disturbance that transfers energy as it travels through matter or space. **(p. 274)**

wavelength: distance between one point on a wave and the nearest point just like it. **(p. 280)**

weak acid: any acid that only partly dissociates in solution. **(p. 684)**

weak base: any base that does not dissociate completely in solution. **(p. 685)**

weather front: zone along which two or more air masses interact, producing certain weather conditions. **(p. 892)**

weathering: process that involves the physical or chemical breakdown of materials on Earth's surface. **(p. 846)**

weight: gravitational force exerted on an object. **(p. 78)**

white dwarf: giant star that has lost its outer layers, leaving behind a hot, dense core that continues to contract under gravity. **(p. 991)**

work: transfer of energy when a force is applied over a distance; measured in joules. **(p. 106)**

capa freática: la parte superior de la zona saturada de un manto acuífero **(p. 864)**

onda: alteración repetitiva que transfiere energía a través de la materia o el espacio. **(p. 274)**

longitud de onda: distancia entre un punto en una onda y el semejante punto más cercano. **(p. 280)**

ácido débil: cualquier ácido que solamente se disocie parcialmente en una solución. **(p. 684)**

base débil: cualquier base que no se disocie completamente en una solución. **(p. 685)**

frente: la zona a lo largo de dos o más masas de aire que interactúan; produce condiciones meteorológicas específicas. **(p. 892)**

meteorización: proceso que incluye el desgaste de los materiales en la superficie de la tierra por causas químicas o físicas. **(p. 846)**

peso: fuerza gravitacional ejercida sobre un objeto. **(p. 78)**

enana blanca: estrella gigante que ha perdido sus capas exteriores, y existe como un núcleo denso y caliente que sigue contraccionando por la gravedad. **(p. 991)**

trabajo: transferencia de energía cuando una fuerza actúa por una distancia y que se mide en julios. **(p. 106)**

English	X	Español

X-ray: electromagnetic wave with wavelength between about 10 billionths of a meter and 10 trillionths of a meter, often used for medical imaging. **(p. 350)**

rayo X: onda electromagnética con longitud de onda entre 10 billonésimas de metro y 10 trillonésimas de metro, la cual se utiliza con frecuencia para producir imágenes de uso médico. **(p. 350)**

Glossary

Bolded page numbers refer to vocabulary terms. The following abbreviations appear after page numbers.

Activities and labs = *act.*
Illustrations/photographs = *illus.*
Problems = *prob.*
Tables = *table*

B

C

T

Photo Credits

COVER i Koji Kitagawa/SuperStock; **v** Laura Doss/CORBIS; **2-3** Darryl Leniuk/Stone/Getty Images; **4-5** Tony Hertz/age fotostock; **6** Olivier Grunewald/Getty Images; **9** NASA; **10** David Herbig/Danita Delimont/Alamy; **11** Terry Vine/Getty Images; **12** Norbert Michalke/age fotostock; **13** Lars Borges/Getty Images; **14** Matt Meadows; **16** Amanita Pictures; **17** (l)Nick Kirk/Alamy Images, (r)Stockbyte; **19** (t to b)CORBIS/age footstock, Photodisc/Getty Images, CORBIS, Comstock Images/PictureQuest, Ingram Publishing/Fotosearch, Matt Meadows, Liane Cary/age fotostock, Matt Meadows, Matt Meadows; **26** Brandi Simons/Getty Images; **27** (t)CORBIS, (c)Stockbyte/Getty Images, (bl)DEA/G. DAGLI ORTI/Getty Images, (br)marka/scataglini/age fotostock; **28** Harry Hook/Getty Images; **29** Nancy Ney/Getty Images; **30** Jeff Kowalsky/Bloomberg via Getty Images; **31** Spencer Grant/age fotostock; **32** IMAGEMORE Co., Ltd./Getty Images; **34 35** Matt Meadows; **36** Sheila Terry/Photo Researchers, Inc.; **38** Amanita Pictures; **41** Stockbyte/Getty Images; **42-43** David Madison/Photodisc/Getty Images; **44** Mark Ransom; **49** Ryan McGinnis/Getty Images; **51** Jonathan Ferrey/Getty Images; **52** Comstock Images/Getty Images; **55** (l)Lorcan/Digital Vision/Getty Images, (r)Sam Clemens/Getty Images; **59** liquidlibrary/PictureQuest; **60** Richard Megna/Fundamental Photographs; **61** Horizons Companies; **62** Rim Light/PhotoLink/Getty Images; **64** AP Photo/Damian Dovarganes; **70-71** Stephen Dalton/Photo Researchers, Inc.; **72** Lawrence Manning/CORBIS; **73** Tim Courlas/Horizons Companies; **74** (t)Michael W. Davidson/Photo Researchers, (b)PhotoDisc; **75** (t)Matt Meadows, (b)Ultimate Group, LLC/Alamy; **79** NASA; **80** Matt Meadows; **81** (l)Lee Strickland/Getty Images, (r)Donald Miralle/Getty Images; **82** (l)Nicolas Russell/Getty Images, (r)Ken Karp/McGraw-Hill Companies; **84** Matt Meadows; **85** Frederic J. Brown/AFP/Getty Images; **86** (l)TRL Ltd./Photo Researchers, Inc., (r)Fstop/Getty Images; **87** Erich Schrempp/Photo Researchers, Inc.; **88** Matt Meadows; **90** Bernard Wolff/Photo Researchers, Inc.; **91** (t)Matt Meadows, (b)David Leah/Getty Images; **92** NASA; **94 95** Mark Ransom; **96** NASA; **102-103** Ashley Cooper/CORBIS; **104-105** Russ Curtis/Photo Researchers, Inc.; **106** Purestock/age fotostock; **108** Dennis Macdonald/age fotostock; **109** (tl)The McGraw-Hill Companies, (tr)Martina Berg/age fotostock, (c)Paul Giamou/Getty Images, (bl)Steve Cole/Photodisc/Getty Images, (bc)Hanson Ng/age fotostock, (br)Martin Poole/Getty Images; **111** (t)Dennis Welsh/Getty Images, (c)Mark Douet/Getty Images, (b)Irina Drazowa-fischer/age fotostock; **113** Matt Meadows; **114** Amoz Eckerson/Visuals Unlimited, Inc.; **115** (l)Comstock Images/Alamy, (tr)Pixel Shepherd/age fotostock, (br)Brand X Pictures/PunchStock; **118** Matt Meadows; **120** Shenval/Alamy Images; **121** Pixtal/age fotostock; **123** The Columbus Dispatch; **129** Matt Meadows; **130** (t)AP Photo/Roland Weihrauch, (b)David McNew/Getty Images; **132** The Columbus Dispatch; **135** Michael Newman/PhotoEdit; **136-137** Birger Lallo/Alamy; **138** The McGraw-Hill Companies; **140** David Chasey/Getty Images; **145** Samuel T. Kendall; **146** (t)Jules Frazier/Getty Images, (b)NASA/Goddard Space Flight Center Scientific Visualization Studio; **148** (l)Paul Souders/Getty Images, (c)CORBIS, (r)age fotostock/SuperStock; **149** Garry Wade/Getty Images; **151** Geostock/Getty Images; **153** (t)Jean-Yves Bruel/Getty Images, (b)Bruce Hands/Getty Images; **154** CORBIS/age fotostock; **155** Getty Images/Flickr RM; **159** Messerschmidt/Getty Images; **161** Tim Courlas/Horizons Companies; **164** CORBIS/age fotostock; **166** David Chasey/Getty Images; **168-169** Mark Karrass/CORBIS; **170** Geoff Butler; **171 173 174** Matt Meadows; **176** Paul Katz/Getty Images; **181** Photodisc/Getty Images; **183** The McGraw-Hill Companies; **184 185** Matt Meadows; **186** (t)The McGraw-Hill Companies, (b)Matt Meadows; **188** (tl)Steve Cole/Getty Images, (tr)Andersen Ross/Photodisc/Getty Images, (c b)Matt Meadows; **192 193** Matt Meadows; **194** Kenneth M. Swezey Papers, Archives Center, National Museum of American History, Smithsonian Institution; **196** Steve Cole/Getty Images;

200-201 Digital Vision/PunchStock; **202** Matt Meadows; **203 204** Cordelia Molloy/Photo Researchers, Inc.; **207** (l)Siede Preis/Getty Images, (r)The McGraw-Hill Companies; **210** The McGraw-Hill Companies; **212** Jorg Grueul/Photographer's Choice/Getty Images; **213** (t)Chase Jarvis/Photographer's Choice RF/Getty Images, (b)Spike Mafford/Getty Images; **215 216** Matt Meadows; **217** The McGraw-hIll Companies; **218** (l)Ingram Publishing/SuperStock, (r)Brand X Pictures; **219** (tl)The McGraw-Hill Companies, (tr)Spike Mafford/Getty Images, (bl)Comstock/PunchStock, (br)Ingram Publishing/SuperStock; **223** Tim Courlas/Horizons Companies; **224 225** Matt Meadows; **226** Steven L. Raymer/National Geographic/Getty Images; **232-233** CYRIL RUOSO/JH EDITORIAL/MINDEN PICTURES/National Geographic Stock; **234** Paul Souders/Getty Images; **236** Matt Meadows; **237** Photodisc/Getty Images; **241** Russell Illig/Getty Images; **242** (t)Photo Researchers, (b)Digital Vision/Getty Images; **245** (l)Igor Kostin/Sygma/CORBIS, (r)Gerd Ludwig/CORBIS; **247** Steve Allen/Brand X Pictures; **248** Creativ Studio Heinemann/Getty Images; **250** Eric Carle/Bruce Coleman, Inc.; **251** (t)ScottishPower/PA Wire URN:6399655/Press Association/AP Images, (b)Russell Illig/Getty Images; **253** Photofusion Picture Library/Alamy; **254** Matt Meadows; **256** Lester Lefkowitz/Taxi/Getty Images; **257** (t)Digital Vision/Getty Images, (b)Comstock Images; **258** (t)Doug Sherman/Geofile, (bl)BananaStock/PunchStock, (br)Digital Vision/Alamy; **259** (t)Peter Arnold, Inc./Alamy, (b)Robert Madden/National Geographic/Getty Images; **261** VisionsofAmerica/Joe Sohm/Photodisc/Getty Images; **262 263** Matt Meadows; **264** ASSOCIATED PRESS; **269** Robert Essel/CORBIS; **270-271** Bryce R. Bradford/Flickr/Getty Images; **272-273** David Clapp/Photolibrary/Getty Images; **274** Thinkstock/Getty Images; **286** Ken Karp/ The McGraw-Hill Companies; **287** Richard Megna/Fundamental Photographs; **288** Matt Meadows; **289** stumayhew/Getty Images; **290** (t)Richard Cooke/Alamy, (b)Richard Megna/Fundamental Photographs; **291** Richard Megna/Fundamental Photographs; **294** Ted Kinsman/Photo Researchers; **295** Comstock Images/Getty Images; **296** Matt Meadows; **304-305** Bob Krist/CORBIS; **307** Stocktrek/age fotostock; **308** Tim Courlas/Horizons Companies; **310** SPL/Photo Researchers, Inc.; **312** Imagestopshop/Alamy; **313** (l)Brand X Pictures/PunchStock, (c)Comstock/PunchStock, (r)Ryan McVay/Getty Images; **316** Christopher Furlong/Getty Images; **317** (l)BLOOMimage/Getty Images, (r)Lew Robertson/Brand X Pictures/Getty Images; **319** Photodisc/Getty Images; **320** Ray Stubblebine/Reuters/CORBIS; **321** (t)Jake Rajs/Photonica/Getty Images, (b)Adalberto Rios Lanz/Sexto Sol/Photodisc/Getty Images; **323** Doug Martin; **324** Rob Melnychuk/Photodisc/Getty Images; **325** Oxford Scientific/Photolibrary/Getty Images; **327** Michael Bradley/Getty Images; **328 329** Matt Meadows; **336-337** NASA; **338** (l)Brand X Pictures/PunchStock, (r)Ken Cavanagh/The McGraw-Hill Companies; **340** Leslie Garland Picture Library/Alamy; **342** fotog/Getty Images; **346** (l)Matthew McVay/Getty Images, (r)CORBIS; **348** Ted Kinsman/Photolibrary/Getty Images; **349** Richard Megna/Fundamental Photographs; **350** Jim Wehtje/Getty Images; **351** Allan Shoemake/Getty Images; **353** Getty Images/Digital Vision; **354** Comstock Images/Alamy; **355** (t)Russell Illig/Getty Images, (b)Matt Meadows; **356** Erik Simonsen/Getty Images; **359** Thinkstock/Comstock Images/Getty Images; **360** SSPL via Getty Images; **366-367** Richard Dirscherl/Visuals Unlimited; **368** Mark Burnett; **370** (l)Creative Crop/Digital Vision/Getty Images, (r)Joe Ginsberg/Getty Images; **371** Matthias Kulka/CORBIS; **372** Kent Wood/Photo Researchers; **375** (t)Steve Allen/Brand X Pictures/Getty Images, (b)Matt Meadows; **376** Matt Meadows; **378** Mike Kemp/Getty Images; **379** The McGraw-Hill Companies; **380** Jason Todd/Digital Vision/Getty Images; **383** Andrew Syred/Photo Researchers; **384** Charles D. Winters/Timeframe Photography/The McGraw-Hill Companies; **385** (l)Yoav Levy/Phototake, (r)PSL Images/Alamy; **387** J Silver/SuperStock; **388** Phil Degginger/Stone/Getty Images; **389** Matt Meadows; **390** (t)Matt Meadows, (b)Dominic Oldershaw; **392** (t)©Jan Woitas/dpa/Corbis, (b)©George Steinmetz/Corbis;

(cr)Joe Raedle/Getty Images, (bl)Marty Honig/Getty Images, (br)Phil Degginger/Getty Images; **745** Dorothy Riess MD/Getty Images; **746** Steven Puetzer/Getty Images; **751** (t to b)Apic/Getty Images, Jerry Cooke/CORBIS, Tim Ridley/Getty Images, SSPL/Getty Images, Getty Images, Hulton Archive/Getty Images, Jupiterimages/Getty Images; **752** Ryan McVay/Getty Images; **753** Karen Moskowitz/Getty Images; **754** (tl)Comstock Images/Alamy, (tr)Ingram Publishing/Alamy, (bl)Paul Conrath/Getty Images, (br)Ingram Publishing/SuperStock; **755** (tl)William Mebane/Getty Images, (tc)Steve Cole/Getty Images, (tr)YOSHIKAZU TSUNO/AFP/Getty Images, (bl)Jeffrey Coolidge/Getty Images, (br)Benne Ochs/Getty Images; **756** (t)Eye of Science/Photo Researchers, Inc., (b)Patrick Bennett/CORBIS; **757** Gary He/The McGraw-Hill Companies; **758** Brand X Pictures/PunchStock; **759** Aaron Haupt; **760** John Humble/Getty Images; **761** Tim Fuller; **762** Courtesy Dupont; **764** David J. Green/Alamy; **766** (l)Getty Images, (c)Chloe Johnson/Alamy, (r)Ingram Publishing/Alamy; **767** (t)Jacques Cornell/The McGraw-Hill Companies, (b)Ken Karp/The McGraw-Hill Companies; **768-769** AVO/NASA; **770-771** Neale Clark/Robert Harding World Imagery/Getty Images; **773** PhotoSpin, Inc/Alamy; **774** NOAA; **781** Mads Nissen/Reportage/Getty Images; **785** (t)David McNew/Getty Images, (b)Code Red/Getty Images; **794** (t)Game McGimsey/Alaska Volcano Observatory/U.S. Geological Survey, (b)Philippe Bourseiller/The Image Bank/Getty Images; **795** Emmanuel LATTES/Alamy; **803** U.S. Geological Survey; **806-807** Chad Ehlers/Stone/Getty Images; **809** DEA PICTURE LIBRARY/Getty Images; **810** (tl tr b)RF Company/Alamy, (tc)Dr. Parvinder Sethi, (c)Doug Sherman/Geofile; **811** (l)Ken Cavanagh/The McGraw-Hill Companies, (r)Doug Sherman/Geofile; **813** (l)Tony Waltham/Robert Harding/Getty Images, (r)CORBIS; **814** (t to b)Gary Ombler/Dorling Kindersley/Getty Images, Phil Degginger/Jack Clark Collection/Alamy, RF Company/Alamy, RF Company/Alamy, Doug Sherman/Geofile, Siede Preis/Getty Images, Doug Sherman/Geofile, Doug Sherman/Geofile, Doug Sherman/Geofile, Siede Preis/Getty Images, Doug Sherman/Geofile, Dorling Kindersley/Getty Images; **816** Fundamental Photographs; **817** Gabbro/Alamy; **819** (l)Ken Cavanagh/The McGraw-Hill Companies, (r)DEA/C. Bevilacqua/Getty Images; **820** (t)Sami Sarkis/Photographer's Choice/Getty Images, (bl)Tony Lilley/Alamy, (br)Dr. Parvinder Sethi; **821** (tl)Dr. Parvinder Sethi, (tc bl)Dirk Wiersma/Photo Researchers, Inc., (tr)Jacques Cornell/The McGraw-Hill Companies, (bc)Photolibrary/age fotostock, (br)Maurice Nimmo/Frank Lane Picture Agency/CORBIS; **823** Emmanuel LATTES/Alamy; **824** CORBIS; **825** Momatiuk-Eastcott/CORBIS; **827** Ingo Jezierski/Getty Images; **828** (l)Virginia Zozaya/Flickr/Getty Images, (r)Peter Adams/Digital Vision/Getty Images; **830** Andreas Strauss/Getty Images; **831** (l)Jacques Cornell/The McGraw-Hill Companies, (r)RF Company/Alamy; **833** Robert Harding Picture Library Ltd/Alamy; **834** (tl)U.S. Geological Survey, (tr)Game McGimsey/AVO/USGS, (b)Christina Neal/AVO/USGS; **836 837** Matt Meadows; **838** Bloomberg via Getty Images; **840** Ken Cavanagh/The McGraw-Hill Companies; **841** Altrendo Nature/Getty Images; **844-845** CORBIS; **846** Dr. Parvinder Sethi; **848** David C Tomlinson/Photographer's Choice/Getty Images; **849** Steve Hamblin/CORBIS; **852** Keren Su/China Span/Getty Images; **854** Doug Sherman/Geofile; **855** Alan Kearney/Digital Vision/Getty Images; **856** (tl)Digital Archive Japan/Alamy, (tr)Bob Nichols, USDA Natural Resources Conservation Service, (b)Alan Morgan; **857** (t)NASA, (b)SPL/Photo Researchers; **859** (t)Austin Post, U.S. Geological Survey, Tacoma, WA, (c)Doug Sherman/Geofile, (b)Wonderlust Industries/Photodisc/Getty Images; **860** Design Pics/PunchStock; **861** Image Source/CORBIS; **862** Mark Reid/USGS; **865** Jim Richardson/National Geographic Stock; **874 875** KS Studios; **876** Stockbyte/Getty Images; **882-883** NOAA; **887** (t b)Stockbyte/Getty Images, (c)Don Farrall/Getty Images; **889** Matt Meadows; **893** J. Schmidt/NPS; **894** NOAA's National Weather Service (NWS) Collection; **895** Digital Vision/PunchStock; **896** Getty Images; **899** Adam Jones/The Image Bank/Getty Images; **902** NASA/Goddard Space Flight Center Scientific Visualization Studio; **906** Jeremy Woodhouse/Getty Images; **908** Ryan McGinnis/age fotostock; **910** Adam Jones/The Images Bank/Getty Images; **914-915** NASA; **916-917** Digital Vision/Getty Images; **918** (l)Roger Ressmeyer/CORBIS, (r)TAKASHI KATAHIRA/amanaimages/CORBIS; **919** NASA; **920** Theo Allofs/Digital Vision/Getty Images; **921** Photodisc/Getty Images; **935** (t)CORBIS, (b)NASA Kennedy Space Center (NASA-KSC); **936** StockTrek/Getty Images; **940** (l)Stockbyte/PunchStock, (tr)StockTrek/Getty Images, (br)NASA Johnson Space Center Collection; **942** NASA; **947** (l)NASA Johnson Space Center Collection, (r)NASA/GSFC; **948-949** NASA/JPL/Space Science Institute; **954** NASA; **955** NASA/Johns Hopkins University Applied Physics Laboratory/Carnegie Institution of Washington; **956** (t)NASA, (b)JPL/NASA; **957** (t)Michele Falzone/Photographer's Choice/Getty Images, (b)Steve Lee (University of Colorado), Jim Bell (Cornell University), Mike Wolff (Space Science Institute), and NASA; **958** (t)NASA/JPL/Cornell/USGS, (b)NASA/JPL/University of Arizona; **959** NASA/JPL/Arizona State University; **960** (t)Mars Exploration Rover Mission, Cornell, JPL, NASA, (c)NASA/JPL/Cornell, (b)NASA/JPL-Caltech/Cornell; **961** (l)JPL/NASA, (r)NASA; **963 964** NASA/JPL; **965** (t)NASA/JPL/Space Science Institute, (b)NASA/JPL/ESA/University of Arizona; **966** (l)NASA/ESA and Erich Karkoschka, University of Arizona, (r)NASA/JPL; **967** StockTrek/Photodisc/Getty Images; **968** Caltech/JPL/NASA; **969** EMORY KRISTOF/National Geographic Stock; **970** (t)Darrell Gulin/The Image Bank/Getty Images, (b)Gerhard Zwerger-Schon/age fotostock; **971** (t)ESA/DLR/FU Berlin (G. Neukum), (b)NASA/JPL/DLR; **972** (t)courtesy Michael Studinger, (c)Science Source/Photo Researchers, Inc., (b)NASA/JPL/University of Arizona; **973** NASA/JPL/University of Arizona/University of Nantes; **974 975** Matt Meadows; **976** NASA/JPL; **978** NASA/JPL/University of Arizona; **980** Photodisc/Getty Images; **981** ESA/DLR/FU Berlin (G. Neukum); **982-983** Bryan Allen/CORBIS; **987** (t)NRAO/AUI, (b)NASA; **991** NASA Goddard Space Flight Center; **992** (l)Getty Images/Digital Vision, (r)NASA; **994** (t)T.Rimmele (NSO), M.Hanna (NOAO)/AURA/NSF, (b)NASA; **995** SOHO (ESA & NASA); **997** NASA/JPL/California Institute of Technology; **998** (t)NASA, ESA, and Hubble Heritage (STScI/AURA); **999** (t)courtesy NRAO/AUI; **999** NASA; **1001** NASA/CXC/MIT/F.K. Baganoff et al.; **1002 1003** NASA; **1004** NASA, ESA, S. Beckwith (STScI) and the HUDF Team; **1005** the 2dF Galaxy Redshift Survey team; **1008** Bettmann/CORBIS; **1010** JPL/NASA; **1012** Bruce Balick et al./HST/NASA; **1014** Getty Images/Photodisc; **1016** The McGraw-Hill Companies; **1017 1018** Tim Fuller; **1023** Digital Vision/Alamy; **1024** (t)Mark Burnett, (b)Dominic Oldershaw; **1025** StudioOhio; **1026** Tim Fuller; **1027** Rob Melnychuk/JupiterImages/Brand X/Alamy.

PERIODIC TABLE OF THE ELEMENTS

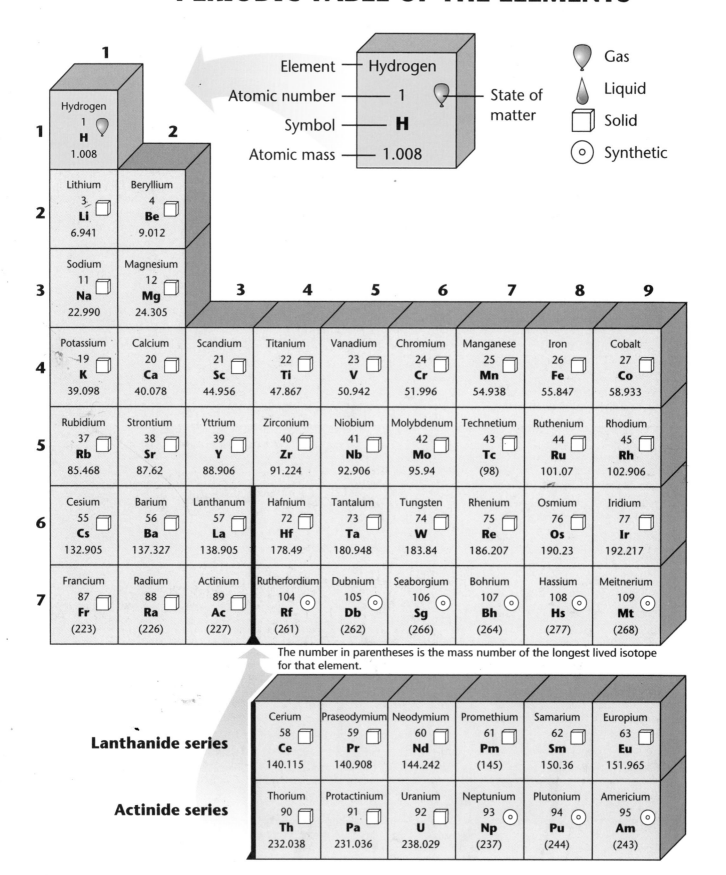

Element — Hydrogen
Atomic number — 1
Symbol — H
Atomic mass — 1.008

State of matter

Gas
Liquid
Solid
Synthetic

The number in parentheses is the mass number of the longest lived isotope for that element.

Lanthanide series

Actinide series